T0418387

Fermented Milk and Dairy Products

FERMENTED FOODS AND BEVERAGES SERIES

Series Editors
M.J.R. Nout and Prabir K. Sarkar

Fermented Milk and Dairy Products (2015)
Editor: Anil Kumar Puniya

Indigenous Fermented Foods of Southeast Asia (2014)
Editor: J. David Owens

Cocoa and Coffee Fermentations (2014)
Editors: Rosane F. Schwan and Graham H. Fleet

Handbook of Indigenous Foods Involving Alkaline Fermentation (2014)
Editors: Prabir K. Sarkar and M.J.R. Nout

Solid State Fermentation for Foods and Beverages (2013)
Editors: Jian Chen and Yang Zhu

Valorization of Food Processing By-Products (2013)
Editor: M. Chandrasekaran

Fermented Foods and Beverages Series

Fermented Milk
and
Dairy Products

Edited by
Anil Kumar Puniya

CRC Press
Taylor & Francis Group
Boca Raton London New York

CRC Press is an imprint of the
Taylor & Francis Group, an **informa** business

CRC Press
Taylor & Francis Group
6000 Broken Sound Parkway NW, Suite 300
Boca Raton, FL 33487-2742

© 2016 by Taylor & Francis Group, LLC
CRC Press is an imprint of Taylor & Francis Group, an Informa business

No claim to original U.S. Government works

Printed on acid-free paper
Version Date: 20150708

International Standard Book Number-13: 978-1-4665-7797-8 (Hardback)

This book contains information obtained from authentic and highly regarded sources. Reasonable efforts have been made to publish reliable data and information, but the author and publisher cannot assume responsibility for the validity of all materials or the consequences of their use. The authors and publishers have attempted to trace the copyright holders of all material reproduced in this publication and apologize to copyright holders if permission to publish in this form has not been obtained. If any copyright material has not been acknowledged please write and let us know so we may rectify in any future reprint.

Except as permitted under U.S. Copyright Law, no part of this book may be reprinted, reproduced, transmitted, or utilized in any form by any electronic, mechanical, or other means, now known or hereafter invented, including photocopying, microfilming, and recording, or in any information storage or retrieval system, without written permission from the publishers.

For permission to photocopy or use material electronically from this work, please access www.copyright.com (http://www.copyright.com/) or contact the Copyright Clearance Center, Inc. (CCC), 222 Rosewood Drive, Danvers, MA 01923, 978-750-8400. CCC is a not-for-profit organization that provides licenses and registration for a variety of users. For organizations that have been granted a photocopy license by the CCC, a separate system of payment has been arranged.

Trademark Notice: Product or corporate names may be trademarks or registered trademarks, and are used only for identification and explanation without intent to infringe.

Visit the Taylor & Francis Web site at
http://www.taylorandfrancis.com

and the CRC Press Web site at
http://www.crcpress.com

Dedicated to dairy farmers,
researchers,
and entrepreneurs.

Anil Kumar Puniya

Contents

SERIES PREFACE — xi
PREFACE — xiii
ACKNOWLEDGMENTS — xvii
EDITOR — xix
CONTRIBUTORS — xxiii

PART I INTRODUCTION, SCOPE, AND SIGNIFICANCE

CHAPTER 1 FERMENTED MILK AND DAIRY PRODUCTS: AN OVERVIEW — 3

ANIL KUMAR PUNIYA, SANJEEV KUMAR, MONICA PUNIYA, AND RAVINDER MALIK

CHAPTER 2 NUTRITIONAL AND THERAPEUTIC CHARACTERISTICS OF FERMENTED DAIRY PRODUCTS — 25

ARUN MALIK, HARNISH, PRIYANKA CHOPRA, STANZIN ANGMO, DINESH KUMAR DAHIYA, SHALINI JAIN, AND HARIOM YADAV

CHAPTER 3 LACTIC ACID BACTERIA AND ASSOCIATED PRODUCT CHARACTERISTICS WITHIN MILK FERMENTATION — 47

CHRISTOPHER BEERMANN AND JULIA HARTUNG

Part II Starter Cultures in Milk Fermentation

Chapter 4 Types and Application of Lactic Starters 83
PRADIP BEHARE, JASHBHAI B. PRAJAPATI, AND RAMESHWAR SINGH

Chapter 5 Metabolic Characteristics of Lactic Starters 109
S. MANJULATA DEVI AND PRAKASH M. HALAMI

Chapter 6 Molecular Biology of Adaptation of Starter Lactic Acid Bacteria to Dairy System 133
RAJESH KUMAR AND JAI KUMAR KAUSHIK

Chapter 7 Preservation of Lactic Starter Cultures 167
CHALAT SANTIVARANGKNA

Part III Dairy Products Based on the Type of Fermentation: Pure Lactic Fermentations (Mesophilic, Thermophilic, and Probiotic Products)

Chapter 8 Natural and Cultured Buttermilk 203
RAVINDER KUMAR, MANPREET KAUR, ANITA KUMARI GARSA, BHUVNESH SHRIVASTAVA, VELUGOTI PADMANABHA REDDY, AND ASHISH TYAGI

Chapter 9 Acidophilus Milks 227
SUJA SENAN AND JASHBHAI B. PRAJAPATI

Chapter 10 Bifidus Milks 269
SREEJA V. AND JASHBHAI B. PRAJAPATI

Chapter 11 Yogurt: Concepts and Developments 311
SARANG DILIP POPHALY, HITESH KUMAR, SUDHIR KUMAR TOMAR, AND RAMESHWAR SINGH

Chapter 12 Technology of Fresh Cheeses 329
YOGESH KHETRA, S.K. KANAWJIA, APURBA GIRI, AND RITIKA PURI

Chapter 13	Dahi, Lassi, and Shrikhand	355
	S.V.N. Vijayendra and M.C. Varadaraj	
Chapter 14	Indonesian Dadih	377
	Ingrid Suryanti Surono	
Chapter 15	Traditional Fermented Dairy Products of Turkey	401
	Ozlem Istanbullu and Bulent Kabak	
Chapter 16	Cheese Whey Fermentation	427
	Carla Oliveira, Giuliano Dragone, Lucília Domingues, and José A. Teixeira	

Part IV Dairy Products Based on the Type of Fermentation: Fungal Lactic Fermentations

Chapter 17	Milk Kefir: Structure and Microbiological and Chemical Composition	461
	Rosane Freitas Schwan, Karina Teixeira Magalhães-Guedes, and Disney Ribeiro Dias	
Chapter 18	Koumiss: Nutritional and Therapeutic Values	483
	Tejpal Dhewa, Vijendra Mishra, Nikhil Kumar, and K.P.S. Sangu	
Chapter 19	Viili	497
	Ming-Ju Chen and Sheng-Yao Wang	

Part V Health Benefits

Chapter 20	Probiotic Dairy Products	519
	Surajit Mandal, Subrota Hati, and Chand Ram	
Chapter 21	Health Benefits of Fermented Probiotic Dairy Products	557
	Arthur C. Ouwehand and Julia Tennilä	

Chapter 22 Validation of Health Claims for Fermented Milks — 571
EDWARD FARNWORTH AND FARAH HOSSEINIAN

Part VI Quality Assurance, Packaging, and Other Prospects

Chapter 23 Quality Assurance of Fermented Dairy Products — 603
H.V. RAGHU, NIMISHA TEHRI, LAXMAN NAIK, AND NARESH KUMAR

Chapter 24 Packaging of Fermented Milks and Dairy Products — 637
P. NARENDER RAJU AND ASHISH KUMAR SINGH

Chapter 25 Xylitol: An Alternative Sweetener for Dairy Products — 673
MAMATHA POTU, SREENIVAS RAO RAVELLA, JOE GALLAGHER, DAVID WALKER, AND DAVID BRYANT

Index — 689

Series Preface

Natural fermentation precedes human history, and since ancient times humans have been controlling the fermentation process. Fermentation, an anaerobic way of life, has attained a wider meaning in biotransformations, resulting in a wide variety of fermented foods and beverages.

Fermented products made with uncontrolled natural fermentations or with defined starter cultures achieve their characteristic flavor, taste, consistency, and nutritional properties through the combined effects of microbial assimilation and metabolite production, as well as from enzyme activities derived from food ingredients.

Fermented foods and beverages span a wide and diverse range of starchy root crops, cereals, pulses, vegetables, nuts, and fruits, as well as animal products such as meats, fish, seafood, and dairy.

The science behind chemical, microbiological, and technological factors and changes associated with the manufacture, quality, and safety of fermented products is progressing and is aimed at achieving higher levels of control over the quality, safety, and profitability of fermented-food manufacture.

Both the producer and the consumer benefit from scientific, technological, and consumer-oriented research. Small-scale production needs to be better controlled and safeguarded. Traditional products need to be characterized and described to establish, maintain, and protect their authenticity. Medium- and large-scale food fermentation requires selected, tailor-made, or improved processes that provide sustainable solutions for the future conservation of energy and water as well as for the responsible utilization of resources and disposal of by-products in the environment.

The scope of CRC Press's *Fermented Foods and Beverages Series* shall include (1) globally known foods and beverages of plant and animal origin (such as dairy, meat, fish, vegetables, cereals, root crops, soybeans, legumes, pickles, cocoa and coffee, wines, beers, spirits, starter cultures, and probiotic cultures), their manufacture, chemical and microbiological composition, processing, compositional and functional modifications taking place as a result of microbial and enzymic effects, their safety, legislation associated with and development of novel products, and opportunities for industrialization; (2) indigenous commodities from Africa, Asia (south, east, and southeast), Europe, Latin America, and the Middle East, their traditional and industrialized processes and their contribution to livelihood; and (3) several aspects of general interest such as valorization of food-processing by-products, biotechnology, engineering of solid-state processes, modern chemical and biological analytical approaches (such as genomics, transcriptomics, and metabolomics), safety, health, and consumer perception.

The sixth book born in the series deals with *fermented milk and dairy products*. This treatise, edited by Dr. Anil Kumar Puniya, deals with the science and technology of fermented dairy products, which are of importance all over the world. The nutritional and health benefits of fermented milk products, as well as their functionality in the context of probiotic cultures, draw international attention. In addition, technological aspects of the manufacture of fermented milk products in a variety of regions of the world, as well as the microbiology and preservation of starter cultures for milk fermentations, are presented. This treatise, with contributions from experts from all over the world, will serve students, teachers, researchers, entrepreneurs, and the general public as a source of information and inspiration.

Preface

The trend of functional foods is increasing and attracting customers globally due to their wide range of therapeutic attributes. Stress-free assimilation, high nutritional value, and simple processing of these products are other leading reasons for their great success. A number of fermented dairy products are getting credibility in the food market with an increase in the overall turnover of industries. Therefore, traditional and innovative fermented dairy products are launched into the market by food manufacturers.

This book envisages an understanding that ranges from the general biology of the fermented milk and dairy products to various microorganisms associated with different types of fermented products available worldwide. The genesis of this book came after feeling a sense of urgency to find literature that included information about different fermented milk and dairy products, their occurrence, the microorganisms involved, nutritional characteristics, and health benefits in one place. However, while the information about several fermented products is already available in the public domain, it is mainly limited and/or incomplete. This book provides the details of manufacturing methods, therapeutic attributes, and social values. The book is the result of utmost desperation to generate literature that brings together researchers of relevant areas and incorporates their recent inputs to enable readers, especially scholars, to learn, realize, and deliver the best available content about fermented milk and dairy products. Thus, the content is suitable for use in a dairy/food microbiology and technology curriculum or as a primary course in food science. To make the book convenient and consistent for reading, it has been divided into smaller units.

The book is divided into 25 chapters and 6 parts. Part I focuses on the scope and significance of fermented milk products, with a brief introduction in Chapter 1 and their major nutritional characteristics in Chapter 2. While keeping in mind the wider importance and application of lactic acid bacteria in milk fermentation, the characteristics of products generated have been highlighted in Chapter 3. In fact, Part II is dedicated to lactic acid bacteria and their huge role in determining the fate of fermented dairy products. Chapters 4 and 5 emphasize the type of different lactic acid bacteria involved and their metabolic characteristics. The material in Chapter 6 is designed to provide a brief background on the molecular biology aspects of lactic acid bacteria to keep readers updated with recent advances and future challenges for genetic manipulations. Chapter 7 focuses on the sustainability of these important starter cultures. Part III is a massive compilation of various types of dairy products available and is further segregated on the basis of the types of fermentation involved. Part III provides deep insights into pure lactic fermentation–based products such as natural and cultural buttermilk; acidophilus milk; bifidus milk; yogurt; fresh cheeses; dahi, lassi, and srikhand; Indonesian dadih; traditional fermented milks of Turkey; and cheese whey fermentation (Chapters 8 through 16). Part IV provides a special emphasis on the description of fungus-mediated lactic acid fermentation and products generated thereof. Chapters 17 through 19 primarily discuss milk kefir, koumiss, and viili. Part V is the most important section of this book, revealing probiotic dairy products, reporting health benefits from these products, and validating health claims, over Chapters 20 through 22. Part VI concludes this book, explaining the methods to ensure the quality assurance, packaging, and role of alternative sweeteners in fermented milk products. In summary, the book is a systematic collection of the knowledge available in the area of fermentation-based milk products. I am sure this compilation, by virtue of its content and continuity, will popularize itself among dairy students, dairy entrepreneurs, and the academic fraternity. I fully acknowledge the experts' contributions and also greatly appreciate the support of CRC Press.

Anil Kumar Puniya
ICAR-National Dairy Research Institute

Acknowledgments

This book came into existence through the enormous contributions of a number of scientists, especially the chapter contributors. I acknowledge their highly determined effort to complete this remarkable task.

I feel privileged to express my heartfelt gratitude to Dr. Anil Kumar Srivastava, Director and Vice-Chancellor, ICAR-National Dairy Research Institute, Karnal, India, who extended full support and greatly encouraged this academic assignment. I also owe my profound thanks to Dr. G. R. Patil, Joint Director (Academic), and Dr. R. K. Malik, Joint Director (Research), for their constructive comments throughout the editing of this book.

I also extend my thanks to all of my professional colleagues in the institute and elsewhere who supported me directly or indirectly in giving final shape to this book. I would also like to acknowledge Dr. A. S. Nanda, Vice-Chancellor, Guru Angad Dev Veterinary and Animal Sciences University, Ludhiana.

I am sincerely thankful to Dr. Ravinder Kumar and Dr. Monica Puniya, who provided their technical help in compiling and editing the chapters. In addition, the scholars who assisted me, namely, Prasanta Choudhury, Nikhil Kumar, Bhuvnesh Shrivastava, Anita Garsa, Manpreet Kaur, Nidhi Alekar, and Manu Verma, deserve full appreciation.

I am also grateful to Dr. Rob Nout and Dr. P. K. Sarkar, the editors of the *Fermented Food and Beverages Series*, and CRC Press for inviting me as the editor of this book.

Words would not suffice to thank my mother, Savitri Devi, wife, Monica, and children, Dharun Vijay and Ishaan Vijay, for their affectionate concern throughout the compilation of this work.

Editor

Dr. Anil Kumar Puniya, Dean of the College of Dairy Science and Technology, Guru Angad Dev Veterinary and Animal Sciences University, Ludhiana, Punjab, India, has significantly contributed in the areas of dairy and rumen microbiology. Prior to this, he served as Principal Scientist (Professor) in the Dairy Microbiology Division of ICAR-National Dairy Research Institute, Karnal. He received his master's degree in microbiology in 1988 from Gurukul Kangri University, Haridwar, India, and a doctorate from ICAR-NDRI, Karnal, in 1994 (over the course of which he was also awarded the prestigious DAAD Fellowship of Germany). He has developed expertise in the field of probiotic science for the control of lifestyle diseases besides the role of rumen microbes as direct-fed microbials for enhanced animal productivity.

Dr. Puniya has published over 120 research papers, reviews, and book chapters of high impact and written 5 lab manuals. He has also served as chairman and keynote/invited speaker for different scientific events, nationally and internationally. He has been involved in 18 externally funded and institutional projects, such as Principal Investigator/Co-Principal Investigator (PI/Co-PI), and has supervised a number of master's and doctoral research scholars. He remained paper coordinator of *Food and Dairy Microbiology* for e-contents for postgraduate subjects (i.e., E-PG Pathshala) of the University Grants Commission, New Delhi, under the National Mission on Education. Additionally, he has developed an e-course, Introductory Dairy Microbiology, for the BTech (Dairy Technology) course under the National Agricultural Innovation Project of the Indian Council of Agricultural Research.

Dr. Puniya was awarded a DSE grant to attend the International Conference of Tropical Veterinary Medicine (1995), Germany, and Department of Science and Technology (DST) & CSIR funding to present a paper at BioMicroWorld 2007, Spain. He chaired the sessions of Food Microbiology during BioMicroWorld 2009, Portugal, and Rumen Microbiology during ANINUE-2012, Thailand. Besides, he has visited Aberystwyth University, United Kingdom, twice, under the Exchange of Scientists Programme of INSA with the Royal Society, London (2007), and as an DBT-CREST awardee (2013). He has also visited Taiwan, in 2010, during an exposure visit of the Department of Science and Technology, Government of India.

Dr. Puniya was a recipient of the Young Scientist Award (1996) of the Association of Microbiologists of India; Kautilya Gold Medal (1998–1999) of Nature Conservators; Best Young Scientist Award (2004) of Eureka Forbes; K.K. Iya Award, 2006; Honorable Mention 2010 of the American Society for Microbiology; and Certificate of Appreciation of ABRCMS, United States, 2012, 2013, and 2014. He has also received a number of Best Poster and Paper awards at different fora. He was a member of the International Scientific Advisory Board of Formatex Research Centre, Spain, and an Advisory Committee member of ICBB-2014, Dubai, United Arab Emirates.

Dr. Puniya underwent the Training Programme for Scientists & Technologists (2007) of DST and other specialized training programs in microbial and molecular biology at Indian Agricultural Research Institute (IARI) and Institute of Microbial Technology (IMTECH), besides the E-course Development program at National Academy of Agricultural Research Management (NAARM), Hyderabad. He was the co-organizing secretary of the 45th Conference of Association of Microbiologists of India (2004) and organized the National Environment Awareness Campaign of Ministry of Environment & Forests at NDRI (2007). He is an editorial board member, a guest editor, and a referee of several reputed journals worldwide. He is life member of a number of renowned scientific bodies, for example, Association of Microbiologists of India (AMI), Animal Nutrition Association (ANA), Animal Nutrition Society of India (ANSI), Biotech Research Society of India (BRSI), Dairy Technology Society of India (DTSI), Indian Dairy Association (IDA), Indian Science Congress Association (ISCA), Probiotic Association of India (PAI), and Society of Biological Chemists (SBC) (India), and is a fellow of the Animal Nutrition Society of India and the National Academy of Dairy Society of India.

Contributors

Stanzin Angmo
National Agri-Food
  Biotechnology Institute
Mohali, India

Christopher Beermann
Department of Biotechnology
Fulda University of Applied
  Science
Fulda, Germany

Pradip Behare
Dairy Microbiology Division
ICAR-National Dairy Research
  Institute
Karnal, India

David Bryant
Institute of Biological and Environmental and Rural Sciences
Aberystwyth University
Aberystwyth, Wales,
  United Kingdom

Ming-Ju Chen
Department of Animal Science
  and Technology
National Taiwan University
Taipei City, Taiwan

Priyanka Chopra
National Agri-Food
  Biotechnology Institute
Mohali, India

Dinesh Kumar Dahiya
Dairy Microbiology Division
ICAR-National Dairy Research
  Institute
Karnal, India

S. Manjulata Devi
Microbiology and Fermentation
  Technology
CSIR-Central Food Technological Research Institute
Mysuru, India

Tejpal Dhewa
Bhaskaraycharya College of
   Applied Science
University of Delhi
New Delhi, India

Disney Ribeiro Dias
Department of Food Science
Federal University of Lavras
Lavras, Brazil

Lucília Domingues
Institute for Biotechnology and
   Bioengineering
Centre of Biological
   Engineering
Universidade do Minho
Braga, Portugal

Giuliano Dragone
Institute for Biotechnology and
   Bioengineering
Centre of Biological
   Engineering
Universidade do Minho
Braga, Portugal

Edward Farnworth
Knowledge Broker: Food,
   Nutrition, Health
Ottawa, Ontario, Canada

Joe Gallagher
Institute of Biological and Environmental and Rural Sciences
Aberystwyth University
Aberystwyth, Wales,
   United Kingdom

Anita Kumari Garsa
Dairy Microbiology Division
ICAR-National Dairy Research
   Institute
Karnal, India

Apurba Giri
Cheese and Fermented Foods
   Laboratory
Dairy Technology Division
ICAR-National Dairy Research
   Institute
Karnal, India

Prakash M. Halami
Microbiology and Fermentation
   Technology
CSIR-Central Food
   Technological Research
   Institute
Mysuru, India

Harnish
National Agri-Food
   Biotechnology Institute
Mohali, India

Julia Hartung
Department of Biotechnology
Fulda University of Applied
   Science
Fulda, Germany

Subrota Hati
Dairy Microbiology
   Department
Anand Agricultural University
Anand, India

Farah Hosseinian
Food Science and Nutrition Program
Department of Chemistry and Institute of Biochemistry
Carleton University
Ottawa, Ontario, Canada

Ozlem Istanbullu
Department of Food Engineering
Hitit University
Corum, Turkey

Shalini Jain
National Institutes of Health
National Institute of Diabetes and Digestive and Kidney Diseases
Bethesda, Maryland

Bulent Kabak
Department of Food Engineering
Hitit University
Corum, Turkey

S.K. Kanawjia
Cheese and Fermented Foods Laboratory
Dairy Technology Division
ICAR-National Dairy Research Institute
Karnal, India

Manpreet Kaur
Dairy Microbiology Division
ICAR-National Dairy Research Institute
Karnal, India

Jai Kumar Kaushik
Animal Biotechnology Centre
ICAR-National Dairy Research Institute
Karnal, India

Yogesh Khetra
Cheese and Fermented Foods Laboratory
Dairy Technology Division
ICAR-National Dairy Research Institute
Karnal, India

Hitesh Kumar
Dairy Microbiology Division
ICAR-National Dairy Research Institute
Karnal, India

Naresh Kumar
Dairy Microbiology Division
ICAR-National Dairy Research Institute
Karnal, India

Nikhil Kumar
Basic and Applied Sciences
National Institute of Food Technology Entrepreneurship and Management
Kundli, India

Rajesh Kumar
Department of Environment
Ministry of Environment and Forests
Chandigarh, India

Ravinder Kumar
Dairy Microbiology Division
ICAR-National Dairy Research
  Institute
Karnal, India

Sanjeev Kumar
Department of Life Science
Assam University
Silchar, India

Karina Teixeira Magalhães-Guedes
Department of Biology
Federal University of Lavras
Lavras, Brazil

Arun Malik
National Agri-Food
  Biotechnology Institute
Mohali, India

Ravinder Malik
Dairy Microbiology Division
ICAR-National Dairy Research
  Institute
Karnal, India

Surajit Mandal
Dairy Microbiology Division
ICAR-National Dairy Research
  Institute
Karnal, India

Vijendra Mishra
Basic and Applied Sciences
National Institute of Food
  Technology Entrepreneurship
  and Management
Kundli, India

Laxman Naik
Dairy Chemistry Division
ICAR-National Dairy Research
  Institute
Karnal, India

Carla Oliveira
Institute for Biotechnology and
  Bioengineering
Centre of Biological
  Engineering
Universidade do Minho
Braga, Portugal

Arthur C. Ouwehand
Active Nutrition
DuPont Nutrition & Health
Kantvik, Finland

Sarang Dilip Pophaly
Department of Dairy
  Microbiology
College of Dairy Technology
Raipur, India

Mamatha Potu
Institute of Biological and
 Environmental and Rural
 Sciences
Aberystwyth University
Aberystwyth, Wales,
 United Kingdom

Jashbhai B. Prajapati
Department of Dairy
 Microbiology
SMC College of Dairy Science
Anand Agricultural University
Anand, India

Monica Puniya
Dairy Microbiology Division
ICAR-National Dairy Research
 Institute
Karnal, India

Ritika Puri
Cheese and Fermented Foods
 Laboratory
Dairy Technology Division
ICAR-National Dairy Research
 Institute
Karnal, India

H.V. Raghu
Dairy Microbiology Division
ICAR-National Dairy Research
 Institute
Karnal, India

P. Narender Raju
Dairy Technology Division
ICAR-National Dairy Research
 Institute
Karnal, India

Chand Ram
Dairy Microbiology Division
ICAR-National Dairy Research
 Institute
Karnal, India

Sreenivas Rao Ravella
Institute of Biological and
 Environmental and Rural
 Sciences
Aberystwyth University
Aberystwyth, Wales,
 United Kingdom

Velugoti Padmanabha Reddy
College of Dairy
 Technology
Sri Venkateswara Veterinary
 University
Tirupati, India

K.P.S. Sangu
Department of Dairy Science
 and Technology
Janta Vedic College
Baraut, India

Chalat Santivarangkna
Institute of Nutrition
Mahidol University
Nakhon Pathom, Thailand

Rosane Freitas Schwan
Department of Biology
Federal University of Lavras
Lavras, Brazil

Suja Senan
Department of Dairy
  Microbiology
SMC College of Dairy
  Science
Anand Agricultural
  University
Anand, India

Bhuvnesh Shrivastava
Dairy Microbiology Division
ICAR-National Dairy Research
  Institute
Karnal, India

Ashish Kumar Singh
Dairy Technology Division
ICAR-National Dairy Research
  Institute
Karnal, India

Rameshwar Singh
Directorate of Knowledge
  Management in
  Agriculture
Indian Council of Agricultural
  Research
New Delhi, India

Ingrid Suryanti Surono
Food Technology
  Department
Bina Nusantara University,
  Alam Sutera Campus
Tangerang, Indonesia

Nimisha Tehri
Dairy Microbiology Division
ICAR-National Dairy Research
  Institute
Karnal, India

José A. Teixeira
Institute for Biotechnology and
  Bioengineering
Centre of Biological
  Engineering
Universidade do Minho
Braga, Portugal

Julia Tennilä
Active Nutrition, DuPont
  Nutrition & Health
Kantvik, Finland

Sudhir Kumar Tomar
Dairy Microbiology Division
ICAR-National Dairy Research
  Institute
Karnal, India

Ashish Tyagi
Rumen Biotechnology Unit
DCN Division
ICAR-National Dairy Research
  Institute
Karnal, India

Sreeja V.
Department of Dairy Microbiology
Sheth MC College of Dairy Science
Anand Agricultural University
Anand, India

M.C. Varadaraj
Microbiology and Fermentation Technology
CSIR-Central Food Technological Research Institute
Mysuru, India

S.V.N. Vijayendra
Microbiology and Fermentation Technology
CSIR-Central Food Technological Research Institute
Mysuru, India

David Walker
Institute of Biological and Environmental and Rural Sciences
Aberystwyth University
Aberystwyth, Wales, United Kingdom

Sheng-Yao Wang
Experimental Farm
National Taiwan University
Taipei City, Taiwan

Hariom Yadav
National Agri-Food Biotechnology Institute
Mohali, India

and

National Institutes of Health
National Institute of Diabetes and Digestive and Kidney Diseases
Bethesda, Maryland

PART I
Introduction, Scope, and Significance

1
FERMENTED MILK AND DAIRY PRODUCTS

An Overview

ANIL KUMAR PUNIYA, SANJEEV KUMAR, MONICA PUNIYA, AND RAVINDER MALIK

Contents

1.1	Introduction	3
1.2	Milk and Its Varieties	4
1.3	Fermented Milk and Characteristics	5
	1.3.1 Starter Cultures	6
	1.3.2 Production of Fermented Milk and Dairy Products	6
	1.3.3 Approaches for the Development of Fermented Dairy Foods	8
1.4	Diversified Fermented Dairy Products	10
1.5	Health Benefits of Fermented Dairy Products	17
1.6	Associated Challenges	18
1.7	Prospects and Newer Approaches	20
1.8	Conclusions	20
References		20

1.1 Introduction

The food industry is the second largest sector after automobiles, and its market is increasing quite dynamically. This sector is catering to the needs and perceptions of the modern society worldwide. The fortunes of the food industry mainly depend on the innovations in foods by adopting the changes in processing parameters, using more efficient machineries, and making novel products as per consumer demands (Tamime 2006; Panesar 2011; Shiby and Mishra 2013; Kaur et al. 2014). With the advancements in technologies, people are shifting fast from traditional foods to therapeutic and nutritional foodstuffs. However, foods with therapeutic benefits for human health have a long history in different cultures for the past more than 2000 years. In recent times, the scientific validations of therapeutic effects of these foods have dramatically changed the spectrum of production and processing of both. Nowadays, foods that claim a number of health benefits are present in the market.

Therapeutic attributes in foods are mainly imparted by fermentation that causes physicochemical changes in the food matrix, which make it usually more nutritious and healthy. During fermentation, the inoculated microbes liberate different health-promoting metabolites and also break down the complex constituents of food (i.e., carbohydrates, proteins, and lipids) into simpler ones that could be bioactive, are easy to digest, and fulfill nutritional aspects (Shiby and Mishra 2013). Fermented foods originated from the dairy, meat, beverage, cereal, and bakery are quite famous throughout the world, and each geographical region has its own specific food, for example, miso in Japan, douche and sufu in China, temph in Indonesia, idli and dosa in India, and yogurt and cheese in Europe and the United States.

Among several fermented foods, dairy products are very popular throughout the world; for example, yogurt is famous in most of the countries. Cheese, leben, viili, kumiss, and kefir, made directly by fermentation of milk, are common in Arabia. The churpi variety of cheese in the Himalayan region, shubat in Kazakhstan, and gariss in Sudan are famous dairy products. Indian fermented milk products such as mistidoi, lassi, dahi, and shrikhand are largely consumed in this subcontinent.

Dairy products are very popular due to their simple formulation, high nutritional value, and therapeutic attributes. Milk and fermented dairy products are important part of traditional diet, which provide sufficient nutrients and modulate various physiological functions (Zubillaga et al. 2001; Tamime 2006; Granato et al. 2010; Panesar 2011). All these attributes make researchers to innovate and design new dairy foods. This chapter is an updated view of the importance of fermented dairy products and futuristic approaches of newer products along with challenges associated.

1.2 Milk and Its Varieties

Milk drawn from the mammary glands of mammals (i.e., goat, buffalo, cattle, sheep, camel) is the primary source of nutrition available for infants before they are able to digest other feeds (Pehrsson et al. 2000). Milk is also considered as a whole food for adults, as it contains almost all essential substances like minerals, vitamins, and easily digestible proteins with balanced amino acids, for human nutrition (Drewnowski 2005; Miller et al. 2007). This nutrition support is not only for newborns or adults but also for microbes (Ebringer et al. 2008).

Energy, water, carbohydrate, fat, protein, vitamins, minerals, and minor biological proteins and enzymes are the major bioactive milk components, which make milk a wholesome food. It contains nearly 87% water; hence, it is a good source of water in the diet. In milk, carbohydrates (i.e., lactose, glucose, and galactose) are the primary source of energy. Milk also contains important fatty acids, namely, linoleic and linolenic acids that cannot be synthesized in body and must be produced (Nagpal et al. 2007a; Puniya et al. 2009) via fermentation or supplemented in diet. It is a rich source of casein and whey proteins. All amino acids that are essential to the body are provided by milk (i.e., vitamins A, D, E, K, riboflavin, niacin, pantothenic acid, and folate). Minerals such as calcium, copper, iron, manganese, magnesium, phosphorus, sodium, and zinc that have different roles in the body including enzyme functions, bone formation, water balance maintenance, and oxygen transport are also found in milk. Other minor proteins and enzymes in milk that are of nutritional interest include lactoferrin and lactoperoxidase (Davoodi et al. 2013).

1.3 Fermented Milk and Characteristics

Fermented milks are produced by the addition of bacterial culture(s) in raw or heat treated milk. In olden times, the fermentation of milk was carried out to preserve or increase the shelf life of the perishable milk. Additionally, the fermentation of milk improves taste and enhances the digestibility of the milk. A portion of the previous-day fermented product was generally used as fresh starter to inoculate the fresh milk, but nowadays well-defined starters, for example, lactobacilli, lactococci, and streptococci, are normally used for the manufacture of fermented products that have excellent nutritional and flavoring characteristics (Tamime 2006; Panesar 2011).

Moreover, fermented milk and its products consist of a number of functional properties, for example, preservation, flavor enhancement, texture enhancement, low caloric content, emulsification, foaming, and nutritional benefits. The fermented milk products used in different countries may be classified into moderately sour with pleasant aroma (i.e., cultured milk), sour and very high sour (i.e., curd and yogurt), and acid-cum alcohol in addition to lactic acid (i.e., kumiss and kefir).

Some of the peculiar characteristics of certain fermented milk products are due to their fermentation profiles that make them more appetizing, and this feature is prevalent if the product is properly stored. These special characteristics may be due to low water activity, production of organic acids, and other chemical compounds during fermentation that are inhibitory to spoilage microflora. Other microbes that produce desirable organoleptic changes are added as starter cultures, which produce organic acids and serve as preservative agents. Sometimes yeasts and molds also participate along with lactic acid bacteria (LAB) (Jan et al. 2002; Tamime 2006; Panesar 2011).

1.3.1 Starter Cultures

Starter cultures are group of microorganisms that are used to produce the fermented milk or dairy product. The use of starters has been positive with respect to the quality of the product, but it has diminished the diversity of fermented dairy products on a global scale. As dairy industry is keen to explore newer possibilities for enhancing the product diversity, there is a growing interest nowadays to look for the potential starter microorganisms that existed during raw milk fermentation. Many new strains have been identified, and their functional characteristics have been explored with reference to a particular disease (Ouwehand et al. 1998; Tamime 2006; Panesar 2011; Downey 2014). Starter cultures form a large group of microorganisms that include bacteria, yeasts, and molds, for example, lactococci, leuconostoc, streptococci, lactobacilli, pediococci, bifidobacteria, brevibacteria, propoinibacteria, and enterococci (Table 1.1).

1.3.2 Production of Fermented Milk and Dairy Products

The details like type of milk, methods of preparation, and storage of fermented milk products are well documented (Marshall and Cole 1985;

Table 1.1 Microorganism Commonly Used as Starter Cultures in Fermented Dairy Products

BACTERIA	YEASTS	MOLDS
• *B. adolescentis, B. brevis, B. bifidum, B. infantis, B. lactis, B. longum* • *Brevibacterium* (Bre.) *linens, Bre. casei* • *Lc. lactis* subsp. *lactis, Lc. lactis* subsp. *cremoris, Lc. lactis* subsp. *lactis* biovar. *diacetylactis* • *Leu. mesenteroides* subsp. *dextranicum, Leu. mesenteroides* subsp. *mesenteroides, Leu. cremoris* • *Lb. delbrueckii* subsp. *bulgaricus, Lb. reuteri, Lb. casei, Lb. fermentum, Lb. plantarum, Lb. helveticus, Lb. acidophilus, Lb. paracasei, Lb. rhamnosus* • *Propoinibacterium* (P.) *freudenreichii* subsp. *freudenreichii, P. freudenreichii* subsp. *shermanii* • *Str. thermophilus*	• *Candida kefir* • *Kluyveromyces marxianus* • *Saccharomyces cerevisiae*	• *Aspergillus oryzae* • *Geotrichum candidum* • *Mucor rasmusen* • *Penicillium roquefortii, P. camemberti*

Tamime and Marshall 1997; Rati Rao et al. 2006; Tamime 2006; Panesar 2011). Such processes involved in all the fermented products is more or less same having some common ingredients like milk and food supplements (i.e., milk solids, sugar and/or stabilizer, colorants, and flavors added after fermentation). Sterilized fruit pulp, whole fruit crunches, nuts, chocolate, herbs, and spices may also be used as additive along with starter cultures.

The milk is pasteurized and transferred to a hygienically clean fermentation or mixing vessel. The temperature is adjusted as per optimal growth temperature (e.g., mesophilic cultures: 24°C–32°C and thermophilic cultures: 35°C–43°C). The appropriate culture mass (either liquid or freeze-dried) is added immediately upon approaching the fermentation temperature and mixed thoroughly. The final fermentation mix containing starter cultures, fermentation substrates, and other ingredients is then transferred to the main container to initiate fermentation. For batch-fermented products, fermentation must be completed prior to the addition of flavors, colorants, and solid foodstuffs. These additives are added only after the fermented product is cooled. The temperature and acid development should be controlled thoroughly to prevent overacidification. Similar trend is to be followed for other flavors or structural contributions. The cooling must be applied to the product before the desired acidity is attained. The products should not be disturbed in the vessel before substantially cooling, so as to prevent the rupturing of coagulum and whey separation. Once the product reached desired cooling, it should only then be stirred and pumped via a mixer to the packaging machine. The packaged products should be stored under refrigeration until distributed (Marshall and Cole 1985; Tamime and Marshall 1997; Rati Rao et al. 2006).

1.3.3 Approaches for the Development of Fermented Dairy Foods

- *Probiotics*: The probiotics (live microorganisms that, when administered in adequate amounts, confer a health benefit on the host) concept was introduced in the early twentieth century. Generally, these are LAB, which are used extensively in the therapeutic preparations and added to foods (Nagpal et al. 2007b; Granato et al. 2010; Aggarwal et al. 2013).

    Lactobacilli and bifidobacterial species possess potential probiotic attributes and are now well-established probiotics (e.g., *Lactobacillus* [*Lb.*] *acidophilus*, *Lb. delbrueckii* subsp. *bulgaricus*, *Lb. lactis*, *Lb. casei*, *Lb. plantarum*, *Lb. rhamnosus*, *Lb. reuteri*, *Lb. paracasei*, *Lb. fermentum*, *Lb. helveticus*, *Bifobacterium* [*B.*] *adolescentis*, *B. longum*, *B. breve*, *B. bifidus*, *B. lactis*, *B. essensis*, *B. infantis*, and *B. laterosporus*). The beneficial effects of probiotics consumption include improvement of intestinal health by the regulation of microbiota, stimulation and development of the immune system, synthesizing and enhancing the bioavailability of nutrients, reducing symptoms of lactose intolerance, and reducing the risk of certain other diseases (Kumar et al. 2011; Nagpal et al. 2012; Garsa et al. 2014).

    Some of probiotic dairy products are present in market produced by several companies, namely, Procter & Gamble, Lallemand, Valio Dairy, Mother Dairy, Amul, Nestle, Britannia, Probi AB, Biocodex, Ganeden Biotech, and Yakult

Danone. Mother Dairy dahi and lassi are available with brand name B-actin. Amul has also entered into the probiotics segment with the introduction of ice cream and dahi. Nestle has launched Active Plus and Yakult Danone, a probiotic drink named Yakult.

- *Prebiotics*: "Non-digestible food ingredients (i.e. prebiotics) that beneficially affects the host by selectively stimulating the growth and/or activity of one or a limited number of bacteria in colon that can improve the host health" (Gibson and Roberfroid 1995, pp. 1401–1412). By using prebiotics, the selective growth of certain indigenous gut bacteria can be improved, thereby any viability problems of orally administered bacteria in upper gastrointestinal tract can be solved. The commonly used prebiotics in dairy foods are fructo-oligosaccharides, xylo-oligosaccharides, and lactose derivatives such as lactulose, lactitol, galacto-oligosaccharides, and soya bean oligosaccharides (Wilder-Smith et al. 2013).

- *Synbiotics*: Food and dairy products that are prepared using probiotics in combination with prebiotics are termed *synbiotics* (Gibson and Roberfroid 1995). It is believed that consumption of synbiotics has greater beneficial effects on the host than probiotics or prebiotics alone (Gmeiner et al. 2000; Kurien et al. 2005; Mandal et al. 2006, 2013). The efficient implantation of probiotics in colonic microbiota is favored by synbiotics, because the prebiotics have a stimulating effect on the growth and activities of exo- and endogenous bacteria.

- *Bioactive peptides*: These are hydrolyzed specific amino acid sequences from the native protein that exert a positive physiological influence on the human body. These are inert within the native protein, but once transformed to active form by microbial or added enzymes and gastrointestinal enzymes during digestion, these exert their beneficial impacts. Fermented dairy products are potential sources of bioactive peptides. An industrial-scale production of biologically active peptides can be achieved by the controlled hydrolysis of milk protein precursor, followed by membrane separation processes (Bargeman et al. 2002; Korhonen and Pihlanto 2003, 2006).

1.4 Diversified Fermented Dairy Products

There is an increasing trend in the consumption of fermented dairy products world over (Table 1.2); some of them are described here.

- *Cheese*: It is one of the most popular fermented dairy products with wide range of applications and is consumed in almost every part of the world. It is produced by milk coagulation, caused by lactic acid and enzyme rennet. The coagulation of casein takes place due to increased lactic acid that leads to drop in pH. If rennet is not supplemented, it will yield soft cheese such as cottage or cream cheeses. Yeasts, molds, and bacteria are all involved in the processes of manufacturing and ripening of different cheeses (Tamime and Marshall 1997).
- *Camel cheese*: It is a fermented product prepared from camel milk (El Zubeir and Jabreel 2008). The low lactose content of this cheese makes it suitable for lactose-intolerant patients. It is high in vitamin C (Farah et al. 1992) and low in cholesterol.

Table 1.2 Important Fermented Dairy Products Along with the Associated Microorganisms

PRODUCT	ORIGIN OF MILK	FERMENTATIVE MICROORGANISM(S)
Acidophilus milk	Cow	*Lb. acidophilus*
Bulgarian buttermilk	Cow	*Lb. delbrueckii* subsp. *bulgaricus*
Cheese	Cow, buffalo, goat, or sheep	*Lc. lactis* subsp. *lactis*, *Lc. lactis* subsp. *cremoris*, *Lb. lactis* subsp. *diacetylactis*, *Str. thermophilus*, *Lb. delbrueckii* subsp. *bulgaricus*, *Priopionibacterium shermanii*, *Penicillium roqueforti*, etc.
Cultured buttermilk	Buffalo or cow	*Str. lactis* subsp. *diacetylactis*, *Str. Cremoris*
Curd	Buffalo or cow	*Lc. lactis* subsp. *Lactis*, *Lb. delbrueckii* subsp. *bulgaricus*, *Lb. plantarum*, *Str. lactis*, *Str. thermophilus*, *S. cremoris*
Kefir	Sheep, cow, goat, or mixed	*Str. lactis*, *Leuconostoc* sp., *Saccharomyces kefir*, *Torula kefir*, *Micrococci*
Kumiss	Mare, camel, or ass	*Lb. acidophilus*, *Lb. delbueckii* subsp. *bulgaricus*, *Saccharomyces*, *Micrococci*
Lassi	Buffalo or cow	*Lb. delbueckii* subsp. *bulgaricus*
Leben	Goat or sheep	*Str. lactis*, *S. thermophilus*, *Lb. delbueckii* subsp. *bulgaricus*, lactose fermenting yeast
Shrikhand	Buffalo or cow	*Str. thermophilus*, *Lb. delbueckii* subsp. *bulgaricus*
Yogurt	Cow	*Lb. acidophilus*, *Str. thermophilus*, *Lb. delbueckii* subsp. *bulgaricus*

Lactobacilli are the dominant microflora that perform lactic fermentation. Among lactobacilli, *Lb. delbrueckii* and *Lb. fermentum* are dominant species followed by *Lb. plantarum* and *Lb. casei*.

- *Gariss*: It is a fermented camel milk product of Sudan and is a full-cream sour milk product prepared by a starter mixture of LAB and yeasts in two leather bags of tanned goat skin. This is produced by semicontinuous or fed-batch fermentation. Usually, a small increment of fresh milk is added, and afterward, the same volume is withdrawn for utilization by consumers.
- *Misti dahi or mistidoi*: It is a sweetened variety of fermented milk product popular in the eastern part of India. Traditionally, it was prepared by the culture of undefined nature and now by the mixed cultures of *Streptococcus* (*Str.*) *lactis*, *Str. diacetylactis*, *Str. cremoris*, and *Leuconostoc* ssp. at the rate of 1% for commercial production. *Str. thermophilus* and *Lb. delbueckii* subsp. *bulgaricus* in the ratio of 3% are used for the production of misti dahi–based functional food, called misti yogurt, and it can be served as a dessert or snack. It is made from buffalo milk or a combination of buffalo and cow milk. Misti dahi has high sugar (13%–19%) and fat (6%–12%) contents. It is prepared by heating milk with sugarcane juice at 60°C–70°C for 6–7 h in an open vessel. Then the cooling starter is added into the cool brownish milk and incubated for lactic fermentation. The final product will be in cream to light brown color due to caramelization of sugar with firm consistency, smooth texture, and pleasant aroma and is stored at low temperature for enhanced shelf life.
- *Shmen*: It is a bright white-colored Algerian butter-type product made from the sour camel milk. Moisture and fat contents are 34%–35% and 49%–56%, respectively (Maurad and Meriem 2008). The bright white color of Shmen is due to the high amount of non-fat components like proteins linked to the fat globules.
- *Shubat*: Also known as *chal*, it is a fermented camel milk product popular in Kazakhstan and Turkmenistan. *Agaran* is fermented cream on the surface of chal. The mixed starter cultures such as *Lb. casei*, *Str. thermophilus*, and lactose-fermenting yeasts

are used for its production. The starter culture is inoculated into camel milk and incubated for 8 h at 25°C followed by 16 h at 20°C. Shubat has digestive as well as therapeutic properties, besides having a high content of vitamins B_1, B_2, and C.

- *Suusac*: It is a traditional Kenyan fermented milk product made from camel milk. Lactate fermentation carried out at 26°C–29°C results in white-colored final product after 1–2 days with distinct smoky flavor and astringent taste (Lore et al. 2005). Two different methods may be used in the preparation of suusac: first is traditional and called *home method*, where fresh camel milk is required, and second is *modified suusac*, where camel milk is heated to 85°C for 30 min and then cooled to 22°C–25°C, and fermentation starts with mesophilic starters at 27°C–30°C for 24 h (Shori 2012).
- *Yogurt*: Similar to cheese, yogurt is another very popular dairy product of modern times. It is produced from whole or skim milk of any origin (i.e., cow, buffalo, goat, sheep, yak). Milk is first standardized and pasteurized followed by the inoculation of starter cultures (*Lb. bulgaricus* and *Str. thermophilus* act symbiotically during fermentation) under standard batch fermentation. These cultures ferment lactose in milk to lactic acid, causing milk to curdle and form yogurt. It may be supplemented with fruits to enhance the flavor. The health-promoting attributes of yogurt containing live and active cultures are well documented (Adolfsson et al. 2004; Patel and Walker 2004; El-Abbadi et al. 2014).
- *Yak milk products*:
  - Qula is a Tibetan yak cheese product for more than 1000 years. It is a type of cheese prepared in the summer season by the addition of yogurt to yak milk. Interestingly, qula can be kept for more than 2–3 years at room temperature without any contamination or degradation; final product is grainy, hard, and yellowish in color.
  - Churpi cheese is a traditional yak milk product of Himalayan region of Darjeeling hills, Sikkim, Ladakh, Nepal, Bhutan, and Tibet. During its preparation, milk is churned in a large wooden drum called *shoptu*. Cheese

blocks are brined and aged for 4–5 months. There are two varieties of churpi: (1) soft variety and (2) hard variety, which is commonly eaten as masticator. Hard variety is made from yak milk, and it is of two types (i.e., chhurpi and dudh chhurpi).
- Churkham is a traditional cheese product formed during the preparation of churpi. Churkham cakes are aged for 2–12 months, and during this fermentation there is an improvement in taste.
- Kurut is a traditional yak milk-fermented product of Qinghai and Tibet of China. Like kefir and koumiss, it is an alcoholic fermented milk product. The fermentation of yak milk occurs in a big container at 10°C–20°C for 7–8 days. The final product is of white color and has alcoholic and acidic sensation. *Lactococcus* (*Lc.*) *lactis* subsp. *lactis*, *Lb. helveticus*, *Str. thermophilus*, *Lb. delbrueckii* subsp. *bulgaricus*, and *Acetobacter* and many other bacteria are reported from kurut by culture-independent techniques.
- *Blaand*: It is a traditional Scottish drink made from the fermented whey with quite an alcoholic fermentation. Whey is then traditionally poured into an oak cask similar to wine and allowed for alcohol fermentation. The fermentation must be monitored closely to avoid vinegar conversion instead of blaand.
- *Dahi*: This is a nutritious fermented milk product also known as curd or Indian yogurt. Dahi is made by lactic fermentation of milk by mixed culture of LAB. A number of mixed species of *Streptococci*, *Lactobacilli*, *Lactococcus*, and *Leuconostoc* have been reported from dahi. It strengthens immune system, improves digestion system, provides benefit during stomach problems and dysentery, and also benefits in osteoporosis and vaginal infections. Probiotic dahi is prepared by the fermentation of milk with probiotics. Dahi could be considered as a functional food, as it possesses many nutritional and therapeutic functions (Sarkar et al. 2008).
- *Kefir*: It is a fermented dairy product made by the combination of lactic acid and alcohol fermentation. Kefir originated from the Caucasus region, its grains are rich in nutrients including

folic acid, and it is linked to the suppression of high blood pressure. Kefir is a cultured milk product of spontaneous fermentation resulting from the introduction of kefir grains into raw milk. Kefir grains are small, spongy, symbiotic colonies of beneficial yeasts and bacteria with an appearance that resembles cottage cheese. Kefir is an exotica sour dairy cultured beverage with little effervescence. The number of lactic acid producers in finished product should not be less than 10^7 CFU/g and yeasts should not be less than 10^4 CFU/g.

- *Koumiss*: It is a fermented dairy product made from the mare's milk originated in nomadic tribes of Central Asia. It is produced from a liquid starter culture that may comprise of *Lactobacillus, Lactococcus, Streptococcus, Leuconostoc,* and *Pediococcus*. To produce koumiss, milk is heated at 90°C–92°C for 5 min and cooled to 26°C–28°C. Then, starter is added and allowed to ferment. After fermentation, it is then cooled to 20°C, while stirring for 1–2 h. The product is bottled and allowed to ripen at 6°C–8°C for 1–3 days. Apart from the nutritional aspects, koumiss has various therapeutic attributes.

- *Matsoni*: It is a cultured dairy food attributed to Bulgaria, Georgia, Armenia, or Russia. Starter cultures have unique LAB, which produce mildly sour yogurt with a syrupy consistency. It can be defined as a milk product of mesophilic fermentation, as it requires ambient temperature after mixing with starters for 2 days or less.

- *Piima*: This is a cultured dairy food of Scandinavian origin. It has a sour flavor with the understated shades of a mild cheese. Unlike other cultured dairy foods, piima has thin uniformity, which is quite much similar to buttermilk. Piima is cultured at room temperature, which makes it an easy and cost-effective process. Only thing required is starter culture beneficial microbiota to inoculate the milk.

- *Quark*: This fermented dairy product is prepared by the inoculation of milk with *Lactobacillus* cultures or in combination of *Lactococcus* cultures and *Str. thermophilus* by the methods of acid, acid-rennet, or thermal-acid coagulation of proteins following with the removal of whey by pressing. The number of

LAB in finished product should not be less than 10^5 CFU/g and protein content not less 14.0%.

- *Ryazhenka*: This is a Russian fermented dairy product produced by the inoculation of *Str. thermophilus* in baked milk. Its number in the finished product should not be less than 10^7 CFU/g.
- *Shrikhand*: It is a semisoft, sweetish sour, whole-milk lactic fermented product prepared from buffalo milk because it is rich in solid not fat and calcium. Shrikhand is manufactured by the fermentation of milk using lactic starters followed by draining of whey of curd. The drained intermediate portion known as *chakka* is later blended with sugar (Kulkarni et al. 2006). Attempts have been made to improve the nutritive and sensory characteristics of shrikhand by adding fruit pulp and other flavoring agents in chakka. Starters contribute the flavor and acidity to the finished product (Nadaf et al. 2012). Shrikhand has high sugar content (33%–39%), which is important to enhance the shelf life, but it has critical health issues due to high sugar content. The shelf life of shrikhand can be enhanced up to 15 days at 35°C and >70 days at 8°C–10°C by heating it at 70°C for 5 min.
- *Viili*: It is a cultured dairy food of Scandinavian origin. It initially hails from Sweden but is now popular in Finland. Yeast and LAB provide a unique ropy, slimy, and sticky texture. It is mildly sour compared to other cultured dairy products making it a good option for children, who are interested in naturally soured probiotic foods. Viili is mainly served with a gooseberry jam or any other sweetener jam jelly.
- *Dairy drinks*: Fermented dairy beverages are products of lactic fermentation of milk by LAB. These can be grouped into nonalcoholic to mild alcoholic products. The nonalcoholic products include lassi, buttermilk, acidophilus milk, sweet acidophilus milk, leben, ayran, chal, and calpis.
  - Lassi or buttermilk is a traditional fermented milk beverage derived by churning of curd and is quite popular in the northern region of India. Lassi is a white to creamy white, viscous sweetish liquid with a rich aroma and has mild to high acidic taste. Conversely, cultured lassi is a

modified product made by blending of curd with water, salt, and spices or sugar (Nair and Thompkinson 2008). To produce lassi, milk is inoculated with lactic acid starter culture and incubated till the formation of curd. The starter cultures used for lassi production consist of mixed cultures of *Lb. acidophilus* and *Str. thermophilus* (Patidar and Prajapati 1998). Use of exo-polysaccharide-producing lactococci as starters provides additional benefits to the final product like controlled whey separation, improved viscosity, flavor, consistency, color, and appearance of lassi (Behare et al. 2010). Variants are generated by blending with fruits, honey, and aloe vera (George et al. 2010, 2012). Buttermilk is sour in taste due to high content of lactic acid produced by LAB fermentation that converts milk sugar into lactic acid by its metabolism. Buttermilk is primarily consumed as a beverage, but it is also used for cooking purposes, especially in bakery items (O'Connell and Fox 2000).

- Fruit lassi is prepared by blending of fruit pulp such as pineapple, mango, and banana pulp with curd or yogurt (Shuwu et al. 2011). Mango lassi is prepared by the addition of mango pulp (7%–8%) and water to yogurt. Pineapple lassi is mildly more acidic as compared to mango and banana lassi.
- Honey lassi is prepared by blending honey syrup with yogurt. Honey is a natural sweetener and contains primarily fructose (38.5%) and glucose (31.3%); blending of honey syrup causes sweetened flavor of lassi. The combined optimum sweetness, honey flavor, and mild acidic taste contribute to the flavor of honey lassi (Shuwu et al. 2011).
- Leben is prepared using LAB; that is, *Lactococcus* species, *Lb. acidophilus*, *Bifidobacteria*, and *Str. thermophilus* are responsible for lactic fermentation of milk (Chammas et al. 2006). Starter cultures contribute to low pH, typical flavor, and texture of the final product. Antimicrobial metabolites produced by the starters enhance the shelf life of the product. The production of traditional leben

involves inoculation of fresh milk with leben of previous batch and fermented up to 18 h (Samet-Bali et al. 2012).
- Ymer is a Danish product with high protein content (5%–6%) and pleasant acidic flavor. It was developed and registered as a sour milk product. Whole milk or pasteurized skim milk is fermented by starter cultures of LAB. Starter cultures used for milk fermentation are mesophilic, generally *Lc. lactis* subsp. *lactis* var. *diacetylactis* and *Leuconostoc* (*Leu.*) *mesenteroides* subsp. *cremoris*. Milk is inoculated with starter culture and incubated at 18°C–20°C until curdling. Starters convert milk lactose into lactic acid and drop the pH to make the product sour. Coagulum attained after fermentation is stirred and heated to remove whey. Cream is added to adjust the fat content about 3.5% before cooling and packaging.

1.5 Health Benefits of Fermented Dairy Products

Fermented milk products are widely distributed and used worldwide (Li et al. 2012). These are modified foods generated by the action of microbes or their enzymes to attain desirable biochemical changes. Dairy products provide a strong foundation for health-promoting innovative ingredients toward the development of functional foods and dietary supplements (Michaelidou and Steijns 2006; Steijns 2008).

Fermented foods are potent detoxifiers, capable of drawing out a wide range of heavy metals from the body. Additionally, fermented foods supplemented with probiotics are rich in essential nutrients like vitamins B_{12}, B_6, K_2, biotin, protein, essential amino acids, and fatty acids that fulfill the body needs. Many of the probiotics produce wide variety of antimicrobial substances, for instance, lactic acid, acetic acid, formic acid, propionic acid, ethanol, diacetyl, acetaldehyde, reutericycline, reuterin, fatty acids, and bacteriocins that are inhibitory to pathogens (Jain et. al. 2009).

Functional dairy foods are, therefore, recommended as an alternative to boost immune system, especially among children and elders (Parvez et al. 2006; Toma and Pokrotnieks 2006). Further, the consumption of functional dairy products with certain LAB helps to lower the cholesterol in blood (Sanders 1998; Parvez et al. 2006).

Lactose intolerance is common that causes diarrhea, bloating, abdominal pain, and flatulence. Probiotics support by secreting lactase to digest lactose. Strains of *Lb. acidophilus* and *Bifidobacteria* improve digestion of lactose (Zubillaga et al. 2001; Levri et al. 2005). Anticancerous property is another attribute of fermented food consisting probiotics. Probiotics inhibit carcinogens and/or pro-carcinogens, inhibit bacteria that convert pro-carcinogens to carcinogens, increase intestinal acidity to alter microbial activity, and reduce bile acid solubility (Zubillaga et al. 2001; Shiby and Mishra 2013).

Probiotics present in functional dairy products also deconjugate bile acids, which are easily excreted. High excretion of bile salts enhances cholesterol consumption, which ultimately reduces cholesterol. Alternate mechanisms of cholesterol assimilation are reported by interaction with cell walls of probiotics. This mechanism is the incorporation of cholesterol into cellular membranes of probiotics during growth and conversion of cholesterol into coprostanol, which is directly excreted in feces (Zubillaga et al. 2001; Kumar et al. 2012).

Functional fermented dairy foods are well known for their use in diarrheal diseases and management of acute viral and bacterial diarrhea, as well as in controlling of antibiotic-associated diarrhea (Benchimol and Mack 2004; Stefano et al. 2008).

Preliminary evidence indicates that probiotics may play a role in blood pressure control. Hypertensive patients, who consumed fermented dairy products containing *Lb. helveticus* and *Saccharomyces cerevisiae*, experienced reduction in systolic and diastolic blood pressure (Nakamura et al. 1995, 1996). Additionally, probiotics present in fermented milk and foods contribute in individual health such as increase in healthy gut flora that helps strengthen digestion for better elimination of toxic wastes. Mental dysfunction problems and diseases can be cured by addressing the gut and probiotics as best alternate.

1.6 Associated Challenges

Fermented dairy foods are contributing to society in various ways, and the consumption of these foods is rising day by day. Yet, there are a number of challenges associated with these foods such as insufficient quantity and quality of milk and poor shelf life of dairy products.

The growth and commercialization of functional food products is complex, is costly, and needs more skills. Various initiative and inputs are mandatory to meet the consumer demands, technical conditions, and legislative regulatory background.

Although dairy products are part of food since the beginning of the civilization, these have a great scope as functional probiotic products. A majority of the population still do not prefer to consume dairy products. The consumers have ethical and health reasons for avoiding dairy products. Low demand and other hurdles discourage the development of value-added dairy foods. Such problems need to be overcome to enhance the production of these fermented dairy products. Most of the countries are lacking suitable regulatory systems for these functional dairy products. There is a need for a clear regulatory system for the production, sales, certification, and advertisement of functional foods and consistent enforcement. Development of institutional capacity in food research, advisory services for manufacturing, and authorities approving health claims for functional foods is essentially required.

The development and marketing of functional foods requires significant research efforts. This involves identifying of new functional components and assessing their physiological effect, bioavailability to humans, and potential health benefits. There is a tremendous scope to conduct clinical trials to understand the product efficacy in order to gain approval for health-enhancing marketing claims by the companies. Consumer awareness is the key to diet and health, and innovation is required to screen the local biodiversity to uncover potential new food sources. Consumer requirement and market will define what regulatory action and science are needed in manufacturing of these products. Producers, processors, and retailers need to be more attentive to understand the domestic as well as export market demand.

It is important to have intellectual property right protection for newly developed products. The successful marketing for functional foods includes consumer awareness about health issues. In spite of existing diversified raw material for producing functional foods, a sustainable management plan for such resources is important to avoid dramatic reduction in resources and also to maintain the constant supply.

1.7 Prospects and Newer Approaches

Trends for demand of functional food indicate a great potential for probiotic dairy products. There is much scope to devise proposals focused on human clinical validation to prove the efficacy of these probiotics and to strengthen the concept of "health and wellness through foods." This will certainly promote probiotics more effectively. Regulatory sectors should assess the opportunities at local, national, and international levels. Further, studies could establish the production system for newer fermented or processed products.

1.8 Conclusions

Undoubtedly, fermented dairy products have attained a very significant position in the human diet. These are important not only from the nutritional but also from the economic point of view. Research in the area of their role in managing lifestyle diseases explains their role in human nutrition. However, it requires careful selection of probiotics, their optimization, and processing strategies that are involved in formulation. It will ensure their safety for consumption and also the confidence of consumers. Overall, these products should occupy a prominent position, as a regular food in our daily diet. Certainly, these fermented foods containing "domesticated microbes" promise the society long-term health benefits to consumers.

References

Adolfsson, O., Meydani, S.N., and Russell, R.M. 2004. Yogurt and gut function. *American Journal of Clinical Nutrition* 80(2): 245–256.

Aggarwal, J., Swami, G., and Kumar, M. 2013. Probiotics and their effects on metabolic diseases: An update. *Journal of Clinical and Diagnostic Research* 7(1): 173–177.

Bargeman, G., Houwing, J., Recio, I., Koops, G.H., and van der Horst, C. 2002. Electro-membrane filtration for the selective isolation of bioactive peptides from anas 2-casein hydrolysate. *Biotechnology and Bioengineering* 80: 599–609.

Behare, P.V., Singh, R., Tomar, S.K., Nagpal, R., Kumar, M., and Mohania, D. 2010. Effect of exopolysaccharides-producing strains of *Streptococcus thermophilus* on technological attributes of fat free lassi. *Journal of Dairy Science* 93(7): 2874–2879.

Benchimol, E.I. and Mack, D.R. 2004. Probiotics in relapsing and chronic diarrhea. *Journal of Pediatric Hematology/Oncology* 26: 515–517.

Chammas, G.I., Saliba, R., Corrieu, G., and Béal, C. 2006. Characterisation of lactic acid bacteria isolated from fermented milk "laban." *International Journal of Food Microbiology* 110: 52–61.

Davoodi, H., Esmaeili, S., and Mortazavian, A.M. 2013. Effects of milk and milk products consumption on cancer: A review. *Comprehensive Reviews in Food Science and Food Safety* 12: 249–264. doi:10.1111/1541-4337.12011.

Downey, M. May 2014. Probiotics provide vital protection against chronic disease. *Life Extension Magazine* 1–9.

Drewnowski, A. 2005. Concept of a nutritious food: Towards a nutrient density score. *American Journal of Clinical Nutrition* 82: 721–732.

Ebringer, L., Ferencik, M., and Krajcovic, J. 2008. Beneficial health effects of milk and fermented dairy products—Review. *Folia Microbiologica* 53(5): 378–394.

El-Abbadi, N.H., Dao, M.C., and Meydani, S.N. 2014. Yogurt: Role in healthy and active aging. *American Journal of Clinical Nutrition* 99(5): 1263S–1270S.

El Zubeir, I.E.M. and Jabreel, S.O. 2008. Fresh cheese from camel milk coagulated with Camifloc. *International Journal of Dairy Technology* 61(1): 90–95.

Farah, Z., Rettenmayer, R., and Atkins, D. 1992. Vitamin content of camel milk. *International Journal of Vitamin and Nutrition Research* 62: 30–33.

Garsa, A.K., Kumariya, R., Kumar, A., Lather, P., Kapila, S., Sood, S.K., and Kapasiya, M. 2014. In vitro evaluation of the probiotic attributes of two pediococci strains producing pediocin PA-1 with selective potency as compared to nisin. *European Food Research and Technology.* doi:10.1007/s00217-014-2243-7.

George, V., Aurora, S., Sharma, V., Wadhwa, B.K., and Singh, A.K. 2012. Stability, physic chemical, microbial and sensory properties of sweetener/sweetener blends in lassi during storage. *Food and Bioprocess Technology* 5: 323–330.

George, V., Aurora, S., Wadhwa, B.K., Singh, A.K., and Sharma, G.S. 2010. Optimization of sweetener blends for the preparation of lassi. *International Journal of Dairy Technology* 63: 256–261.

Gibson, G.R. and Roberfroid, M.B. 1995. Dietary modulation of the human colonic microbiota: Introducing the concept of prebiotics. *Journal of Nutrition* 125: 1401–1412.

Gmeiner, M., Kneifel, W., Kulbe, K.D., Wouters, R., De Boever, P., Nollet, L., and Verstraete, W. 2000. Influence of a synbiotic mixture consisting of *Lactobacillus acidophilus* 74-2 and a fructooligosaccharide preparation on the microbial ecology sustained in a simulation of the human intestinal microbial ecosystem. *Applied Microbiology and Biotechnology* 53: 219–223.

Granato, D., Branco, G.F., Cruz, A.G., Faria, J.d.A.F., and Shah, N.P. 2010. Probiotic dairy products as functional foods. *Comprehensive Reviews in Food Science and Food Safety* 9: 455–470.

Jain, S., Yadav, H., and Sinha, P.R. 2009. Probiotic dahi containing *Lactobacillus casei* protects against *Salmonella enteritidis* infection and modulates immune response in mice. *Journal of Medicinal Food* 12(3): 576–583.

Jan, T.M.W., Ayad, H.E.E., Hugenholtz, J., and Smit, G. 2002. Microbes from raw milk for fermented dairy products. *International Dairy Journal* 12: 91–109.

Kaur, M., Kumar, H., Kumar, N., and Puniya, A.K. 2014. Recent trends in production of fermented dairy and food products. In: *Dairy and Food Processing Industry Recent Trend*, ed. B.K. Mishra, Part 1, vol 9, 141–161. New Delhi, India: Biotech Books.

Korhonen, H. and Pihlanto, A. 2003. Bioactive peptides: Novel applications for milk proteins. *Applied Biotechnology and Food Science Policy* 1:133–144.

Korhonen, H. and Pihlanto, A. 2006. Bioactive peptides: Production and functionality. *International Dairy Journal* 16: 945–960.

Kulkarni, C., Belsare, N., and Lele, A. 2006. Studies on shrikhand rheology. *Journal of Food Engineering* 74: 169–177.

Kumar, M., Nagpal, R., Kumar, R., Hemalatha, R., Verma, V., Kumar, A., Chakraborty, C. et al. 2012. Cholesterol-lowering probiotics as potential biotherapeutics for metabolic diseases. *Experimental Diabetes Research*. doi:10.1155/2012/902917.

Kumar, M., Verma, V., Nagpal, R., Kumar, A., Behare, P.V., Singh, B., and Aggarwal, P.K. 2011. Anticarcinogenic effect of probiotic fermented milk and chlorophyllin on aflatoxin-B1 induced liver carcinogenesis in rats. *British Journal of Nutrition* 107: 1006–1016.

Kurien, A., Puniya, A.K., and Singh, K. 2005. Selection of prebiotic and *Lactobacillus acidophilus* for synbiotic yoghurt preparation. *Indian Journal of Microbiology* 45: 45–50.

Levri, K.M., Ketvertis, K., Deramo, M., Merenstein, J.H., and D'Amico, F. 2005. Do probiotics reduce adult lactose intolerance? A systematic review. *Journal of Family Practice* 54(7): 613–620.

Li, W., Yang, Z., and Talashek, T. 2012. Blends for fermented milk products. US Patents 20130059032 A1.

Lore, T.A., Mbugua, S.K., and Wangoh, J. 2005. Enumeration and identification of microflora in suusac, a Kenyan traditional fermented camel milk product. *LWT-Food Science and Technology* 38: 125–130.

Mandal, S., Hati, S., Puniya, A.K., Singh, R., and Singh, K. 2013. Development of synbiotic milk chocolate using encapsulated *Lactobacillus casei* NCDC 298. *Journal of Food Processing and Preservation* 37(5): 1031–1037.

Mandal, S., Puniya, A.K., and Singh, K. 2006. Effect of alginate concentrations on survival of microencapsulated *Lactobacillus casei* NCDC-298. *International Dairy Journal* 16: 1190–1195.

Marshall, V.M. and Cole, W.M. 1985. Methods for making kefir and fermented milks based on kefir. *Journal of Dairy Research* 52: 451–456.

Maurad, K. and Meriem, K. 2008. Probiotic characteristics of *Lactobacillus plantarum* strains from traditional butter made from camel milk in arid regions (Sahara) of Algeria. *Grasas Y Aceites* 59: 218–224. doi:10.3989/gya.2008.v59.i3.511.

Michaelidou, A. and Steijns, J. 2006. Nutritional and technological aspects of minor bioactive components in milk and whey: Growth factors, vitamins and nucleotides. *International Dairy Journal* 16: 1421–1426.

Miller, G.D., Jarvis, J.K., and McBean, L.D. 2007. Contribution of dairy foods to health throughout the life cycle. In: *Handbook of Dairy Foods and Nutrition* (3rd edition), eds. G.D. Miller, J.K. Jarvis, and L.D. McBean, 339–399. Boca Raton, FL: CRC Press.

Nadaf, N.Y., Patil, R.S., and Zanzurne, C.H. 2012. Effect of addition of gulkand and rose petal powder on chemical composition and organoleptic properties of Shrikhand. *Recent Research in Science and Technology* 4: 52–55.

Nagpal, R., Behare, P.V., Kumar, M., Mohania, D., Yadav, M., Jain, S., Menon, S. et al. 2012. Milk, milk products, and disease free health: An updated overview. *Critical Reviews in Food Science and Nutrition* 52: 321–333.

Nagpal, R., Yadav, H., Puniya, A.K., Singh, K., Jain, S., and Marotta, F. 2007a. Conjugated linoleic acid: Sources, synthesis and potential health benefits—An overview. *Current Topics in Nutraceutical Research* 5: 55–66.

Nagpal, R., Yadav, H., Puniya, A.K., Singh, K., Jain, S., and Marotta, F. 2007b. Potential of probiotics and prebiotics for synbiotic functional dairy foods: An overview. *International Journal of Probiotics and Prebiotics* 2: 75–84.

Nair, K. and Thompkinson, D. 2008. Optimization of ingredients for the formulation of a direct acidified whey based lassi-like beverage. *International Journal of Dairy Technology* 61: 199–205.

Nakamura, Y., Masuda, O., and Takano, T. 1996. Decrease of tissue angiotensin-1 converting enzyme activity upon feeding sour milk in spontaneously hypertensive rats. *Bioscience, Biotechnology, and Biochemistry* 60:488–489.

Nakamura, Y., Yamamoto, N., Sakal, K., and Takano, T. 1995. Antihypertensive effect of sour milk and peptides isolated from it that are inhibit ORS to angiotensin-1-converting enzyme. *Journal of Dairy Science* 78: 1253–1257.

O'Connell, J.E. and Fox, P.F. 2000. Heat stability of buttermilk. *Journal of Dairy Science* 83(8): 1728–1732.

Ouwehand, A.C. and Salminen, S. 1998. The health effects of cultured milk products with viable and non-viable bacteria. *International Dairy Journal* 8: 749–758.

Panesar, P.S. 2011. Fermented dairy products: Starter cultures and potential nutritional benefits. *Food and Nutrition Sciences* 2:47–51.

Parvez, S., Malik, K.A., Ah Kang, S., and Kim, H.Y. 2006. Probiotics and their fermented food products are beneficial for health. *Journal of Applied Microbiology* 100: 1171–1185.

Patel, D. and Walker, M. 2004. Semi-solid cultured dairy products: Principles and applications. In: *Handbook of Food Products Manufacturing* (2nd edition), eds. Y.H. Hui, R.C. Chandan, S. Clak, N. Cross, J. Dobbs, W.J. Hurst, L.M. Nollet et al., 113–124. New York: John Wiley & Sons, Inc.

Patidar, S.K. and Prajapati, J.B. 1998. Standardisation and evaluation of lassi prepared using *Lactobacillus acidophilus* and *Streptococcus thermophilus*. *Journal of Food Science and Technology* 35: 428–431.

Pehrsson, P.R., Haytowitz, D.B., Holden, J.M., Perry, C.R., and Beckler, D.G. 2000. USDA's National food and nutrient analysis program: Food sampling. *Journal of Food Composition and Analysis* 13(4): 379–389.

Puniya, A.K., Reddy, C.S., Kumar, S., and Singh, K. 2009. Influence of sunflower oil on conjugated linoleic acid production by *Lactobacillus acidophilus* and *Lactobacillus casei*. *Annals of Microbiology* 59: 505–507.

Rati Rao, E., Vijayendra, S.V.N., and Varadaraj, M.C. 2006. Fermentation biotechnology of traditional foods of the Indian subcontinent. In: *Food Biotechnology* (2nd edition), eds. K. Shetty, G. Paliyath, A. Pometto, and R.E. Levin, 1759–1794. Boca Raton, FL: CRC Press/Taylor & Francis.

Samet-Bali, M.A., Ayadi, O., and Attia, H. 2012. Development of fermented milk "Leben" made from spontaneous fermented cow's milk. *African Journal of Biotechnology* 1:1829–1837.

Sanders, M.E. 1998. Overview of functional foods: Emphasis on probiotic bacteria. *International Dairy Journal* 8: 341–347.

Sarkar, S. 2008. Innovations in Indian fermented milk products—A review. *Food Biotechnology* 22:78–97.

Shiby, V.K. and Mishra, H.N. 2013. Fermented milks and milk products as functional foods—A review. *Critical Reviews in Food Science and Nutrition* 53(5):482–496.

Shori, A.B. 2012. Comparative study of chemical composition, isolation and identification of micro-flora in traditional fermented camel milk products: Gariss, Suusac, and Shubat. *Journal of the Saudi Society of Agricultural Sciences* 11: 79–88.

Shuwu, M.P., Ranganna, B., Suresha, K.B., and Veena, R. 2011. Development of value added lassi using honey. *The Mysore Journal of Agricultural Sciences* 45(4): 757–763.

Stefano, G. 2008. Probiotics for children with diarrhea: An update. *Journal of Clinical Gastroenterology* 42: 53–S57.

Steijns, J.M. 2008. Dairy products and health: Focus on their constituents or on the matrix? *International Dairy Journal* 18: 425–435.

Tamime, A.Y. 2006. *Fermented Milks*. Oxford: Blackwell Science Ltd.

Tamime, A.Y. and Marshall, V.M.E. 1997. Microbiology and technology of fermented milk. In: *Microbiology and Biochemistry of Cheese and Fermented Milk* (2nd edition), ed. B.A. Law, 57–152. London: Blackie Academic and Professional.

Toma, M.M. and Pokrotnieks, J. 2006. Probiotics as functional food: Microbiological and medical aspects. *Acta Universitatis Latviensis* 710: 117–129.

Wilder-Smith, C.H., Materna, A., Wermelinger, C., and Schuler, J. 2013. Fructose and lactose intolerance and malabsorption testing: The relationship with symptoms in functional gastrointestinal disorders. *Alimentary Pharmacology and Therapeutics* 37: 1074–1083.

Zubillaga, M., Weill, R., Postaire, E., Goldman, C., Caro, R., and Boccio, J. 2001. Effect of probiotics and functional foods and their use in different diseases. *Nutrition Research* 21: 569–579.

2
Nutritional and Therapeutic Characteristics of Fermented Dairy Products

ARUN MALIK, HARNISH, PRIYANKA CHOPRA, STANZIN ANGMO, DINESH KUMAR DAHIYA, SHALINI JAIN, AND HARIOM YADAV

Contents

2.1	Introduction	26
2.2	Milk Lipids	27
2.3	Milk Sterols	27
2.4	Milk Proteins	28
2.5	Milk Sugars	28
2.6	Milk Salts	29
2.7	Milk Foods with Health-Promoting Properties	29
2.8	Therapeutic Potential of Functional Milk Foods	30
2.9	Health Benefits of Consuming Milk-Based Foods	33
	2.9.1 Improving Lactose Tolerance	34
	2.9.2 Effects on Blood Pressure	35
	2.9.3 Role of Fermented Dairy Products on Heart Health	35
	2.9.4 Dairy Foods and Obesity	36
	2.9.5 Dental Benefits from Dairy Foods	36
	2.9.6 Role in Blood Sugar Management	36
	2.9.7 Antimicrobial Potential	37
	2.9.8 Anticancer Activities	37
	2.9.9 Appetite Suppressors	38
	2.9.10 Anti-HIV Activity	39

2.9.11 Relief against Rotaviral Diarrhea 39
2.9.12 Milk-Derived Bioactive Peptides 40
2.9.13 Conclusions and Future Perspectives 41
References 43

2.1 Introduction

Milk is a complete food for all ages of individuals as it contains ideal balanced nutrients for growth and development. Varieties of milk are marketed under different brand names like cow's milk, sheep milk, camel's milk, and milk obtained from plant sources such as almond, oat, soya, and rice milk (Zivkovic and Barile 2011). The market value of all these kinds of milk is due to their beneficial role in daily lifestyle of humans. Moreover, products derived from milk have abundant source of proteins and vitamins and are generally prepared from low-fat milk. Cheese and yogurt, food products prepared from milk, also contain same beneficial nutrients as present in milk. They are also very rich source of calcium, which is an important mineral and helps to build strong muscles, bones, and teeth (Abrams 2013). For the growth of children, calcium is utmost necessary, which helps to grow strong bones and teeth. Various kinds of milk food products that have high nutritional value are available in the market in a variety of forms like clotted cream, kefir, powdered milk, whole-milk products, buttermilk products, skim milk, whey products, ice cream, high milk fat and nutritional products, khoa (milk tofu), infant formula, ghee (butter), and evaporated and baked milk.

Components of milk are necessary to determine the right balance of nutrients. Milk is about 87% water, and other components of milk are distributed in many forms. Main milk components like lipids, proteins, sugars, and salts are also available (Figure 2.1). The distinctive biomolecules present in milk differ largely on the basis of age and type of breed under study, lactation period, and overall

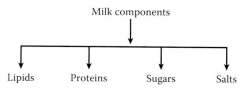

Figure 2.1 Major components of milk.

physiological condition of milch animals. This chapter includes the nutritional composition of dairy products as well as their therapeutic attributes.

2.2 Milk Lipids

Among various biomolecules present in milk, fat is a major key macromolecule and is modified in a number of ways to get the desired final product. Milk fat is mainly composed of complex mixture of lipids. Lipids are triglycerides and secreted as milk fat globules. The combinations of triglycerides constitute the major proportion of fat micelles, about 98%. It also encompasses few vitamins (fat soluble) and pigments like carotenes that signify the yellowish appearance of butter, waxes, and sterols. Milk lipids help mammals for accumulating fat in adipose tissues in the initial days after birth and act as a major reservoir of essential and nonessential fatty acids in diet. No doubt they act as a solubilizing medium for fat-soluble vitamins A, D, E, and K. Milk lipids provide energy to body by the oxidation of fat that yields 9 calories/g. All three types of fatty acids, saturated, mono, and polyunsaturated that differ in nature of bonds present between the carbon atoms, are present in milk fat, and this distribution varies between animals and mammals (Livingstone et al. 2012). On average, the composition of human milk fat ranges from 3% to 5%, while cow milk contains somewhat higher percent of fat (3.6%). In general, higher amounts of palmitic and oleic acids are present in human milk fat than bovine milk fat. Different types of fatty acids such as butyric, caproic, caprylic, capric, lauric, myristic, palmitic, stearic, arichidonic, oleic, linoleic, and linolenic acids are present in milk fat in conjunction with glycerol molecules/moieties (Livingstone et al. 2012).

2.3 Milk Sterols

Sterols are a form of cholesterol, and they constitute the major part of milk. They are the descendants of various steroidal hormones and fat-soluble vitamins. Importantly, they help in the formation of bile salts in the body. Cholesterol is an important sterol and is very crucial for the overall growth and maintenance of young ones. As evident, its deficiency leads to serious neurodegenerative diseases in humans due to lack of its proper supply to nerve tissues for functioning (Koletzko et al. 2011).

2.4 Milk Proteins

Milk contains approximately 3.3% of protein and is considered as complete protein balance required for humans because it comprises of nine essential amino acids in its composition. Important essential amino acids present in milk are a pool of simple acidic (lysine and histidine), hydrophobic (leucine, isoleucine, valine, phenylalanine, tryptophan, methionine), and polar (theronine) amino acids (Kotler et al. 2013). Milk proteins have approximately 82% caseins, which help to maintain proper growth and development of young ones. In addition to casein, all other important biological milk proteins are whey proteins (18%–20% protein in milk), but most of these whey proteins are not digestible in human intestine.

Casein proteins are of four different kinds, namely, alpha S-1, beta S-1, beta, and kappa casein. Whey proteins are of four various types, namely, beta-lactoglobulin, beta-lactalbumins, immunoglobulin, and serum albumin. Milk also contains important enzymes other than caseins. The important enzymes present in milk are lipases, peroxidases, proteases, and phosphates that play significant roles in different catalysis.

2.5 Milk Sugars

The major abundant sugar present in milk is lactose, a disaccharide that is made up of two monomeric units of glucose and galactose. Minute amounts of other oligosaccharides are also present in milk (Rudloff and Kunz 2012). Glucose is considered as a paramount requirement to the mammary gland for milk synthesis and is absorbed through intestine. Also it is the major and readily available source of metabolic energy and for short carbon chains. Moreover, many popular dairy products like yogurt are manufactured only due to the presence of this lactose sugar, which is the main carbon source available for many lactic acid bacteria (LAB) to act, which thereby allows production of lactic acid from lactose (De Vos and Vaughan 1994). Some other carbohydrates are also freely found in milk that include amino sugars, sugar phosphate, neutral and acid oligosaccharides, and nucleotide sugars.

2.6 Milk Salts

Milk salts contribute a small proportion to milk. Depending upon the type of ions, they are diffusible or partially associated with casein molecules in milk to form large colloidal molecules. They are mainly responsible for the buffering capacity of milk, thereby maintaining the pH, ionic strength, and osmotic pressure of milk. As reported, calcium and phosphorus are the major minerals found in milk, which help in the development of strong bones and soft tissues of the body. Other minerals such as iron, zinc, copper, manganese, cobalt, and molybdenum contribute to the overall mineral pool, and upon consumption they will be utilized as enzyme cofactors. The composition of iron in milk is low and that too is bounded to lactoferrin, transferrin, xanthine oxidase, and some of the caseins. In case of cow's milk, the minutely present zinc is found associated with casein, whereas molybdenum is associated to xanthine oxidase, manganese in conjunction with milk fat, and cobalt to vitamin B_{12}.

2.7 Milk Foods with Health-Promoting Properties

Increase in population literacy rate and reach of social media around the world have created a huge awareness in population regarding their health, especially in developing and underdeveloping countries. This is evident by many recent studies flooded in scientific domain, wherein both consumers and medical practitioners at first preferred preventive approach over curative therapy in treating diseases. Several health benefits of dairy foods, especially probiotic products, have been well documented after their scheduled intake. With growing scientifically proven health benefits, these products are nowadays sold under different names like immunobiotics, biopharma foods, designer foods, nutritional foods, super foods, and functional foods (Childs and Poryzees 1998).

2.8 Therapeutic Potential of Functional Milk Foods

Conventionally, the healthiness of a food is measured on the basis of nutritional factors it have such as content of fat, fibers, salts, and vitamins. Besides this traditional healthiness, certain foods have single bioactive components that exert a positive impact on our well-being by their therapeutic value. As described earlier, such food products are called *functional foods*. Japan is a pioneer in the field of functional foods where they were first launched under a specialized category of foods called FOSHU (foods for specific health use) constituted in 1991; its primary goal is to reduce the increased burden of health maintenance costs. In this category, a milk drink fermented by a suitable probiotic culture popularly called *Yakult* was widely available and consumed in Japan since 1935 (Karimi and Pena 2003).

According to Functional Food Science in Europe concerted action project coordinated by ILSI (International Life Sciences Institute), "a food can be regarded as functional if it has been satisfactorily demonstrated to affect beneficially one or more target functions in the body beyond adequate nutritional effects in a way that is relevant to either an improved state of health and well-being and/or a reduction of risk of disease" (ILSI Europe 2002, p. S6). Instead of scientifically improving the various aspects of human and animal health, these products should behave like natural products so that they can be equally liked, picked, and included in one's daily diet. As per Diplock et al. (1999, p. S26) definition, a "functional food must remain food and it must demonstrate its effects in amounts that can normally be expected to be consumed in the diet: it is not a pill or a capsule, but it is a part of the normal food pattern."

Very recent studies have now well proved it that the food we consume in our daily diet is not merely a source of nutrition but full of bioactive ingredients that in one or another manner positively modulate many important functions and keep us healthy in this world of disease. This makes direct relationship with food and added health attributes. Initially, it was thought that these domesticated foods are only liked in some regions of the world, but as this science grows in the past few years and continues its pace, nowadays they are equally accepted in the United States after Japan. We can easily find the involvement of these domesticated tiny bugs in a variety of food products ranged from the conventional to designer medical foods. Dairy-based products are probably the best mode available for the delivery of these bacteria to host in question. However, more research in this area and likely investments of food and pharmaceutical industries in this field bring newer technologies that make it feasible to prepare and supply these probiotics in drug formates (capsules) and in combination with medicinal foods (to enhance their effect) that are stable for longer duration. One can easily find a yogurt, lassi, flavored milk, ice creams, and to some extent cheese varieties prepared with probiotic bacteria—all of which are the routinely consumed dairy products. The basis reason as described earlier is the utilization of milk sugars and other constituents present in milk and the formation of lactic acid as end product in fermentation. Second, they were naturally presented in traditional foods and in their likely natural environment. In fact, most of the probiotics we use today in manufacturing of dairy products are descendants isolated from such foods. Importantly, we consumed these foods from ancient times as our ancestors taught their symbiotic benefits and subsequently got habitual of them (Sanders 2000). These products were consumed for decades due to their presumptive health benefits in some developed countries. This is evident from Childs and Poryzees (1990) studies where they realized that more than 40% of Americans prefer to buy and ingest foods that have an effective role in establishing good health and the prevention of disease.

Generally, these foods are consumed when one is in good health, but some specially designed (spores of lactobacillus) foods are prescribed in special cases, such as to those suffering from gastrointestinal (GI) disorders. To get their intended beneficial effect, these foods must be supplied in sufficient numbers and that too (as believed) not surpassing their natural amounts present in normal diet (Korhonen 2002). On the basis of the types of ingredients (posing health attributes) present, functional foods may be categorized into 1) probiotics based, that are prepared from the combination of probiotics, prebiotics and certain other dietary fibers, and 2) antioxidants, that are fortified with essential vitamins, various polyphenols, sterols mostly obtained from plant, important fatty acids like conjugated linoleic acids (CLAs), and rarer minerals. The primary and major sites for effective function of these functional foods are the GI tracts, and obviously, it is the area where they bring about their maximum effects due to large surface area available to act and direct association with the host mucosal epithelia. In particular, they interfere with the bioavailability of various nutrients (Roberfroid 2000). However, they exert their effect directly or indirectly on different organelles and parts of the system, either by modulating the laid metabolic and physiological processes, through interaction with toll-like receptors, or by producing bioactive components like CLAs and phenolic compounds. These bioactive molecules have many functions: they aid in preventing the progression of cancer by inducing apoptosis in cancerous cells, boost immunity of proliferation of immune cells, and control obesity by acting on satiety hormones and adipose tissue-specific transcriptional factors. Sometimes probiotic influx the supply of macronutrients like amino acids, carbohydrates, and fatty acids. In recent times, dairy-based functional foods in the form of probiotic product are gaining much popularity. Probiotics, more often called *friendly bacteria*, are live microbial strains of one species or a consortium of more than one genera or species that when ingested in standardized colony-forming units altogether improve the health and life quality of an individual. One thing is for sure that commercially prepared and sold probiotic-based foods must have a living probiotic microorganism in amounts that are claimed on their composition leaflet, so that later on after their prescribed consumption, one might feel the suggested advantage (De Vrese and Schrezenmeir 2001). To be considered a microorganism-suitable probiotic candidate, it should have some of the following general criteria:

- It is easily grown, subcultured, and maintained for longer passages under standard laboratory conditions.
- It lacks pathogenic islands, has antibiotic-resistant genes, and more importantly, is barred of proteins responsible for horizontal gene transfer.
- It should have external appendages to make better and everlasting adhesion to host epithelial mucosa.
- It should tolerate high concentrations of acid and biles prevalent inside host GI tract.
- It should have better stability so that it can survive for longer times during processing and storage. Bacterial probiotics

mostly used in the formulation of functional products include (1) species of lactobacilli such as *Lactobacillus johnsonii, Lb. acidophilus, Lb. delbrueckii* subsp. *bulgaricus, Lb. casei, Lb. brevis, Lb. reuteri, Lb. fermentum, Lb. plantarum, Lb. cellobiosus,* and *Lb. curvatus*; (2) species from streptococci, lactococci, and enterococci group such as *Streptococcus salivarius* subsp. *thermophilus, Str. intermedius, Str. diacetylactis, Lactococcus lactis* subsp. *cremoris,* and *Enterococcus faecium*; and (3) important bifidobacteria species include strains of *Bifobacterium infantis, B. animalis, B. adolescentis, B. bifidum,* and *B. longum* (Mercenier et al. 2002). Instead of these bacteria, species of other genera may also be used in probiotics food formulation that include cultures of *Propionibacterium* ssp., unicellular yeast (*Saccharomyces boulardii*), and *Bacillus subtilis* (Jan et al. 2001).

The advancement in probiotic science has led to genesis of more targeted probiotics called *second-generation probiotics/genetically modified microorganisms* by the involvement of modern biotechnology and bioengineering, where the host can be provided with some vital components like interleukins. The earlier discussion makes a clear indication that a probiotic food may be conventional dairy food (cheese, buttermilk, and yogurt) or in the form of nutritional supplements, drugs in the form of tablets, lyophilized capsules, and powders. Of that, some are consumed with a motive to obtain energy and other to solve some specific medicinal purpose (Ross 2000; Temmerman et al. 2002).

As of today, the probiotic products available in market are prepared just from a single bacterial culture (e.g., Yakult, a popular Japanese drink prepared from *Lb. casei* shirota strain), while other products contain a mixture of the same genera (e.g., Bacilac, a Belgium product prepared by adding *Lb. acidophilus* and *Lb. rhamnosus* strains) or a consortium of two different genera (VSL#3, a product from Italy is prepared by adding eight LAB species and is available in market in capsular form).

2.9 Health Benefits of Consuming Milk-Based Foods

The main objective behind milk-based prolonged investigations and deeply designed studies is to check and obtain biotherapeutically active milk components, and other derived dairy/functional foods, that in all manner sum up the life expectancy of population not only from nutrition delivery point of view but also from medicinal and economic points. Here, in this section, the text is emphasized on health-alleviating effects of functional dairy products (Figure 2.2).

2.9.1 Improving Lactose Tolerance

Lactose, generally called *milk sugar*, is hydrolyzed in the small intestine by the catalytic activity of the enzyme lactase. A major proportion of the population mostly in their later phase of life gets devoid of this lactase activity due to many reasons and in turn cannot digest the ingested milk. The intact lactose is later acted on by the microorganisms sited in large intestine and here starts the symptoms of gas formation, uneasiness, bloating, and diarrhea—a condition known as *lactose intolerance*. It may be hereditary that an individual is inborn with lactase deficiency or it develops during due course of his or her life (Heyman 2006).

Several studies had proved that a regular intake of dairy products have reduced this condition to a major extent (Nagpal et al. 2007). This is due to the activity of microbial lactase present in products that utilized most of the available lactose. Second, microbial lactases are

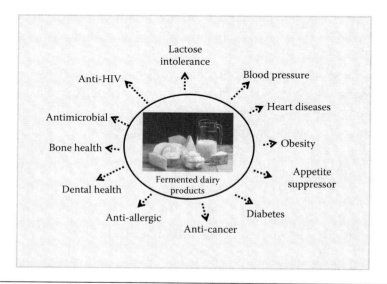

Figure 2.2 Possible health benefits of consumption of fermented dairy products.

added to host pool enzymes that help in digesting lactose. De Vrese et al. (2001) suggested that probiotic bacteria supplied the beta-galactosidase enzyme, as observed from the colonic ileum after taking fermented dairy food and is credited for lactose hydrolysis. Similar reasons were suggested with most of fermented dairy products, mainly after yogurt/dahi (Indian yogurt) consumption. As milk foods are considered as the best source of natural nutrients, there are always chances of building some sort of deficiencies in lactose-intolerant persons. It is recommended that these persons should not take these products at once but instead consume them after regular intervals in prescribed amounts, that is, a half bowl of yogurt in the morning, a little bit of ice cream and milk in the evening, and so on. In fact they should consume milk products in conjunction with other foods rich in lactase activity (Talhouk et al. 1996; Salminen et al. 1999).

2.9.2 Effects on Blood Pressure

Hypertension, a problem of the modern world, is predisposed by many factors with the most important being molecular (genetic) and surrounding environment. Those who are already suffering from abnormal blood pressures are at high risk of getting other lethal diseases like cardiac arrest and brain stroke. Many studies have already well-established the more pronounced positive effect of consuming milk on systolic pressures than calcium can have alone. Consuming milk-derived products on a regular basis provides three important nutrients that have antihypertensive active (calcium, whey-derived peptides, and phosphopeptides of casein) and certain other biological active peptides that are responsible for lowering blood pressure (Patel and Renz-Schanen 1998; Ashar and Chand 2004a, 2004b; Marshall 2004).

2.9.3 Role of Fermented Dairy Products on Heart Health

As described earlier for hypertensive activity (Section 2.9.2), these fermented foods serve as a medium to supply bioactive components—calcium, some antioxidants, isomers of CLAs, and probiotic bacteria—that have been intervened as positive controllers in many heart-associated diseases. However, they should be supplied in controlled ratio of cholesterol, so that no signs of hypercholesterolemia developed. For the various hypocholesteromic components, CLAs are the first choice of researchers and consumers because of their multifunctional application. However, much research was also devoted to supply functional foods with probiotics, as they housed some important cholesterol-utilizing enzymes (Gilliland et al. 1984; James et al. 1999; Xiao et al. 2003).

2.9.4 Dairy Foods and Obesity

Dairy foods formulated with low calories-based formula are helpful to manage weight in obese persons. The active dairy component strikes on many important regulators of obesity. Summarily, they help to prolong or prevent the maturation of pre-adipocytes into mature adipocytes inside adipose tissue; they interfere with the fat depository mechanism in adipose cells. However, recent studies have shown that probiotics incorporated in dairy foods may be helpful in reducing the brown fat deposits of the body. This altogether fat reduction brought about by various means is helpful in reducing the waist circumference of the body. Moreover, it is proposed that less fat deposition around the abdominal part reduces the risk of diabetes and associated heart diseases (Zemel 2005).

2.9.5 Dental Benefits from Dairy Foods

Calcium, phosphorus, and protein constitute the key components of dairy foods, which are very important for the formation and maintenance of healthy bones and teeth (Bowen 2002; Johansson 2002). Moreover, dairy foods and nutrients present in milk somehow help to reduce the deposition of secretary acids on teeth, which develops as a microbial fermentation of carbohydrate-based foods. The combination of various ingredients present in cheese, such as phosphorus, calcium, and casein, may help to re-mineralize teeth and also reduce the risk of getting dental plaques.

2.9.6 Role in Blood Sugar Management

It is evident that a daily intake of standardized or low-fat milk minimizes the risk of type-2 diabetes mellitus development. Therefore, it confers that besides drugs and exercise, the type of food we consume and our living style can modulate various factors responsible for diabetes. Some authors had reported that consumption of dairy foods, like dahi, may have insulinotropic effects and thereby lowers or disfavors diabetes development (Hyon et al. 2005; Yadav et al. 2006). However, there are numerous contradictions to these evidences, and the exact hypothesis is still to be reframed.

2.9.7 Antimicrobial Potential

Milk components impart antimicrobial potential, for example, milk is rich source of immunoglobulins present in colostrum and provides immunity against various infections, resist pathogens to colonize in intestinal epithelia, reduce absorption of food antigens and promotes phagocytosis of pathogens (Walzem, 2001). In addition, lactoferrin, a milk protein that has strong affinity for iron present in milk, removes freely present iron and renders it unavailable for most of the pathogenic enzymes and proteins to function properly (Reiter 1985; Marshall 2004). Another important defensive component of milk is lysozyme, which hydrolyzes the peptidoglycan subunits of pathogens cell wall. Besides, lactoperoxidase of milk reacts with microorganism-derived hydrogen peroxide and generates some highly reactive intermediary compounds that have cell wall–degrading potential of invading microorganisms (Ballongue 1998).

2.9.8 Anticancer Activities

Continuous milk intake manifests anticancerous effects, especially of colorectal and breast cancer. Cis-9, trans-11 CLA is the principal biomolecule of milk lipids responsible for this activity and was found to be effective against breast cancer in animal models. Moreover, this activity is also strengthened by other components of milk like calcium ions, polyphenols, and peptides generated by probiotic LAB and vitamin D (Marshall 2004; Rehmeyer 2006), as depicted in Figure 2.3. The overall concentration of CLA's isomer is significantly higher in whole milk in comparison to processed milk or milk low in fat. The daily inclusion of yogurt in diet has been linked to prevent or reduce

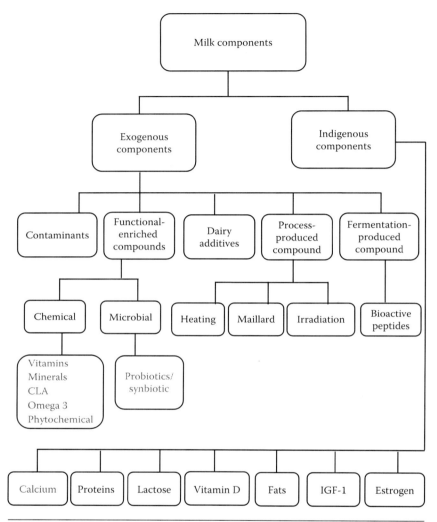

Figure 2.3 Anticancer activity of milk components and products.

the progression of colon cancer in distinctive population groups, which is attributed due to the formation of various potent peptides through incubation (Ganjam et al. 1997).

2.9.9 Appetite Suppressors

Whey is a rich source of many nutritious components. The watery fluid left over in whey following rennet action in milk can satisfy your hunger by activating various pancreatic hormones, which promote contraction and mobility of the gall bladder and bowel and further maintain gastric activity (Walzem 2001).

2.9.10 Anti-HIV Activity

Whey proteins are found beneficial for HIV-infected individuals. Inclusion of isolated whey proteins in diet increases the amounts and activity of various glutathione-based antioxidative enzymes that improve the overall treatment. Micke et al. (2001, 2002) supplemented diet with isolated whey proteins for a period of 2 weeks and found significant fold increase in plasma glutathione levels. It has been demonstrated with 2 weeks' trial that the commercial source of protein did influence the extent of increase in plasma. These whey proteins remarkably strengthen the muscles in females suffering from HIV infection (Agin et al. 2001). Milk lactoferrin is also found to play a preventive role against HIV infection (Berkhout et al. 2003). In fact, some authors had screened the activity of various bovine milk proteins against HIV-1 enzymes that are crucial in establishing the infectious life cycle in the host and obtained significant protective effects (Ng et al. 2001).

2.9.11 Relief against Rotaviral Diarrhea

Many previously conducted studies had shown that selective intake of some probiotic strains of lactobacilli like *Lb. casei* shirota, *Lb. reuteri*, and *Lb. delbrueckii* subsp. *bulgaricus* and bifidobacteria like *B. lactis* Bb12 shortens the life cycle of this virus-developed diarrhea probably by 1 day (Kaila et al. 1992; Saavedra et al. 1994; Shornikova et al. 1997; Boudraa et al. 2001; Chandra 2002). Intake of probiotic strain *Lb. rhamnosus* GG strongly cut short the duration of rotavirus diarrhea and is probably the best-studied evidence. However, other strains were also reported with this activity—the intake of *Lb. acidophilus* shortens the infectious cycle of the disease (Simakachorn et al. 2000). The suggested mechanism for this activity is the overall increase in the amount of serum IgA antibody levels and other immune effectors molecules to rotavirus that boost immunity and prevent recurrence of infection (De Roos and Katan 2000; McFarland 2000).

2.9.12 Milk-Derived Bioactive Peptides

Milk is a rich source of various bioactive constituents and known for numerous biological actions, which not only extends mothers' health benefits but also transmits healthy effects from mother to child. Bioactive peptides are produced during fermentation, processing and digestion, and breakdown for milk proteins (Brantl et al. 1979; Zioudrou et al. 1979; Loukas et al. 1983). Hence, milk proteins are the major source of bioactive peptides in dairy products. Bioactive peptides have been known for numerous bioactivities including immunomodulatory, hypotensive, opiate, heart protecting, and enhanced calcium and other mineral utilization efficacy. Interestingly, few of the bioactive peptides have shown potential activities to regulate glucose homeostasis by influencing insulin secretion and/or enhancing insulin sensitivity (Daniel et al. 1990). Bioactive peptide generation is a common phenomenon of the human digestion process, where various digestive enzymes (i.e., proteases, chymotrypsin, pepsin, etc.) digest milk proteins in small pieces of peptides and free amino acids. The release of bioactive peptides during the digestion process of milk proteins helps in GI tract development and function. Interestingly, these bioactive peptides are generated in close proximity of mucosal immune cells, and they modulate the functioning and secretory properties of immune cells that carry the signals of immune action to the whole body. Milk-derived bioactive peptides and other related peptides like prolactin, growth factors, and rosuposin play a crucial role in infants' overall development, immune maturation, as well as microbial activities.

Although the presence of bioactive peptides in numerous types of processed and fermented foods is well known, their exact physiological role in the human body is completely unknown. Normally, the notion is eating a diet enriched with various types of bioactive peptides may help maintaining a homeostatic function of human immune, digestive, nervous, and metabolic function. This notion basically establishes the fact that food enriched with particular types of bioactive peptides and their biofunctionalities will be of higher importance to design functional foods via personalized nutrition. But these facts have to be proven scientifically. Peptides derived from casein (one of the major proteins in milk) have been very well known for various applications including dietary supplements, that is, source of phosphopeptides, and also used in various pharmaceutical products, that is, casomorphins and phospopeptides. We always have to remember that effectiveness and safety of those bioactive peptides in human and animals are not well established and are yet to be proven. Recently, the research trend shows that bioactive peptides derived from dairy products exhibit activities of angiotensin-converting-enzyme inhibitors and phosphopeptidases, and these areas are considered to be significant in near future research. These areas will generate not only unique information but also industrial and economic application by developing new dietary formulations for human health.

2.9.13 Conclusions and Future Perspectives

Human health is affected by various factors in which diet, environment, and genetics are three major important factors that largely play a role in maintaining normal functionalities of the human body. Commonly, foods are designed to fulfill the nutritional requirements of the human body, while medicines are designed to treat various health ailments. But now this notion is changing very fast, and food is considered a very important component of your health; in some cases, foods and their ingredients are considered as important as medicine to treat various health ailments. It is said very clearly that what our health today is largely depends on what and how we eat and what types of genes we carry. The importance of food, especially milk/dairy, in human health has been generated in the late 1980s and particularly during the 1990s, and it has grown substantially till date. In the current scenario, there are various foods that have been designed to have biological activities, and these foods are called *functional foods* (i.e., foods that carry health beneficial properties to consumers beyond nutritional values). In the recent few years, the health consciousness around the society has spread at a faster rate; therefore, consumption of these functional foods has increased tremendously. It is estimated that the functional food market is going to be a major market around the world for business and economic importance. Although the function related to health benefits and biological activities is becoming more obvious for these functional foods, their exact biological ingredients and efficacy in humans have yet to be illustrated in well-controlled studies. Another fact is that the activity of functional food also depends on the quantity of bioactive ingredients present in that particular type of functional food at the time of consumption. Nowadays, terms like *functional foods* and *probiotics* are becoming very popular, but a very small fraction of the global population is known by these terms. Milk products are the most favorable vehicles for carrying health beneficial bacteria, that is, bifidobacteria and lactobacillus, and provide the appropriate environment for proper growth of these probiotic organisms. Milk products such as yogurt, cheese, and others are known not only for favoring the growth of probiotics in them but also for favoring efficient and maximum delivery of these bacteria into the human intestine by buffering GI pH (i.e., low pH in stomach and higher pH in duodenum). In addition, health beneficial biofunctionalities of probiotics go very well along with the bioactivities of milk-based foods. The immunomodulatory function of probiotics is one of the famous arenas studied in the scientific community, and the effects of probiotics are known in various health anomalies including allergies and autoimmune diseases. The effects of probiotics are known by various ways including stimulatory and functional modulation of various immune cell type population and their cytokine and antibody secretory functionalities. Although various facts have been established in this line, more rigorous and well-planned studies have to be carried out to make these claims robust and applicable for human application.

Although milk and other dairy products carry various health benefits for the human body, improper handling and malpractices in manufacturing of these products also threaten to carry various pathogenic bacteria. The origin of these pathogenic bacteria can be from milch animals or environmental contamination occurring during collection and storage of milk products, which is commonly found in developing countries (Bhatnagar et al. 2007; Le-Jeune and Schultz 2009). Therefore, it is very critical to initiate and spread public awareness to educate milch producers and consumers for proper milk collection, storage and processing, and awareness information enriched with scientific background.

References

Abrams, S.A. 2013. Calcium and vitamin D requirements of eternally fed preterm infants. *Pediatrics* 131: 1676–1683.

Agin, D., Gallagher, D., Wang, J., Heymsfield, S.B., Pierson, R.N., and Kottler, D.P. 2001. Effect of whey protein and resistant exercise of body cell mass, muscle strength and quality of life in women with HIV. *AIDS* 15: 2431–2440.

Ashar, M.N. and Chand, R. 2004a. Fermented milk containing ACE-inhibitory peptides reduces blood pressure in middle aged hypertensive subjects. *Milchwissenschaft* 59: 363–366.

Ashar, M.N. and Chand, R. 2004b. Antihypertensive peptides purified from milks fermented with *Lactobacillus delbrueckii* ssp. *bulgaricus*. *Milchwissenschaft* 59: 14–17.

Ballongue, J. 1998. Bifidobacteria and probiotic action. In: *Lactic Acid Bacteria*, eds. S. Salminen and A. von Wright, 519–587. Hong Kong, China: Marcel Dekker Inc.

Berkhout, B., van Wamel, J., Beljaars, L.B., Leoni, M., Dirk, K.F., Visser, S., and Floris, R. 2003. Characterization of the anti-HIV effects of native lactoferrin and other milk proteins and protein-derived peptides. *Antiviral Research* 55: 341–355.

Bhatnagar, P., Khan, A.A., Jain, M., Kaushik, S., and Jain, S.K. 2007. Microbiological study of khoa sold in Chambal region (Madhya Pradesh): A case study. *Indian Journal of Microbiology* 47: 263–266.

Boudraa, G., Benbouabdellah, M., Hachelaf, W., Boisset, M., Desjeux, J.F., and Touhami, M. 2001. Effect of feeding yogurt versus milk in children with acute diarrhea and carbohydrate malabsorption. *Journal of Pediatric Gastroenterology and Nutrition* 33: 307–313.

Bowen, W.H. 2002. Effects of dairy products on oral health. *Scandinavian Journal of Nutrition* 46: 178–179.

Brantl, V., Teschemacher, H., Henschen, A., and Lottspeich, F. 1979. Novel opioid peptides derived from casein (α-casomorphins). Isolation from bovine casein peptone. *Hoppe-Seyler's Zeitschrift für Physiologische Chemie* 360: 1211–1216.

Chandra, R.K. 2002. Nutrition and the immune system from birth to old age. *European Journal of Clinical Nutrition* 56: S73–S76.

Childs, N.M. and Poryzees, G.H. 1998. Foods that help prevent disease: Consumer attitudes and public policy implications. *British Food Journal* 9: 419–426.

Daniel, H., Vohwinkel, M., and Rehner, G. 1990. Effect of casein and α-casomorphins on gastrointestinal motility in rats. *Journal of Nutrition* 120: 252–257.

De Roos, N.M. and Katan, M.B. 2000. Effects of probiotic bacteria on diarrhea, lipid metabolism, and carcinogenesis: a review of papers published between 1988 and 1998. *American Journal of Clinical Nutrition* 71: 405–411.

De Vos, W.M. and Vaughan, E.E. 1994. Genetics of lactose utilization in lactic acid bacteria. *FEMS Microbiology Reviews* 15: 217–237.

De Vrese, M. and Schrezenmeir, J. 2001. Pro and prebiotics. *Innovative Food Technology* May: 49–55.

Diplock, A.T., Aggett, P.J., Ashwell, M., Bornet, F., Fern, E.B., and Roberfroid, M.B. 1999. Scientific concepts of functional foods in Europe: Consensus document. *British Food Journal* 81: 11–27.

Ganjam, L.S., Thornton, W.H., Marshall, R.T., and MacDonald, R.S. 1997. Anti-proliferative effects of yogurt fractions obtained by membrane dialysis on cultured mammalian intestinal cells. *Journal of Dairy Science* 80: 2325–2329.

Gilliland, S.E., Staley, T.E., and Busl, L.J. 1984. Importance of bile tolerance of *Lactobacillus acidophilus* used as dietary adjunct. *Journal of Dairy Science* 67: 3045–3051.

Heyman, M.B. 2006. Lactose intolerance in infants, children, and adolescents. *Pediatrics* 118: 1279–1286.

Hyon, K.C., Walter, C.W., Meir, J.S., Eric, R., and Frank, B.H. 2005. Dairy consumption and risk of type 2 diabetes mellitus in men. *Archives of Internal Medicine* 165: 997–1003.

ILSI Europe. 2002. Scientific concepts of functional foods in Europe: Consensus document. *British Journal of Nutrition* 81: 1S–27S.

James, W., Anderson, M.D., and Gilliland, S.E. 1999. Effect of fermented milk (yogurt) containing *Lactobacillus acidophilus* L1 on serum cholesterol in hypercholesterolemia in humans. *Journal of the American College of Nutrition* 18: 43–50.

Jan, G., Leverrier, P., Pichereau, V., and Boyaval, P. 2001. Changes in protein synthesis and morphology during acid adaptation of *Propionibacterium freudenreichii*. *Applied and Environmental Microbiology* 67: 2029–2036.

Johansson, I. 2002. Milk and dairy products: Possible effects on dental health. *Scandinavian Journal of Nutrition* 46: 119–122.

Kaila, M., Isolauri, E., Soppi, E., Virtanen, E., Laine, S., and Arvilommi, H. 1992. Enhancement of the circulating antibody secreting cell response in human diarrhea by a human *Lactobacillus* strain. *Pediatric Research* 32: 141–144.

Karimi, O. and Pena, A.S. 2003. Probiotics: Isolated bacterial strains or mixtures of different strains? *Drugs Today* 39: 565–597.

Koletzko, B., Agostoni, C., Bergmann, R., Ritzenthaler, K., and Shamir, R. 2011. Physiological aspects of human milk lipids and implications for infant feeding: A workshop report. *Acta Paediatrica* 100: 1405–1415.

Korhonen, H. 2002. Technology options for new nutritional concepts. *International Journal of Dairy Technology* 55: 79–88.

Kotler, B.M., Kerstetter, J.E., and Insogna, K.L. 2013. Claudins, dietary milk proteins, and intestinal barrier regulation. *Nutrition Review* 71: 60–65.

Le-Jeune, J.T. and Rajala-Schultz, P.J. 2009. Food safety: Unpasteurized milk: A continued public health threat. *Clinical Infectious Diseases* 48: 93–100.

Livingstone, K.M., Lovegrove, J.A., and Givens, D.I. 2012. The impact of substituting SFA in dairy products with MUFA or PUFA on CVD risk: Evidence from human intervention studies. *Nutrition Research Reviews* 25: 193–206.

Loukas, S., Varoucha, D., Zioudrou, C., Streaty, R., and Klee, W.A. 1983. Opioid activities and structures of α-casein derived exorphins. *Biochemistry* 22: 4567–4573.

Marshall, K. 2004. Therapeutic applications of whey protein. *Alternative Medicine Review* 9: 136–156.

McFarland, L.V. 2000. Beneficial microbes: Health or hazard? *European Journal of Gastroenterology and Hepatology* 12:1069–1071.

Mercenier, A., Pavan, S., and Pot, B. 2002. Probiotics as bio-therapeutic agents: Present knowledge and future prospects. *Current Pharmaceutical Design* 8: 9–110.

Micke, P., Beeh, K.M., and Buhl, R. 2002. Effect of long term supplementation with whey proteins on plasma glutathione levels of HIV infected patients. *European Journal of Nutrition* 41: 12–18.

Micke, P., Beeh, K.M., Schlaak, J.F., and Buhl, R. 2001. Oral supplementation with whey proteins increases plasma glutathione levels of HIV infected patients. *European Journal of Clinical Investigation* 31: 171–178.

Nagpal, R., Yadav, H., Puniya, A.K., Singh, K., Jain, S., and Marotta, F. 2007. Potential of probiotics and prebiotics for synbiotic functional dairy foods: An overview. *International Journal of Probiotics and Prebiotics* 2: 75–84.

Ng, T.B., Lam, T.L., Au, T.K., Ye, X.Y., and Wan, C.C. 2001. Inhibition of human immunodeficiency virus type 1 reverse transcriptase, protease and integrase by bovine milk proteins. *Life Science* 69: 2217–2223.

Patel, R.S. and Renz-Schanen, A. 1998. Dietary calcium and its role in disease prevention. *Indian Dairyman* 50: 49–51.

Rehmeyer, J.J. 2006. Milk therapy: Breast-milk compounds could be a tonic for adult ills. *Science News* 170: 376–378.

Reiter, B. 1985. The biological significance of the non-immunoglobulin protective proteins in milk: Lysozyme, lactoferrin and lactoperoxidase. In: *Developments in Dairy Chemistry* (3rd edition), ed. P.F. Fox, 281–336. New York: Elsevier Applied Science Publishers.

Roberfroid, M.B. 2000. Prebiotics and probiotics: Are they functional foods? *American Journal of Clinical Nutrition* 71: 1682S–1687S.

Ross, S. 2000. Functional foods: Food and drug administration prospective. *American Journal of Clinical Nutrition* 71: 1735S–1738S.

Rudloff, S. and Kunz, C. 2012. Milk oligosaccharides and metabolism in infants. *Advances in Nutrition* 3: 398S–405S.

Saavedra, J.M., Bauman, N.A., Oung, I., Perman, J.A., and Yolken, R.H. 1994. Feeding of *Bifidobacterium bifidum and Streptococcus thermophilus* to infants in hospital for prevention of diarrhoea and shedding of rotavirus. *Lancet* 344: 1046–1049.

Salminen, S., Ouwehand, A.C., Benno, Y., and Lee, Y.K. 1999. Probiotics: How should they be defined? *Trends in Food Science & Technology* 10: 107–110.

Sanders, M.E. 2000. Symposium: Probiotic bacteria—Implications for human health. Considerations for use of probiotic bacteria to modulate human health. *Journal of Nutrition* 130: 384S–390S.

Shornikova, A.V., Casas, I.A., Isolauri, E., Mykkänen, H., and Vesikari, T. 1997. *Lactobacillus reuteri* as a therapeutic agent in acute diarrhea in young children. *Journal of Pediatric Gastroenterology and Nutrition* 24: 399–404.

Simakachorn, N., Pichaipat, V., Rithipornpaisarn, P., Kongkaew, C., Tongpradit, P., and Varavithya, W. 2000. Clinical evaluation of the addition of lyophilized, heat-killed *Lactobacillus acidophilus* LB to oral rehydration therapy in the treatment of acute diarrhea in children. *Journal of Pediatric Gastroenterology and Nutrition* 30: 68–72.

Talhouk, R.S., Abdo, R.A., and Saad, A. 1996. Lactose intolerance. Diagnosis and dietary treatment with milk substitutes. *Journal Medical Libanais* 44: 36–40.

Temmerman, R., Pot, B., Huys, G., and Swings, J. 2002. Identification and antibiotic susceptibility of bacterial isolates from probiotic products. *International Journal of Food Microbiology* 81: 1–10.

Walzem, R.L. 2001. Health enhancing properties of whey proteins and whey fractions. In: *Applications Monograph-Nutritional and Beverages*, 1–8. Arlington, VA: US Dairy Export Council.

Xiao, J.Z., Kondo, S., Takahashi, N., Miyaji, K., Oshida, K., Hiramatsu, A., Iwatsuki, K., Kokubo, S., and Hosono, A. 2003. Effects of milk products fermented by *Bifidobacterium longum* on blood lipids in rats and healthy adult male volunteers. *Journal of Dairy Science* 86: 2452–2461.

Yadav, H., Jain, S., and Sinha, P.R. 2006. Effect of skim milk and dahi (yogurt) on blood glucose, insulin, and lipid profile in rats fed with high fructose diet. *Journal of Medicinal Food* 9: 328–335.

Zemel, M.B. 2005. Dairy augmentation of total and central fat loss in obese subjects. *International Journal of Obesity* 29: 391–397.

Zioudrou, C., Streaty, R.A., and Klee, W.A. 1979. Opioid peptides derived from food proteins. *Journal of Biological Chemistry* 254: 2446–2449.

Zivkovic, A.M. and Barile, D. 2011. Bovine milk as a source of functional oligosaccharides for improving human health. *Advances in Nutrition* 2: 284–289.

3
Lactic Acid Bacteria and Associated Product Characteristics within Milk Fermentation

CHRISTOPHER BEERMANN
AND JULIA HARTUNG

Contents

3.1	Introduction	48
3.2	LAB with Distinct Acidification Properties	54
3.3	Specific Characteristics of Lactobacilli Co-Cultures	55
3.4	Acidification, Proteolysis, and Other Metabolic Properties of Lactococci	56
3.5	Metabolic Peculiarities of *Streptococcus thermophilus* in Yogurt and Other Milk Products	57
3.6	*Enterococcus*, *Leuconostoc*, and *Pediococcus* Species for Product Perfection	58
3.7	Metabolic Incompatibility of Bifidobacteria with Milk Products	59
3.8	Coexistence of LAB and Yeasts in Kefir and Similar Milk Products	60
3.9	Non-Starter LAB and Strain Selection Strategies for New LAB Starter Cultures	62
3.10	Flavor Development in Milk Products by LAB	63
3.11	EPS and the Texture of Fermented Milk Products	64
3.12	Importance of the Proteolytic System of LAB in Milk Fermentation	67
3.13	Proteolytic Systems of LAB in Cheese Production	69
3.14	Debittering of Milk Products by Proteolytic LAB	70
3.15	Conclusion	71
References		72

3.1 Introduction

The traditional fermentation processes of milk from different animals have been known for centuries around the world. Various microbial fermentation processes with different milk qualities lead to a huge diversity of known products, for instance, kefir, koumiss, yogurt, dahi, and cheese. Dissimilar to traditional spontaneous acidification, today's fermentation processes are determined by microbiologically defined starter cultures (Beermann and Hartung 2012). For milk fermentation, principally lactic acid bacteria (LAB), occasionally in combination with yeasts and molds, are applied. Aside from common acidification properties, specific proteolytic activities of selected bacteria species and associated enzyme systems are necessary to achieve the defined textures, flavor development, and support ripening processes of milk products (Beermann and Hartung 2013; Sfakianakis and Tzia 2014). To date, distinctively selected LAB from the milk germ pool and milk product-associated non-starter LAB are discussed to broaden the portfolio of milk fermenting organisms in industrial applications. Further, new insights into genetic engineering, strain selection strategies, and target-adjusted application procedures might lead to exact tools to products with defined characteristics. Table 3.1 gives an overview of the most relevant microorganisms in milk fermentation.

Table 3.1 Lactic Acid Bacteria and Related Microorganism (Yeasts) of Fermented Milk Products

MICROORGANISM(S)	PRODUCT	REFERENCES
GRAM-POSITIVE LACTIC ACID BACTERIA		
Lactobacilli	Cheese (Pecorino Sardo cheese, Genestoso cheese), yogurt, kefir fermented milk (nunu, dahi)	Akabanda et al. (2010), Arias et al. (2013), Kourkoutas et al. (2007), Maqsood et al. (2013)
Lb. acidophilus	Fermented milk (dahi, kurut, koumiss, Suero Costeño), probiotic dairy products (cheese, Gouda cheese), yogurt, cheese	Al-Awwad et al. (2009), Bergamini et al. (2009), El-Tanboly et al. (2010), Hassan and Amjad (2010), Maqsood et al. (2013), Movsesyan et al. (2010), Sun et al. (2010)

(*Continued*)

Table 3.1 (*Continued*) Lactic Acid Bacteria and Related Microorganism (Yeasts) of Fermented Milk Products

MICROORGANISM(S)	PRODUCT	REFERENCES
Lb. brevis	Cheese (ewe cheese, Pecorino Umbro, Pecorino Sardo, Pecorino Toscano, Pecorino Romano, Fossa, Canestrato Pugliese), fermented milk (kurut, koumiss, gariss, Suero Costeño), kefir	Ashmaig et al. (2009), Khalid and Marth (1990), Sun et al. (2010)
Lb. buchneri	Cheese, fermented milk (koumiss)	Khalid and Marth (1990), Sun et al. (2010)
Lb. parabuchneri	Fermented milk (koumiss)	Sun et al. (2010)
Lb. casei	Cheese (soft cheese, Swiss-type cheese), fermented milk (koumiss, kurut, dahi), yogurt, probiotic dairy products (cheese)	Bergamini et al. (2009), Khalid and Marth (1990), Kourkoutas et al. (2006), Maqsood et al. (2013), Sun et al. (2010)
Lb. casei subsp. *casei*	Cheese (ewe cheese, Pecorino Umbro, Pecorino Sardo, Pecorino Toscano, Pecorino Romano, Fossa, Canestrato Pugliese), yogurt	Song et al. (2010)
Lb. paracasei	Fermented milk (gariss), yogurt, cheese (Danbo cheese, Ganestoso cheese)	Antonsson et al. (2003), Arias et al. (2013), Ashmaig et al. (2009), Song et al. (2010)
Lb. paracasei subsp. *paracasei*	Cheese (Beyaz cheese, ewe cheese, Pecorino Umbro, Pecorino Sardo, Pecorino Toscano, Pecorino Romano, Fossa, Canestrato Pugliese), probiotic dairy products, fermented milk (kurut, koumiss, Suero Costeño)	Durlu-Ozkaya et al. (2001), Sun et al. (2010)
Lb. casei subsp. *pseudoplantarum*	Cheese (ewe cheese, Pecorino Umbro, Pecorino Sardo, Pecorino Toscano, Pecorino Romano, Fossa, Canestrato Pugliese), fermented milk (dahi)	Maqsood et al. (2013)
Lb. crustorum	Fermented milk (kurut, koumiss)	Sun et al. (2010)
Lb. curvatus	Cheese (Danbo cheese, ewe cheese, Pecorino Umbro, Pecorino Sardo, Pecorino Toscano, Pecorino Romano, Fossa, Canestrato Pugliese), fermented milk (koumiss)	Antonsson et al. (2003), Sun et al. (2010)
Lb. delbrueckii	Cheese (Swiss-type cheese), fermented milk (Kurut, Dahi)	Maqsood et al. (2013), Sun et al. (2010)

(*Continued*)

Table 3.1 (*Continued*) Lactic Acid Bacteria and Related Microorganism (Yeasts) of Fermented Milk Products

MICROORGANISM(S)	PRODUCT	REFERENCES
Lb. delbrueckii subsp. *Bulgaricus*	Fermented milk (koumiss, kurut, dahi), yogurt, kefir, cheese	Atta et al. (2009), François et al. (2007), Georgala et al. (1995), Hassan and Amjad (2010), Khalid and Marth (1990), Liu et al. (2009), Maqsood et al. (2013), Movsesyan et al. (2010), Omae et al. (2008), Sun et al. (2010), Urshev et al. (2008)
Lb. delbrueckii subsp. *Delbrueckii*	Fermented milk (Suero Costeño)	Maqsood et al. (2013)
Lb. delbrueckii subsp. *lactis*	Yogurt, fermented milk (dahi, kurut)	Maqsood et al. (2013), Sun et al. (2010)
Lb. fermentum	Fermented milk (kurut, koumiss, gariss, dahi), probiotic dairy products (yogurt, cheese, kefir), cheese (Ras cheese, ewe cheese, Pecorino Umbro, Pecorino Sardo, Pecorino Toscano, Pecorino Romano, Fossa, Canestrato Pugliese)	Ashmaig et al. (2009), El-Ghaish et al. (2010), Mikelsaar and Zilmer (2009), Sun et al. (2010)
Lb. gasseri	Fermented milk (gariss, koumiss)	Ashmaig et al. (2009), Sun et al. (2010)
Lb. helveticus	Yogurt, cheese, kefir, fermented milk (kurut, koumiss, dahi)	Khalid and Marth (1990), Maqsood et al. (2013), Song et al. (2010), Sun et al. (2010)
Lb. kefiranofaciens	Kefir	Chen et al. (2009)
Lb. kefiranofaciens subsp. *Kefirgranum*	Fermented milk (koumiss, kurut)	Sun et al. (2010)
Lb. kefiri	Kefir, fermented milk (kurut, koumiss)	Chen et al. (2009), Sun et al. (2010)
Lb. parakefiri	Fermented milk (koumiss)	Sun et al. (2010)
Lb. pentosus	Cheese (ewe cheese, Pecorino Umbro, Pecorino Sardo, Pecorino Toscano, Pecorino Romano, Fossa, Canestrato Pugliese), fermented milk (koumiss, Suero Costeño)	Sun et al. (2010)
Lb. plantarum	Cheese (Beyaz cheese, artisan-produced fresh cheese, soft cheese, ewe cheese, Pecorino Umbro, Pecorino Sardo, Pecorino Toscano, Pecorino Romano, Fossa, Canestrato Pugliese,	Arias et al. (2013), Ashmaig et al. (2009), Durlu-Ozkaya et al. (2001), Khalid and Marth (1990), Sun et al. (2010)

(*Continued*)

Table 3.1 (*Continued*) Lactic Acid Bacteria and Related Microorganism (Yeasts) of Fermented Milk Products

MICROORGANISM(S)	PRODUCT	REFERENCES
	Danbo cheese, soft Kareish-type cheese, Ganestoso cheese), fermented milk (gariss, kurut, oumiss, Suero Costeño)	
Lb. paraplantarum	Fermented milk (koumiss)	Sun et al. (2010)
Lb. rhamnosus	Fermented milk (Suero Costeño, koumiss, gariss), cheese (ewe cheese, Pecorino Umbro, Pecorino Sardo, Pecorino Toscano, Pecorino Romano, Fossa, Canestrato Pugliese, Danbo cheese)	Antonsson et al. (2003), Ashmaig et al. (2009), Movsesyan et al. (2010), Sun et al. (2010)
Lb. salivarius	Cheese	Movsesyan et al. (2010)
Lb. alimentarium, Lb. animalis, Lb. Divergens	Fermented milk (gariss)	Ashmaig et al. (2009)
Lb. amylolyticus, Lb. amylovorus, Lb. crispatus, Lb. gallinarum	Fermented milk (koumiss, kurut)	Sun et al. (2010)
Lb. hilgardii, Lb. suntoryeus	Fermented milk (kurut)	Sun et al. (2010)
Lb. amylophilus, Lb. amylotrophicus, Lb. diolivorans, Lb. johnsonii, Lb. pontis, Lb. reuteri, Lb. zeae	Fermented milk (koumiss)	Sun et al. (2010)
Lactococci	Cheese (Pecorino Sardo ewes´milk cheese, Ganestoso cheese), fermented milk (nunu, Iben, dahi), kefir	Akabanda et al. (2010), Arias et al. (2013), Bekkali et al. (2013), Kourkoutas et al. (2007), Maqsood et al. (2013)
Lc. Garvieae	Fermented milk (kurut)	Sun et al. (2010)
Lc. lactis	Fermented milk (dahi), cheese (soft cheese, artisan-produced fresh cheese, Pecorino Sardo ewes´milk cheese)	Maqsood et al. (2013), Omae et al. (2008)
Lc. lactis subsp. cremoris	Fermented milk (kurut, Iben), cheese (soft Kareish-type cheese)	Antunes et al. (2007), Bekkali et al. (2013), Sun et al. (2010)
Lc. lactis subsp. diacetylactis	Cheese (soft Kareish-type cheese)	Sun et al. (2010)
Lc. lactis subsp. lactis biovar. *Diacetylactis*	Fermented milk	Antunes et al. (2007)

(*Continued*)

Table 3.1 (*Continued*) Lactic Acid Bacteria and Related Microorganism (Yeasts) of Fermented Milk Products

MICROORGANISM(S)	PRODUCT	REFERENCES
Lc. lactis subsp. *lactis*	Fermented milk (kurut, Iben, dahi, Suero Costeño), cheese (brie, yak, soft Kareish-type cheese, Beyaz cheese, Ganestoso cheese), kefir	Antunes et al. (2007), Arias et al. (2013), Bekkali et al. (2013), Durlu-Ozkaya et al. (2001), Sun et al. (2010)
Lc. raffinolactis	Fermented milk (Gariss, Dahi)	Ashmaig et al. (2009)
Enterococci	Fermented milk (nunu), cheese (Pecorino Sardo ewes´milk cheese, Ganestoso cheese)	Akabanda et al. (2010), Arias et al. (2013)
E. durans	Fermented milk (kurut), cheese (Beyaz cheese), yogurt	Durlu-Ozkaya et al. (2001), Jamaly et al. (2010), Sun et al. (2010)
E. faecalis	Fermented milk (kurut), cheese (Beyaz cheese, Ganestoso cheese)	Arias et al. (2013), Durlu-Ozkaya et al. (2001), Sun et al. (2010)
E. faecium	Fermented milk (dahi, kurut), cheese (Beyaz cheese)	Durlu-Ozkaya et al. (2001), Sun et al. (2010)
E. hirae	Fermented milk (kurut), cheese (Beyaz cheese)	Durlu-Ozkaya et al. (2001), Sun et al. (2010)
Leuconostoc	Fermented milk (nunu), kefir, cheese (Ganestoso cheese)	Akabanda et al. (2010), Arias et al. (2013), Kourkoutas et al. (2007)
Leu. Citreum	Fermented milk (kurut)	Sun et al. (2010)
Leu. Lactis	Fermented milk (kurut, Suero Costeño)	Sun et al. (2010)
Leu. mesenteroides	Fermented milk (kurut), cheese (artisan-produced fresh cheese, soft cheese, Ganestoso cheese), kefir, yogurt	Arias et al. (2013), Chen et al. (2009)
Leu. mesenteroides subsp. *Cremoris*	Fermented milk (Suero Costeño)	Antunes et al. (2007)
Leu. mesenteroides subsp. *Dextranicum*	Fermented milk (dahi)	Sun et al. (2010)
Leu. mesenteroides subsp. *Mesenteroides*	Fermented milk (kurut, dahi)	Sun et al. (2010)
Leu. pseudomesenteroides	Fermented milk (kurut), cheese (Ganestoso cheese)	Arias et al. (2013), Sun et al. (2010)
Streptococcus	Fermented milk (nunu), cheese, yogurt	Akabanda et al. (2010)
Str. Bovis	Fermented milk (dahi)	Sun et al. (2010)
Str. Cremoris	Fermented milk (dahi)	Maqsood et al. (2013)
Str. lactis subsp. *diacetylactis*	Fermented milk (dahi)	Maqsood et al. (2013)

(*Continued*)

Table 3.1 (*Continued*) Lactic Acid Bacteria and Related Microorganism (Yeasts) of Fermented Milk Products

MICROORGANISM(S)	PRODUCT	REFERENCES
Str. salivarius subsp. Thermophilus	Fermented milk (dahi, kurut), cheese (Swiss-type cheese), kefir, yogurt	Atta et al. (2009), François et al. (2007), Georgala et al. (1995), Khalid and Marth (1990), Liu et al. (2009), Maqsood et al. (2013), Omae et al. (2008), Sun et al. (2010), Urshev et al. (2008), Vaningelgem et al. (2004)
Pediococcus		
Pediococcus pentosaceus	Fermented milk (dahi), cheese	Khalid and Marth (1990)
OTHER GRAM-POSITIVE BACTERIA		
Propionibacteria		
P. shermanii	Cheese	Khalid and Marth (1990)
Bifidobacteria	Probiotic dairy products, yogurt	Antunes et al. (2007)
B. angulatum	Probiotic dairy products	Corbo et al. (2001)
B. animalis subsp. lactis	Probiotic dairy products, fermented milk, yogurt	Antunes et al. (2007)
B. bifidum	Probiotic dairy products (Canestrato Pugliese hard cheese)	Corbo et al. (2001)
B. breve	Probiotic dairy products	Corbo et al. (2001)
B. infantis	Probiotic dairy products	Al-Awwad et al. (2009)
B. longum	Probiotic dairy products (Canestrato Pugliese hard cheese)	Corbo et al. (2001)
GRAM-NEGATIVE BACTERIA		
Gluconobacter oxidans	Yogurt	Kourkoutas et al. (2007)
Yeasts		
Candida	Kefir	Kourkoutas et al. (2007)
C. kefyr	Fermented milk (nunu)	Akabanda et al. (2010)
C. stellata	Fermented milk (nunu)	Akabanda et al. (2010)
Kluyveromyces	Kefir	Kourkoutas et al. (2007)
K. marxianus	Fermented milk (nunu, viili), kefir	Akabanda et al. (2010), Chen et al. (2009), Wang et al. (2008)
Pichia	Kefir	Kourkoutas et al. (2007)
Pichia fermentans	Fermented milk (viili), kefir	Chen et al. (2009), Wang et al. (2008)
Saccharomyces	Kefir	Wang et al. (2008)
S. cerevisiae	Fermented milk (nunu)	Akabanda et al. (2010)
S. pastorianus	Fermented milk (nunu)	Akabanda et al. (2010)
S. turicensis	Kefir	Chen et al. (2009), Wang et al. (2008)

(*Continued*)

Table 3.1 (*Continued*) Lactic Acid Bacteria and Related Microorganism (Yeasts) of Fermented Milk Products

MICROORGANISM(S)	PRODUCT	REFERENCES
S. unisporus	Fermented milk (viili)	Wang et al. (2008)
Yarrowia		
Yarrowia lipolytica	Fermented milk (nunu)	Akabanda et al. (2010)
Zygosaccharomyces		
Zygosaccharomyces bisporus	Fermented milk (nunu)	Akabanda et al. (2010)
Zygosaccharomyces rouxii	Fermented milk (nunu)	Akabanda et al. (2010)

3.2 LAB with Distinct Acidification Properties

LAB can be subclassified into seven phylogenetic classes, *Lactococcus*, *Lactobacillus*, *Enterococcus*, *Pediococcus*, *Streptococcus*, *Leuconostoc*, and *Oenococcus* (O'Sullivan et al. 2009), and contain post-pasteurization contaminants of food starter bacteria for fermented milk products. Contrary to other starter bacteria, which autolyze in milk and release intracellular enzymes, lactobacilli generally persist and grow in the product. Additionally, commensal gastrointestinal LAB exert probiotic health benefits, like the improvement of epithelial barrier function and restoration of microbial homeostasis through microbe–microbe interactions. Further, aspects broadly discussed here are supportive of immune system functions and suppression of pathogens (Bergamini et al. 2009; El-Tanboly et al. 2010; Zalán et al. 2010).

Lactic acid fermentation is part of the anaerobic energy metabolic pathway of pro- and eukaryotic cells. Yeasts, molds, and also LAB are able to form lactic acid from different hexoses, such as glucose or fructose, and also from pentoses, such as xylose or ribose (McSweeney and Sousa 2000). The genera *Streptococcus*, *Enterococcus*, *Lactococcus*, and *Pediococcus* and some strains of *Lactobacillus* directly degrade hexoses to lactate by the aldolase-dependent homofermentative pathway. Other homolactic lactobacilli such as *Lactobacillus plantarum* and *Lb. casei* are non-starter lactobacilli (NSLAB), which sometimes cause product quality problems, like calcium lactate crystals on cheese surfaces. In Swiss-type cheese toxic amines, decarboxylase activity by NSLAB has been described (Khalid and Marth 1990). On the other hand, this group of LAB is an interesting germ pool for screening unknown organisms for new milk fermentation procedures.

LAB without aldolase activity, like the genus *Leuconostoc* and specific strains of the genus *Lactobacillus*, are heterofermentative, with less lactate production than homofermentative LAB. The phosphoketolase-dependent *Bifidobacterium* fermentation pathway of the gut-associated genus *Bifidobacterium*, a strict anaerobe, degrades glucose to lactate and acetate at a ratio of 3:2 and generates several side products, like ethanol, CO_2, and acetate (Haug et al. 2007). In order to modulate the acidification rate of LAB, genetic engineering concerning aldolase and phosphoketolase activities is conceivable.

3.3 Specific Characteristics of Lactobacilli Co-Cultures

Lb. delbrueckii and the three subspecies *Lb. delbrueckii* subsp. *bulgaricus*, *Lb. delbrueckii* subsp. *delbrueckii*, and *Lb. delbrueckii* subsp. *lactis* are used as a component of starter cultures for milk fermentation, predominantly combined with *Streptococcus thermophilus* and *Lb. helveticus* for yogurt and cheese. During milk fermentation, these bacteria control the pH value of the product, influence the growth of other organisms, and generate product-specific flavor compounds along proteo- and lipolysis. *Lb. helveticus* is a thermophilic starter. Although multiple amino acid auxotrophs have been described for *Lb. helveticus*, an efficient proteolytic system can cope with cheese and other complex protein matrices. However, in combination with *Str. thermophilus* and *Lb. delbrueckii* subsp. *bulgaricus*, *Lb. helveticus* is relevant for cheese production, particularly Swiss-type and Italian-type cheeses (Genay et al. 2009; Khalid and Marth 1990). As a flavor-adjunct culture, this bacterial combination reduces bitterness and accelerates flavor development. *Lb. helveticus* strains are among the most nutritionally fastidious lactobacilli.

Lb. delbrueckii subsp. *bulgaricus* belongs to the acidophilus complex, which is a group of lactobacilli with probiotic properties and related to *Lb. acidophilus*, *Lb. johnsonii*, and *Lb. gasseri* (Van de Guchte et al. 2006). The probiotic potential of *Lb. acidophilus* has been extensively discussed in several studies (Al-Awwad et al. 2009). Hassan and Amjad (2010) proposed *Lb. acidophilus* as a starter culture for yogurt. This homofermenter metabolizes hexoses to long polymeric carbohydrates like inulin fructo-oligosaccharides, which have been described to support the growth of probiotic bacteria. Other strains with probiotic properties, like *Lb. plantarum*, *Lb. rhamnosus*, and *Lb. casei*, are also applied in fermented milk products (Klaenhammer et al. 2002). For *Lb. fermentum* ME-3, applicable in yogurt, kefir, or cheese, antimicrobial activity against intestinal pathogens and antioxidative activity have been described (Mikelsaar and Zilmer 2009). A new strain, *Lb. fermentum* IFO 3956, isolated from Egyptian Ras cheese, showed proteolytic activity on β-casein and whey β-lactoglobulin, which is attractive for the production of hypoallergenic milk products (El-Ghaish et al. 2010).

3.4 Acidification, Proteolysis, and Other Metabolic Properties of Lactococci

Lactococci are mesophilic, metabolically consistent LAB, and most of the milk fermentative enzymes are encoded by plasmid DNA (Gitton et al. 2005; Klaenhammer et al. 2002). Though lactococci tend to amino acid auxotrophy, milk is an optimal culture medium (Hansen and Martinussen 2009). *Lactococcus lactis* is widely applied for the acidification and proteolysis in milk fermentation and cheese manufacturing. The *Lc. lactis* subsp. *lactis* biovar. *diacetylactis* is incorporated into starter cultures for diacetyl aroma development. Lactococci synthesize several specific food additives. Among these, an important metabolite class is exopolysaccharides (EPS), which improve texture and syneresis in yogurt. Also, lactococci synthesize several bacteriocins, such as nisin. Interestingly, cheese ripening could be accelerated by release of intracellular enzymes of sensitive bacteriocin-lysed starter strains (Oliveira et al. 2009; Pogačić et al. 2013).

In milk, lactococci produce a vast negative redox potential (E_h), which affects the microbial and sensorial qualities of fermented products. On the one hand, growth of spoilage microorganisms is suppressed, but on the other hand, the metabolic activity of secondary flora in cheese, the viability of probiotics in yogurt, and the stability of aroma compounds are affected (Tachon et al. 2010).

Not only milk-derived *Lc. lactis* strains but also plant-derived strains that are genetically related to milk-derived strains are applicable for milk fermentation. These bacteria ferment numerous additional kinds of carbohydrates and reveal great stress tolerance during production processes (Rademaker et al. 2007).

3.5 Metabolic Peculiarities of *Streptococcus thermophilus* in Yogurt and Other Milk Products

Str. thermophilus is a Gram-positive, homofermentative LAB that is primarily utilized in yogurt production. This coccus rapidly converts lactose into lactate and also produces molecules that contribute to flavor and texture (Arioli et al. 2009; Klaenhammer et al. 2002). Aside from the anaerobic nature of *Str. thermophilus*, CO_2 and bicarbonate are growth-stimulating factors. For example, fixation of CO_2 for aspartic acid biosynthesis is essential for the growth of *Str. thermophilus* in milk (Arioli et al. 2009). Most *Str. thermophilus* strains are β-galactosidase-negative. In the production of Mozzarella cheese, non-metabolized galactose leads to intolerable brown coloring during cooking at high temperatures. Concerning this, genetically engineered recombinant β-galactosidase-positive strains have been discussed (Robitaille et al. 2007).

In yogurt, *Str. thermophilus* is synergistically associated with *Lb. delbrueckii* subsp. *bulgaricus* with high acidification rates (Herve-Jimenez et al. 2009). Generally, *Streptococcus* shows high proteolytic activity. However, not only mutual supplementation with milk protein–derived amino acids and peptides but also growth stimulatory factors like CO_2, pyruvate, and folate are the basis of this supportive relationship (Herve-Jimenez et al. 2009). *Str. thermophilus* co-incubated with *Lb. bulgaricus* also upregulates several cellular peptide and amino acid transporters, notably for sulfur amino acids, during the late stage of milk fermentation, and several enzymes involved in the metabolism of various sugars are additionally activated. Liu et al. (2009) suggested that agonistic genetically co-evolutional processes along this coexistence probably improved synergistic growth strategies. As the genomes of both strains have been sequenced completely, putative horizontal gene transfer studies might be helpful in decoding these genetic aspects.

Concerning product characteristics, organoleptic attributes, acidity, and the concentration of free fatty acids, ascorbic acid, and total solids of yogurt are significantly influenced by the amount and ratio of both starter strains (Atta et al. 2009). In cheese, intra- and extracellular LAB proteases hydrolyze milk proteins, for example, $α_{S1}$-casein or β-casein, resulting in the absence of these proteins. In cheese, *Str. thermophilus* and *Propionibacterium shermanii* release specific amino acids from casein, which are characteristic cheese flavor molecules (O'Sullivan et al. 2009).

3.6 *Enterococcus*, *Leuconostoc*, and *Pediococcus* Species for Product Perfection

Starter cultures of enterococci descend from the intestinal microbiota and are occasionally splendid components of milk fermentation cultures. Prominent species are *Enterococcus faecalis*, *E. faecium*, and *E. durans*, which are especially applied in production of cheeses, such as Cheddar or Mozzarella. Due to their specific citrate utilization and high lipolytic activity, these bacteria contribute to typical sensorial and flavor cheese characteristics. For example, cheese manufactured with enterococci possesses high concentrations of free amino acids, volatile or long-chain free fatty acids, diacetyl, and acetoin. Several studies have demonstrated that *E. durans* enhances the growth of lactococci and streptococci and, therefore, might be supportive for fermentations with complex protein matrices. Enterococci produce bacteriocins and EPS to protect the fermentation process and to develop creamy or stiff product textures. Finally, in current application, several enterococci are probiotic adjunct cultures applied in Cheddar cheese (Jamaly et al. 2010).

Other LAB applicable in milk fermentation are *Leuconostoc* strains, which are traditionally applied in kefir. This group of starters are famous synthesizers of complex oligosaccharides, which develop texture probabilities of milk products and are prebiotic factors for the intestinal gut flora. A study from Seo Mi et al. (2007) demonstrated that *Leuconostoc citreum* co-cultured with *Lb. casei*, *Lb. delbrueckii* subsp. *bulgaricus*, and *Str. thermophilus* synthesizes complex oligosaccharides. Dextransucrase of *Leu. citreum* catalyzes the transglycosylation reaction of sugars. Often, sucrose acts as a glucose donor and lactose or maltose as acceptor molecules. For example, glycosyl lactose is synthesized when sucrose is added to the fermentation process. By applying sucrose and maltose together, panose and other iso-malto-oligosaccharides are formed. Interestingly, the production of oligosaccharides does not affect lactate fermentation or growth patterns. In addition, *Leu. mesenteroides* synthesizes conjugated linoleic acid (Abd El-Salam et al. 2010). For this typical milk fat fatty acid, several health benefits have been described. Thus, milk fermentation with combined *Leuconostoc* strains might provide pre-, probiotic, and conjugated linolenic acid features in one product.

3.7 Metabolic Incompatibility of Bifidobacteria with Milk Products

Bifidobacterium bifidum, *B. longum* subsp. *infantis*, *B. longum*, and *B. animalis* subsp. *lactis* are well-established probiotics in cheese, bifidus milk, and yogurt (Raeisi et al. 2013). In order to understand the bacterial physiology, *B. longum* was the first bifidobacterium to have its genome totally sequenced (Roy 2005; Sela et al. 2008). Several milk-specific physiological benefits of bifidobacteria have been described, like the improvement of lactose digestibility and absorption of calcium, as well as a reduction of total serum cholesterol levels. Bifidobacteria produce a range of B-complex vitamins, such as thiamine (B_1), and also generate nicotinic acid and conjugated linoleic acid. Antimicrobial activities of bifidobacteria are mainly based on acidification properties and on bacteriocins (Barboza et al. 2009).

In order to optimize selection criteria and product incorporation strategies, the growth, viability, and stability during production processes and storage as well as the enzymatic activities of bifidobacteria in milk, camel milk, acidified milk, and fermented products have been examined (Antunes et al. 2007; Kabeir et al. 2009). Bifidobacteria are able to ferment lactose, galactose, raffinose, sucrose, and also oligosaccharides (Barboza et al. 2009). In milk, the growth of bifidobacteria is limited, probably due to low proteolytic activities. Consequently, in milk products the application of bifidobacteria is often recommended, in combination with LAB with high proteolytic activity, such as *Str. thermophilus* or *Lb. acidophilus* as feeding bacteria. However, a competitive growth situation and hampered bifidobacterial growth due to organic acids, bacteriocins, and hydrogen peroxide derived from LAB limit this feed-culture constellation in milk. High inocula of bifidobacteria and the addition of growth-promoting nitrogen sources like amino acids or casein hydrolysates stabilize this co-culture (Roy 2005).

A study from Petschow and Talbott (1990) showed that specific fractions of acid whey, acid casein, and rennin casein from human and bovine milk promoted the growth of bifidobacterial strains isolated from stools of human infants. Another possibility for stimulating the growth of bifidobacteria is combined application with propionic acid bacteria (PAB), which are commonly utilized for typical Swiss-type cheese ripening (Warminska-Radyko et al. 2002). Falentin et al. (2010) studied the complete genome of *Propionibacterium freudenreichii* CIRM-BIA1^T, with new insights into mechanisms to support bifidobacteria. For example, 1,4-dihydroxy-2-naphtohoic acid produced by *P. freudenreichii* stimulates the growth of bifidobacteria (Falentin et al. 2010). In addition, mouse studies have demonstrated that PAB influence the composition and metabolic activities of the intestinal microflora. At least, PAB possess good credentials to be probiotic itself (Pérez Chaia et al. 1999). Therefore, countless probiotic strains exist and await scientific characterization to fulfill the increasing demands of the functional food market (Movsesyan et al. 2010; Schäffer et al. 2010; Song et al. 2010).

Compared with milk and yogurt, cheese is a preferable product matrix for bifidobacteria. The less acidic pH value and high lipid content of cheese and the anaerobic environment within the cheese body are supportive for bifidobacterial growth during the ripening process. Additionally, bifidobacteria never affect the typical cheese characteristics. For example, supplementation of Canestrato Pugliese hard cheese with bifidobacteria led to a hard ewe cheese body with a considerable number of viable cells, up to 6 log^{10} CFU/g after 90 days of ripening, without affecting the main microbiological, chemical, and sensory characteristics of the cheese (Corbo et al. 2001). The production of functional cheeses might be a suitable alternative to yogurt or fluid milk products.

3.8 Coexistence of LAB and Yeasts in Kefir and Similar Milk Products

Yeasts play an important role in the preparation of fermented dairy products. These provide essential growth nutrients such as amino acids and vitamins, produce ethanol and CO_2, and influence the pH value of the milk product. β-Galactosidase-positive and -negative yeast species are present in sour milk products like kefir, viili, and koumiss grains (Farnworth 2005). Kefir is a probiotic fermented milk beverage with a slight lactic acid buttery taste, carbon dioxide effervescence, and an ethanol concentration of <2% vol/vol (Farnworth 2005; Wang et al. 2008). Kefir grains contain lemon-shaped or long filamentous yeast cells growing in close association with LAB like *Candida friedrichii*, *C. humilis*, *C. kefyr*, *Saccharomyces exiguous*, *S. cerevisiae*, *S. pastorianus*, *Klyveromyces marxianus*, and *K. lactis* var. *lactis* (Farnworth 2005; Lopitz-Otsoa et al. 2006). Kefir grains are an explicit example of symbiosis between eukaryotic yeast and prokaryotic LAB in fermented milk products. The yeasts and bacteria are enclosed by a flexible protein, lipid, and kefiran—EPS matrix, which forms the grains. Kefiran is produced by *Lb. kefiranofaciens*, formerly known as *Lb. kefirgranum* (Vancanneyt et al. 2004). Wang et al. (2008) studied yeast communities in kefir grains and viili starters in Taiwan and found strains of *K. marxianus*, *S. turicensis*, and *Pichia fermentans*, with a distribution of 76%, 22%, and 2%, respectively, in kefir.

Multifaceted interactions between yeasts and LAB influence product characteristics and quality. In kefir, yeasts are the ethanol and carbon dioxide producers. The ability of yeasts to metabolize galactose, lactate, or citrate is the basis for coexistence with galactose-releasing LAB. Yeasts produce several vitamins and release free amino acids and fatty acids from the milk matrix that support the growth of LAB. Further, lactate assimilation of yeasts and lactate production of LAB lead to a metabolic cycle (Lopitz-Otsoa et al. 2006). Characteristic end products of kefir fermentation are lactic acid, acetaldehyde, acetoin, diacetyl, ethanol, and CO_2 (Lopitz-Otsoa et al. 2006). Guzel-Seydim et al. (2000) found that the average pH value of kefir was constant during refrigerated storage periods, though the concentrations of lactic, orotic, and citric acid increased, while pyruvic and hippuric acid, both prominent components during fermentation, vanished during storage. The flavoring acetaldehyde content doubled and acetoin decreased. These findings illustrate the intricate metabolic system of the microorganisms involved in the fermentation.

The microbiological profile of traditional kefir grains alters along the production and storage phases and is hard to imitate for industrial processes. Several studies attempted to produce kefir with desirable pure cultures for industrial production lines, for instance, by combining yogurt cultures and yeasts (Farnworth 2005). Chen et al. (2009) suggested a kefir production with immobilized starter cultures isolated from traditional kefir grains containing *K. marxianus*, *S. turicensis*, *P. fermentans*, *Lb. kefiranofaciens*, *Lb. kefiri*, and *Leu. mesenteroides*. In the final product, the fundamental microbiological and physicochemical characteristics of traditional kefir were reproduced. Starter cultures containing freeze-dried LAB and yeasts from kefir grains are now commercially available and are also valuable for cheese ripening (Farnworth 2005; Kourkoutas et al. 2007).

In contrast to kefir, viili is a complex fermentation with surface-growing, yeast-like fungus *Geotrichum candidum* and several EPS-producing LAB strains, mainly *Lc. lactis* subsp. *cremoris*. Most traditional viili cultures also contain yeasts (Wang et al. 2008). Although viili starters never form grains, it has been suggested that the yeasts support EPS-synthesizing LAB (Wang et al. 2008). New combinations of different microorganisms might expand the range of possible product applications with new sensory characteristics.

3.9 Non-Starter LAB and Strain Selection Strategies for New LAB Starter Cultures

Non-starter LAB (NSLAB) are the non-inoculated and uncontrolled developing part of the microflora of cheese and other milk products. Primarily, adventitious mesophilic lactobacilli like *Lb. casei* subsp. *casei*, *Lb. casei* subsp. *pseudoplantarum*, *Lb. paracasei* subsp. *paracasei*, and *Lb. plantarum*, but also pediococci or micrococci, which are hardly detectable in fresh curd, dominate the microflora of mature cheese (Antonsson et al. 2003). Compared with starter lactobacilli, NSLAB adapt more efficiently to the altering growth conditions of cheese during ripening, which, at the endpoint, is characterized by reduced accessible carbohydrates and nitrogen sources, total moisture of 32%–39%, with 4%–6% salt in moisture, and a pH value of 4.9–5.3 at storage temperatures of between 5°C and 13°C (Antonsson et al. 2003).

Raw milk-derived NSLAB accelerate cheese ripening and increase the level of free amino acids, peptides, and free fatty acids, which intensify the specific flavor notes of the cheese body. Therefore, several studies screened NSLAB in order to select the most suitable strains for application in cheese manufacturing (Antonsson et al. 2003; Sgarbi et al. 2013). NSLAB of Swiss-type cheese are dominated by facultative heterofermentative lactobacilli, namely, *Lb. plantarum*, *Lb. rhamnosus*, *Lb. brevis*, and *Lb. paracasei* (Daly et al. 2010). In Swiss-type cheese, these bacteria can control the activity of PAB and reduce secondary fermentation processes during ripening. Competitive mechanisms related to bacterial citrate metabolism are involved. Thus, PAB more readily develop in cheeses with citrate-negative NSLAB strains (Daly et al. 2010).

The criteria of NSLAB strain selection for production are product compatibility and metabolic properties of the bacteria, considering the desired characteristics of the final product (Arias et al. 2013; François et al. 2007). Interesting sources for NSLAB are traditional fermented milk products such as kurut, koumiss, and gariss (Akabanda et al. 2010; Ashmaig et al. 2009; Sun et al. 2010). Several NSLAB strains were isolated from beyaz cheese made from raw ewe's (Durlu-Ozkaya et al. 2001), dromedary, or goat's milk (Hassaïne et al. 2008; Marokki et al. 2011), from dahi and from lben, which are both traditionally fermented dairy products (Bekkali et al. 2013; Maqsood et al. 2013).

The search for effective strains with specific technological properties always has to consider compatibility with the raw materials and production processes. For example, freeze-dried kefir co-culture does not always fit into feta-type cheese production (Kourkoutas et al. 2007). In yogurt, microbial acidification rate, flavor development properties, and EPS production for increasing gel thickness and reducing syneresis are product-specific selection criteria for NSLAB (François et al. 2007). For instance, typical starter cultures for bovine milk yogurt show advanced growth, organic acid production, and proteolytic activity in camel milk, but the consistency of the product is watery, fragile, and poorly structured (Rahman et al. 2009). In the near future, NSLAB strain selection from the pool in native milk might become an essential tool in the development of fermented dairy products.

3.10 Flavor Development in Milk Products by LAB

The quality of fermented milk products significantly depends on sensory perception. Starter and non-starter bacteria, added and indigenous milk enzymes, together with non-enzymatic, chemical conversions, form the specific product flavor, whereas the compositional balance of distinct aroma and taste molecules, organic acids, lipids, and protein fragments is designed to avoid off-flavors (Smit et al. 2005). The flavor compounds are categorized by the associated metabolic activity, namely, proteolysis and peptidolysis, transaminase pathway, lyase pathway, and non-enzymatic conversion (Smit et al. 2005).

In cheese ripening, glycol-, lipo-, and proteolytic activities, especially casein degradation, as well as the conversion of methionine and several aromatic and branched-chain amino acids are responsible for flavor formation (Smit et al. 2005). Aside from these, fermentative byproducts like organic acids, acetaldehyde, diacetyl, ethanol, and others are of importance. For example, the phosphotransferase system of LAB synthesizes aromatic aldehydes, alcohols, and carboxylic acids. In yogurt, diacetyl, acetoin, acetaldehyde, and acetic acid are responsible for the typical flavor, and succinic acid gives the typical sour taste (Omae et al. 2008). Although lipolysis of LAB is less important, diverse PAB, molds, and yeasts possess vast lipolytic activities during surface ripening of cheese and release several flavor compounds, like free fatty acids, methylketones, secondary alcohols, esters, and lactones.

In flavor development, combinations of LAB with other microorganisms and enzymes influence the sensorial outcome of fermented milk. For instance, *Str. thermophilus* and *Lc. lactis* produce less sour yogurt compared with other LAB combinations (Omae et al. 2008). Different combinations of *Str. thermophilus* and *Lb. delbrueckii* subsp. *bulgaricus* strains in ewe's milk yogurt resulted in dissimilar concentrations of acetaldehydes, acetone, diacetyl, and ethanol (Georgala et al. 1995). In order to produce milk products, new combinations of LAB with NSLAB, yeasts, and other microorganisms might offer alternative aromatizing and flavoring potential, with sufficient acidifying and proteolytic activities (Golić et al. 2013; Price et al. 2014).

3.11 EPS and the Texture of Fermented Milk Products

EPS are a diverse group of polysaccharides with product-specific texture and potential to produce rheological characteristics. EPS-secreting ropy LAB are classified into four groups: group I includes capsule-forming ropy strains and ropy strains secreting unattached EPS, capsule-forming non-ropy strains secreting unattached EPS belong to group II, non-capsule-forming ropy LAB strains are represented in group III, and group IV contains strains producing no or undetectable EPS (Hassan 2008).

In general, EPS are classified into homo- and heteropolysaccharides. Homopolysaccharides contain only D-glucopyranose or D-fructofuranose, and heteropolysaccharides represent complex structures with different monosaccharides, monosaccharide derivatives, or substituted monosaccharides, mostly D-galactose and L-rhamnose (Hassan 2008; Jolly et al. 2002). Several screening methods for EPS secretors have been established, illustrating the variety of LAB and secreted structures (Borgio et al. 2009; Urshev et al. 2008). For example, Frengova et al. (2002) described kefiran composed of glucose and galactose in a 1.0:0.94 ratio. EPS from *Str. thermophilus* are primarily composed of galactose, glucose, and rhamnose, but also contain N-acetyl-galactosamine, fucose, and acetylated galactose (Vaningelgem et al. 2004).

The biosynthesis of heteropolysaccharides involves several steps directly connected to the cell's carbon metabolism (Jolly et al. 2002). Genes of EPS biosynthesis are organized in different gene clusters with similar operon structures coding the carbohydrate chain length, the composition of the repeating units, and the degree of polymerization. Sugar residues from activated sugar donors are sequentially transferred onto a cellular lipophilic carrier, and specific glycosyltransferases assemble the oligosaccharide-repeating unit accordingly. In order to generate EPS with defined characteristics, the structure, composition, and chain length can be altered by genetic engineering. Direct modification of EPS genes leads to different repeating units and chain lengths. For instance, gene shuffling and heterologous gene expression of glycosyltransferase genes influence the type and position of sugars attached to the EPS core structure (Jolly et al. 2002). Stack et al. (2010) expressed the pediococcal glycosyltransferase gene in probiotic *Lb. paracasei* NFBC 338 in order to produce health beneficial β (1, 3) D-glucans. Further, EPS synthesis depends on growth and is influenced by bacterial culture conditions (Audy et al. 2010; Hassan 2008; Jolly et al. 2002). Vaningelgem et al. (2004) observed that in cultures based on milk with an uncontrolled pH value, EPS production by *Str. thermophilus* was limited, while a constant pH value of 5.5–6.6 and adding hydrolyzed whey protein to the medium increased the production rate. Maximal EPS production was observed in the temperature range of between 32°C and 42°C. These results underline that control processes are indispensable for large-scale EPS production.

In general, milk curdling and gelation are normally initiated by a pH value below 5.5. Casein micelles destabilize and the protein matrix irreversibly precipitates. At this pH value, LAB produce adequate amounts of EPS and influence the gelation process. Scanning electron and confocal scanning laser microscopy approaches nicely illustrate the network of bacterial carbohydrate chains and globular milk protein (Ayala-Hernandez et al. 2008). Charged and hydrophobic sides of EPS intercalate into the milk protein matrix and affect texture formation (Hassan 2008; Jolly et al. 2002). Importantly, rheological characteristics of the resulting gels depend on the ratio between EPS and total protein content (Hassan 2008).

The textural and rheological alterations of fermented milk by capsular and ropy EPS have been examined in various studies (Amatayakula et al. 2006). For instance, EPS from Lb. helveticus contribute to water retention in cheese and other milk products (Hassan 2008, Jolly et al. 2002). Ayala-Hernádez et al. (2009) demonstrated that viscosity and viscoelastic properties of solutions containing >2% whey protein were increased by adding EPS. Another study found that pre-fermentation of milk with immobilized EPS-producing LAB improved product texture (Grattepanche et al. 2007). EPS capsules disrupt the continuity of the milk protein network. A smoother product body with higher viscosity and less syneresis is typical for milk products fermented with ropy strains of LAB compared with milk products fermented with non-ropy LAB, like Scandinavian viili and langmjolk (Hassan 2008). On the other hand, yogurt produced with EPS-encapsulated ropy bacteria also shows a milk protein gel structure with decreased syneresis (Jolly et al. 2002). Further, those products indicate a high degree of protein-gel breakdown and hysteresis loop.

EPS are also new important tools for the production of low-fat fermented dairy products. EPS bind water to the product matrix and increase the moisture in the non-fat portion. In cheese, for instance, EPS reduce the rigidity of the protein network and increase the viscosity of the serum phase of cheese (Hassan 2008; Lynch et al. 2014). A disadvantage of EPS in reduced-fat cheese is that after 2–3 months of ripening, bitterness occurs, due to increased remaining chymosin activity. However, Agrawal and Hassan (2007) demonstrated that ultra-filtration of raw milk successfully reduced this bitterness in cheese later on. However, EPS in low-fat or fat-free yogurt help to prevent textural defects and contribute to creaminess and to firmness in mouth feel. Additionally, various studies have demonstrated that EPS increase water holding capacity and lower syneresis of low-fat yogurts (Robitaille et al. 2009).

To date, several studies have focused on health beneficial effects of EPS. For instance, the influence of EPS on the adhesion properties of probiotics and pathogens to intestinal mucus as well as chemopreventive, immunomodulating, and cholesterol-lowering properties of EPS have been investigated (Purohit et al. 2009). In this respect, human *Lactobacillus* and *Bifidobacterium* strains produce EPS with viscosifying properties and enable the combination of probiotic and EPS-associated health claims in milk products (Salazar et al. 2009).

3.12 Importance of the Proteolytic System of LAB in Milk Fermentation

The proteolytic system of LAB consists of cell wall-bound proteinases and several intracellular peptidases that successively degrade casein to long-chain oligopeptides, peptides, and amino acids. Generally, the degradation of casein by LAB is initiated by one type of cell envelope serine protease. Only *Lb. helveticus* and *Lb. bulgaricus* possess two different types of cell envelope proteinases, which differ in substrate specificity and proteolytic activity (Griffiths and Tellez 2013, Savijoki et al. 2006).

Several cell envelope proteinases from LAB have been characterized and categorized, including PrtH from *Lb. helveticus*, PrtR from *Lb. rhamnosus*, PrtS from *Str. thermophilus*, PrtB from *Lb. bulgaricus*, and PrtP from *Lc. lactis* and *Lb. paracasei* (Savijoki et al. 2006). In LAB, the gene loci of proteinases differ from strain to strain. In lactococci, the *prtP* genes are either plasmid- or genome-encoded, whereas in lactobacilli these genes are only genome-encoded (Savijoki et al. 2006). Both extra- and intragenomic gene loci are accessible for genetic engineering transformation techniques.

PrtPs are divided into PI- and PIII-type proteinases. PI and PIII digest preferentially β-casein and $α_{S1}$-casein. κ-casein is a substrate of PIII. Several studies considered substrate specificities of various cell envelope proteinases with regard to β-casein or α_{S1}-casein. For example, the affinity of PI-type proteinases is due to positively charged peptide residues, whereas the PIII-type recognizes negatively and positively charged protein structures (Savijoki et al. 2006; Visser 1993). PrtPs are subclassified into seven groups (a, b, c, d, e, f, and g), in accordance with the cleavage specificities toward the α_{S1}-casein fragment (f1–23). Hebert et al. (2008) identified two new cleavage sites, at the α_{S1}-casein (f1–23) and (Glu_{14}-Val_{15} and Glu_{18}-Asn_{19}) fragments, for the cell envelope proteinase of *Lb. delbrueckii* subsp. *lactis* CRL 581.

After extracellular digestion, peptides released from casein are actively taken up by the bacteria using the oligopeptide transport (Opp) system, which includes adenosine triphosphate-binding cassette transporters (Savijoki et al. 2006). Then, absorbed peptides are degraded by the N-terminal aminopeptidases PepC and PepN, as well as the X-prolyl dipeptidyl aminopeptidase PepX. These metallopeptidases preferentially cleave α_{S1}-casein f1–23 and/or β-casein f193–209 (Savijoki et al. 2006). PepC and PepN of the peptidase system of *Lb. helveticus* possess minor substrate specificity and PepX removes proline-containing dipeptides from the N-terminus of the amino acid chain (Christensen et al. 2003). Other endopeptidases (PepE and PepO) hydrolyze internal peptide bonds independently of the N-terminal amino acid residues. Though the proteolytic activity of bifidobacteria is limited, Janer et al. (2005) investigated the enzymatic ability of *Bb. animalis* subsp. *lactis* to hydrolyze milk proteins and identified a similar PepO intracellular endopeptidase. The peptide hydrolysis profile of this PepO showed that the N-terminal side of phenylalanine residues of oligopeptides and post-proline peptide linkages were predominantly cleaved. Also, peptide hydrolysis profiles of *Lb. helveticus* indicated an unidentified endopeptidase with specificity for peptide bonds C-terminal to proline residues, which might be important for initiating the hydrolysis of α_{S1}-casein (f1–9) and β-casein (f193–209) (Christensen et al. 2003). In contrast, a tripeptidase (PepT) isolated from the same LAB strain cleaves hydrophobic tripeptides, with the highest activity for Met-Gly-Gly (Savijoki and Palva 2000).

Interestingly, growth phase and growth medium affect peptidase activities. Simova and Beshkova (2007) evaluated the proteolytic activities of amino-, di-, tri-, and endopeptidases of *Lb. casei*, *Lb. helveticus*, *Lb. bulgaricus*, and *Str. thermophilus* toward 19 substrates along the distinct phases of microbial growth kinetics. The highest specific activities of peptidases were significantly associated with the late-log phase.

3.13 Proteolytic Systems of LAB in Cheese Production

During cheese ripening, the proteolytic systems of LAB catalyze disintegration of the casein matrix. Texture alterations of the cheese body are attributed to partial digestion of α_{S1}-casein, which leads to the release of flavoring peptides and amino acids. Proteolytic activity of LAB on cheese is strain dependent and affected by technological enzymes such as chymosin. These interactions could result in a higher degree of proteolytic activity, the formation of bitter peptides and unwanted off-flavors, and extensive loss of protein and yield (Savijoki et al. 2006; Visser 1993).

Peptidases of thermophilic LAB like *Lb. helveticus*, *Lb. delbrueckii*, and *Str. thermophilus* predominantly contribute to proteolysis in Swiss-type cheese. Deutsch et al. (2000) compared the proteolytic activity of intracellular thermophilic peptidases. The enzyme system of *Str. thermophilus* was not able to release free proline from the β-casein hydrolysate that is associated with the specific sweet flavor of Swiss-type cheese. In contrast, *Lb. helveticus* flavoring proteases were highly active. In case of all tested LAB, phosphorylated peptides remained undigested. A further important prerequisite for flavor formation in cheese is the release of cytoplasmic peptidases from LAB into the curd by autolysis during fermentation. Autolysis rate of LAB differs markedly and depends on the fermentation conditions. Increased autolysis of starter accelerates ripening and improves flavor development and sensory properties of cheese (Savijoki et al. 2006). In consequence, the selection of starter microorganisms directly influences the flavor characteristics of cheese during production and ripening.

In order to accelerate proteolysis and flavor development of cheese, the proteolytic system of LAB can be genetically modified. For example, recombinant expression of different peptidases derived from *Lb. helveticus* or *Lb. delbrueckii* subsp. *lactis* controlled by constitutive or inducible promoters enhances the proteolytic activity of *Lc. lactis* (Desfossés-Foucault et al. 2013). Overexpression of peptidase gene *pepN* or *pepC* in *Lc. lactis* subsp. *cremoris* NM1 increased the level of specific free amino acids in cheese and improved flavor (McGarry et al. 1994). The formation of flavoring and aromatic molecules depends not only from the proteolytic or peptidolytic activities of enzymes but also from the content of free amino acids in the product, which limits the possibilities to accelerate the ripening process by genetic engineering (McGarry et al. 1994; Savijoki et al. 2006).

3.14 Debittering of Milk Products by Proteolytic LAB

Chemical and biocatalytic proteolysis of milk protein matrices always carry the risk of bittering, which is a serious quality concern, especially in Gouda and Cheddar cheese production. For instance, accumulation of hydrophobic peptides like proline peptides causes bitterness. Cell envelope proteinases of particular LAB starter strains influence the formation of bitter peptides. For instance, less bitterness is generated from casein by the *Lc. lactis* subsp. *cremoris* AM_1 (PIII-type) proteinase compared with the *Lc. lactis* subsp. *cremoris* HP (PI-type) proteinase. The PIII-type proteinase initially cleaves large C-terminal fragments, mainly the β-casein fragment (f53–209), whereas the PI-type proteinase activity directly leads to short-chain and bitter tasting C-terminal casein fragments (Savijoki et al. 2006; Visser 1993).

On the other hand, bitter peptides can be degraded by sequential action of peptidases, especially PepN, PepX, PepO2, and PepO3. For example, recombinant *Lc. lactis* expressing *Lb. helveticus* CNRZ32 PepO2 and PepO3 in combination with PepN is a useful tool to reduce bitterness in cheese (Savijoki et al. 2006). Degradation of bitter peptides derived from β-casein by LAB peptidases was recently examined by Shimamura et al. (2009). The ability of different lactococcal cheese starters to degrade the bitter peptide Gly-Pro-Phe-Pro-Ile-Ile-Val varied depending on the type of LAB. The highest debittering capacity was shown by *Lc. lactis* subsp. *lactis* 527. Sridhar et al. (2005) identified three putative debittering endopeptidase genes, *pepO3*, *pepF*, and *pepE2*, from *Lb. helveticus* CNRZ32. The potential of using transgenetically expressed CNRZ32 endopeptidases PepE, PepE2, PepF, PepO, PepO2, and PepO3 to hydrolyze the specific β-casein bitter peptide (f193–209) and $α_{S1}$-casein peptide (f1–9) in cheese was examined in the same study. PepO3 hydrolyzed both peptides, and PepE and PepF had distinctive peptide-cleavage specificities toward $α_{S1}$-casein (f1–9) and β-casein (f193–209) fragments, whereas PepE2 and PepO did not hydrolyze either peptide at all.

Debittering potential of *Lb. helveticus* CNRZ32 endopeptidases PepO2 and PepO3 expressed in *Lc. lactis* has been proven in Cheddar cheese serum during ripening. Another study verified that X-prolyl dipeptidyl peptidase of *Lb. casei* subsp. *casei* LLG degraded two bitter peptides (f53–97 and f203–209). Also, this peptidase debitters cheese by degrading hydrophobic peptides with Ala-Pro-Phe-Pro-Glu-Val and Phe-Leu-Leu residues (Habibi-Najafi and Lee 2007). Further, the salt concentration of cheese often influences the activity of proteolytic debittering peptidase systems and the formation of hydrophobic bitter peptides complexes, especially in the case of the effects of C-terminal β-casein fragments (Visser 1993).

3.15 Conclusion

In conclusion, LAB offer a broad spectrum of functional properties for milk fermentation that determine the appearance, quality, and safety of the final product. Distinctively selected LAB from the milk germ pool and milk-associated NSLAB, as well as new combinations of LAB with molds, yeasts, and enzymes, will lead to novel acidification or efficient proteolytic strategies and compatible tools for flavor and texture development of fermented milk products. In this context, oligosaccharides secreted by LAB that interact with the milk protein matrix will become more important. Finally, complex proteinase systems of LAB combined with other bacteria and microorganisms are available to enhance ripening processes of milk products, develop new product-specific flavors, and prevent bittering. This open field of research might establish the starting point of a new era of product design for fermented milk products.

References

Abd El-Salam, M.H., El-Shafei, K., Sharaf, O.M., Effat, B.A., Asem, F.M., and El-Aasar, M. 2010. Screening of some potentially probiotic lactic acid bacteria for their ability to synthesis conjugated linoleic acid. *International Journal of Dairy Technology* 63(1): 62–69.

Agrawal, P. and Hassan, A.N. 2007. Ultrafiltered milk reduces bitterness in reduced-fat Cheddar cheese made with an exopolysaccharide-producing culture. *Journal of Dairy Science* 90: 3110–3117.

Akabanda, F., Owusu-Kwarteng, J., Glover, R.L.K., and Tano-Debrah, K. 2010. Microbiological characteristics of Ghanaian traditional fermented milk product, *Nunu*. *Nature and Science* 8(9): 178–187.

Al-Awwad, N.J., Haddadin, M.S., and Takruri, H.R. 2009. The characteristics of locally isolated *Lactobacillus acidophilus* and *Bifidobacterium infantis* isolates as probiotics strains. *Jordan Journal of Agricultural Sciences* 5(2): 192–206.

Amatayakula, T., Halmosb, A.L., Sherkatb, F., and Shah, N.P. 2006. Physical characteristics of yoghurts made using exopolysaccharide-producing starter cultures and varying casein to whey protein ratios. *International Dairy Journal* 16: 40–51.

Antonsson, M., Molin, G., and Ardö, Y. 2003. *Lactobacillus* strains isolated from Danbo cheese as adjunct cultures in a cheese model system. *International Journal of Food Microbiology* 85: 159–169.

Antunes, A.E.C., Grael, E.T., Moreno, I., Rodrigues, L.G., Dourado, F.M., Saccaro, D.M., and Lerayer, A.L.S. 2007. Selective enumeration and viability of *bifidobacterium animalis* subsp. *lactis* in a new fermented milk product. *Brazilian Journal of Microbiology* 38: 173–177.

Arias, L.G., Fernández, D., Sacristán, N., Arenas, R., Fresno, J.M., and Tornadijo, E. 2013. Enzymatic activity, surface hydrophobicity and biogenic amines production in lactic acid bacteria isolated from an artisanal Spanish cheese. *African Journal of Microbiology Research* 7(19): 2114–2118.

Arioli, S., Roncada, P., Salzano, A.M., Deriu, F., Corona, S., Guglielmetti, S., Bonizzi, L., Scaloni, A. and Mora, D. 2009. The relevance of carbon dioxide metabolism in *Streptococcus thermophilus*. *Microbiology* 155: 1953–1965.

Ashmaig, A., Hasan, A., and El Gaali, E. 2009. Identification of lactic acid bacteria isolated from traditional Sudanese fermented camel's milk (*Gariss*). *African Journal of Microbiology Research* 3(8): 451–457.

Atta, M.S., Hashim, M.M., Zia, A., and Masud, T. 2009. Influence of different amounts of starter cultures on the quality of yoghurt prepared from buffalo milk. *Pakistan Journal of Zoology Supplementary Series* 9: 129–134.

Audy, J., Labrie, S., Roy, D., and LaPointe, G. 2010. Sugar source modulates exopolysaccharide bio-synthesis in *Bifidobacterium longum* subsp. *longum* CRC 002. *Microbiology* 156: 653–664.

Ayala-Hernandez, I., Goff, H.D., and Corredig, M. 2008. Interactions between milk proteins and exo-polysaccharides produced by *Lactococcus lactis* observed by scanning electron microscopy. *Journal of Dairy Science* 91: 2583–2590.

Ayala-Hernández, I., Hassan, A.N., Goff, H.D., and Corredig, M. 2009. Effect of protein supplementation on the rheological characteristics of milk permeates fermented with exopolysaccharide-producing *Lactococcus lactis* subsp. *cremoris*. *Food Hydrocolloids* 23: 1299–1304.

Barboza, M., Sela, D.A., Pirim, C., LoCascio, R.G., Freeman, S.L., German, J.B., Mills, D.A., and Lebrilla, C.B. 2009. Glycoprofiling bifidobacterial consumption of galacto-oligosaccharides by mass spectrometry reveals strain-specific, preferential consumption of glycans. *Applied and Environmental Microbiology* 75(23): 7319–7325.

Beermann, C. and Hartung, J. 2012. Current enzymatic fermentation processes for novel milk product applications. *European Food Research Technology* 235(1): 1–12.

Beermann, C. and Hartung, J. 2013. Physiological properties of milk ingredients released by fermentation. *Food and Function* 4(2): 185–199.

Bekkali, N., El Amraoui, A., Hammoumi, A., Poinsot, V., and Belkhou, R. 2013. Use of *Lactococci* isolated from Moroccan traditional dairy product: Development of a new starter culture. *African Journal of Biotechnology* 12(38): 5662–5669.

Bergamini, C.V., Hynes, E.R., Candioti, M.C., and Zalazar, C.A. 2009. Multivariate analysis of pro-teolysis patterns differentiated the impact of six strains of probiotic bacteria on a semi-hard cheese. *Journal of Dairy Science* 92: 2455–2467.

Borgio, J.F., Bency, B.J., Ramesh, S., and Amuthan, M. 2009. Exopolysaccharide production by *Bacillus subtilis* NCIM 2063, *Pseudomonas aeruginosa* NCIM 2862 and *Streptococcus mutans* MTCC 1943 using batch culture in different media. *African Journal of Biotechnology* 9(20): 5454–5457.

Chen, T.H., Wang, S.Y., Chen, K.N., Liu, J.R., and Chen, M.J. 2009. Microbiological and chemical properties of kefir manufactured by entrapped microorganisms isolated from kefir grains. *Journal of Dairy Science* 92: 3002–3013.

Christensen, J.E., Broadbent, J.R., and Steele, J.L. 2003. Hydrolysis of casein-derived peptides αS1-casein(f1-9) and β-casein(f193-209) by *Lactobacillus helveticus* peptidase deletion mutants indicates the presence of a previously undetected endo-peptidase. *Applied and Environmental Microbiology* 69(2): 1283–1286.

Corbo, M.R., Albenzio, M., De Angelis, M., Sevi, A., and Gobbetti, M. 2001. Microbiological and biochemical properties of Canestrato Pugliese hard cheese supplemented with bifidobacteria. *Journal of Dairy Science* 84: 551–561.

Daly, D.F.M., Mc Sweeney, P.L.H., and Sheehan, J.J. 2010. Split defect and secondary fermentation in Swiss-type cheeses—A review. *Dairy Science and Technology* 90: 3–26.

Desfossés-Foucault, E., LaPointe, G., and Roy, D. 2013. Dynamics and rRNA transcriptional activity of lactococci and lactobacilli during Cheddar cheese ripening. *International Journal of Food Microbiology* 166(1):117–124.

Deutsch, S.M., Molle, D., Gagnaire, V., Piot, M., Atlan, D., and Lortal, S. 2000. Hydrolysis of sequenced β-casein peptides provides new insight into peptidase activity from thermophilic lactic acid bacteria and highlights intrinsic resistance of phosphopeptides. *Applied and Environmental Microbiology* 66(12): 5360–5367.

Durlu-Ozkaya, F., Xanthopoulos, V., Tunail, N., and Litopoulou-Tzanetaki, E. 2001. Technologically important properties of lactic acid bacteria isolates from Beyaz cheese made from raw ewes' milk. *Journal of Applied Microbiology* 91: 861–870.

El-Ghaish, S., Dalgalarrondo, M., Choiset, Y., Sitohy, M., Ivanova, I., Haertlé, T., and Chobert, J.M. 2010. Characterization of a new isolate of *Lactobacillus fermentum* IFO 3956 from Egyptian Ras cheese with proteolytic activity. *European Food Research and Technology* 230: 635–643.

El-Tanboly, E.S., El-Hofi, M., Abd-Rabou, N.S., and El-Desoki, W. 2010. Contribution of mesophilic starter and adjunct lactobacilli to proteolysis and sensory properties of semi hard cheese. *The Journal of American Science* 6(9): 697–703.

El-Tanboly, E.S., El-Hofi, M., Youssef, Y.B., El-Desoki, W., and Jalil, R.A. 2010. Influence of freezes-hocked mesophilic lactic starter bacteria and adjunct lactobacilli on the rate of ripening Gouda cheese and flavour development. *The Journal of American Science* 6(11): 465–471.

Falentin, H., Deutsch, S.M., Jan, G., Loux, V., Thierry, A., Parayre, S., Maillard, M.B. et al. 2010. The complete genome of *Propionibacterium freudenreichii* CIRM-BIA1^T, a hardy actinobacterium with food and probiotic applications. *PLoS ONE* 5(7): e11748.

Farnworth, E.R. 2005. Kefir—A complex probiotic. *Food Science and Technology Bulletin: Functional Foods* 2(1): 1–17.

François, Z.N., El Hoda, N., Florence, F.A., Paul, M.F., Félicité, T.M., and El Soda, M. 2007. Biochemical properties of some thermophilic lactic acid bacteria strains from traditional fermented milk relevant to their technological performance as starter culture. *Biotechnology* 6(1): 14–21.

Frengova, G.I., Simova, E.D., Beshkova, D.M., and Simov, Z.I. 2002. Exopolysaccharides produced by lactic acid bacteria of kefir grains. *Zeitschrift für Naturforschung* 57c: 805–810.

Genay, M., Sadat, L., Gagnaire, V., and Lortal, S. 2009. PrtH2, not *prtH*, is the ubiquitous cell wall proteinase gene in *Lactobacillus helveticus*. *Applied and Environmental Microbiology* 75(10): 3238–3249.

Georgala, A.I.K., Tsakalidou, E., Kandarakis, I., and Kalantzopoulos, G. 1995. Flavour production in ewe's milk and ewe's milk yoghurt, by single strains and combinations of *Streptococcus thermophilus* and *Lactobacillus delbrueckii* subsp. *bulgaricus*, isolated from traditional Greek yoghurt. *Lait* 75: 271–283.

Gitton, C., Meyrand, M., Wang, J., Caron, C., Trubuil, A., Guillot, A., and Mistou, M.Y. 2005. Proteomic signature of *Lactococcus lactis* NCDO763 cultivated in milk. *Applied and Environmental Microbiology* 71(11): 7152–7163.

Golić, N., Čadež, N., Terzić-Vidojević, A., Šuranská, H., Beganović, J., Lozo, J., Kos, B., and Raspor, J.S.P. 2013. Evaluation of lactic acid bacteria and yeast diversity in traditional white pickled and fresh soft cheeses from the mountain regions of Serbia and lowland regions of Croatioa. *International Journal of Food Microbiology* 166(2): 294–300.

Grattepanche, F., Audet, P., and Lacroix, C. 2007. Enhancement of functional characteristics of mixed lactic culture producing Nisin Z and exo-polysaccharides during continuous prefermentation of milk with immobilized cells. *Journal of Dairy Science* 90: 5361–5373.

Griffiths, M.W. and Tellez, A.M. 2013. *Lactobacillus helveticus*: The proteolytic system. *Frontiers in Microbiology* 4: 30.

Guzel-Seydim, Z., Seydim, A.C., and Greene, A.K. 2000. Organic acids and volatile flavour components evolved during refrigerated storage of kefir. *Journal of Dairy Science* 83: 275–277.

Habibi-Najafi, M.B. and Lee, B.H. 2007. Debittering of tryptic digests from β-casein and enzyme modified cheese by x-prolyl dipeptidylpeptidase from *Lactobacillus casei* ssp. *casei*. LLG. *Iranian Journal of Science and Technology, Transaction A* 31(A3): 263–270.

Hansen, S.W. and Martinussen, J. 2009. Strains of *Lactococcus lactis* with a partial pyrimidine requirement show sensitivity toward aspartic acid. *Dairy Science and Technology* 89: 125–137.

Hassaïne, O., Zadi-Karam, H., and Karam, N.E. 2008. Phenotypic identification and technological properties of lactic acid bacteria isolated from three breeds dromedary raw milks in south Algeria. *Emirates Journal of Food and Agriculture* 20(1): 46–59.

Hassan, A. and Amjad, I. 2010. Nutritional evaluation of yoghurt prepared by different starter cultures and their physiochemical analysis during storage. *African Journal of Biotechnology* 9 (20):2913-2917.

Hassan, A.N. 2008. *ADSA Foundation Scholar Award:* possibilities and challenges of exopolysaccharide-producing lactic cultures in dairy foods. *Journal of Dairy Science* 91: 1282–1298.

Haug, A., Høstmark, A.T., and Harstad, O.M. 2007. Bovine milk in human nutrition—A review. *Lipids in Health and Disease* 6: 25.

Hebert, E.M., Mamone, G., Picariello, G., Raya, R.R., Savoy, G., Ferranti, P., and Addeo, F. 2008. Characterization of the pattern of α_{s1}- and β-casein breakdown and release of a bioactive peptide by a cell envelope proteinase from *Lactobacillus delbrueckii* subsp. *lactis* CRL 581. *Applied and Environmental Microbiology* 74(12): 3682–3689.

Herve-Jimenez, L., Guillouard, I., Guedon, E., Boudebbouze, S., Hols, P., Monnet, V., Maguin, E., and Rul, F. 2009. Postgenomic analysis of *Streptococcus thermophilus* cocultivated in milk with *Lactobacillus delbrueckii* subsp. *bulgaricus*: Involvement of nitrogen, purine, and iron metabolism. *Applied and Environmental Microbiology* 75(7): 2062–2073.

Jamaly, N., Benjouad, A., Comunian, R., Daga, E., and Bouksaim, M. 2010. Characterization of *Enterococci* isolated from Moroccan dairy products. *African Journal of Microbiology Research* 4(16): 1768–1774.

Janer, C., Arigoni, F., Lee, B.H., Peláez, C., and Requena, T. 2005. Enzymatic ability of *Bifidobacterium animalis* subsp. *lactis* to hydrolyze milk proteins: Identification and characterization of endo-peptidase O. *Applied and Environmental Microbiology* 71(12): 8460–8465.

Jolly, L., Vincent, S.J.F., Duboc, P., and Neeser, J.R. 2002. Exploiting exopolysaccharides from lactic acid bacteria. *Antonie van Leeuwenhoek* 82: 367–374.

Kabeir, B.M., Yazid, A.M., Hakim, M.N., Khahatan, A., Shaborin, A., and Mustafa, S. 2009. Survival of *Bifidobacterium pseudocatenulatum* G4 during the storage of fermented peanut milk (PM) and skim milk (SM) products. *African Journal of Food Sciences* 3(6): 150–155.

Khalid, N.M. and Marth, E.H. 1990. Lactobacilli—Their enzymes and role in ripening and spoilage of cheese: A review. *Journal of Dairy Science* 73: 2669–2684.

Klaenhammer, T., Altermann, E., Arigoni, F., Bolotin, A., Breidt, F., Broadbent, J., Cano, R. et al. 2002. Discovering lactic acid bacteria by genomics. *Antonie van Leeuwenhoek* 82: 29–58.

Kourkoutas, Y., Bosnea, L., Taboukos, S., Baras, C., Lambrou, D., and Kanellaki, M. 2006. Probiotic cheese production using *Lactobacillus casei* cells immobilized on fruit pieces. *Journal of Dairy Science* 89: 1431–1151.

Kourkoutas, Y., Sipsas, V., Papavasiliou, G., and Koutinas, A.A. 2007. An economic evaluation of freeze-dried kefir starter culture production using whey. *Journal of Dairy Science* 90: 2175–2180.

Liu, M., Siezen, R.J., and Nauta, A. 2009. *In silico* prediction of horizontal gene transfer events in *Lactobacillus bulgaricus* and *Streptococcus thermophilus* reveals protocooperation in yoghurt manufacturing. *Applied and Environmental Microbiology* 75(12): 4120–4129.

Lopitz-Otsoa, F., Rementeria, A., Elguezabal, N., and Garaizar, J. 2006. Kefir: A symbiotic yeasts-bacteria community which alleged healthy capabilities. *Revista Iberoamericana de Micologia* 23: 67–74.

Lynch, K.M., McSweeney, P.L.H., Arendt, E.K., Uniacke-Lowe, T., Galle, G., and Coffey, A., 2014. Isolation and characterization of exopolysaccharide-producing *Weisella* and *Lactobacillus* and their application as adjunct cultures in Cheddar cheese. *International Dairy Journal* 34(1): 125–134.

Maqsood, S., Hasan, F., and Masud, T. 2013. Characterization of lactic acid bacteria isolated from indigenous dahi samples for potential source of starter culture. *African Journal of Biotechnology* 12(33): 5226–5231.

Marokki, A., Zúñiga, M., Kihal, M., and Pérez-Martínez, G. 2011. Characterization of *Lactobacillus* from Algerian goat's milk based on phenotypic, 16S sequencing and their technological properties. *Brazilian Journal of Microbiology* 42: 158–171.

McGarry, A., Law, J., Coffey, A., Daly, C., Fox, P.F., and Fitzgerald, G.F. 1994. Effect of genetically modifying the lactococcal proteolytic system on ripening and flavour development in Cheddar cheese. *Applied and Environmental Microbiology* 60(12): 4226–4233.

McSweeney, P.L.H. and Sousa, M.J. 2000. Biochemical pathways for the production of flavour compounds in cheeses during ripening: A review. *Lait* 80: 293–324.

Mikelsaar, M. and Zilmer, M. 2009. *Lactobacillus fermentum* ME-3—An anti-microbial and antioxidative probiotic. *Microbial Ecology in Health and Disease* 21: 1–27.

Movsesyan, I., Ahabekyan, N., Bazukyan, I., Madoyan, R., Dalgalarrondo, M., Chobert, J., Popov, Y., and Haertlé, T. 2010. Properties and survival under simulated gastrointestinal conditions of lactic acid bacteria isolated from Armenian cheeses and matsuns. *Biotechnology & Biotechnological Equipment* 24(Special Edition): 444–449.

Oliveira, M.N., Almeida, K.E., Damin, M.R., Rochat, T., Gratadoux, J.J., Miyoshi, A., Langella, P., and Azevedo, V. 2009. Behavior and viability of spontaneous oxidative stress-resistant *Lactococcus lactis* mutants in experimental fermented milk processing. *Genetics and Molecular Research* 8(3): 840–847.

Omae, M., Maeyama, Y., and Nishimura, T. 2008. Sensory properties and taste compounds of fermented milk produced by *Lactococcus lactis* and *Streptococcus thermophilus*. *Food Science and Technology Research* 14(2): 183–189.

O'Sullivan, O., O'Callaghan, J., Sangrador-Vegas, A., McAuliffe, O., Slattery, L., Kaleta, P., Callanan, M., Fitzgerald, G.F., Ross, R.P., and Beresford, T. 2009. Comparative genomics of lactic acid bacteria reveals a niche-specific gene set. *BMC Microbiology* 9: 50.

Pérez Chaia, A., Zárate, G., and Oliver, G. 1999. The probiotic properties of propionibacteria. *Lait* 79: 175–185.

Petschow, B.W. and Talbott, R.D. 1990. Growth promotion of *Bifidobacterium* species by whey and casein fractions from human and bovine milk. *Journal of Clinical Microbiology* 28(2): 287–292.

Pogačić, T., Mancini, A., Santarelli, M., Bottari, B., Lazzi, C., Neviani, E., and Gatti, M. 2013. Diversity and dynamic of lactic acid bacteria strains during aging of a long ripened hard cheese produced from raw milk and undefined natural starter. *Food Microbiology* 36(2): 207–215.

Price, J.E., Linforth, R.S.T., Dodd, C.E.R., Phillips, C.A., Hewson, L., Hort, J., and Gkatzionis, K. 2014. Study of the influence of yeast inoculums concentration (*Yarrowia lipolytica* and *Kluyveromyces lactis*) on blue cheese aroma development using microbiological models. *Food Chemistry* 145: 464–472.

Purohit, D.H., Hassan, A.N., Bhatia, E., Zhang, X., and Dwivedi, C. 2009. Rheological, sensorial, and chemopreventive properties of milk fermented with exopolysaccharide-producing lactic cultures. *Journal of Dairy Science* 92: 847–856.

Rademaker, J.L.W., Herbet, H., Starrenburg, M.J.C., Naser, S.M., Gevers, D., Kelly, W.J., Hugenholtz, J., Swings, J., and van Hylckama Vlieg, J.E.T. 2007. Diversity analysis of dairy and nondairy *Lactococcus lactis* isolates, using a novel multilocus sequence analysis scheme and (GTG) 5-PCR fingerprinting. *Applied and Environmental Microbiology* 73(22): 7128–7137.

Raeisi, S.N., Ouoba, L.I.I., Farahmand, N., Sutherland, J., and Ghoddusi, H.B. 2013. Variation, viability and validity of bifidobacteria in fermented milk products. *Food Control* 34(2): 691–697.

Rahman, I.E.A., Dirar, H.A., and Osman, M.A. 2009. Microbiological and biochemical changes and sensory evaluation of camel milk fermented by selected bacterial starter cultures. *African Journal of Food Science* 3(12): 398–405.

Robitaille, G., Moineau, S., St-Gelais, D., Vadeboncoeur, C., and Britten, M. 2007. Galactose metabolism and capsule formation in a recombinant strain of *Streptococcus thermophilus* with a galactose-fermenting phenotype. *Journal of Dairy Science* 90: 4051–4057.

Robitaille, G., Tremblay, A., Moineau, S., St-Gelais, D., Vadeboncoeur, C., and Britten, M. 2009. Fat-free yoghurt made using a galactose-positive exopolysaccharide-producing recombinant strain of *Streptococcus thermophilus*. *Journal of Dairy Science* 92: 477–482.

Roy, D. 2005. Technological aspects related to the use of bifidobacteria in dairy products. *Lait* 85: 39–56.

Salazar, N., Prieto, A., Leal, J.A., Mayo, B., Bada-Gancedo, J.C., de los Reyes-Gavilán, C.G., and Ruas-Madiedo, P. 2009. Production of exopolysaccharides by *Lactobacillus* and *Bifidobacterium* strains of human origin, and metabolic activity of the producing bacteria in milk. *Journal of Dairy Science* 92: 4158–4168.

Savijoki, K., Ingmer, H., and Varmanen, P. 2006. Proteolytic systems of lactic acid bacteria. *Applied Microbiology and Biotechnology* 71: 394–406.

Savijoki, K. and Palva, A. 2000. Purification and molecular characterization of a tripeptidase (PepT) from *Lactobacillus helveticus*. *Applied and Environmental Microbiology* 66(2): 794–800.

Schäffer, B., Keller, B., Daróczi, L., and Lőrinczy, D. 2010. Examination of growth of probiotic microbes by an isoperibolic calorimetry. *Journal of Thermal Analysis and Calorimetry* 102: 9–12.

Sela, D.A., Chapman, C., Adeuya, A., Kim, J.H., Chen, F., Whitehead, T.R., Lapidus, A. et al. 2008. The genome sequence of *Bifidobacterium longum* subsp. *infantis* reveals adaptations for milk utilization within the infant microbiome. *Proceedings of the National Academy of Sciences* 105(48): 18964–18969.

Seo Mi, D., Kim, S.Y., Eom, H.J., and Han, N.S. 2007. Synbiotic synthesis of oligosaccharides during milk fermentation by addition of *Leuconostoc* starter and sugars. *Journal of Microbiology and Biotechnology* 17(11): 1758–1764.

Sfakianakis, P. and Tzia, C. 2014. Conventional and innovative processing of milk for yogurt manufacture, development of texture and flavour: A review. *Foods* 3: 176–193.

Sgarbi, E., Lazzi, C., Tabanelli, G., Gatti, M., Neviani, E., and Gardini F. 2013. Nonstarter lactic acid bacteria volatilomes produced using cheese components. *Journal of Dairy Science* 96(7): 4223–4234.

Shimamura, T., Nishimura, T., Iwasaki, A., Odake, S., and Akuzawa, R. 2009. Degradation of a bitter peptide derived from casein by lactic acid bacterial peptidase. *Food Science and Technology Research* 15(2): 191–194.

Simova, E. and Beshkova, D. 2007. Effect of growth phase and growth medium on peptidase activities of starter lactic acid bacteria. *Lait* 87: 555–573.

Smit, G., Smit, B.A., and Engels, W.J.M. 2005. Flavour formation by lactic acid bacteria and biochemical flavour profiling of cheese products. *FEMS Microbiology Reviews* 29: 591–610.

Song, T.S., Kim, J.Y., Kim, K.H., Jung, B.M., Yun, S.S., and Yoon, S.S. 2010. In vitro evaluation of probiotic lactic acid bacteria isolated from dairy and non-dairy environments. *Food Science and Biotechnology* 19(1): 19–25.

Sridhar, V.R., Hughes, J.E., Welker, D.L., Broadbent, J.R., and Steele, J.L. 2005. Identification of endo-peptidase genes from the genomic sequence of *Lactobacillus helveticus* CNRZ32 and the role of these genes in hydrolysis of model bitter peptides. *Applied and Environmental Microbiology* 71(6): 3025–3032.

Stack, H.M., Kearney, N., Stanton, C., Fitzgerald, G.F., and Ross, R.P. 2010. Association of beta-glucan endogenous production with increased stress tolerance of intestinal lactobacilli. *Applied and Environmental Microbiology* 76(2): 500–507.

Sun, Z., Liu, W., Zhang, J., Yu, J., Zhang, W., Cai, C., Menghe, B., Sun, T., and Zhang, H. 2010. Identification and characterization of the dominant lactobacilli from Koumiss in China. *Journal of General and Applied Microbiology* 56: 257–265.

Tachon, S., Brandsma, J.B., and Yvon, M. 2010. NoxE NADH oxidase and the electron transport chain are responsible for the ability of *Lactococcus lactis* to decrease the redox potential of milk. *Applied and Environmental Microbiology* 76(5): 1311–1319.

Urshev, Z.L., Dimitrov, Z.P., Fatchikova, N.S., Petrova, I.G., and Ishlimova, D.I. 2008. Partial characterization and dynamics of synthesis of high molecular mass exo-polysaccharides from *Lactobacillus delbrueckii* ssp. *bulgaricus* and *Streptococcus thermophilus*. *World Journal of Microbiology and Biotechnology* 24: 171–179.

Vancanneyt, M., Mengaud, J., Cleenwerck, I., Vanhonacker, K., Hoste, B., Dawyndt, P., Degivry, M.C., Ringuet, D., Janssens, D., and Swings, J. 2004. Reclassification of *Lactobacillus kefirgranum* Takizawa et al. 1994 as *Lactobacillus kefiranofaciens* subsp. *kefirgranum* subsp. nov. and emended description of *L. kefiranofaciens* Fujisawa et al. 1988. *International Journal of Systematic and Evolutionary Microbiology* 54: 551–556.

Van de Guchte, M., Penaud, S., Grimaldi, C., Barbe, V., Bryson, K., Nicolas, P., Robert, C. et al. 2006. The complete genome sequence of *Lactobacillus bulgaricus* reveals extensive and ongoing reductive evolution. *Proceedings of the National Academy of Sciences* 103(24): 9274–9279.

Vaningelgem, F., Zamfir, M., Adriany, T., and De Vuyst, L. 2004. Fermentation conditions affecting the bacterial growth and exopolysaccharide production by *Streptococcus thermophilus* ST 111 in milk-based medium. *Journal of Applied Microbiology* 97:1257–1273.

Visser, S. 1993. Proteolytic Enzymes and their relation to cheese ripening and flavour: An overview. *Journal of Dairy Science* 76:329–350.

Wang, S.Y., Chen, H.C., Liu, J.R., Lin, Y.C., and Chen, M.J. 2008. Identification of yeasts and evaluation of their distribution in Taiwanese kefir and viili starters. *Journal of Dairy Science* 91: 3798–3805.

Warminska-Radyko, I., Laniewska-Moroz, L., and Babuchowski, A. 2002. Possibilities for stimulation of *Bifidobacterium* growth by propionibacteria. *Lait* 82: 113–121.

Zalán, Z., Hudáček, J., Štětina, J., Chumchalová, J., and Halász, A. 2010. Production of organic acids by *Lactobacillus* strains in three different media. *European Food Research and Technology* 230: 395–404.

PART II
STARTER CULTURES IN MILK FERMENTATION

4

Types and Application of Lactic Starters

PRADIP BEHARE, JASHBHAI B. PRAJAPATI, AND RAMESHWAR SINGH

Contents

4.1	Introduction	84
	4.1.1 Functions of Starter Cultures	85
4.2	Characteristics of Dairy Starter Cultures	85
	4.2.1 Bacteria	87
	4.2.2 Yeasts	92
	4.2.3 Molds	92
4.3	Types of Starter Cultures	93
	4.3.1 Composition of Starter Flora	93
	4.3.1.1 Single Strain Starter	93
	4.3.1.2 Mixed Strain Starters	93
	4.3.1.3 Paired Compatible Starters	93
	4.3.1.4 Multiple Strain Starters	93
	4.3.2 Growth Temperature of Starter Culture	94
	4.3.2.1 Mesophilic Starters	94
	4.3.2.2 Thermophilic Starters	94
	4.3.3 Production of End Product by the Starter Culture	94
	4.3.3.1 Lactic Starters	94
	4.3.3.2 Non-Lactic Starters	94
	4.3.4 Physical Forms of Starter Culture	94
	4.3.4.1 Liquid Starter	94
	4.3.4.2 Frozen Starter Culture	95
	4.3.4.3 Frozen Concentrated Starter Culture	95
	4.3.4.4 Dried Starter Cultures	95
	4.3.4.5 Dried Concentrated Cultures	95
	4.3.5 Product for Which Starters Used	96
4.4	Propagation of Starters	96
4.5	Problems Associated with Production of Starter Culture	98

4.6 Application of Starter Cultures	99
4.7 Conclusion	105
References	106

4.1 Introduction

Starter cultures are used in the manufacturing of different types of fermented milk and dairy products including yogurt, dahi, cultured buttermilk, sour cream, quarg, kefir, koumiss, and cheese. The use of these cultures for the preparation of products has been practiced since time immemorial. Traditionally, the common method was to use the previous-day's product (i.e., milk, dahi, whey, buttermilk, etc.) as an inoculum to produce the fresh batches of fermented product. Such methods were not reliable and often resulted in off-flavor, inconsistent products, and product failure due to undesirable fermentation. These drawbacks were mainly due to the lack of scientific knowledge of starter culture technology. The ideal of pure culture for making good-quality fermented milk became more visible in the mid-nineteenth century, where starters were widely studied and their metabolisms were well established. This leads to manufacturers handling large volumes of milk started selection of appropriate starters to obtain uniform quality product. This made the selection of appropriate starters even for manufacturers handling large volumes of milk to obtain uniform quality of product. Thereafter, companies started producing pure cultures for commercial application.

Starter cultures may be defined as the carefully selected group of microorganisms that are deliberately added to milk and milk products to bring desirable fermentative changes. These have multifunctional role in dairy fermentation; the primary one is to produce lactic acid, hence, popularly called as lactic acid bacteria (LAB). Besides production of lactic acid, certain cultures perform secondary functions such as the production of acetic, propionic, and folic acids, CO_2, H_2O_2, ethanol, bacteriocins, exopolysaccharides (EPS), and so on (Cintas et al. 2001; Padalino et al. 2012; Yang et al. 2012). This chapter reviews the characteristics of different starters, their types, scaling up, problems associated with starters, and their application.

4.1.1 Functions of Starter Cultures

Different functions of starter cultures are to

- Produce lactic acid and other metabolites (i.e., alcohol, CO_2, propionic acid, acetic acid, etc.).
- Produce aromatic compounds like diacetyl, acetaldehyde, and acetoin.
- Control the growth of pathogens and spoilage causing microorganisms.
- Produce certain vitamins (i.e., folic acid, vitamin B_{12}, niacin, etc.).
- Bring proteolytic and lipolytic activities.
- Improve body and texture of certain products by producing EPS.
- Assist in overall acceptability of the final product.

4.2 Characteristics of Dairy Starter Cultures

Starter cultures form a large group of microorganisms that include bacteria, yeasts, and molds (Table 4.1) for particular fermented milk products. Although starter cultures are genetically diverse, the common characteristics of these groups include Gram-positive, non-spore

Table 4.1 Microorganisms Used as Starter Cultures

MICROORGANISM(S)	MAJOR CHARACTERISTICS
A. BACTERIA	
Lc. lactis subsp. lactis, Lc. lactis subsp. cremoris, Lc. lactis subsp. lactis biovar. diacetylactis	Cocci shaped, production of lactic acid, exopolysaccharides, riboflavin, bacteriocins, diacetyl, etc.
Leuconostoc mesenteroides subsp. dextranicum, Leu. mesenteroides subsp. mesenteroides, Leu. cremoris	Cocci shaped, production of lactic acid, exopolysaccharides (homopolysaccharides, dextran), diacetyl, CO_2, etc.
Streptococcus thermophilus	Cocci shaped, production of lactic acid, exopolysaccharides (homo- or heteropolysaccharides), bacteriocins, folate, etc.
Lactobacillus delbrueckii subsp. bulgaricus, Lb. reuteri, Lb. casei, Lb. fermentum, Lb. plantarum, Lb. helveticus, Lb. acidophilus, Lb. paracasei, Lb. rhamnosus	Rod shaped, production of lactic acid, exopolysaccharides (homo- or heteropolysaccharides), riboflavin, bacteriocins, etc. Few species are used as probiotic.

(Continued)

Table 4.1 (Continued) Microorganisms Used as Starter Cultures

MICROORGANISM(S)	MAJOR CHARACTERISTICS
P. acidilactici	Cocci shaped, forms tetrads, production of lactic acid, bacteriocins, etc.
Bifidobacterium adolescentis, B. brevis, B. bifidum, B. infantis, B. lactis, B. longum	Anaerobic heterofermentative, non-spore forming rods, production of two molecules of lactate and three molecules of acetate. Species are also used as probiotic.
Brevibacterium linens, B. casei	Rods, pleomorphic, obligate aerobes, impart reddish-orange color in cheeses.
Propionibacterium freudenreichii subsp. freudenreichii, P. freudenreichii subsp. shermanii	Rods, pleomorphic, production of propionic acid, acetic acid, and CO_2.
Enterococcus faecium, E. faecalis, E. durans	Cocci shaped, production of lactic acid, bacteriocins.
B. YEASTS	
Candida kefir	Short ovoid to long ovoid, budding yeast-like cells or blastoconidia, production of ethanol and CO_2.
Kluyveromyces marxianus	Formation of pseudomycelium, fermentation of lactose and inulin, produces aroma compounds such as fruit esters, carboxylic acids, ketones, furans, alcohols, monoterpene alcohols, and isoamyl acetate.
Saccharomyces cerevisiae	Forms blastoconidia (cell buds), produces ascospores, production of alcohol and CO_2.
Torulospora delbrueckii	Forms buds, produces ethanol and CO_2.
C. MOLDS	
Penicillium roquefortii, P. camemberti	Production of asexual spores in phialides with a distinctive brush-shaped configuration, production of mycotoxins like roquefortine and PR toxin.
Geotrichum candidum	Produce chains of hyaline, smooth, one-celled, subglobose to cylindrical, slimy arthroconidia (ameroconidia) by the holoarthric fragmentation of undifferentiated hyphae. Septate hyphae that disarticulate into arthroconidia and do not form budding yeast cells. Contributes to an aroma.
Aspergillus oryzae	Filamentous fungus, highly aerobic and are found in almost all oxygen-rich environments.
Mucor rasmusen	Spores or sporangiospores can be simple or branched and form apical, globular sporangia that are supported and elevated by a column-shaped columella.

forming, non-pigmented, and unable to produce catalase and cytochrome, growing anaerobically but are aero-tolerant and obligatorily ferment sugar with lactic acid as the major end product. The nutritional requirement of these cultures varies from species to species. Most of the cultures are nutritionally fastidious, often requiring specific amino acids, vitamin B, and other growth factors, while unable to use complex carbohydrates.

4.2.1 Bacteria

Lactococci have been widely used for manufacturing a variety of fermented milk products. So far, five species are recognized but only *Lactococcus lactis* is used as starter culture that has practical significance in dairy fermentations. There are two subspecies, *Lc. lactis* subsp. *lactis* and *Lc. lactis* subsp. *cremoris,* and one variant, *Lc. lactis* subsp. *lactis* biovar. *diacetylactis*, which are commonly used as single or in mixed cultures. Lactococci are homofermentative and mesophilic, and when grown in milk, more than 95% of their end product is lactic acid, L(+) isomer. However, being weakly proteolytic, they can use milk proteins and grow at 10°C but not at 45°C. *Lc. lactis* subsp. *lactis* is more heat and salt tolerant than other subspecies. It ferments maltose, grows at 40°C, and in pH 9.5, produces ammonia from arginine, whereas *Lc. lactis* subsp. *cremoris* did not show these characteristics. *Lc. lactis* biovar. *diacetylactis* shows a close relationship with *Lc. lactis* subsp. *lactis* but differs by exhibiting citrate positive ability and does not produce as much lactic acid in milk as the latter. Nisin and diplococin are produced by *Lc. lactis* subsp. *lactis* and *Lc. lactis* subsp. *cremoris*, respectively, while bacteriocins produced by *Lc. lactis* subsp. *lactis* biovar. *diacetylactis* are not named. Some lactococci can produce EPS and improve textural properties of cultured dairy products (Cerning 1990; Behare et al. 2009a).

Leuconostoc ssp. occur in pairs and chains of cocci and are often ellipsoidal. It is difficult to differentiate *Leuconostoc* from lactococci. Both are catalase negative and form chains of coccal to oval shaped cells. However, a useful method by which one can fairly distinguish these two is by growing them in litmus milk. Lactococci reduce litmus before coagulation, whereas leuconostocs do not. Fundamentally, leuconostocs are heterofermentative and produce D-lactate, and with the exception of *Leuconostoc lactis*, these show no change in litmus milk

(Garvie 1960). These also do not hydrolyze arginine and require various B vitamins for growth. *Leuconostoc* ssp. grows at 10°C but not at 40°C, and can ferment lactose, galactose, fructose, and ribose. The end product produced includes diacetyl, carbon dioxide, and acetoin from citrate. The species that are widely used as dairy starters include *Leu. mesenteroides* subsp. *cremoris* (previously referred as *Leu. cremoris* or *Leu. citrovorum*), *Leu. mesenteroides* subsp. *mesenteroides*, *Leu. mesenteroides* subsp. *dextranicum*, and *Leu. lactis*. They are primarily used as flavor producers in butter, cheeses, and flavored milks. These cultures are often used in combination with other fast growing lactic cultures.

Streptococcus thermophilus used as a starter is fairly close with *Str. salivarius*, a common inhabitant of the mouth. Earlier, *S. thermophilus* was combined with *Str. salivarius*, but analysis of DNA hybridization data indicated these two as different and *S. thermophilus* got separate species status (Axelsson 1993). *S. thermophilus* can be easily distinguished from lactococci and *Leuconostocs* by sugar fermentation profile and growth temperature. It can grow at 45°C (but no growth at 10°C), while lactococci and leuconostocs cannot grow during these conditions. Additionally, *S. thermophilus* strains differ in their ability to utilize galactose. Use of non-galactose fermenting strains will result in high levels of this reducing sugar in products. Since galactose and other reducing sugars react with amino acids in the Maillard reaction, it is usual to only select galactose-utilizing strains to reduce the probability of undesirable color changes in heated products. Most strains of *S. thermophilus* hydrolyze aesculin, lactose, and saccharose. It is one of the fastidious starter cultures that coagulate milk in very short time. Several dairy products subjected to high temperatures (>40°C) during fermentation are acidified by the combined use of *S. thermophilus* and *Lactobacillus* ssp. Streptococci are generally isolated from milk and milk products. But *Enterococcus* species, which also resembles *Str. thermophilus* by growing at 45°C and having similar morphological features, often created confusion for isolation of *S. thermophilus* from milk and milk products. However, most enterococci can grow at 10°C in 6.5% NaCl and at pH 9.6 and contain the lancified group D-antigen while *S. thermophilus* does not show these characteristics.

Lactobacillus consists of a genetically and physiologically diverse group of rod shaped LAB. These are generally found in milk, cheeses, butter, and traditional fermented milk products (Bettache et al. 2012). Orla-Jensen (1931) classified lactobacilli into three groups: *thermobacterium*, *streptobacterium*, and *betabacterium*. *Lactobacillus* was further divided into three main groups, I, II, and III resembling the classification of Orla-Jensen (1931), but designating them as subgeneric taxa (Kandler and Weiss 1986). Lactobacilli are the most acid tolerant of the starter cultures, liking to initiate growth at acidic pH (5.5–6.2) and lowering the pH of milk to below 4.0. Some species are homofermentative, while others are heterofermentative. While some species produce mainly L-lactate from glucose, others produce D-lactate. Since certain strains exhibit significant racemase activity and a racemase is an isomerase, D/L lactic acid is also produced (Table 4.2). In pure

Table 4.2 Selected Characteristics of *Lactobacillus* ssp.

LACTOBACILLI	LACTIC ACID ISOMER	GROWTH AT		CARBOHYDRATE UTILIZATION							
		15°C	45°C	ESC	AMY	ARA	CEL	GLU	GAL	MAN	XYL
GROUP I (THERMOBACTERIUM)—OBLIGATE HOMOFERMENTATIVE											
Lb. acidophilus, Lb. gasserie	DL	−	+	+	+	−	+	+	+	−	−
Lb. delbrueckii subsp. *bulgaricus*	D(−)	−	+	−	−	−	−	+	−	−	−
Lb. helveticus	DL	−	+	−	−	−	−	+	+	−	−
Lb. johnsonii	DL	+	+	ND	+	ND	+	+	+	−	−
Lb. kefiranofaciens	D(L)	−	−	ND	−	ND	−	+	+	−	−
GROUP II (STREPTOBACTERIUM)—FACULTATIVE HETEROFERMENTATIVE											
Lb. paracasei subsp. *paracasei*	L/DL	+	d	+	+	−	+	+	+	+	−
Lb. rhamnosus	L(+)	+	+	+	+	d	+	+	+	+	−
Lb. plantarum	DL	+	−	+	d	+	+	+	+	+	−
GROUP III (BETABACTERIUM)—OBLIGATE HETEROFERMENTATIVE											
Lb. brevis	DL	+	−	d	−	+	−	+	d	−	−
Lb. fermentum, Lb. reuteri	DL	−	+	−	−	d	d	+	+	−	−
Lb. viridescens	DL	+	−	−	−	−	−	−	+	−	−

ESC, aesculin; AMY, amygdalin; ARA, arabinose; CEL, cellobiose; GLU, glucose; GAL, galactose; MAN, mannitol; XYL, xylose; d, 11%–89% strains showed positive reaction; ND, not detected.

cultures, many lactobacilli are slow growers, and due to this reason, these are generally combined with other fast growing cultures. Few lactobacilli are used as probiotics for treatment of various disorders (Delzenne et al. 2011).

Pediococci divide to form tetrads that differentiate these morphologically from other LAB. Only *Pediococcus pentosaceus* and *P. acidilactici* are used as dairy starters but these are less important than other LAB. However, *P. acidilactici* is used with therapeutic properties along with other starter cultures *P. acidilactici, Lb. acidophilus,* and *Bifidobacterium bifidum* in the ratio of 1.0:0.1:1.0 (Tamime and Marshall 1997). The important characteristics of *P. acidilactici* and *P. pentosaceus* are shown in Table 4.3.

Only six (*B. adolescentis, B. breve, B. bifidum, B. infantis, B. lactis,* and *B. longum*) out of 30 *Bifidobacterium* species are used in the dairy industry. These produce lactic acid and acetic acid in the ratio of 2:3. These are catalase negative, Gram-positive, irregularly shaped (pleomorphic) rods, many of which form branched cells. These are anaerobic in nature and can grow poorly in milk possibly because of the lack of small peptidases. The optimum temperature for growth is 37°C–41°C, while no growth occurs below 20°C and above 46°C. Growth at 45°C seems to discriminate between animal and human strains. The optimum pH is between 6.5 and 7.0 and no growth is recorded at pH lower than 4.5. These can utilize lactose, galactose, fructose, maltose, and sucrose. Bifidobacteria differ with other LAB by possessing the enzyme fructose-6-phosphate phosphoketolase, the key enzyme of the bifid-shunt. Some of the members of these groups

Table 4.3 Selected Characteristics of Pediococci

CHARACTERISTICS	P. PENTOSACEUS	P. ACIDILACTICI
Lactic acid isomer	D	DL
Growth at 10°C	−	−
45°C	−	+
Growth in 10% NaCl	V	−
Acid from		
Arabinose	+	V
Xylose	V	+
Maltose	+	−
Trehalose	+	V

V, Variable; +, positive reaction; −, negative reaction.

are also used as probiotics as these find a suitable environment in the human host and provide beneficial health effects by improving intestinal disorders like diarrhea, constipation, and irritable bowel syndrome (Grandy et al. 2010; Guglielmetti et al. 2011). Bifidobacteria are the first organisms to establish in new borne babies.

Brevibacterium linens and *Brevibacterium casei* are used in cheeses to impart a distinctive reddish-orange color to the rind or cause the fermentation of smear on brick and Limburger cheese (Olson 1969; Reps 1993). The bacteria are Gram positive, pleomorphic rods, and obligate aerobes, and optimum growth temperature is 20°C–25°C (*B. linens*) or 30°C–37°C (*B. casei*). These do not use lactose or citrate but can grow on the lactate produced during cheese fermentation. These organisms are salt tolerant and non-motile and produce no endospore.

Propionibacterium ssp. are non-spore forming, pleomorphic, Gram-positive rods that produce large amounts of propionic and acetic acids, and carbon dioxide from sugars and lactic acid. These are anaerobic to aerotolerant mesophiles and are closely related to coryneforms in the *Actinomycetaceae* group. The most useful species for dairy is *P. freudenreichii* (subsp. *freudenreichii* and *shermanii*) widely used in Swiss cheeses (i.e., Emmental and Gruyere) predominantly for producing large gas holes in cheese during maturation. Propionibacteria grow on lactic acid produced during cheese fermentation. Lactate is oxidized to pyruvate, which is then converted to acetate and carbon dioxide or propionate. The other species in dairy products are *P. jensenii*, *P. thoenii*, and *P. acidipropionici*.

Enterococci are generally found in milk and milk products like other LAB. Based on 16SRNA sequencing within *Enterococcus*, three species, that is, *E. faecium*, *E. faecalis*, and *E. durans*, are revealed. These are Gram positive, catalase negative cocci, and produce L(+) lactic acid from glucose. These are normal inhabitants of the human intestinal tract and, hence, are indicators of fecal contamination, and some species are pathogenic. Enterococci are not widely used as starter cultures due to the role of some species in causing food borne illness. However, in some of the Southern European countries, these are used as starter culture in some cheese varieties and fermented milk products (Tamime and Marshall 1997). In addition, selected *Enterococcus* ssp. are commercially available as probiotics for the prevention and control of intestinal disorders.

4.2.2 Yeasts

The presence of yeasts in dairy products is unacceptable and considered as contaminants (IDF 1998). These are quite common in the environment and are often isolated from sugar-rich materials. These are Gram positive, short to long ovoid, form blastoconidia and produce ethanol and CO_2 during fermentation of lactose. Yeasts are chemo-organotrophs, as these use organic compounds as a source of energy and do not require sunlight to grow. Carbon is obtained mostly from glucose and fructose, or disaccharides (i.e., sucrose and maltose). Few species can metabolize pentose sugars like ribose alcohols and organic acids. Yeast species either require oxygen for aerobic respiration or are anaerobic, but also have aerobic methods of energy production (i.e., facultative anaerobes). Unlike bacteria, there are no known yeasts that grow only anaerobically. These grow best in a neutral or slightly acidic environment. Although yeasts are commonly used in bread and wine production, few species are used in the manufacture of kefir and koumiss, where these carry out yeasty-lactic fermentation. The important species of yeasts, which are used in fermented milk products, include *Saccharomyces cerevisiae*, *Saccharomyces boulardii*, *Torulaspora delbrueckii*, *Kluyveromyces marxianus*, and *Candida kefyr*.

4.2.3 Molds

Molds are used in the cheese industry for making some semisoft cheese varieties. The major role of molds is to enhance the flavor and aroma and modify the body and texture of the curd slightly. *Penicillium camemberti* and *P. roquefortii* are used in mold ripened cheeses. *P. camemberti*, also called white mold, grows on the surface of the cheese (Camembert, Brie, and similar varieties), while *P. roquefortii*, called blue mold, grows in the interior of the blue-veined cheeses (Roquefort, Stilton, Gorgonzola). Both species are lipolytic and proteolytic and produce methyl ketones and free fatty acids that impart distinctive flavor and aroma to the cheeses. Other molds have limited application but are used in some parts of the world including *Mucor rasmusen* in Norway for ripened skim milk cheese (Kosikowski and Mistry 1997) and *Geotrichum candidum* in viili, a fermented milk product of Finland. *G. candidum* is cosmopolitan in distribution (i.e., air, water, plants, and milk and milk products). Lipases and proteases of *G. candidum* release fatty acids and peptides that can be metabolized by subsequent microbial populations and contribute to the development of distinctive flavors (Litthauer et al. 1996; Holmquist 1998).

4.3 Types of Starter Cultures

Basically, starters can be grouped as lactic or non-lactic starters. The starter cultures used in the manufacture of cheese can be classified into different groups.

4.3.1 Composition of Starter Flora

4.3.1.1 Single Strain Starter It consists of only one type of microorganism and a lactic acid producer is commonly used to achieve the desirable changes. However, use of such culture is always at risk if a culture fails due to inherent or external factors, for example, *Lc. lactis* subsp. *lactis* or *Lc. lactis* subsp. *cremoris* or *S. thermophilus*.

4.3.1.2 Mixed Strain Starters These consist of two or more strains in an unknown proportion. The advantage of using a mixed strain starter is if one strain fails due to any reason the other performs. It also gives a wider tolerance to other factors like temperature and pH changes. However, mixed starters are difficult to maintain as in repeated transfer one strain may become dominant over another. Consequently, it is advisable to use such cultures in a correct proportion or distinctly grow those and mix just before the inoculation of milk, for example, *Lc. lactis* subsp. *lactis*, *Lc. lactis* subsp. *cremoris*, and *Leuconostoc* ssp.

4.3.1.3 Paired Compatible Starters Two single strains are used in a desired ratio to have a better performance. Compatible strains are selected by monitoring their growth in a liquid medium. However, none of the strains should produce an inhibitory substance that may inhibit the growth of another, for example, *Lc. lactis* subsp. *lactis* and *Lc. lactis* subsp. *cremoris*.

4.3.1.4 Multiple Strain Starters These are a mixture of known compatible, non-phage related, carefully selected strains that give consistent product. Starters can be used for extended periods as the numbers of strains are known, for example, *Lc. lactis* subsp. *cremoris* and *Lc. lactis* subsp. *lactis* biovar. *diacetylactis*.

4.3.2 Growth Temperature of Starter Culture

4.3.2.1 Mesophilic Starters These have an optimum growth temperature between 20°C and 30°C and comprise mainly of *Lactococcus* and *Leuconostoc* species. Apart from the production of lactic acid, certain cultures also produce diacetyl. These are widely used in the making of dahi, culture buttermilk, butter, lassi, and so on.

4.3.2.2 Thermophilic Starters These cultures exhibit a higher optimum growth temperature that lies between 37°C and 45°C and are useful in the manufacture of products that require acidification at higher temperature (>40°C). These are used in the making of yogurt, dahi, acidophilus milk, and high scalded cheeses. The examples of thermophilc starters are *S. thermophilus*, *Lb. delbrueckii* subsp. *bulgaricus*, and *Lb. fermentum*. *S. thermophilus* produces lactic acid at a faster rate and are therefore generally combined with other thermophilic starters.

4.3.3 Production of End Product by the Starter Culture

4.3.3.1 Lactic Starters Starter cultures that produce lactic acid as the principal end product from lactose. *Lc. lactis* subsp. *lactis*, *Lc. lactis* subsp. *cremoris*, and *S. thermophilus* are the typical examples of lactic cultures.

4.3.3.2 Non-Lactic Starters Non-lactic starters produce other end products like acetic acid, carbon dioxide, ethanol, and propionic acid. Such cultures can also produce lactic acid as one of the end products. The examples of non-lactic starters include *B. bifidum*, *P. freudenreichii* subsp. *freudenreichii*, and some lactobacilli.

4.3.4 Physical Forms of Starter Culture

4.3.4.1 Liquid Starter This form of culture is more popular and handled in a dairy plant on routine basis. Cultures are available in fluid form and normally preserved in small volumes for a few days but their quantities can be scaled up as per the requirement. These cultures have a limited shelf life and require periodic transfer to maintain them active. Commonly, starter cultures are made in sterile reconstituted skim milk or litmus chalk milk.

4.3.4.2 Frozen Starter Culture These are made in a frozen state by deep-freezing (–20°C to –40°C) or freezing in liquid nitrogen (–196°C). Cultures have more shelf life than liquid state and can be used for few months. However, at low temperature mechanical process affects the performance of cultures that can be sorted out by adding cryoprotective agents (like sucrose, gelatin, and glycerin).

4.3.4.3 Frozen Concentrated Starter Culture Cells are first concentrated to have a high number of cell population approximately 10^{10}–10^{13} CFU/g followed by rapid freezing that is done in liquid nitrogen. The protection to cells is given by appropriate cryoprotective agents. These are preserved for longer duration and used for large quantities of milk. These cultures are often used as direct-vat-set (DVS) cultures.

4.3.4.4 Dried Starter Cultures Starter cultures are dried to increase their shelf-life for longer times and even more than a year. The dried form of the culture is achieved by vacuum-drying, spray-drying, and freeze-drying. In all types of drying, viability of the culture is most important. Drying should not affect the characteristics of the cultures. Spray-drying and vacuum-drying give much lower survivability of the cultures than the freeze-drying, hence freeze-drying becomes the method of choice. In freeze-drying, the culture of interest is grown overnight in an appropriate medium, and the cell pellet is obtained by centrifugation. The cell pellet is resuspended in a minimal quantity of milk containing a cryoprotective agent. The cell mass is then subjected to freeze-drying (–40°C). The entire process takes 8–10 h.

4.3.4.5 Dried Concentrated Cultures These are the cultures usually dried by lyophilization after concentration of cells by techniques like high speed centrifugations or diffusion culture. These cultures contain active cells in a range of 10^{11}–10^{14} CFU/g and can be used as DVS cultures.

Table 4.4 Starter Cultures Named as Per the Intended Use

STARTERS/PRODUCTS	CULTURES EMPLOYED
Yogurt culture	S. thermophilus + Lb. delbrueckii subsp. bulgaricus
Dahi culture	Lactococci, Leuconostoc ssp., Lb. delbrueckii subsp. bulgaricus, Lb. acidophilus, S. thermophilus
Cheddar cheese culture	Lactococci
Cottage cheese culture	Leu. mesenteroides subsp. cremoris, Lb. casei
Swiss cheese culture	Thermophilic lactobacilli, S. thermophilus, P. freudenreichii subsp. shermanii
Acidophilus culture	Lb. acidophilus
Bifidus culture	Bifidobacterium bifidum
Yakult culture	Lb. casei
Koumiss culture	Lb. delbrueckii subsp. bulgaricus, Lb. acidophilus, Kluyveromyces fragilis, K. marxianus
Kefir culture	Kefir grains (contain lactobacilli, yeasts, lactococci and acetic acid bacteria)
Brick cheese culture	Lactococci, Brevibacterium linens
Roquefort cheese culture	Lactococci, Penicillium roquefortii
Camembert cheese culture	Penicillium camemberti
Cultured buttermilk culture	Lactococci, Leuconostoc ssp.
Bulgarian milk culture	Lb. delbrueckii subsp. bulgaricus
Leben/Labneh culture	Lc. lactis subsp. lactis, S. thermophilus, Lb. delbrueckii subsp. bulgaricus, lactose fermenting yeasts
Probiotic cultures	Lb. acidophilus, B. bifidum, B. longum, B. infantis

4.3.5 Product for Which Starters Used

Starter cultures can also be categorized on the basis of intended use (Table 4.4). The cultures are named after the product for which they are meant, for example, yogurt culture, dahi culture, kefir culture, and cheese culture.

4.4 Propagation of Starters

Propagation is the process of multiplication of pure or mixed starter cultures. This is essentially required to prepare large quantities of product commercially. A typical flow diagram for propagation of starters in dairy is given in Figure 4.1.

For the manufacture of fermented milk products, starters are grown in heat-treated milk or milk-based media. Starters are added to the milk medium at a predetermined rate so as to allow conventional stages of manufacture. Strict asepsis is required to maintain the purity of

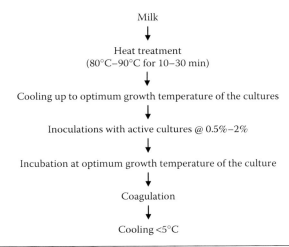

Figure 4.1 Methods of propagation of culture.

starters during propagation. Similarly, the medium (milk) should be free from antibiotic residues or any other substances harmful to the starter. Control of temperatures during incubation and cooling are also important factors affecting the activity of the culture. The conventional method of starter preparation includes a stepwise increase in volume of starters from stock culture (0.4 ml) (liquid, dried, or frozen) to mother culture (40 ml), mother culture to working culture (4 liters), and working culture to bulk culture (200 liters).

Preparation of mother culture is a very important step in the production of bulk starter for commercial application. The methods of bulk starter production are illustrated in Figure 4.2. The traditional method is time consuming, laborious, requires skilled personnel, and is more prone to contamination. The other method makes use of concentrated cultures sufficient to inoculate a large quantity of milk for bulk starter production or directly for product manufacture.

Several new techniques have been developed in the recent past to prevent contamination during starter propagation and bulk starter production, especially from bacteriophage. These include the Lewis system, the Jones system, and the Alfa-Laval system, which employ mechanical and chemical control measures in the design of equipment and bulk starter tank to prevent the entry of contaminants during propagation. Starter propagation in specially designed media, devoid of calcium

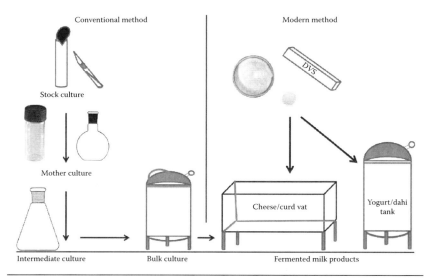

Figure 4.2 Production of starter cultures.

ions, also helps in controlling phage infection. Such media are called phage-inhibitory media or phage-resistant media.

Many professional laboratories now supply DVS or DVI (direct-vat-inoculation) cultures, which are highly concentrated cultures, having cell population of about 10^{10} to 10^{11} cells per gram which are used directly for setting the vat or for the preparation of bulk culture. These cultures omit the need of culture propagation at the factory.

4.5 Problems Associated with Production of Starter Culture

Starter cultures may show a number of defects with respect to growth and performance. Among these, the defects called weak acid, no coagulation, flat flavor, and thin body are due to slowness or sluggishness of the starter. This gives inadequate performance during product manufacture leading to a delay in manufacture, poor quality product, and economic loss. The slowness in starters comes due to a number of factors (Table 4.5) that can adversely affect the growth, leading to reduced rate of acid production or sometimes complete cessation of acid production.

By far the most common reason for slowness is found to be antibiotic residues in milk. When animals are given antibiotics for certain diseases, they are likely to be secreted into the milk. The starter

Table 4.5 Problems Associated with Starter Cultures and Their Remedies

PROBLEMS	REMEDIES
PROBLEM WITH STARTER ITSELF	
Spontaneous loss of vitality	Replace the culture
Strain variation	Check the balance
Physiological state of cells	Inoculate at the correct stage
INCOMPETENCE IN THE CONTROL OF STARTER	
Starter contaminated	Prevent contamination
Too infrequent sub-culturing	Sub-culture regularly at right stage
Use of unsuitable media (milk)	Use high quality reconstituted skim milk for propagation
Culturing at wrong temperature	Ensure proper temperature
PROBLEM WITH MILK	
Abnormal milk, for example, mastitic, colostrum, late lactation	Never allow mixing with normal milk
Antibiotic residues in milk	Never mix milk from antibiotic-treated animal for 3 days
Milk affected by feeds, seasonal factors, aeration, and so on.	Never use such milk or use after treatment
Inhibitory substances in milk, for example, preservative, detergent and sanitizer residues	Avoid such milk
PROBLEMS ASSOCIATED WITH THE PRODUCTION METHODS	
Changes in ripening time and temperature	Ensure proper adjustment of time and temperature
Cooking and clotting temperature in cheese	Use resistant cultures, don't change temperatures
Bacteriophage action	Observe strict asepsis, use starter rotation system, phage inhibitory media or phage resistant media, environmental control, use of modified cultures.

organisms are inhibited depending upon the concentration of antibiotic and the type of strain. *S. thermophilus* is most sensitive to penicillin residues in milk. The other reason contributing to slowness is bacteriophage infection.

4.6 Application of Starter Cultures

Starter cultures are used in manufacturing of variety of fermented milk products. The properties that are required by the lactic cultures for industrial use may differ from product to product (Panesar 2011). Apart from the production of useful metabolites in the product, certain

Table 4.6 Role of Exopolysaccharides Producing Starter Cultures in Fermented Milk Products

EPS PRODUCING STARTER CULTURES	PRODUCTS	EFFECT(S)	REFERENCE
Lc. lactis subsp. lactis	Cream cheese	Better firmness, consistency and melting properties	Nauth and Hayashi (2004)
Lb. kefir	Kefir	Improved texture	Micheli et al. (1999)
Lc. lactis subsp. cremoris JFR1	Cheddar cheese	Increased moisture retention, improved sensory, textural and melting properties	Awad et al. (2005)
Lc. lactis subsp. lactis B6	Dahi	Improved rheological and sensory properties, less syneresis	Behare et al. (2009b)
S. thermophilus (EPS−) and Lb. delbrueckii subsp. bulgaricus (EPS+)	Stirred yogurt	Increased viscosity	Marshall and Rawson (1999)
S. thermophilus (EPS+) and Lb. delbrueckii subsp. bulgaricus (EPS−)	Yogurt	Ropiness, low serum separation, higher viscosity	Folkenberg et al. (2005)
S. thermophilus (EPS+) and Lb. delbrueckii subsp. bulgaricus MR-1R (EPS+)	Mozzarella cheese	Retained moisture, better melting properties	Perry et al. (1997)
S. thermophilus IG16	Lassi	Improved consistency and sensory attributes, higher viscosity	Behare et al. (2010)

cultures are known to produce EPS, which improves textural, rheological, and sensory properties of fermented milk products (Table 4.6). Some starters can also produce bacteriocins that have potential significance as biopreservatives for food application. Bacteriocins are the proteins that are inhibitory to self or closely related species. However, some bacteriocins of starter cultures have broad spectrum activities, especially against spoilage causing and pathogens (Table 4.7). The application of starters in some of the fermented milk products is as explained below.

Yogurt is made by the combination of *S. thermophilus* and *Lb. delbrueckii* subsp. *bulgaricus* that shows symbiotic association during their growth. *Lb. delbrueckii* subsp. *bulgaricus* produces certain amino acids from casein that stimulates the growth of *S. thermophilus*, whereas *S. thermophilus* stimulates the growth of *Lb. delbrueckii* subsp. *bulgaricus* by removing oxygen, lowering pH, and producing formic acid. One of the major problems of yogurt making is excessive sourness produced

Table 4.7 Bacteriocins of Lactic Acid Bacteria and Their Inhibitory Spectra

BACTERIOCINS	PRODUCER	INHIBITORY SPECTRUM
Acidolin	Lb. acidophilus	Broad spectrum. Spore formers, Enteric pathogens
Acidophilin	Lb. acidophilus	Lactic acid bacteria, Spore formers, Salmonella ssp., E. coli, S. aureus, Pseudomonas
Bavaricin A	Lb. bavaricus	Lactobacillus ssp., Lactococcus ssp., Enterococcus ssp., Listeria monocytogenes
Brevicin	Lb. brevis	Pediococci, Lactobacilli, Leuconostocs, N. coralina
Carnocin	Leu. carnosum	Lactobacilli, Pediococci, Enterococci, Leuconostocs, Carnobacteria, Listeria
Curvacin A	Lb. curvatus	Lactobacillus ssp., Carnobacterium ssp., Listeria monocytogenes
Diplococcin	Lc. lactis subsp. cremoris	Lc. lactis subsp. cremoris, Lc. lactis subsp. lactis
Fermenticin	Lb. fermentum	Lactobacilli
Helveticin J	Lb. helveticus	Lb. delbrueckii subsp. bulgaricus, Lb. lactis, Lb. helveticus
Lactacin B	Lb. acidophilus	Lactobacilli
Lactocidin	Lb. acidophilus	Broad spectrum, effective against Gram +ve and Gram –ve bacteria
Lactococcin	Lc. lactis subsp. cremoris	Lactococci, S. aureus, B. cereus, S. typhi
Mesenerocin	Leu. mesenteroides	L. monocytogenes, B. linens, E. faecalis, P. pentosaceus
Nisin	Lc. lactis	Lactococci, Bacilli, Clostridia, Micrococci, S. aureus
Pediocin AcH	P. acidilactici	Lactobacilli, S. aureus, Leuconostocs, P. putida, L. monocytogenes, Cl. perfringens
Reuterin	Lb. reuteri	Salmonella, Shigella, Clostridia, Staphylococci, Listeria, Candida, Trypanosoma
Sakacin 674	Lb. sake	L. monocytogenes

by the starter itself. Careful selection of yogurt starters is essential to overcome the problem of over acidification due to dominance of any single strain. Usually, lactobacilli dominate during the storage and make a very sour product. Therefore, the culture selection, rods:cocci ratio (1:1), and fermentation control are required to be maintained to ensure the better product. The ideal yogurt flavor is a balanced blend of acidity and acetaldehyde. The main source of acetaldehyde is from the conversion of threonine to acetaldehyde catalyzed by threonine aldolase of *Lb. delbrueckii* subsp. *bulgaricus* (Hickey et al. 1983). Yogurt cultures also improve the texture of the yogurt due to the complex interaction that occurs between milk proteins, acid and EPS produced by the starter culture (Ariga et al. 1992).

Cultured buttermilk is generally prepared by mesophilic starters, which has the ability to acidify the substrate and produce flavor and aroma compounds. *Lc. lactis* subsp. *lactis* or *Lc. lactis* subsp. *cremoris* is combined with citrate fermenting bacteria such as *Leu. mesenteroides* subsp. *cremoris* or *Lc. lactis* subsp. *lactis* var. *diacetylactis*. Diacetyl is the major flavoring compound produced in this product.

Dahi is a popular fermented milk product of India resembling yogurt (Behare and Prajapati 2007) that can be prepared from both mesophilic and thermophilic starters. Most commonly used starters for dahi include *Lc. lactis* subsp. *lactis*, *Lc. lactis* subsp. *cremoris*, *S. thermophilus*, *Lb. delbreuckii* subsp. *bulgaricus*, *Lb. lactis*, and *Lb. casei*, and *Lecuonostoc* ssp. However, mesophilic lactococci, leuconostocs, and thermophilic streptococci are usually preferred over lactobacilli due to the production of high acidity, slow fermentation, and lengthy incubation period by the latter. In southern India, dahi making still relies on *Lactobacillus* cultures that produce sour curd. *S. thermophilus* is a fastidious culture generally used in industries to make dahi. Also where typical flavor is required, mesophilic cultures comprised of *Lc. lactis* subsp. *lactis* var. *diacetylactis* and *Leuconostoc* ssp. are used.

Lassi, a yogurt-like drinking product in the Indian subcontinent is prepared from milk having 1.5%–4.5% fat, after making a set curd (i.e., Dahi, followed by vigorous stirring and addition of sugar syrup plus optional flavor). It is a white to creamy white viscous liquid with a sweetish rich aroma and mild to medium acidic taste (Behare et al. 2010). Similar starter cultures can be used that are used in the manufacture of dahi.

The process of cheese manufacture is governed by the rate at which the starter grows and a number of changes it brings about. The starter has several functions to carry out in cheese making, and hence, careful selection and proper maintenance of starter cultures is essential in the cheese industry. Mesophilic starters containing lactococci and leuconostocs are used in Cheddar, Cottage, mold ripened, and some other cheese varieties, while thermophilic starters containing *S. thermophilus*, *Lb. helveticus*, and *Lb. delbrueckii* are used in Swiss types and Italian cheeses.

Molds are important in the ripening of wide range of cheeses. Mold-ripened cheeses are divided into those which are ripened by *P. roquefortii*, which grows and forms blue veins within cheese (Roquefort, Gorgonzola, Stilton and Danish blue), and those which are ripened with *P. camemberti*, which grows on the surface of cheese (Camembert and Brie). Molds are also associated with other cheese varieties. The surface of the French cheeses, St. Nectaire and Tome de Savoie, is covered by a complex fungal flora containing *Penicillium*, *Mucor*, *Cladosporium*, *Geotrichum*, *Epicoccum*, and *Sporotrichum*, while *Penicillium* and *Mucor* have been reported on the surface of the Italian cheese Taleggio and *Geotrichum* on that of Robiola (Gripon 1993). Mucor is sprayed on the surface of the Norwegian cheese Gammelost (Oterholm 1984). *P. roquefortii* has also been reported to develop within Gammelost cheese; in such situations, *P. roquefortii* is introduced to the cheese at piercing (Gripon 1993). Interior- or surface-mold-ripened cheeses have different appearances, and the high biochemical activities of these molds produce typical aroma and taste.

Acidophilus products are made by fermentation with *Lb. acidophilus* alone or in combination with some other bacteria or yeasts. *Lb. acidophilus* is a natural inhabitant of the human intestinal tract and this property is helpful to combat intestinal pathogens and cure several intestinal disorders. The culture is incorporated in different probiotic foods and feeds for the improvement of growth and performance of the host. When different forms of products are made, the culture must be resistant to processing conditions and should also survive conditions of gastrointestinal tract. Traditionally, acidophilus milk is prepared by inoculating *Lb. acidophilus* culture at the rate of 2%–5% in autoclaved or severally boiled milk. The milk is incubated at 37°C till it attains 1%–1.5% of acidity. This milk gives a cooked flavor and unappetizing flat taste. Hence, it could not become popular and was consumed as medicine, when needed. Looking at the limitation of traditional acidophilus milk, attempts were made to prepare several other products, which can supply large numbers of viable acidophilus cells in acceptable forms. These attempts were directed toward the supplementation of *Lb. acidophilus* with other flavor-producing cultures, concentration and drying of the products, or dispensing cell concentrates into other popular dairy products like pasteurized milk or ice cream. The recommended dose of *Lb. acidophilus* is 10^7–10^8 cells per day to derive maximum therapeutic benefits.

Kefir is a foamy, effervescent milk product resulting from mixed lactic acid and alcoholic fermentation of milk by kefir grains. It is an old product from Caucasian mountains in Russia. Kefir grains are gelatinous white or cream colored irregular grains of varying size (0.5–2.0 cm dia), similar to grains of wheat. They are made of a polysaccharide called *kefiran* produced by *Lb. kefir* and are insoluble in water. Bacteria and yeasts reside in symbiotic relationship within the folds of the granules. Kefir grains are reported to contain at least six functionally different groups of organisms, namely (1) mesophilic homofermentative lactic streptococci, (2) mesophilic heterofermentative streptococci, (3) thermophilic lactobacilli, (4) mesophilic lactobacilli, (5) yeasts, and (6) acetic acid bacteria. All these organisms grow in association during kefir making and produce lactic acid (0.91%–1%), alcohol (0.5%–1.0%), and CO_2 (0.03%–0.07%) as major end products. Before making kefir, the grains are added to pasteurized milk and incubated at 18°C–20°C for 15–16 h with intermittent agitation. Then the inoculation rate varies from 2%–7%. The first incubation is done at 25°C for 8–12 h to allow acid production and subsequently packed in crown-capped bottles/cartons and incubated further at 8°C–10°C for 10–12 h for accumulation of CO_2 and alcohol.

Koumiss is a product similar to kefir made by acid and alcoholic fermentation of milk. It is very popular in Russia and Central Asia, and traditionally made from Mare's milk. However, due to increased demand of Koumiss and short supply of Mare's milk, the product is now made from cow milk after some modifications. Mare's milk has low casein and high whey protein content; hence, it does not form a firm curd but remains liquid. Additionally, it has a high lactose content and is rich in albumin, peptone, and certain vitamins. Hence, to make the product from cow milk resembling Mare's milk, certain modifications are required to be done for adjusting the composition of milk. Koumiss culture consists of *Lb. delbrueckii* subsp. *bulgaricus*, *Lb. acidophilus*, and *Kluyveromyces lactis* or *K. marxianus* subsp. *marxianus*. Yeasts and lactobacilli grow in association in this product and produce 1%–1.5% lactic acid, 1%–2% alcohol, and 0.5%–0.9% carbon dioxide. The koumiss is graded as weak, medium, and strong based on acidity and alcohol contents. Koumiss is used for the treatment of pulmonary tuberculosis in Russia.

This is yogurt like fermented milk of Middle East. The sour milk is stored in porous earthenware or hanged in cloth bags to allow its concentration. In some areas, it is rolled into balls and sun dried. The product contains mixed microflora consisting of *Lc. lactis*, *S. thermophilus*, *Lb. bulgaricus*, *Lb. plantarum*, *Lb. brevis*, *Lb. casei*, and yeasts.

Yakult is a therapeutic fermented milk popular in Japan. It is cultured with a special strain of *Lb. casei* strain shirota, which has shown several therapeutic benefits. To make the product rich in proteins, a variety called chlorella Yakult is also introduced. It makes use of an algae—chlorella.

Viili is a traditional Finnish fermented milk made from pasteurized milk with slime-forming starters consisting of lactic streptococci and a mold, *Geotrichum candidium*. The mold grows on the top and forms a velvet-like layer. The mold and cocci are symbiotic in nature.

The milk cultured with Bifidobacteria is called bifidus milk. Bifidiobacteria are normal inhabitants of the intestinal tract of a newborn. These possess special therapeutic properties and increase the resistance of the infant to several disorders. True bifidobacteria are strict anaerobes and slow growers in milk; hence, it is difficult to prepare bifidus milk. However, the product is made from severely heat-treated milk taking greater aseptic precautions and using a high rate of inoculum. The species regularly employed are *B. bifidum* or *B. longum*. However, use of aerotolerant strains of *B. adolescentis* gives a better product.

4.7 Conclusion

Starter cultures have a long history of safe use and therefore received a generally recognized as safe status. The successful production of fermented milk products depends upon the quality of starter cultures. The fermented milk industry aims for the multifunctional strains of starter that contribute to the organoleptic, technological, nutritional, and health properties of products. The production of organic acids, antimicrobial compounds, flavoring compounds, EPS, and vitamin-producing cultures are the examples of functional starters. However, a careful selection of specific strains combined with proper production and handling procedures is essential to ensure that the desired benefits are given to the consumers.

References

Ariga, H., Urashima, T., Michihata, E., Ito, M., Morizono, N., Kimura, T., and Takahashi, S. 1992. Extracellular polysaccharide from encapsulated *Streptococcus salivarius* subsp. *thermophilus* OR 901isolated from commercial yoghurt. *Journal of Food Science* 57: 625–628.

Awad, S., Hassan, N., and Muthukumarappan, K. 2005. Application of exopolysaccharides producing cultures in reduced fat cheddar cheese: Texture and melting properties. *Journal of Dairy Science* 88: 4204–4213.

Axelsson, L.T. 1993. Lactic acid bacteria: Classification and physiology. In: *Lactic Acid Bacteria*, ed. S. Salmen and A. von Wright, 1–64. New York: Marcel Dekkar.

Behare, P.V. and Prajapati, J.B. 2007. Thermization as method for enhancing the shelf-life of cultured buttermilk. *Indian Journal of Dairy Science* 60: 86–93.

Behare, P.V., Singh, R., Kumar, M., Prajapati, J.B., and Singh, R.P. 2009a. Exopolysaccharides of lactic acid bacteria: A review. *Journal of Food Science and Technology* 46: 1–11.

Behare, P.V., Singh, R., and Singh, R.P. 2009b. Exopolysaccharide producing mesophilic lactic cultures for preparation of reduced fat dahi-An Indian fermented milk. *Journal of Dairy Research* 76: 90–97.

Behare, P.V., Singh, R., Tomar, S.K., Nagpal, R., Kumar, M., and Mohania, D. 2010. Effect of exopolysaccharides-producing strains of *Streptococcus thermophilus* on technological attributes of fat-free lassi. *Journal of Dairy Science* 93: 2874–2879.

Bettache, G., Fatma, A., Miloud, H., and Mebrouk, K. 2012. Isolation and identification of lactic acid bacteria from dhan, a traditional butter and their major technological traits. *World Applied Sciences Journal* 17: 480–488.

Cerning, J. 1990. Extracellular polysaccharides produced by lactic acid bacteria. *FEMS Microbiology Review* 87: 113–130.

Cintas, L.M., Casaus, M.P., Herranz, C., Nes, I.F., and Hernández, P.E. 2001. Review: Bacteriocins of lactic acid bacteria. *Food Science Technology International* 7: 281–305.

Delzenne, N.M., Neyrinck, A.M., Backhed, F., and Cani, P.D. 2011. Targeting gut microbiota in obesity: Effects of prebiotics and probiotics. *Nature Reviews Endocrinology* 7: 639–646.

Folkenberg, D.M., Dejmek, P., Skriver, A., and Ipsen, R. 2005. Relation between sensory texture properties and exopolysaccharide distribution in set and in stirred yoghurts produced with different starter cultures. *Journal of Texture Studies* 36: 174–189.

Garvie, E.L. 1960. The genus *Leuconostoc* and its nomenclature. *Journal of Dairy Research* 27: 283–292.

Grandy, G., Medina, M., Soria, R., Teran, C.G., and Arya, M. 2010. Probiotics in the treatment of acute rotavirus diarrhea. A randomized, double-blind, controlled trial using two different probiotic preparations in Bolivian children. *BMC Infectious Disease* 10: 1–7.

Gripon, J.C. 1993. Mould ripened cheeses. In: *Cheese: Chemistry, Physics and Microbiology*, ed. P.F. Fox, 111–136. London: Chapman & Hall.
Guglielmetti, S., Mora, D., Gschwender, M., and Popp, K. 2011. Randomized clinical trial: *Bifidobacterium bifidum* MIMBb75 significantly alleviates irritable bowel syndrome and improves quality of life-a double-blind, placebo-controlled study. *Alimentary and Pharmacology Therapeutics* 33: 1123–1132.
Hickey, M.W., Hillier, A.J., and Jago, G.R. 1983. Enzymatic activities associated with lactobacilli in dairy products. *Australian Journal of Dairy Technology* 38: 154–157.
Holmquist, M. 1998. Insights into the molecular basis for fatty acyl specificities of lipases from *Geotrichum candidum* and *Candida rugosa*. *Chemistry and Physics of Lipid* 93: 57–65.
IDF. 1998. *Yeasts in Dairy Industry: Positive and Negative Aspects*. Special issue No. 9801. Brussels, Belgium: International Dairy Federation.
Kandler, O. and Weiss, N. 1986. Genus *Lactobacillus* Beijerinck 1901, 212AL. In: *Bergey's Manual of Systematic Bacteriology* (volume 2), ed. P.E. Sneath, N.S. Mair, M.E. Sharpe, and J.G. Holt, 1209–1234. Baltimore, MD: Williams & Wilkins.
Kosikowski, F.V. and Mistry, V.V. 1997. *Cheese and Fermented Milk Foods* (volume I). Westport, CT: F. V. Kosikowski LLC.
Litthauer, D., Louw, C.H., and Du Toit, P. 1996. *Geotrichum candidum* P-5 produces an intracellular serine protease resembling chymotrypsin. *International Journal of Biochemistry and Cell Biology* 28: 1123–1130.
Marshall, V.M. and Rawson, H.L. 1999. Effects of exopolysaccharide-producing strains of thermophilic lactic acid bacteria on the texture of stirred yoghurt. *International Journal of Food Science and Technology* 34: 137–143.
Micheli, L., Uccelletti, D., Palleschi, C., and Creasenzi, V. 1999. Isolation and characterization of a ropy *Lactobacillus* strains producing the exopolysaccharide kefiran. *Applied Microbiology Biotechnology* 53: 69–74.
Nauth, R.K. and Hayashi, D. 2004. Methods for manufacture of fat-free cream cheese. US Patent 6,689,402.
Olson, N.F. 1969. *Ripened Semisoft Cheeses-Pfizer Cheese Monograph* (volume IV). New York: Chas. Pfizer and Co.
Orla-Jensen, S. 1931. *Dairy Bacteriology* (2nd edition), translated by P.S. Arup. London: J and A. Churchill.
Oterholm, A. 1984. Cheese making in Norway. *IDF Bulletin* 171: 21–29.
Padalino, M., Perez-Conesa, D., Lopez-Nicolas, R., and Frontela-Saseta, C. 2012. Effect of fructooligosaccharides and galactooligosaccharides on the folate production of some folate-producing bacteria in media cultures or milk. *International Dairy Journal* 27: 27–33.
Panesar, P.S. 2011. Fermented dairy products: Starter cultures and potential nutritional benefits. *Food and Nutrition Sciences* 2: 47–51.
Perry, D.B., McMahon, D.J., and Oberg, C.J. 1997. Effect of exopolysaccharide producing cultures on moisture retention in low-fat Mozzarella cheese. *Journal of Dairy Science* 80: 799–805.

Reps, A. 1993. Bacterial surface ripened cheeses. In: *Cheese: Chemistry, Physics and Microbiology*, ed. P.F. Fox, 137–172. London: Chapman & Hall.

Tamime, A.Y. and Marshall, V.M.E. 1997. Microbiology and technology of fermented milks. In: *Microbiology and Biochemistry of Cheese and Fermented Milk* (2nd edition), ed. B.A. Law, 57–152. London: Blackie Academic and Professional.

Yang, E., Fan, L., Jiang, Y., Doucette, C., and Fillmore, S. 2012. Antimicrobial activity of bacteriocin-producing lactic acid bacteria isolated from cheeses and yogurts. *AMB Express* 2: 48. http://www.amb-express.com/content/2/1/48.

5
Metabolic Characteristics of Lactic Starters

S. MANJULATA DEVI AND PRAKASH M. HALAMI

Contents

5.1	Introduction	109
5.2	Metabolic Pathways of LAB	110
	5.2.1 Homofermentative Pathway	111
	5.2.2 Heterofermentative Pathway	112
5.3	Metabolic Pathways for Flavor Enhancers	112
	5.3.1 Glycolysis and Citrate Metabolism	113
	5.3.2 Protein Metabolism	115
	5.3.3 Lipid Metabolism	116
5.4	Flavor Compounds in Fermented Milk Products	117
	5.4.1 Carbonyl Compounds	120
	5.4.2 Acids and Alcoholic Compounds	121
	5.4.3 Ester Compounds	122
	5.4.4 Sulfur Compounds	123
5.5	Other Miscellaneous Metabolites	124
5.6	Future Perspectives	125
5.7	Conclusions	126
References		127

5.1 Introduction

Fermentation of food is a desirable process because of the modification in metabolites generated by certain microorganisms, which impart beneficial attributes. For fermentation of milk and its products, lactic acid bacteria (LAB) are considered to be the major workhorses due to their genetic and metabolic properties (Pedersen et al. 2002). LAB are used worldwide in the manufacture of fermented dairy foods, because they facilitate lactic acid and exopolysaccharide (EPS) production, exhibit antimicrobial activity, and enhance flavor and texture (Papagianni 2012). Generally, LAB are fastidious in nature and possess limited biosynthetic apparatus due to their small genome size (~2–3 Mb) and their ability to utilize simple sugars for their metabolism. Moreover, LAB carry several plasmids with technological traits, such as ability to utilize carbohydrates (lactose, sucrose, raffinose, etc.), citrate metabolism, production of EPS, bacteriocin, and antibiotic resistance, all of which are advantages that can be exploited for metabolic engineering purposes (Ayad 2008; Papagianni 2012).

LAB are used in the dairy industry in the fermentation of milk, as they not only use lactose as the major carbon source but also use other mono- and disaccharides as substrates. The major product during milk fermentation is lactic acid, which acts as a natural preservative. Moreover, other metabolites, like carbohydrates, alcohols, aldehydes, ketones, esters, and sulfur-containing compounds, that are produced during the fermentation improve the flavor, texture, taste, storage, and safety of the end products (Buruleanu et al. 2010; Cheng 2010). Species of *Lactococcus*, *Lactobacillus*, and *Streptococcus* are the most widely used LAB in the manufacture of dairy products like yogurt, buttermilk, and cheese.

Flavor remains a major criterion among consumers purchasing a food. The quality of fermented dairy products is mainly determined by sensory perception. The major basic flavors recognized by tongue are bitter, acid, sweet, salt, and umami, whereas the nose recognizes a large variety of volatile flavor compounds (Smit et al. 2005). The flavor in fermented dairy products is a consequence of several biochemical pathways in which LAB play a significant role as starters. Glycolysis and pyruvate metabolism (conversion of lactose and citrate), lipolysis (conversion of fat), and proteolysis (conversion of casein) are the three main pathways in the fermentation, leading to the production of compounds of different flavors in dairy products (Kranenburg et al. 2002; Ayad 2008).

5.2 Metabolic Pathways of LAB

LAB are Gram positive, catalase negative, non-spore forming, non-aerobic, fastidious, acid tolerant, and fermentative and produce mainly lactic acid from sugar fermentation (Axelsson 1993; Suskovic et al. 2010). They are mainly used as starters in the fermentation of several dairy, meat, vegetable, and cereal products. LAB attract to use this microorganism in fermented dairy products such as anti-mutagenicity, immune stimulation and anti-tumor activity (Harutoshi 2013). Most of the LAB drive their energy via different systems, in which fermentation of carbohydrates plays a major role. Two major pathways are recognized during sugar fermentation in LAB: homofermentative and heterofermentative pathways.

5.2.1 Homofermentative Pathway

In the glycolytic pathway (Embden–Meyerhof pathway [EMP]), lactic acid is the main end product obtained from the metabolism of hexoses (Figure 5.1). Some species of *Pediococcus*, *Enterococcus*, *Lactococcus*,

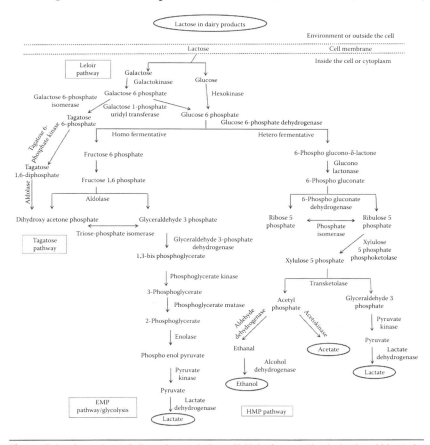

Figure 5.1 General metabolic pathways during milk/dairy fermentation by lactic acid bacteria. (Modified from McSweeney, P.L.H. and Sousa, M.J., *Lait*, 80, 293–324, 2000; Cheng, H., *Curr. Rev. Food Sci. Nutr.*, 50, 938–950, 2010; Khalid, K., *Int. J. Biosci.*, 1, 1–13, 2011; Papagianni, M., *Comput. Struct. Biotechnol. J.*, 3, 1–8, 2012.)

and *Lactobacillus* metabolize more than 90% of the sugar, which is converted to lactic acid. The most important group of bacteria that are usually associated with the dairy industry, like *Streptococcus thermophilus, Lactobacillus delbrueckii* subsp. *bulgaricus*, and *Lactococcus lactis*, follow the EMP to metabolize the carbohydrates inside the microbial cell. The lactose in milk gets converted to glucose and galactose via the Leloir pathway with galactokinase as the first metabolic enzyme. This pathway was observed in *Lb. acidophilus, St. thermophilus*, and *Lb. delbrueckii* subsp. *bulgaricus*, when glucose is depleted in a cell (Khalid 2011; Papagianni 2012).

5.2.2 Heterofermentative Pathway

In the phospho-ketolase pathway, lactose and glucose are fermented and several end products such as acetic acid, propionic acid, carbon dioxide and lactic acid are released (Figure 5.1). Several species of *Lactobacillus*, mainly *Lb. casei, Lb. paracasei,* and *Lb. plantarum, Leuconostoc, Oenococcus, Weisella* follow this pathway and transport lactose into the cell by permease, which in turn is hydrolyzed to glucose and galactose. These LAB utilize only 50% of the substrate and are converted to lactic acid. *Bifidobacterium* sp. ferment glucose to lactate and acetate. These end products contribute to the rapid acidification and improvement of flavor, texture, and nutrition of the food products (Khalid 2011; Kockova et al. 2011; Papagianni 2012).

5.3 Metabolic Pathways for Flavor Enhancers

LAB are used in the fermentation of several dairy products like milk, curd, yogurt, buttermilk, and cheese, because of their contribution to the production of compounds of several flavors. These flavor compounds in dairy products are derived from several biochemical pathways, which mainly metabolize sugars (lactose), milk proteins (caseins), and milk fat (fatty acids) as the precursors (Marilley and Casey 2004). The proteolytic enzymes from LAB degrade the caseins, peptides, and fatty acids, releasing several flavor compounds, for example, keto-acids, aldehydes, amino acids, acids, alcohol, and esters, which contribute to taste and aroma (Irygoyen et al. 2007; Ilicic et al. 2012).

5.3.1 Glycolysis and Citrate Metabolism

The metabolism of lactate/lactose/pyruvate and/or citrate may be referred to erroneously as glycolysis, where, first, the lactose gets converted to lactate and further releases pyruvate. During the fermentation process, one molecule of glucose releases two molecules of pyruvate to yield two adenosine triphosphate molecules. In most of the LAB, pyruvate plays an essential role in the regeneration of nicotinamide adenine dinucleotide (NAD^+) in order to perform the fermentation process. Glycolysis is determined by a series of 10 reactions with 10 intermediate enzymes. This pathway occurs in most aerobic and anaerobic bacteria. The final product of glycolysis, that is, pyruvate, gets reduced to lactic acid, which is involved in the flavor formation during the fermentation process. Several cultures of *Lactobacillus*, *Oenococcus*, and *Leuconostoc* metabolize glucose by the pentose phosphate pathway leading to heterolactic fermentation, which releases lactate, ethanol, and acetic acid as end products (Quintans et al. 2008).

Pyruvate and lactate are produced from several substrates, which include carbohydrates, amino acids, and organic acids, produced by different species of LAB (Liu 2003). The lactose in milk and milk-containing products is hydrolyzed by several starter cultures of LAB (*Lactococcus* and *Lactobacillus* sp.) to glucose and galactose. Glucose is further oxidized to pyruvate by the glycolysis pathway (EMP). The galactose present in the cell environment is converted into glucose 6-phospate or glyceraldehyde 3-phosphate by lactococci or leuconostocs by the Leloir pathway or tagatose pathway, respectively (Figure 5.1) (Cogan and Hill 1993; Liu 2003).

The metabolism of pyruvate from citrate by LAB leads to the formation of several end products such as lactate, acetate, formate, and ethanol, and aromatic compounds like diacetyl, acetoin, and butanediol (Figure 5.2a). The presence of these aromatic compounds in dairy products such as butter, acid cream, and cottage cheese imparts a buttery aroma and has a positive effect on the fermented end product. During citrate fermentation, a variety of metabolic products are produced by a bacterium depending on the growth conditions. In most of the cases, the transport of citrate takes place through a membrane protein and gets converted into acetate and oxaloacetate. The oxaloacetate gets converted into pyruvate and releases several aromatic compounds. Some strains of *Lactococcus* such as *Lc. lactis* and

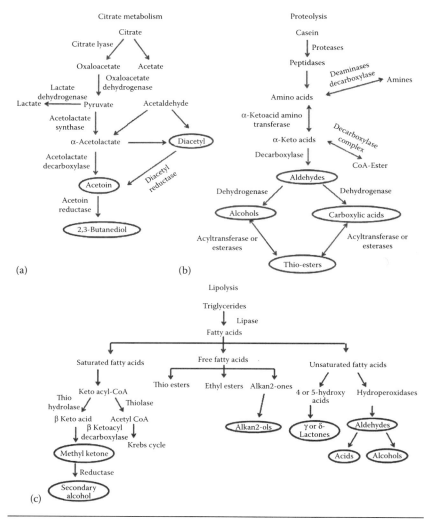

Figure 5.2 Schematic representation of the flavor-enhancing metabolic pathway: (a) citrate metabolism, (b) proteolysis, and (c) lipolysis (Modified from Marilley, L. and Casey, M.G., *Int. J. Food Microbiol.*, 90, 139–159, 2004; Smit, G. et al., *FEMS Microbiol. Rev.*, 29, 591–610, 2005; Quintans, N.G. et al., Citrate metabolism and aroma compound production in lactic acid bacteria. In: *Molecular Aspects of Lactic Acid Bacteria for Traditional and New Applications*, ed. B. Mayo et al., 65–88. Kerala, India: Research Signpost, 2008.)

Lc. diacetylactis, *Leuconostoc*, and *Weisella* ssp. are generally used in diacetyl production and applied in the dairy industry. Additionally, some species of *Enterococcus* are widely used in the development of aroma in cheese products (Quintans et al. 2008).

The citrate lyase, which is responsible for the conversion of citrate to oxaloacetate, was found in several of the LAB like *Leuconostoc mesenteroides*, *Weisella paramesenteroides*, *Lc. diacetylactis*, *Lb. plantarum*, and *Oenococcus oeni*. The metabolism of citrate in *Enterococcus* sp. is still not clear. However, it was observed that *E. faecalis*, *E. durans*, and *E. faecium* encode the citrate-producing genes with strain-to-strain variation. Moreover, *E. faecalis* and *E. faecium* can co-metabolize with lactose and not with glucose (Sarantinopoulos et al. 2001, 2003; Rea and Cogan 2003). Citrate is mainly metabolized to produce flavor compounds like diacetyl, acetoin, and 2,3-butanediol during fermentation of dairy products (Marilley and Casey 2004).

5.3.2 Protein Metabolism

The degradation of casein and peptides by proteolytic enzymes of LAB leads to the production of free amino acids, which in turn enhances flavor in yogurt and cheese (Hannon et al. 2007; Irygoyen et al. 2007; Ilicic et al. 2012). LAB are the most studied microorganisms for proteolytic systems, because of their physiological traits, which contribute toward the development of organoleptic properties of fermented dairy products (Savijoki et al. 2006). *Lactococcus lactis* is the most extensively studied organism for proteolysis, transport, and regulation. Additionally, several LAB starter cultures like *St. thermophilus*, *Lb. delbrueckii* subsp. *bulgaricus*, and *Lb. helveticus* are also extensively used in the manufacture of fermented dairy products for economic and beneficial importance. Proteolytic enzymes play a major role in the fermentation of dairy products (Figure 5.2b) (Savijoki et al. 2006; Ilicic et al. 2012).

In amino acid metabolism, the transamination pathway is being followed for the degradation of branched chain (leucine, valine, and isoleucine), aromatic (tryptophan, phenylalanine, tyrosine), and sulfur-containing amino acids (methionine, cysteine), which ultimately lead to the formation of aromatic flavor compounds like aldehydes, alcohols, and esters (Smit et al. 2005; Ardo 2006). Different aminotransferases are involved in the transamination of amino acids, which include branched chain aminotransferases (BcAT) and aromatic aminotransferases. According to Yvon et al. (2000), most of the lactococcal and streptococcal strains produce BcAT, whereas several of the lactobacilli strains (*Lb. sakei*, *Lb. reuteri*, *Lb. johnsonii*) are devoid of this aminotransferases; hence, these organisms were unable to produce 2-methyl butanoate and isobutyric acid flavor enhancing compounds. It was observed that deamination is often linked to a transamination reaction in several genomes of *Lb. salivarius*, *Lb. plantarum*, and *St. thermophilus*, where the α-keto acid aminotransferases and glutamate dehydrogenases are the key components in flavor formation, especially in cheese (Tanous et al. 2002; McSweeney and Sousa 2004; Liu et al. 2008). α-Keto acid decarboxylases are involved in oxidative decarboxylation, which was found to be an active compound in several of the lactococcal strains. Further, the acyltransferases/esters involved in the amino acid degradation pathways give a fruity flavor in several of the cheese products and is well characterized in *Lb. helveticus*, *Lb. casei* and several lactococci strains (Fenster et al. 2000; Fernandez et al. 2000; Yebra et al. 2004).

5.3.3 Lipid Metabolism

Triglycerides are the other main components of milk fat and are usually involved in the manufacture of flavor compounds. A series of several enzymatic reactions are involved directly/indirectly in the production of volatile compounds. The lipolytic pathway followed by most of the LAB during cheese ripening is summarized in Figure 5.2c. Lipase is the main enzyme required for the breakdown of milk fat/triglycerides/tributyrin. Several genera of LAB like *Lactococcus*, *Lactobacillus*, *Leuconostoc*, and *Enterococcus* are isolated from milk products like cheese. Katz et al. (2002) reported that not all strains of LAB have the ability to utilize the milk fat substrate, and the esterase and lipase activities were found to be species and/or strain specific.

Lipolysis of milk fat leads to the release of saturated, unsaturated, and free fatty acids. The saturated fatty acids release methyl ketones and secondary alcohols (pentan-2-ol, heptan-2-ol) through a series of enzymatic reactions. The unsaturated fatty acids release lactones, aldehydes, alcohols, and acids, which are usually found in cheese ripening. Lactones are cyclic compounds and found to enhance the flavor of cheese, yogurt, and milk products. The alcohols released during lipolysis by LAB can subsequently combine with free fatty acids to release aromatic esters like ethyl acetate and butyl acetate (Marilley and Casey 2004). The lipases/esterase producing LAB were found to be intracellular and a number of bacteria have been characterized and identified. The cultures of *Lb. helveticus*, *Lb. acidophilus*, and *Lb. delbrueckii* subsp. *bulgaricus* and subsp. *lactis* showed high levels of esterase activity compared to lipase activity (Hofi et al. 2011). These organisms are usually used as starters in cheese making, which subsequently play a major role during cheese ripening.

5.4 Flavor Compounds in Fermented Milk Products

The role of fermented milk has become an important component of a nutritional diet of consumers. Several LAB are widely used in the fermentation of different milk products to improve the health status of humans. Table 5.1 lists different LAB used in the fermentation of various dairy products. The sensory importance of any food product does not completely depend on the volatile compounds. However, the flavor compounds produced by microbial strains are relatively well studied. The aroma compounds are classified into several families according to their physiochemical, chemical structure, or sensory properties. The aroma of different compounds produced by LAB during milk-based fermentation is summarized in Table 5.2. The major families associated in flavor development are carbonyl compounds,

Table 5.1 Different Milk Products Fermented by Lactic Acid Bacteria

DAIRY PRODUCT	TYPE OF MILK USED	STARTER CULTURES INVOLVED	MAJOR AROMATIC COMPOUNDS OBSERVED
Buttermilk	Cow's/buffalo's milk	*St. lactis* subsp. *diacetylactis, St. cremoris*	Carbonyl compounds, acids
Cheese	Cow's/buffalo's/goat's milk	*Lc. lactis* subsp. *lactis, Lc. lactis* subsp. *cremoris, Lc. lactis* subsp. *diacetylactis, St. thermophilus, Lb. delbrueckii* subsp. *bulgaricus*	Carbonyl compounds, alcohol, acids, ketones, esters, sulfur compounds
Curd	Cow's/buffalo's milk	*Lc. lactis* subsp. *lactis, Lb. delbrueckii* subsp. *bulgaricus, Lb. plantarum, St. cremoris, St. thermophilus, Lb. acidophilus*	Carbonyl compounds, acids, esters
Lassi	Cow's/buffalo's milk	*Lb. bulgaricus*	Carbonyl compounds, acids
Shrikand	Cow's/buffalo's milk	*St. thermophilus, Lb. bulgaricus*	Carbonyl compounds, acid
Yogurt	Cow's milk	*Lb. acidophilus, Lb. bulgaricus, Lb. plantarum, St. thermophilus*	Carbonyl compounds, acids, esters, alcohol

Source: Ayad, E.H.E., *Alexanderia J. Food Sci. Technol.*, Sp.v, 31–43, 2008; Panesar, P.S., *Food Nutr. Sci.*, 2, 47–51, 2011.

Table 5.2 Different Aromatic Compounds Produced by LAB During Fermentation of Dairy Products

FLAVOR COMPOUND	DESCRIPTION	PRECURSOR REQUIRED	ENZYME(S) RESPONSIBLE	METABOLISM	STARTER ORGANISMS
(A) CARBONYL COMPOUNDS (ALDEHYDE AND KETONES)					
Acetaldehyde	Green apple, buttery, pleasant	Pyruvate, threonine	Pyruvate decarboxylase, theronine aldolase	Pyruvate, amino acid	Lb. bulgaricus and St. thermophilus
Acetoin	Buttery, fruity	Diacetyl	Diacetyl reductase	Citrate/pyruvate	Leuconostoc sp., Lc. lactis subsp. lactis, Lc. lactis subsp. lactis biovar. diacetylactis
Acetone	Sweet, fruity	Acetyl phosphate	Acetokinase	Citrate/pyruvate	Lc. lactis subsp. lactis var. diacetylactis, Lb. bulgaricus, St. thermophilus
Diacetyl (2,3-butanedione)	Buttery, strong	α-Acetolactate	α-acetolactate decarboxylase	Citrate/aspartic acid	Lc. lactis, St. thermophilus, Leu. mesenteroides, Lc. lactis subsp. lactis var. diacetylactis
(B) ACIDS					
Acetic acid	Acidic, pungent, vinegar	Pyruvate	Lactate dehydrogenase	Glycolysis	Lb. acidophilus, Lb. delbruckii, Lb. helveticus, Lb. rhamnosus, Ped. pentosaceus, E. faecium
Lactic acid	Acidic, pungent	Pyruvate	Lactate dehydrogenase	Glycolysis	Lb. acidophilus, Lb. delbruckii, Lb. helveticus, Lb. rhamnosus, Ped. pentosaceus, E. faecium
Propionic acid	Pungent, vinegar, sour milk	Pyruvate	Lactate dehydrogenase	Glycolysis	Lb. acidophilus, Lb. delbruckii, Lb. helveticus, Lb. rhamnosus, Ped. pentosaceus, E. faecium
(C) ALCOHOL					
Ethanol	Mild, ether	Pyruvate, amino acids	Alcohol dehydrogenase	Glycolysis, proteolysis	Lc. lactis, St. thermophilus, Pediococcus sp.

(Continued)

Table 5.2 (Continued) Different Aromatic Compounds Produced by LAB During Fermentation of Dairy Products

FLAVOR COMPOUND	DESCRIPTION	PRECURSOR REQUIRED	ENZYME(S) RESPONSIBLE	METABOLISM	STARTER ORGANISMS
(D) ESTERS					
Butyl acetate	Pineapple/fruity	Ethyl butanoate and ethyl isovalerate	Esterase	Proteolysis, lipolysis	*Lb. helveticus, Lc. lactis, Lb. salivarius, Lb. casei*
Ethyl acetate	Solvent-like, fruity	Ethyl butanoate and ethyl isovalerate	Esterase	Proteolysis, lipolysis	*Lb. helveticus, Lc. lactis, Lb. salivarius, Lb. casei, Lb. acidophilus*
Methyl acetate	Solvent-like, fruity, sweet, nail polish	Ethyl butanoate and ethyl isovalerate	Esterase	Proteolysis, lipolysis	*Lb. helveticus, Lc. lactis, Lb. salivarius, Lb. casei*
Methyl formate,	Fruity, mild alcoholic	Ethyl butanoate and ethyl isovalerate	Esterase	Proteolysis, lipolysis	*Lb. helveticus, Lc. lactis, Lb. salivarius, Lb. casei*
(E) SULFUR COMPOUNDS					
Dimethyl sulfide	Lactone-like, intense, sulfurous	Methionine, cysteine	Aminotransferases	Alternative metabolism	*Lb. delbrueckii* subsp. *bulgaricus, Lb. plantarum, St. thermophilus, Lc. lactis* subsp. *cremoris*
Dimethyl trisulfide	Sulfurous	Methionine, cysteine	Aminotransferases	Alternative metabolism	*Lb. delbrueckii* subsp. *bulgaricus, Lb. plantarum, St. thermophilus, Lc. lactis* subsp. *cremoris*
Dimethyldisulfide	Boiled cabbage/cauliflower, garlic	Methionine, cysteine	Aminotransferases	Alternative metabolism	*Lb. delbrueckii* subsp. *bulgaricus, Lb. plantarum, St. thermophilus, Lc. lactis* subsp. *cremoris*
Methional	Cooked cabbage/potato	Methionine, cysteine	Aminotransferases	Alternative metabolism	*Lb. delbrueckii* subsp. *bulgaricus, Lb. plantarum, St. thermophilus, Lc. lactis* subsp. *cremoris*

Source: McSweeney, P.L.H., *Int. J. Dairy Technol.* 57, 127–144, 2004; Liu, M. et al., *Appl. Environ. Microbiol.* 74, 4590–4600, 2008; Routray, W. and Mishra, H.N., *Comprehensive Rev. Food Sci. Food Safety,* 10, 208–220, 2011.

acids, alcohols, esters, fatty acids, and ketones. Further, the flavor compounds produced by several starter cultures of LAB may be divided into four main categories. First, non-volatile acids include acids such as lactic acid, pyruvic acid, oxalic acid, and succinic acid. The second group consists of volatile acids such as formic acid, acetic acid, propionic acid, and/or butyric acid followed by carbonyl compounds, for example, acetaldehyde, acetone, acetoin, or diacetyl. The last group includes some miscellaneous compounds that are derived from amino acids, fat, protein, and lactose.

5.4.1 Carbonyl Compounds

Carbonyl compounds are one of the major aromatic compounds of dairy products (Pourahmad and Assadi 2005; Cheng 2010). Acetaldehyde produced by sugar metabolism produces a very pleasant fruity aroma at very low concentrations. It was reported that *Lb. bulgaricus* and *St. thermophilus* impart a characteristic apple and/or nutty flavor in yogurt (Cheng 2010). Not all the LAB utilize the same pathway for the production of acetaldehyde. For example, more acetaldehyde production was achieved by *St. thermophilus* through amino acid metabolism, where threonine aldolase converts the threonine to acetaldehyde (Chaves et al. 2002). However, through genetically modified LAB, it was observed that *Lc. lactis* produces acetaldehyde by pyruvate metabolism from lactose (Bongers et al. 2005). Dana et al. (2011) reported that *Lactobacillus* strains utilize the acetoin degradation pathway for the production of acetaldehyde, where acetoin dehydrogenase is the main enzyme responsible for flavor formation during milk-based fermentation.

Diacetyl is another major aromatic compound in many dairy products and gives a buttery taste. This compound is produced by LAB utilizing citric acid in milk; α-acetolactate decarboxylase is the enzyme responsible for the high production of diacetyl in dairy products. *Lc. lactis*, *St. thermophilus*, and *Leu. mesenteroides* are usually involved in citrate metabolism and produce aromatic compounds (Cheng 2010). Hugenholtz and Starrenburg (1992) reported that *Lc. lactis* subsp. *lactis* var. *diacetylactis* strain Ru4 produces high amounts of diacetyl in dairy products and utilize citrate in milk for the production of this metabolite. Another carbonyl compound responsible for flavor in dairy products is acetoin, which produces a mild creamy, sweet, and buttery taste. The enzyme responsible for the production of acetoin is diacetyl reductase, produced by the breakdown of diacetyl. A rich perception of pleasant buttery taste is obtained in yogurts when diacetyl is mixed with acetoin. Several *Leuconostoc* sp., *Lc. lactis* subsp. *lactis*, and *Lc. lactis* subsp. *lactis* biovar. *diacetylactis* produce different end products such as diacetyl, acetoin, acetaldehyde, 2,3-butanediol during the fermentation of dairy products (buttermilk, sour cream, ripened cream butter, yogurt, etc.) (Hugenholtz and Starrenburg 1992; Cogan and Jordan 1994; Rattray et al. 2003).

Acetone, a carbonyl ketonic compound, is another aroma substance found usually in fermented dairy products. Acetone was found to be one of the volatile organic flavor compounds in amasi (a Zimbabwean naturally fermented milk). Moreover, several LAB species like *Lc. lactis* subsp. *lactis* Lc39 and Lc261, *Lb. paracasei* Lb11, and *Lc. lactis* subsp. *lactis* biovar. *diacetylactic* C1 were used as starters in the fermentation of milk (Gadaga et al. 2007). Wang et al. (2012) reported a significant concentration of acetone in cheese made with *Str. thermophilus* SP 1.1. Additionally, high levels of 3,5-octadien-2-one and acetophenone were also observed. The presence of several volatile carbonyl compounds (acetaldehyde, acetone, acetoin, and other organic acids) in high amounts during the fermentation of laban (Lebanese traditional fermented milk) by *Lb. delbrueckii* subsp. *bulgaricus* and *St. thermophilus* has also been shown (Chammas et al. 2006).

5.4.2 Acids and Alcoholic Compounds

Lactic acid and other acids like propionic acid and butyric acid play a major role in the enhancement of flavor and aroma of several dairy products like yogurt, cheese, and kefir. Lactic acid is widely used as a preservative, acidulant, and flavoring agent in several foodstuffs, pharmaceuticals, and cosmetics and in the chemical industry (Wee et al. 2006). Several of the LAB strains produce lactic acid by glycolysis or EMP pathway from pyruvate by lactate dehydrogenase enzyme. It was reported that enhanced production of lactic acid was observed by batch fermentation with *Lb. rhamnosus* (Berry et al. 1999). Lactic acid production from lactose by *Lb. helveticus* was observed in concentrated cheese whey (Schepers et al. 2002). The cultures of

LAB such as *Lb. delbruckii*, *Lc. lactis* subsp. *lactis*, *Lb. acidophilus*, and *Pediococcus pentosaceus* are commonly utilized in the production of lactic acid (Escamilla-Hurtado et al. 2005). Lactic and butyric acids have also been used in the food, chemical, and pharmaceutical industries as a flavoring, varnishing, disinfectant agent. The cultures of LAB in dairy products produce butyric acid by lipolysis of milk (Van Immerseel et al. 2010). Vaseji et al. (2012) used *Lb. acidophilus*, *Bifidobacterium*, *St. thermophilus*, and *Lb. delbruckii* subsp. *bulgaricus* as probiotic starter cultures and observed the fermentation of milk and production of butyric acid.

Alcohol was found to be another volatile aromatic compound in yogurt and cheese. Ethanol was found to be one of the essential compounds released during the breakdown of glucose and catabolism of amino acids (Marilley and Casey 2004; Guler 2007; Cheng 2010). The primary alcoholic compounds 2-methyl propanol and 3-methyl butanol were observed during the end of the storage in slated yogurt. Molina et al. (1999) observed high levels of ethanol production in goat's milk when compared to cow's and sheep's milk. However, type of milk may also alter ethanol production (Guler 2007). Though ethanol is considered to be a major aromatic compound, it was reported that ethanol provides a complementary flavor; the overall aroma and flavor in yogurt and cheese products fermented with LAB are not clear (Vedamuthu 2006). Several of the LAB such as *Lc. lactis*, *St. thermophilus*, and *Pediococcus* sp. were found to produce this organic compound.

5.4.3 Ester Compounds

Ester comprises an important volatile flavoring compound most commonly found in cheese. Ester contributes not only to the taste but also to the aroma of dairy products. In general, low concentrations of esters are desirable for overall flavor balance, which includes a fruity flavor. In general, biosynthesis of ester takes place either by esterification, wherein the formation of esters from alcohol and carboxylic acid takes place, or by alcoholysis, also known as transferase reaction, wherein the production of ester takes place from alcohol and acylglycerol. In several dairy products, ester contributes a vital flavoring agent. In milk and cheese, ethyl esters of fatty acids are found. In cheese, a range of ethyl esters are identified. LAB are known to possess ester biosynthesis mechanism by virtue of their ability to produce enzymes like esterase/lipase/alcohol acyltransferase. In cheese and fermented milk products, LAB are being used as a starter or starter adjuvant that essentially contributes to the glycolysis, proteolysis, and lipolysis process during cheese ripening. *Lactococcus lactis* subsp. *lactis* are known to induce a fruity off-flavor in Cheddar cheese due to ester production (Nardi et al. 2002). Esterase produced by LAB is capable of producing ester by esterification of ethanol with butanoic and hexanoic acids. Several strains of mesophilic LAB are known to possess esterase activity. For instance, *Str. cremoris* strains are known to possess ethyl butanoate synthesizing activity. The cultures of *Str. thermophilus* and *Lc. lactis* subsp. *cremoris* are known to possess esterases activity that catalyzes both hydrolysis, liberating free fatty acids, and alcoholysis of the substrates. In an aqueous environment, ester synthesis through alcoholysis gives a much higher ester concentration than esterification, and this process is likely mainly responsible for ester formation in semi-hard cheese types. Despite this enzymatic potential, in semi-hard type cheeses such as Cheddar and Gouda, ethanol or other alcohols are generally the limiting factor in ester synthesis, as acid(-precursor) and enzymes are usually present (Liu et al. 2004).

During fermentation, esters that are formed due to the hydrolysis of straight chain esters include ethyl acetate, ethyl butyrate, pentyl acetate, propyl butyrate, ethyl hexanoate, octanoate, and dodecanoate. Branched esters are generally formed by the degradation of protein and carbohydrate. These include methyl-2-methylbutanoate, ethyl isobutanoate, 3-methylbutyl acetate, ethyl-2-methylbutanoate, ethyl-3-methylbutanoate, isobutyl butanoate, 3-octyl acetate, and 2-methylbutyl hexanoate. A few aromatic esters are also formed through protein sources, for example, ethyl benzoate and phenethyl acetate.

5.4.4 Sulfur Compounds

Sulfur compounds are associated with flavor and aroma of a majority of fermented foods, in which they can play an attractive or a repulsive role, depending on their identity and concentration. Sulfur compounds in the form of volatile sulfur compounds (VSC) essentially arise from common sulfur-bearing amino acids, methionine, and cysteine catabolism.

Since these amino acids are present in limited amounts in milk, their biosynthesis and catabolic pathways are the only source of production. Methanethiol (MTL), a common precursor of some VSC, gets generated through catabolism of L-methionine by microorganisms in cheese and subsequently other flavorings are formed, such as methionol, dimethyl sulfide, dimethyl trisulfide, carbonyl sulfide, and hydrogen sulfide. Most of these sulfur-containing compounds have a great impact on sensory profiles on dairy products, especially on cheese (Weimer et al. 1999).

Catabolism of MTL is found to be due to the activity of aminotransferase from 4-methylthio-2-oxobutyric acid. Subsequently, due to de-methiolating activity, methionethiol is formed. In another step, cystathione by lysis is known to catalyze the elimination of methionine to MTC. The cultures of LAB are known to display considerably low activity toward methionine, thus leading to the limited access for the VSC production. The biosynthesis of methionine is known to be complex due to its multiple alternative pathways and enzymes involved. In LAB, it has been observed that there is a large difference in the distribution of required enzymes, suggesting high variability in the biosynthetic pathways. In the recent past, attempts have been made to evaluate the role of C-S lyse activities in VSC production through methionine, which could in turn lead to enhancement of cheese sulfur aroma compound (Bustos et al. 2011).

5.5 Other Miscellaneous Metabolites

Some LAB secrete polymers called EPS and are found to improve the functional properties of food as well as health of the humans (Welman and Maddox 2003). Such LAB improve the texture and rheology of fermented milk products, simultaneously imparting enhancement of flavor and taste to the end product (Duboc and Mollet 2001). Some LAB, like *Lec. mensenteroides* subsp. *mesenteroides*, *Lec. mesenteroides* subsp. *dextranicum*, *Lb. curvatus*, *St. mutans*, *Lb. plantarum*, *Lb. fermentum*, *Lb. delbrueckii*, *Lb. sanfranciscensis*, and *Lc. lactis*, play a major role in the industrial production of yogurt, cheese, and milk-based products (Harutoshi 2013). These LAB utilize mono- and/or disaccharides as the sole energy source and finally release several homo-EPS (α-D-glucans, β-D-glucans, fructans) and hetero-EPS (D-glucose,

D-galactose, L-rhamnose), subsequently improving the texture and taste of the fermented milk product (Vijayendra et al. 2009; Harutoshi 2013).

LAB are known to produce antimicrobial metabolites, besides organic acids, hydrogen peroxide, inorganic substances, and many low molecular weight peptides called bacteriocins. Such low molecular weight antimicrobial peptides are found to exhibit antagonistic activity against closely related organisms and food-borne pathogens. These substances are known to ensure safety and improve the shelf life of dairy products by inhibiting the growth of spoilage bacteria. Nisin, an antibiotic containing bacteriocin produced by a milk bacterium *Lc. lactis*, was commonly used for the preservation of dairy products, especially to inhibit *Clostridium* spores (Cotter et al. 2005). Other class II group of bacteriocins have received considerable attention in food preservation due to their anti-listerial activity, and these include pediocin and enterocin (Halami 2010; Devi and Halami 2011). Though most of the bacteriocins of LAB are active against Gram-positive bacteria, a synergistic effect of enterocin in combination with lipase was found to inhibit Gram-negative bacteria such as *E. coli* (Vrinda et al. 2012b). Several of these bacteriocin-producing cultures can be directly used as starter cultures or as protective cultures, where the shelf life of the finished product can be extended (Devi et al. 2012; Vrinda et al. 2012a). In addition, bacteriocin-producing culture with phytate degrading ability can be used for functional-food formulations (Raghavendra et al. 2011). Some of the bacteriocin-producing LAB have very limited usage in dairy milk products due to non-lactose fermenting ability; in such cases the bacteriocin-producing ability can be mobilized into a starter culture and allowed to improve the safety and shelf life of the fermented end product (Somkuti and Steinberg 2010; Devi and Halami 2013).

5.6 Future Perspectives

LAB are widely used for the industrial production of dairy products where starter cultures play an important role. This chapter provides requisite knowledge based on flavor compounds that improve the quality and sensory properties of fermented products. Compounds like acetaldehyde, diacetyl, and acetone are the most commonly found aromatic compounds in different milk-based fermented products. Phylogenetically closely related LAB follow common metabolic pathways for sugar fermentation, production of lactic acid, and flavor compounds. However, considerable genetic evolution might occur while adapting to diverse environmental ecological niches. This may provide an important view in the metabolic pathways of different strains. The biodiversity between different strains of LAB offers a good possibility for flavor diversification and novel product development. The knowledge of genomic data provides information in prediction and designing of attractive or selected flavor compounds, which can be applied for industrially potent cultures to meet consumer demands. Genomics may assist in faster selection of novel strains with interesting sensory aromatic flavor compounds.

Recent genome sequence data of several LAB associated with fermentation can provide detailed insight into the protein metabolites and enzymes involved in flavor-forming reaction. Thus, genomics could play an important role in designing starter cultures for their specific flavor compounds. In addition, there is a need to develop techniques that facilitate the development of novel starter cultures with flavor and aromatic compounds.

5.7 Conclusions

Metabolic activity of lactic starters is an important aspect in flavor development during fermentation. The principal metabolic activity of a starter culture includes conversion of sugar through glycolysis and pyruvate metabolism, besides metabolism of fats and proteins. Pyruvate acts as a precursor for several aroma and flavor compounds, which include diacetyl, acetaldehyde, and acetoin; their end products are strain and product specific. Though in lactic cultures lipolysis is limited, it contributes in flavor development in curd, yogurt, buttermilk, and cheese. Similarly, the proteolytic property of LAB contributes to flavor and texture formation. Detailed knowledge of metabolic activity of starter cultures and factors deciphering their regulation could contribute to the improvement of fermented product. While developing starter cultures for industrial production of dairy products, it is important to understand the defined regulatory mechanisms associated with in situ enrichment. The production of specific flavor and aroma compounds will have greater impact on finished product and also depends on selection of defined starter culture.

References

Ardo, Y. 2006. Flavour formation by amino acid catabolism. *Biotechnology Advances* 24: 238–242.

Axelsson, L.T. 1993. Lactic acid bacteria: Classification and physiology. In: *Lactic Acid Bacteria*, eds. S. Salminen and A. von Wright, 1–64. New York: Marcel Decker.

Ayad, E.H.E. 2008. Biodiversity of lactococci in flavour formation for dairy products innovation. *Alexanderia Journal of Food Science and Technology* Sp.v: 31–43.

Berry, A.R., Franc, C.M.M., Zhang, W., and Middelberg, A.P.J. 1999. Growth and lactic acid production in batch culture of *Lactobacillus rhamnosus* in a defined medium, *Biotechnology Letters* 21: 163–167.

Bongers, R.S., Hoefnagel, M.H.N., and Kleerebezem, M. 2005. High-level acetaldehyde production in *Lactococcus lactis* by metabolic engineering. *Applied and Environmental Microbiology* 71: 1109–1113.

Buruleanu, L., Nicolescu, C.L., Bratu, M.G., Manea, I., and Avram, D. 2010. Study regarding some metabolic features during lactic acid fermentation of vegetable juices. *Romanian Journal of Biotechnology Letters* 15: 5177–5188.

Bustos, I., Martinez-Bartolome, M.A., Achemchem, F., Palez, C., Requena, T., and Cuesta, M. 2011. Volatile sulphur compounds-forming abilities of lactic acid bacteria: C-S lyase activities. *International Journal of Food Microbiology* 148: 121–127.

Chammas, G.I., Saliba, R., Corrieu, G., and Beal, C. 2006. Characterisation of lactic acid bacteria isolated from fermented milk "laban." *International Journal of Food Microbiology* 1: 52–61.

Chaves, A.C.S.D., Fernandez, M., Lerayer, A.L.S., Mierau, I., Kleerebezem, M., and Hugenholtz, J. 2002. Metabolic engineering of acetaldehyde production by *Streptococcus thermophilus*. *Applied and Environmental Microbiology* 68: 5656–5662.

Cheng, H. 2010. Volatile flavour compounds in Yogurt: A review. *Current Review of Food Science and Nutrition* 50: 938–950.

Cogan, T.M. and Hill, C. 1993. Cheese starter cultures. In: *Cheese: Chemistry, Physics and Microbiology*, ed. P.F. Fox, 193–255. London: Chapman & Hall.

Cogan, T.M. and Jordan, K.N. 1994. Metabolism of *Leuconostoc* bacteria. *Journal of Dairy Science* 77: 2704–2717.

Cotter, P.D., Hill, C., and Ross, R.P. 2005. Bacteriocins: Developing innate immunity for food. *Nature Reviews Microbiology* 3: 777–788.

Dana, M.G., Yakhchali, B., Salmanian, A.H., and Jazil, F.R. 2011. High-level acetaldehyde production by an indigenous *Lactobacillus* strain obtained from traditional dairy products of Iran. *African Journal of Microbiology* 5: 4398–4405.

Devi, S.M., Asha, M.R., and Halami, P.M. 2012. In situ production of pediocin PA-1 like bacteriocin by different genera of lactic acid bacteria in soymilk fermentation and evaluation of sensory properties of the fermented soy curd. *Journal of Food Science and Technology*. doi:10.1007/s13197-012-0870-1.

Devi, S.M. and Halami, P.M. 2011. Detection and characterization of pediocin PA-1/AcH like bacteriocin producing lactic acid bacteria. *Current Microbiology* 62: 181–185.

Devi, S.M. and Halami, P.M. 2013. Conjugal transfer of pediocin PA-1 bacteriocin from different genera of lactic acid bacteria by in vitro and in situ methods. *Annals of Microbiology*. doi:10.1007/s13213-013-0624-y.

Duboc, P. and Mollet, B. 2001. Applications of exopolysaccharides in the dairy industry. *International Dairy Journal* 11: 759–768.

Escamilla-Hurtado, M.L., Valdes-Martinez, S.E., Soriano-Santos, J., Gomez-Pliego, R., Verde-Calvo, J.R., Reyes-Dorantes, A., and Tomasini-Campocosio, A. 2005. Effect of culture conditions on production of butter flavor compounds by *Pediococcus pentosaceus* and *Lactobacillus acidophilus* in semisolid maize-based cultures. *International Journal of Food Microbiology* 105: 305–316.

Fenster, K.M., Parkin, K.L., and Steele, J.L. 2000. Characterization of an arylesterase from *Lactobacillus helveticus* CNRZ32. *Journal of Applied Microbiology* 88: 572–583.

Fernandez, L., Beerthuyzen, M.M., Brown, J., Siezen, R.J., Coolbear, T., Holland, R., and Kuipers, O.P. 2000. Cloning, characterization, controlled overexpression, and inactivation of the major tributyrin esterase gene of *Lactococcus lactis*. *Applied and Environmental Microbiology* 66: 1360–1368.

Gadaga, T.H., Viljoen, B.C., and Narvhus, J.A. 2007. Volatile organic compounds in naturally fermented milk and milk fermented using yeasts, lactic acid bacteria and their combinations as starter cultures. *Food Technology and Biotechnology* 45: 195–200.

Guler, Z. 2007. Changes in salted yoghurt during storage. *International Journal of Food Science and Technology* 42: 235–245.

Halami, P.M. 2010. Isolation and characterization of a nitrosoguanidine-induced *Enterococcus faecium* MTCC 5153 mutant defective in enterocin biosynthesis. *Research in Microbiology* 161: 590–594.

Hannon, J.A., Kilcawley, K.N., Wilkinson, M.G., Delahunty, C.M., and Beresford, T.P. 2007. Flavour precursor development in Cheddar cheese due to lactococcal starters and the presence and lyses of *Lactobacillus helveticus*. *International Dairy Journal* 17: 316–327.

Harutoshi, T. 2013. Exopolysaccharides of lactic acid bacteria for food and colon health applications. In: *Lactic Acid Bacteria—R & D for Food, Health and Livestock Purposes*, ed. M. Kongo, 515–538. Rijeka, Croatia: InTech Europe.

Hofi, M.E., Tanboly, E.S.E., and Rabou, N.S.A. 2011. Industrial application of lipases in cheese making: A review. *Internet Journal of Food Safety* 13: 293–302.

Hugenholtz, J. and Starrenburg, M.J.C. 1992. Diacetyl production by different strains of *Lactococcus lactis* subsp. *lactis* var. *diacetylactis* and *Leuconostoc* spp. *Applied Microbiology and Biotechnology* 38: 17–22.

Ilicic, M.D., Milanovic, S.D., Caric, M.D., Kanuric, K.G., Vukic, V.R., Hrnjez, D.V., and Ranogajec, M.I. 2012. Volatile compounds of functional dairy products. *APTEFF* 43: 11–19.

Irygoyen, A., Ortigosa, M., Juansaras, I., Oneca, M., and Torre, P. 2007. Influence of an adjunct culture of *Lactobacillus* on the free amino acids and volatile compound in a Roncal-type ewe's milk cheese. *Food Chemistry* 100: 71–80.

Katz, M., Medina, R., Gonzalez, S., and Oliver, G. 2002. Esterolytic and lipolytic activities of lactic acid bacteria isolated from Ewe's milk and cheese. *Journal of Food protection* 65: 1997–2001.

Khalid, K. 2011. An overview of lactic acid bacteria. *International Journal of Bioscience* 1: 1–13.

Kockova, M., Gerekova, P., Petrulakova, Z., Hybenove, E., Sturdik, E., and Valik, L. 2011. Evaluation of fermentation properties of lactic acid bacteria isolated from sourdough. *Acta Chimica Slovaca* 2: 78–87.

Kranenburg, R., Kleerebezem, M., Hylckama Vlieg, J., Ursing, B.M., Jos Smit, B.A., Ayad, E.H.E., Smit, G., and Siezen, R.J. 2002. Flavour formation from amino acids by lactic acid bacteria: Predictions from genome sequence analysis. *International Dairy Journal* 12: 111–121.

Liu, M., Nauta, A., Francke, C., and Siezen, R.J. 2008. Comparative genomics of enzymes in flavour forming pathways from amino acids in lactic acid bacteria. *Applied and Environmental Microbiology* 74: 4590–4600.

Liu, S.Q. 2003. Practical implications of lactate and pyruvate metabolism by lactic acid bacteria in food and beverage fermentations. *International Journal of Food Microbiology* 83: 115–131.

Liu, S.Q., Holland, R., and Crow, V.L. 2004. Esters and their biosynthesis in fermented dairy products: A review. *International Dairy Journal* 14: 923–945.

Marilley, L. and Casey, M.G. 2004. Flavours of cheese products: Metabolic pathways, analytical tools and identification of producing strains. *International Journal of Food Microbiology* 90: 139–159.

McSweeney, P.L.H. 2004. Biochemistry of cheese ripening. *International Journal of Dairy Technology* 57: 127–144.

McSweeney, P.L.H. and Sousa, M.J. 2000. Biochemical pathways for the production of flavour compounds in cheeses during ripening: A review. *Lait* 80: 293–324.

Molina, E., Ramos, M., Alonso, L., and Lopez-Fandino, R. 1999. Contribution of low molecular weight water soluble compounds to the taste of cheeses made of cows', ewes' and goats' milk. *International Dairy Journal* 9: 613–621.

Nardi, M., Fiez-Vandal, C., Tailliez, P., and Monnet, V. 2002. The EstA esterase is responsible for the main capacity of *Lactococcus lactis* to synthesize short chain fatty acid esters *in vitro*. *Journal of Applied Microbiology* 93: 994–1002.

Panesar, P.S. 2011. Fermented dairy products: Starter cultures and potential nutritional benefits. *Food and Nutrition Science* 2: 47–51.

Papagianni, M. 2012. Metabolic engineering of lactic acid bacteria for the production of industrially important compounds. *Computational and Structural Biotechnology Journal* 3: 1–8.

Pedersen, M.B., Jensen, P.R., Janzen, T., and Nilsson, D. 2002. Bacteriophage resistance of a *thyA* mutant of *Lactococcus lactis* blocked in DNA replication. *Applied and Environmental Microbiology* 68: 3010–3023.

Pourahmad, R. and Assadi, M.M. 2005. Yoghurt production by Iranian native starter cultures. *Nutrition and Food Science* 35: 410–415.

Quintans, N.G., Blancato, V., Repizo, G., Magni, C., and Lopez, P. 2008. Citrate metabolism and aroma compound production in lactic acid bacteria. In: *Molecular Aspects of Lactic Acid Bacteria for Traditional and New Applications*, eds. B. Mayo, P. Lopez, and G. Pérez-Martínez, 65–88. Kerala, India: Research Signpost.

Raghavendra, P., Usha-Kumari, S.R., and Halami, P.M. 2011. Phytate-degrading *Pediococcus pentosaceus* CFR R123 for application in functional foods. *Beneficial Microbes* 2: 57–61.

Rattray, F.P., Myling-Petersen, D., Larsen, D., and Nilson, D. 2003. Plasmid-encoded diacetyl (acetoin) reductase in *Leuconostoc pseudomesenteroides*. *Applied and Environmental Microbiology* 69: 304–311.

Rea, M.C. and Cogan, T.M. 2003. Glucose prevents citrate metabolism by enterococci. *International Journal of Food Microbiology* 88: 201–206.

Routray, W. and Mishra, H.N. 2011. Scientific and technical aspects of yogurt aroma and taste: A review. *Comprehensive Reviews in Food Science and Food Safety* 10: 208–220.

Sarantinopoulos, P., Kalantzopoulos, G., and Tsakalidou, E. 2001. Citrate metabolism by *Enterococcus faecalis* FAIR-E 229. *Applied and Environmental Microbiology* 67: 5482–5487.

Sarantinopoulos, P., Lefteris Makras, L., Vaningelgem, F., Kalantzopoulos, G., De Vuyst, L., and Tsakalidou, E. 2003. Growth and energy generation by *Enterococcus faecium* FAIR-E 198 during citrate metabolism. *International Journal of Food Microbiology* 84: 197–206.

Savijoki, K., Ingmer, H., and Varmanen, P. 2006. Proteolytic systems of lactic acid bacteria. *Applied Microbiology and Biotechnology* 71: 394–406.

Schepers, A.W., Thibault, J., and Lacroix, C. 2002. *Lactobacillus helveticus* growth and lactic acid production during pH-controlled batch cultures in whey permeate/yeast extract medium. Part I. Multiple factor kinetic analysis. *Enzyme and Microbial Technology* 30: 176–186.

Smit, G., Smit, B.A., and Engels, W.J. 2005. Flavour formation of lactic acid bacteria and biochemical flavour profiling of cheese products. *FEMS Microbiology Reviews* 29: 591–610.

Somkuti, G.A. and Steinberg, D.H. 2010. Pediocin production in milk by *Pediococcus acidilactici* in co-culture with *Streptococcus thermophilus* and *Lactobacillus delbrueckii* subsp. *bulgaricus*. *Journal of Industrial Microbiology and Biotechnology* 37: 65–69.

Suskovic, J., Kos, B., Benanovic, J., Pavunc, A.L., Habjanic, K., and Matosic, S. 2010. Antimicrobial activity—The most important property of probiotic and starter lactic acid bacteria. *Food Technology and Biotechnology* 43: 296–307.

Tanous, C., Kieronczyk, A., Helinck, S., Chambellon, E., and Yvon, M. 2002. Glutamate dehydrogenase activity: A major criterion for the selection of flavour-producing lactic acid bacteria strains. *Antonie van Leeuwenhoek* 82: 271–278.

Van Immerseel, F., Ducatelle, R., De Vos, M., Boon, N., Van De Wiele, T., Verbeke, K., Rutgeerts, P., Sas, B., and Louis, P. 2010. Butyric acid-producing anaerobic bacteria as a novel probiotic treatment approach for inflammatory bowel disease. *Journal of Medical Microbiology* 59: 141–143.

Vaseji, N., Mojgani, N., Amirinia, C., and Iranmanesh, M. 2012. Comparison of butyric acid concentrations in ordinary and probiotic yogurt samples in Iran. *Iranian Journal of Microbiology* 4: 87–93.

Vedamuthu, E.R. 2006. Starter cultures for yogurt and fermented milk. In: *Manufacturing Yogurt and Fermented Milks*, eds. R.C. Chandan, C. White, A. Kilara, and Y.H. Hui, 89–116. Ames, IA: Blackwell Publishing Professional.

Vijayendra, S.V.N., Palanivel, G., Mahadevamma, S., and Tharanathan, R.N. 2009. Partial characterization of a new heteropolysaccharide produced by a native isolate of heterofermentative *Lactobacillus* sp. CFR 2182. *Archives of Microbiology* 191: 301–310.

Vrinda, R., Goveas, L.C., Prakash, M., Halami, P.M., and Narayan, B. 2012a. Optimization of conditions for probiotic curd formulation by *Enterococcus faecium* MTCC 5695 with probiotic properties using response surface methodology. *Journal of Food Science and Technology*. doi:10.1007/s13197-012-0821-x.

Vrinda, R., Narayan, B., and Halami, P.M. 2012b. Combined effect of enterocin and lipase from *Enterococcus faecium* NCIM5363 against food-borne pathogens: Mode of action studies. *Current Microbiology* 65: 162–169.

Wang, W., Zhang, L., and Li, Y. 2012. Production of volatile compounds in reconstituted milk reduced-fat cheese and the physicochemical properties as affected by exopolysachharide-producing strain. *Molecules* 17: 14393–14408.

Wee, Y.J., Kim, J.N., and Ryu, H.W. 2006. Biotechnological production of lactic acid and its recent applications. *Food Technology and Biotechnology* 44: 163–172.

Weimer, B., Seefeldt, K., and Dias, B. 1999. Sulfur metabolism in bacteria associated with cheese. *Antonie van Leeuwenhoek* 76: 247–261.

Welman, A.D. and Maddox, I.S. 2003. Exopolysachharides in lactic acid bacteria: Perspectives and challenges. *Trends in Biotechnology* 21: 269–274.

Yebra, M.J., Viana, R., Monedero, V., Deutscher, J., and Perez-Martinez, G. 2004. An esterase gene from *Lactobacillus casei* cotranscribed with genes encoding a phosphoenolpyruvate: Sugar phosphotransferase system and regulated by a LevR-like activator and σ^{54} factor. *Journal of Molecular Microbiology and Biotechnology* 8: 117–128.

Yvon, M., Chambellon, E., Bolotin, A., and Roudot-Algaron, F. 2000. Characterization and role of the branched-chain aminotransferase (BcaT) isolated from *Lactococcus lactis* subsp. *cremoris* NCDO 763. *Applied and Environmental Microbiology* 66: 571–577.

6

Molecular Biology of Adaptation of Starter Lactic Acid Bacteria to Dairy System

RAJESH KUMAR AND JAI KUMAR KAUSHIK

Contents

6.1	Introduction	134
6.2	Genome Characteristics of LAB	135
	6.2.1 Lactococci	137
	6.2.2 Lactobacilli	138
	6.2.3 *Leuconostoc*	143
	6.2.4 *Streptococcus thermophilus*	144
6.3	Stress-Tolerance by LAB in GIT	144
6.4	Adaptation of LAB to Milk Environment	146
6.5	Sugar Transport and Glycolysis in LAB	148
6.6	Respiration in *Lc. lactis*	150
6.7	Citrate Utilization and Flavor Generation by LAB	151
6.8	Proteolytic System of LAB	152
	6.8.1 Cell Envelope Proteinases	153
	6.8.2 Peptide Uptake Systems	154
	6.8.3 Regulation of the Proteolytic Systems	155
	6.8.4 Bioactive Peptides	156
6.9	Restriction/Modification Systems Combating Bacteriophage Attack among LAB	156
References		157

6.1 Introduction

Microbial cultures used for making fermented dairy foods are called dairy starter cultures. These cultures possess a long history in the preparation of fermented milks dating back to 1000 years and have evolved along with food systems. Lactic acid bacteria (LAB) are widely used in the production of fermented dairy foods and constitute the major part of the commercial dairy starter cultures. These are arguably the most valuable group of bacteria for human health and have tremendous economic importance in world food systems.

LAB are relatively heterogeneous group of Gram-positive cocci, bacilli, and coccobacilli, which inhabit a broad range of ecological niches. Yet, these are related by different defining characteristics, including low (<55 mol%) G + C content, acid tolerant, non-sporulating, non-aerobic, but aerotolerant, nutritionally fastidious, unable to synthesize porphyrins, and strictly fermentative metabolism with lactic acid as the major metabolic end product (Hansen 2002; Broadbent and Steele 2005). The core group of LAB comprises several species of *Lactobacillus, Lactococcus, Leuconostoc, Pediococcus,* and *Streptococcus,* which have also been commercially exploited to serve as the starters for the production of cheese and fermented milks (Broadbent and Steele 2005). LAB are fastidious microorganisms and commonly present in nutrient-rich environments like animal GIT (gastrointestinal tract), plants, and dairy and meat products. Even if these grow in a variety of habitats, they rely on fermentable carbohydrates, amino acids, fatty acids, salts, and vitamins for their growth (Cogan et al. 2007). Specific LAB strains are also often included in probiotics that on consumption are thought to impart beneficial effect to the host. LAB can be regarded as homo- or heterofermentative, depending on how these ferment hexoses under non-limited growth conditions (Hansen 2002; Broadbent and Steele 2005). The homofermentative LAB use the glycolytic pathway, resulting in the production of lactic acid as the major end product, whereas the heterofermentative LAB use the 6-phosphogluconate/phosphoketolase pathway (6-PG/PK) to produce lactic acid, carbon dioxide, and ethanol (or acetic acid) as the major end products. However, differentiation of homo- and heterofermentative LAB does not solely depend on the production of certain fermentation products, since few LAB species could be facultative heterofermenters. These species are homofermenters in the presence of hexose, but switch to heterofermenters, when pentose is the available carbon source by inducing the 6-PG/PK pathway (Axelsson 2004). LAB not only contribute to the preservation of dairy foods due to acidification but also contribute to the development of product flavor and texture. Understanding the role of genes and genetics of biochemical pathways in LAB might help to control their starter performance more precisely for their exploitation in the dairy industry. Understanding of metabolic pathways and their adaptation to specific dairy environments might help in 'imparting starter-specific characteristics like flavor, aroma, texture, and functional properties to the development of novel designer foods.

6.2 Genome Characteristics of LAB

LAB possess average genome size of 2 Mbp, with a coding strength of approximately 1600–3000 genes (Klaenhammer et al. 2005; Makarova et al. 2006). At present, complete genome sequences of many LAB strains of different species have been published at NCBI (Table 6.1). All typical bacterial genomes display architectural features of chromosome like co-orientation between DNA replication and gene transcription and an asymmetric trend in nucleotide assemblage of leading and lagging DNA strands (McLean et al. 1998). Klappenbach et al. (2000) suggested that rRNA operon copy number reflects ecological strategies of bacteria to influence the structure of natural microbial communities. On the other hand, tRNA is considered as the workhorse of protein synthesis, and it accepts amino acids and donates them to the growing polypeptide chain at the ribosome. The secret of involvement of tRNA in many biochemical steps of the translation remains hidden in its sequence and structural features (Saks and Conery 2007). Among various LAB genomes, few genetic characteristics seem to be universally conserved, comprising enzymes associated with glycolysis (McLean et al. 1998; Klaenhammer et al. 2005). Genetic phenomena such as horizontal gene transfer (HGT), mutation, gene duplication, gene decay, gene loss, and genome rearrangements have been presumed to contribute to the structure, shape and adaptation of LAB. Factual adaptation to nutrient-enriched environments (e.g., milk, vegetation, mammalian gut) stimulates gene decay and gain of vital genes

Table 6.1 Typical Genome Characteristics of Common Lactic Acid Bacteria (LAB)

GENUS/SPECIES/STRAINS	GENOME SIZE (MBP)	% GC CONTENT	NO. OF PROTEINS	NO. OF rRNA OPERONS	NO. OF tRNAS	NCBI GENBANK ACCESSION NO.	REFERENCE
Lb. acidophilus 30SC	2.08	38	2138	12	63	CP002559-61	Oh et al. (2011)
Lb. acidophilus NCFM	1.99	34	1864	4	61	NC_006814	Altermann et al. (2005)
Lb. casei LC2W	3.03	46	3121	5	64	CP002616	Chen et al. (2011)
Lb. casei LOCK919	3.11	46	3092	5	60	CP005486	Koryszewska-Baginska et al. (2013)
Lb. casei Zhang	2.86	46	2804	5	59	CP001084	Zhang et al. (2010)
Lb. citreum KM20	1.79	39	1820	4	69	NC_010471	Kim et al. (2008)
Lb. delbrueckii subsp. bulgaricus ATCC BAA365	1.85	49	1725	9	98	NC_008529	Mayo et al. (2008)
Lb. delbrueckii subsp. bulgaricus ND02	2.13	49	2012	9	94	CP002341	Sun et al. (2011a)
Lb. delbrueckii subsp. bulgaricus ATCC11842	1.86	49	1562	9	95	NC_008054	Van de Guchte et al. (2006)
Lb. helveticus DPC4571	2.08	38	1848	4	73	NC_010080	Callanan et al. (2008)
Lb. johnsonii N6.2	1.88	34	1728	4	55	CP006811	Leonard et al. (2014)
Lb. johnsonii NCC533	1.99	34	1821	6	79	NC_005362	Pridmore et al. (2004)
Lb. lactis subsp. cremoris MG 1363	2.53	35	2436	6	62	NC_009004	Wegmann et al. (2007)
Lb. lactis subsp. cremoris SK11	2.64	35	2509	6	62	NC_008527	Makarova et al. (2006)
Lb. lactis subsp. lactis biovar. diacetylactis IL1403	2.36	35	2310	6	62	NC_002662	Bolotin et al. (2001)
Lb. plantarum WCFS1	3.31	44	3052	5	62	NC_004567	Kleerebezem et al. (2003)
Lb. plantarum WCFS1	3.31	44	3042	5	70	AL935263	Siezen et al. (2012)
Leu. mesenteroides subsp. mesenteroides ATCC8293	2.07	37	2009	4	71	NC_008531	Makarova et al. (2006)
Str. thermophilus CNRZ1066	1.79	39	1915	6	67	NC_006449	Bolotin et al. (2004)
Str. thermophilus LMD9	1.86	39	1710	6	67	NC_008532	Makarova et al. (2006)
Str. thermophilus LMG13811	1.79	39	1890	6	67	NC_006448	Bolotin et al. (2004)
Str. thermophilus MTCC5461	1.73	39	1764	13	59	ALIL00000000	Prajapati et al. (2013)

through HGT (Bolotin et al. 2004; Altermann et al. 2005). Among the majority of sequenced LAB, genome decay was noticed, especially in genes associated in uptake and utilization of carbohydrates and proteins (de Vos and Vaughan 1994; Kleerebezem et al. 2003; de Vos 2011). Genome simplification and decay of metabolic routes were revealed by the genome analysis of the yogurt starter cultures—*Str. thermophilus* and *Lb. delbrueckii* subsp. *bulgaricus* (Altermann et al. 2005; Van de Guchte et al. 2006), as well as of the cheese starter culture *Lb. helveticus* (Callanan et al. 2008). Out of ~2100–2200 genes, common ancestor of Lactobacillales had lost 600–1200 genes (~25%–30%) and acquired ~100 genes after the genetic divergence from the Bacilli ancestor, having genome size of ~2700–3700 genes. Among these bacteria, approximately 10%–12% of the coding genes are predicted to be pseudogenes. Hence, numerous genes encoding for the biosynthesis of cofactors like heme, molybdenum coenzyme, and pantothenate were lost, and vice-versa, certain cofactor transporters, such as nicotinamide mononucleotide transporter, were acquired. Deprecating various metabolic pathways required to synthesize various amino acids and the acquisition of diverse peptidases by LAB resulted in genome evolution for survival in the protein-rich environments like milk and gut (Savijoki et al. 2006; Liu et al. 2010). Ancestors of Lactobacillales were obligate anaerobe as reflected by the absence of genes encoding characteristic enzymes of aerobic bacteria in their genome such as heme/copper-type cytochrome/quinol oxidase-related genes and catalase (Mayo et al. 2008). Numerous genes of *Str. thermophilus* were vanished, while several new pseudogenes were acquired through an active and ongoing process of genome decay (Marchesi and Shanahan 2007; Mayo et al. 2008). Among *Lactobacillus* ssp. with larger genomes, such as *Lb. plantarum* and *Lb. casei*, the deficit of ancestral genes was equalized by the emergence of several new genes via duplication and HGT (Kleerebezem et al. 2003). A brief description of the characteristic features of the genome of common LAB genera and species used as dairy starters is described below.

6.2.1 *Lactococci*

Lactococci are AT-rich, Gram-positive LAB, which were formerly included in the genus *Streptococcus* group N1 (Garvie 1986). These are homofermentative, non-motile, non-sporulating, and catalase-negative cocci that inhabit natural niches such as dairy environment and plant material. *Lactococcus lactis* subsp. *lactis/cremoris* is the workhorse of dairy starter cultures because of its extensive use in the production of buttermilk and cheeses, imparting acidification and flavor through their proteolytic activity (Hols et al. 1999; Smit et al. 2005). *Lc. lactis* has the generally recognized as safe (GRAS) status and studied extensively as is the model organism for novel biotechnological applications, such as heterologous expression of proteins, generation of nutraceuticals, and in vaccine delivery (Nouaille et al. 2003). Louis Pasteur described *Lc. lactis* in 1850s during his studies on soured milks, and it was the first bacterium to be isolated as a pure culture from a mixed population by Joseph Lister (Lister 1878; O'Sullivan 2006). The genome size of *Lc. lactis* strains varies in the range of 2.3–2.6 Mbp, comprising 2300–2500 protein encoding genes. A number of insertion sequence (IS) elements (about 43–130) and certain prophages (4–6) were encountered among *Lc. lactis* strains. Although distribution of IS elements does not seem to be random, however yet, one-fifth of the 71 IS elements have been observed to be concentrated in an integration hotspot region spanning a distinct 56-kb genome region of the *Lc. lactis* MG1363 strain (Makarova and Koonin 2007; Mayo et al. 2008; O'Sullivan et al. 2009). This integration hotspot carries genes that are typically associated with lactococcal plasmids and a repeat sequence specifically found on *Lc. lactis* plasmids and in the so-called lateral gene transfer hotspot of the *Str. thermophilus* genome (Bolotin et al. 2004). Among all sequenced *Lc. lactis* strains, existence of a late competence gene was noticed, which indicates that *Lc. lactis* could develop the natural competence ability under appropriate physiological conditions (Wydau et al. 2006; Mayo et al. 2008). For extracellular proteolytic degradation of casein, *Lc. lactis* acquired cell envelope proteinases (CEPs) PrtP and PrtM and many intracellular aminopeptidases (PepC, PepN, PepA, PepX, PepV, PepQ, PepO, PepP, and PepT) (Siezen 1999; Savijoki et al. 2006).

6.2.2 Lactobacilli

The genus *Lactobacillus* is extremely heterogeneous, Gram-positive, non-sporulating rods, or coccobacilli with 32%–53% GC content of the chromosomal DNA. These are arranged into various groups based on differences in sugar metabolism because of the presence or absence of fructose-1, 6-bisphosphate aldolase and phosphoketolase (Axelsson 1998).

To date, more than 100 species of *Lactobacillus* have been identified. Various *Lactobacillus* species used as probiotics include *Lactobacillus GG* (*Lb. rhamnosus* or *Lb. casei* subsp. *rhamnosus*), *Lb. acidophilus* NCFM, *Lb. casei* strain Shirota, *Lb. paracasei*, *Lb. johnsonii* 100-100, *Lb. johnsonii* LA1, *Lb. ruteri*, *Lb. helveticus*, *Lb. jensenii*, and *Lb. plantarum*. Lactobacilli are homofermenters or heterofermenters, microaerophilic, aciduric or acidophilic and exhibit complex nutritional requirements because of their adaptation to diverse habitats (i.e., plants, dairy, meat, and mammalian gut) typically enriched in carbohydrates and proteins (Axelsson 2004; Broadbent and Steele 2005). Several *Lactobacillus* ssp. are excellent starters and adjunct cultures for fermenting milk into variety of dairy products such as yakult, yogurt, and cheese (Broadbent and Steele 2005; O'Sullivan 2006). The autochthonous population of lactobacilli inhabiting mammalian gut exerts beneficial influence on the maintenance of healthy gut microbial structure. Gut colonization, antagonistic nature, immunomodulation, and cholesterol assimilation are some of the probiotic attributes of lactobacilli (Chin et al. 2000; FAO/WHO 2002; Kumar et al. 2011a, 2012).

Lb. bulgaricus is another homofermentative and acidophilic LAB species, which has been employed in association with *Str. thermophilus* for the commercial production of yogurt worldwide. Complete genome of *Lb. bulgaricus* is constituted of 1.8–2.0 Mbp with 49.7% of GC content and is characterized by the presence of a 47.5-kbp IR, depicting an exceptional structure in bacterial genomes. Evolutionary size reduction of its genome is evidenced by the existence of comparatively high number of rRNA operons (9), tRNA genes (94–98), and pseudogenes (Van de Guchte et al. 2006; Mayo et al. 2008; Sun et al. 2011a). The *Lb. bulgaricus* genome is quickly gaining higher GC content as suggested by the presence of much higher GC content at the third position of the codon than overall GC composition (Van de Guchte et al. 2006; Mayo et al. 2008; Sun et al. 2011a). This variation could be the consequence of *Lb. bulgaricus* adaptation to milk from a plant-associated habitat, as evidenced by the presence of phosphotransferase systems (PTSs), other sugar transport systems, and hydrolytic enzymes. In metabolic cooperation with *Str. thermophilus*, the loss of excess functions could be advantageous in the lactose and casein enriched milk environment (Bolotin et al. 2004; Van de Guchte et al. 2006).

Lb. helveticus is a homofermentative LAB, which is commonly used as a starter or adjunct culture to lower the bitterness and enhance the flavor in cheese (Fox et al. 2000; Tompkins et al. 2012). The genome size of *Lb. helveticus* is ≥2.0 Mbp with GC content of 37.73% (Callanan et al. 2008; Mayo et al. 2008; Tompkins et al. 2012). It encodes ~1600 proteins, 4 rRNA operons, 73 tRNA genes, and numerous pseudogenes. Genes encoding 16S rRNA of *Lb. helveticus* and *Lb. acidophilus* species exhibit >98% similarity (Felis and Dellaglio 2007). Surprisingly, 213 IS elements of different classes were identified in the genome of *Lb. helveticus* DPC4571 (Callanan et al. 2008). A chromosomal region of 100 kbp with GC content of 42% and flanked by IS elements and unique 12 bp direct repeats observed in the genome of *Lb. helveticus* DPC4571 could be an adaptation-associated genomic island. The gut inhabiting *Lb. helveticus* strains have been observed to be highly proteolytic and possessing elaborate proteolytic systems that have been associated with the adaptation of microorganisms to protein-rich environment (Callanan et al. 2008). Genome of *Lb. helveticus* R0052 strain contained genes encoding mucus-binding proteins similar to those identified in *Lb. acidophilus*, which probably help in the adhesion to intestinal epithelia (Tompkins et al. 2012). Moreover, certain *Lb. helveticus* strains are known to generate bioactive peptides with antihypertensive and immunomodulatory properties by milk fermentation. The *Lb. helveticus* genome encodes aminopeptidases like PepC, PepE, PepN, PepX, PepD, PepV, PepI, PepQ, and PepR, which play an important role in the generation of bioactive peptides by breaking down milk protein casein (Christensen et al. 1999; Luoma et al. 2001; Takano 2002; Savijoki et al. 2006; Griffiths and Tellez 2013).

Lb. acidophilus is a homofermentative *Lactobacillus* that can survive transit through GIT, where low pH and high bile concentration to toxic level for microbes are prevalent (Liong and Shah 2005). *Lb. acidophilus* NCFM is a potential probiotic strain that has been utilized commercially for over 25 years in a variety of probiotic products (Sanders and Klaenhammer 2001; Altermann et al. 2005). *Lb. acidophilus* contained a genome size of around 2 Mbp with GC content of 34.71%. *Lb. acidophilus* genome encodes an array of ATP-binding cassette (ABC) transporters for amino acids, oligopeptides, and several permeases for amino acids. The genome comprises up to 1864 predicted open reading frames (ORF), of which 72.5% ORFs have been functionally classified. In addition, 4 rRNA operons and 61 tRNA genes were also found in *Lb. acidophilus* genome (Altermann et al. 2005; Stahl and Barrangou 2013). *Lb. acidophilus* NCFM strain can synthesize only cysteine, serine, and aspartic acid among the amino acids (Sanders and Klaenhammer 2001). In addition, the genome of strain NCFM possesses several gene clusters encoding for the utilization of oligosaccharides such as fructooligosaccharide and raffinose, which are transcriptionally controlled by genes belonging to the *lacI* family (Altermann et al. 2005). On the cost of loss of function for biosynthetic pathway enzymes for amino acids, *Lb. acidophilus* has acquired more than 20 putative peptidases (PepX) and proteases for the extensive proteolytic breakdown and oligopeptide transporters for peptide utilization (Christensen et al. 1999). The gut inhabiting *Lb. acidophilus* also contained genes encoding surface-binding proteins like mucus-binding protein (*mub*) and fibronectin-binding protein (*fnb*), which probably play a crucial role in their adhesion to intestinal epithelia (Buck et al. 2005).

Among LAB, *Lb. plantarum* has the largest genome size of around 3.3 Mbp with GC content of 44.5% and contained up to 3042 predicted protein encoding genes, 5 rRNA operons, and 70 tRNA genes (Siezen et al. 2012). *Lb. plantarum* is a facultative heterofermentative LAB, and genes encoding all enzymes associated with glycolysis and phosphoketolase pathways are present in its large genome, which also contains genes for biosynthetic pathways of most of the amino acids with the exception of branched-chain amino acids (BCAAs) (Kleerebezem et al. 2003). *Lb. plantarum* inhabits variety of environmental niches (Kumar et al. 2014). *Lb. plantarum* adapted to gut system has been observed to possess genes encoding collagen-binding protein (*cbp*), 04 bile salt hydrolase (*bsh*) proteins, *mub* proteins (Kaushik et al. 2009; Kumar et al. 2011b, 2014; Yadav et al. 2013), and resistance against low pH and bile salts (Kaushik et al. 2009; Duary et al. 2010). The presence of numerous genes encoding regulatory, transport, and stress-related proteins in its genome reflects the ecological flexibility of this species (Kleerebezem et al. 2003). Majority of the genes associated with broad range of sugars transport and utilization, and genes encoding extracellular functions are clustered in a region near to the origin of replication (Kleerebezem et al. 2003). The deviation in the base composition of many genes of this region compared to the rest of the genome suggests that these genes are acquired from horizontal transfers (Molenaar et al. 2005; Kumar et al. 2014). Based on these facts, this chromosomal segment was designated as the so-called lifestyle adaptation island (Kleerebezem et al. 2003).

Lb. johnsonii is a member of the acidophilus group of intestinal lactobacilli that has been comprehensively studied for probiotic attributes such as adhesion to intestinal epithelia, pathogen inhibition, and immunomodulation. *Lb. johnsonii* genome can be ≥1.88 Mbp in size with an average GC content of 34.5% and may contain up to 1728 protein-coding ORFs, 4–6 rRNA operons, 55–79 tRNAs, 14 full IS elements, and two complete prophages (Pridmore et al. 2004; Leonard et al. 2014). Surprisingly, its genome lacks genes encoding biosynthetic pathways for amino acids, purine nucleotides, and most cofactors. On the other side, *Lb. johnsonii* acquired numerous amino acid permeases, peptidases, and phosphoenolpyruvate-PTSs, indicating its strong dependency for nutrients on the host and other intestinal microbes. Hence, competition appears unlikely among *Lb. johnsonii*, bifidobacteria, and bacteroides; instead, this species should adapt better in the upper GIT part, where amino acids, peptides, and mono- and oligosaccharides are abundant. Genome analysis predicted numerous surface-binding proteins, a characteristic feature to enhance GIT survival and colonization (Pridmore et al. 2004). In addition, genes encoding *bsh* and bile acid transporters could help in the survival of *Lb. johnsonii* in hostile GIT environment (Elkins et al. 2001). Comparative genomic comparison revealed a high degree of genetic relatedness among *Lb. johnsonii*, *Lb. taiwanensis*, and *Lb. gasseri* (Sarmiento-Rubiano et al. 2010).

Lb. casei is a homofermentative gut inhabiting LAB, commonly used as a starter culture in the production of Yakult and other fermented milk products. Specific *Lb. casei* strains like Shirota and Zhang are mainly used in the preparation of Yakult and Koumiss. The complete genome size of *Lb. casei* may be in the range of 2.86–3.11 Mbp with 46.2%–46.5% chromosomal GC content. It may encode up to 2804–3121 coding sequences, 59–64 tRNAs, and 5 rRNA operons (Zhang et al. 2010; Chen et al. 2011; Koryszewska-Baginska et al. 2013). *Lb. casei* LC2W can synthesize several fractions of exopolysaccharide (EPS) from skim milk as its chromosomal genome contained 17 EPS-related genes. These genes are involved in the regulation, polymerization, and chain length termination and export of the EPS (Chen et al. 2011). Comparative genome analysis confirmed *Lb. casei* LOCK919 genome as the largest genome among the fully sequenced genomes of this species. In addition, strain LOCK919 also contained proteins with putative roles in the persistence, adhesion, and colonization in the human gut (Koryszewska-Baginska et al. 2013).

6.2.3 Leuconostoc

The genus *Leuconostoc* is a heterofermentative, catalase negative, and slime-forming Gram-positive cocci (Garvie 1986). *Leuconostoc mesenteroides* subsp. *cremoris* and *Leu. lactis* are industrially important species and are the common starters/adjunct cultures in the production of butter and cheeses. *Leuconostoc*s propagate slowly, especially in milk, and hence, they are less important for lactic acid fermentation, but they are excellent producer of flavoring compounds such as diacetyl through citrate utilization (Hugenholtz 1993). *Leu. citreum*, a dextran-producing species, is generally encountered in fermented foods and feeds of plant and dairy origin, such as cheese, pickles, sauerkraut, and cabbage (Garvie 1986). A potential starter strain KM20 was originally identified as dominant bacteria in Kimchi, a popular fermented Korean commodity made from a variety of spicy vegetables (Choi et al. 2003). The complete genome of *Leuconostoc* consists of about 1.79–2.01 Mbp with GC content in the range of 37.77%–39% and may contain 1702–1942 protein-coding genes, 4 rRNA operons, and 67 tRNA genes (Kim et al. 2008; Jung et al. 2012). Genome of *Leu. citreum* KM20 lacked complete phages, but contained IS*3*- and IS30-related IS elements. In addition, a complete gene set for heterolactic fermentation *via* the phosphoketolase pathway, with an incomplete tricarboxylic acid cycle and a limited biosynthetic ability for various amino acids and cofactors, was found within the genome of *Leu. citreum* KM20. Multiple genes for dextransucrases and alternansucrases and an array of carbohydrate hydrolases and transporter genes were also identified within *Leu. citreum* genome. These genes could have contributed to the potential of *Leuconostoc* for medical and food applications. Furthermore, five mucus-binding domains were embedded in a plasmid encoded putative cell wall-anchored protein of *Leu. citreum*. *Mub* proteins are supposed to help a bacterium in gut colonization, and hence, the probiotic potential of *Leuconostoc* strains should be explored (Kim et al. 2008).

6.2.4 Streptococcus thermophilus

Str. thermophilus is the only species of the genus *Streptococcus*, which has GRAS status. *Str. thermophilus* is the second most valuable dairy starter following *Lc. lactis* and is being utilized for fermenting milk into yogurt, hard cooked cheeses of the Italian and Swiss types as well as soft cheeses (Fox et al. 2000). The genome size of sequenced *Str. thermophilus* strains is around 1.83 Mbp with 39.1% GC content. The genome also contains 56–67 tRNA genes and 5–6 rRNA operons (O'Sullivan et al. 2009; Sun et al. 2011b). *Str. thermophilus* still share certain physiological and metabolic features with its pathogenic relatives. This organism has acquired numerous pseudogenes, which indicate regressive evolution mainly through the loss of virulence genes that are no longer required in the dairy niche (Bolotin et al. 2004; O'Sullivan et al. 2009). During the course of evolution, sugar catabolic abilities of *Str. thermophilus* are degenerated, while it has acquired a well-developed nitrogen metabolism. Symbiotic cooperation occurs between *Str. thermophilus* and *Lb. bulgaricus* as both of these organisms share similar ecological niche. Numerous adaptation genomic island acquired by HGT imparted important adaptive traits, which are of industrial significance such as EPS biosynthesis, bacteriocin production, or oxygen tolerance (O'Sullivan et al. 2009) to *Str. thermophilus*. Six large insertion islands formed by the genes encoding transposase, glutamate decarboxylase, acetyltransferase, glycosyltransferase, polysaccharide biosynthesis protein, and the EPS biosynthesis gene cluster were found in the genome of *Str. thermophilus* ND03 strain. EPS-gene cluster contained *epsA-G*, *epsI*, *epsJ*, and *epsP* genes, which were involved in the regulation, polymerization, chain length determination, and export of the EPS (Sun et al. 2011b).

6.3 Stress-Tolerance by LAB in GIT

To impart health-promoting properties to the host, a dairy starter LAB should be able to overcome hostile conditions prevalent in GIT, including highly acidic condition in stomach and high bile concentration in intestine. Advances in genomics and proteomics have led to the identification of relevant genes encoding F_1F_0-ATPase (*atp* operon), conjugated bile salt hydrolases (*cbsH*), and conjugated bile salt transporters (*cbsT*), which could be associated with various stress responses by LAB (Cotter and Hill 2003; Duary et al. 2010; Kumar et al. 2011a). The acid tolerance of LAB depends on the equilibrium between extra- and intra-cellular pH and when the internal pH reaches a threshold value, cellular functions are inhibited and the cells die (Cotter and Hill 2003; Duary et al. 2010). Low pH appears to limit the growth of starter bacteria during fermentation and GIT transit period. The F_0F_1-ATPase is a known factor, which offers protection to LAB against acidic conditions (Cotter and Hill 2003; Duary et al. 2010). The F_0F_1-ATPase consisted of a catalytic portion (F_1) constituted of α, β, γ, δ, and ε subunits for adenosine triphosphate (ATP) hydrolysis and an integral membrane portion (F_0) including a, b, and c subunits, which function as a membranous channel for proton translocation (Cotter and Hill 2003). The F_1F_0-ATPase is chiefly involved in pumping out of protons (H^+) from the bacterial cytoplasm by proton motive force (PMF) and hence helps bacterium to tolerate acidic pH (Cotter and Hill 2003; Duary et al. 2010).

Following the stomach passage, the small intestine is the second major barrier in GIT. Although pH of the small intestine is more favorable toward bacterial survival, the presence of high bile concentration might be deleterious. Bile salts are secreted as bile into the duodenum in the form of *N*-acyl compounds conjugated with glycine or taurine, and enhance the emulsification and absorption of dietary lipids apart from undergoing enterohepatic circulation. In addition to their function in the intestine as natural emulsifiers, bile salts possess some detergent-like antimicrobial properties. To survive in intestine, many LAB species have evolved mechanisms to resist the detergent action of bile salts by enzymatic hydrolysis. Bsh, an intracellular enzyme commonly found in certain intestinal bacteria, helps in degradation of conjugated bile. The Conjugated bile acid hydrolase (CBAH) proteins (EC 3.5.1.24) and penicillin V amidases (EC 3.5.1.11) belong to the choloylglycine hydrolase family of enzymes and have been classified as N-terminal nucleophilic hydrolases with an N-terminal cysteine residue. CBAH catalyses the hydrolysis of glycine or taurine-conjugated bile acids into the amino acid residue and deconjugated bile acid. Free bile salts are less effective detergents than their conjugated counterparts and less efficient in forming mixed micelle and hence excreted through feces from the intestinal tract (Kumar et al. 2011a, 2012, 2013).

In high bile conditions, expression of *cbsh* encoding genes by lactobacilli could offer bile resistance. However, the presence and genetic organization of *bsh* genes in lactobacilli are highly variable. Among lactobacilli, a single or multiple *bsh* genes have been reported and characterized (Elkins et al. 2001; McAuliffe et al. 2005; Lambert et al. 2008; Zhang et al. 2009; Kumar et al. 2013, 2014). Elkins et al. (2001) identified a locus encoding BSHβ, a partial *cbsT1* and a complete *cbsT2* in *Lb. johnsonii* strain 100-100 and *Lb. acidophilus* strain KS-13. Analyses of genomes suggest that bile exposure could have driven the dissemination and evolution of *bsh* genes in the human intestinal microbial metagenome (Jones et al. 2008). According to Kurdi et al. (2000), *Lc. lactis* negotiates the high bile concentration by active extrusion of cholic acid from the cell in an ATP-dependent manner, whereas several *Lactobacillus* ssp. accumulate cholic acid into cell membrane upon glucose energization. The mechanism of cholic acid accumulation is not transporter mediated, but depends on cholic acid diffusion across the bacterial cell according to the transmembrane proton gradient.

6.4 Adaptation of LAB to Milk Environment

LAB utilize lactose in milk as carbon source and casein as a source of amino acids. Genome and transcriptome analyses explain adaptation of LAB in the milk environment by acquiring specialized systems for transporting and metabolizing lactose, oligopeptides, and amino acids along with several stress response mechanisms (Makarova et al. 2006; Mayo et al. 2008; O'Sullivan et al. 2009). Adaptation to the milk environment by LAB had resulted in the extensive loss of ancestral genes and the gain of numerous pseudogenes as revealed by the phylogenetic comparison of current and ancestral gene sets (Kleerebezem et al. 2003; Bolotin et al. 2004; Pridmore et al. 2004; Altermann et al. 2005; Makarova et al. 2006; Mayo et al. 2008; O'Sullivan et al. 2009). Gene loss and metabolic simplification are accompanied with lineage-specific duplication and acquisitions of key genes via HGT. A small percentage (≈5% of the total number of genes) of plasmid-encoded genes in *Lc. lactis* are essential for growth of LAB species in specific environments (Siezen et al. 2005). Additionally, genomes of LAB encode transposons and IS elements, which contribute significantly to the genetic diversity and adaptation of microbes (Mayo et al. 2008; O'Sullivan et al. 2009). Comparative genomic analysis of two *Lc. lactis* plant isolates to those of dairy adapted strains provided a first view of the molecular basis of adaptation of this bacterium to plant and milk environments. A high synteny was observed between the genomes of dairy and plant *Lc. lactis* isolates, but numerous genes were lost in dairy *Lc. lactis* strains during the course of evolutionary adaptation in comparison to plant isolates (Siezen et al. 2005). Certain dairy practices also might have induced specific adaptation among LAB to particular fermented milks. Strains of *Lb. delbrueckii*, *Lb. helveticus* and *Str. thermophilus*, whose closest relatives are represented by commensal as well as pathogenic bacteria from the human and animal GIT, have adapted to stringent conditions faced during processing of dairy products like yogurt and cooked cheeses (Beresford et al. 2001; Wouters et al. 2002). During the adaptation, *Str. thermophilus* acquired new set of genes for efficient exploitation of milk's nutrients by HGT and at the same time lost dispensable genes and amino acid biosynthesis ability for its growth in milk by genome miniaturization (Bolotin et al. 2004; Sun et al. 2011b). The absence of specific lactose-synporter from the pathogenic relatives of *Str. thermophilus* is a classic example. Similarly, the adaptation of *Lb. helveticus* to dairy niches was accompanied by the loss of genes for amino acid biosynthesis, transport proteins, and energy metabolism. The lateral gene transfer events helped *Lb. helveticus* to gain functions for fatty acid biosynthesis, DNA restriction and modification, and amino acid biosynthesis (Callanan et al. 2008). A gain of 17-kbp DNA region containing multiple copies of IS1191 and a gene cluster by *Lb. bulgaricus* and *Lb. lactis* enabled them to synthesize essential nutrients from milk components (Kok et al. 2005). In this 17-kbp DNA region, a unique gene was found, which endowed LAB to synthesize methionine, a rare amino acid in milk. Several LAB strains had acquired 2–300 kbp size plasmids, which were reckoned as their tools of adaptation to dairy system. However, many LAB species sustained chromosomal regions dedicated to the incorporation of adaptive DNA, which were called as life style island (O'Sullivan et al. 2009).

6.5 Sugar Transport and Glycolysis in LAB

LAB have acquired genes for many carbohydrate-degrading enzymes to utilize various mono- and disaccharides like glucose, fructose, galactose, lactose, maltose, mannose, and cellobiose as carbon source. However, noticeable variations occurred in the number and types of glycosyl hydrolases in LAB species during the course of adaptation to diverse nutrient environments (O'Sullivan et al. 2009; Mayo et al. 2010). Among LAB, carbohydrate-metabolizing genes generally consist of three intact sugar hydrolases, namely, β-galactosidase, sucrose-6-phosphate hydrolase, and intracellular α-amylase, in addition to (1) three specific PTS for mannose/fructose, cellobiose, and sucrose transport; (2) three distinguished sugar ABC transporters of unknown substrates; and (3) two permeases for the predictive uptake of glucose and lactose (de Vos and Vaughan 1994; Kleerebezem et al. 2000). The *lac-gal* gene clusters are conserved in the genome of *Lb. lactis*, *Lb. acidophilus*, *Lb. johnsonii*, *Lb. plantarum*, and *Lb. salivarius*; however, *Lb. bulgaricus* genome contained only a *lacSZ* operon. Utilization of milk lactose by LAB mainly depends on genes encoding β-galactosidases, β-phospho-galactosidases, and β-glucosidases (de Vos and Vaughan 1994). During the intracellular import or transport, sugars are phosphorylated via the phosphoenolpyruvate (PEP)-dependent PTS (PEP-PTS) that is an active process for sugar transport since the event is directly coupled to sugar phosphorylation at the loss of one ATP (Figure 6.1). The PEP-PTS system transfers the phosphate group from PEP to the importing sugar via two cytoplasmic phosphocarrier proteins: enzyme I and HPr. Protein HPr plays a crucial role in the regulation of lactose utilization at both levels, namely, sugar transport and gene transcription (Jamal et al. 2013). In *Lc. lactis*, a catabolite control protein CcpA regulates the expression of numerous genes and operons, which are under the catabolite repression, and additionally, CcpA affects glycolytic flux through *las* operon activation to control lactose metabolism (Luesink et al. 1998; Kleerebezem et al. 2000). *Lc. lactis* utilizes lactose as the main carbon source in various dairy fermentation processes, and hence, the lactose metabolizing pathways are

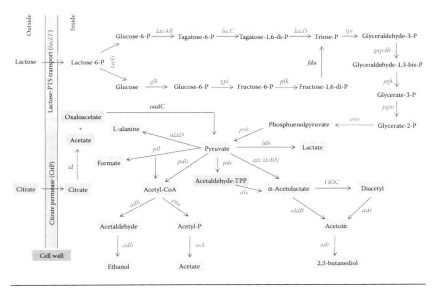

Figure 6.1 Schematic representation of the genes involved in lactose and citrate metabolism pathways in *Lc. lactis*. Enzymes are indicated by their encoded genes, using abbreviations based on established genetic nomenclature. Lactose-PTS transport (lacEF), citrate permease (*CitP*), phospho β-galactosidase (*lacG*), citrate lyase (*cl*), galactose-6-phosphate isomerase (*lacAB*), tagatose-6-phosphate kinase (*lacC*), tagatose-1,6-diphosphate aldolase (*lacD*), glucokinase (*glk*), glucose-phosphate isomerase (*gpi*), phosphofructo kinase (*pfk*), fructose diphosphate aldolase (*fda*), triosephosphate isomerase (*tpi*), glyceraldehydes 3-phosphate dehydrogenase (*gapdh*), phosphate glycerate kinase (*pgk*), phosphogluco mutase (*pgm*), enolase (*eno*), pyruvate kinase (*pyk*), lactate dehydrogenase (*ldh*), oxaloacetate decarboxylase (*oadc*), pyruvate formate lyase (*pfl*), L-alanine dehydrogenase (*alaD*), pyruvate dehydrogenase (*pdh*), pyruvate decarboxylase (*pdc*), acetohydroxy acid synthase (*ilvBN*), alcohol dehydrogenase (*adh*), α-acetolactate synthase (*als*), non-enzymatic oxidative decarboxylation (ODC), phosphate acetyltransferase (*pta*), acetate kinase (*ack*), acetolactate decarboxylase (*aldB*) and acetoin diacetyl reductase (*adr*).

well understood. Genes encoding lactose-PTS transport, lactose-6P hydrolysis, and the tagatose-6P pathway are plasmid encoded among many lactococci (Figure 6.1). Transcriptional repressor protein encoded by *lacR* gene controls the expression of these *las*-operon genes, and tagatose-6P, an intermediate of the tagatose-6P pathway, relieves the repression of *lac* gene expression (Kleerebezem et al. 2000; Van Rooijen et al. 1993). Moreover, CcpA protein activates the transcription of three key enzymes associated in glycolysis and lactate production, phosphofructo-kinase, pyruvate kinase, and lactate dehydrogenase encoded on *las* operon (Luesink et al. 1998). Therefore, *CcpA* gene inactivation of *Lc. lactis* may result in the decreased expression levels of the *lac* operon genes (Luesink et al. 1998).

6.6 Respiration in *Lc. lactis*

Respiratory capability of LAB was overlooked and initial reports were based on whole genome sequence analysis of *Lc. lactis* IL1403 (Sijpesteijn 1970; Duwat et al. 2001). Interestingly, few LAB respire in the presence of oxygen and exogenous heme, resulting in the production of extremely decreased level of lactic acid (Gaudu et al. 2002). Heme, an iron containing porphyrin and an essential cofactor of the cytochrome oxidase system, acts as a limiting factor of the respiratory ability of *Lc. lactis*, since this bacterium lacks a functional biosynthetic pathway for heme, and hence, the exogenous heme supplementation in growth medium is essential to allow respiration in the presence of oxygen (Bolotin et al. 2001; Duwat et al. 2001; Gaudu et al. 2002). Under respiratory conditions, carbon sources are more efficiently converted into biomass, resulting in higher cell yields and survival. Acetate, acetoin, and diacetyl are produced from pyruvate at the expense of lactic acid that was shown by the upregulation of pyruvate dehydrogenase complex, acetolactate synthase, and α-acetolactate decarboxylase genes of *Lc. lactis* (Duwat et al. 2001). Under respiratory conditions, *ygfC* gene encoding a putative regulatory protein was the most upregulated, while the expression of pyruvate formate lyase and alcohol dehydrogenase genes were downregulated by 2.5- and 50-folds, respectively (Vido et al. 2004). These evidences were the milestone toward the development of a patented process for the industrial production of LAB starter cultures by Chr. Hansen (Pedersen et al. 2005). Respiratory ability of *Lc. lactis* strains were also studied in the absence of exogenous heme and numerous genes, which showed differential expression under different conditions (Pedersen et al. 2005). However, high cell yield could not be obtained in the presence of aeration and exogenous heme in case of *Str. thermophilus*, *Lb. bulgaricus*, and *Lb. helveticus*. Complete genome sequence analysis of these species could not confirm the presence of genes encoding cytochrome oxidase and enzymes involved in the biosynthesis of quinones. These genetic features are recognized to be essential for respiration (Bolotin et al. 2004; Van de Guchte et al. 2006; Callanan et al. 2008; Mayo et al. 2008).

6.7 Citrate Utilization and Flavor Generation by LAB

The industrial value of LAB strains is mainly due to their ability to convert carbohydrates into lactic acid, while few LAB species like *Lc. lactis* and *Leu. mesenteroides* produce C4 flavor compounds diacetyl and acetoin (Figure 6.1) through citrate metabolism (Garcia-Quintans et al. 2008; Pudlik and Lolkema 2011, 2012). Intracellular import of citrate is mediated by citrate permease, which develops a membrane potential upon electrogenic exchange of divalent citrate and monovalent lactate (Garcia-Quintans et al. 2008; Pudlik and Lolkema 2011, 2012). Three types of genetic organization associated with citrate metabolism have been identified in LAB genome (Drider et al. 2004). Citrate lyase cleaves internal citrate into acetate and oxaloacetate, and oxaloacetate is decarboxylated into pyruvate by an oxaloacetate decarboxylase, resulting into a transmembrane pH gradient due to scalar proton consumption (Marty-Teysset et al. 1996; Drider et al. 2004). Citrate metabolism in *Lc. lactis* and *Leu. mesenteroides* generates a secondary PMF by secondary transporters, malate permease and citrate permease (*CitP*), which translocate net negative charges across the membrane and catalyze a heterologous exchange of two structurally related precursor and product molecules malate/lactate and citrate/lactate, respectively (Marty-Teysset et al. 1996; Drider et al. 2004). In *Lactococcus* and *Leuconostoc* ssp., the citrate metabolic pathway is a precursor-product exchange system. The presence of citrate in medium could stimulate the growth rate and enhance *citP* expression level in *Leuconostoc* ssp., while the similar observations could not be observed in *Lc. lactis* (López de Felipe et al. 1995; Marty-Teysset et al. 1995; Garcia-Quintans et al. 1998; Drider et al. 2004; Pudlik and Lolkema 2012). During citrolactic fermentation, citrate serves as the main precursor of lactate in *Leuconostoc*, and *CitP* catalyzes a precursor-product exchange reaction, while in case of *Lactococcus*, lactate is produced via homofermentative pathway (Pudlik and Lolkema 2011, 2012). Ability of *Leuconostoc* to ferment glucose and citrate offers a growth advantage in comparison with the growth on glucose substrate alone due to the metabolic shift in the heterofermentative pathway of glucose (Marty-Teysset et al. 1996).

The ability to maintain redox balance is vital to all organisms as these are central to both anabolic and catabolic metabolism. Van Hoek and Merks (2012) found that maintaining redox balance is the key to explain that some microbes decrease the flux through the high-yield pathway, while other microbes like *Lc. lactis* use overflow-like low-yield metabolism. In *Leu. mesenteroides*, redox balance of the phosphoketolase pathway is maintained by the ethanol produced from acetyl phosphate that is formed from glucose in the absence of citrate. However, in the presence of citrate, redox balance is shuttled to pyruvate, which further generates lactate and acetyl phosphate. Later acetate is produced from acetyl phosphate via the acetate kinase pathway, which derives the formation of one ATP per mole of acetyl phosphate (Marty-Teysset et al. 1996; Drider et al. 2004). Genetic organization of CitP has been well studied in *Lc. lactis* subsp. *diacetylactis* and *Leu. mesenteroides* subsp. *cremoris* (López de Felipe et al. 1995; Garcia-Quintans et al. 1998; Drider et al. 2004). Genes encoding *citP*, regulatory protein (*citR*), and leader peptide (*citQ*) remain under the control of citQRP operon encoded by lactococcal plasmid pCIT264 (López de Felipe et al. 1995; Drider et al. 2004). Transcriptome analysis suggested that acid stress induced *citQRP* operon as the mRNA expression was 14-fold more abundant at acidic pH of 4.5 than at 6.5 (Garcia-Quintans et al. 1998; Drider et al. 2004). At low pH, intracellular lactate accumulation via citrate metabolic pathway confers higher resistance on *Lc. lactis* subsp. *diacetylactis* cells and relieves the growth inhibition caused by lactate toxicity (Magni et al. 1999). Gene (*citP*) is located within a cluster of eight genes called *citCDEFGOP* on the chromosome of *Leu. mesenteroides* subsp. *cremoris*. Two transcripts of 5.2 kb (*citCDEFG*) and 4 kb (*citGOP*) were detected when cells were grown in the presence of citrate (Bekal et al. 1998, 1999; Drider et al. 2004). However, in *Lc. lactis* subsp. *diacetyllactis*, *citP* gene was found on plasmid-encoded *citQRP* operon that is transcribed from the pH-dependent P1 promoter (Garcia-Quintans et al. 2008). Hence, at low pH transcriptional activation of the promoters controlling *cit* operon might offer the starter bacteria an adaptive response under acidic stress (Mayo et al. 2010).

6.8 Proteolytic System of LAB

LAB are fastidious organisms that depend upon exogenous sources of peptides or amino acids. A protein-enriched environment like milk and its proteolytically-broken-down products provide essential amino acids.

The proteolytic activity of LAB enables their successful adaptation in the milk environment. CEPs of LAB initiate the breakdown of casein by degrading it into smaller oligopeptides, which are subsequently imported inside cells via specific peptide transport systems (Kunji et al. 1996; Christensen et al. 1999; Griffiths and Tellez 2013). Inside the cytoplasm, various intracellular peptidases further degrade these oligopeptides into smaller peptides and eventually into amino acids and various flavoring compounds. These peptides, amino acids, and their derivatives contribute to the characteristic flavor and texture of the fermented milks (Figure 6.2). LAB also possess stress-inducible proteolytic enzymes (Savijoki et al. 2006).

6.8.1 Cell Envelope Proteinases

In LAB, the CEPs are typically synthesized as preproproteins of some 2000 residues containing several distinct functional domains. Starting from the N terminus, the CEPs include the prepro domain matching to a signal pro-sequence, catalytic serine protease domain, an insert domain, which might be responsible for the substrate specificity of

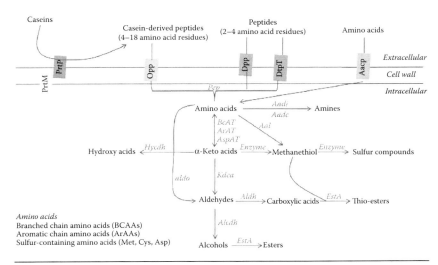

Figure 6.2 Simplified presentation of the genes mediating flavor generating pathways. Enzymes are indicated by their encoding genes: peptidase (*Pep*), branched-chain aminotransferase (*BcAT*), aromatic aminotransferase (*ArAT*), aspartate aminotransferase (*AspAT*), amino acid deaminase (*Aadi*), amino acid decarboxylase (*Aadc*), amino acid lyase (*Aal*), aldolase (*aldo*), keto acid decarboxylase (*kdcA*), hydroxy acid dehydrogenase (*Hycdh*), aldehyde dehydrogenase (*Aldh*), alcohol dehydrogenase (*Alcdh*), esterase (*EstA*).

CEPs, A domain of unknown function, B domain likely involved in stabilizing the CEP activity or specificity, the helix domain associated with the extracellular positioning of A and B domains, and a hydrophilic W domain functioning as a cell wall spacer (Fernandez-Espla et al. 2000). In *Lc. lactis*, a divergently transcribed gene encoding a membrane-bound lipoprotein (PrtM) was found essential for the autocatalytic maturation of PrtP (Haandrikman et al. 1989, 1991). *Lb. paracasei* possessed same genetic organization for these two genes, whereas *prtM* gene could not be identified in the flanking regions of CEPs encoding gene of *Lb. helveticus*, *Lb. bulgaricus*, and *Str. thermophilus*. However, it was observed that in case of *Lb. bulgaricus*, PrtM-like chaperone was not required for the maturation process of CEPs (Holck and Naes 1992; Gilbert et al. 1996; Siezen 1999; Fernandez-Espla et al. 2000; Germond et al. 2003). The PrtR (*prtR*) CEP of *Lb. rhamnosus* BGT10 was quite different from CEPs of starter LABs as its B domain was smaller and different. This B domain lacked helix and insert domains, while the W domain of PrtR was more homologous to certain cell surface antigens of oral and vaginal streptococci (Pastar et al. 2003).

6.8.2 Peptide Uptake Systems

Casein proteolysis mediated by CEPs generates oligopeptides, which are transported into the cell via oligopeptide-binding protein (Opp) system. The Opp proteins belong to a superfamily of highly conserved ATP-binding cassette transporters, which allow the uptake of casein-derived oligopeptides (Drider et al. 2004). In the genome of *Lc. lactis* MG1363, genes encoding oligopeptide-binding protein (OppA), two integral membranes proteins (OppB and OppC), and two nucleotide-binding proteins (OppD and OppF) constitute an operon (Mayo et al. 2010). This Opp system can transport up to 18-amino acid long peptides; however, the kinetics of transport can be significantly affected by the nature of peptides to be transported (Detmers et al. 1998; Juillard et al. 1998). The Opp systems could not be widely investigated among other LAB but in *Str. thermophilus* and *Lb. bulgaricus*, the composition of the Opp system was homologous to *Lactococcus* (Garault et al. 2002; Peltoniemi et al. 2002). *Str. thermophilus* comprises three paralogs of genes encoding OppA proteins, associated in oligopeptide import (Garault et al. 2002). Additionally, in *Lc. lactis* strains, other peptide transporters comprised a PMF-driven dipeptide/tripeptide DtpT and an ATP-driven Dpp system. Dpp can transport di-, tri-, and tetra-peptides containing relatively hydrophobic BCAAs and exhibits the highest affinity for tripeptides, whereas DtpT displayed a preference toward hydrophilic and charged di/tri-peptides. A gene encoding a PMF-driven DtpT could also be identified in *Lb. helveticus* and its transport specificity resembles that of *Lc. lactis* (Savijoki et al. 2006).

6.8.3 Regulation of the Proteolytic Systems

Regulation of the proteolytic system of LAB ensures adequate intracellular nitrogen balance by sensing the changes in the intracellular nitrogen pool. It was reported that di- or tri-peptides with hydrophobic BCAAs act as effector molecules in the transcriptional regulation of the Opp system and consequently the whole proteolytic system of *Lc. lactis* (Mayo et al. 2010). Addition of casein hydrolysate containing 80% peptides and 20% amino acids in growth medium downregulated the expression of six transcriptional units including *prtP*, *prtM*, *opp–pepO1*, *pepD*, *pepN*, *pepC*, and *pepX* by 5- to 150-folds. However, under nitrogen-limiting conditions, the downregulated expression was relieved (Savijoki et al. 2006; Mayo et al. 2010). In a more similar proteomics approach, upregulation of Opp, PepO1, PepN, PepC, PepF, and the putative substrate-binding protein (OptS) of a second oligopeptide ABC transporter system could be noticed in *Lc. lactis*, which was propagated in a medium lacking free amino acids and peptides (Gitton et al. 2005). Expression of several components of the proteolytic system of *Lc. lactis* remained under the negative regulation of transcriptional regulator CodY in the presence of intracellular pool of BCAAs such as isoleucine, leucine, and valine (Guédon et al. 2001b; Petranovic et al. 2004). Under *in vitro* conditions, CodY binds upstream of the *opp*-operon and the binding was stimulated by BCAAs (Den Hengst et al. 2005b). In a combined approach of whole genome expression profiling and bioinformatic analysis with classical DNA–protein interaction studies, Den Hengst et al. (2005a, 2005b) discovered the conserved target sequence of CodY, the CodY box (AATTTTCWGAAAATT) in *Lc. lactis*. They also observed that CodY regulates its own synthesis and requires isoleucine, leucine, or valine to bind to the CodY box present at its promoter. The proteolytic system was repressed by CodY in nitrogen-enriched medium and the limiting concentration of BCAAs relieved the expression (Guédon et al. 2001b; Den Hengst et al. 2005a). In *Lc. lactis*, *pepP* expression could be affected by the carbon source in the growth medium (Guédon et al. 2001a). Gene encoding *pepP* is preceded by several potential catabolite-responsive element boxes, indicating the direct control by a CcpA-like regulator (Guédon et al. 2001b). Furthermore, stress response regulator CtsR, which binds to a direct-repeat heptad located upstream of target genes, regulates the expression of the stress-inducible proteolytic enzymes like proteolytic subunit of Clp protease, ClpP, and regulatory subunits, ClpC, ClpE, and ClpB (Derre et al. 1999; Varmanen et al. 2000; Savijoki et al. 2006). In *Lc. lactis*, *trmA* gene inactivation partially complements a heat sensitive *clpP* mutation by stimulating Clp-independent degradation of non-native proteins. Hence, TrmA could be another negative regulator of proteolysis (Frees et al. 2001). Therefore, at least three regulators, CodY, CtsR, and TrmA, may regulate the proteolytic activity in *Lc. lactis* (Savijoki et al. 2006; Mayo et al. 2010).

6.8.4 Bioactive Peptides

In modern times, milk proteins are considered the chief and safe source of bioactive peptides. A number of bioactive peptides formed in fermented milks and milk protein hydrolysates by the extensive proteolysis of milk proteins. Upon oral intake, these bioactive peptides may have beneficial impact on various functions of the human body. Bioactive efficacy of these peptides largely depends on their amino acid sequence. Peptides with various physiological functions like immunomodulation, antimicrobial nature, and antihypertensive property have been identified and isolated. The ability of *Lb. helveticus* to produce antihypertensive peptides has been associated with its CEPs; however, till date any relationship could not be established between antihypertensive effect and proteinase specificity (Griffiths and Tellez 2013).

6.9 Restriction/Modification Systems Combating Bacteriophage Attack among LAB

Restriction/modification (R/M) enzymes enable the digestion of foreign DNA, if encountered in the cytoplasm; however, the host DNA remains safe. These enzymes can be subclassified into three groups: type I, type II, and type III. Type I enzymes contain three subunits, which are responsible for modification (M), restriction (R), and specificity (S) and have been denoted as Hsd (host specificity determinant). Three type I R/M enzymes of *Lb. helveticus* were dairy microorganism specific, hsdR (lhv_1031), hsdS1 (lhv_1152), and hsdR (lhv_1978). Additionally, there was one dairy specific type III R/M enzyme mod (lhv_0028), and a major basic local alignment search tool (BLAST) search confirmed that these genes only occurred in organisms capable of surviving in a dairy environment with homologues in *Pediococcus, Ruminococcus,* and *Clostridia* species (O'Sullivan et al. 2009). It could not be clarified why these R/M proteins are specific to LAB of dairy origin and not present in the organism of gut environment. One possible explanation might be that generally higher bacterial populations are present in the dairy environment and these could be more susceptible to phage attacks and hence need more R/M pathways. Propagation of starter strains in the dairy environment could enhance their population dramatically in comparison with the population achieved by similar starter strains in other environmental niches. Besides, in dairy fermentations the same starter strains are often employed repeatedly over extended periods of time. Hence, using starter cultures in this manner can allow the development of bacteriophage sensitive starter populations. In milk fermentations, use of mixed starter strains provided the advantage of various R/M systems that would offer some protection against bacteriophage attack. Prior to the development of the modern strain selection techniques, the use of back slopping could ensure that only strains from successful fermentations were propagated in further fermentations. Hence, long history of milk fermentations explains the strong selective pressure toward picking phage-resistant strains even before the bacteriophage existence was known (O'Sullivan et al. 2009).

References

Altermann, E., Russell, W.M., Azcárate-Peril, M.A., Barrangou, R., Buck, B.L., McAuliffe, O., Souther, N. et al. 2005. Complete genome sequence of the probiotic lactic acid bacterium *Lactobacillus acidophilus* NCFM. *Proceedings of the National Academy of Sciences of the United States of America* 102: 3906–3912.

Axelsson, L. 1998. Lactic acid bacteria: Classification and physiology. In: *Lactic Acid Bacteria: Microbiology and Functional Aspects* (2nd edition), ed. S. Salminen and A. Von Wright, 1–72. New York: Marcel Dekker Inc.

Axelsson, L. 2004. Lactic acid bacteria: Classification and physiology. In: *Lactic Acid Bacteria, Microbiological and Functional Aspects* (3rd edition), ed. S. Salminen, and A. Von Wright, 1–66. New York: Marcel Dekker Inc.

Bekal, S., Diviès, C., and Prévost, H. 1999. Genetic organization of the *citCDEF* locus and identification of *mae* and *clyR* genes from *Leuconostoc mesenteroides*. *Journal of Bacteriology* 181: 4411–4416.

Bekal, S., Van Beeumen, J., Samyn, B., Garmyn, D., Henini, S., Diviès, C., and Prévost, H. 1998. Purification of *Leuconostoc mesenteroides* citrate lyase and cloning and characterization of the *citCDEFG* gene cluster. *Journal of Bacteriology* 180: 647–654.

Beresford, T.P., Fitzsimons, N.A., Brennan, N.L., and Cogan, T.M. 2001. Recent advances in cheese microbiology. *International Dairy Journal* 11: 259–274.

Bolotin, A., Quinquis, B., Renault, P., Sorokin, A., Ehrlich, S.D., Kulakauskas, S., Lapidus, A. et al. 2004. Complete sequence and comparative genome analysis of the dairy bacterium *Streptococcus thermophilus*. *Nature Biotechnology* 22: 1554–1558.

Bolotin, A., Wincker, P., Mauger, S., Jaillon, O., Malarme, K., Weissenbach, J., Ehrlich, S.D., and Sorokin, A. 2001. The complete genome sequence of the lactic acid bacterium *Lactococcus lactis* ssp. *Lactis* IL1403. *Genome Research* 11: 731–753.

Broadbent, J.R. and Steele, J.L. 2005. Cheese flavour and the genomics of lactic acid bacteria. *ASM News* 71: 121–128.

Buck, B.L., Altermann, E., Svingerud, T., and Klaenhammer, T.R. 2005. Functional analysis of putative adhesion factors in *Lactobacillus acidophilus* NCFM. *Applied and Environmental Microbiology* 71: 8344–8351.

Callanan, M., Kaleta, P., O'Callaghan, J., O'Sullivan, O., Jordan, K., McAuliffe, O., Sangrador-Vegas, A. et al. 2008. Genome sequence of *Lactobacillus helveticus*, an organism distinguished by selective gene loss and insertion sequence element expansion. *Journal of Bacteriology* 190: 727–735.

Chen, C., Ai, L., Zhou, F., Wang, L., Zhang, H., Chen, W., and Guo, B. 2011. Complete genome sequence of the probiotic bacterium *Lactobacillus casei* LC2W. *Journal of Bacteriology* 193: 3419–3420.

Chin, J., Turner, B., Barchia, I., and Mullbacher, A. 2000. Immune response to orally consumed antigens and probiotic bacteria. *Immunology and Cell Biology* 78: 55–66.

Choi, I.K., Jung, S.H., Kim, B.J., Park, S.Y., Kim, J., and Han, H.U. 2003. Novel *Leuconostoc citreum* starter culture systems for the fermentation of Kimchi, a fermented cabbage product. *Antonie van Leeuwenhoek* 84: 247–253.

Christensen, J.E., Dudley, E.G., Pederson, J.A., and Steele, J.L. 1999. Peptidases and amino acid catabolism in lactic acid bacteria. *Antonie van Leeuwenhoek* 76: 217–246.

Cogan, T.M., Beresford, T.P., Steele, J., Broadbent, J., Shah, N.P., and Ustunol, Z. 2007. Invited review: Advances in starter cultures and cultured foods. *Journal of Dairy Science* 90: 4005–4021.

Cotter, P.D. and Hill, C. 2003. Surviving the acid test: Responses of gram-positive bacteria to low pH. *Microbiology and Molecular Biology Reviews* 67: 429–453.
De Vos, W.M. 2011. System solutions by lactic acid bacteria: From paradigms to practice. *Microbial Cell Factories* 10(Suppl 1): S2.
De Vos, W.M. and Vaughan, E.E. 1994. Genetics of lactose utilization in lactic acid bacteria. *FEMS Microbiology Reviews* 15: 217–237.
Den Hengst, C.D., Curley, P., Larsen, R., Buist, G., Nauta, A., van Sinderen, D., Kuipers, O.P., and Kok, J. 2005a. Probing direct interactions between CodY and the *oppD* promoter of *Lactococcus lactis*. *Journal of Bacteriology* 187: 512–521.
Den Hengst, C.D., van Hijum, S.A., Geurts, J.M., Nauta, A., Kok, J., and Kuipers, O.P. 2005b. The *Lactococcus lactis* CodY regulon: Identification of a conserved cis-regulatory element. *The Journal of Biological Chemistry* 280: 34332–34342.
Derre, I., Rapoport, G., and Msadek, T. 1999. CtsR, a novel regulator of stress and heat shock response, controls *clp* and molecular chaperone gene expression in gram-positive bacteria. *Molecular Microbiology* 31: 117–131.
Detmers, F.J., Kunji, E.R., Lanfermeijer, F.C., Poolman, B., and Konings, W.N. 1998. Kinetics and specificity of peptide uptake by the oligopeptide transport system of *Lactococcus lactis*. *Biochemistry* 37: 16671–16679.
Drider, D., Bekal, S., and Prevost, H. 2004. Genetic organization and expression of citrate permease in lactic acid bacteria. *Genetics and Molecular Research* 3: 273–281.
Duary, R.K., Batish, V.K., and Grover, S. 2010. Expression of *atpD* gene in putative indigenous 480 probiotic *L. plantarum* strains under in vitro acidic conditions by RT-qPCR. *Research in Microbiology* 161: 399–405.
Duwat, P., Sourice, S., Cesselin, B., Lamberet, G., Vido, K., Gaudu, P., Le Loir, Y., Violet, F., Loubiere, P., and Gruss, A. 2001. Respiration capacity of the fermenting bacterium *Lactococcus lactis* and its positive effects on growth and survival. *Journal of Bacteriology* 183: 4509–4516.
Elkins, C.A., Moser, S.A., and Savage, D.C. 2001. Genes encoding bile salt hydrolases and conjugated bile salt transporters in *Lactobacillus johnsonii* 100-100 and other *Lactobacillus* species. *Microbiology* 147: 3403–3412.
FAO/WHO. 2002. *Guidelines for the Evaluation of Probiotics in Food*. Report of a joint FAO/WHO working group on drafting guidelines for the evaluation of probiotics in food, London.
Felis, G.E. and Dellaglio, F. 2007. Taxonomy of lactobacilli and bifidobacteria. *Current Issues in Intestinal Microbiology* 8: 44–61.
Fernandez-Espla, M.D., Garault, P., Monnet, V., and Rul, F. 2000. *Streptococcus thermophilus* cell wall-anchored proteinase: Release, purification, and biochemical and genetic characterization. *Applied and Environmental Microbiology* 66: 4772–4778.
Fox, P.F., Guinee, T.P., Cogan, T.M., and McSweeney, P.L.H. 2000. *Fundamentals of Cheese Science*. Gaithersburg, MA: AN Aspen Publishers Inc.

Frees, D., Varmanen, P., and Ingmer, H. 2001. Inactivation of a gene that is highly conserved in gram-positive bacteria stimulates degradation of non-native proteins and concomitantly increases stress tolerance in *Lactococcus lactis. Molecular Microbiology* 41: 93–103.

Garault, P., Le Bars, D., Besset, C., and Monnet, V. 2002. Three oligopeptide-binding proteins are involved in the oligopeptide transport of *Streptococcus thermophilus. The Journal of Biological Chemistry* 277: 32–39.

Garcia-Quintans, N., Blancato, V.S., Repizo, G.D., Magni, C., and Lopez, P. 2008. Citrate metabolism and aroma compound production in lactic acid bacteria. In: *Molecular Aspects of Lactic Acid Bacteria for Traditional and New Applications*, eds. B. Mayo, P. López, and G. Pérez-Martínez, 65–88. Kerala, India: Research Signpost.

Garcia-Quintans, N., Magni, C., de Mendoza, D., and López, P. 1998. The citrate transport system of *Lactococcus lactis* subsp. *lactis* biovar *diacetylactis* is induced by acid stress. *Applied and Environmental Microbiology* 64: 850–857.

Garvie, E.I. 1986. Genus *Leuconostoc*. In: *Bergey's Manual of Systematic Bacteriology* Vol.2, eds. P.H.A Sneath, N.S. Mair, M.E. Sharpe, J. G. Holt. Baltimore, MD: Williams & Wilkins.

Gaudu, P., Vido, K., Cesselin, B., Kulakauskas, S., Tremblay, J., Rezaiki, L., Lamberret, G., Sourice, S., Duwat, P., and Gruss, A. 2002. Respiration capacity and consequences in *Lactococcus lactis. Antonie van Leeuwenhoek* 82: 263–269.

Germond, J.E., Delley, M., Gilbert, C., and Atlan, D. 2003. Determination of the domain of the *Lactobacillus delbrueckii* subsp. *bulgaricus* cell surface proteinase PrtB involved in attachment to the cell wall after heterologous expression of the *prtB* gene in *Lactococcus lactis. Applied and Environmental Microbiology* 69: 3377–3384.

Gilbert, C., Atlan, D., Blanc, B., Portalier, R., Germond, G.J., Lapierre, L., and Mollet, B. 1996. A new cell surface proteinase: sequencing and analysis of the *prtB* gene from *Lactobacillus debrueckii* subsp. *bulgaricus. Journal of Bacteriology* 178: 3059–3065.

Gitton, C., Meyrand, M., Wang, J., Caron, C., Trubuil, A., Guillot, A., and Mistou, M.Y. 2005. Proteomic signature of *Lactococcus lactis* NCDO763 cultivated in milk. *Applied and Environmental Microbiology* 71: 7152–7163.

Griffiths, M.W. and Tellez, A.M. 2013. *Lactobacillus helveticus*: The proteolytic system. *Frontiers in Microbiology*. doi:10.3389/fmicb.2013.00030.

Guédon, E., Renault, P., Ehrlich, D., and Delorme, C. 2001a. Transcriptional pattern of genes coding for the proteolytic system of *Lactococcus lactis* and evidence for coordinated regulation of key enzymes by peptide supply. *Journal of Bacteriology* 183: 3614–3622.

Guédon, E., Serror, P., Ehrlich, S.D., Renault, P., and Delorme, C. 2001b. Pleiotropic transcriptional repressor CodY senses the intracellular pool of branched-chain amino acids in *Lactococcus lactis. Molecular Microbiology* 40: 1227–1239.

Haandrikman, A., Kok, J., Laan, H., Soemitro, S., Ledeboer, A., Konings, W., and Venema, G. 1989. Identification of a gene required for maturation of an extracellular lactococcal serine proteinase. *Journal of Bacteriology* 171: 2789–2794.

Haandrikman, A., Kok, J., and Venema, G. 1991. Lactococcal proteinase maturation protein PrtM is a lipoprotein. *Journal of Bacteriology* 173: 4517–4525.

Hansen, E.B. 2002. Commercial bacterial starter cultures for fermented foods of the future. *International Journal of Food Microbiology* 78: 119–131.

Holck, A. and Naes, H. 1992. Cloning, sequencing and expression of the gene encoding the cell-envelope-associated proteinase from *Lactobacillus paracasei* subsp. *paracasei* NCDO151. *Journal of General Microbiology* 138: 1353–1364.

Hols, P., Kleerebezem, M., Schanck, A.N., Ferain, T., Hugenholtz, J., Delcour, J., and de Vos, W.M. 1999. Conversion of *Lactococcus lactis* from homolactic to homoalanine fermentation through metabolic engineering. *Nature Biotechnology* 17: 588–592.

Hugenholtz, J. 1993. Citrate metabolism in lactic acid bacteria. *FEMS Microbiology Reviews* 12: 165–178.

Jamal, Z., Miot-Sertier, C., Thibau, F., Dutilh, L., Lonvaud-Funel, A., Ballestra, P., Marrec, C.L., and Dols-Lafargue, M. 2013. Distribution and functions of phosphotransferase system genes in the genome of the lactic acid bacterium *Oenococcus oeni*. *Applied and Environmental Microbiology* 79: 3371–3379.

Jones, B.V., Begley, M., Hill, C., Gahan, C.G., and Marchesi, J.R. 2008. Functional and comparative metagenomic analysis of bile salt hydrolase activity in the human gut microbiome. *Proceedings of the National Academy of Sciences of the United States of America* 105: 13580–13585.

Juillard, V., Guillot, A., Le Bars, D., and Gripon, J.C. 1998. Specificity of milk peptide utilization by *Lactococcus lactis*. *Applied and Environmental Microbiology* 64: 1230–1236.

Jung, J.Y., Lee, S.H., and Jeon, C.O. 2012. Complete genome sequence of *Leuconostoc mesenteroides* subsp. *mesenteroides* strain J18, isolated from kimchi. *Journal of Bacteriology* 194: 730–731.

Kaushik, J.K., Kumar, A., Duary, R.K., Mohanty, A.K., Grover, S., and Batish, V.K. 2009. Functional and probiotic attributes of an indigenous isolate of *Lactobacillus plantarum*. *PLoS One*. doi:10.1371/journal.pone.0008099.

Kim, J.F., Jeong, H., Lee, J.S., Choi, S.H., Ha, M., Hur, C.G., Kim, J.S. et al. 2008. The complete genome sequence of *Leuconostoc citreum* KM20. *Journal of Bacteriology* 190: 3093–3094.

Klaenhammer, T.R., Barrangou, R., Buck, B.L., Azcaráte-Peril, M.A., and Altermann, E. 2005. Genomic features of lactic acid bacteria effecting bioprocessing and health. *FEMS Microbiology Reviews* 29: 393–409.

Klappenbach, J.A., Dunbar, J.M., and Schmidt, T.M. 2000. rRNA operon copy number reflects ecological strategies of bacteria. *Journal of Bacteriology* 66: 1328–1333.

Kleerebezem, M., Boekhorst, J., van Kranenburg, R., Molenaar, D., Kuipers, O.P., Leer, R., Tarchini, R. et al. 2003. Complete genome sequence of *Lactobacillus plantarum* WCFS1. *Proceedings of the National Academy of Sciences of the United States of America* 100: 1990–1995.

Kleerebezem, M., Hols, P., and Hugenholtz, J. 2000. Lactic acid bacteria as a cell factory: Rerouting of carbon metabolism in *Lactococcus lactis* by metabolic engineering. *Enzyme and Microbial Technology* 26: 840–848.

Kok, J., Buist, G., Zomer, A.L., van Hijum, S.A., and Kuipers, O.P. 2005. Comparative and functional genomics of lactococci. *FEMS Microbiology Reviews* 29: 411–433.

Koryszewska-Baginska, A., Aleksandrzak-Piekarczyk, T., and Bardowski, J. 2013. Complete genome sequence of the probiotic strain *Lactobacillus casei* (formerly *Lactobacillus paracasei*) LOCK919. *Genome Announcements* 1: e00758-13.

Kumar, R., Grover, S., and Batish, V.K. 2011a. Hypocholesterolemic effect of dietary inclusion of two putative probiotic bile salt hydrolase (bsh) producing *Lactobacillus plantarum* strains in sprague-dawley rats. *British Journal of Nutrition* 105: 561–573.

Kumar, R., Grover, S., and Batish, V.K. 2011b. Molecular identification and typing of putative probiotic indigenous *Lactobacillus plantarum* strain lp91 of human origin by specific primed-PCR assays. *Probiotics and Antimicrobial Proteins* 3: 186–193.

Kumar, R., Grover, S., and Batish, V.K. 2012. Bile salt hydrolase (Bsh) activity screening of lactobacilli: In vitro selection of indigenous *Lactobacillus* strains with potential bile salt hydrolyzing and cholesterol lowering ability. *Probiotics and Antimicrobial Proteins* 4: 162–172.

Kumar, R., Grover, S., Kaushik, J.K., and Batish, V.K. 2014. *IS30*-related transposon mediated insertional inactivation of bile salt hydrolase (bsh1) gene of *Lactobacillus plantarum* strain Lp20. *Microbiological Research* 169: 553–560.

Kumar, R., Hemalatha, R., Kumar, M., Varikuti, S.R., Athimamula, R., Shujauddin, M., Ramagoni, R., and Kondapalli, N. 2013. Molecular cloning, characterization and heterologous expression of bile salt hydrolase (Bsh) from *Lactobacillus fermentum* NCDO394. *Molecular Biology Reports* 40: 5057–5066.

Kunji, E.R.S., Mierau, I., Hagting, A., Poolman, B., and Konings, W.N. 1996. The proteolytic systems of lactic acid bacteria. *Antonie van Leeuwenhoek* 70: 187–221.

Kurdi, P., van Veen, H.W., Tanaka, H., Mierau, I., Konings, W.N., Tannock, G.W., Tomita, F., and Yokota, A. 2000. Cholic acid is accumulated spontaneously, driven by membrane Δ pH, in many lactobacilli. *Journal of Bacteriology* 182: 6525–6528.

Lambert, J.M., Siezen, R.J., de Vos, W.M., and Kleerebezem, M. 2008. Improved annotation of conjugated bile acid hydrolase superfamily members in Gram-positive bacteria. *Microbiology* 154: 2492–2500.

Leonard, M.T., Valladares, R.B., Ardissone, A., Gonzalez, C.F., Lorca, G.L., and Triplett, E.W. 2014. Complete genome sequences of *Lactobacillus johnsonii* strain N6.2 and *Lactobacillus reuteri* strain TD1. *Genome Announcements* 2: e00397-14.

Liong, M.T. and Shah, N.P. 2005. Acid and bile tolerance and cholesterol removal ability of lactobacilli strains. *Journal of Dairy Science* 88: 55–66.

Lister, J. 1878. On lactic fermentation and its bearing on pathology. *Transactions Pathological Society of London* 29: 425–467.

Liu, M., Bayjanov, J.R., Renckens, B., Nauta, A., and Siezen, R.J. 2010. The proteolytic system of lactic acid bacteria revisited: A genomic comparison. *BMC Genomics* 11: 36.

López de Felipe, F., Magni, C., de Mendoza, D., and López, P. 1995. Citrate utilization gene cluster of the *Lactococcus lactis* biovar *diacetylactis*: Organization and regulation of expression. *Molecular and General Genetics* 246: 590–599.

Luesink, E.J., van Herpen, R.E., Grossiord, B.P., Kuipers, O.P., and de Vos, W.M. 1998. Transcriptional activation of the glycolytic *las* operon and catabolite repression of the *gal* operon in *Lactococcus lactis* are mediated by the catabolite control protein CcpA. *Molecular Microbiology* 30: 789–798.

Luoma, S., Peltoniemi, K., Joutsjoki, V., Rantanen, T., Tamminen, M., Heikkinen, I., and Palva, A. 2001. Expression of six peptidases from *Lactobacillus helveticus* in *Lactococcus lactis*. *Applied and Environmental Microbiology* 67: 1232–1238.

Magni, C., de Mendoza, D., Konings, W.N., and Lolkema, J.S. 1999. Mechanism of citrate metabolism in *Lactococcus lactis*: Resistance against lactate toxicity at low pH. *Journal of Bacteriology* 181: 1451–1457.

Makarova, K., Slesarev, A., Wolf, Y., Sorokin, A., Koonin, E., Pavlov, A., Pavlova, N. et al. 2006. Comparative genomics of the lactic acid bacteria. *Proceeding of National Academy of Sciences of the United States of America* 103: 15611–15616.

Makarova, K.S. and Koonin, E.V. 2007. Evolutionary genomics of lactic acid bacteria. *Journal of Bacteriology* 189: 1199–1208.

Marchesi, J. and Shanahan, F. 2007. The normal intestinal microbiota. *Current Opinion in Infectious Diseases* 20: 508–513.

Marty-Teysset, C., Lolkema, J.S., Schmitt, P., Diviès, C., and Konings, W.N. 1995. Membrane potential-generating transport of citrate and malate catalyzed by CitP of *Leuconostoc mesenteroides*. *The Journal of Biological Chemistry* 270: 25370–25376.

Marty-Teysset, C., Posthuma, C., Lolkema, J.S., Schmitt, P., Divies, C., and Konings, W.N. 1996. Proton motive force generation by citrolactic fermentation in *Leuconostoc mesenteroides*. *Journal of Bacteriology* 178: 2175–2185.

Mayo, B., Aleksandrzak-Piekarczyk, T., Fernández, M., Kowalczyk, M., Álvarez-Martín, P., and Bardowski, J. 2010. Updates in the metabolism of lactic acid bacteria. In: *Biotechnology of Lactic Acid Bacteria: Novel Applications*, eds. F. Mozzi, R.R. Raya, and G.M. Vignolo. Oxford: Wiley-Blackwell. doi:10.1002/9780813820866.ch1.

Mayo, B., van Sinderen, D., and Ventura, M. 2008. Genome analysis of food grade lactic acid producing bacteria: From basics to applications. *Current Genomics* 9: 169–183.

McAuliffe, O., Cano, R.J., and Klaenhammer, T.R. 2005. Genetic analysis of two bile salt hydrolase activities in *Lactobacillus acidophilus* NCFM. *Applied and Environmental Microbiology* 71: 4925–4929.

McLean, M.J., Wolfe, K.H., and Deine, K.M. 1998. Base composition skews, replication orientation, and gene orientation in 12 prokaryote genomes. *Journal of Molecular Evolution* 47: 691–696.

Molenaar, D., Bringel, F., Schuren, F.H., de Vos, W.M., Siezen, R.J., and Kleerebezem, M. 2005. Exploring *Lactobacillus plantarum* genome diversity by using microarrays. *Journal of Bacteriology* 187: 6119–6127.

Nouaille, S., Ribeiro, L.A., Miyoshi, A., Pontes, D., Le Loir, Y., Oliveira, S.C., Langella, P., and Azevedo, V. 2003. Heterologous protein production and delivery systems for *Lactococcus lactis*. *Genetics and Molecular Research* 2: 102–111.

Oh, S., Roh, H., Ko, H., Kim, S., Kim, K., Lee, S.E., Chang, I., Kim, S., and Choi, I. 2011. Complete genome sequencing of *Lactobacillus acidophilus* 30SC, isolated from swine intestine. *Journal of Bacteriology* 193: 2882–2883.

O'Sullivan, D.J. 2006. Genetics of dairy starter culture. In: *Food Biotechnology* (2nd edition), eds. K. Shetty, G. Paliyath, A. Pometto, and R.E. Levin, 242–266. New York: Taylor & Francis.

O'Sullivan, O., O'Callaghan, J., Sangrador-Vegas, A., McAuliffe, O., Slattery, L., Kaleta, P., Callanan, M., Fitzgerald, G.F., Ross, R.P., and Beresford, T. 2009. Comparative genomics of lactic acid bacteria reveals a niche-specific gene set. *BMC Microbiology* 9: 50.

Pastar, I., Tonic, I., Golic, N., Kojic, M., van Kranenburg, R., Kleerebezem, M., Topisirovic, L., and Jovanovic, G. 2003. Identification and genetic characterization of a novel proteinase, PrtR, from the human isolate *Lactobacillus rhamnosus* BGT10. *Applied and Environmental Microbiology* 69: 5802–5811.

Pedersen, M.B., Iversen, S.L., Sorensen, K.I., and Johansen, E. 2005. The long and winding road from the research laboratory to industrial applications of lactic acid bacteria. *FEMS Microbiology Reviews* 29: 611–624.

Peltoniemi, K., Vesanto, E., and Palva, A. 2002. Genetic characterization of an oligopeptide transport system from *Lactobacillus delbrueckii* subsp. *bulgaricus*. *Archives of Microbiology* 177: 457–467.

Petranovic, D., Guedon, E., Sperandio, B., Delorme, C., Ehrlich, D., and Renault, P. 2004. Intracellular effectors regulating the activity of the *Lactococcus lactis* CodY pleiotropic transcription regulator. *Molecular Microbiology* 53: 613–621.

Prajapati, J.B., Nathani, N.M., Patel, A.K., Senan, S., and Joshi, C.G. 2013. Genomic analysis of dairy starter culture *Streptococcus thermophilus* MTCC5461. *Journal of Microbiology and Biotechnology* 23: 459–466.

Pridmore, R.D., Berger, B., Desiere, F., Vilanova, D., Barretto, C., Pittet, A.C., Zwahlen, M.C. et al. 2004. The genome sequence of the probiotic intestinal bacterium *Lactobacillus johnsonii* NCC 533. *Proceedings of the National Academy of Sciences of the United States of America* 101: 2512–2517.

Pudlik, A.M. and Lolkema, J.S. 2011. Mechanism of citrate metabolism by an oxaloacetate decarboxylase—Deficient mutant of *Lactococcus lactis* IL 1403. *Journal of Bacteriology* 193: 4049–4056.

Pudlik, A.M. and Lolkema, J.S. 2012. Substrate specificity of the citrate transporter CitP of *Lactococcus lactis*. *Journal of Bacteriology* 194: 3627–3635.

Saks, M.E. and Conery, J.S. 2007. Anticodon-dependent conservation of bacterial tRNA gene sequences. *RNA* 13: 651–660.

Sanders, M.E. and Klaenhammer, T.R. 2001. Invited review: The scientific basis of *Lactobacillus acidophilus* NCFM functionality as a probiotic. *Journal of Dairy Science* 84: 319–331.

Sarmiento-Rubiano, L.A., Berger, B., Moine, D., Zunigal, M., Perez-Martinez, G., and Yebral, M.J. 2010. Characterization of a novel *Lactobacillus* species closely related to *Lactobacillus johnsonii* using a combination of molecular and comparative genomics methods. *BMC Genomics* 11: 504.

Savijoki, K., Ingmer, H., and Varmanen, P. 2006. Proteolytic systems of lactic acid bacteria. *Applied Microbiology and Biotechnology* 71: 394–406.

Siezen, R.J. 1999. Multi-domain, cell-envelope proteinases of lactic acid bacteria. *Antonie van Leeuwenhoek* 76: 139–155.

Siezen, R.J., Francke, C., Renckens, B., Boekhorst, J., Wels, M., Kleerebezem, M., and van Hijum, S.A.F.T. 2012. Complete resequencing and reannotation of the *Lactobacillus plantarum* WCFS1 Genome. *Journal of Bacteriology* 194: 195.

Siezen, R.J., Renckens, B., van Swam, I., Peters, S., van Kranenburg, R., Kleerebezem, M., and de Vos, W.M. 2005. Complete sequences of four plasmids of *Lactococcus lactis* subsp.*cremoris* SK11 reveal extensive adaptation to the dairy environment. *Applied and Environmental Microbiology* 71: 8371–8382.

Sijpesteijn, A.K. 1970. Induction of cytochrome formation and stimulation of oxidative dissimilation by hemin in *Streptococcus lactis* and *Leuconostoc mesenteroides*. *Antonie van Leeuwenhoek* 36: 335–348.

Smit, G., Smit, B.A., and Engels, W.J. 2005. Flavor formation by lactic acid bacteria and biochemical flavor profiling of cheese products. *FEMS Microbiology Reviews* 29: 591–610.

Stahl, B. and Barrangou, R. 2013. Complete genome sequence of Probiotic strain *Lactobacillus acidophilus* La-14. *Genome Announce* 1: e00376-13.

Sun, Z., Chen, X., Wang, J., Zhao, W., Shao, Y., Guo, Z., Zhang, X. et al. 2011a. Complete genome sequence of *Lactobacillus delbrueckii* subsp. *bulgaricus* strain ND02. *Journal of Bacteriology* 193: 3426–3427.

Sun, Z., Chen, X., Wang, J., Zhao, W., Shao, Y., Wu, L., Zhou, Z. et al. 2011b. Complete genome sequence of *Streptococcus thermophilus* strain ND03. *Journal of Bacteriology* 193: 793–794.

Takano, T. 2002. Anti-hypertensive activity of fermented dairy products containing biogenic peptides. *Antonie van Leeuwenhoek* 82: 333–340.

Tompkins, T.A., Barreau, G., and Broadbent, J.R. 2012. Complete genome sequence of *Lactobacillus helveticus* R0052, a commercial probiotic strain. *Journal of Bacteriology* 194: 6349.

Van de Guchte, M., Penaud, S., Grimaldi, C., Barbe, V., Bryson, K., Nicolas, P., Robert, C. et al. 2006. The complete genome sequence of *Lactobacillus bulgaricus* reveals extensive and ongoing reductive evolution. *Proceedings of the National Academy of Sciences of the united States of America* 103: 9274–9279.

Van Hoek, M. and Merks, R. 2012. Redox balance is key to explaining full vs. partial switching to low-yield metabolism. *BMC Systems Biology* 6: 22.

Van Rooijen, R.J., Dechering, K.J., Wilmink, C.M.J., and de Vos, W.M. 1993. Lysines 72, 80, 213, and aspartic acid 210 of the *Lactococcus lactis* LacR repressor are involved in the response to the inducer tagatose-6-phosphate leading to induction of lac operon expression. *Protein Engineerings* 6: 208–215.

Varmanen, P., Ingmer, H., and Vogensen, F.K. 2000. *ctsR* of *Lactococcus lactis* encodes a negative regulator of *clp* gene expression. *Microbiology* 146: 1447–1455.

Vido, K., Le Bars, D., Mistou, M.Y., Anglade, P., Gruss, A., and Gaudu, P. 2004. Proteome analyses of heme-dependent respiration in *Lactococcus lactis*: Involvement of the proteolytic system. *Journal of Bacteriology* 186: 1648–1657.

Wegmann, U., O'Connell-Motherway, M., Zomer, A., Buist, G., Shearman, C., Canchaya, C., Ventura, M. et al. 2007. Complete genome sequence of the prototype lactic acid bacterium *Lactococcus lactis* subsp. *cremoris* MG1363. *Journal of Bacteriology* 189: 3256–3270.

Wouters, J.T.M., Ayad, E.H.E., Hugenholtz, J., and Smit, G. 2002. Microbes from raw milk for fermented dairy products. *International Dairy Journal* 12: 91–109.

Wydau, S., Dervyn, R., Anba, J., Dusko Ehrlich, S., and Maguin, E. 2006. Conservation of key elements of natural competence in *Lactococcus lactis* ssp. *FEMS Microbiology Letters* 257: 32–42.

Yadav, A.K., Tyagi, A., Kaushik, J.K., Saklani, A.C., Grover, S., and Batish, V.K. 2013. Role of surface layer collagen binding protein from indigenous *Lactobacillus plantarum* 91 in adhesion and its anti-adhesion potential against gut pathogen. *Microbiology* 168: 639–645.

Zhang, W., Yu, D., Sun, Z., Wu, R., Chen, X., Chen, W., Meng, H., Hu, S., and Zhang, H. 2010. Complete genome sequence of *Lactobacillus casei* Zhang, a new probiotic strain isolated from traditional homemade Koumiss in inner Mongolia, China. *Journal of Bacteriology* 192: 5268–5269.

Zhang, W.Y., Wu, R.N., Sun, Z.H., Sun, S.T., Meng, H., and Zhang, H.P. 2009. Molecular cloning and characterization of bile salt hydrolase in *Lactobacillus casei* Zhang. *Annals of Microbiology* 59: 721–726.

7

Preservation of Lactic Starter Cultures

CHALAT SANTIVARANGKNA

Contents

7.1 Introduction	168
7.2 Cultivation of Lactic Starter Cultures	169
7.3 Harvest of Cells	172
7.4 Addition of Protectants	173
7.4.1 Cryoprotectants	173
7.4.2 Dehydration Protectants	174
7.5 Preservation of Starter Cultures	175
7.5.1 Short-Term Preservation of Liquid Cultures	175
7.5.2 Long-Term Preservation by Freezing	176
7.5.2.1 Freezing of Lactic Starter Cultures	177
7.5.2.2 Freezing of Concentrated and DVS Lactic Starter Cultures	178
7.5.2.3 Frozen Storage and Thawing	178
7.5.3 Long-Term Preservation by Drying	181
7.5.3.1 Drying of Lactic Starter Cultures	182
7.5.3.2 Drying of Concentrated and DVS Lactic Starter Cultures	183
7.5.3.3 Storage and Rehydration of Powder of the Cultures	184
7.6 Other Factors Affecting Viability during Preservation	187
7.6.1 Intrinsic Factors	187
7.6.2 Stress Induction	187
7.6.3 Encapsulation	188
7.6.4 Genetic Modification	189
7.7 Concluding Remarks	189
References	190

7.1 Introduction

Conventionally, lactic starter cultures are freshly prepared by the successive propagation of cells from a small amount of stock culture until a large volume of bulk inoculum with approximately 10^{8-9} viable cells/ml is obtained. The bulk inoculum will be added to milk to get approximately 10^{6-7} viable cells/ml of milk. Fermented milk plants may have their own culture collection for the preparation of stock cultures or purchase stock cultures from starter producing labs. This practice is used especially by plants, where they produce local special products and where the regular supply of stock cultures is easy. The successive propagation of starter cultures has disadvantages: (1) it is very laborious and labor intensive, (2) several stepwise propagations necessitate a high degree of production planning, (3) it has a high risk of bacterial contamination or phage infection, and (4) the repeated subculture carries a potential for the loss of some properties due to mutation and auto-selection.

Advances in the biomass production such as neutralization and cell concentration processes enable the production of starter cultures in frozen and dried concentrated forms (Figure 7.1).

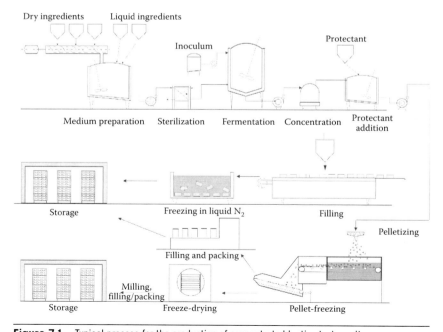

Figure 7.1 Typical process for the production of concentrated lactic starter cultures.

Nowadays, lactic starter cultures are increasingly produced in the concentrated forms for an inoculation into bulk starter media and for a direct inoculation into milk. The latter form of concentrated cultures is designated to as direct vat inoculation (DVI) cultures or direct vat set (DVS). It is a prerequisite that concentrated cultures are stable during storage to ensure a high enough viability to resume fermentation activities immediately after inoculation. Viability of concentrated lactic starter cultures and DVS cultures is typically ca. 10^{10}–10^{12} cells/ml or g. Although the concentrated starter cultures and DVS cultures reduce or eliminate the drawbacks of the successive starter culture propagation, these comparatively also have some disadvantages: (1) these are more expensive, (2) dried starter cultures have a longer lag period before the growth and acid production, (3) not all lactic starter cultures retain high viability and activities after the concentration and preservation processes, and (4) frozen concentrated cultures require low temperature storage facilities at the dairy plants and during the transportation.

Preservation methods are closely related to the practices in the utilization of lactic starter cultures. The preserved forms of lactic starter cultures and their using practices are shown in Figure 7.2.

This chapter covers the different methods and forms of preservation that are used to preserve lactic starter cultures at different levels of scaling along with the associated advantages and disadvantages. The content will flow in accordance with the major steps in the production and preservation processes of lactic acid starter cultures, that is, cultivation of the cultures, harvest of cells, addition of protectants, and preservation and storage methods. Important alternative preservation methods and other factors affecting viability of preserved starter cultures will be discussed as well.

7.2 Cultivation of Lactic Starter Cultures

Growing of cells is the first step for the preparation of starter cultures. The culture media and conditions can influence viability of the starters during preservation. Culture media for lactic starter cultures in laboratories are undefined rich media such as MRS, LAPT, M17, and Elliker media. Lactic starter cultures grow generally better in the rich media than in milk, but the media are too expensive for the industrial scale preparation of bulk starters and may impart off-flavors

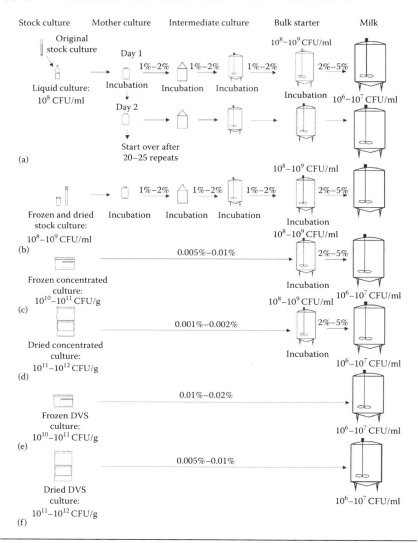

Figure 7.2 Different preserved forms of lactic starter cultures in accordance with their using practices in fermented dairy product manufactures: the preparation of bulk starter by successive subcultures (a and b), the preparation of bulk starter by the inoculation with concentrated cultures (c and d), and the use of DVS cultures (e and f). The inoculum size of the bulk starter, concentrated and DVS cultures is generally varying among products and the manufacturers.

in the products. Skim milk and whey-based media are more common and economical for the industries. Nutrients such as yeast extract, and specific vitamins and amino acids usually are added to promote the growth. Some nutrients improve tolerances of lactic starters to stresses occurred during the preservation. For example, in a study with *Streptococcus thermophilus*, the addition of oleic acid in the whey-based medium increased the concentration of unsaturated fatty acids and decreased the concentration of saturated fatty. As a result, the recovery of the acidification activity of cells was improved after frozen storage at −20°C for 8 weeks (Beal et al. 2001).

The industrial scale cultivation of lactic starter cultures is free-cell batch fermentation. After inoculation of the cultures, growth is promoted by maintaining the pH at optimal ranges for the cultures. The lactic acid produced is neutralized with, for example, NH_4OH, Na_2CO_3, KOH, NaOH, and gaseous NH_3. The ammonium salt from the neutralization with gaseous NH_3 and NH_4OH can be additionally served as the nitrogen source for the cells (Peebles et al. 1969). Little information is available on the effects of the neutralizing agents on preservation of lactic starter cultures. Calcium hydroxide was shown to improve the viability of *Lactobacillus bulgaricus* after freezing and freeze-drying (Wright and Klaenhammer 1983). Furthermore, growth of lactic starter cultures is less inhibited by calcium lactate than ammonium lactate or sodium lactate, and the dried powder the cultures obtained is less hygroscopic (Stadhouders et al. 1969). To obtain maximal biomass, other parameters that must be optimized for lactic starter cultures are temperature, agitation rate, and oxygen content in media. Recently, a change from the traditional anaerobic fermentation to an aerobic respiration process has been introduced for *Lactococcus* ssp. The lactic acid bacteria exhibit a respiratory lifestyle in the presence of oxygen and heme. Consequences of the respiration are the higher biomass and lower amount of lactic acid. Under the optimal conditions, cell concentration are tenfold or more concentrated than the cell concentration of a normal bulk starter. However, the aeration and heme do not influence the biomass yield of *Str. thermophilus, Lb. delbrueckii* subsp. *bulgaricus*, and *Lb. helveticus* (Pedersen et al. 2005).

In addition to the batch fermentation, continuous fermentation such as membrane bioreactor (Taniguchi et al. 1987a, 1987b) and immobilized cell systems (Doleyres et al. 2004) were also investigated. In these systems, cells are retained in a bioreactor, with continuous feeding of fresh medium, by the membrane filtration and the entrapment of cells in food-grade biopolymer gel matrices, respectively. The continuous fermentation processes give higher yields and productivity, but a disadvantage is that the process is more susceptible to contamination and cell characteristics can be lost over time (Lacroix and Yildirim 2007). So far, the continuous fermentation is still mainly for experimental purposes.

In the presence of some sugars, lactic acid bacteria produce or increase the production of metabolites that may affect viability of cells during the preservation. For example, it has been suggested that exopolysaccharides (EPS) protect cells from desiccation (Cerning 1990; Roberts 1996; Schwab et al. 2007), and *Weisella cibaria, Lb. plantarum,* and *Pediococcus pentosaceus* (Di et al. 2006) produced EPS when sucrose was used as a carbon source. Lactose, fructose, and galactose increased the amounts of EPS produced by *Lb. helveticus* (Torino et al. 2005), *Lb. delbrueckii* subsp. *bulgaricus* (Carvalho et al. 2004), *Str. salivarius* subsp. *thermophilus* (Gancel and Novel 1994), and *Lb. casei* (Mozzi et al. 2001), respectively.

7.3 Harvest of Cells

In batch cultivation, exponential growth of cells is limited by the depletion of essential nutrients, accumulation of fermentation products, and non-optimal growth conditions. Harvest of cells at the stationary phase yields a higher amount of biomass and increases the tolerance of cells to stresses during the preservation. The higher stress tolerance is because of the production of general stress proteins. (Van de Guchte et al. 2002). For instance, the cryotolerance of stationary phase cells of *Lb. lactis* subsp. *lactis* (Kim et al. 1999; Wouters et al. 1999), *Lb. bulgaricus* and *Str. thermophilus* (Fonseca et al. 2001), and *Lb. lactis* subsp. *diacetylactis* (Lee 2004) was significantly higher than in the log phase cells. Similarly, higher viability was found in stationary phase cells of *Lb. rhamnosus* (Corcoran et al. 2004) and *Lb. delbrueckii* subsp. *bulgaricus* (Teixeira et al. 1995a) after spray drying, compared with log phase cells.

Following the cultivation, cells must be harvested and concentrated before the production of frozen and dried concentrated lactic starter cultures. When milk-based media are employed, it is recommended to add trisodium citrate (e.g., 1% w/v) to clarify the media before the concentration of cells (Sandine 1996). Centrifugation is conventionally employed to harvest and concentrate cells. The separation is based on the difference of density between cells and the medium. A high viscosity medium needs stronger centrifugal conditions to harvest cells. The concentration rates from the centrifugation are between 10 and 40 times. The drawbacks of the centrifugation are high production and maintenance costs. Furthermore, in some strains, it is difficult to obtain a complete recovery of cells by centrifugation. For example, in a study with four *Bifidobacterium longum* strains, *B. longum* S9 and *B. longum* III produced firm cell pellets, whereas *B. longum* ATCC 15707 and *B. longum* II produced soft cell pellets. The soft pellet is an indicative of difficulty in recovering cells by centrifugation (Reilly and Gilliland 1999).

Alternatively, a cross-flow membrane filtration can be used to concentrate cells. The cell suspension flows tangentially to a membrane, which selectively separates the cells (retentate) from the fermented medium (permeate). An advantage of the cross-flow filtration is a possibility to obtain the complete recovery of cells and the reduced costs (Van Reis and Zydney 2001). Disadvantages of membrane filtration are, for example, the longer separation time and the formation of filter cake, which reduces the membrane performance. Performance of a membrane depends on several factors such as pH and protein content of the culture media. The filtration affects cell physiology, depending on the operating conditions. Following the filtration, *Lb. bulgaricus* CFL1 had the increased unsaturated to saturated and cyclic to saturated fatty acid ratios. The change in membrane fatty acid composition enhanced membrane fluidity and resulted in the improved cryotolerance of cells to freezing and frozen storage (Streit et al. 2011).

7.4 Addition of Protectants

7.4.1 Cryoprotectants

Cryoprotectants prevent excessive dehydration and the formation of large ice crystals within cells (see Section 7.5.2 for injuries of cells during freezing). These can be roughly classified based on the ability to diffuse through the cell membrane into penetrating compounds such as glycerol and dimethylsulfoxide (DMSO, Me_2SO) and non-penetrating compounds such as sugars, proteins, and polymers like maltodextrin. The penetrating compounds are more effective against injuries during slow freezing, whereas the non-penetrating compounds are more effective against injuries during rapid freezing. The common cryoprotectants used for lactic starters are sugars, glycerol, and skim milk (Thunell et al. 1984; Beal and Corrieu 1994; Beal et al. 2001). Examples of the sugars are lactose, sucrose, and trehalose (Baumann and Reinbold 1966; Chavarri et al. 1988; De Antoni et al. 1989; Carcoba and Rodriguez 2000; Panoff et al. 2000). Glycerol can be prepared as a 20% glycerol stock and mixed with the cells suspension in a 1:1 ratio to get the final glycerol concentration of 10%. Skim milk (10%–20% total solid) is a favorite suspending medium and protectant, either used it alone or in a combination with the aforementioned protectants (Carcoba and Rodriguez 2000; Juárez et al. 2004; Modesto et al. 2004; Volkert et al. 2008). Dimethylsulfoxide is used less often with lactic starter cultures. DMSO cannot be sterilized conveniently by autoclavation but by filtration. Moreover, it has unpleasant garlic flavor, which may cause off-flavor in the products. It is advisable to allow the protectants to equilibrate the intracellular solutes before freezing. The incubation time of cells with the cryoprotectants is typically 10–60 min at 0°C–10°C (Hubalek 2003). The incubation at a higher temperature and for longer time should be avoided. The incubation of *Lb. delbrueckii* subsp. *bulgaricus* with glycerol at 42°C for 1 h ascribed the higher susceptibility of cells to freezing and thawing (Panoff et al. 2000).

7.4.2 Dehydration Protectants

The addition of protectants before drying prevents damage of cell membrane, which is believed to be the major site of dehydration inactivation during drying and rehydration (Brennan et al. 1986; Santivarangkna et al. 2008a, 2008b). During the freeze- and spray drying, some protectants such as trehalose and sucrose also prevent cells from cryo- (Carcoba and Rodriguez 2000) and thermal injuries (Kilimann et al. 2006; Tymczyszyn et al. 2007a). The effectiveness of a protectant seems to vary largely and be interdependent between the starter cultures and the drying processes (Santivarangkna et al. 2007a). Sugars are the favorite protectants because of their relatively low prices, chemically innocuous nature, and common use in food industries. These sugars are, for example, trehalose (Conrad et al. 2000), sorbitol (Linders et al. 1997; Santivarangkna et al. 2006), mannitol (Efiuvwevwere et al. 1999), sucrose (Carvalho et al. 2003), lactose

(Higl et al. 2007), and mannose (Carvalho et al. 2004). Among them, one of the most extensively investigated sugars is trehalose. However, its high price may confine the use in lactic starter culture preservation at the industrial scale.

The protectants can also be complex components and polymers used as suspending medium or carrier, for example, skim milk, gelatin, soluble starch, and gum acacia. At outlet air temperatures of 100°C–105°C, *Lb. paracasei* spray dried with 10% gum acacia survived 10-fold better than a control sample dried with 20% reconstituted skim milk (Desmond et al. 2002a). In a study with bifidobacteria spray dried with various carriers (10% of skim milk, gum arabic, soluble starch, and gelatin), skim milk gave the highest viability (Lian et al. 2002). The viability of bifidobacteria is highly dependent on carriers and varies with strains. *B. infantis* strains survived better with gum arabic followed by gelatin and soluble starch, whereas *B. longum* strains survived better with gelatin followed by gum arabic and soluble starch. In addition to the dehydration protectants, antioxidants, such as ascorbate (Kurtmann et al. 2009a; Martin-Dejardin et al. 2013) and tocopherol (Ying et al. 2010), are often incorporated into the drying media to protect cells from oxidation during storage of the dried cells.

It is likely that an interconnection exists between protective effects of the sugars in growth media and the sugars in drying media (Carvalho et al. 2004). The protective effects of sucrose in drying media were higher if cells were previously grown in MRS media supplemented with sucrose, compared with cells grown in the same osmolality of MRS medium supplemented with polyethylene glycol or in standard MRS media (Tymczyszyn et al. 2007b).

7.5 Preservation of Starter Cultures

7.5.1 Short-Term Preservation of Liquid Cultures

Fermented milk plants conventionally employ stock cultures from starter culture suppliers or from their own labs for the preparation of mother cultures. The stock cultures can be in liquid, frozen, and freeze-dried forms. The liquid form is more customary but can maintain high viability and activities at a refrigerated temperature only up to a week, or few weeks at most. Reconstituted skim milk (e.g., 10%)

is often used for the preparation of liquid cultures. The starter cultures are inoculated and incubated before cooling and stored at refrigerated temperatures. Calcium carbonate and yeast extract are often added to increase storage stability. Stability during the chill storage is strain dependent. The viability of *Lb. casei* strain Shirota, and *Lb. paracasei* NCIMB 8001 and NCIMB 9709 remained higher than 90% in the cultured milk at 4°C after storage for 28 days. In contrast, the viability of *Lb. rhamnosus* GG, *Lb. acidophilus* NCIMB 8690, and *Str. thermophilus* NCIMB 10387 at the same storage conditions was only 50%, 1.9%, and 0.25%, respectively (Lee 2008). Some lactic starter cultures are greatly sensitive to the chill storage. After one day of storage at 7°C, the number of viable cells of *Lb. paracasei* CRL 1289 in milk-yeast extract medium (MYE medium, 13% non-fat milk, 0.5%, yeast extract, and 1% glucose) decreased by about 1 log cycle, and the survival rate was only 17% (Juárez et al. 2004).

7.5.2 Long-Term Preservation by Freezing

Freezing removes water from a solution in the form of ice crystals. The cytoplasm of a cell generally remains unfrozen at −10°C to −15°C, and hence, extracellular ice crystals form first during freezing. If the rate of freezing is very low, cells will lose water rapidly enough to maintain the osmotic equilibrium between the inside and the outside of them. Therefore, ice formation outside harms cells virtually similar to dehydration. In contrast, if the cooling rate is too high for cells to lose water rapidly enough, the osmotic equilibrium is maintained by frozen water inside the cells. In this case, viability-threatening intracellular ice formation (IIF) will occur. Therefore, the cooling rate is considered as one of the most important freezing parameters (Fonseca et al. 2001; Roos 2004; Schoug et al. 2006). Theoretically, an optimal cooling rate should be low enough to avoid IIF but high enough to minimize the dehydration effects. An optimal cooling rate depends on factors such as water permeability of cell membrane and the presence of a cell wall (Dumont et al. 2004), membrane fatty acid composition (Goldberg and Eschar 1977; Beal et al. 2001), as well as cell type and cell size. Practically, slow freezing of concentrated cell suspension is realized by placing the suspension in a freezer or deep freezer. The freezing with a higher cooling rate can be obtained with a liquid bath containing coolants such as ethanol cooled by dry ice or refrigerated coils (−78°C), and with liquid N_2 (−196°C). To avoid intracellular ice formation during freezing, a two-step method was recommended (Valdez 2001). According to this method, lactic starter cultures should be cooled at a rate of about 5°C/min in a freezer or cooling bath to a temperature between −20°C and −40°C to allow sufficient efflux of water. Cells are subsequently frozen and stored in liquid N_2.

In addition to these conventional freezing, innovative rapid freezing methods were explored by Volkert et al. (2008). The methods are spray freezing and pressure shift freezing. In the spray freezing, a cell suspension is sprayed in air blast freezer to get droplets (e.g., size of ca. 5–30 μm) with a high surface area to volume ratio. In the pressure shift freezing, a cell suspension is pressurized in the unfrozen state and cooled under pressure to a minimum of −21°C at 210 MPa. Subsequently, the freezing is triggered by instance pressure release, leading to a sudden super cooling of water with respect to the atmospheric freezing point of water at 0°C. Between the two methods, the spray freezing could retain higher viability of *Lb. rhamnosus* (87.4% and 90.5%, spraying with in PBS buffer and 20% skim milk, respectively).

The robustness to freezing and frozen storage is specie or strain dependent (Champagne et al. 1991; Sanders et al. 1999; Rault et al. 2007). For example, viability of *Lb. plantarum* is quite high over the wide range of the freezing. There was no significant change in its viability between low (5°C/min) and very high freezing rates (30,000°C/min) (Dumont et al. 2004). In a study with eight *Lb. acidophilus* strains, there was no significant difference in viability for five strains frozen by rapid freezing (in liquid N_2) and by slow freezing (in cold air at −30°C), whereas the rest three strains did (Foschino et al. 1996). Similarly, in a study with four *Lb. delbrueckii* strains, results showed different behavior to freezing and storage at −80°C. One strain was resistant to both freezing and storage steps, one strain was very sensitive to the freezing process, and two strains were sensitive to frozen storage (Rault et al. 2007).

7.5.2.1 Freezing of Lactic Starter Cultures Stock and mother cultures can be preserved in the frozen form by placing in freezers, cooled liquid bath, or liquid N_2. The cultures are usually prepared in skim milk and dispensed in metal cans or plastic containers. The frozen cultures must be incubated before subculturing into a larger volume of media. For very long-term storage of lactic starter cultures in a culture collection, there are two common containers used, that is, a heavy-wall borosilicate ampoule and plastic cryovial. The glass ampoule needs to be sterilized before use and sealed properly with hot flame. The glass ampoule is inconvenient, and there is a potential risk of injuries during unsealing the ampoule. A suitable plastic cryovial consists of a polypropylene cryogenic tube with a high density polyethylene closure. They customarily have two varieties: those with an internal thread and silicone gasket, and those with an external thread. The silicone gasket provides an excellent seal, but over-tightening of the vial cap can distort the seal. In other words, the plastic cryovial will leak if the seal is too loose or too tight.

7.5.2.2 Freezing of Concentrated and DVS Lactic Starter Cultures The concentrated and DVS cultures can be filled in ring-pull metal cans and frozen in liquid N_2. Alternatively, the cultures can be produced as frozen pellets by adding the cultures dropwise on trays in a conventional industrial freeze-dryer or by dripping the cell concentrate into liquid N_2. For high output and uniform quality, fully automated sterilizable pelletizing units are available for the industrial production. Cell concentrate is distributed into liquid N_2 by a designed liquid dispenser, and the droplets congeal instantly into pellets after a few seconds. The pellets are then conveyed out of the immersion with a mesh-belt conveyor. The pellet-frozen lactic starter cultures are available in a laminated paper board containers and plastic pouches. The pellets are free-flowing and convenient for weighing and dispensing. However, if the storage temperature is higher than the ice melting temperature (Tm') of a pellet-frozen lactic starter culture, individual pellets will stick together and form larger clumps. The Tm' can be routinely measured by differential scanning calorimetry techniques (Roos 1995).

7.5.2.3 Frozen Storage and Thawing Storage temperatures down to $-20°C$ and $-40°C$ can be maintained by common and deep freezers, respectively. Ultra-low storage temperatures can be maintained by a specialized electric freezer or by a liquid N_2 freezer. A general ultra-low temperature electrical freezer can maintain the temperature down to −85°C, whereas some ultra-low temperature freezers can maintain the exceptionally low temperature down to −152°C. The electrical storage system is more practical and does not require low temperature storage solution. However, the ultra-low temperature freezer should be equipped with a CO_2 or liquid N_2 back-up system for the power outage in areas, where the power supplies are not reliable.

There are two basic liquid N_2 storage systems, that is, immersing containers in the liquid or holding containers in the vapor phase above the liquid. The liquid-phase system requires less maintenance. However, it should be aware that the liquid-to-gas expansion ratio of nitrogen is ca. 1:700 at 20°C. Therefore, a container for the freezing in liquid N_2 should be sturdy and leak-proof. The leakage may cause not only contamination, but also injuries due to the explosion when the container is subjected to a warmer temperature. To avoid the explosion, it is also recommended to loosen the closures immediately upon removal of the containers from liquid N_2. The storage of starter cultures in vapor phase of liquid N_2 has an advantage that the contamination risk from a leaking sample and container surface is minimized. However, the temperature in vapor phase storage system varies and depends upon the amount of liquid N_2. Therefore, it is necessary to maintain sufficient levels of liquid N_2 to ensure low temperatures at the top of the container.

Shelf life of frozen cultures depends on the storage temperatures. The lower the storage temperatures, the longer shelf life is. For this reason, liquid N_2 provides the best storage stability (Baumann and Reinbold 1966; Gibson et al. 1966), and only the small reduction in viability of cells occurs during storage at −70°C to −80°C for about 1 or 2 years (Foschino et al.1996; Juárez et al. 2004). Storage at the temperatures of −20°C to −45°C is common in fermented milk industries, and the reduction in viability is more pronounced at −20°C (Thunell et al. 1984; Juárez et al. 2004). Therefore, commercial frozen concentrated cultures should be stored at a temperature lower than −45°C to get the shelf life for at least 12 months (Bylund 2003). The stability of frozen starter cultures depends also on strains and the freezing media. For instance, in a study with six *Lactobacillus* strains, *Lb. salivarius* CRL 1328 showed a slightly decrease in viability during storage up to 18 months at −20°C. *Lb. paracasei* CRL 1251 and CRL 1289, *Lb. acidophilus* CRL 1259 and CRL 1294, and

Lb. crispatus CRL 1266 showed the progressive decrease in viability between 3 and 24 months during storage at −20°C. This decrease differed between the cryoprotective media used. After 24 months of the storage, viable counts for the strains CRL 1251, CRL 1266, and CRL 1289 in LAPTg-glycerol media (1.5% peptone, 1% tryptone, 1% glucose, 1% yeast extract, 0.1% Tween 80) were 1 or 2 log cycles lower than in MYE (13% non-fat milk, 0.5% yeast extract, and 1% glucose) (Juárez et al. 2004).

Notably, in a study with three *Lb. acidophilus* strains by Gilliland and Lara (1988), there was not a significant interaction between the frozen storage at −196°C and the subsequent storage at 5°C. The survival during chill storage was similar independently, whether the cells were freshly harvested or previously frozen and stored in liquid N_2 for 28 days. However, a contrary result was observed in a study with four *B. longum* strains that were frozen, stored in liquid N_2 for 28 days, and subsequently stored at 5°C. *B. longum* survived variably during the chill storage, depending on the strains, growth pH, and duration of the storage (Reilly and Gilliland 1999). In light of these results, it means that for some lactic starter cultures, cells can be stored at the chilled temperature without reduced viability for a certain period of time. This may give more flexibility in the production planning of fermented milk and for a sudden change in the production plan.

During storage at an increased temperature, fluctuation of the storage temperature, and thawing, the small ice crystals can reorganize into the lethally large crystals. For this reason, the storage temperature should be well controlled, and the rate of thawing should be rapid to confine the ice recrystallization. The frozen lactic starter cultures stored in a small cryogenic ampoule or cryogenic vial in liquid N_2 can be rapidly thawed by plunging it into a water bath at 37°C–40°C. Ice will be melted in 40–60 s with moderate agitation (Nakamura 1996). The vial or ampoule can be decontaminated by 70% ethanol before a subsequent inoculation. In industries, thawing and inoculation accomplish simultaneously by pouring the pellets of frozen concentrated cultures directly into milk or pouring the pellets through the automatic and in-line inoculation systems. Thus, the thawing occurs rapidly due to the small pellet size and the large volume of milk being inoculated. However, in the case of frozen concentrated lactic starter cultures that are filled in a can, the cultures can be detached from the can by placing in water containing 25–50 ppm available chlorine at 20°C–22°C for 20 min. (Sandine 1996).

7.5.3 Long-Term Preservation by Drying

Water molecules contribute to the stability of proteins, DNA, and lipids. It confers structural order of cells. On the one hand, removal of water inactivates the cells, and the amount of water remaining after drying affects the viability. The dehydration inactivation of *Lb. helveticus* was reported to be pronounced at water content below around 0.3–0.4 (Selmer-Olsen et al. 1999) and 0.3–0.5 g H_2O.g dry weight^{-1} (Santivarangkna et al. 2007b). This implies that the removal of free water has less detrimental effects on the viability of cells (Santivarangkna et al. 2007b). On the other hand, lactic starters should be dried to water content low enough to slow down metabolic processes and stabilize cellular structures and functions during storage, for example, below 0.1 g H_2O.g dry weight^{-1} (Aguilera and Karel 1997).

Freeze-drying or lyophilization is widely used and is the established process for the preservation of lactic starter cultures. The typical freeze-drying process consists of three steps: freezing, primary drying or sublimation, and secondary drying or desorption. The freezing stage can be conducted with the procedures mentioned previously (Section 7.5.2). The ice crystals formed during freezing affect mass and heat transfer and determine the morphology and distribution of the pores that are created during the removal of ice crystals in the subsequent primary drying stage. The inactivation of the freeze-dried starter cultures is mostly attendant on the freezing step (Tsvetkov and Brankova 1983; To and Etzel 1997). In a study with three different lactic acid bacteria species, 60%–70% of cells that survived the freezing step survived through the dehydration step (To and Etzel 1997). The primary drying stage represents most of the freeze-drying time. During removal of ice crystals, the process temperature in the primary drying stage should not be higher than the so-called collapse temperature (T_c, the maximum temperature preventing the structure of the dried product from macroscopic collapse) to avoid ice melting back (Fonseca et al. 2004a). The collapse temperature depends on cell type and concentration (Fonseca et al. 2004b). The collapsed structure not only lacks elegance but also causes reduced viability. The melted products loss porous structure, and hence, the water vapor cannot be removed easily. The products generally have high residual water contents and lengthy reconstitution times. At the end of the primary drying, there is principally no ice or only small amount present.

In addition to free water, cells contain bound water, which is not frozen and removed during primary drying. This bound water is removed by desorption during the secondary drying. During the secondary drying, the shelf temperature is much higher than the temperature during the primary drying to promote desorption of water. At the end of the secondary drying, the sample should have water content that is optimal for storage. This is generally at 1% or less for a long-term shelf life (Nakamura 1996). However, the moisture content of 4% was also suggested to be enough in practice (Gardiner et al. 2000).

7.5.3.1 Drying of Lactic Starter Cultures The freeze-dried stock cultures cannot be directly used for the propagation of bulk starter because the cultures have too low viable cells. The dried cultures must be reactivated and grown in the larger volume of media for the preparation of mother and intermediate cultures. The dried stock cultures have advantages over the liquid and frozen cultures that they have longer shelf life, and these can be distributed to distant areas without a cold chain. The frequently used containers for freeze-drying of stock cultures are crimped glass vial (closed with a rubber stopper and a metal cap) and glass ampoule. The glass vials can be easily dried on temperature-controlled shelves, which permit a better control of the heat transfer to avoid collapse of the product structure. Rehydration of the dried product is convenient and can be done directly in the vial. However, the vial is less hermetic. The gradual release or exchange of air and moisture through the rubber stopper can occur during a long storage period. In contrast, glass ampoules are attached to the freeze-dryer manifold's nipples. The product temperature increases to the ambient temperature of its own accord, and for this reason, the product structure cannot be well controlled. At the end of the drying, the ampoules can be sealed with a flame air/gas torch. Pull-seal method (stem of ampoules is flamed and drawn down during melted until a seal is formed) is preferable to tip seal (ampoule is flamed until a seal is formed) because the pinhole leakage is minimized. Advantages of the ampoules are that they can be sealed tightly under a vacuum, which renders better storage stability, for example, for storage of lactic starter cultures in a culture collection.

7.5.3.2 Drying of Concentrated and DVS Lactic Starter Cultures The concentrated and DVS culture powder is produced by freeze-drying of the frozen pellets in trays of the dryer. The freeze-dried starter culture granules may be ground to ensure uniformity, and the powder is dispensed in aluminum pouches. Commonly, freeze-dried concentrated cultures have a longer lag phase than that of the frozen ones. In addition to the conventional freeze-drying process, alternative drying processes are of industrial interest because of their advantages over freeze-drying such as less capital and energy costs, shorter processing time, more reliable process monitoring, enhanced stability at room temperature, and possible continuous processing mode (Santivarangkna et al. 2007a). Furthermore, they may allow better viability for lactic starter cultures that are sensitive to freezing such as *Lb. bulgaricus* (Beal and Corrieu 1994; Fonseca et al. 2000) and *Lb. sanfranciscensis* (De Angelis and Gobbetti 2004). The major alternative drying processes are spray, fluidized bed, and vacuum drying (Santivarangkna et al. 2007b).

Spray drying is a prevalent process in dairy industries. It can produce dry powder with a relatively lower cost than freeze-drying. For spray drying, the outlet air temperature is the major drying parameter affecting the viability. The outlet temperature is inversely related to the viability of lactic starter cultures (Kim and Bhowmik 1990; Bielecka and Majkowska 2000; Desmond et al. 2002b; Ananta et al. 2005). Nevertheless, a too low outlet air temperature causes too high residual moisture content for long-term storage. Reduction in the viability caused by increased outlet air temperature varies with the carrier used. For example, the greatest reduction was observed in bifidobacteria dried in soluble starch and a lesser reduction in gelatin, gum arabic, and skim milk (Lian et al. 2002). It should be noted that a considerably longer delay in lactic acid production was observed in spray-dried *Lb. lactis* subsp. *cremoris*, *Lb. casei* subsp. *pseudoplantarum*, and *Str. thermophilus*, especially at high outlet air temperatures (To and Etzel 1997).

Fluidized bed drying has comparable costs to those of spray drying, and the process is also viable for large-scale continuous production.

The drawback is that only granulatable materials can be dried, and therefore, cells must be entrapped or encapsulated in support materials such as skim milk (Roelans and Taeymans 1990), potato starch (Linders et al. 1997), alginate (Selmer-Olsen et al. 1999), and casein (Mille et al. 2004). It is not suitable for oxygen sensitive lactic starter cultures because in the process cells will be suspended and dried in the air stream for quite long time.

The major advantage of vacuum drying is that oxidation reactions during drying can be minimized for oxygen sensitive lactic starter cultures. However, the drying time of the process is much longer than that of spray and fluidized bed drying. The vacuum drying system is conventionally a batch process. Nevertheless, Hayashi et al. (1983) have successfully developed a continuous vacuum dryer whose cost is one-third of that of freeze-drying. Studies reported the comparable viable rates between lactic starter cultures dried by controlled low temperature vacuum drying and freeze-drying in *Lb. acidophilus* (King and Su 1993) and *Lb. paracasei* subsp. *paracasei* (Higl et al. 2008). It was reported also that with an appropriate protectant, vacuum-dried *Lb. paracasei* F19 was more stable at non-refrigerated temperatures than the freeze-dried cells (Foerst et al. 2012).

7.5.3.3 Storage and Rehydration of Powder of the Cultures Although the stability of cells during storage is believed to be specie dependent, the tentative correlation exists between morphology of lactic starter cultures and their storage stabilities. In a study by Miyamoto-Shinohara et al. (2008), it was remarked that for *Lactobacillus*, species with glycerol teichoic acid (GTA) in the cell wall and cell membrane (*Lb. helveticus*), and in the cell membrane (*Lb. fermentum*) had higher viability than that had GTA in the cell wall (*Lb. brevis* and *Lb. delbrueckii*). Furthermore, species with ribitol teichoic acid (*Lb. plantarum*) or polysaccharides (*Lb. casei* and *Lb. salivarius*) in the cell wall had higher viability than species with GTA.

Stability of lactic starters is inversely related to the storage temperature (King et al. 1998; Andersen et al. 1999; Gardiner et al. 2000; Silva et al. 2002) and the moisture content (Clementi and Rossi 1984; Higl et al. 2007). Dried cells should be kept under low relative humidity environment and at low temperatures. It is recommended that the storage temperature of dried cells should be lower than their glass transition temperatures (T_g) or a temperature below which materials exhibit extremely high viscosity and gives them solid-like properties at the corresponding moisture content (Aguilera and Karel 1997). The T_g is not an absolute threshold of bacterial stability during storage because not every inactivation process is diffusion limited, for example, free radical reactions. However, the diffusion of oxygen into the glassy matrix is slow, and this will decrease the production rate of the free radicals (Lievense and van't Riet 1994). In general, freeze-dried products have vastly expanded dry surface area and therefore are increasingly susceptible to oxygen and moisture.

The storage temperature of 4°C is recommended to be suitable for the storage of freeze-dried lactic starter cultures (Wang et al. 2004; Coulibaly et al. 2009). A minimum shelf life of 1 year can be obtained for commercial freeze-dried starter for the preparation of mother culture at a temperature below 5°C. For longer storage stability, it is recommended that freeze-dried cells should be stored at a temperature lower than −20°C to reduce the activity of water molecules (Miyamoto-Shinohara et al. 2006). For the shelf life at least a year, a temperature below −18°C is required for freeze-dried powder for the preparation of bulk starter and freeze-dried DVS (Bylund 2003).

In addition to a low temperature, low water activity gives rise to the glassy state of dried starter cultures. Nevertheless, studies have shown that too low water activity may reduce the storage stability of dried cells. At very low value of water activity (<0.1), a slight increase in the degradation rate can probably occur by oxidative mechanisms and slow Maillard reaction (Kurtmann et al. 2009b; Passot et al. 2012). In studies with *Lb. bulgaricus* CFL1, it was shown that optimal water activity for the storage of the dried cells is at ca. 0.1–0.2 (Teixeira et al. 1995b; Passot et al. 2012). An increase in storage water activity from 0.11 to 0.22 was shown to have relatively less negative impact on the viability than did an increase in water activity from 0.22 to 0.32 (Kurtmann et al. 2009a). Taken together, the storage of dried starter cultures at water activity of ca. 0.1 should be practically low enough for long-term storage. Due to the deteriorative browning reaction, a non-reducing sugar is the preferable protectant to the reducing sugar (Kurtmann et al. 2009b; Strasser et al. 2009).

Since the major damage during storage of lactic starter cultures is due to the lipid oxidation (Teixeira et al. 1996; Andersen et al. 1999;

Yao et al. 2008; Coulibaly et al. 2009), antioxidants are often added. However, it should be used with care. Sodium ascorbate, which is a customary antioxidant, in the freeze-drying medium (up to 10%) improved the storage stability of *Lb. acidophilus* at 30°C (Kurtmann et al. 2009a). However, ascorbic acid together with monosodium glutamate protected *Lb. delbrueckii* subsp. *bulgaricus* only during storage at 4°C. The death rate of the culture was even higher at the storage temperature of 20°C (Teixeira et al. 1995b). Similarly, the incorporation of Na-ascorbate alone or in the combination with tocopherol decreased viability of microencapsulated *Lb. rhamnosus* GG (LGG) spray-dried powders during storage at 4 and 25°C, whereas tocopherol alone improved the viability at both storage temperatures (Ying et al. 2010). These controversial effects are supposedly due to the differences in abilities of various lactic starter cultures to catabolize ascorbate and form products that are detrimental to cell viability (Ying et al. 2010).

Packages for the storage of dried cells should be moisture and air impermissible. Aluminum laminated pouch is superior to glass and polyester (PET) bottles for the storage of lactic starter cultures (Wang et al. 2004). For the very long-term storage of original lactic starter cultures in a culture collection, a double ampoule with desiccant added in the outer ampoule is used by culture collections. The survival of cells dried in ampoules is strongly influenced by the degree of vacuum. Ampoules sealed under vacuum <1 Pa render better survival than those sealed under vacuum of 7 Pa (Miyamoto-Shinohara et al. 2006).

Rehydration of cells is an inevitable step for the use of dried cells. Viability of cells seems to depend on rehydration media (Champagne et al. 1991). However, it is difficult to have a universal rehydration medium for dried lactic starter cultures. Variations were found among species and strains (Sinha et al. 1974; De Valdez et al. 1985; Zhao and Zhang 2005). Still, it is quite clear that viability of the dried cultures correlates directly with the rehydration temperatures. Higher viability is obtained when cells are rehydrated with the warm media (Teixeira et al. 1995a; Mille et al. 2004; Wang et al. 2004). Cells dried with alternative drying processes often require long rehydration time. For example, Ca-alginate gel beads with entrapped *Lb. helveticus* need lag time of almost 3 h to reach the original bead volume during the rehydration (Selmer-Olsen et al. 1999).

7.6 Other Factors Affecting Viability during Preservation

7.6.1 Intrinsic Factors

Robustness to freezing and drying is peculiar to each strain or specie of lactic bacteria. Some studies suggested that viability is correlated to cell morphology, that is, cell size and shape. Bozoglu et al. (1987) reported that small spherical streptococci cells were more resistant to freeze-drying than long rod lactobacilli cells because of the lower surface area to volume ratio. This explanation disagrees with a study with *Lb. reuteri* grown with various elapsed times in the stationary phase. The bacterium differed in viability after freeze-drying even if shape of the cells is proved the same (Palmfeldt and Hahn-Hagerdal 2000). The mechanism behind this difference is still unclear, and this poses a difficulty in adopting preservation conditions from one given strain to other strains. A strategy that can be used to obtain a robust strain is the isolation of survivors from numerous cycles of preservation-related stresses, for example, repeated freezing and thawing cycles (Monnet et al. 2003).

7.6.2 Stress Induction

Exposing to a mild stress can trigger protective mechanisms of cells to subsequent lethal stresses during the preservation. For example, *Lb. paracasei* cells incubated at 0.3 M NaCl for 30 min had viabilities 16-fold higher than unstressed cells after spray drying at outlet air temperatures of 100°C–105°C (Desmond et al. 2002b). The pre-incubation of cells at temperatures lower than the optimal temperatures before freezing improved viability after freezing of *Lb. acidophilus* (Baati et al. 2000), *Lb. bulgaricus* (De Urraza and De Antoni 1997; Panoff et al. 2000), *Lb. helveticus*, *Lb. lactis* subsp. *lactis* (Wouters et al. 1999), *Lb. lactis* subsp. *cremoris*, and *P. pentosaceus* (Kim and Dunn 1997). Similarly, the pre-incubation improved viability of *Lb. lactis* subsp. *cremoris* and *Lb. lactis* subsp. *lactis* after freeze-drying (Broadbent and Lin 1999). At the end of the pH-controlled fermentation (pH 6.0), *Lb. delbrueckii* subsp. *bulgaricus* was acidified and held at pH 5.25 for 30 min. The higher viability was observed in acid-shocked cells after freezing (42.9%), as compared to the control cells (4.7%) (Streit et al. 2008).

Notably, stress induction during cultivation often compromises the growth rate and yield of cells. Therefore, it should be sure that increased viability after preservation will clearly outweigh the decrease in cell number due to the stress induction. For example, viability of *Lb. johnsonii* was improved by 19% when heat shocking at 55°C for 45 min, but the number of cells was reduced by 1.5-log unit for this heating duration (Walker et al. 1999). In other words, in this study the heat shock had actually a negative impact on net viability after the preservation. Effects of the stress induction depend also on the nature and degree of the stress applied. The induction can also negatively affected cell viability after the preservation (Schoug et al. 2008; Mozzetti et al. 2013). For the screening of the optimal sublethal stress conditions, a two-stage continuous culture tool was investigated by Mozzetti et al. (2013). The first reactor was operated under fixed conditions to produce cells with controlled physiology, and the second reactor was used to test the stress pre-treatment combinations. In this setup, the effects of stress pre-treatments with up to four different conditions can be examined at once per day.

7.6.3 Encapsulation

Encapsulation is a process of coating or entrapping of an active component or a mixture of active components within the matrix (encapsulant). Encapsulation of lactic starter cultures consists of the mixing of a cell suspension with an appropriate encapsulant to form the capsule or gel globule. Alginates, carrageen, cellulose derivatives (acetate, phthalate), chitosan, gelatin, and starch are among the most common encapsulants. The best-known encapsulation method is the dropwise addition of a sodium alginate solution containing lactic starter cultures to a calcium chloride solution. Encapsulation may also mutually be carried out during the spray drying process (O'Riordan et al. 2001; Ying et al. 2010). Although studies on the encapsulation have largely focused on its roles on the survival and stability of probiotics in food matrices and its roles during the passage of cells through the gastrointestinal tract, some studies also reported protective effects of the encapsulation during the preservation and storage of lactic starter cultures. For instance, the encapsulation of *Lb. plantarum* in alginate beads increased the survival of cells by approximately 30% after freeze-drying (Kearney et al. 1990). *Bifidobacterium longum* encapsulated in beads containing 3% alginate and 0.15% xanthan gum retained the viability during storage in acidified milk at 4°C for 10 days (Woo et al. 1999), whereas viability of the control free cells decreased progressively. A comprehensive review on the encapsulation of probiotics and lactic starter cultures including its influence on their survival was well described in Krasaekoopt and Bhandari (2012).

7.6.4 Genetic Modification

Tolerances of lactic starter cultures can be specifically modified by the genetic manipulation. Genetically modified *Lb. paracasei* strain, overproducing the chaperone protein GroESL, which refolds denatured protein during heat stress, had viabilities after spray drying and freeze-drying ca. 10-fold and twofold higher than those of parent cells, respectively (Desmond et al. 2004). In addition, some lactic starter cultures were modified so that cells could accumulate solutes that were able to protect cells. The nisin-controlled expression system was used to direct the heterologous expression of the listerial betaine uptake system BetL in *Lb. salivarius* UCC118 (Sheehan et al. 2006). Following nisin induction, the UCC118-BetL$^+$ survived significantly better than UCC118-BetL$^-$ after freeze- and spray drying. Similarly, the trehalose biosynthesis genes *ots* BA from *E. coli* was introduced in *Lb. lactis* NZ9000 (Termont et al. 2006). Following nisin induction, the intracellular accumulation of trehalose reached maxima at approx. 50 mg/g wet cell weight after 2–3 h. The viability of induced NZ9000 was markedly higher than that of the non-induced NZ9000 after freeze-drying with skim milk (94.0% vs. 56.7%) and without skim milk (ca. 100% vs. 20%). The induced cells had also higher storage stability and retained almost 100% viability for at least 1 month when being stored at 8°C and 10% relative humidity.

7.7 Concluding Remarks

The preservation of lactic starter cultures must be considered in relation to their applications and practices. For very long-term storage of the original stock cultures in a culture collection, hundreds of culturable cells, for example, 300 CFU/0.1 ml of suspension prepared from an ampoule, can be satisfactory (Nakamura 1996). In contrast, preservation of the cultures for bulk culture preparation and DVS cultures should maintain, for example, billion and trillion of active cells to continue successive subculturing and to resume fermentation immediately after the inoculation. A troublesome finding reported by several studies is that many technological traits and responses of lactic starter cultures upon the preservation processes and conditions are intrinsic to each species or strain. It means that on the one hand, optimal preservation processes and conditions need to be determined on a case-to-case basis, and the process optimization may be necessary for a new lactic starter culture. On the other hand, this is a challenge for fermented milk industries and starter culture specialists to identify as many as possible the common principles that can be applied to most starter cultures. The principles will help the industries and culture specialists keeping up with the increasing number of food and dairy products fermented or supplemented with novel lactic starter cultures. It should be noted also that these dairy products often employ the cultures with probiotic properties, and some studies reported changes in these properties after the preservation (Golowczyc et al. 2011; Paez et al. 2013). Therefore, the preservation of lactic starter cultures should be redefined, and its purpose should include the maintenance of essential probiotic properties of lactic starter cultures, in addition to the maintenance of viability and fermentation activities.

References

Aguilera, J.M. and Karel, M. 1997. Preservation of biological materials under desiccation. *Critical Reviews in Food Science and Nutrition* 37(3): 287–309.

Ananta, E., Volkert, M., and Knorr, D. 2005. Cellular injuries and storage stability of spray-dried *Lactobacillus rhamnosus* GG. *International Dairy Journal* 15(4): 399–409.

Andersen, A.B., Fog-Peterson, M.S., Larsen, H., and Skibsted, L.H. 1999. Storage stability of freeze-dried starter cultures (*Streptococcus thermophilus*) as related to physical state of freezing matrix. *Food Science and Technology* 32: 540–547.

Baati, L., Fabre-Gea, C., Auriol, D., and Blanc, P.J. 2000. Study of the cryotolerance of *Lactobacillus acidophilus*: Effect of culture and freezing conditions on the viability and cellular protein levels. *International Journal of Food Microbiology* 59(3): 241–247.

Baumann, D.P. and Reinbold, G.W. 1966. Freezing of lactic cultures. *Journal of Dairy Science* 49(3): 259–264.

Beal, C. and Corrieu, G. 1994. Viability and acidification activity of pure and mixed starters of *Streptococcus salivarius* ssp. *thermophilus*-404 and *Lactobacillus delbrueckii* ssp.*bulgaricus*-398 at the different steps of their production. *Food Science and Technology-Lebensmittel-Wissenschaft & Technologie* 27(1): 86–92.

Beal, C., Fonseca, F., and Corrieu, G. 2001. Resistance to freezing and frozen storage of *Streptococcus thermophilus* is related to membrane fatty acid composition. *Journal of Dairy Science* 84(11): 2347–2356.

Bielecka, M. and Majkowska, A. 2000. Effect of spray drying temperature of yoghurt on the survival of starter cultures, moisture content and sensoric properties of yoghurt powder. *Nahrung-Food* 44(4): 257–260.

Bozoglu, T.F., Ozilgen, M., and Bakir, U. 1987. Survival kinetics of lactic acid starter cultures during and after freeze-drying. *Enzyme and Microbial Technology* 9(9): 531–537.

Brennan, M., Wanismail, B., Johnson, M.C., and Ray, B. 1986. Cellular-damage in dried *Lactobacillus acidophilus*. *Journal of Food Protection* 49(1): 47–53.

Broadbent, J.R. and Lin, C. 1999. Effect of heat shock or cold shock treatment on the resistance of *Lactococcus lactis* to freezing and lyophilization. *Cryobiology* 39(1): 88–102.

Bylund, G. 2003. *Dairy Processing Handbook*. Lund, Sweden: Tetra Pak Processing Systems AB.

Carcoba, R. and Rodriguez, A. 2000. Influence of cryoprotectants on the viability and acidifying activity of frozen and freeze-dried cells of the novel starter strain *Lactococcus lactis* ssp. *lactis* CECT 5180. *European Food Research and Technology* 211(6): 433–437.

Carvalho, A.S., Silva, J., Ho, P., Teixeira, P., Malcata, F.X., and Gibbs, P. 2003. Effects of addition of sucrose and salt, and of starvation uponthermotolerance and survival during storage of freeze-dried *Lactobacillus delbrueckii* ssp. *bulgaricus*. *Journal of Food Science* 68: 2538–2541.

Carvalho, A.S., Silva, J., Ho, P., Teixeira, P., Malcata, F.X., and Gibbs, P. 2004. Effects of various sugars added to growth and drying media upon thermotolerance and survival throughout storage of freeze-dried *Lactobacillus delbrueckii* ssp. *bulgaricus*. *Biotechnology Progress* 20(1): 248–254.

Cerning, J. 1990. Exocellular polysaccharides produced by lactic acid bacteria. *FEMS Microbiological Review* 7(1–2): 113–130.

Champagne, C.P., Gardner, E., Brochu, N., and Beaulieu, Y. 1991. The freeze-drying of lactic-acid bacteria—A review. Canadian *Institute* of *Food Science* and *Technology Journal* 24(3–4): 118–128.

Chavarri, F.J., De Paz, M., and Nunez, M. 1988. Cryoprotective agents for frozen concentrated starters from non-bitter *Streptococcus lactis* strains. *Biotechnology Letter* 10(1): 11–16.

Clementi, F. and Rossi, J. 1984. Effect of drying and storage conditions on survival of *Leuconostocoenos*. *American Journal of Enology and Viticulture* 35(3): 183–186.

Conrad, P.B., Miller, D.P., Cielenski, P.R., and de Pablo, J.J. 2000. Stabilization and preservation of *Lactobacillus acidophilus* in saccharide matrices. *Cryobiology* 41: 17–24.

Corcoran, B.M., Ross, R.P., Fitzgerald, G.F., and Stanton, C. 2004. Comparative survival of probiotic lactobacilli spray-dried in the presence of prebiotic substances. *Journal of Applied Microbiology* 96(5): 1024–1039.

Coulibaly, I., Amenan, A.Y., Lognay, G., Fauconnier, M.L., and Thonart, P. 2009. Survival of freeze-dried *Leuconostoc mesenteroides* and *Lactobacillus plantarum* related to their cellular fatty acids composition during storage. *Applied Biochemistry and Biotechnology* 157(1): 70–84.

De Angelis, M. and Gobbetti, M. 2004. Environmental stress responses in *Lactobacillus*: A review. *Proteomics* 4(1): 106–122.

De Antoni, G.L., Perez, P., Abraham, A., and Anon, M.C. 1989. Trehalose, a cryoprotectant for *Lactobacillus bulgaricus*. *Cryobiology* 26: 149–153.

De Urraza, P. and De Antoni, G. 1997. Induced cryotolerance of *Lactobacillus delbrueckii* supsp. *bulgaricus* LBB by preincubation at suboptimal temperatures with a fermentable sugar. *Cryobiology* 35: 159–164.

De Valdez, G.F., Degiori, G.S., Holgado, A.P.D., and Oliver, G. 1985. Effect of the rehydration medium on the recovery of freeze-dried lactic acid bacteria. *Applied and Environmental Microbiology* 50(5): 1339–1341.

Desmond, C., Fitzgerald, G.F., Stanton, C., and Ross, R.P. 2004. Improved stress tolerance of GroESL-overproducing *Lactococcus lactis* and probiotic *Lactobacillus paracasei* NFBC 338. *Applied and Environmental Microbiology* 70(10): 5929–5936.

Desmond, C., Ross, R.P., O'Callaghan, E., Fitzgerald, G., and Stanton, C. 2002a. Improved survival of *Lactobacillus paracasei* NFBC 338 in spray-dried powders containing gum acacia. *Journal of Applied Microbiology* 93(6): 1003–1011.

Desmond, C., Stanton, C., Fitzgerald, G.F., Collins, K., and Ross, R.P. 2002b. Environmental adaptation of probiotic lactobacilli towards improvement of performance during spray drying. *International Dairy Journal* 12(2–3): 183–190.

Di Cagno, R., De Angelis, M., Limitone, A., Minervini, F., Carnevali, P., Corsetti, A., Gaenzle, M., Ciati, R., and Gobbetti, M. 2006. Glucan and fructan production by sourdough *Weissella cibaria* and *Lactobacillus plantarum*. *Journal of Agricultural and Food Chemistry* 54(26): 9873–9881.

Doleyres, Y., Fliss, I., and Lacroix, C. 2004. Continuous production of mixed lactic starters containing probiotics using immobilized cell technology. *Biotechnology Progress* 20(1): 145–150.

Dumont, F., Marechal, P.A., and Gervais, P. 2004. Cell size and water permeability as determining factors for cell viability after freezing at different cooling rates. *Applied and Environmental Microbiology* 70(1): 268–272.

Efiuvwevwere, B.J.O., Gorris, L.G.M., Smid, E.J., and Kets, E.P.W. 1999. Mannitol-enhanced survival of *Lactococcus lactis* subjected to drying. *Applied Microbiology and Biotechnology* 51: 100–104.

Foerst, P., Kulozik, U., Schmitt, M., Bauer, S., and Santivarangkna, C. 2012. Storage stability of vacuum-dried probiotic bacterium *Lactobacillus paracasei* F19. *Food and Bioproducts Processing* 90(2): 295–300.

Fonseca, F., Beal, C., and Corrieu, G. 2000. Method of quantifying the loss of acidification activity of lactic acid starters during freezing and frozen storage. *Journal of Dairy Research* 67(1): 83–90.

Fonseca, F., Beal, C., and Corrieu, G. 2001. Operating conditions that affect the resistance of lactic acid bacteria to freezing and frozen storage. *Cryobiology* 43: 189–198.

Fonseca, F., Passot, S., Cunin, O., and Marin, M. 2004a. Collapse temperature of freeze-dried *Lactobacillus bulgaricus* suspensions and protective media. *Biotechnology Progress* 20(1): 229–238.

Fonseca, F., Passot, S., Lieben, P., and Marin, M. 2004b. Collapse temperature of bacterial suspensions: The effect of cell type and concentration. *CryoLetter* 25(6): 425–434.

Foschino, R., Fiori, E., and Galli, A. 1996. Survival and residual activity of *Lactobacillus acidophilus* frozen cultures under different conditions. *Journal of Dairy Research* 63(2): 295–303.

Gancel, F. and Novel, G. 1994. Exopolysaccharide production by *Streptococcus salivarius* ssp. *thermophilus* cultures. 1. Conditions of production. *Journal of Dairy Science* 77(3): 685–688.

Gardiner, G.E., O'Sullivan, E., Kelly, J., Auty, M.A., Fitzgerald, G.F., Collins, J.K., Ross, R.P., and Stanton, C. 2000. Comparative survival rates of human-derived probiotic *Lactobacillus paracasei* and *L. salivarius* strains during heat treatment and spray drying. *Applied and Environmental Microbiology* 66(6): 2605–2612.

Gibson, C.A., Landerkin, G.B., and Morse, P.M. 1966. Effects of additives on the survival of lactic streptococci in frozen storage. *Applied Microbiology* 14(4): 665–669.

Gilliland, S.E. and Lara, R.C. 1988. Influence of storage at freezing and subsequent refrigeration temperatures on beta-Galactosidase activity of *Lactobacillus acidophilus*. *Applied and Environmental Microbiology* 54(4): 898–902.

Goldberg, I. and Eschar, L. 1977. Stability of lactic acid bacteria to freezing as related to their fatty acid composition. *Applied and Environmental Microbiology* 33(3): 489–496.

Golowczyc, M.A., Silva, J., Teixeira, P., De Antoni, G.L., and Abraham, A.G. 2011. Cellular injuries of spray-dried *Lactobacillus* spp. isolated from kefir and their impact on probiotic properties. *International Journal of Food Microbiology* 144(3): 556–560.

Hayashi, H., Kumazawa, E., Saeki, Y., and Ishioka, Y. 1983. Continuous vacuum dryer for energy saving. *Drying Technology* 1(2): 275–284.

Higl, B., Kurtmann, L., Carlsen, C.U., Ratjen, J., Forst, P., Skibsted, L.H., Kulozik, U., and Risbo, J. 2007. Impact of water activity, temperature, and physical state on the storage stability of *Lactobacillus paracasei* ssp. *paracasei* freeze-dried in a lactose matrix. *Biotechnology Progress* 23(4): 794–800.

Higl, B., Santivarangkna, C., and Foerst, P. 2008. Bewertung und Optimierung von Gefrier-und Vakuumtrocknungsverfahren in der Herstellung von mikrobiellen Starterkulturen. *Chemie Inginieur Technik* 80(8): 1157–1164.

Hubalek, Z. 2003. Protectants used in the cryopreservation of microorganisms. *Cryobiology* 46(3): 205–229.
Juárez, T., Silvina, M., Ocaña, V.S., and Nader-Macías, M.E. 2004. Viability of vaginal probiotic lactobacilli during refrigerated and frozen storage. *Anaerobe* 10(1): 1–5.
Kearney, L., Upton, M., and Mc Loughlin, A. 1990. Enhancing the viability of *Lactobacillus plantarum* inoculum by immobilizing the cells in calcium alginate beads incorporating cryoprotectants. *Applied and Environmental Microbiology* 56(10): 3112–3116.
Kilimann, K.V., Doster, W., Vogel, R.F., Hartmann, C., and Ganzle, M.G. 2006. Protection by sucrose against heat-induced lethal and sublethal injury of *Lactococcus lactis*: An FT-IR study. *Biochimica et Biophysica Acta* 1764(7): 1188–1197.
Kim, S.S. and Bhowmik, S.R. 1990. Survival of lactic acid bacteria during spray drying of plain yogurt. *Journal of Food Science* 55(4): 1008–1011.
Kim, W.S. and Dunn, N.W. 1997. Identification of a cold shock gene in lactic acid bacteria and the effect of cold shock on cryotolerance. *Current Microbiology* 35(1): 59–63.
Kim, W.S., Ren, J., and Dunn, N.W. 1999. Differentiation of *Lactococcus lactis* subspecies *lactis* and subspecies cremoris strains by their adaptive response to stresses. *FEMS Microbiology Letter* 171(1): 57–65.
King, V.A.E., Lin, H.J., and Liu, C.F. 1998. Accelerated storage testing of freeze-dried and controlled low-temperature vacuum dehydrated *Lactobacillus acidophilus*. *Journal of General and Applied Microbiology* 44(2): 161–165.
King, V.A.E. and Su, J.T. 1993. Dehydration of *Lactobacillus acidophilus*. *Process Biochemistry* 28(1): 47–52.
Krasaekoopt, W. and Bhandari, B. 2012. Properties and applications of different probiotic delivery systems. In: *Encapsulation Technologies and Delivery Systems for Food Ingredients and Nutraceuticals*, eds. N. Garti and D.J. McClements, 541–594. Cambridge: Woodhead Publishing Limited.
Kurtmann, L., Carlsen, C.U., Risbo, J., and Skibsted, L.H. 2009a. Storage stability of freeze–dried Lactobacillus acidophilus (La-5) in relation to water activity and presence of oxygen and ascorbate. *Cryobiology* 58(2): 175–180.
Kurtmann, L., Skibsted, L.H., and Carlsen, C.U. 2009b. Browning of freeze-dried probiotic bacteria cultures in relation to loss of viability during storage. *Journal of Agricultural and Food Chemistry* 57(15): 6736–6741.
Lacroix, C. and Yildirim, S. 2007. Fermentation technologies for the production of probiotics with high viability and functionality. *Current Opinion in Biotechnology* 18(2): 176–183.
Lee, K. 2004. Cold shock response in *Lactococcus lactis* ssp. *diacetylactis*: A comparison of the protection generated by brief pre-treatment at less severe temperatures. *Process Biochemistry* 39(12): 2233–2239.
Lee, Y.K. 2008. Selection and maintenance of probiotic microorganisms. In: *Handbook of Probiotics and Prebiotics*, eds. Y.K. Lee and S. Salminen, 177–187. Hoboken, NJ: John Wiley & Sons, Inc.

Lian, W.C., Hsiao, H.C., and Chou, C.C. 2002. Survival of bifidobacteria after spray-drying. *International Journal of Food Microbiology* 74(1–2): 79–86.

Lievense, L.C. and van't Riet, K. 1994. Convective drying of bacteria. II. Factors influencing survival. *Advances in Biochemical Engineering/Biotechnology* 51: 71–89.

Linders, L.J.M., Wolkers, W.F., Hoekstra, F.A., and Vantriet, K. 1997. Effect of added carbohydrates on membrane phase behavior and survival of dried *Lactobacillus plantarum*. *Cryobiology* 35(1): 31–40.

Martin-Dejardin, F., Ebel, B., Lemetais, G., Nguyen Thi Minh, H., Gervais, P., Cachon, R., and Chambin, O. 2013. A way to follow the viability of encapsulated *Bifidobacterium bifidum* subjected to a freeze-drying process in order to target the colon: Interest of flow cytometry. *European Journal of Pharmaceutical Sciences* 49(2): 166–174.

Mille, Y., Obert, J.P., Beney, L., and Gervais, P. 2004. New drying process for lactic bacteria based on their dehydration behavior in liquid medium. *Biotechnology and Bioenggineering* 88(1): 71–76.

Miyamoto-Shinohara, Y., Sukenobe, J., Imaizumi, T., and Nakahara, T. 2006. Survival curves for microbial species stored by freeze-drying. *Cryobiology* 52(1): 27–32.

Miyamoto-Shinohara, Y., Sukenobe, J., Imaizumi, T., and Nakahara, T. 2008. Survival of freeze-dried bacteria. *Journal of General* and *Applied Microbiology* 54(1): 9–24.

Modesto, M., Mattarelli, P., and Biavati, B. 2004. Resistance to freezing and freeze-drying storage processes of potential probiotic bifidobacteria. *Annals of Microbiology* 54(1): 43–48.

Monnet, C., Beal, C., and Corrieu, G. 2003. Improvement of the resistance of *Lactobacillus delbrueckii* ssp. *bulgaricus* to freezing by natural selection. *Journal of Dairy Science* 86(10): 3048–3053.

Mozzetti, V., Grattepanche, F., Berger, B., Rezzonico, E., Arigoni, F., and Lacroix, C. 2013. Fast screening of *Bifidobacterium longum* sublethal stress conditions in a novel two-stage continuous culture strategy. *Beneficial Microbes* 4(2): 167–178.

Mozzi, F., Rollan, G., de Giori, G.S., and Font de Valdez, G. 2001. Effect of galactose and glucose on the exopolysaccharide production and the activities of biosynthetic enzymes in *Lactobacillus casei* CRL 87. *Journal of Applied Microbiology* 91(1): 160–167.

Nakamura, L.K. 1996. Preservation and maintenance of eubacteria. In: *Maintaining Cultures for Biotechnology and Industry*, eds. J.C. Hunter-Cevera and A. Belt, 65–84. London: Academic Press, Inc.

O'Riordan, K., Andrews, D., Buckle, K., and Conway, P. 2001. Evaluation of micro encapsulation of a *Bifidobacterium* strain with starch as an approach to prolonging viability during storage. *Journal of Applied Microbiology* 91(6): 1059–1066.

Paez, R., Lavari, L., Audero, G., Cuatrin, A., Zaritzky, N., Reinheimer, J., and Vinderola, G. 2013. Study of the effects of spray-drying on the functionality of probiotic lactobacilli. *International Journal of Dairy Technology* 66(2): 155–161.

Palmfeldt, J. and Hahn-Hagerdal, B. 2000. Influence of culture pH on survival of *Lactobacillus reuteri* subjected to freeze-drying. *International Journal of Food Microbiology* 55(1–3): 235–238.

Panoff, J.M., Thammavongs, B., and Gueguen, M. 2000. Cryoprotectants lead to phenotypic adaptation to freeze-thaw stress in *Lactobacillus delbrueckii* ssp. *bulgaricus* CIP 101027T. *Cryobiology* 40(3): 264–269.

Passot, S., Cenard, S., Douania, I., Tréléa, I.C., and Fonseca, F. 2012. Critical water activity and amorphous state for optimal preservation of lyophilised lactic acid bacteria. *Food Chemistry* 132(4): 1699–1705.

Pedersen, M.B., Iversen, S.L., Sørensen, K.I., and Johansen, E. 2005. The long and winding road from the research laboratory to industrial applications of lactic acid bacteria. *FEMS Microbiology Reviews* 29(3): 611–624.

Peebles, M.M., Gillilan, S.E., and Speck, M.L. 1969. Preparation of concentrated lactic streptococcus starters. *Applied Microbiology* 17(6): 805–810.

Rault, A., Beal, C., Ghorbal, S., Ogier, J. C., and Bouix, M. 2007. Multiparametric flow cytometry allows rapid assessment and comparison of lactic acid bacteria viability after freezing and during frozen storage. *Cryobiology* 55(1): 35–43.

Reilly, S.S., and Gilliland, S.E. 1999. *Bifidobacterium longum* survival during frozen and refrigerated storage as related to pH during growth. *Journal of Food Science* 64(4): 714–718.

Roberts, I.S. 1996. The biochemistry and genetics of capsular polysaccharide production in bacteria. *Annual Review of Microbiology* 50: 285–315.

Roelans, E. and Taeymans, D. 1990. Effect of drying conditions on survival and enzyme activity of microorganims. In: *Engineering and Food Volume 3 Advanced Processes*, eds. W.E.L. Spiess and H. Schubert, 559–569. London: Elsevier Applied Science.

Roos, Y.H. 1995. Methodology. In: *Phase Transitions in Foods*, ed. Y.H. Roos, 49–71. San Diego, CA: Academic Press.

Roos, Y.H. 2004. Phase and state transitions in dehydration of biomaterials and foods. In: *Dehydration of Products of Biological Origin*, ed. A.S. Mujumdar, 3–22. Enfield, NH: Science Publishers, Inc.

Sanders, J.W., Venema, G., and Kok, J. 1999. Environmental stress responses in *Lactococcuslactis*. *FEMS Microbiology Reviews* 23(4): 483–501.

Sandine, W.E. 1996. Commercial production of dairy starter cultures. In: *Dairy Starter Cultures*, eds. T.M. Cogan and J.P. Accolas, 191–206. New York: VCH.

Santivarangkna, C., Higl, B., and Foerst, P. 2008a. Protection mechanisms of sugars during different stages of preparation process of dried lactic acid starter cultures. *Food Microbiology* 25(3): 429–441.

Santivarangkna, C., Kulozik, U., and Foerst, P. 2006. Effect of carbohydrates on the survival of *Lactobacillus helveticus* during vacuum drying. *Letters in Applied Microbiology* 42: 271–276.

Santivarangkna, C., Kulozik, U., and Foerst, P. 2007a. Alternative drying processes for the industrial preservation of lactic acid starter cultures. *Biotechnology Progress* 23(2): 302–315.

Santivarangkna, C., Kulozik, U., and Foerst, P. 2008b. Inactivation mechanisms of lactic acid starter cultures preserved by drying processes. *Journal of Applied Microbiology* 105(1): 1–13.

Santivarangkna, C., Wenning, M., Foerst, P., and Kulozik, U. 2007b. Damage of cell envelope of *Lactobacillus helveticus* during vacuum drying. *Journal of Applied Microbiology* 102(3): 748–756.

Schoug, A., Fischer, J., Heipieper, H.J., Schnurer, J., and Hakansson, S. 2008. Impact of fermentation pH and temperature on freeze-drying survival and membrane lipid composition of *Lactobacillus coryniformis* Si3. *Journal of Industrial Microbiology and Biotechnology* 35(3): 175–181.

Schoug, S., Olsson, J., Carlfors, J., Schnurer, J., and Hakansson, S. 2006. Freeze-drying of *Lactobacillus coryniformis* Si3-effects of sucrose concentration, cell density, and freezing rate on cell survival and thermophysical properties. *Cryobiology* 53(1): 119–127.

Schwab, C., Vogel, R., and Ganzle, M.G. 2007. Influence of oligosaccharides on the viability and membrane properties of *Lactobacillus reuteri* TMW1.106 during freeze-drying. *Cryobiology* 55: 108–114.

Selmer-Olsen, E., Sorhaug, T., Birkeland, S.E., and Pehrson, R. 1999. Survival of *Lactobacillus helveticus* entrapped in Ca-alginate in relation to water content, storage and rehydration. *Journal of Industrial Microbiology & Biotechnology* 23(2): 79–85.

Sheehan, V.M., Sleator, R.D., Fitzgerald, G.F., and Hill, C. 2006. Heterologous expression of BetL, a betaine uptake system, enhances the stress tolerance of *Lactobacillus salivarius* UCC118. *Applied and Environmental Microbiology* 72(3): 2170–2177.

Silva, J., Carvalho, A.S., Teixeira, P., and Gibbs, P.A. 2002. Bacteriocin production by spray-dried lactic acid bacteria. *Letters in Applied Microbiology* 34(2): 77–81.

Sinha, R.N., Dudani, A.T., and Ranganathan, B. 1974. Protective effect of fortified skim milk as suspending medium for freeze drying of different lactic acid bacteria. *Journal of Food Science* 39: 641–642.

Stadhouders, J., Janson, L.A., and Hup, G. 1969. Preservation of starters and mass production of starter bacteria. *Netherland Milk and Dairy Journal* 23: 182–199.

Strasser, S., Neureiter, M., Geppl, M., Braun, R., and Danner, H. 2009. Influence of lyophilization, fluidized bed drying, addition of protectants, and storage on the viability of lactic acid bacteria. *Journal of Applied Microbiology* 107(1): 167–177.

Streit, F., Athes, V., Bchir, A., Corrieu, G., and Beal, C. 2011. Microfiltration conditions modify *Lactobacillus bulgaricus* cryotolerance in response to physiological changes. *Bioprocess and Biosystems Engineering* 34(2): 197–204.

Streit, F., Delettre, J., Corrieu, G., and Beal, C. 2008. Acid adaptation of *Lactobacillus delbrueckii* subsp. *bulgaricus* induces physiological responses at membrane and cytosolic levels that improves cryotolerance. *Journal of Applied Microbiology* 105(4): 1071–1080.

Taniguchi, M., Kotani, N., and Kobayashi, T. 1987a. High-concentration cultivation of lactic acid bacteria in fermentor with cross-flow filtration. *Journal of Fermentation Technology* 65(2): 179–184.

Taniguchi, M., Kotani, N., and Kobayashi, T. 1987b. High concentration cultivation of *Bifidobacterium longum* in fermenter with cross-flow filtration. *Applied Microbiology and Biotechnology* 25(5): 438–441.

Teixeira, P., Castro, H., and Kirby, R. 1995a. Spray drying as a method for preparing concentrated cultures of *Lactobacillus bulgaricus*. *Journal of Applied Bacteriology* 78(4): 456–462.

Teixeira, P., Castro, H., and Kirby, R. 1996. Evidence of membrane lipid oxidation of spray-dried *Lactobacillus bulgaricus* during storage. *Letters in Applied Microbiology* 22(1): 34–38.

Teixeira, P.C., Castro, M.H., Malcata, F.X., and Kirby, R.M. 1995b. Survival of *Lactobacillus delbrueckii* ssp. *bulgaricus* following spray-drying. *Journal of Dairy Science* 78(5): 1025–1031.

Termont, S., Vandenbroucke, K., Iserentant, D., Neirynck, S., Steidler, L., Remaut, E., and Rottiers, P. 2006. Intracellular accumulation of trehalose protects *Lactococcus lactis* from freeze-drying damage and bile toxicity and increases gastric acid resistance. *Applied and Environment Microbiology* 72(12): 7694–7700.

Thunell, R.K., Sandine, W.E., and Bodyfelt, F.W. 1984. Frozen starters from internal-pH-control grown cultures. *Journal of Dairy Science* 67(1): 24–36.

To, B.C.S. and Etzel, M.R. 1997. Spray drying, freeze drying, or freezing of three different lactic acid bacteria species. *Journal of Food Science* 62(3): 576–585.

Torino, M.I., Hebert, E.M., Mozzi, F., and Font de Valdez, G. 2005. Growth and exopolysaccharide production by *Lactobacillus helveticus* ATCC 15807 in an adenine-supplemented chemically defined medium. *Journal of Applied Microbiology* 99(5): 1123–1129.

Tsvetkov, T. and Brankova, R. 1983. Viability of micrococci and lactobacilli upon freezing and freeze drying in the presence of different cryoprotectants. *Cryobiology* 20(3): 318–323.

Tymczyszyn, E.E., Del Rosario, D.M., Gomez-Zavaglia, A., and Disalvo, E.A. 2007a. Volume recovery, surface properties and membrane integrity of *Lactobacillus delbrueckii* subsp. *bulgaricus* dehydrated in the presence of trehalose or sucrose. *Journal of Applied Microbiology* 103(6): 2410–2419.

Tymczyszyn, E.E., Gomez-Zavaglia, A., and Disalvo, E.A. 2007b. Effect of sugars and growth media on the dehydration of *Lactobacillus delbrueckii* ssp. *bulgaricus*. *Journal of Applied Microbiology* 102(3): 845–851.

Valdez, G.F. 2001. Maintenance of lactic acid bacteria. In: *Food Microbiology Protocols*, eds. J.F.T. Spencer and A.R. Spencer, 163–171. Totowa, NJ: Humana Press.

Van de Guchte, M., Serror, P., Chervaux, C., Smokvina, T., Ehrlich, S.D., and Maguin, E. 2002. Stress responses in lactic acid bacteria. *Antonie van Leeuwenhoek* 82(1–4): 187–216.

Van Reis, R. and Zydney, A. 2001. Membrane separations in biotechnology. *Current Opinion in Biotechnology* 12(2): 208–211.

Volkert, M., Ananta, E., Luscher, C., and Knorr, D. 2008. Effect of air freezing, spray freezing, and pressure shift freezing on membrane integrity and viability of *Lactobacillus rhamnosus* GG. *Journal of Food Engineering* 87(4): 532–540.

Walker, D.C., Girgis, H.S., and Klaenhammer, T.R. 1999. The groESL chaperone operon of *Lactobacillus johnsonii*. *Applied and Environmental Microbiology* 65(7): 3033–3041.

Wang, Y.C., Yu, R.C., and Chou, C.C. 2004. Viability of lactic acid bacteria and bifidobacteria in fermented soymilk after drying, subsequent rehydration and storage. *International Journal of Food Microbiology* 93(2): 209–217.

Woo, C., Lee, K., and Heo, T. 1999. Improvement of *Bifidobacterium longum* stability using cell-entrapment technique. *Journal of Microbiology and Biotechnology* 9(2): 8.

Wouters, J.A., Rombouts, F.M., De Vos, W.M., Kuipers, O.P., and Abee, T. 1999. Cold shock proteins and low-temperature response of *Streptococcus thermophilus* CNRZ302. *Applied and Environmental Microbiology* 65(10): 4436–4442.

Wright, C.T. and Klaenhammer, T.R. 1983. Survival of *Lactobacillus bulgaricus* during freezing and freeze-drying after growth in the presence of calcium. *Journal of Food Science* 48(3): 773–777.

Yao, A.A., Coulibaly, I., Lognay, G., Fauconnier, M.L., and Thonart, P. 2008. Impact of polyunsaturated fatty acid degradation on survival and acidification activity of freeze-dried *Weissella paramesenteroides* LC11 during storage. *Applied Microbiology and Biotechnology* 79(6): 1045–1052.

Ying, D.Y., Phoon, M.C., Sanguansri, L., Weerakkody, R., Burgar, I., and Augustin, M.A. 2010. Microencapsulated *Lactobacillus rhamnosus* GG Powders: Relationship of powder physical properties to probiotic survival during storage. *Journal of Food Science* 75(9): E588–E595.

Zhao, G. and Zhang, G. 2005. Effect of protective agents, freezing temperature, rehydration media on viability of malolactic bacteria subjected to freeze-drying. *Journal of Applied Microbiology* 99(2): 333–338.

PART III
Dairy Products Based on the Type of Fermentation

Pure Lactic Fermentations (Mesophilic, Thermophilic, and Probiotic Products)

8
Natural and Cultured Buttermilk

RAVINDER KUMAR, MANPREET KAUR, ANITA KUMARI GARSA, BHUVNESH SHRIVASTAVA, VELUGOTI PADMANABHA REDDY, AND ASHISH TYAGI

Contents

8.1	Introduction	203
8.2	Fermented or Cultured Milk: An Overview	205
8.3	Buttermilk	207
	8.3.1 Chemical Composition of Buttermilk	209
	8.3.2 Processing and Drying of Buttermilk	210
8.4	Cultured Buttermilk	210
	8.4.1 Starter Cultures Used for Cultured Buttermilk	210
	8.4.2 Production of Cultured Buttermilk	213
8.5	Uses of Buttermilk	215
8.6	Health Benefits of Buttermilk	220
8.7	Summary	222
References		222

8.1 Introduction

Milk has been a part of food since the dawn of civilization and also considered as a complete food for human beings. The fermentation of milk is also an ancient technique for the preservation of milk. It is largely used as a means to preserving highly perishable products like milk apart from imparting other benefits to the finished product. This process is carried out by the normal microbiota, while some of the fermentation is intentionally done by using specific microbes for a particular purpose. But the actual process and role of milk fermentation is yet to be completely understood. The inoculation of fresh milk with fermented milks was the process to maintain routine cultures for further use (Kerr and McHale 2001). But now, well-established starter cultures (i.e., lactobacilli, streptococcus, etc.)

are commercially available for the production of various fermented dairy foods, which are discussed elsewhere in this book.

Fermentation is not only used for the preservation of food but also imparts various flavors or tastes, forms, and desired sensory ambiances. Sometime, it also confers several therapeutic and nutritional properties in the finished product. Gradually, consumers have started to recognize the therapeutic and nutritional aspects of fermented foods, which increased the consumption as well as popularity of these foodstuffs (Kaur et al. 2014). The validation of health benefits (i.e., anti-obesity, anti-diabetes, anti-cholesterol level, anticancer, immune modulation, etc.) of some of the fermented milks and its products, for instance, yogurt, has changed the vision of consumers, thereby causing a concomitant rise in its production (Shiby and Mishra 2013).

Nowadays, several functional dairy foods (i.e., probiotic dahi, yogurt, yakult, low cholesterol milk, milk omega-3-milk, etc.), which have a beneficial effect on lifestyle diseases and disorders, are very common in the market. Apart from all these products, buttermilk is one of the classical examples of such products. Buttermilk has a fresh, piquant taste imparted by lactic acid bacteria (LAB), which remains as an integral part of buttermilk even after fermentation. It is not necessary that LAB present in buttermilk elicit probiotic attributes always, as sometimes microflora may also belong to the non-probiotics category. Biochemically, these microbes utilize sugar and yield acids, which results in sourness in buttermilk and also leads to a decrease in pH. Further, the decrease in the pH of milk affects the casein content, which causes the thickening of milk. Both sourness and thickening of buttermilk are imparted by lactic acid produced during the fermentation of milk. Traditional as well as cultured buttermilk has remained an excellent source of nutrition as it consists of good amounts of potassium, phosphorus, vitamin B_{12}, riboflavin, enzymes, protein, and calcium (Conway et al. 2014b).

Buttermilk has wide food applications and can be used for drinking purposes; it may be supplemented to produce sour cream or cultured butter. The name, however, is a bit misleading because it does not contain butter. It is a milky liquid leftover after the churning of cream, which is processed for the preparation of butter. The fat content in buttermilk remains very low. Buttermilk is often used in baking because of its special properties (for instance, sourness, enzymes, and microflora present in this by-product). It can enhance the flavor of various preparations, that is, dips and cakes.

Further, buttermilk can be used as a battering agent for frying of chops of pork and chicken (Sharma et al. 1998; Raval and Mistry 1999; Shukla et al. 2004; Patel and Gupta 2008). Various studies have also shown the therapeutic importance of the consumption of buttermilk (Larsson et al. 2008; Conway et al. 2013a, 2013b; Fuller et al. 2013; Fu et al. 2014). The objective of this chapter is to cover the different aspects of buttermilk production, its nutritional importance, its health benefits, and its use in different products.

8.2 Fermented or Cultured Milk: An Overview

Cultured milk is produced by the addition or fermentation of LAB and there are wide ranges of fermented milks, which have a number of common characteristics (Driessen and Puhan 1988). The cultures used in fermented milk have their own optimum temperature range for growth. To meet the requirements of cultured milk, that is, mildly acidic and slightly pricking, the content of lactic acid and carbohydrate has to be controlled during the manufacturing of the product. By cooling milk at a specific temperature and at a proper time, acidification can be limited, whereas the excess of carbon dioxide in the product at the end of fermentation can be removed by stirring or deaeration by vacuum. The final taste of cultured milk is the result of a mixture of compounds (e.g., diacetyl flavor is associated with butter and buttermilk) present in a certain ratio in the finished product.

Fermented milk production involves various steps, that is, pasteurization of milk, standardization and homogenization of milk, inoculation culture, breakdown of coagulum, cooling, and packaging. The pasteurization of milk used for the production of cultured milk is carried out to inhibit the pathogens and deactivate the native substances, which are inhibitory to LAB. Additionally, heat treatment denatures whey proteins, which improves the texture of the final milk product. Whey proteins should be denatured, which results in a coagulum that can be stirred easily to realize a smooth and viscous product (Snoeren et al. 1981). This can be achieved by heating milk at 80°C–85°C for 2–5 min (Hillier and Lyster 1979).

Ultra-high-temperature-sterilized milk and other high-temperature-treated milks result in lower viscosity, age-thickening, and a cooked flavor in cultured milks. The lowering of pasteurization temperature of milk results in decreased firmness and retarded acidification. Therefore, homogenization of milk at 55°C and 20 MPa is sufficient and good for distribution of fat. Decreasing solids and not fat content in milk results in a taste difference as the product may turn *flat* and *watery*; buttermilk with this defect is said to be *astringent*. Increasing milk solids-not-fat (SNF) leads to a *full* taste, higher viscosity, and a stable cultured milk without wheying-off during storage. The inoculum added causes production of acid, which leads to the formulation of coagulum. The coagulum is generally stirred to get smooth fermented milk. The detailed steps involved in fermented milk production are shown in Figure 8.1. Cultured milks consist of all required nutrition with easy digestibility

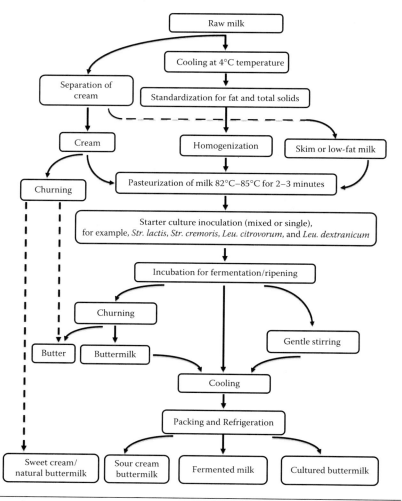

Figure 8.1 Production of fermented milk, cultured buttermilk, and natural buttermilk.

and also impart health benefits. This chapter covers the various aspects of buttermilk, one of the popular forms or examples of cultured milk.

8.3 Buttermilk

Natural buttermilk is a leftover liquid by-product made during the churning of butter. It is a very famous fermented drink in India as well as in other Middle Eastern countries. Cultured buttermilk is probably the easiest fermented milk product to produce but still the exact quantity of production of buttermilk is not assessed. However, the quantity of buttermilk production can be estimated on the basis of production of butter. Approximately, 6.5%–7.0% of total milk produced worldwide is used for the preparation of butter that yields high amounts (around 3.2 million tons/annum) of buttermilk as a by-product. In India, sour buttermilk (lassi) is also obtained during the preparation of butter (makkhan) and curd (dahi). The utility of buttermilk solids was unnoticed and untouched for a long time. However, following a shift in focus of investigation, the consumption of buttermilk, its functions, and its therapeutic attributes are currently explored worldwide. Still, this by-product needs more attention for its many uses; buttermilk-derived health-beneficial products can be developed as it is present in bulk amounts all over the world.

Usually, four types of buttermilk are produced, as shown in Table 8.1 and Figure 8.1. Natural buttermilk is very low in fat (since most of the fat goes to the butter part). It can be consumed as it is or added to recipes in place of water for a nutritional boost.

Cultured buttermilk is another type of buttermilk obtained by intentional acidification of skim milk done by buttermilk starter cultures. It is similar to yogurt in the sense that it is cultured using live beneficial bacteria and can be consumed as a thick and creamy beverage (Conway et al. 2014a, 2014b).

Sweet cream buttermilk (SCBM) is produced from churning of cream. Churning of cream results in the separation of butter and an aqueous phase called SCBM. Generally, cream is not ripened in this case. SCBM has high fat content as compared to skim milk, which can be decreased by centrifuging it or by ultrafiltration (Conway et al. 2014a, 2014b). SCBM also consists of huge amounts of proteins, which are drawn by churning from the fat globule–milk serum interface

Table 8.1 Different Types of Buttermilk and Their Properties

SWEET CREAM BUTTERMILK	SOUR CREAM BUTTERMILK	CULTURED BUTTERMILK	COMMERCIAL BUTTERMILK
Result of the churning process of cream separated from milk.	End product of churning process of ripened cream.	Produced by culturing of skim milk.	Commercial buttermilk is most widely available in grocery stores.
Produced from fresh, or sweet, milk, which is converted to cream that convert into butter and buttermilk and has a taste similar to regular skim milk.	Made from raw, unpasteurized sour milk. The milk is allowed to sour naturally prior to churning.	Produced by addition of bacterial culture, which provide a rich, fuller and tangier flavor.	Milk is not used to make butter, instead manufactures add bacterial cultures to skim or low-fat cow's milk and let it mature.
Exact taste of the sweet cream milk depends on the flavor of the original milk. For example, goat's milk has a naturally more pungent taste than cow's milk.	Sour cream buttermilk has a tart taste, similar to yogurt or sour cream. As with sweet cream milk, the source of the milk provides slight variations in taste.	The resulting buttermilk has a similar tangy flavor.	Over time, the milk thickens and develops its characteristic sour taste. Commercial manufacturers also make powdered buttermilk. They use the same process as with wet milk, then remove the liquid.

(King 1955). Apart from their ability to release biologically active peptides (Roesch and Corredig 2002), these proteins also contribute as a mixture of glycol-phospholipids in buttermilk. The phospholipids content in SCBM is around nine times greater as compared to skim milk (Table 8.2). SCBM lacks short-chain fatty acids, principally consisting of C_{16} (palmitic) and higher acids. Around 40% of fatty acids

Table 8.2 Composition and Physiochemical Properties of Sweet Cream Buttermilk and Skim Milk

CHARACTERISTIC	SWEET CREAM BUTTERMILK	SKIM MILK
Ash (%)	0.73	0.80
Fat (%)	0.60	0.09
Lactose (%)	4.84	5.25
Titratable acidity	0.13	0.15
Total proteins (%)	3.70	4.30
Total solid (%)	9.75	10.80
Phospholipids (mg)	78.60	8.50
pH	6.85	6.70

present in buttermilk are saturated by dry weight. Non-conjugated di- to penta-unsaturated acids make up the rest of the volume (Garton 1963). Additionally, the phospholipids content of buttermilk comprises cephalin, lecithin, and sphingomyelin in same amounts along with small quantities of cerebrosides. Buttermilk also varies in physical and chemical properties as compared to skim milk (Table 8.2). For instance, it has low acidity and high viscosity in comparison with skim milk. Such a difference in properties makes buttermilk suitable for numerous applications in the dairy sector, but the chemical composition of buttermilk and skim milk remain almost the same when produced under standard parameters or conditions.

8.3.1 Chemical Composition of Buttermilk

Buttermilk has emulsion and flavor-enhancing ability, which makes it a key dairy component in several food applications. The composition of sweet and cultured buttermilk is similar to skim milk. Additionally, the composition of whey buttermilk is also similar to whey. But the fat content is high in buttermilk (6%–20%) compared to skim milk (0.3%–0.4%) or whey (Sodini et al. 2006). The chemical composition of buttermilk depends largely on the amount of water added to cream. However, buttermilk is almost similar to skim milk in composition, as mentioned earlier, when produced under standard conditions (Table 8.2). The major difference between sour buttermilk and SCBM pertains to its titratable acidity. The titratable acidity is higher in sour buttermilk (>0.15%); it may be more than 1% sometimes, whereas in SCBM it lies between 0.10% and 0.15%. On the other hand, natural buttermilk has wider variations in its composition as it varies with milk quality used for the preparation of curd and the amount of water added in between the churning. Yet, on the average, it consists of total solids (4%), lactose (3%–4%), lactic acid (1.2%), protein (1.3%), and fat (0.8%).

Buttermilk contains high amounts of calcium, which contributes significantly to its health benefits. The human body requires 1,000 mg of calcium per day. Low-fat buttermilk contains around 28% calcium. Consumption of 500 ml buttermilk can fulfill the calcium requirement of the body. Riboflavin is another crucial content of buttermilk. Additionally, buttermilk also boosts protein intake.

8.3.2 Processing and Drying of Buttermilk

SCBM can be used in dry form for various food applications. SCBM is more suitable for processing as it has higher heat stability and its constituent composition is similar to skim milk (Bratland 1972). SCBM remains similar in processes of separation, clarification, pasteurization, concentration, and drying at high temperatures. The processes of spray drying and concentration for SCBM are similar to those of skim milk powder (SMP). Spray drying of buttermilk is generally carried out at 185°C–195°C, and the drying process to concentrate the buttermilk is carried out till the 40%-50% solid in end product has been achieved. The major difference between SCBM and SMP is the concentration of total lipid and density. The total lipid consisting of phospholipids content remained high bulk in SCBM than SMP, while bulk density is found low. It is generally observed that during storage, high lipid or fat concentration can decrease the shelf life of milk powder. But the high phospholipids content present in SCBM reduces the chances of oxidation in powder.

8.4 Cultured Buttermilk

Cultured buttermilk is the sour end product obtained after fermentation of skim or partially skim milk inoculated with LAB cultures.

8.4.1 Starter Cultures Used for Cultured Buttermilk

Microorganisms that are intentionally supplemented into milk for desired fermentation to produce fermented milk products under controlled conditions are called starter cultures. The use of starters has been tremendously important as it decides the quality and nutritional value of the desired end product. But on the other hand, it has diminished the diversity of fermented dairy products (Kaur et al. 2014). Buttermilk starters contain certified organic milk and live active cultures. Large portion of these active cultures (e.g., *Lactococcus, Lactobacillus, Streptococcus,* and *Leuconostocs*) belong to LAB (Table 8.3). Additionally, non-lactic starters can also be used as a co-inoculant with LAB for the production of buttermilk.

Table 8.3 Microorganisms Used as Starter Cultures for Preparation of Buttermilk

MICROORGANISM(S)	TEMPERATURE FOR GROWTH (°C)
HETEROFERMENTATIVE	
Leu. mesenteroides	25
Lb. brevis	30
Lb. kefir	32
HOMOFERMENTATIVE	
Lb. lactis subsp. lactis biovar. diacetylactis	25
Lb. casei	30
Lb. lactis subsp. cremoris	30
Lb. lactis subsp. lactis	30
Lb. acidophilus	37
Lb. delbueckii subsp. lactis	40
Str. thermophilus	40
Lb. helveticus	42
Lb. delbueckii subsp. bulgaricus	45

Starter cultures can be used as single, mixed, and multiple strains depending upon the type of products to be prepared for a specific purpose. The purity and activity of starter cultures define their ability to perform functions efficiently. An ideal starter culture should have some characteristics, for example, should be quick and steady in acid production, should produce a product with fine and clean lactic flavor, and should not produce any pigments, gas, off-flavor, and bitterness in the finished products.

The major role of starter cultures during the fermentation of milk are the production of lactic acid and a few other organic acids, for example, formic acid and acetic acid, changes in body and texture in final products followed by coagulation of milk, production of flavoring compounds, such as diacetyl, acetoin, and acetaldehyde, and production of antibacterial substances in the finished product.

Generally, buttermilk products (i.e., sour and cultured buttermilk) are produced by different types of starter cultures. These cultures are classified on the basis of their temperature and fermentation of glucose for growth purposes, for instance, mesophilic, thermophilic, homofermentative, and heterofermentative bacteria.

Products made by use of mesophilic lactic starter cultures (optimal temperature 30°C–40°C) may use one of the following types of microorganisms: O (homofermentative *Lactococcus lactis* subsp. *cremoris* and *Lc. lactis* subsp. *lactis*), D (microbes of O types and *Streptococcus lactis* subsp. *lactis* var. diacetylactis), L (in addition to the O type bacteria, it contains *Leuconostoc mesenteroides* subsp. *mesenteroides*), and LD (combination of *Str. lactis* subsp. *lactis* var. *diacetylactis* and *Leu. mesenteroides* subsp. *mesenteroides*).

Thermophilic bacteria work over an optimal temperature range of 50°C–60°C and digesters are usually operated as close as possible to 55°C. This offers the advantages of faster reaction rates and better pathogen-killing power/capacity as compared to mesophilic digestion, which leads to shorter retention times.

Homofermentative bacteria consume or ferment glucose that yields lactic acid as the primary end metabolite. In various dairy culture applications, *Lactococcus* ssp. is commonly used as a starter culture, when the quick lactic acid production or low pH is desirable. The sugar fermentative pathway of homofermentative bacteria is shown in Figure 8.2a.

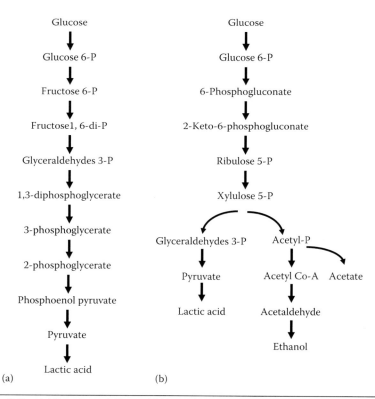

Figure 8.2 (a) Homofermentative and (b) heterofermentative pathways of lactic acid bacteria.

Heterofermentative bacteria also consume glucose; common end products are lactic acid, ethanol, and carbon dioxide (Figure 8.2b). The common identification of heterofermentative bacteria is generally carried out by evolved gases. The application of heterofermentative LAB as starter culture in dairy products is not common. Yet, these are not rare in dairy products.

8.4.2 Production of Cultured Buttermilk

Skim or low-fat milk is used for the production of buttermilk. First, milk is pasteurized at 82°C–85°C for 2–5 min as per need. The major objective of pasteurization is killing of bacteria that are present naturally in milk and denaturing the milk protein to decrease the wheying off. Afterward, milk is allowed to cool to 22°C followed by the addition of required starter cultures, for example, *Lc. lactis* subsp. *lactis*, *Lc. lactis* subsp. *cremoris*, *Leu. citrovorum*, and *Leu. dextranicum*. These cultures are desirable candidates for imparting buttermilk's unique sour flavor by increasing acidity if butter or buttermilk is produced at an industrial or commercial level (Figure 8.3). The starter cultures *Lc. lactis* subsp. *cremoris*, *Leu. citrovorum*, and *Leu. dextranicum* are typically used to generate the flavor in butter, while *Lc. lactis* subsp. *lactis* is associated with the production of lactic acid, which causes the typical tangy flavor of cultured buttermilk. Hence, LAB are common and dominating groups of bacteria, which are required for fermentation of milk to produce buttermilk. Additionally, LAB are also naturally accepted and generally recognized as safe for human consumption (Aguirre and Collins 1993). During fermentation of milk, growth and metabolic activities of LAB cause a few changes in milk, which results in chemical and physical modifications in milk as shown in Table 8.4. After the fermentation, coagulum is stirred, packed, and stored at refrigerated temperatures. The production scheme of buttermilk is shown in Figures 8.1 and 8.3.

Moreover, probiotics are very popular throughout the world due to their health benefits, so adding probiotics to starter cultures or after the production of buttermilk is often carried out these days in addition to the conventional method (El-Fattah and Ibrahim 1998; Rodas et al. 2002; El-Shafei 2003).

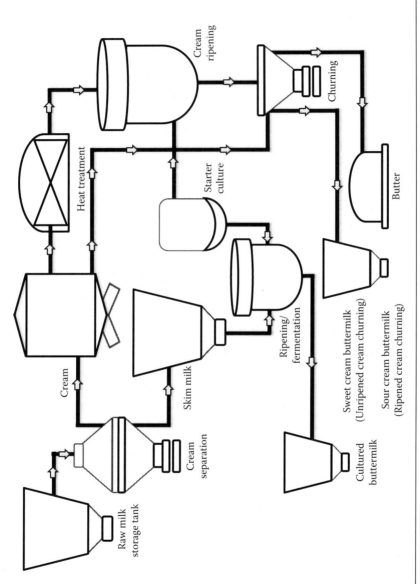

Figure 8.3 Large-scale production scheme of buttermilk.

Table 8.4 Changes in the Constituents during Milk Fermentation

COMPONENTS	FORMULATION OF COMPONENTS
Breakdown of fat	Flavored compounds and acetic, propionic, butyric, isovaleric, caproic, caprylic, and capric acids generated due to breakdown of fats
Breakdown of protein, for example, casein	Generation of amino acids (serine, glutamic acid, proline, valine, leucine, isoleucine, and tyrosine) due to breakdown of proteins
Breakdown of vitamins, carotenoids	Simple vitamins, B_2, B_6, and B_{12}
Lactose lytic	Lactic acid, galactose, and glucose produced due to breakdown of lactose

8.5 Uses of Buttermilk

Apart from its use for consumption purposes, buttermilk has various applications in several food product formulations, especially with reference to the fermented products. Traditional buttermilk is itself a popular drink in many countries (Patel and Gupta 2008). Yet, addition of fruit juices or fruit pulps is an attractive avenue to utilize buttermilk as well to increase the earning of the associated sector. Several types of refreshing drinks of buttermilk have been prepared using litchi, apple, banana, mango, and guava pulp (Shukla et al. 2004). Furthermore, the addition of cashew (15%) and kokum syrup (10%) has been employed to prepare flavored buttermilk (Kankhare et al. 2005; Patel and Gupta 2008). Rao and Kumar (2005) developed a spray dried buttermilk powder with mango pulp. A product with desired sensory effect was obtained when buttermilk (80%) was blended with mango pulp (20%).

Moreover, large-scale dairy plants and industries blend pure SCBM in whole milk for fluid milk supply or in skim milk for processing, that is, drying. The blending of SCBM in buffalo milk to prepare toned milk usually increases the various properties of end products such as palatability, heat stability, viscosity, and decreased curd tension (Pal and Mulay 1983; Patel and Gupta 2008). Flavor milks and beverages can also be prepared with use of SCBM. Moreover, skim milk and SCBM powder blends are commonly used for reconstitution applications.

Buttermilk can be used in many food formulations in place of SMP. Generally, buttermilk is used to enhance the texture, thickness, and reduction of fats. Several examples are discussed here. For instance, the thick and smooth texture of yogurt, which is a highly desirable characteristic, can be achieved by increasing total solids in the yogurt mix. Conventionally, to achieve the required total solid content, the milk is boiled till it reduces by up to two-thirds of its original volume. Boiling of milk is not regarded as a suitable approach, and hence, blending of SMP in milk is the best alternative to increase total solids. SCBM has been used as a substitute of SMP. It has been demonstrated that substitution of SMP up to 50% by buttermilk for low-fat yogurt preparation yielded the end product which was closely similar to control product (Vijayalakshmi et al. 1994). Further, addition of 4.8% SCBM in low-fat yogurt imparted a smooth soft texture (Trachoo and Mistry 1998; Patel and Gupta 2008). Moreover, the addition of SCBM also reduces syneresis in yogurt as compared to other constituents (Guinee et al. 1995). Romeih et al. (2014) added buttermilk powder with reduced fat content. The addition of buttermilk powder caused a decline in pH while it improved the functionality of the yogurt gel in terms of water-holding capacity and exhibited the most desirable organoleptic attributes. The investigation proved that buttermilk powder can be used as a valuable alternative in fat-free yogurt production with a source of extra protein.

Buttermilk also plays a potential role in cheese formulations. The process of formulation of hard cheeses, for example, Cheddar and Gouda carried out with low-fat milk as high fat content imparts soft texture to cheese body. Generally, skim milk is added to bulk milk to adjust the fat content so that hard body cheese can be obtained. In various dairy applications, SCBM has been used to replace the skim milk. Therefore, SCBM has also been used to prepare hard cheese in place of skim milk. The major problem with the supplementation of SCMB is the presence of the high fat membrane material that yields soft body cheese. However, the blending of SCBM up to some extent can result in the desirable hardness in Cheddar cheese (Joshi and Thakar 1996a, 1996b). To overcome the problem of high fat content, ultrafiltrated SCBM is generally used. The ultrafiltered SCBM supplementation is reported to give improved quality (body and texture) to cheese in comparison with the control (Mistry et al. 1996). Using the same strategy, low-fat Cheddar cheese was also produced. Interestingly, this cheese consisted of less fat (14.5%) as compared to control (15.1%), when milk was blended with 5% ultrafiltered SCBM (Raval and Mistry 1999;

Patel and Gupta 2008). Further, ultrafiltered buttermilk was used to prepare reduced fat Mozzarella cheese (Poduval and Mistry 1999). SCBM (30%–40%) was also utilized to prepare cottage cheese by replacing whole milk; the end product was an enhanced flavored soft cheese (Shodjaodini et al. 2000). Moreover, Gokhale et al. (1999) successfully blended SCBM up to 75% to lower down the water content in processed cheese and no significant alteration was observed in cheese composition. Concentrated buttermilk (CBM) was employed for making processed cheese, and some of its physicochemical properties were also studied. The observations found that CBM can be used to make functional processed cheese and can also be supplemented as an ingredient in other dairy products (Doosh et al. 2014).

Paneer or Indian cheese is generally prepared with addition of acids (e.g., citric acid) in milk, which causes coagulum of milk; that integrated mass of milk coagulum is known as paneer. Prior to preparation of paneer, standardization of milk is required, which is commonly done with the addition of skim milk. But it has been observed that addition of SCBM in buffalo milk in place of skim milk enhanced the yield (around 1%) of paneer without causing any changes in texture and sensory properties (Pal and Garg 1989; Patel and Gupta 2008). Additionally, buttermilk (10%) has been used in place of buffalo milk for the preparation of fried paneer, which showed that the uptake of oil was high and the end product revealed soft texture compared to the control (Sharma et al. 1998). On the other hand, it was observed that up to 50% cultured buttermilk incorporation in paneer whey can result in a good quality drink. The chemical composition of blending results in around 8.31% total solids, 2.20% protein, 1.10% fat, 4.40% lactose, and 0.55 ash. This cultured buttermilk was accepted very well and its shelf life was 5 days under refrigerated conditions (Ghanshyambhai et al. 2014).

Buttermilk powder can be used along with milk for the preparation of curd. Curd or dahi (Indian curd) is prepared from milk with the addition of a mixture of starter cultures. Generally, buffalo milk is used for this preparation. Nevertheless, the addition of 1%–2% SMP with milk enhances the texture body of dahi prepared from buttermilk (Shreshtha and Gupta 1979; Patel and Gupta 2008). The higher amount of SCBM in buffalo milk yields soft body curd.

The preparation of various Indian sweets involves buttermilk and buttermilk powder. For example, chhana is unripened curd cheese prepared from buffalo milk and it is used for formulation of several sweets, for example, rasgulla. Sometimes, compositional changes in buffalo milk yield hard and greasy chhana, which are undesirable defects in this product. To reduce such defects, several suggestions, for instance, addition of salts (e.g., sodium citrate), dilution of milk (20%–30% water), low temperature coagulation, and homogenization, have been proposed (Rajorhia and Sen 1988; Patel and Gupta 2008). Along the same lines, the addition of SCBM (60%; on the total solid basis) buffalo milk showed acceptable preparation of chhana (Kumar 2006). Similarly, the formulation of rasgulla needs chhana prepared from cow or buffalo milk. Usually the soft body and texture of chhana is desirable for rasgulla, but buffalo milk yields hard body chhana. To overcome this, buttermilk continues to be the favorite constituent to increase the softness of chhana. It has been seen that the blending of SCBM in buffalo milk at a ratio of 60:100 with other constituents (i.e., arrowroot, maida, and semolina) produced better quality rasgulla (Kumar 2006; Patel and Gupta 2008).

Sandesh is another Indian sweet that is made from chhana. This sweet is made from cow milk chhana due to its soft body and texture (Sen and Rajorhia 1990). However, trials have been carried out with chhana prepared from buffalo milk. Similar to rasgulla, addition of SCBM in buffalo milk yields better quality sandesh (Kumar 2006). The blending ratio of SCBM and buffalo remained 60:100. Moreover, the preparation of basundi can be carried out with the help of SCBM solids in place of whole buffalo milk solids. However, the 100% replacement of buffalo milk solids with SCBM powder causes a decrease in the lactose and ash contents in end product. It also showed adverse effects on the physicochemical properties (Patel and Upadhyay 2004; Patel and Gupta 2008). Nevertheless, the preparation of basundi with 25% of SCBM yielded a good product, which shows the potential role of SCBM in the formulation of basundi.

Chakka is made from dahi prepared from buffalo milk. Basically, chakka is an intermediate product that is used to prepare shrikhand. When whey is drained from dahi, the leftover semi-solid mass is known as chakka, which is sweetened with sugar and named as shrikhand. The formulation of chakka and shrikhand has also been carried out with the supplementation of SCBM to enhance the properties of the end product. It has been reported that the addition of SCBM (50%)

imparted enhanced flavor, body, and texture, and the end product did not show any alteration in chemical composition. The addition of 15% SCBM yielded the same quality shrikhand as with buffalo skim milk (Karthikeyan et al. 1999).

The production of ice cream and other frozen desserts also involves the use of buttermilk powder. SCBM has been used as a better alternate for SMP. In a study, it has been shown that the mixture of spray dried whey and dried SCBM powder at a ratio of 50:50 yielded good quality ice cream as compared to control (Tirumalesha and Jayaprakasha 1998).

Probiotics are popular foods and their market is increasing day-by-day worldwide. The same is true for buttermilk as well. Trials on probiotic-based buttermilk successfully demonstrated the utility of this drink. The health benefits of such probiotic buttermilk have been validated in Japanese quail hens (El-Fattah and Ibrahim 1998). In this study, buttermilk fermented with *Lb. acidophilus* was fed to hens for 100 days; there was a significant decrease in low density cholesterol levels in serum and liver, whereas there was an increase observed in high density lipoproteins. In another investigation, buttermilk supplemented with *Lb. reuteri* (1%) did not show any change in composition and sensory attributes (Rodas et al. 2002). Other probiotic strains have also been used successfully for the preparation of probiotic buttermilk with potential health attributes (El-Shafei 2003).

Apart from its traditional dairy uses, buttermilk powder can be used to enhance the functional, nutritional, and organoleptic properties of bread. Büşra Madenci and Bilgiçli (2014) added buttermilk powder in leavened and unleavened flat bread dough at different levels, and the supplementation of buttermilk powder enhanced the dough properties (i.e., resistance to extension and maximum resistance and dough stability). Additionally, it also improves the protein and mineral contents with better taste and odor.

Buttermilk improves the flavor and aroma of cakes and dips. It also works as a battering agent for frying of chicken and chops of pork (Raval and Mistry 1999; Shukla et al. 2004). In the preparation of some of the Indian dishes, buttermilk is also added to initiate fermentation in raw materials of bhature, jalebi, and idli.

Moreover, it has been noticed that buttermilk powder has heat resistance ability and in combination with glucose syrup improves oxidative stability. On pH adjustment and heat treatment, buttermilk revealed superior encapsulating properties compared to skim milk for fish oils (Augustin et al. 2014).

8.6 Health Benefits of Buttermilk

The health benefits of buttermilk have been evidenced by many studies. Clinical trials studying the effects of buttermilk on various diseases (e.g., cholesterol reduction, blood pressure reduction, antiviral effects, and anticancer) have showed positive effects.

Lactose intolerance or inability to digest lactose in adults is common worldwide. People with lactose intolerance are not able to consume whole milk as consumption may cause abdominal pain, bloating and diarrhea, and flatulence. Inability to digest lactose in adults occurs due to the absence or presence of lactase in low levels in small intestine. However, the population suffering from lactose intolerance can consume buttermilk as an alternative for whole milk. The microflora present in buttermilk have the ability to metabolize lactose (Figure 8.2). The metabolic activity of microbes converts lactose in lactic acid, which is easy to digest. Additionally, buttermilk also plays an assistive role in the removal of undesired stomach acid (which causes indigestion and heartburn) by creating a layer on the lining of the stomach and moving the acid through esophagus.

Moreover, the interaction of buttermilk with resveratrol has been shown to be helpful in the delivery of this component (Ye et al. 2013). In fact, the binding ability of whole buttermilk with resveratrol or a complex formulation enhances the aqueous solubility of resveratrol, which is quite helpful in the delivery of this component in system via the natural route. Further, it has also been suggested that hydrolyzed proteins of buttermilk are a rich source of natural antioxidants (Conway et al. 2013b).

Buttermilk edible films are also nowadays used, and the effect of incorporating different ratios of both non-heated and heated (95°C) buttermilk to corn starch films was analyzed in terms of its structural, mechanical, barrier, optical, and bioactive properties. It was observed that only films formulated with heated buttermilk exhibited antioxidant activity, probably due to the release of the antioxidant peptides during thermal treatment of proteins (Moreno et al. 2014).

Turmeric is good source of phenolic components, for example, curcuminoids (curcumin, demethoxycurcumin, and bisdemethoxycurcumin), which have anti-inflammatory and antioxidant activities and other properties of clinical importance. Sometimes, poor stability of these phenolic components in foodstuffs discourages their potential health attributes. However, it has been seen that buttermilk has a good capability to stabilize these phenolic components compared to buffers. So, buttermilk can be used to deliver these turmeric components in the system to impart health benefits (Fu et al. 2014).

Further, it has been observed that consumption of cultured buttermilk can decrease the chances of bladder cancer (Larsson et al. 2008). It has also been suggested that buttermilk intake can reduce cholesterol levels by inhibiting the assimilation of cholesterol in the intestinal tract (Conway et al. 2013a).

Consumption of buttermilk can reduce blood pressure in moderately hypercholesterolemic individuals. It was seen that arterial blood pressure and systolic blood pressure get reduced with continous buttermilk consumption. The concentration of angiotensin-I converting enzyme also decreased in plasma, while levels of angiotensin-II and aldosterone remained unchanged (Conwey et al 2014a). In another study, Fuller et al. (2013) showed anti-rotavirus activity using the milk fat globule membrane obtained from buttermilk.

Buttermilk is also quite helpful in losing or controlling weight as it is low in calories when compared with whole milk. It is estimated that buttermilk has only 50% of the calories and fat found in milk. The routine consumption of buttermilk leads to a decrease in weight. Individuals can replace whole milk with buttermilk to gain health attributes (Conway et al. 2014b). Although the health benefits of fermented milk products are great, research is still going on to investigate the health benefits of products that are either indigenous or commercial. In this regard, buttermilk is one of the most important cultured products. The cultured or fermented products also have additional advantages as compared to the non-fermented products because fermentation itself imparts the nutritional, therapeutic, functional, and other qualities to a particular product, which is after all the final requirement. Nowadays, many human problems related to health are solved by these cultured products along with cultured buttermilk, discussed above. Hence, cultured buttermilk can be used as a promising key to overcome some health-related problems.

8.7 Summary

Buttermilk is a traditional by-product and famous as a dairy drink, which is used in various formulations. Yet this product is not well explored for different roles in dairy and other associated food sectors at commercial level. Many workers are investigating to find out the role of buttermilk as health foods; for example, the functions of buttermilk phospholipids have been related with many health benefits. However, the validation of buttermilk benefits (i.e., anticancerous and anticholestrolic activities) is needed in detail. To enhance buttermilk quality and health attributes, the appropriate selection of microbes, production strategies, and suitable storage protocols have to be adopted or developed. Moreover, innovations in buttermilk, for example, probiotic buttermilk, are ongoing; yet, more concerted efforts are required to prove the potential of this dairy product.

References

Aguirre, M. and Collins, M.D. 1993. Lactic acid bacteria and human clinical infection. *Journal of Applied Bacteriology* 75(2): 95–107.

Augustin, M.A., Bhail, S., Cheng, L.J., Shen, Z., Øiseth, S., and Sanguansri, L. 2014. Use of whole buttermilk for microencapsulation of omega-3 oils. *Journal of Functional Foods*. doi:10.1016/j.jff.2014.02.014.

Bratland, A. 1972. Production of milk. *Dairy Science Abstract* 40: 1294.

Büşra Madenci, A. and Bilgiçli, N. 2014. Effect of whey protein concentrate and buttermilk powders on rheological properties of dough and bread quality. *Journal of Food Quality* 37(2): 117–124.

Conway, V., Couture, P., Gauthier, S., Pouliot, Y., and Lamarche, B. 2014a. Effect of buttermilk consumption on blood pressure in moderately hypercholesterolemic men and women. *Nutrition* 30: 116–119.

Conway, V., Couture, P., Richard, C., Gauthier, S.F., Pouliot, Y., and Lamarche, B. 2013a. Impact of buttermilk consumption on plasma lipids and surrogate markers of cholesterol homeostasis in men and women. *Nutrition, Metabolism and Cardiovascular Diseases* 23(12): 1255–1262.

Conway, V., Gauthier, S.F., and Pouliot, Y. 2013b. Antioxidant activities of buttermilk proteins, whey proteins, and their enzymatic hydroly sates. *Journal of Agricultural and Food Chemistry* 61(2): 64–72.

Conway, V., Gauthier, S.F, and Pouliot, Y. 2014b. Buttermilk: Much more than a source of milk phospholipids. *Animal Frontiers* 4(2): 44–51.

Doosh, K.S., Alhusyne, L.A., and Almosawi, B.N. 2014. Utilization of concentrated buttermilk in functional processed cheese manufacturing and studying some of its physicochemical properties. *Pakistan Journal of Nutrition* 13(1): 33–37.

Driessen, F.M. and Puhan, Z. 1988. Technology of mesophilic fermented milk. *Bulletin of IDF* 227: 75–80.

El-Fattah, A.M.A. and Ibrahim, F.A.A. 1998. Effect of feeding fresh and fermented buttermilk on serum and egg yolk cholesterol of Japanese quail hens. *Bulletin of Faculty of Agriculture, University of Cairo* 49(3): 355–368.

El-Shafei, K. 2003. Production of probiotic cultured buttermilk using mixed cultures containing polysaccharide producing *Leuconostoc mesenteroides*. *Egyptian Journal of Dairy Science* 31(1): 43–60.

Fu, S., Shen, Z., Ajlouni, S., Ng, K., Sanguansri, L., and Augustin, M.A. 2014. Interactions of buttermilk with curcuminoids. *Food Chemistry* 149: 47–53.

Fuller, K.L., Kuhlenschmidt, T.B., Kuhlenschmidt, M.S., Jiménez-Flores, R., and Donovan, S.M. 2013. Milk fat globule membrane isolated from buttermilk or whey cream and their lipid components inhibit infectivity of rotavirus in vitro. *Journal of Dairy Science* 96(6): 3488–3497.

Garton, G.A. 1963. The composition and biosynthesis of milk lipids. *Journal of Lipid Research* 4: 237–254.

Ghanshyambhai, M.R., Balakrishnan, S., and Aparnathi, K.D. 2014. Standardization of the method for utilization of paneer whey in cultured buttermilk. *Journal of Food Science and Technology*. doi:10.1007/s13197-014-1301-2.

Gokhale, A.J., Pandya, A.J., and Upadhyay, K.G. 1999. Effect of substitution of water with sweet cream buttermilk on quality of processed cheese spread. *Indian Journal of Dairy Science* 52(4): 256–261.

Guinee, T.P., Mullins, C.G., Reville, W.J., and Cotter, M.P. 1995. Physical properties of stirred-curd unsweetened yoghurts stabilized with different dairy ingredients. *Milchwissenschaft* 50(4): 196–200.

Hillier, R.M. and Lyster, R.L.J. 1979. Whey protein denaturation in heated milk and cheese whey. *Journal of Dairy Research* 46: 95–102.

Joshi, N.S. and Thakar, P.N. 1996a. Utilization of buttermilk in manufacture of buffalo Cheddar cheese—Standardizing the conditions for curd setting. *Indian Journal of Dairy Science* 49(5): 350–352.

Joshi, N.S. and Thakar, P.N. 1996b. Utilization of buttermilk Cheddar cheese made using butter milk solids in processed cheese manufacture. *Indian Journal of Dairy Science* 49(5): 353–355.

Kankhare, D.H., Joshi, S.V., Toro, V.A., and Dandekar, V.S. 2005. Utilization of fruit syrups in the manufacture of flavoured buttermilk. *Indian Journal of Dairy Science* 58(6): 430–432.

Karthikeyan, S., Desai, H.K., and Upadhyay, K.G. 1999. Effect of varying levels of total solids in sweet cream buttermilk on the quality of fresh shrikhand. *Indian Journal of Dairy Science* 52(2): 95–99.

Kaur, M., Kumar, H., Kumar, H., and Puniya, A.K. 2014. Recent trends in production of fermented dairy and food products. In: *Dairy and Food Processing Industry Recent Trends-Part I*, ed. B.K. Mishra, 141–161. Delhi, India: Biotech Books, Chawla Offset Printers.

Kerr, T.J. and McHale, B.B. 2001. *Applications in General Microbiology: A Laboratory Manual*. Winston-Salem, NC: Hunter Textbooks.

King, N. 1955. *The Milk Fat Globule Membrane*. Farnham Royal: Commonwealth Agricultural Bureau.

Kumar, J. 2006. *Admixture of Buttermilk to Buffalo Milk for Production of Chhana and Chhana Based Sweets.* PhD dissertation. Karnal, India: National Dairy Research Institute.

Larsson, S.C., Andersson, S.O., Johansson, J.E., and Wolk, A. 2008. Cultured milk, yogurt, and dairy intake in relation to bladder cancer risk in a prospective study of Swedish women and men. *American Journal of Clinical Nutrition* 88: 1083–1087.

Mistry, V.V., Metzger, L.E., and Maubois, J.L. 1996. Use of ultrafiltered sweet buttermilk in the manufacture of reduced fat Cheddar cheese. *Journal of Dairy Science* 79(7): 1137–1145.

Moreno, O., Pastor, C., Muller, J., Atarés, L., González, C., and Chiralt, A. 2014. Physical and bioactive properties of corn starch—Buttermilk edible films. *Journal of Food Engineering* 141: 27–36.

Pal, D. and Garg, F.C. 1989. Studies on utilization of sweet cream buttermilk in the manufacture of paneer. *Journal of Food Science and Technology* 26: 259–264.

Pal, D. and Mulay, C.A. 1983. Influence of buttermilk solids on the physicochemical and sensory properties of market milks. *Asian Journal of Dairy Research* 2: 129–135.

Patel, A.A. and Gupta, V.K. 2008. *Course Compendium: Technological Advances in the Utilization of Dairy By-Products.* Karnal, India: National Dairy Research Institute. http://www.dairyprocessingcaft.com/wp-content/uploads/2012/05/Byproducts-2008.pdf. Accessed August 18, 2014.

Patel, H.G. and Upadhyay, K.G. 2004. Substitution of milk solids by sweet cream buttermilk solids in the manufacture of basundi. *Indian Journal of Dairy Science* 57(4): 272–275.

Poduval, V.S. and Mistry, V.V. 1999. Manufacture of reduced fat Mozzarella cheese using ultrafiltered sweet buttermilk and homogenized cream. *Journal of Dairy Science* 82(1): 1–9.

Rajorhia, G.S. of and Sen, D.C. 1988. Technology of chhana—A review. *Indian Journal Dairy Science* 41(2): 141–148.

Rao, H.G.R. and Kumar, A.H. 2005. Spray drying of mango juice-butter milk blends. *Lait* 85(4/5): 395–404.

Raval, D.M. and Mistry, V.V. 1999. Application of ultrafiltered sweet buttermilk in the manufacture of reduced fat process cheese. *Journal of Dairy Science* 82(11): 2334–2343.

Rodas, B.A., Angulo, J.O., Cruz, J.De-La., and Garcia, H.S. 2002. Preparation of probiotic buttermilk with *Lactobacillus reuteri*. *Milchwissenschaft* 57(1): 26–28.

Roesch, R.R. and Corredig, M. 2002. Production of buttermilk hydrolyzates and their characterization. *Milchwissenschaft* 57(7): 376–380.

Romeih, E.A., Abdel-Hamid, M., and Awad, A.A. 2014. The addition of buttermilk powder and trans-glutaminase improves textural and organoleptic properties of fat-free buffalo yogurt. *Dairy Science & Technology* 94(3): 297–309.

Sen, D.C. and Rajorhia, G.S. 1990. Production of soft grade sandesh from cow milk. *Indian Journal of Dairy Science* 43(3): 419–427.

Sharma, H.K., Singhal, R.S., and Kulkarni, P.R. 1998. Characteristics of fried paneer prepared from mixtures of buffalo milk, skimmed milk, soy milk and buttermilk. *International Journal of Dairy Technology* 51(4): 105–107.

Shiby, V.K. and Mishra, H.N. 2013. Fermented milks and milk products as functional foods—A review. *Critical Review in Food Science and Nutrition* 53(5): 482–496.

Shodjaodini, E.S., Mortazavi, A., and Shahidi, F. 2000. Study of cottage cheese production from sweet buttermilk. *Agricultural Science Technology* 14(2): 61–70.

Shreshtha, R.G. and Gupta, S.K. 1979. Dahi from sweet cream buttermilk. *Indian Dairyman* 31: 57–60.

Shukla, F.C., Sharma, A., and Singh, B. 2004. Studies on the preparation of fruit beverages using whey and buttermilk. *Journal of Food Science and Technology* 41(1): 102–105.

Snoeren, R.H.M., Damman, A.J., and Klok, H.J. 1981. De viscositet van ondermelk concentraat. *Zuivelzicht.* 73: 887–889.

Sodini, I., Morin, P., Olabi, A., and Jiménez-Flores, R. 2006. Compositional and functional properties of buttermilk: A comparison between sweet, sour, and whey buttermilk. *Journal of Dairy Science* 89: 525–536.

Tirumalesha, A. and Jayaprakasha, H.M. 1998. Effect of admixture of spray dried whey protein concentrate and dried butter milk powder on physico-chemical and sensory characteristics of ice cream. *Indian Journal of Dairy Science* 51(1): 13–19.

Trachoo, N. and Mistry, V.V. 1998. Application of ultrafiltered sweet buttermilk and sweet buttermilk powder in the manufacture of nonfat and low-fat yoghurts. *Journal of Dairy Science* 81(12): 3163–3171.

Vijayalakshmi, R., Khan, M. M.H., Narasimhan, R., and Kumar, C.N. 1994. Utilization of butter milk powder in preparation of low-fat yoghurt. *Cheiron* 23(6): 248–254.

Ye, J.H., Thomas, E., Sanguansri, L., Liang, Y.R., and Augustin, M.A. 2013. Interaction between whole buttermilk and resveratrol. *Journal of Agricultural and Food Chemistry* 61(29): 7096–7101.

9
ACIDOPHILUS MILKS

SUJA SENAN AND JASHBHAI B. PRAJAPATI

Contents

9.1	Introduction	227
9.2	Characteristics of *Lactobacillus acidophilus*	229
9.3	Technology and Microbiology of Acidophilus Milks	229
	9.3.1 Acidophilus Milk (Sour)	230
	9.3.2 Acidophilus Milk (Sweet)	236
	9.3.3 Acidophilus Yeast Milk	237
	9.3.4 Acidophilus Whey	238
	9.3.5 Acidophilus Cream Culture	238
	9.3.6 Acidophilus Ice Cream	239
	9.3.7 Acidophilus Cheese	241
	9.3.8 Acidophilus Paste	242
	9.3.9 Acidophilus Powder	242
	9.3.10 Acidophilin	245
	9.3.11 Acidophilus Yogurt	246
9.4	Nutritional Benefits of Acidophilus Milk	250
	9.4.1 Improved Bioavailability of Minerals	251
	9.4.2 Improved Vitamin Supplementation	252
9.5	Therapeutic Benefits of Acidophilus Milk	253
9.6	Conclusions	258
References		259

9.1 Introduction

Lactobacillus acidophilus, which has almost grown synonymous to probiotics, was initially isolated by Ernst Moro (Frost and Butterworth 1931) from the feces of breast-fed infants and named *Bacillus acidophilus*, meaning *acid loving*. Elie Metchnikoff suggested that humans should consume milk fermented with *Bacillus bulgaricus* to prolong life, while Rettgerin (1915) showed the same results of sweet or sour milk fermented with the bacillus of Metchnikoff on the growth and mortality of rats and fowls. However, Kopeloff's published book on *Lb. acidophilus* appeared in 1926. Thus, two scientifically validated views were established: (1) lactose diet exerted an important influence on the intestinal bacteria; and (2) *Lb. acidophilus* was considered to be the most likely species to fulfill the base criteria expected of probiotics, survival through gut, bile tolerance, acid tolerance, and antimicrobial production. Thus, it was a logical progression to blend milk and acidophilus termed as *acidophilus milk*. These acidophilus products utilize large numbers of *Lb. acidophilus* alone or in combinations with other types of microbes to get nutritional and therapeutic benefits.

Acidophilus milk products are usually fermented but because of the acid flavor, consumer's acceptance declined. To make it more palatable, other forms and, especially, non-fermented or sweet acidophilus milk were developed. The different types of milk products containing *Lb. acidophilus* are acidophilus milk, sweet acidophilus milk, acidophilus yeast milk, acidophilin, acidophilus paste, acidophilus ice cream, acidophilus cheese, and so on. In addition to food products containing acidophilus, there are various pharmaceutical preparations that fall in the category of drugs/supplements in the form of freeze-dried bacteria that are used for gastrointestinal disturbances.

The work covering a century indicates the importance of preparations, which contain large numbers of live lactobacilli. The viability should be maintained or destruction of lactobacilli slowed down to ensure that the therapeutic minimum of cells should be maintained till the end of shelf life of the product. Large chunks of the work involving probiotics stability are centered on encapsulation of the microbe. The technology includes encapsulation with nano-structured polyelectrolyte layers that display high protection to the cells due to the impermeability of polyelectrolyte nano-layers to large enzyme molecules like pepsin and pancreatin and to the stability of the polyelectrolyte nano-layers in gastric and intestinal pH.

The knowledge-driven modern day consumers can be influenced to consume acidophilus products, only if they are shown proof of concept of the functionality by clinical studies. For example, *Lb. acidophilus* NCFM probiotic available in conventional foods and dietary supplements in the United States since mid-1970s is predicated on its safety, amenability to commercial manipulation, biochemical, and physiological attributes presumed to be important to human probiotic functionality. In the manufacturing of acidophilus milk, prime importance is given to the therapeutic properties of the product followed by microbiological and technological aspects. This chapter aims to compile the works done on the various aspects of acidophilus milks and their variants.

9.2 Characteristics of *Lactobacillus acidophilus*

Lb. acidophilus is a Gram-positive rod (0.6–0.9 × 1.5–6.0 μm) with rounded ends. These are non-flagellated, non-motile, and non-spore forming, and intolerant to salt. They lack cytochromes and are microaerophilic, so surface growth on solid media is generally enhanced by anaerobiosis or reduced oxygen pressure and 5%–10% CO_2. Most strains of *Lb. acidophilus* can ferment amygdalins, cellobiose, fructose, galactose, glucose, lactose, maltose, mannose, salicin, sucrose, trehalose, and aesculin. Apart from lactose, *Lb. acidophilus* has been reported to utilize sucrose. This could be due to the activity of β-fructofuranosidase, a constitutive enzyme, and β-galactosidase, which is induced in *Lb. acidophilus* (Nielsen and Gilliland 1992). The yield of lactic acid is 1.8 mol/mol glucose, accompanied by minor amounts of other compounds. Acetaldehyde, a carbonyl flavoring molecule, is a by-product of acidophilus activity in milk. Growth of *Lb. acidophilus* may occur at a high temperature of 45°C, but optimum growth occurs within 35°C–40°C. Its acid tolerance varies from 0.3% to 1.9% titratable acidity, with an optimum pH lying at 5.5–6.0. Further information on the biological, biochemical, technological, and therapeutic properties of *Lb. acidophilus* relevant for use as probiotics was reviewed by Gomes and Malcata (1999).

9.3 Technology and Microbiology of Acidophilus Milks

Before probiotics are chosen on the basis of their safety and functional characteristics, these must be able to be manufactured at industrial scale. Furthermore, these have to survive and retain their functionality during storage, and also in the foods into which these are incorporated without producing off-flavors. Factors related to the technological and sensory aspects of probiotic food production are of utmost importance since only by satisfying the demands of the consumers can the food industry succeed in promoting the consumption of functional probiotic products in future. There are several technological factors related to probiotics like oxygen tolerance, acid and bile tolerance, heat tolerance, and types of food carrier, which are affecting the manufacturing of probiotic food products. As probiotics are mainly incorporated in fermented milk products, whose consumption may be limited by allergies, intolerances, or by low-cholesterol diets, new carriers of probiotic acidophilus products like fruit and vegetable products, which may ensure a regular consumption of beneficial microorganisms, are developed.

In the following passage, there would be a collection of technological and microbiological aspects of acidophilus milks and its variants. The authors have written the chapter with the intention of reporting relevant work exclusively on *Lb. acidophilus* containing products (Table 9.1). Apart from the milk-based acidophilus products, *Lb. acidophilus* fermented pear juice has been suggested recently as a new strategy for anti-hyperglycemia and antihypertensive therapy that reduce the oxidative stress associated with type-2 diabetes and its complications (Ankolekar et al. 2012). A novel probiotic product, oblea (i.e., wafer-type dehydrated traditional Mexican dessert) was developed using sweet goat whey fermented with *B. infantis* or *Lb. acidophilus* and maintained above the minimum concentration required in a probiotic product (Santiago et al. 2012). In Italy, research institutions and food companies are working together in developing a probiotic vegetable line bringing the probiotic benefits in a range of traditional foods, for example, seasoned table olives, artichokes, and salads (Lavermicocca et al. 2010). In a functional bread, combining the microencapsulation and starch-based coatings showed that the microencapsulated *Lb. acidophilus* survived after baking and storage time, although reduction was higher in sandwich treatment (i.e., starch solution/sprayed microcapsules/starch solution) (Altamirano-Fortoul et al. 2012).

9.3.1 Acidophilus Milk (Sour)

Acidophilus milk was the first dietary product developed containing large numbers of *Lb. acidophilus*. It was natural that milk should be used

Table 9.1 Variants Acidophilus Milk Products Available in the Global Market

PRODUCT	PLACE	ORGANISM(S)
A-38	Denmark	A + B + mesophilic LD culture
A-B yogurt	France	Lb. acidophilus, B. bifidum
ABC ferment	Germany	A + B + L. casei
Acidophilus buttermilk	United States	Lb. acidophilus, B. bifidum, Leu. mesenteroides ssp. cremoris, mesophilic lactococci
Acidophilus milk	Sweden	A + B + mesophilic LD culture
Acidophilus yeast milk	Former USSR	Lb. acidophilus, S. fragilis, S. cerevisiae
Acidophilus yogurt	Several countries	Lb. acidophilus, Lb. delbrueckii ssp. bulgaricus, Str. thermophilus
AKTIFITplus	Switzerland	A + B + L. casei GG + S. thermophilus
BAlive	United Kingdom	A + B + yogurt culture
B-active	France	Lb. acidophilus, Lb. delbrueckii ssp. bulgaricus, Str. thermophilus, B. bifidum
BIO	France	A + B + yogurt culture
Biobest	Germany	B. bifidum or B. longum + yogurt culture
Biogarde	Germany	A + B + Str. thermophilus
Bioghurt	Germany	A + B + Str. thermophilus
Biokys (=Femilact)	Czechoslovakia	A + B + Pediococcus acidilactici
Biomild	Several countries	Ibidem
Biomild	Germany	A + B
Cultura	Denmark	A + B
Kyr	Italy	A + B + yogurt culture
LA-7 plus	Bauer	A + B
LC-1	Germany	Lb. acidophilus

as a carrier because soured or cultured milk has been an important compound in the diet of many people. In 1928, the preparation of acidophilus milk was first standardized on a laboratory scale and concluded that acidophilus milk has its best flavor at 0.8%–1.0% lactic acid with one teacupful of old culture to inoculate one quart of diluted evaporated milk, starting the fermentation in a thermos bottle at 100°F–102°F and allowing the fermentation to proceed for 13–17 h (Rice 1928). Study on the viability of *Lb. acidophilus* in acidophilus milk suggested that a satisfactory viability of *Lb. acidophilus* for more than a week can be achieved if the number of foreign bacteria must always remain negligible and second, rigid precaution against excessive acidity must be observed followed by the storage temperature (between 12°C and 16°C), when the initial acidity is 0.65% (Kulp 1931).

An economical method of manufacturing acidophilus milk was soon designed, which involved heating skim milk to as near the boiling point for 30 min, quick cooling at 37°C–40°C, immediate inoculation with an actively growing culture of *Lb. acidophilus* and holding at that temperature till coagulation takes place or a few hours thereafter. This method had the merit of cutting into half the time required to get the skim milk ready for inoculation (Knaysi 1932). Johnson et al. (1987) made an exclusive study on the selection criteria of strains for acidophilus products that included fermentation of 8 carbohydrates, growth at 15°C and 45°C, resistance to 0.2% oxgall, lysis by lysozyme, or sensitivity to 17 antibiotics. They concluded that for use in dietary adjuncts, *Lb. acidophilus* strains should be selected based on β-galactosidase activity, presence of S layer, and capable of growing in lactose. Production of high-quality fermented milk products containing *Lb. acidophilus* is a major challenge to dairy plants owing to the sensitive character of microbes in milk (Gomes and Malcata 1999). Growth of *Lb. acidophilus* in milk has been a concern owing to its slow propagation in milk. These hardly remain competitive in the presence of other microbes and will thus be easily outnumbered. Consequently, careful selection and application of *L. acidophilus* strains should be based on the growth rate, proteolytic activity, and flavor promotion for success in multiple industrial applications.

The procedure for the preparation of acidophilus sour milk with 1.5%–2.0% lactic acid had been standardized (Figure 9.1). The products obtained were too sour with total acidity ranging from 1.2% to 1.5% lactic acid (pH 3.7–4.0). The total viable cells of *Lb. acidophilus* ranged from 10×10^8 to 43×10^8 cells/g of the product. It possessed a satisfactory antimicrobial effect against *E. coli*, as well as other pathogenic and undesirable bacteria of the intestine. The product could be preserved up to 10 days below 5°C, without loss of antibacterial activity and taste (Gandhi and Nambudripad 1979).

Even though acidophilus milk is very good for health, it never gained popularity because flavors developed in its manufactures caused it to be unappetizing (Speck 1980). As a result acidophilus milk was used primarily as a therapeutic agent. Efforts to mask the undesirable flavor ranged from the addition of extra lactic acid to the use of different flavors, but none seemed to result in providing a more

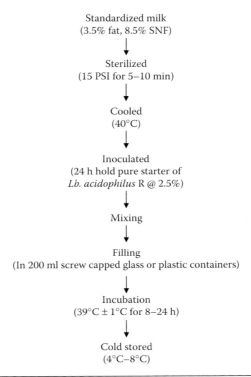

Figure 9.1 Manufacturing of acidophilus sour milk. (Data from Gandhi, D.N. and Nambudripad, V.K.N., *Indian Dairyman*, 31, 813–818, 1979.)

acceptable beverage. In Yugoslavia, acidophilus milk is produced on a large scale. A popular acidophilus product was produced in a Zagreb dairy from good quality milk with a minimum of 8.5% solids-not-fat (SNF), homogenized and pasteurized at 90°C–95°C with 5–10 min holding. The milk was then cooled to 40°C–45°C, inoculated with 2%–5% of a mixed culture of *Lb. acidophilus* strains, filled into 200 ml plastic containers, and incubated at about 38°C for 5–6 h. After cooling, milk was aged for more than 10–15 h at 4°C–8°C and sold as soon as possible to contain more than 2,000–3,000 million viable *Lb. acidophilus* per ml (Lang and Lang 1975).

Milk cultures of *Lb. acidophilus* show little difference in viability in acidophilus milk kept for several days at refrigerator temperature, 9°C and at 22°C (Black 1931). It is also known that cultures of *Lb. acidophilus* are unable to initiate rapid growth in milk, thereby allowing other microbes to successfully overgrow. As a result, the thermal treatment given to milk was so severe that the milk had a marked cooked flavor. This treatment also produces lacunose, a growth factor for lactobacilli, as a degradation product of milk sugar. Usually use of two different strains of *Lb. acidophilus* is made in acidophilus milk manufacture, that is, a mucilaginous and a normal strain. Mucilaginous strains imparted a viscous consistency, but they are weak acid producers. Acidophilus milk prepared only with the mucilaginous strain was excessively viscous and had an insufficiently developed flavor, while the non-mucilaginous strains were strong acid giving excessively sour taste. Solution lay in using them at different ratios (Lang and Lang 1975). The generation time of the acidophilus strains was shorter in glucose, as compared to sucrose or lactose medium (Srinivas et al. 1990), which could be due to the utilization patterns of *Lb. acidophilus* in the order: glucose > fructose > sucrose > lactose > galactose and presence of lactose hydrolyzing enzymes β-galactosidase and β-phosphogalactosidase. Addition of tomato juice to skimmed milk resulted in higher viable counts of *Lb. acidophilus*, shorter generation time, and improved sugar utilization with more acid production (Babu et al. 1992). Attempts to increase the growth rates and acid production of *Lb. acidophilus* by adding milk hydrolyzates did not positively affect *Lb. acidophilus* due to its self-sufficient proteolytic system generating its own nitrogen source. Nevertheless, co-culture with *B. lactis* led to enhanced rates of growth and acidification, when compared with single strain, suggesting some degree of symbiosis between the strains (Gomes and Malcata 1999). Response surfaces analysis was put to use to select optimum conditions for enhanced positive organoleptic traits and for improved survival of *Lb. acidophilus* strains (Gardini et al. 1999). To improve the growth of *Lb. acidophilus* unable to grow in pure milk, a mixture of 20 amino acids was added to skim milk, and alanine, serine, isoleucine, and cysteine are the most important amino acids for growth in milk. Supplementation of milk with four identified amino acids, four ribonucleosides, and one source of iron was able to support the growth of the probiotic *Lb. acidophilus* strain in milk with cell counts and final pH similar to that obtained by the addition of yeast extract (Elli et al. 1999). A combined effect of milk supplementation and culture composition on acidification, textural properties, and microbiological stability of fermented milks containing *Lb. acidophilus* LA-5 was that acidifying activity was enhanced with mixed cultures resulting in a shorter time to reach pH 4.5. The acidifying activity was greatly improved with casein hydrolysate, with a reduction of fermentation time by about 55%. The texture of the fermented products was not dependent on culture composition but strongly dependent on milk supplementation because sweet whey supplementation gave products with lower firmness and viscoelasticity than products supplemented with casein hydrolysates (Oliveira et al. 2001).

Carbonation was considered as an alternative to reduce oxidative stress of lactobacilli. A carbonated acidophilus milk was made with raw milk supplemented with 2% skim milk powder. The mix was pasteurized at 85°C for 30 min, cooled to 4°C, carbonated with food-grade CO_2 to pH 6.3. The carbonated pasteurized milk was inoculated (2%v/v) with a mixture of *Str. thermophilus* St73 and *Lb. acidophilus* LaA3 to produce AT (*Str. thermophilus/Lb. acidophilus*) fermented milk. Inoculated carbonated milk was packed in sterile glass bottles and incubated without stirring at 42°C. After a pH of 5 was reached, the fermented milk was stored at 4°C to avoid an excessive post-acidification during the first few days of refrigerated storage that could lead to a reduction of the viability of probiotics (Vinderola et al. 2000). Shah and Prajapati (2013) formulated a carbonated probiotic drink using indigenous fully sequenced strains to give a viable lactobacilli and streptococci count of 8.94 and 8.87 log cfu/mL at the end of 28 days of storage at 5°C. Different amounts of metabolic products like volatile compounds, organic acids and carbon dioxide vary according to fermentation time and strain used; hence, controlling the fermentation time is of utmost importance in acidophilus milk manufacture (Ostlie et al. 2003). While selecting a strain for the preparation of acidophilus milks, its survivability at the storage temperature or its cryotolerance is of paramount importance. Better cryotolerance was obtained in cells grown at 30°C or at pH 5. These cells showed no loss in acidification activity during freezing and a low rate of loss in acidification activity during frozen storage, while the cells grown at 42°C or at pH 4.5 displayed poor cryotolerance. High resistance during frozen storage was related to a high cyclic C19:0 concentrations, while a low cryotolerance was explained by a low C18:2 content (Wang et al. 2005). Whole genome microarray was employed to study the growth pattern and monitor the gene expression of *Lb. acidophilus* NCFM cells propagated in 11%

total solids (TS) skim milk during early, mid-, and late logarithmic phase, and stationary phase. Expression of the genes involved in lactose utilization increased immediately upon exposure to milk, while genes of the proteolytic enzyme system increased overtime. The expression of a 2-component regulatory system shown to regulate acid tolerance and proteolytic activity increased during the early log and early stationary phases of growth (Azcarate-Peril et al. 2009). Information on gene expression patterns can suggest new strategies for strain improvement to help the lactobacilli to withstand stresses and give good technological stability.

9.3.2 Acidophilus Milk (Sweet)

Sweet acidophilus was introduced in 1975 in the United States. This non-fermented milk was made by mixing low fat pasteurized milk with *Lb. acidophilus* concentrate. The product had gained much popularity within a very short time, because it had overcome the main drawbacks of fermented acidophilus milk (Speck 1980). The *Lb. acidophilus* culture used was a human isolate possessing bile resistance and grown in a milk medium. It was later harvested, re-suspended in milk, and frozen at $-196\,°C$. The culture in the form of a concentrate was added to cold pasteurized low-fat milk in a tank, mixed well, packaged, and then maintained below $40\,°F$, and distributed in normal channels used for milk. Several million viable and bile-resistant cells of *Lb. acidophilus* were present per ml, and the population was maintained for 2–3 weeks with proper refrigeration. The flavor of milk remained unaltered by the culture. Another non-fermented acidophilus milk product called *digest* was made from pasteurized, homogenized low-fat milk with added *Lb. acidophilus* (8×10^6/ml) and enriched with vitamins A and B (Lang and Lang 1975).

Sinha et al. (1979) formulated a non-fermented acidophilus milk prepared by adding live *Lb. acidophilus* cells ($4.05–4.55 \times 10^6$/ml of milk) to low-fat milk. Lactobacilli showed 80% viability after 10 days storage. Feeding trials showed the product to suppress fecal coliforms and increase lactobacilli. In rats, weight gains were 8.6 versus 6.3 g/day in control rats fed low-fat milk. Blood serum cholesterol was also lower in experimental rats than in control.

9.3.3 Acidophilus Yeast Milk

The poor survival of probiotics added to yogurts is mainly attributed to the low pH of the product environment. Since yeasts have the ability to metabolize organic acids, resulting in a decrease in acidity, the inclusion of yeasts as part of the normal microflora, in association with probiotics has been suggested with the intention to assure better survival of the probiotics in bio-yogurts.

Little is known regarding the technology of acidophilus yeast milk that is only produced in former USSR. Whole of skimmed milk is heated to about 90°C–95°C for 10–15 min, cooled to 35°C to be inoculated with 3%–5% mixture of *Lb. acidophilus* and *Saccharomyces lactis*. The milk was bottled and first stage of fermentation took place at 35°C until acidity reached 0.8% lactic acid, followed by second stage fermentation at 10°C–17°C for 6–12 h (Eller 1971; Koroleva 1991). *Lb. acidophilus* produced the acid, while the yeasts produced ethanol and CO_2. The final product was stored at <8°C until consumed. Acidophilus yeast milk could be prepared in large fermentation tanks (Eller 1971), provided the CO_2 content may be reduced due to pumping before packaging. The product thus formed was described as viscous, slightly acidic, and sharp with a yeasty taste. Subramanian and Shankar (1985) achieved high viability of *Lb. acidophilus* in the presence of lactose-fermenting *S. fragilis* and *Candida pseudotropicalis*. The milk was heated to 90°C for 20 min and coagulation achieved in less than 20 h at 33°C or 37°C. The consistency of the coagulum was improved by fortification of the milk with 1.5% skim milk powder and 0.5% agar. The latter compound prevented the breakup of the curd caused by the production of CO_2. The packaging of yeast lactic products was done in hermetically sealed laminated paperboard cartons or plastic containers that pose the problem of consumer rejection due to swollen packages because of CO_2 formation. Now a breathing membrane, which allows CO_2 to escape, has been developed (Tamime and Robinson 1988). A communalistic association between yeasts and lactic acid bacteria (LAB) exists as yeasts possess stability-enhancing effects on LAB. The specific effects of yeasts on LAB stability vary with yeasts. *Williopsis saturnus* var. *saturnus* enhanced the survival of *Lb. acidophilus*, *Lb. rhamnosus*, and *Lb. reuteri* to 10^6 cfu/g but the same yeast failed to improve the survival of *Lb. johnsonii*, *Str. thermophilus*, and *Lb. bulgaricus* in fermented milk (Liu and Tsao 2009).

9.3.4 Acidophilus Whey

Attempts have been done to convert surplus nutritive whey into a palatable, refreshing, and economical acidophilus whey drink named *acido whey*. Filtered fat-free whey was inoculated with pure and active culture of *Lb. acidophilus* R. The product was found to have pH in the range of 3.25–3.50 and exhibited antibacterial activity against several test microbes (Gandhi and Nambudripad 1979). Air on fortified whey drink was formulated with whole milk (58.0%), sweet whey (25.0%), sucrose (10.0%), mango pulp (7.0%), culture of *Str. thermophilus* and *Lb. bulgaricus* (0.8%), culture of probiotic *Lb. acidophilus* (1.2%), and iron amino acid chelate (0.0002%). The probiotic human strain of *Lb. acidophilus* provided 10^8 cfu/ml with 3 mg of iron per 80-mL dose (Silva et al. 2001).

9.3.5 Acidophilus Cream Culture

Acidophilus cultures (i.e., having favorable dietetic and therapeutic effects), cream cultures (i.e., consisting of *Lc. lactis* and *Lc. cremoris*), and aroma producing leuconostocs (i.e., *Leu. citrovorum* + *Leu. dextranicum* or *Lc. diacetylactis*) to impart improved organoleptic properties (Lang and Lang 1975) were mixed to produce the acidophilus cream culture. At a ripening temperature close to the optimum for *Lb. acidophilus* (37°C), the product was sharply acidic with a tendency for over acidification and possessed a coarse consistency. The finished product had good organoleptic qualities but *Lb. acidophilus* flora showed very weak multiplication and had a low viability.

A new process of acidophilus milk manufacturing was introduced based on the separate incubation of milk with an acidophilus culture and separate incubation of milk with cream culture and then mixing at a ratio of 1:9 after ripening for 16–18 h. The resultant coagulum was thick and has a typical sharply acid flavor and acidity in the range of 85–100° Soxhlet Henkel degrees (SH). After the completion of the ripening of both milks in their separate container, the coagulum was gently stirred and pumped with a positive pump into the ripening tank with gentle mixing. Then, the coagulum is incubated with the culture, cooled to 8°C–10°C, and kept in a cold store at 8°C–10°C until next day. Preferably, the milk should contain 3.6% fat, 8% SNF, and have an acidity between 36 and 50°SH. The coagulum should be milky white to creamy, the consistency sufficiently thick, flavor and texture characteristics of the culture used, that is, cream like with pleasant lactic odor and have refreshing, clean, and aromatic taste. The acidophilus milk produced according to the new technology has organoleptic properties similar to those of kefir milk, because the ratio of streptococci to lactobacilli (9:1) is about the same in both the product (Lang and Lang 1975).

9.3.6 Acidophilus Ice Cream

During probiotic ice cream production, each process stage ought to be optimized, aiming at an increased survival of probiotics, so as to guarantee the product functional properties. This means that the challenges involved in the production of conventional ice cream must be taken into consideration during the development of probiotic ice cream. These challenges include the contributions for the microstructure and colloidal properties provided by the ingredients and/or components present in the formulation, the knowledge and control of the ice crystallization, the choice of appropriate stabilizers, and, finally, the understanding and control of the fat destabilization and the emulsifier functionality (Goff 2008). Overall, the steps of probiotic ice cream processing are reception/weighing of the ingredients involved (i.e., milk, emulsifiers, stabilizers, milk powder, sugar), mixing, pasteurizing, cooling to a temperature of around 37°C–40°C, for the addition of the freeze-dried starter cultures (i.e., usually yogurt cultures) and the probiotic cultures, subsequent fermentation to a pH of 4.8–4.7, or the addition of a previously fermented inoculum containing both types of lactic cultures, cooling to 4°C and keeping the mixture at this temperature for 24 h for maturation. The processing steps up to this point lead to the production of the ice cream mix. The mix is subsequently beaten/frozen, in order to produce the final product, which is packaged and maintained frozen throughout transport, commercial distribution and maintenance, and storage for consumption. During all these steps after freezing, the temperature of the frozen product should be strictly controlled in case of frozen products like ice cream; this implies overcoming intrinsic hurdles, which take place in the processing of ice creams, such as the beating step, where air is incorporated and storage under freezing temperatures, which affects survival of probiotics during storage and also the way in which the inoculum is added to the product. In addition, great attention should be given to the correct choice of the other ingredients to be used in the product, especially any fruit pulp/juice, which will give the final flavor to the product (Cruz et al. 2009).

Monitoring the survival of the *Lb. acidophilus* and β-galactosidase activity in a standard ice cream revealed that after freezing of the fermented mix, counts were 1.5×10^8 for *L. acidophilus*, while these decreased to 4×10^6 cfu/ml (Hekmat and McMahon 1992) after storage. *Sweet acidophilus ice cream* was formulated by the direct incorporation of a culture of *Lb. acidophilus* in form of a cell concentrate into the ice cream mix at the time of storage, mixing, and freezing. Organoleptic evaluation and physical characteristics of the product were equally acceptable, as the ice cream was prepared without microbes. Feeding trial with human volunteers revealed that the ingestion of *Lb. acidophilus* ice cream led to a perceptible decline in the blood cholesterol (Hagen and Narvhus 1999). Duthic et al. (1981) also described ice cream as the ideal vehicle for *Lb. acidophilus* culture, when added to the mix and frozen. During freezing, large destruction was observed, while during storage, there were no more fluctuations in counts of *Lb. acidophilus*, which was maintained at >2 million/ml.

In a study, acidophilus cultures were grown at 37°C for 12 h in ultra-high temperature (UHT) skimmed milk, and the fermented milk was added to ice cream mix at 10% w/w to derive a final probiotic count of 10^6 cfug/L (Akin et al. 2007). A chocolate flavored probiotic ice cream was made from goat's milk using probiotics comprising *Lb. acidophilus* LA, *B. animalis* subsp. *lactis* BB-12, and novel probiotic *Propionibacterium jensenii* 702 with viable numbers of all probiotics maintained to 10^8 cfu/g up to 52 weeks at −20°C (Ranadheera et al. 2013). A novel probiotic ice cream is developed, where finger millet or ragi (*Eleusine coracana*) is being studied as a feasible additive in probiotic ice cream to increase its functionality (Chaudhary 2013). Ragi is rich in calcium, contains an amino acid called tryptophan that lowers appetite, and helps in keeping weight in control in addition to being a very good source of natural iron.

9.3.7 Acidophilus Cheese

The studies on acidophilus addition in cheese have also been researched to substantial lengths. Turkish white cheese proved to be a good vehicle for probiotic acidophilus, when the ripening was done in vacuum packed bags. *Lb. acidophilus* survived to numbers more than 10^7 cfu g^{-1}, which is necessary for positive effects on health (Kasimoglu et al. 2004). The treatment given to milk has had an impact of the viability of *Lb. acidophilus* H5. It was seen that pasteurized milk plus probiotics cheese showed a cell load decrease of about 1 log cfu/g, while high-pressure homogenized milk plus probiotic cheese positively affected the viability during the refrigerated storage of the probiotics (Burns et al. 2008). *Lb. acidophilus* LA-5 added solely or in co-culture with a starter culture of *Str. thermophilus*, for the production of Minas fresh cheese, resulted in a good quality product, with a small rate of post-acidification, indicating that traditional yogurt culture could be employed in co-culture with LA-5 to improve the quality of this cheese (Souza and Saad 2009). Comparing the favorable forms of addition of adjunct culture to cheese-making milk between lyophilized powder dispersed in milk, or within a substrate composed of milk and milk fat, it was deciphered that direct addition of the probiotics performed best, however the pre-incubation presented some advantages such as an increased population of lactobacilli in the initial inoculum (Bergamini et al. 2005). The effect of the supplementation of increasing amounts of *Lb. acidophilus* strains indicated some negative sensory effects in probiotic cheese processing (Gomes et al. 2011). Milk microencapsulation in calcium alginate gel and resistant starch was able to increase the survival rate of *Lb. acidophilus* LA-5 in Iranian white brined cheese after 6 months of storage (Mirzaei et al. 2012). Bacterial interaction was the single most factor other than high inoculation level of *Lb. acidophilus* in cheese-based French onion dip to obtain greater than 6 log of individual population at the end of shelf life (Tharmaraj and Shah 2004). Probiotic caprine coalho cheese was proposed as an excellent functional cheese, when prepared using milk enriched in conjugated linoleic acid (CLA) and incorporation of *Lb. acidophilus* LA-5 giving the final population of *Lb. acidophilus* around 7.5 log cfug^{-1} with no influence of CLA content on the probiotic viability (Santos dos et al. 2012). For making acidophilus Pecorino cheese, lamb rennet paste containing encapsulated *Lb. acidophilus* helped retain the probiotic viability for 4–5 days and then showed a fast reduction, although the ability of *Lb. acidophilus* to produce CLA in the cheese matrix translated into a functional cheese (Santillo et al. 2012).

9.3.8 Acidophilus Paste

Acidophilus paste is a concentrated cultural product obtained by partial elimination of whey from acidophilus milk. Its manufacture consists of adding to sterilized or pasteurized milk (cooled to 45°C) a starter, which contains pure cultures of mucilaginous and non-mucilaginous strains of *Lb. acidophilus*. The coagulum was cut into cubes of 2 cm in size and transferred into sterilized bags, which are arranged on a draining table for spontaneous pressing and partial removal of whey. The finished acidophilus paste had the consistency of thick cream and contained 8% fat and 80% moisture with an acidity of 180–200°T (Lang and Lang 1975).

The product can be made from sterilized, concentrated (20% TS) milk also. Then, it was inoculated with *Lb. acidophilus* culture. An acidity of 1.8%–2.0% is obtained within 5–6 h. The coagulum is then stirred to a homogeneous consistency and mixed with 66% sugar syrup. The resultant paste contained 40% TS including 21% fat, 6% protein, 10% lactose, and 2% sucrose (Gandhi and Nambudripad 1979). Acidophilus paste could be flavored with sugar, cocoa, or canned fruits, and it is considered suitable for feeding infants.

9.3.9 Acidophilus Powder

Probiotics are gaining increasing importance as health food supplements, but are usually sold in high moisture dairy products like yogurts and drinks. However, these products have a limited shelf life. One solution is to spray dry the bacteria. Survival of *Lb. acidophilus* in spray drying has also been discussed by Espina and Packard in 1979. Korobkina and coworkers in 1981 developed a

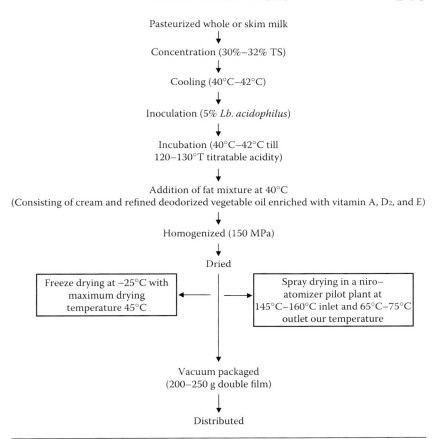

Figure 9.2 Manufacturing of acidophilus infant milk powder. (Data from Korobkina, G.S. et al., *Voprosy Pitaniya*, 1, 51–54, 1981.)

dried acidophilus milk formula of the *Malyutta* and *Malysh* type (Figure 9.2). As early in 1982, a Russian formula for acidophilus dry products was standardized for nutrition of athletes engaged in speed skating and weight lifting (Brents et al. 1982). Efforts have been made for preparation of spray dried acidophilus powder with an ability to ferment milk within 12 h at 40°C when used at the level of 1%–3% (Gandhi and Kumar 1980). A unique blend of banana, tomato juice, and sugar to acidophilus milk and spray drying the mixture was successfully carried out by Prajapati et al. (1986) shown in Figure 9.3. The blend of acidophilus milk with banana and tomato juice gave a final lactobacilli count of 21×10^7 cfu/g with 28.71% protein. The nutritional value of the blend was also higher due to its folic acid content 1056 µg/g compared to unfermented skim milk

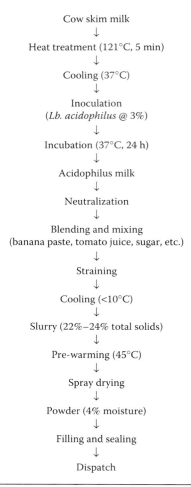

Figure 9.3 Manufacturing of spray dried acidophilus banana powder. (Data from Prajapati, J.B. et al., *Cultured Dairy Products Journal*, 21, 16–21, 1986.)

banana tomato powder that supplied only 800 µg/g. A nutritionally superior spray dried acidophilus malt powder was formulated that gave a lactobacilli count of 106 cfu/g and 17% total protein (Shah et al. 1987). As a means to increase the storage stability of *Lb. acidophilus* strain, Chan and Zhang (2002) devised a method of compressing microbial cell containing powders into a pellet, which was then encapsulated with a coating material of a combination of sodium alginate and hydroxypropyl cellulose by further compression. However, not specific to *Lb. acidophilus*, there are many published papers that relate to the selection of an appropriate drying medium for probiotics encapsulation, such as carbohydrates (e.g., starches, maltodextrins, and sugars), hydrocolloids, and proteins in order to increase their survival rate after thermal processing (Lian et al. 2002; Ananta et al. 2005; Rodríguez-Huezo et al. 2007). Corcoran et al. 2004 have tested cells in different growth phases and found that mid-log stage cells showed less viability after spray drying than stationary phase cells. Zamora et al. (2006) tested 12 different strains of LAB and found minimal losses directly after spray drying but significant falls after storage which varied with cell type.

Espina spray drying was performed in a tall-form co-current spray dryer, and atomization was performed by a twin-fluid nozzle, using compressed air as the atomizing gas. The outlet air temperatures ranged from 60°C and 80°C. After spray drying with maltodextrin either with or without broth components, it was found that early stationary cells of both *E. coli* K12 and *Lb. acidophilus* showed significantly greater survival rates than mid-log cells including during storage (Espina and Packard 1979). Lower survival rates were recorded as the outlet air temperature was increased (particularly above 80°C), although bacteria surviving higher temperatures were more likely to survive storage. *Lb. acidophilus* generally showed better survival than *E. coli* and this is attributed to the thicker cell wall of the Gram-positive cell. Atomization without drying produced negligible reductions in cell viability and so all losses can be attributed to the drying stage. High temperatures detrimental to probiotics and can be circumvented by sub-lethal thermal shock (50°C for *Lb. acidophilus*) and by increasing the microencapsulating material concentration increased the survival rate of the probiotics (Anekella and Orsat 2013).

9.3.10 Acidophilin

Acidophilin was made from pasteurized milk using a mixture of *Lb. acidophilus*, *Str. lachi*, and kefir fungi. This product differs from acidophilus milk in the composition of microflora, flavor, and therapeutic properties. The starter containing three cultures in equal proportion is generally used at the rate of 6%–9% of the weight of milk. The milk was fermented after cooling to 28°C–32°C, at 18°C–25°C. At this temperature, *Lb. acidophilus* does not grow well. Acidophilin, which had been fermented at 20°C–26°C, constituted not more than 2% of the total microflora, streptococci 97% and yeast cells 1%. Thus, streptococci predominated in acidophilin. However, by raising the temperature of incubation to 32°C–36°C, the relative content of lactobacilli increased. At 41°C–42°C, the growth of the lactobacilli was intensified and product acquired flavor typical of acidophilus milk (Lang and Lang 1975). The product was packed in bottles of 0.25 or 0.5 L. The acidity of the product was 0.75%–1.30%. Sharma and Gandhi (1981) prepared acidophilin with four strains of *Lb. acidophilus*, namely, R, CH_2, I, and H along with *Str. lactis* C10 and kefir grains, which gave a product of desirable acidity and maximum antibacterial activity against *M. flavus*, *E. coli*, *S. aureus*, and *B. subtilis*.

9.3.11 Acidophilus Yogurt

Yogurts are produced from pasteurized milk fermented by *Str. thermophilus* and *Lb. delbrueckii* ssp. *bulgaricus*. Probiotic *Lb. acidophilus* is not required in the production, but are often added to carry out co-fermentation with the starter cultures for added functionality. The application of probiotics to yogurt is commonly referred as bio-yogurt. Yogurt to include live strains of *Lb. acidophilus* and species of *Bifidobacterium*, in addition to the conventional yogurt microbes, is known as AB cultures. To translate the functionality of the strain into the product, the focus is given on the therapeutic minimum count (10^8–10^{10} CFU/day) of the probiotic to be maintained till the end of the shelf life (Kailasapathy and Chin 2000). Studies have shown low viability of probiotics in yogurt and fermented milks (Gilliland and Speck 1977; Iwana et al. 1993; Shah and Lankaputhra 1997; Gardini et al. 1999; Schillinger 1999; Vinderola et al. 2000). To obtain the count and viability, one should be aware of the factors affecting the survival of *Lb. acidophilus* in yogurt.

Manipulating the conditions in the manufacture and storage of yogurt could increase the survival of LAB and bifidobacteria. The various approaches have been reviewed by Kailasapathy and Chin (2000), who summed it up vividly in the following points:

- Terminating fermentation at a higher pH (>5).
- Enrichment of yogurt mix with whey protein concentrate (increases the buffering capacity of yogurt, retards decrease in pH, and prevents pH change during storage of yogurt).

- Application of hydrostatic pressure (200–300 MPa for 10 min) to yogurt prevents after acidification and hence maintains initial number of viable LAB.
- Heat shock (58°C for 5 min) of yogurt (prevents excess acid production and acidity remains constant during storage).
- Lowering the incubation temperature to 37°C favors growth of bifidobacteria and increases incubation time.
- Lowering the storage temperature to less than 3°C–4°C increases LAB culture (*Lb. acidophilus* and bifidobacteria) survival.

Rational selection and design of probiotics remains an important challenge and will require a platform of basic information about the physiology and genetics of strains relevant to their intestinal roles, functional activities, and interactions with other resident microbes. In this context, genetic characterization of probiotics is essential to unequivocally define their contributions to the intestinal microbiota and ultimately identify the genotypes that control any unique and beneficial properties. To be useful as a health adjunct, *Lb. acidophilus* should be able to resist physiological stresses such as acid and bile. Chou and Weimer (1999) proposed the use of acid and bile tolerant isolates that retained their ability to grow at pH 3.5 with 0.3% bile after the selective pressure was removed and reapplied. It was observed that the isolated isolates possessed growth advantages over that of the parents under stress conditions and hence poses as better candidates for probiotics. Matsumura et al. (1999) further added the binding activity of surface layer protein to rat colonic mucin, which contains sugar chains similar to those in human colonic mucin, should be added to the selection criterion for *Lb. acidophilus* to be used as a probiotic yogurt starter. Probiotics added to yogurt cultures are commonly in the form of direct-vat-set cultures containing strains in high numbers, which alone can guarantee the desired performance in commercial manufacturing of fermented milk bio-products (Hoier 1992).

The possible interactions among the strains selected to manufacture a probiotic fermented dairy product like stimulation, delay, complete inhibition of growth, and no effects should be taken into account, when choosing the best combination(s) to optimize their performance in the process and their survival in the products during storage. The composition of the species participating in the fermentation affects the survival of *Lb. acidophilus* and *Bifidobacterium* ssp. The increase in probiotics during manufacture and the viability of these microbes during storage were dependent on the species and strain of associative yogurt microbes. Metabolic products secreted by other microbes influence the viability of *Lb. acidophilus* (Gilliland and Speck 1977). Between the two starter culture species, *Lb. delbrueckii* ssp. *bulgaricus* exerted a greater detrimental effect on the survival of some *Lb. acidophilus* strains perhaps by producing inhibitory metabolites like hydrogen peroxide. Therefore, the exclusion of this species in yogurt production has been suggested to eliminate the antagonistic effect (Gilliland and Speck 1977; Villegas and Gilliland 1998; Lourens-Hattingh and Viljoen 2001; Talwalkar and Kailasapathy 2003). The solution was in using modified or ABT (acidophilus-bifidus-thermophilus)-yogurt starter cultures (i.e., fermented with *Lb. acidophilus, B. bifidum*, and *Str. thermophilus*) (Kim and Gilliland 1983). The proteolytic strains of *Lb. delbrueckii* subsp. *bulgaricus* further improved the viability of probiotics in yogurt (Shihata and Shah 2002). Another approach to increase the viability was to use ruptured yogurt bacterial cells to release their intracellular β-galactosidase and reduce their viable counts, and to contain less hydrogen peroxide during fermentation (Shah and Lankaputhra 1997). *Lb. acidophilus* was the sole species that was inhibited by *Lb. casei* and *Bifidobacterium* in a co-culturing (Vinderola et al. 2002).

In a study to understand the relationship *of Lb. acidophilus* with yogurt starter activity, Ng et al. (2011) revealed that *Lb. acidophilus* strains exhibited good survival at low pH brought about by glucono delta-lactone that released gluconic acid gradually at a rate comparable to organic acids produced by the starter cultures, yogurts made with starters without probiotics, and killed starters did not affect the survival of *Lb. acidophilus* NCFM. *Lb. acidophilus* is also more tolerant to acidic conditions than *B. bifidum* (Lankaputhra and Shah 1997). *Over-acidification* or *post-production acidification* occurs after fermentation and during storage at refrigerated temperature, which is mainly due to the uncontrollable growth of strains of *Lb. bulgaricus* at low pH values and refrigerated temperatures. Over-acidification can be prevented in bio-yogurts by applying *good manufacturing practices* and by using cultures with reduced *over-acidification* (Kneifel et al. 1993).

The higher acidity of carbonated milk (i.e., production of carbonic acid) enhanced growth and metabolic activity of the starter during fermentation and was the reason for reduction in incubation period (Vinderola et al. 2000). Yogurts prepared with microencapsulated cultures presented lower values for post-acidification and greater stability compared to the product prepared with the addition of the free culture (Shoji et al. 2013). Incorporation of free and encapsulated probiotics did not substantially change the overall sensory properties of yogurts, while greatly enhanced the survival of probiotics *Lb. acidophilus* ATCC 4356 against an artificial human gastric digestive system (Ortakci and Sert 2012). Incorporation of 0.5% (w/v) of xanthan gum or the 1% (w/v) of cellulose acetate phthalate within the 3% (w/v) of alginate solution for bead formation increased the survival of the probiotics. *Lb. acidophilus* LA14 grown in acidic conditions displayed increased viability from 63% of the freeze-dried bacteria up to 76%, while smaller microcapsules increased the survival of the probiotics after acid incubation to 91% (Albertini et al. 2010). Spray-chilling using fat as carrier was again an innovative technology for the protection, application, and delivery of probiotics (Pedroso et al. 2012).

The effect of refrigerated storage temperature was studied at 2°C, 5°C, and 8°C on the viability of probiotics in ABY (*Lb. acidophilus*, *B. lactis* BB-12, and yogurt bacteria). After 20 days, storage at 2°C resulted in the highest viability of *Lb. acidophilus*. The effect of cold storage on the viability of probiotics in fermented milks has been the subject of different studies (Centeno-de-Laro 1987; Nighswonger et al. 1996; Bolin et al. 1998; Han-Seung et al. 2000).

Apart from the application of prebiotics to increase the viability of acidophilus strains, many alternatives have also been investigated. Supplementation of apple and banana fibers to skim milk yogurts increased the viability of *Lb. acidophilus* L10 during shelf life (Espírito Santo et al. 2010). Although passion fruit fiber did not show any clear effect on probiotic counts, Açai pulp favored an increase in *Lb. acidophilus* L10 (Espírito Santo et al. 2012). Exhaustive studies like the growth patterns of 24 strains of LAB (*Str. thermophilus*, *Lb. delbrueckii* subsp. *bulgaricus*, and *Lc. lactis*) and 24 strains of probiotics (*Lb. acidophilus*, *Lb. casei*, *Lb. paracasei*, *Lb. rhamnosus*, and *bifidobacteria*) in liquid media containing different substances were assessed. The substances used were salts (NaCl and KCl), sugars (sucrose and lactose), sweeteners (acesulfame and aspartame), aroma compounds (diacetyl, acetaldehyde, and acetoin), natural colorings for fermented milk (red, yellow, and orange colorings), flavoring agents (strawberry, vanilla, peach, and banana essences), flavoring, coloring agents (strawberry, vanilla, and peach), nisin, natamycin, and lysozyme were carried out (Vinderola et al. 2002). Inulin was used as a prebiotic to improve quality of skim milk fermented by pure cultures of *Lb. acidophilus*, binary co-cultures with *Str. thermophilus*, or a combination containing all. Inulin supplementation to pure cultures lowered the generation time, with particular concern *Lb. acidophilus*: mono-cultures, co-cultures, and combination (Oliveira et al. 2011). A cyanobacterial (*Spirulina platensis*) biomass had a beneficial effect on the survival of ABT starter bacteria regardless of storage temperature (Varga et al. 2002). The survival of *Lb. acidophilus* increased by more than 1.5 $\log_{10}$ cycle, upon addition of cereal extracts (Charalampopoulos et al. 2003). *Lb. acidophilus* survival in simulated gastrointestinal fluids during the *in vitro* assays was improved by the addition of inulin, whey protein concentrate, and freezing in synbiotic guava mousse (Buriti et al. 2010). Probiotic *Lb. acidophilus* in milk supplemented with soy and pulse ingredients: pea protein, chickpea flour, lentil flour, pea fiber, soy protein concentrate, and soy flour had no negative effect on acidification of the fermented milks and the highest effects were obtained with lentil and soy flour (Zare et al. 2012). An impressive proteomic change was also demonstrated through the micro-2DE system. Prebiotics induced a stronger resistance to gastrointestinal juices than glucose. Prebiotics induced a more effective production of butyrate than glucose (Nazzaro et al. 2012). Green lentils selectively enhanced the number of probiotics in yogurt in the initial stages of storage and maintained an overall count of starter culture and probiotics over a 28-day storage (Agil et al. 2013).

9.4 Nutritional Benefits of Acidophilus Milk

The nutritional benefits of milk-based products fermented with *Lb. acidophilus* may add or replenish the constituents present in milk. The biological value of acidophilus milk is definitely high, because of the predigested proteins present and available essential amino acids of milk proteins and microbial cell protein. Functionality added depends on the type of milk used, on the type of microbes added, and on the manufacturing process employed. They are characterized by lower levels of free amino acids and certain vitamins than non-fermented milks. *Lb. acidophilus* has been reported to synthesize folic acid, niacin, thiamine, riboflavin, pyridoxine, and vitamin K, which are slowly absorbed by the body. The vitamins of the B-complex are frequently obtained as natural ingredients in foods, so addition of *Lb. acidophilus* to the diet will more effectively help to meet those requirements. The bio-availabilities to the host of such minerals as calcium, zinc, iron, manganese, copper, and phosphorus may also be enhanced upon consumption of fermented dairy products and improve the digestibility of the proteins (Tamime and Robinson 1988).

Magdi et al. (2010) studied biochemical changes that occur in amino acids, water soluble vitamins, soluble sugars, and organic acids occurring during fermentation (43°C for 6 h) of camel milk inoculated with *Str. thermophilus* 37, *Lb. delbrueckii* subsp. *bulgaricus* CH2, and *Lc. lactis. Lb. acidophilus* and mixed yogurt culture (*Str. thermophiles* and *Lb. bulgaricus* @ 1:1) report revealed that amino acids slightly increase in alanine, leucine, histidine, lysine, and arginine, while valine, methionine, and tyrosine were slightly decreased. Vitamin C content decreased, while no significant change observed in riboflavin and thiamine content, and organic acids like lactic acid, formic acid, and acetic acid increased during fermentation. Bioactive peptides are also released after protein degradation due to proteolysis by acidophilus strains during fermentation. These bioactive peptides are the main source of a range of biologically active peptides such as casomorphins, casokinins, immunopeptides, lactoferrin, lactoferricin, and phosphopeptides. In fermented milk, the available lysine was reduced by 40%, when skim milk was fermented with *Lb. acidophilus* (Rao and Shahani 1987). Milk fermentation results in a complete solubilization of calcium, magnesium, and phosphorus and a partial solubilization of trace minerals. *Lb. acidophilus* was also found to increase folic acid levels in skim milk (Deeth and Tamime 1981; Lin et al. 2000).

9.4.1 Improved Bioavailability of Minerals

Acidophilus and iron fortified yogurt could be considered as a low cost and wide reaching alternative to address iron deficiency anemia. The reason for increased bioavailability of iron could be due to the fermentation or fortification or both. Clarification was sought by a study with rats, where a higher hemoglobin gain in the animals fed on a diet with *Lb. acidophilus* was observed than in those on a similar diet without microorganisms (Oda et al. 1994). CLA have several beneficial health effects including reduced risk of carcinogenesis, atherosclerosis, obesity, improved hyperinsulinemia, and prevention of catabolic effects of the immune system (Ogawa et al. 2005). With this knowledge, and the need to supply natural *inproduct* CLA enriched food, *Lb. acidophilus* NCDC14 added dahi was prepared with increased production of free fatty acids by lipolysis of milk fat and increased CLA content by using internal linoleic acid, compared to control dahi during fermentation and 10 days of storage at 4°C (Yadav et al. 2007a). Preschool children fed with iron-fortified fermented milk beverage added with a probiotic improved their iron status, and a positive correlation between iron intake and hemoglobin was established (Silva et al. 2008). A significant effect of fermentation by probiotics *Lb. helveticus* MTCC 5463 was seen in amount of soluble calcium that increased from 46.65 in milk to 153.25 mg/100 g product). Volatile fatty acids synthesized by *Lb. helveticus* MTCC 5463 were 6.80 mg/g lactic acid, 0.02 mg/g acetic acid, and 0.11 mg/g butyric acid (Goswami 2012).

9.4.2 Improved Vitamin Supplementation

During fermentation, probiotic LAB produce a range of secondary metabolites, some of which have been associated with health-promoting properties, among these are the B-vitamins and bioactive peptides released from milk proteins. Fermentation causes changes in vitamin content by several mechanisms. These include synthesis of vitamins by fermenting microbes or loss of vitamins due to their metabolism, loss of vitamins by chemical reactions not directly related to the fermentation, increase or decrease in stability of vitamins due to pH changes, and heating losses associated with preparation of raw materials prior to or after fermentation. *Lb. helveticus* MTCC 5463 strains were found to contain folate and biotin metabolism genes during its whole genome sequencing (Prajapati et al. 2011). Goswami (2012) reported that *Lb. helveticus* MTCC 5463 was found to be the highest producer of folic acid as estimated by HPLC, and the folic acid concentration was 123 μg/L. Milk had 0.18 μg/g of biotin, which increased to 0.23μg/g by *Lb. helveticus* MTCC 5463.

9.5 Therapeutic Benefits of Acidophilus Milk

The tremendous increase in scientific publications on probiotics in the past 30 years resulted in (1) important advances in the understanding of the fundamental mechanisms by which probiotics may confer beneficial effects to the host and (2) the ability to verify the attributed health benefits of probiotics. The understanding of the mode of action of probiotics constitutes a base to develop new strategies for the prevention and treatment of specific human diseases.

Clinical studies have verified some of the anecdotal reports on the beneficial effects of consuming certain cultures, or their metabolites, or both, including reduction in fecal enzymes implicated in cancer initiation, treatment of diarrhea associated with travel, antibiotic therapy, control of rota virus and *Clostridium difficile*, control of ulcers related to *Helicobacter pylori*, antagonism against foodborne pathogens and tooth decay microbes, and amelioration of lactose malabsorption symptoms.

Consumption of non-fermented milk containing *Lb. acidophilus* by healthy adult men increased facultative lactobacilli in their feces with nonsignificant effects on numbers of coliforms or anaerobic lactobacilli including bifidobacteria (Gilliland et al. 1977). Acido yeast milk was successfully applied for the treatment of certain intestinal diseases. *S. lactis* used in the product possessed an antibacterial effect against *Mycobacterium tuberculosis* (Koroleva 1991). *Lb. acidophilus* SBT 2074, when administered to *E. coli* infected rats resulted in significant inhibition of coliforms and anaerobes complemented with decreased fecal enzyme β-glucuronidase activity (Sreekumar and Hosono 2000). Yogurt supplemented with *B. animalis* subsp. *lactis* (BB-12) and *Lb. acidophilus* (LA-5) showed that the probiotics remained active during gut transit and were instrumental in the reduction in potential pathogens. Proper standardization of the dose is essential to reinforce lactobacilli without altering the balanced microbial ecosystem of healthy individuals (Larsen et al. 2006). A study showed using real-time polymerase chain reaction (PCR) that *Lb. acidophilus* (LA-5) (dose of 10^9 cfu/100 g of yogurt supplemented with green tea) remained active during gut transit and were associated with an increase in beneficial bacteria and a reduction in potential pathogens (Savard et al. 2011). *Lb. acidophilus* LAP5, isolated from swine was successful in inhibiting the invasion of *Salmonella choleraesuis* to human Caco-2 cell line (Lin et al. 2008). The inhibition effect appears to be multifactorial that includes the adhesion to host intestine epithelium, production of organic acids and bacteriocin by lactobacilli cells. MALDI-TOF/TOF technique was employed to understand the effects of key components of *Lb. acidophilus* L-92 that affect adhesion to Caco-2 cells. It was stated that surface layer protein A might play a key role in its attachment and in the release of IL-12 from dendritic cells (Ashida et al. 2011).

For therapeutic value of acidophilus milk on constipation, experiments were conducted on 124 persons over a period of 18 months. Out of 74 persons, who finished these experiments, 43 were constipated. Approximately two-thirds of those persons experiencing intestinal difficulties were benefitted by acidophilus therapy. A majority of the non-constipated persons reported themselves in a much better physical condition, while they were drinking acidophilus milk (Stark et al. 1934). Treatment of diarrhea by administering living or dried bacteria to restore a disturbed intestinal microflora has a long tradition. The mechanisms and the efficacy of a probiotic effect often depend on interactions with the specific microflora of the host or immunocompetent cells of the intestinal mucosa. *Lb. acidophilus* providing immunostimulatory properties or the alleviation of symptoms and shortening of acute infections are perhaps the best-documented probiotic effects validated by clinical studies. Administration of yogurt made with starters containing the conventional yogurt bacteria *Lb. bulgaricus* and *Str. thermophilus* supplemented with *Lb. acidophilus*, *B. bifidum*, and *B. infantis* to rats showed enhanced mucosal and systemic IgA responses to the cholera toxin immunogen than yogurt that was manufactured with starters containing only conventional yogurt bacteria (Tejada-Simon et al. 1999). Undernutrition impaired the ability of the lactobacillus supplement to prevent children diarrhea (Río et al. 2004). *Lb. acidophilus* was recommended as adjuvant therapy in combination with olsalazine to achieve more effective treatment for ulcerative colitis (Abdin and Saeid 2008). An abnormal gut bacterial colonization has been proposed as a reason of necrotizing enterocolitis (NEC) in newborns. As an alternate therapy against NEC prophylactic, 250 million live

Lb. acidophilus and 250 million *B. infantis* strains were orally administered to newborn infants requiring intensive care hospitalization. The results were encouraging with cases decreasing to 34 in probiotic group compared to 85 NEC cases in control (Hoyos 1999).

In an era, where consumers are on the look for clean labels that are devoid of artificial ingredients and preservatives, antimicrobial peptides have been gaining attention as antimicrobial alternatives to chemical food preservatives and commonly used antibiotics. The quantity of the antimicrobial substances produced by *Lb. acidophilus* strains depends significantly on the composition of the nutrient medium, pH of the medium at the end of the fermentation, presence of bile salts, and buffering agent (Fernandes et al. 1988). Joseph et al. (1997) reported that *Lb. acidophilus* strains produced a substance active against strains of *Lb. delbrueckii* sp. *bulgaricus*. *Lb. acidophilus* 30SC produced a heat-stable antimicrobial compound that inhibited a number of Gram-positive bacteria including *Listeria ivanovii* (Oh et al. 2000). A bacteriocin produced by *Lb. acidophilus* LA-1 called acidophilicin LA-1 was active against *Lb. delbrueckii* sp. *bulgaricus*, *Lb. helveticus*, *Lb. jugurti*, and *Lb. casei*. Shahani et al. (1977) developed methods to purify the factor responsible for antibiosis produced by *Lb. acidophilus*. The antibiotic thus isolated has been termed *acidophilin*. The production of 50,000 Da acidophilicin LA-1 increased on addition of β-glycerophosphate and was highest at pH 5.5–6.0. *Lb. acidophilus* n.v. Er 317/402 strain Narine produces acidocin LCHV, that has remarkable heat stability (90 min at 130°C), active over a wide pH range, and has a broad spectrum of activity both against methicillin-resistant *S. aureus* (MRSA) and *Clostridium difficile* (Mkrtchyan et al. 2010). Karska-Wysocki et al. (2010) revealed that the direct interaction of *Lb. acidophilus* CL1285 and *Lb. casei* LBC80R as pure cultures and MRSA in liquid medium led to the elimination of 99% of the MRSA cells after 24 h of their incubation at 37°C.

The development of symptoms of lactose intolerance depends mainly on the dose of lactose ingested. Lactose loads of 15 g or greater invariably produce symptoms in lactase-deficient persons, while with a dose of 12 g lactose, symptoms can be minimal or absent. Tolerance up to 20 g of lactose in acidophilus milk and yogurt has been suggested and is thought to be due to a low lactose content or *in vivo* autodigestion by microbial β-galactosidase in yogurt is tolerated well by lactase-deficient persons (Savaiano et al. 1984). Milks inoculated with 10^{10} cells of *Lb. acidophilus* or with a yogurt cultured played decreased symptoms compared with uninoculated milk in lactose maldigesting children (Montes et al. 1995). While selecting strains for the purpose of alleviating lactose intolerance, bile and acid tolerance of the strain are the crucial factors to be considered. It was observed that *Lb. acidophilus* N1 exhibiting lower β-galactosidase activity and lactose transport but higher bile and acid tolerance among the various strains studied was most effective in improving lactose tolerance in human subjects (Mustapha et al. 1997).

Lb. acidophilus La1 is known to induce changes in intestinal flora by persisting in the gastrointestinal tract, thereby acts as adjuvant to the humoral immune response (Link-Amster et al. 1994). The probiotic potential of *Lb. helveticus* MTCC 5463 (earlier known as *Lb. acidophilus* V3) has been extensively studied under *in vitro* and *in vivo* conditions. The strain has also shown positive immunomodulating effects in a chick model (Patidar and Prajapati 1999). Immunofluorescence and immunosorbent assays were used to collect evidence on the enhanced gut and systemic immune responses in mice fed with cultures of either *Lb. acidophilus* or *Lb. paracasei* at 10^8 CFU/50 µl per day for 14 days. The results indicated an increase in the number of IgA producing cells, IL-10 and IFN-γ cytokine producing cells in small intestine with enhanced secretion of anti-inflammatory cytokine and pro-inflammatory cytokine (Paturi et al. 2007).

Lb. acidophilus of human origin, which assimilates cholesterol, grows well in the presence of bile, produces bacteriocins, and will have an advantage over another strain that is not effective in establishing and assimilating cholesterol in the intestinal tract (Gilliland and Walker 1990). The hypocholesterolemic effect of *Lb. acidophilus* MTCC 5463 had been verified in human volunteers (Ashar and Prajapati 2000). Feeding of acidophilus milk resulted in reduction of total cholesterol by 11.7%, 21.0%, 12.4%, and 16.4% in volunteer group A1 (40–60 years), C2 (200–220 mg/dl initial cholesterol), C3 (220–250 mg/dl initial cholesterol), and H1 (normal health), respectively. Remarkable positive effects on mice fed with commercial rodent chow plus yogurt made from milk inoculated with a 0.01% (w/v) freeze-dried culture of *Str. thermophilus* plus *Lb. acidophilus* were observed on the weight gain, serum cholesterol, high-density lipoprotein cholesterol, low-density lipoprotein cholesterol, triglycerides, and the numbers of fecal lactobacilli and coliforms (Akalin et al. 1997). Studies claim that certain lactobacilli or bifidobacteria decrease the concentration of the blood stream cholesterol in humans (Mann and Spoerry 1974; Agerbaek et al. 1995), swine (Gilliland et al. 1985), and rat (Taranto et al. 1998). Consumption of non-fermented forms like synbiotic capsules containing *Lb. acidophilus* CHO-220 and inulinal so improved the plasma, total and LDL-cholesterol levels among hypercholesterolemic men and women by modifying the interconnected pathways of lipid transporters and without contributing to bile-related toxicity (Ooi et al. 2010).

Consumption of a diet of low-fat (2.5%) dahi containing probiotic *Lb. acidophilus* and *Lb. casei* significantly delayed the onset of glucose intolerance, hyperglycemia, hyperinsulinemia, dyslipidemia, and oxidative stress in high fructose-induced diabetic rats (Yadav et al. 2007b). Supplementation of *Lb. acidophilus* with *dahi* cultures increased the efficacy of dahi to suppress treptozotocin-induced diabetes in rats by inhibiting depletion of insulin, as well as preserving diabetic dyslipidemia, inhibiting lipid peroxidation and nitrite formation. This may empower antioxidant system of β-cells and may slow down the reduction of insulin and elevation of blood glucose levels (Yadav et al. 2008). Feeding probiotic yogurt containing *Lb. acidophilus* LA-5 and *Bifidobacterium lactis* Bb12 significantly decreased atherogenic indices, improved total cholesterol and LDL-C concentrations in type 2 diabetic people, leading to the improvement of cardiovascular disease risk factors (Ejtahed et al. 2011).

The hypothesis that probiotics may promote obesity by altering the intestinal flora (Raoult 2009) remains controversial, like a study performed on 15 Indian obese female children had shown that *Bacteroides, Prevotella, Eubacterium rectale, Bifidobacterium* ssp., or *Lb. acidophilus* were equivalent between lean and obese subjects (Balamurugan et al. 2010). Million et al. in 2012 compared the obese and lean subjects by focusing on *Firmicutes, Bacteroidetes, Methanobrevibacter smithii, Lc. lactis, B. animalis,* and seven species of *Lactobacillus* by quantitative PCR and *Bifidobacterium* genera at species level. The gut microbiota associated with human obesity is depleted in *M. smithii*. Some *Bifidobacterium* or *Lactobacillus* species were associated with normal weight (*B. animalis*), while others (*Lb. reuteri*) were associated with obesity. Therefore, gut microbiota composition at the species level is related to body weight and obesity, which might be of relevance for further studies and the management of obesity.

Biochemical effects of probiotics include the reduction of fecal enzymes that can convert procarcinogens to carcinogens in the gastrointestinal system. *Lb. acidophilus* has been found to reduce the levels of fecal enzymes such as β-glucuronidase, azoreductase, nitro-reductase, and urease responsible for catalyzing the conversion of carcinogenic amines. Reconstituted non-fat dry milk was fermented by *Lb. acidophilus* BG2FO4, freeze-dried, extracted in acetone, dissolved in dimethyl sulfoxide, and assayed for antimutagenicity using Ames test against N-methyl, N-nitro, N-nitroso-guanidine, and 3,2′dimethyl-4-aminobiphenyl. Dose-dependent activity of the strain was found to be significant against both mutagens, while the maximal inhibitory activity against mutagens was more than double than exhibited by extracts of unfermented milk even when added with 2% L-lactic acid (Nadathur et al. 1994). Natural killer (NK) cells play a critical role in innate immune response and induce spontaneous killing of tumor cells and virus-infected cells, and heat-killed La205 directly stimulated NK cytolytic activity in dose- and time-dependent manners, through enhancement of granule exocytosis, and granulysin may be a critical mediator in heat-killed La205-induced granule exocytosis (Cheon et al. 2011).

9.6 Conclusions

With growing demand of consumers to balance diet and drug, various carriers of probiotic acidophilus products of dairy and non-dairy origin are being subjected to innovative technological advances. Primarily, the efforts should be directed to select safe and effective strains of *Lb. acidophilus* with known phenotypic and genotypic characteristics for precise identification and enumeration. The strains stability is another challenge that hinges on three key factors: heat and moisture stability down the supply chain, across all delivery systems. Further research, in form of controlled human studies, is needed to determine which probiotics and dosages are associated with the highest efficacy. Information regarding the interaction between bacteria and dairy is focused on the growth and survival of probiotics during production, storage, and gastric transit, and therefore, further research is needed to determine the effect of food on metabolic activities of probiotics associated with their beneficial effects.

References

Abdin, A.A. and Saeid, E.M. 2008. An experimental study on ulcerative colitis as a potential target for probiotic therapy by *Lactobacillus acidophilus* with or without "olsalazine." *Journal Crohns Colitis* 2: 296–303.

Agerbaek, M., Gerdes, L.U., and Richelsen, B. 1995. Hypocholesterolaemic effect of a new fermented milk product in healthy middle aged men. *European Journal Clinical Nutrition* 49:346–352.

Agil, R., Gaget, A., Gliwa, J., Avis, T.J., Willmore, W.G., and Hosseinian, F. 2013. Lentils enhance probiotic growth in yogurt and provide added benefit of antioxidant protection. *Lebensmittel-Wissenschaftund-Technologie* 50:45–49.

Akalin, A.S., Gonc, S., and Duzel, S. 1997. Influence of yogurt and acidophilus yogurt on serum cholesterol levels in mice. *Journal of Dairy Science* 80:2721–2725.

Akin, M.B., Akin, M.S., and Kirmaci, Z. 2007. Effects of inulin and sugar levels on the viability of yogurt and probiotic bacteria and the physical and sensory characteristics in probiotic ice cream. *Food Chemistry* 104: 93–99.

Albertini, B., Vitali, B., Paaserini, N., Cruciani, F., Di Sabatino, M., Rodriguez, L., and Brigidi, P. 2010. Development of microparticulate systems for intestinal delivery of *Lactobacillus acidophilus* and *Bifidobacterium lactis*. *European Journal of Pharmaceutical Sciences* 40: 359–366.

Altamirano-Fortoul, R., Moreno-Terrazas, A., Quezada-Gallo, C., and Rosell, M. 2012. Viability of some probiotic coatings in bread and its effect on the crust mechanical properties. *Food Hydrocolloids* 29: 166–174.

Ananta, E., Volkert, M., and Knorr, D. 2005. Cellular injuries and storage stability of spray dried *Lactobacillus rhamnosus* GG. *International Dairy Journal* 15: 399–409.

Anekella, K. and Orsat, V. 2013. Optimization of microencapsulation of probiotics in raspberry juice by spray drying. *Food Science and Technology* 50: 17–24.

Ankolekar, C., Marcia, P., Duane, G., and Kalidas, S. 2012. In vitro bioassay based screening of antihyperglycemia and antihypertensive activities of *Lactobacillus acidophilus* fermented pear juice. *Innovative Food Science and Emerging Technologies* 13: 221–230.

Ashar, M.N. and Prajapati, J.B. 2000. Verification of hypocholesterolemic effect of fermented milk on human subjects with different cholesterol level. *Folia Microbiologia* 45:263–268.

Ashida, N., Sae, Y., Tadashi, S., and Naoyuki, Y. 2011. Characterization of adhesive molecule with affinity to Caco-2 cells in *Lactobacillus acidophilus* by proteome analysis. *Journal of Bioscience and Bioengineering* 112: 333–337.

Azcarate-Peril, M.A., Tallon, R., and Klaenhammer, T.R. 2009. Temporal gene expression and probiotic attributes of *Lactobacillus acidophilus* during growth in milk. *Journal of Dairy Science* 92: 870–886.

Babu, V., Mital, B.K., and Garg, S.K. 1992. Effect of tomato juice addition on the growth and activity of *Lactobacillus acidophilus*. *International Journal of Food Microbiology* 17: 67–70.

Balamurugan, R., George, G., Kabeerdoss, J., Hepsiba, J., Chandragunasekaran, A.M., and Ramakrishna, B.S. 2010. Quantitative differences in intestinal *Faecalibacterium prausnitzii* in obese Indian children. *Brazilian Journal of Nutrition* 103, 335–338.

Bergamini, C.V., Hynes, E.R., Quiberoni, A., and Zalazar, C.V. 2005. Probiotic bacteria as adjunct starters: Influence of the addition methodology on their survival in a semi-hard Argentinean cheese. *Food Research International* 38: 597–604.

Black, L.A. 1931. Viability of *Lactobacillus acidophilus* and *Lactobacillus bulgaricus* cultures stored at various temperatures. *Journal of Dairy Science* 14:59–72.

Bolin, Z., Libudzisz, Z., and Moneta, J. 1998. Viability of *L. acidophilus* in fermented milk products during refrigerated storage. *Polish Journal of Food and Nutrition Science* 7:465–472.

Brents, M., Slavgorodskaia, I.P., Kalinina, N.N., Verzhinskaia, M.F., and Luchkina, O.I. 1982. Specialized acidophilus products in the nutrition of athletes. *VoprPitan* 2:18–21.

Buriti, C.A., Castro, I.A., and Saad, M.I. 2010. Viability of *Lactobacillus acidophilus* in synbiotic guava mousses and its survival under in vitro simulated gastrointestinal conditions. *International Journal of Food Microbiology* 137: 121–129.

Burns, P., Patrignani, F., Serrazanetti, D., Vinderola, G.C., Reinheimer, J.A., Lanciotti, R., and Guerzoni, M.E. 2008. Probiotic Crescenza cheese containing *Lactobacillus casei* and *Lactobacillus acidophilus* manufactured with high-pressure homogenized milk. *Journal of Dairy Science* 91: 500–512.

Centeno-de-Laro, R. 1987. Effect of frozen and subsequent refrigerated storage on beta-galactosidase activity, viability and bile resistance of *L. acidophilus*. *Dissertation Abstracts International B* 48: 113.

Chan, E.S. and Zhang, Z. 2002. Encapsulation of probiotic bacteria *Lactobacillus acidophilus* by direct compression. *Food and Bioproducts Processing* 80: 78–82.

Charalampopoulos, D., Pandiella, S.S., and Webb, C. 2003. Evaluation of the effect of malt, wheat and barley extracts on the viability of potentially probiotic lactic acid bacteria under acidic conditions. *International Journal Food Microbiology* 82:133–141.

Chaudhary, M.G. 2013. *Development of Probiotic Ice Cream Supplemented with Finger Millet (Ragi)*. MTech dissertation. Anand, India: Anand Agricultural University.

Cheon, E.C., Khashayarsha, K. and Khan, M.W., Strouch, M.J., Krantz, S.B., Phillips, J., Blatner, N.R. et al. 2011. Mast cell 5-lipoxygenase activity promotes intestinal polyposis in APCDelta468 mice. *Cancer Research* 71:1627–1636.

Chou, L. and Weimer, B. 1999. Isolation and characterization of acid- and bile-tolerant isolates from strains of *Lactobacillus acidophilus*. *Journal of Dairy Science* 82:23–31.

Corcoran, B.M., Ross, R.P., Fitzgerald, G.F., and Stanton, C. 2004. Comparative survival of probiotic lactobacilli spray-dried in the presence of prebiotic substances. *Journal of Applied Microbiology* 96: 1024–1039.

Cruz, A.G., Antunes, A.E.C., Sousa, O.P., Faria, A.F., and Saad, S.M.I. 2009. Ice-cream as a probiotic food carrier. *Food Research International* 42: 1233–1239.

Deeth, H.C. and Tamine, A.Y. 1981. Yogurt: Nutritive and therapeutic aspects. *Journal of Food Protection* 44: 78–86.

Duthic, A.H., Duthic, A.E., Nilson, K.M., and Atherton, H.V. 1981. An Ideal vehicle for *Lactobacillus acidophilus*. *Dairy Field* 164: 139–140.

Ejtahed, H.S., Mohtadi-Nia, J., Homayouni-Rad, A., Niafar, M., Asghari-Jafarabadi, M., Mofid, V., and Akbarian-Moghari, A. 2011. Effect of probiotic yogurt containing *Lactobacillus acidophilus* and *Bifidobacterium lactis* on lipid profile in individuals with type 2 diabetes mellitus. *Journal of Dairy Science* 94: 3288–3294.

Eller, H. 1971. *The Technology of Sour Milk Products*. Tallinn, Estonia: The Ministry of Meat and Milk Industry of the Estonian S.S.R.

Elli, M., Ralf, Z., Roberto, R., and Lorenzo, M. 1999. Growth requirements of *Lactobacillus johnsonii* in skim and UHT milk. *International Dairy Journal* 9: 507–513.

Espina, F. and Packard, V.S. 1979. Survival of *Lactobacillus acidophilus* in a spray-drying process. *Journal of Food Protection* 42: 149–152.

Espírito Santo, A.P.d., Cartolano, N.S., Silva, T.F., Soares, F.A., Gioielli, L.A., Perego, P., Converti, A., and Oliveira, M.N. 2012. Fibers from fruit by-products enhance probiotic viability and fatty acid profile and increase CLA content in yoghurts. *International Journal of Food Microbiology* 154: 135–144.

Espírito Santo, A.P.d., Silva, R.C., Soares, F.A.S.M., Anjos, D., Gioielli, L.A., and Oliveira, M.N. 2010. Açai pulp addition improves fatty acid profile and probiotic viability in yoghurt. *International Dairy Journal* 20: 415–422.

Fernandes, C.F., Shahani, M., and Amer, M.A. 1988. Effect of nutrient media and bile salts on growth and antimicrobial activity of *Lactobacillus acidophilus*. *Journal of Dairy Science* 71: 3222–3229.

Frost, W.D. and Butterworth, T.H. 1931. Present status of acidophilus milk. *American Journal of Public Health* 21:862–866.

Gandhi, D.N. and Kumar, A. 1980. Spray-dried acidophilus milk powder. *Indian Dairyman* 32: 599–600.

Gandhi, D.N. and Nambudripad, V.K.N. 1979. An introduction to different acidophilus milk products and its concentrates. *Indian Dairyman* 31: 813–818.

Gardini, F., Lanciott, R., Guerzoni, M.E., and Torriani, S. 1999. Evaluation of aroma production and survival of *Streptococcus thermophilus*, *Lactobacillus delbrueckii* subsp. *bulgaricus* and *Lactobacillus acidophilus* in fermented milks. *International Dairy Journal* 9: 125–134.

Gilliland, S.E., Nelson, C.R., and Maxwell, C. 1985. Assimilation of cholesterol by *Lactobacillus acidophilus*. *Applied Environmental Microbiology* 49: 377–381.

Gilliland, S.E. and Speck, M.L. 1977. Antagonistic action of *Lactobacillus acidophilus* toward intestinal and food-borne pathogens in associative cultures. *Journal of Food Protection* 40: 820–823.

Gilliland, S.E. and Speck, M.L. 1977. Instability of *Lactobacillus acidophilus* in yogurt. *Journal of Dairy Science* 60: 1394–1398.

Gilliland, S.E. and Walker, D.K. 1990. Factors to consider when selecting a culture of Lactobacillus acidophilus as a dietary adjunct to produce a hypocholesterolemic effect in humans. *Journal of Dairy Science* 73: 905–911.

Goff, D. 2008. 65 years of ice-cream science. *International Dairy Journal* 18: 754–758.

Gomes, A.P., Cruz, A.G., Cadena, R.S., Celeghini, R.M., Faria, J.A., Bolini, H.M., Pollonio, M.A., and Granato, D. 2011. Low sodium Minas fresh cheese manufacture: Effect of the partial replacement of NaCl with KCl. *Journal of Dairy Science* 94: 2706–2011.

Gomes, M.P. and Malcata, F.X. 1999. *Bifidobacterium* spp. and *Lactobacillus acidophilus*: Biological, biochemical, technological and therapeutical properties relevant for use as probiotics. *Trends in Food Science and Technology* 10: 139–157.

Goswami, R. 2012. Comparative study of *Lactobacillus helveticus* MTCC 5463 for folate and biotin production in fermented milk. MSc thesis. Anand, India: Anand Agricultural University.

Hagen, M. and Narvhus, J.A. 1999. Production of ice cream containing probiotic bacteria. *Milchwissenschaft* 54:265–268.

Han-Seung, J.L., Pestka, J.J., and Ustunol, Z. 2000. Viability *of Bifidobacterium lactis* in commercial dairy products during refrigerated storage. *Journal of Food Protection* 63: 327–331.

Hekmat, S. and McMahon, D.J. 1992. Survival of *Lactobacillus acidophilus* and *Bifidobacterium bifidum* in ice cream for use as a probiotic food. *Journal of Dairy Science* 75:1415–1422.

Hoier, E. 1992. Use of probiotic starter cultures in dairy products. *Food Australia* 44: 418–420.

Hoyos, A.B. 1999. Reduced incidence of necrotizing enterocolitis associated with enteral administration of *Lactobacillus acidophilus* and *Bifidobacterium infantis* to neonates in an intensive care unit. *International Journal of Infectious Diseases* 3: 197–202.

Iwana, H., Masuda, H., Fujisawa, T., Suzuki, H., and Mitsuoka, T. 1993. Isolation of *Bifidobacterium* spp. in commercial yoghurts sold in Europe. *Bifidobacteria Microflora* 12: 39–45.

Johnson, M.C., Ray, B., and Bhowmik, T. 1987. Selection of *Lactobacillus acidophilus* strains for use in "acidophilus products." *Antonie van Leeuwenhoek* 53(4):215–231.

Joseph, P., Dave, R.I., and Shah, N.P. 1997. Antagonism between yoghurt bacteria and probiotic bacteria isolated from commercial starter cultures, commercial yoghurts and a probiotic capsule. *Food Australia* 50:20–23.

Kailasapathy, K. and Chin, J. 2000. Survival and therapeutic potential of probiotic organism with reference to *Lactobacillus acidophilus* and *Bifidobacterium* spp. *Immunology Cell Biology* 78: 80–88.

Karska-Wysocki, B., Bazo, M., and Smoragiewicz, W. 2010. Antibacterial activity of *Lactobacillus acidophilus* and *Lactobacillus casei* against methicillin-resistant *Staphylococcus aureus* (MRSA). *Microbiology Research* 165: 674–686.

Kasimoglu, A., Göncüglu, M., Akgün, S. 2004. Probiotic white cheese with *Lactobacillus acidophilus*. *International Dairy Journal* 14: 1067–1073.

Kim, H.S. and Gilliland, S. 1983. Lactobacillus acidophilus as a dietary adjunct for milk to aid lactose digestion in humans. *Journal of Dairy Science* 66: 959–966.

Knaysi, G. 1932. An economical method of producing acidophilus milk. *Journal of Dairy Science* 15: 71–72.

Kneifel, W., Jaros, D., and Erhard, F. 1993. Microflora and acidification properties of yogurt and yogurt-related products fermented with commercially available starter cultures. *International Journal of Food Microbiology* 18: 179–189.

Kopeloff, N. 1926. *Lactobacillus acidophilus*. Baltimore, MD: Williams & Wilkins.

Korobkina, G.S., Brents, M.Y., Kalinina, N.N., Vorob'eva, V.M., and Shamanova, G.P. 1981. Development of dried acidophilus milk formulae for infant feeding. *Voprosy Pitaniya* 1: 51–54.

Koroleva, N.S. 1991. Products prepared with lactic acid bacteria and yeasts. In: *Therapeutic Properties of Fermented Milks*, ed. R.K. Robinson, 159–179. London: Elsevier Applied Science.

Kulp, W.L. 1931. Studies on the viability of *L. acidophilus* in "Acidophilus Milk." *American Journal of Public Health and Nations Health* 21: 873–883.

Lang, F. and Lang, A. 1975. Acidophilus milk products: Little known cultured milks of great potential. *The Milk Industry* 77: 4–6.

Lankaputhra, W. E. V. and Shah, N. P. 1995. Survival of *Lactobacillus acidophilus* and *Bifidobacterium* species in the presence of acid and bile salts. *Cultured Dairy Products Journal*, 30: 113–118.

Larsen, C., Nielsen, S., Kaestel, P., and Brockmann, E. 2006. Dose–response study of probiotic bacteria *Bifidobacterium animalis* subsp. *lactis* BB-12 and *Lactobacillus paracasei* subsp. *paracasei* CRL-341 in healthy young adults. *European Journal of Clinical Nutrition* 60: 1284–1293.

Lavermicocca, P., Rossi, M., Russo, F., and Srirajaskanthan, R. 2010. Table olives: A carrier for delivering probiotic bacteria to humans. In: *Olives and Olive Oil in Health and Disease Prevention*, eds. V.R. Preedy and R.R. Vatson, 735–747. London: Academic Press.

Lian, W.C., Hsiao, H.C., and Chou, C.C. 2002. Survival of bifidobacteria after spray drying. *International Journal of Food Microbiology* 74: 79–86.

Lin, C.K., Tsai, H.C., Lin, P.P., Tsen, H.Y., and Tsai, C.C. 2008. *Lactobacillus acidophilus* LAP5 able to inhibit the *Salmonella choleraesuis* invasion to the human Caco-2 epithelial cell. *Anaerobe* 14: 251–255.

Lin, M.Y. and Young, C.M. 2000. Folate levels in cultures of lactic acid bacteria. *International Dairy Journal* 10:409–414.

Link-Amster, H., Rochat, F., Saudan, K.Y., Mignot, O., and Aeschlimann, J.M. 1994. Modulation of a specific humoral immune response and changes in intestinal flora mediated through fermented milk intake. *FEMS Immunology and Medical Microbiology* 10: 55–63.

Liu, S.Q. and Tsao, M. 2009. Enhancement of survival of probiotic and non-probiotic lactic acid bacteria by yeasts in fermented milk under non-refrigerated conditions. *International Journal of Food Microbiology* 30: 34–38.

Lourens-Hattingh, A. and Viljoen, B.C. 2001. Yoghurt as probiotic carrier food. *International Dairy Journal* 11: 1–17.

Magdi, A.O., Ibrahim, E.A.R., and Hamid, A.D. 2010. Biochemical changes occurring during fermentation of camel milk by selected bacterial starter cultures, *African Journal of Biotechnology* 9: 7331–7336.

Mann, G.V. and Spoerry, A. 1974. Studies of a surfactant and cholesteremia in the Maasai. *American Journal of Clinical Nutrition* 27:464–469.

Matsumura, A., Saito, T., Arakuni, M., Kitazawa, H., Kawai, Y., and Itoh, T. 1999. New binding assay and preparative trial of cell-surface lectin from *Lactobacillus acidophilus* Group Lactic Acid Bacteria. *Journal of Dairy Science* 82: 2525–2529.

Million, M., Maraninchi, M., Henry, M., and Armougom, F. 2012. Obesity-associated gut microbiota is enriched in *Lactobacillus reuteri* and depleted in *Bifidobacterium animalis* and *Methanobrevibacter smithii*. *International Journal of Obesity* 36: 817–825.

Mirzaei, H, Pourjafar, H., and Homayouni, A. 2012. Effect of calcium alginate and resistant starch microencapsulation on the survival rate of *Lactobacillus acidophilus* La5 and sensory properties in Iranian white brined cheese. *Food Chemistry* 132:1966–1970.

Mkrtchyan, H., Gibbons, S., Heidelberger, S., and Zloh, M. 2010. Purification, characterization and identification of acidocin LCHV, an antimicrobial peptide produced by *Lactobacillus acidophilus* n.v. Er 317/402 strain Narine. *International Journal of Antimicrobial Agents* 35:255–260.

Montes, R.G., Bayless, T.M., Saavedra, J.M., and Perman, J.A. 1995. Effect of milks inoculated with *Lactobacillus acidophilus* or a yogurt starter culture in lactose mal-digesting children. *Journal of Dairy Science* 78: 1657–1664.

Mustapha, A., Jiang, T., and Savaiano, D.A. 1997. Improvement of lactose digestion by humans following ingestion of unfermented acidophilus milk: Influence of bile sensitivity, lactose transport, and acid tolerance of *Lactobacillus acidophilus*. *Journal of Dairy Science* 80: 1537–1545.

Nadathur, A.R., Gould, S.J., and Bakalinsky, A.T. 1994. Antimutagenicity of fermented milk. *Journal of Dairy Science* 77:3287–3295.

Nazzaro, F., Fratianni, F., Nicolaus, B., and Poli, A. 2012. The prebiotic source influences the growth, biochemical features and survival under simulated gastrointestinal conditions of the probiotic *Lactobacillus acidophilus*. *Anaerobe* 18: 280–285.

Ng, E.W., Yeung, M., and Tong, P.S. 2011. Effects of yogurt starter cultures on the survival of *Lactobacillus acidophilus*. *International Journal of Food Microbiology* 145:169–175.

Nielsen, J.W. and Gilliland, S.E. 1992. The lactose hydrolyzing enzyme from *Lactobacillus acidophilus*. *Cultured Dairy Products Journal* 27: 20–24.

Nighswonger, B.D., Brashears, M.M., and Gilliland, S.E. 1996. Viability of *L. acidophilus* and *L. casei* in fermented milk products during refrigerated storage. *Journal of Dairy Science* 79: 212–219.

Oda, T., Kado-Oka, Y., and Hashibara, H. 1994. Effect of Lactobacillus on iron bioavailability in rats. *Journal of Nutrition Science and Vitaminology* 40: 613–616.

Ogawa, J., Kishino, S., Ando, A., Sugimoto, S., Mihara, K., and Shimizu, S. 2005. Production of conjugated fatty acids by lactic acid bacteria. *Journal of Bioscience and Bioengineering* 100: 355–364.

Oh, S., Kim, S.H., and Worobo, R.W. 2000. Characterization and purification of a bacteriocin produced by a potential probiotic culture, *Lactobacillus acidophilus* 30SC. *Journal of Dairy Science* 83: 2747–2752.

Oliveira, M.N., Sodini, I., Remeuf, F., and Corrieu, G. 2001. Effect of milk supplementation and culture composition on acidification, textural properties and microbiological stability of fermented milks containing probiotic bacteria. *International Dairy Journal* 11: 935–942.

Oliveira, R.P.S., Perego, P., Oliveira, M.N., and Converti, A. 2011. Growth and survival of mixed probiotics in nonfat fermented milk: The effect of inulin. *Chemical Engineering Transactions* 24: 457–462.

Ooi, L.G., Ahmad, R., and Yuen, K.H. 2010. *Lactobacillus gasseri* CHO-220 and inulin reduced plasma total cholesterol and low-density lipoprotein cholesterol via alteration of lipid transporters. *Journal of Dairy Science* 93:5048–5058.

Ortakci, F. and Sert, S. 2012. Stability of free and encapsulated *Lactobacillus acidophilus* ATCC 4356 in yogurt and in an artificial human gastric digestion system. *Journal of Dairy Science* 95: 6918–6925.

Ostlie, H., Hell, M.H., and Narvhus, J. 2003. Growth and metabolism of probiotics in fermented milk. *International Journal of Food Microbiology* 87: 17–27.

Patidar, S.K. and Prajapati, J.B. 1999. Effect of feeding lactobacilli on serum antibody titre and faecal flora in chicks. *Microbiologie, Aliments, Nutrition* 17:145–154.

Paturi, G., Phillips, Jones, M., and Kailasapathy, K. 2007. Immune enhancing effects of Lactobacillus acidophilus LAFTI L10 and *Lactobacillus paracasei* LAFTI L26 in mice. *International Journal of Food Microbiology* 115:115–118.

Pedroso, D.L., Thomazini, M., Heinemann, R.J.B., and Favaro-Trindade, C.S. 2012. Protection of *Bifidobacterium lactis* and *Lactobacillus acidophilus* by microencapsulation using spray-chilling. *International Dairy Journal* 26:127–132.

Prajapati, J.B., Khedkar, C.D., Chitra, J., Suja, S., Mishra, V., Sreeja, V., Patel, R.K. et al. 2011. Whole genome shotgun sequencing of Indian origin strain of *Lactobacillus helveticus* MTCC 5463 with probiotic potential. *Journal of Bacteriology* 5:449–511.

Prajapati, J.B, Shah, R.K., and Dave, J.M. 1986. Nutritional and therapeutical benefits of a blended-spray dried acidophilus. *Cultured Dairy Products Journal* 21: 16–21.

Ranadheera, C.S., Evans, C.A., Adams, M.C., and Baines, S.K. 2013. Production of probiotic ice cream from goat's milk and effect of packaging materials on product quality. *Small Ruminant Research* 112:174–180.

Rao, D.R. and Shahani, K.M. 1987. Vitamin content of cultured milk products. *Cultured Dairy Products Journal* 22: 6–10.

Raoult, D. 2009. Probiotics and obesity: A link? *Nature Reviews Microbiology* 7: 616.

Rice, F.E. 1928. The preparation of acidophilus milk. *American Journal of Public Health and the Nations Health* 18: 105–1108.

Río, M.E., Zago, L.B., Garcia, H., and Winter, L. 2004. Influence of nutritional status on the effectiveness of a dietary supplement of live lactobacillus to prevent and cure diarrhoea in children. ArchivosLatinoamericanos de Nutrición 54:287–292.

Rodríguez-Huezo, M.E., Durán Lugo, R., Prado Barragán, L., Cruz-Sosa, F., Lobato-Calleros, C., Alvarez-Ramírez, J., and Vernon-Carter, E.J. 2007. Pre-selection of protective colloids for enhanced viability of *Bifidobacterium bifidum* following spray-drying and storage, and evaluation of aguamiel as thermoprotective prebiotic. *Food Research International* 40: 1299–1306.

Santiago, T., Sáenz-Collins, G., and Rojas-de Gante, C.P. 2012. Elaboration of a probiotic oblea from whey fermented using *Lactobacillus acidophilus* or *Bifidobacterium infantis*. *Journal of Dairy Science* 95: 6897–6904.

Santillo, A., Albenzio, M., Bevilacqua, A., Corbo, M.R., and Sevi, A. 2012. Encapsulation of probiotic bacteria in lamb rennet paste: Effects on the quality of Pecorino cheese. *Journal of Dairy Science* 95: 3489–3500.

Santos dos, K.M.O., Bomfim, M.A.D., Vieira, A.D.S., Benevides, S.D., Saad, S.M.I., Buriti, F.C.A., and Egito, A.S. 2012. Probiotic caprine coalho cheese naturally enriched in conjugated linoleic acid as a vehicle for Lactobacillus acidophilus and beneficial fatty acids. *International Dairy Journal* 24: 107–112.

Savaiano, D.A., Anouar, A.A., Smith, D.E., and Levitt, M.D. 1984. Lactose malabsorption from yoghurt, pasteurized yoghurt, sweet acidophilus milk, and cultured milk in lactase-deficient individuals. *American Journal of Clinical Nutrition* 40: 1219–1223.

Savard, P., Lamarche, B., Paradis, M.P., and Thiboutot, H. 2011. Impact of *Bifidobacterium animalis* subsp. *lactis* BB-12 and, *Lactobacillus acidophilus* LA-5-containing yoghurt, on fecal bacterial counts of healthy adults. *International Journal of Food Microbiology* 149: 50–57.

Schillinger, U. 1999. Isolation and identification of lactobacilli from novel-type probiotic and mild yoghurts and their stability during refrigerated storage. *International Journal of Food Microbiology* 47: 79–87.

Shah, N. and Prajapati, J.B. 2013. Effect of carbon dioxide on sensory attributes, physico-chemical parameters and viability of probiotic *L. helveticus* MTCC 5463 in fermented milk. *Journal of Food Science and Technology*. doi: 10.1007/s13197-013-0943-9.

Shah, N.P. and Lankaputhra, W.E.V. 1997. Improving viability of *Lactobacillus acidophilus* and *Bifidobacterium* ssp. in yogurt. *International Dairy Journal* 7: 349–359.

Shah, R.K., Prajapati, J.B., and Dave, J.M. 1987. Packaging materials to store a spray dried a acidophilus malt preparation. *Indian Journal of Dairy Science* 40: 287–291.

Shahani, J.M., Vakil, J.R., and Kilara, A. 1977. Natural antibiotic activity of *L. acidophilus* and *bulgaricus*. II. Isolation of acidophilin from *L. acidophilus*. *Cultured Dairy Products Journal* 12(2): 8.

Sharma, N. and Gandhi, D.N. 1981. Preparation of acidophilin,-1. Selection of the starter culture. *Cultured Dairy Products Journal* 16: 6–10.

Shihata, A. and Shah, N.P. 2002. Influence of addition of proteolytic strains of *Lactobacillus delbrueckii* subsp. *bulgaricus* to commercial ABT starter cultures on texture of yoghurt, exopolysaccharide production and survival of bacteria. *International Dairy Journal* 12: 765–772.

Shoji, A.S., Oliveira, A.C., Balieiro, J.C.C., Freitas, O., Thomazini, M., Heinemann, R.J.B., Okuro, P.K., and Favaro-Trindade, C.S. 2013. Viability of *L. acidophilus* microcapsules and their application to buffalo milk yoghurt. *Food and Bioproducts Processing* 91: 83–88.

Silva, M.R., Dias, G., Ferreira, C.L.L.F., Franceschini, S.C.C., and Costa, N.M.B. 2008. Growth of preschool children was improved when fed an iron-fortified fermented milk beverage supplemented with *Lactobacillus acidophilus*. *Nutrition Research* 28: 226–232.

Silva, M.R., Dias, G., Ferreira, L.L.C., Franceschini, S.C., and Costa, M.B. 2001. Growth of preschool children was improved when fed an iron-fortified fermented milk beverage supplemented with *Lactobacillus acidophilus*. *Nutrition Research* 28: 226–232.

Sinha, D.K., Dam, R., and Shahani, K.M. 1979. Evaluation of the properties of a nonfermented acidophilus milk. *Journal of Dairy Science* 62: 52–56.

Souza, H.B. and Saad, M.I. 2009. Viability of *Lactobacillus acidophilus* La-5 added solely or in co-culture with a yoghurt starter culture and implications on physico-chemical and related properties of Minas fresh cheese during storage. *LWT—Food Science and Technology* 42: 633–640.

Speck, M.L. 1980. Preparation of lactobacilli for dietary uses. *Journal of Food Protection* 42: 65–67.

Sreekumar, O., Hosono, A. 2000. Immediate effect of *Lactobacillus acidophilus* on the intestinal flora and fecal enzymes of rats and the *in vitro* inhibition of *Escherichia coli* in co-culture. *Journal of Dairy Science* 83: 931–939.

Srinivas, D., Mital, B.K., and Garg, S.K. 1990. Utilization of sugars by *Lactobacillus acidophilus* strains. *International Journal of Food Microbiology* 10: 51–57.

Stark, C.N., Gordon, R., Mauer, J.C., Curtis, L.R., and Schubert, J.H. 1934. Studies on acidophilus milk. *American Journal of Public Health* 24: 470–488.

Subramanian, P. and Shankar, P.A. 1985. Commensalistic interaction between *Lactobacillus acidophilus* and lactose-fermenting yeasts in the preparation of acidophilus-yeast milk. *Cultured Dairy Products Journal* 20: 17–26.

Talwalkar, A. and Kailasapathy, K., 2003. Metabolic and biochemical responses of probiotics bacteria in oxygen. *Journal of Dairy Science* 86: 2537–2546.

Tamime, A.Y. and Robinson, R.K. 1988. Fermented milks and their future trends: II. *Journal of Dairy Research* 55: 281–307.

Taranto, M.P., Medici, M., Perdigon, G., Ruiz Holgado, A.P., and Valdez, G.F. 1998. Evidence for hypocholesterolemic effect of *Lactobacillus reuteri* in hypercholesterolemic mice. *Journal of Dairy Science* 81:2336–2340.

Tejada-Simon, M.V. and Pestka, J.J. 1999. Pro-inflammatory cytokine and nitric oxide induction in murine macrophages by cell wall and cytoplasmic extracts of lactic acid bacteria. *Journal of Food Protection* 62: 1435–1444.

Tharmaraj, N. and Shah, N.P. 2004. Survival of *Lactobacillus acidophilus*, *Lactobacillus paracasei subsp. paracasei*, *Lactobacillus rhamnosus*, *Bifidobacterium animalis* and *Propionibacterium* in cheese-based dips and the suitability of dips as effective carriers of probiotic bacteria. *International Dairy Journal* 14: 1055–1066.

Varga, L. Szigeti, J., Kovacs, R., and Foldes, T. 2002. Influence of a *Spirulina platensis* biomass on the microflora of fermented ABT Milks during storage. *Journal of Dairy Science* 85: 1031–1038.

Villegas, E. and Gilliland, S.E., 1998. Hydrogen peroxide production by *Lactobacillus delbrueckii* ssp. *lactis* at 5°C. *Journal of Food Science* 63: 1070–1074.

Vinderola, C.G., Bailo, N., and Reinheimer, J.A. 2000. Survival of probiotic microflora in Argentinian yoghurts during refrigerated storage. *Food Research International* 33: 97–102.

Vinderola, C.G., Mocchiutti, P., and Reinheimer, J.A. 2002. Interactions among lactic acid starter and probiotic bacteria used for fermented dairy products. *Journal of Dairy Science* 85:721–729.

Wang, Y., Corrieu, G., and Béal, C. 2005. Fermentation pH and temperature influence the cryotolerance of *Lactobacillus acidophilus* RD758. *Journal of Dairy Science* 88: 21–29.

Yadav, H., Jain, S., and Sinha, P.R. 2007a. Production of free fatty acids and conjugated linoleic acid in probiotic dahi containing *Lactobacillus acidophilus* and *Lactobacillus casei* during fermentation and storage. *International Dairy Journal* 17: 1006–1010.

Yadav, H., Jain, S., and Sinha, P.R. 2007b. Antidiabetic effect of probiotic dahi containing *Lactobacillus acidophilus* and *Lactobacillus casei* in high fructose fed rats. *Nutrition* 23:62–68.

Yadav, J.P., Saini, S., Kalia, A.N., and Dangi, A.S. 2008. Hypoglycemic activity of ethanolic extract of *Salvadoraoleoides* in normal and alloxan—Induced diabetes rats. *Indian Journal of Pharmacology* 40: 23–27.

Zamora, L.M., Carretero, C., and Parés, D. 2006. Comparative survival rates of lactic acid bacteria isolated from blood, following spray-drying and freeze-drying. *Food Science and Technology International* 12: 77–84.

Zare, F., Champagne, C.P., Simpson, B.K., and Orsat, V. 2012. Growth of starter microorganisms and acidification trend in yogurt and probiotic media supplemented with pulse ingredients. *LWT—Food Science and Technology* 45: 155–160.

10
Bifidus Milks

SREEJA V. AND JASHBHAI B. PRAJAPATI

Contents

10.1	Introduction	270
10.2	History	271
10.3	Bifidobacteria	272
10.4	Classification of Bifidus Products	278
10.5	Limitations of Bifidus Products Production	278
	10.5.1 Growth of Bifidobacteria in Milk	279
	10.5.2 Cultures	280
	10.5.3 Interaction with Starter Bacteria	281
	10.5.4 Acid Tolerance	282
	10.5.5 Presence of Oxygen	283
10.6	Products Containing Bifidobacteria	283
	10.6.1 Bifidus Milk	284
	10.6.2 Bifidus Milk with Yogurt Flavor	286
	10.6.3 Bifidus Active	287
	10.6.4 Bifidus Baby Foods	287
	10.6.5 Bifidus Yogurt	288
	10.6.6 Mil-Mil	288
	10.6.7 Sweet Bifidus Milk	289
	10.6.8 Sweet Acidophilus Bifidus Milk	289
	10.6.9 BRA Sweet Milk	289
	10.6.10 Tarag	289
	10.6.11 Progurt	289
	10.6.12 Acidophilus Milk Containing *B. bifidum*	290
	10.6.13 Diphilus Milk	291
	10.6.14 Bifidogene	291
	10.6.15 Bifider	291

10.6.16	Bio-Spread	291
10.6.17	Milk- and Water-Based Cereal Puddings	291
10.7	Measures to Improve Viability of Bifidobacteria in Bifidus Products	292
10.7.1	Encapsulation	292
10.7.2	Use of Prebiotics	293
10.7.3	Exploiting Cellular Stress Responses	293
10.7.4	Genetic Characterization	293
10.8	Therapeutic Aspects of Bifidus Products	295
10.9	Requirements for Marketing Bifidus Products	299
10.10	Conclusions	300
References		301

10.1 Introduction

The growing awareness about probiotic intestinal microbes and their various perceived health benefits has caught the attention of food industry resulting in a newer category of functional foods. Such foods are meant to exploit the beneficial effects of probiotics on intestinal metabolism and thus have a positive effect on human health. Probiotic dairy products, such as yogurts containing *Lactobacillus acidophilus* and *Bifidobacterium* ssp., constitute a significant part among the commercially available probiotic foods (Reid et al. 2003). Bifidus milk belongs to the category of functional milks containing mainly bifidobacteria. This is one of the most important groups of probiotic intestinal microbes. Usually, bifidobacteria are used in combination with other lactic acid bacteria (LAB) for product making owing to their slow growth in milk. The consumption of bifidobacteria through foods may affect the composition of indigenous microflora and may have several beneficial effects on human health, such as maintenance of balanced intestinal microflora, immune modulation, alleviation of lactose intolerance, resistance to enteric pathogens, anti-diarrheal, anti-hypertensive, and anti-carcinogenic effects. Bifidus milk and products are very popular in developed countries such as Japan, the European Union, and the United States. A comprehensive review of different milk products containing bifidobacteria is presented in this chapter.

10.2 History

The process of milk fermentation and the viable microbes responsible for it have received great attention over the past decades especially after discovering the importance of these microbes in conferring a wide range of therapeutic benefits to humans. The first scientific work to investigate the beneficial effects of fermented milk for human health had been done by Metchnikoff in the beginning of 20th century. Since then numerous scientific studies have been published describing the health benefits associated with the consumption of fermented dairy products (Modler et al. 1990; Shin et al. 2000b). LAB associated with such fermentations usually do not inhabit the human and animal gastrointestinal tract (GIT), nor do these survive passing through the digestive system (Klaver et al. 1993). Therefore, to get the beneficial health effects associated in the fermented dairy products, there has been an increased interest in the incorporation of the beneficial intestinal species, mainly *Lactobacillus acidophilus* and *Bifidobacterium* species in the fermented milk products (Reuter 1990; Shah and Lankaputhra 1997; Shah 2001). The importance of the microflora in the GIT of warm-blooded animals was revealed in the late nineteenth century when microbiologists realized that the microflora in the GIT of healthy animals differed from those found in diseased animals. Research for the isolation and characterization of these microbes revealed their role in maintaining the microbial balance in the GIT and thus human health. The growing awareness about probiotics and their numerous perceived health benefits (Table 10.1) has caught the attention of the food industry (Saarela et al. 2002; Salminen and Gueimonde 2004). Food companies are increasingly manufacturing foods with incorporated probiotics, which fall under the category of probiotic/synbiotic products. Bifidobacteria being a major part of the human intestinal microflora play an important role in maintaining health (Tannock 1999a). A significant increase in the number of foods and other dietary products containing live bifidobacteria (Isolauri et al. 2000; Kailasapathy and Chin 2000; Lievin et al. 2000; Crittenden et al. 2001; Ishibashi and Yamazaki 2001) is seen. Bifidobacteria were first used for the manufacture of baby foods by Mayer in 1948. Later, in 1968 Schuler-Malyoth and co-workers suggested the first large-scale commercial process for making fermented

Table 10.1 Potential Health Benefits of Functional Foods Prepared with Probiotic Bacteria

BENEFICIAL EFFECT	POSSIBLE CAUSES AND MECHANISMS
Improved digestibility	Partial breakdown of proteins, fats, and carbohydrates
Improved nutritional value	Higher levels of B vitamins and certain free amino acids, namely, methionine, lysine, and tryptophan
Improved lactose utilization	Reduced lactose in product and further availability of lactase
Antagonistic action toward enteric pathogens	Disorders, such as functional diarrhea, antibiotic-associated diarrhea, rotavirus diarrhea in children, and gastroenteritis in infants and adults ulcerative colitis. Production of antimicrobial substance and prevention of pathogen adhesion or pathogen activation
Colonization in gut	Survival in gastric acid, resistance to lysozyme and low surface tension of intestine, adherence to intestinal mucosa, multiplication in the intestinal tract
Immune modulation	Enhancement of macrophage formation, stimulation of production of suppressor cells and γ-interferon, suppress lymphocyte proliferation
Anti-mutagenic and anticarcinogenic effect	Conversion of potential pre-carcinogens into less harmful compounds, inhibitory action toward some types of cancer, in particular cancers of the gastrointestinal tract by binding and degradation of pre-carcinogens, reduction of carcinogen-promoting enzymes and stimulation of the immune system
Hypocholesterolemic action	Production of inhibitors of cholesterol synthesis. Use of cholesterol by assimilation and precipitation with deconjugated bile salts
Vaccine vehicle	Naturally occurring or rDNA vaccine epitopes

Sources: Gomes, A.M.P. and Malcata, F.X., *Trends Food Sci. Technol.*, 10, 139–157, 1999; Saarela, M. et al., *J. Biotechnol.*, 84, 197–215, 2000; Lin, D.C., *Nutr. Clin. Pract.*, 18(6), 497–506, 2003; Sanders, M.E. and Marco, M.L., *Annu. Rev. Food Sci. Technol.*, 1, 65–85, 2010.

milks containing bifidobacteria (Kurmann et al. 1992). As the years went by, a number of products have been commercialized containing different species of bifidobacterium, in particular *B. bifidum* and *B. longum*, often with other LAB as well.

10.3 Bifidobacteria

Bifidobacteria were discovered and described by Henry Tissier during the period 1899–1900. Henry Tissier was a French pediatrician, studying infantile diarrhea. During his study, he observed that an irregular Y shaped bacterium was present in abundance in the feces of breast-fed infants, but was absent in the feces of infants suffering from diarrhea. He called these bifid bacteria *Bacillus bifidus* (Leahy et al. 2005; Lee and O'Sullivan 2010). A historical overview of bifidobacteria is presented in Table 10.2. From 1923 to 1934, in the 1st to 4th editions of *Bergey's Manual of Systematic Bacteriology*, these bacteria were classified as *Bacteroides bifidus*. From 1939 to 1957, in the 5th to 7th editions of *Bergey's Manual*, these were called as *Lactobacillus bifidus*, because of their rod-like shapes and obligate fermentative

Table 10.2 Important Milestones in the History of Bifidobacteria

PERIOD	MILESTONES
1899–1900	Discovery of bifidobacteria.
1920	Holland named the Tissier strain *Lb. bifidus*.
1924	Orla-Jensen recognized the existence of the genus *Bifidobacterium* as a separate taxon.
1957	Dehnart first realized the existence of multiple biotypes of *Bifidobacterium* and proposed a scheme for the differentiation of these bacteria based on their carbohydrate fermentation patterns.
1963	Reuter recognized and named seven species of *Bifidobacterium*, in addition to the known *B. bifidum*.
1965–1967	Scardovi and Trovatelli, De Vries et al. clarified the characteristic metabolic pathway of hexose fermentation in bifidobacteria. Identified the key enzyme as fructose-6-phosphate phosphoketolase, which splits the hexose phosphate to erythrose-4-phosphate and acetyl phosphate.
1970	Scardovi et al. used DNA–DNA filter hybridization procedure to assess the validity of the bifidobacterial species previously described and to recognize new DNA homology groups among the strains they were isolating in large numbers from diverse ecological niches.
1974	8th edition of *Bergey's Manual of Determinative Bacteriology* classified bifidobacteria in the genus *Bifidobacterium*. The genus comprised eight species, and it was included in the family of Actinomycetaceae of the order Actinomycetales.
1982	Biavati et al. brought in other correction to the classification after the introduction of the electrophoresis of soluble cell proteins on polyacrylamide gel as a criterion of species identification
1986	Scardovi made rearrangements to previous classification and new species identification resulting in recognition of 24 species
1997	Stackebrand and colleagues proposed a novel hierarchical structure collecting the genus *Bifidobacterium* with the genus *Gardnerella* into the single family of Bifidobacteriaceae in the order of Bifidobacteriales based on 16S rRNA analysis.
1998	37 species, with four taxa (*B. longum, B. pseudolongum, B. animalis, B. thermacidophilum*).

Sources: Biavati, B. et al., *Ann. Microbiol.*, 50, 117–132, 2000; Leahy, S.C. et al., *J. Appl. Microbiol.*, 98, 1303–1315, 2005; Lee, J.H. and O'Sullivan, D.J., *Microbiol. Mol. Biol. Rev.*, 74(3): 378–416, 2010; Turroni, F. et al., *Int. J. Food Microbiol.*, 149, 37–44, 2011.

characteristics similar to that of the genus *Lactobacillus*. In the 8th edition of *Bergey's Manual of Determinative Bacteriology* (Rogosa 1974), bifidobacteria were reclassified in the genus *Bifidobacterium*, initially adopted by Orla-Jensen. The genus *Bifidobacterium* belongs to phylum Actinobacteria, order Bifidobacteriales, and family Bifidobacteriaceae. Phylum Actinobacteria include Gram-positive microbes with a high G + C DNA content ranging between 51% and 70% (Turroni et al. 2011). Genus *Bifidobacterium* consisted of only 11 species. Later, this was updated to 24 species in 1986, and currently, there are 37 species divided into four taxa, namely, *Bifidobacterium longum*, *B. pseudolongum*, *B. animalis*, *B. thermacidophilum*, and further subdivided into subspecies, exhibiting >93% identity of their 16S rDNA sequences (Miyake et al. 1998) and comprising isolates from the intestines of humans, animals, and insects, and also from human dental caries and raw milk (Table 10.3). Three new species, isolated from the bumblebee digestive tracts, that is, *B. actinocoloniiforme*, *B. bohemicum*, and *B. bombi*, have been added to the existing bifidobacteria (Killer et al. 2009, 2011; Turroni et al. 2011).

Bifidobacteria are directly or indirectly associated with the human/animal intestinal environment, which is represented by the animal intestine (i.e., humans, cow, rabbit, mice, chicken and insect), oral cavity,

Table 10.3 *Bifidobacterium* ssp. Isolated from Human and Milk

BIFIDOBACTERIUM SP.	ISOLATED FROM
B. breve	Infant feces and vagina
B. infantis	Infant feces and vagina
B. bifidum	Infant and adult feces and vagina
B. longum	Infant and adult feces and vagina
B. catenulatum	Infant and adult feces and vagina
B. pseudocatenulatum	Infant feces
B. adolescentis	Infant and adult feces, appendix, dental caries and vagina
B. angulatum	Adult feces
B. denticolens	Human dental caries
B. dentium	Human dental caries, oral cavity and adult feces
B. inopinatum	Human dental caries
B. catenulatum	Infant and adult feces and vagina
B. gallicum	Adult feces
B. lactis	Fermented milk

Sources: Biavati, B. et al., *Ann. Microbiol.*, 50, 117–132, 2000; Lee, J.H. and O'Sullivan, D.J., *Microbiol. Mol. Biol. Rev.*, 74(3): 378–416, 2010; Kumar, C.L.P. et al., *World Appl. Sci. J.*, 17(11), 1454–1465, 2012.

sewage, blood, and food (Ventura et al. 2007). In the intestinal tract of animals and humans, bifidobacteria coexist with a large variety of bacteria and are believed to play a pivotal role in maintaining a healthy GIT. The distribution pattern of bifidobacteria in feces of infants, in feces of adults, in human vagina, and in dental caries (Biavati et al. 2000) shows the differences in the adaptation of the *Bifidobacterium* species of human origin in the different habitats (Table 10.3).

Bifidobacteria are Gram-positive, catalase-negative, non-motile, non-spore forming, non-gas producing bacteria with a G + C content of 55%–67%. These pleomorphic rods are somewhat irregular, V or Y shaped resembling branches, spatulate, or club shaped. Their ability to form branches depends on the medium used for cultivation and this ability is said to be strain dependent (Tannock 1999b; Lee and O'Sullivan 2010). Ability of bifidobacteria to ferment hexoses by a fructose-6-phosphate phosphoketolase shunt, often termed the *bifidus pathway*, serves as a biochemical test to differentiate bifidobacteria from morphologically similar genera such as *Lactobacillus, Actinomyces, Propionibacterium,* and *Eubacterium*. Bifidobacteria are the only intestinal bacteria known to utilize this fermentation pathway. Glucose fermentation through bifidus pathway produces acetic and lactic acids in a theoretical ratio of 3:2 (Tannock 1999b). Considered to be strict anaerobes, ability of bifidobacteria to tolerate and survive in the presence of oxygen depends on the species or strain and composition of the culture medium (Kailasapathy and Champagne 2011). Bifidobacteria are saccharolytic bacteria and are believed to play an important role in carbohydrate fermentation in the colon (Pokusaeva et al. 2011). All characterized strains have the ability to ferment glucose, galactose, and fructose. These acid-tolerant microbes have their optimum pH for growth between 6.5 and 7.0. Some strains such as *B. lactis* and *B. animalis* can survive exposure at pH 3.5 (Matsumoto et al. 2004). But *Bifidobacterium* strains cannot survive above pH 8.5 (Biavati et al. 2000). Most human strains of bifidobacteria can grow at an optimum temperature of $37 \pm 1°C$, whereas for the animal strains, the optimum temperature is around $42 \pm 1°C$ (Kailasapathy and Champagne 2011).

Bifidobacteria are fastidious in their nutritional requirement and require a number of growth factors for their growth, which are called bifidogenic factors. These are defined as compounds, usually of a carbohydrate nature, that survive direct metabolism by the host and reach the large bowel or cecum, where these are preferentially metabolized by bifidobacteria as source of energy. Bifidogenic factors may be classified under the new prebiotic concept. Prebiotics are non-digestible food ingredients that beneficially affect the host by selectively stimulating the growth and/or activity of one, or a limited number of, bacteria in the colon, and which may thus improve the health of the host (Gibson and Roberfroid 1995). Table 10.4 summarizes the bifidogenic factors and their effects. Non-digestible oligosaccharides have been used in the diet specifically to increase relative numbers of bifidobacteria in the gut microflora (Wang and Gibson 1993; Gibson and Wang 1994a, 1994b; Manning and Gibson 2004). Promising results have been achieved

Table 10.4 Bifidogenic Factors for *Bifidobacterium* ssp., Their Origin, and Possible Effects

BIFIDOGENIC FACTORS	ORIGIN	STRAINS	EFFECTS
Casein bifidus factor (CBF)	Bovine casein subjected to acid and enzyme (papain and pepsin) hydrolysis	*B. bifidum*	Growth promotion, CBF supplies both peptides and carbohydrates necessary for growth
Human κ-casein and glycomacropeptide derived there from	Intact κ-casein from human milk and hydrolyzed with chymosin and pepsin	*B. infantis* S12	Small growth-promoting activity of intact κ-casein enhanced upon enzymatic hydrolysis
κ-casein enzymatic digest	Bovine milk casein digested by trypsin	*Bifidobacterium* ssp. (*B. bifidum*, *B. longum*)	Growth-promoting activity in fully synthetic medium associated with disulfide/sulfhydryl (cysteine) residues and a certain tryptic peptide-unidentified bifidogenic factor
Casein macropeptide (CMP)	Bovine milk	*Bifidobacterium* ssp.	Growth-promoting activity
N-acetylneuraminic acid (Neu-Ac)-containing substances (NeuAc, sialyl lactose, glycomacropeptide)	Bovine milk	*Bifidobacterium* ssp. and lactobacilli	Growth-promoting activity of *B. breve*, *B. bifidum*, *B. infantis*
Transgalactosylated oligosaccharides galactosyl galactose, galactosyl glucose	Transgalactosylation enzymatic reaction on lactose	*Bifidobacterium* ssp.	Growth promoting, decreased levels of toxic short chain fatty acids, fecal pH, fecal ammonia

(*Continued*)

Table 10.4 (*Continued*) Bifidogenic Factors for *Bifidobacterium* ssp., Their Origin, and Possible Effects

BIFIDOGENIC FACTORS	ORIGIN	STRAINS	EFFECTS
Fructans	Ash-free white powder from tubers of Jerusalem artichoke	*Bifidobacterium* ssp. (*B. infantis* ATCC 15697, *B. adolescentis* ATCC 15703, *B. longum* ATCC 15707)	Growth-promoting activity
Xylooligosaccharides (xylobiose)	Wheat bran, corn cobs aspen, wheat straw	*Bifidobacterium* ssp.	Increase of intestinal bifidobacteria with 1–2 g/d
Oligosaccharides	Onion, garlic, chicory root, asparagus, Jerusalem artichoke, soybeans, wheat bran	*Bifidobacterium* ssp.	Proliferation of bifidobacteria and suppression of putrefactive bacteria Prevention of constipation and pathogenic diarrhea
Fructooligosaccharides (inulin, oligofructose)	Onion, Garlic, Burdock	*Bifidobacterium* ssp.	Growth- and acid-promoting activity, good carbon and energy source
Lactulose	Lactose	*Bifidobacterium* ssp.	Growth-promoting activity, activation of immune response and infection limitation
Ginseng extract	Ginseng	*Bifidobacterium* ssp.	Growth-promoting activity

Sources: Kanbe, M., Uses of intestinal lactic acid bacteria and health. In: *Functions of Fermented Milk*, eds. Y. Nakazawa and A. Hosono, Elsevier applied Science publications, London, 289–304, 1992; Gomes, A.M.P. and Malcata, F.X., *Trends Food Sci. Technol.*, 10, 139–157, 1999.

in vitro using inulin and fructooligosaccharides (FOS). Among the different types of (linear and branched) FOS studied, oligofructose, a linear chain molecule comprising glucose and fructose at a degree of polymerization of four, is found to have the highest bifidogenic effect. All FOS studied are oligosaccharides that occur in nature, mainly of plant origin. These compounds are resistant to the strong hydrolytic capacities of the human intestinal enzymes and, therefore, reach the colon essentially unaltered, where they behave like soluble dietary fiber and are digested by the colonic flora (Roberfroid et al. 1993). The transgalactosylated oligosaccharides synthesized from lactose by β-galactosidase transfer reaction act as a growth factor for bifidobacteria. In addition to oligosaccharides, lactulose has been found to be a growth promoter for bifidobacteria (Kanbe 1992).

10.4 Classification of Bifidus Products

Bifidus products can be classified based on the technologies adapted to make these products more palatable and acceptable to the consumers. Cultured milks containing bifidobacteria possess many nutritional and technological advantages, such as limited postproduction acidification, mild taste, the formation of the physiologically desirable L(+) lactic acid, and the therapeutic benefits that follow the consumption of *live* bifidobacteria, but there still remains many problems associated with the manufacture of such products. These limitations come in the way of popularity of bifidus milk and products. However, increasing number of reports on the health benefits provided by bifidobacteria tempted the scientists to develop alternate technologies to overcome the limitations, and the approaches tried are as follows:

- Not to ferment—growth and flavor problems were coming when the milk was fermented with bifidobacteria. Therefore, non-fermented products containing viable cells of these microbes in chilled milk were developed.
- Blending with other bacteria—to enhance the flavor, it was thought to supplement additional microflora as starter cultures that give good flavor.
- Concentration and drying—this changed the form of consumption and improved the shelf life.
- Carry bifidus in other popular products such as yogurt.

A classification of the bifidus milk and products based on the different approaches is shown in Figure 10.1.

10.5 Limitations of Bifidus Products Production

In order to be functional, bifidus milk and products should contain viable bifidobacteria in large numbers during its production and storage. However, maintaining the viability of bifidobacteria during processing and refrigerated storage has been a challenge to

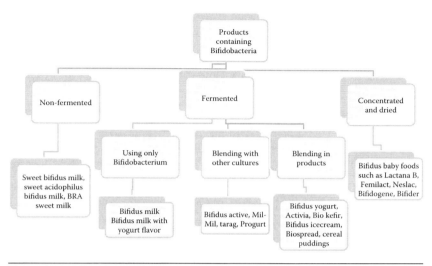

Figure 10.1 Classification of products containing bifidobacteria.

dairy processors. Ability of the bifidobacteria to grow in milk and survive during product manufacture, their tolerance to oxygen, acid, bile and heat, their ability to metabolize prebiotic substances, and survival during storage are some of the very important parameters that can influence the quality and functionality of bifidus products.

10.5.1 Growth of Bifidobacteria in Milk

Although milk contains essential nutrients for the growth of bifidobacteria, the level of amino acids and peptides is insufficient to provide the ideal condition for their rapid growth and survival (Klaver et al. 1993). As a consequence, bifidobacteria have been used only to limited extent in commercial fermented milks (Gomes and Malcata 1999; Heller 2001) in spite of the fact that fermented milk matrix is the preferred carrier for bifidobacteria. Cow milk and buffalo milk, which are devoid of bifidus factor, are poor substrates for the growth of bifidobacteria resulting in slow growth of bifidobacteria in these milks. Because of this reason, the initial milk used in the preparation of bifidus products is severely heated to make it almost sterile, so that the chance of survivors overgrowing bifidobacteria during incubation can be avoided. But the severe heating causes development of cooked type of flavor in natural bifidus milk. Another problem is the prolonged fermentation time due to slow acid production in milk. This results in certain defects in the resultant products, such as whey separation, sandy or slimy texture, too mild taste/yeasty or vinegary taste, or too little aroma (Rasic and Kurmann 1983; Samona et al. 1996).

10.5.2 Cultures

The species of bifidobacteria that are being employed as the starter culture in the preparation of bifidus milk are *B. bifidium*, *B. infantis*, *B. longum*, *B. breve*, *B. adolenscentis*, and *B. animalis*. Among these, *B. bifidium* and *B. longum* are the most widely used strains. Some commercial bifidobacteria strains are shown along with their producers/suppliers (Table 10.5). Like most commercial probiotic culture preparations, bifidobacteria cultures are supplied in highly concentrated form, usually meant for direct vat set (DVS) application. Use of these highly concentrated DVS cultures avoids the need for propagating this culture at the production site. The DVS cultures are supplied either as highly concentrated frozen cultures or as freeze-dried cultures. The cultures should be filled in gas and light proof containers to protect the cultures

Table 10.5 Partial List of Commercially Popular Bifidobacterial Strains

STRAIN	COMPANY
B. longum BB-536	Morinaga Milk Industry
B. longum SBT-2928	Snow Brand Milk Products
B. breve	Yakult
B. breve strain Yakult	
B. lactis LaftiTM B94	DSM
B. longum UCC35624	UCC Cork
B. lactis DR-10/Howaru, *B. lactis* HN019, *B. lactis* Bi-07	Danisco
B. animalis ssp. *lactis*, BB-12	Chr. Hansen
B. bifidum Bb-11	Chr. Hansen
B. longum ssp. *longum* BB-46	Chr. Hansen, Inc. (Denmark)
B. longum strains BL46 (DSM 14583) AND BL2C (DSM 14579)	DSM
Bifidobacterium Rosell-71 and *Bifidobacterium* Rosell-33	Institut Rosell-Lallemand, Canada
B. longum UCC 35624	UCCork
B. lactis FK120	Fukuchan milk (Japan)
B. lactis LKM512	Fukuchan milk (Japan)

Sources: Soccol, C.R. et al., *Food Technol. Biotechnol.*, 48(4), 413–434, 2010; Pokusaeva, K. et al., *Genes Nutr.*, 6, 285–306, 2011.

against light and humidity. Most often aluminum-foil-coated cartons or pouches are used. As the cultures are sensitive, it is important to handle and store these according to the manufacturers' instructions (Saarela et al. 2000). The cell concentration per gram of product varies with the culture and the type of organisms used. Usually, deep-frozen cultures contain more than 10^{10} cfu/g, whereas freeze-dried cultures typically contain more than 10^{11} cfu/g (Oberman and Libudzisz 1998).

In fermented bifidus products, it is important that the bifidobacteria culture used contributes to good sensory properties. Therefore, it is quite common to use bifidobacteria mixed together with other types of bacteria suited for the fermentation of the specific product. For milk-based products, the bifidobacteria strains are often mixed with *Str. thermophilus* and *Lb. delbrueckii* to achieve the desired flavor and texture. In many cases, the consumers find products fermented with *Lb. delbrueckii* too acidic and with too heavy acetaldehyde flavor (i.e., yogurt flavor). Therefore, bifidus cultures have been developed to bring out the preferred flavors in the products in which these are used. Examples of such cultures are fermented milk with *Lb. acidophilus* and *Bifidobacterium* ssp., *Lb. acidophilus*, *Bifidobacterium* ssp. and *Lb. casei*, and *Lb. acidophilus*, *Bifidobacterium*, and *Str. thermophilus*. Tamime et al. (1995) have done an extensive review on the manufacturing technology for producing fermented milks containing *Bifidobacterium* strains. A fermentation temperature of 37°C–40°C is usually recommended for producing bifido-containing fermented products since most human strains of bifidobacteria multiply well in this temperature range. Bifidobacteria strains can be an integral part of the starter and grow in a symbiotic relation with the other strains composing the culture. In such cases, the combination of strains is selected taking into consideration their optimum fermentation temperature. Sometimes bifidobacteria strains are added to the fermented milk after fermentation. In special sweet bifidus milk products, bifidobacteria are added to the product in such a way that they will retain their viability but are prevented from multiplying by cooling the product.

10.5.3 Interaction with Starter Bacteria

The probiotic-starter interaction during processing and storage has an impact on the quality of the finished product. Selecting a suitable starter-probiotic combination ensures functional fermented dairy products with excellent sensory properties and good survival of probiotics (Saarela et al. 2000). For this, a screening process evaluating the impact of different starters on the sensory properties and on the survival of the probiotic strain should be done. The most suitable starters for probiotic lactobacilli and bifidobacteria may include *Str. thermophilus*, yogurt cultures, and mesophilic starters with different combinations of *Lactococcus* strains. For *Bifidobacterium* ssp., a thermophilic starter may be preferred over mesophilic as the optimum growth temperature of the former lies in the thermophilic range (37°C–42°C). Also sometimes, the starter might improve the growth conditions of probiotic by producing substances favorable to the growth of the probiotic or by reducing the oxygen pressure. For example, an increased survival of *B. longum*, which are quite sensitive to oxygen, was achieved by selecting specific strains of *Str. thermophilus* (Ishibashi and Shimamura 1993). Amino acids such as valine, glycine, and histidine released by the proteolytic activity of *Lb. delbrueckii* subsp. *bulgaricus* increased the growth of bifidobacteria (Misra and Kulia 1994). The negative impact of starter on probiotic survival *in vitro* and *in vivo* should also be taken into consideration while selecting a suitable starter. Metabolites formed by the starter such as lactic acid, hydrogen peroxide, and bacteriocins might influence the survival of probiotics. Samona et al. (1996) reported that bifidobacteria strains were unable to grow in the presence of yogurt cultures but by choosing the right combination of probiotic and starter strains this decline in bifidobacteria counts could be prevented.

10.5.4 Acid Tolerance

Acid tolerance, particularly in the fermented product and during gastrointestinal transit, is an important parameter that can affect the viability of bifidobacteria strains and ultimately the potential health benefits of the product. In general, bifidobacteria cultures are less acid tolerant than lactobacilli cultures and this is reflected by their reduced tolerance to human gastric juice (Dunne et al. 1999). The tolerance of *Bifidobacterium* ssp. to acidic conditions is strain-specific (Mättö et al. 2006). *B. longum* has shown better survival in acidic conditions and bile concentrations compared to *B. infantis*, *B. adolsecentis*, and *B. bifidum*. Furthermore, *B. longum* was found to grow more easily in milk compared to other strains, while *B. animalis* ssp. *animalis* has good survival properties in fermented milks. But this species is not of human origin (Lankaputhra and Shah 1996).

10.5.5 Presence of Oxygen

Oxygen toxicity is an important and critical problem when dealing with an anaerobe like bifidobacteria. Better survival and viability of bifidobacteria in deaerated milk has been observed by Klaver et al. (1993). To exclude oxygen during the production of probiotic milk products, special equipment is required to provide an anaerobic environment. Oxygen affects probiotic cultures in two ways. First, it is directly toxic to the cells, certain probiotic cultures are sensitive to oxygen, and they die in its presence. Second, in the presence of oxygen, certain cultures, particularly *Lb. delbrueckii* subsp. *bulgaricus*, produce peroxide. A synergistic inhibition of probiotic cultures due to acid and hydrogen peroxide has been reported (Lankaputhra and Shah 1996), for this reason, removal of *Lb. delbrueckii* ssp. *bulgaricus* from some starter cultures (i.e., *Lb. acidophilus*, *Bifidobacterium*, and *Str. thermophilus* starter cultures) has achieved some success in improving the survival of probiotics. Studies have focused on the prevention of detrimental effects of oxygen on probiotic cultures, including the use of antioxidants or oxygen scavengers (Dave and Shah 1997; Talwalkar and Kailasapathy 2004). To protect bifidobacteria from the toxicity of oxygen in yogurts, measures such as incorporation of ascorbic acid, L-cysteine, and a high-oxygen consuming strain of *Str. thermophilus* have been tried (Kailasapathy and Champagne 2011). Oxygen can also enter the product through packaging materials during storage. Packaging materials such as polyethylene and polystyrene are gas permeable and allow the diffusion of oxygen into the product during storage (Ishibashi and Shimamura 1993).

10.6 Products Containing Bifidobacteria

The reputation of bifidobacteria as biotherapeutic agent for prevention or treatment of gastrointestinal diseases and the fact that these bacteria constitute the larger and more stable microbial populations of human intestinal tract as well as these are the predominant colonic microbiota of breast-fed infants have led to their approval as a valuable functional ingredient for the development of a variety of functional foods and dietary supplements. Some of the popular bifidus products are described hereunder and others are listed in Table 10.6.

10.6.1 Bifidus Milk

This product has its origin in Germany and is named according to the bacteria (former name *Lactobacillus bifidus*) used in fermentation. Bifidus milk is produced in small quantities in some of the European

Table 10.6 Some Milk-Based Bifidus Products

NAME	PHYSICAL TYPE	COUNTRY OF ORIGIN	MICROFLORA
AB–Yogurt	Gel	Denmark	*Lb. acidophilus* + *B. bifidum* + yogurt culture
Akult		Japan	*Lb. acidophilus, B. bifidum, B. breve, Lb. casei* subsp. *casei*
Bifighurt	Gel	FRG	*Lb. acidophilus* + *Str. thermophilus, B. bifidum*
Bio kefir		Poland	*Bifidobacterium* and/or *Lb. acidophilus*
Biogarde	Liquid	West Germany	*Lb. acidophilus* + *B. bifidum, Str. thermophilus*
Biogarde Ice Cream	Frozen	FRG	*Lb. acidophilus* + *B. bifidum*
Biokys	Stirred gel	Czech Republic	*B. bifidum, Lb. acidophilus, Pediococcus acidilactici*
Biomild		Several countries	*Bifidobacterium, Lb. acidophilus*
Cultura	Liquid	Denmark	*Lb. acidophilus* + *B. bifidum*
Life start original	Freeze-dried powder	United States	*B. infantis*
Life start Two	Freeze-dried powder	United States	*B. bifidum*
Liobif	Freeze-dried powder	Yugoslavia	*B. bifidum*
Lunebest	Gel/set type	Germany	*Lb. acidophilus* + *B. bifidum* + yogurt culture
Lyobifidus	Freeze-dried powder	France	*B. bifidum*
Mil-Mil	Gel	Japan	*Lb. acidophilus, B. bifidum, B. breve*
Mil–mil E	Gel	Japan	*Lb. acidophilus* + *B. bifidum* + yogurt culture

(Continued)

Table 10.6 (*Continued*) Some Milk Based Bifidus Products

NAME	PHYSICAL TYPE	COUNTRY OF ORIGIN	MICROFLORA
Miru-Miru/Bifiru	Liquid	Japan	*Lb. acidophilus* + *Lb. casei* + *B. breve*
NuTrish a/B	Liquid	United States	*Lb. acidophilus* + *Bifidobacteria* ssp.
Ofilus nature	Gel/set type	France	*Str. thermophilus* + *Lb. acidophilus* + *B. bifidum*
Omniflora	Freeze-dried powder	Germany	*Lb. acidophilus* + *B. longum* + a saprophytic *E. coli*
Procult		Germany	*B. longum* BB536
Vitality		United Kingdom	*Bifidobacterium.*, *Lb. acidophilus*
NON-MILK-BASED BIFIDUS PRODUCTS			
Soy flour plus WPC	Liquid/gel	Italy	*Lb. acidophilus* + yogurt culture, *B. bifidum*

Sources: Compiled from Kurmann, J.A. et al., *Encyclopedia of Fermented Fresh Milk Products*, Van Nostrand Reinhold, New York, 200–281, 1992; Tannock, G.W., Internal renewal: The potential for modification of the normal microflora. In: *Normal Microflora: An Introduction to Microbes Inhabiting the Human Body*, Chapman & Hall, London, 100, 1995; Muir, D.D. et al., *Int. J. Dairy Technol.*, 52, 129–134, 1999; Farnworth, E.R. and Mainville, I., Kefir: A fermented milk product. In: *Handbook of Fermented Functional Foods*, ed. E.R. Farnworth, CRC Press LLC, Boca Raton, FL, 77–111, 2003; Gurakan, G.C. et al., Probiotic dairy beverages: Microbiology and technology. In: *Development and Manufacture of Yoghurt and Other Functional Dairy Products*, ed. F. Yildiz, CRC Press, Boca Raton, FL, 181–183, 2010; Shah, N.P., *From Bulgarian Milks to Probiotics Fermented Milks*. Summilk, IDF World Dairy Summit, 2011.

countries and its consumption is linked to the dietetic and therapeutic values rather than the sensory properties. Bifidus milk can be prepared from skimmed milk, partially skimmed milk or whole milk. Human strains of *B. bifidum* or *B. longum* are used as culture. The manufacturing process is depicted in Figure 10.2. The final set product has a pH of 4.3–4.7 and contains 10^8–10^9/ml viable bifidobacteria. The count of bifidobacteria decreased by 2 logs during refrigerated storage for 1–2 weeks. Bifidus milk can be produced as stirred product also. This product is easily digestible than the milk from which it is made. It has been used as a protective means against imbalances in the gut microflora, in the treatment of liver diseases, chronic constipation, and also as an aid in the therapy of gastrointestinal disorders (Kurmann et al. 1992).

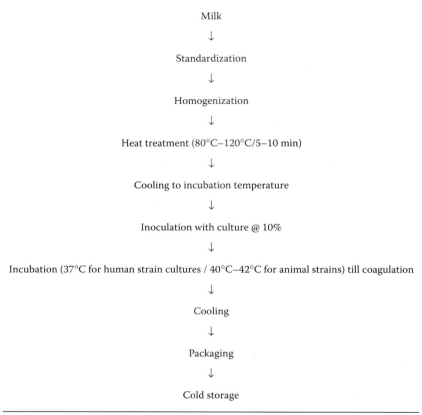

Figure 10.2 Manufacturing process for bifidus milk.

10.6.2 Bifidus Milk with Yogurt Flavor

This dietetic product from specially formulated milk has a pleasant yogurt like flavor, which may be modified by adding fruit. It is prepared from cow's milk and belongs to set product category. *B. bifidum* or *B. longum* is used as culture. Equal volumes of the retentate of ultrafiltered sweet whey and ultrafiltered skim milk are mixed, the mixture is then pasteurized at 80°C for 30 min and then cooled to 37°C, followed by addition of threonine (0.1%) and starter (2%). Incubation is carried out at 37°C for 24 h, followed by cooling to 4°C and subsequent cold storage. The viable bifidobacteria count declined from 10^9/ml to 10^7–10^6/ml during storage of the product at 4°C for 21 days (Kurmann et al. 1992). Final product composition is shown in Table 10.7.

Table 10.7 Composition of Bifidus Milk with Yogurt Flavor

PRODUCT CONSTITUENTS	LEVEL
Total solids	15%
Protein	7.3%
Fat	1.3%
pH	4.7
Acetaldehyde concentration	29–39 ppm
Viable bifidobacteria	10^9/ml

10.6.3 Bifidus Active

This nontraditional fermented milk from France is prepared from cow's milk. It is a set type plain, natural product. It may be added with mandarin orange or mint flavors and vitamin C. Starter culture consists of yogurt culture and human strain of *B. longum*. The product has protein 3.4%, fat 3.0%, calcium 128 mg/100 g, 238 kJ (57 kcal) (Kurmann et al. 1992).

10.6.4 Bifidus Baby Foods

These nontraditional products from France and Germany are being made using cow's milk. Starter culture consists of *B. bifidum* alone or in combination with *Lactobacillus acidophilus* and *Pediococcus acidilactici*. Some examples are Lactana-B, Bifiline, and Femilact. *Lactana-B* is a dried formula product containing prebiotic lactulose and viable *B. bifidum*. It is produced from modified milk and developed in Germany.

Bifiline, developed in USSR, is a liquid formula containing viable bifidobacteria. A milk formula called *Malutka* is used for making bifiline. The heat treated, homogenized milk formula is fermented with 5% starter culture containing 0.5% corn extract at 37°C for 8–10 h until coagulation followed by cooling. The final product has about 0.6% titratable acidity and 10^7–10^9/g viable bifidobacteria.

Femilact is a dried formula developed in Czechoslovakia. It is made by fermenting heat-treated cream (12% fat) with 2%–5% mixed culture consisting of *B. bifidum*, *Lb. acidophilus*, and *P. acidilactici* in 1:0.1:1 ratio at 30°C till desired acidity followed by cooling. To this, heat-treated vegetable oil, lactose, whey proteins, and vitamins are added and the mixture is homogenized and spray dried. The final reconstituted product has a titratable acidity of 0.25% and contains 10^8–10^9/ml viable culture (Kurmann et al. 1992).

Neslac is a commercialized milk powder containing *B. animalis* subsp. *lactis* BB-12. It is meant for older infants (Playne et al. 2003, Chouraqui et al. 2004). The most often used bifidobacterium species in the baby food preparations is *B. bifidum*. It acts alone or together with its *in vivo* growth-promoting substances to modify the gut flora of bottle-fed babies and infants and thus provides protection against enteric infection or side effects of antibiotic therapy and hence acts as an aid in the treatment of intestinal disorders and enteric infections.

10.6.5 Bifidus Yogurt

This nontraditional fermented milk prepared from cow's milk is popular in Germany, the United States, Japan, and several other countries. It is basically yogurt with bifidobacteria. Product is characterized by mild acid flavor and firm body. Its flavor may be masked or modified by fruit flavoring. Culture comprises of yogurt starter (*Str. thermophilus* and *Lb. delbrueckii* subsp. *bulgaricus*) and *B. bifidum* or *B. longum*. The product can be prepared either by mixing a desirable ratio of separately cultured bifidus milk with cultured yogurt or by a simultaneous fermentation. The processing steps include milk standardization, homogenization, heat treatment, cooling to incubation temperature followed by inoculation with starter @ 5%–10% (*B. bifidum* + yogurt starter) and incubation (37°C for human strain cultures/40°C–42°C for animal strains) till coagulation. The product is then subjected to cooling, stirring, packaging, and cold storage (Kurmann et al. 1992).

Activia is a yogurt containing *B. animalis* DN 173 010. The product is owned by Danone and is marketed as a beneficial health product, with claim on improved intestinal motility. It is marketed under the trade names *Bifidus regularis*, *Bifidus actiregularis*, *Bifidus digestivum*, and *B. lactis*. It comes under the functional food category designed to improve digestive health.

10.6.6 Mil-Mil

This nontraditional fermented milk beverage developed by Yakult Honsha Company from Japan uses *B. bifidum*, *B. breve,* and *Lb. acidophilus* as starters. The product is sweetened with small quantities of glucose and/or fructose and is colored with carrot juice. Carrot juice provides provitamin A. This product is sold as a soup as well (Kurmann et al. 1992).

10.6.7 Sweet Bifidus Milk

This is a non-fermented, nontraditional, sweet milk product made from cow's milk sold in Germany and Japan. Product is made by inoculating dried sweet milk or sterilized milk with cultures/freeze-dried powders of bifidobacteria (Kurmann et al. 1992).

10.6.8 Sweet Acidophilus Bifidus Milk

This is a non-fermented sweet milk product from Japan, added with *Lb. acidophilus* and *B. longum*.

Concentrated suspensions of cultures are added to cold pasteurized milk. After mixing, the milk is packaged and refrigerated. Cow milk is used for its production. Finished product has a flavor similar to that of freshly pasteurized milk (Kurmann et al. 1992).

10.6.9 BRA Sweet Milk

This contains *B. infantis*, *Lb. reuteri*, and *Lb. acidophilus*. The product is similar to acidophilus milk. The microorganisms are added to pasteurized cold milk before packaging (Rothschild 1995).

10.6.10 Tarag

This traditional product of Mongolia is made using milk of ewe/goat/cow/yak. It is a set type product with viscous consistency and fresh acidic taste. A mixed starter culture consisting of *Bifidobacterium* sp., *Lb. bulgaricus,* and kefir grains in a 1:0.5:0.5 proportion is used. This mixed starter has been shown to be inhibitory to a strain of *E. coli* and *Salmonella sonnei* (Kurmann et al. 1992).

10.6.11 Progurt

This is a protein enriched cultured product developed in Chile. Cow's milk is used for its preparation. The starter culture consists of mesophilic lactococci, namely, *Lactococcus lactis* subsp. *diacetylactis* and *Lc. lactis* subsp. *cremoris* in 1:1 ratio, or includes the addition of *Lb. acidophilus* and/or *B.bifidum* (Kurmann et al. 1992) as shown in Figure 10.3. Finished product has 5% fat, 6.2% protein, 3% lactose, 0.7% ash, and pH 4.4–4.5. No data are available on the survival of culture bacteria (Schacht and Syrazinski 1975).

10.6.12 Acidophilus Milk Containing B. bifidum

Strains of *B. bifidum* NDRI and *Propionibacterium freudenreichii* subsp. *shermanii* MTCC 1371 were incorporated in plain acidophilus milk to further enhance its prophylactic characteristics. The product retained its dietetic characteristics up to 7 days of storage at refrigerated temperature and its consumption may be advantageous over

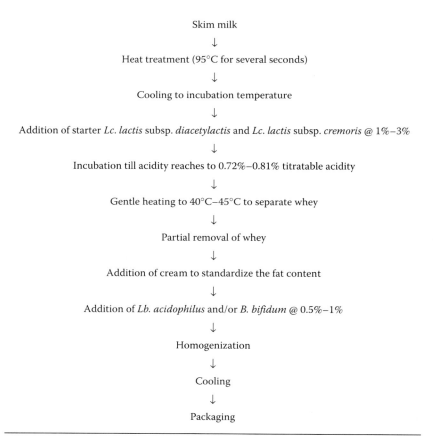

Figure 10.3 Method for production of progurt.

plain acidophilus milk due to its enhanced prophylactic features (Sarkar and Misra 2010).

10.6.13 Diphilus Milk

This is a fermented milk reported by Prevot in 1971. *B. bifidum* and *Lb. acidophilus* are used as starter cultures (Kanbe 1992).

10.6.14 Bifidogene

This is a pharmaceutical preparation made from cow's milk to treat gastrointestinal disorders resulting from antibiotic therapy. It is a freeze-dried preparation from France and contains viable bifidobacteria (Kurmann et al. 1992).

10.6.15 Bifider

This is a pharmaceutical preparation from Japan for use as food in special dietary regimens and in the treatment of gastrointestinal disorders. It is prepared in freeze-dried form using cow's milk and contains viable cells of *B. bifidum* (Kurmann et al. 1992).

10.6.16 Bio-Spread

A reduced-fat edible table bio-spread was made from milk fat and soy oil containing mixed cultures of *Lb. casei* ACA-DC 212.3 and *B. infantis* ATCC 25962. The process was modified to enhance the viability of the probiotics (i.e., both cultures showed 1 log decline after processing). The rate of decline in the viability of bifidobacteria during shelf life was greater than that of the lactobacilli (Charteris et al. 2002).

10.6.17 Milk- and Water-Based Cereal Puddings

Lb. rhamnosus GG, *Lb. acidophilus* LA-5 and 1748, and *B. animalis* subsp. *lactis* BB-12 were successfully used for the production of milk–water-based puddings with and without prebiotics (e.g., polydextrose and Litesse TM). All strains showed good growth and survival in milk-based puddings (e.g., viable counts ranged between 8 and 9.1 log cfu/g) (Helland et al. 2004).

10.7 Measures to Improve Viability of Bifidobacteria in Bifidus Products

Survival of cultures during food processing and storage is a must for successful delivery of probiotics in foods. Studies have demonstrated that probiotic viability can be significantly protected via encapsulation in a variety of carriers including milk proteins and prebiotics. Induction of stress responses in probiotic strains can be applied to improve the ability of cultures to survive processing, such as freeze-drying, spray drying, and gastric transit. The addition of various protective compounds to probiotic cultures can also improve their viability during manufacture. Some measures to improve viability of bifidobacteria are discussed hereunder.

10.7.1 Encapsulation

Encapsulation is being increasingly used in the probiotic food industry as a means of protecting live cells from extremes of heat or moisture, such as those experienced during drying and storage (O'Riordan et al. 2001). This technique allows the active core ingredient, or substrate, to be separated from its environment by a protective film or coating. This separation occurs until the release of the functional ingredient is desired. Several methods of microencapsulation of probiotics are reported and include spray drying, extrusion, emulsion, and phase separation (Kailasapathy 2002). *B. bifidum* cells encapsulated in gel beads composed of alginate, pectin, and whey proteins and surrounded by two membranes exhibited good survival at pH 2.5 for up to 2 h, while free cells did not survive under these conditions, and furthermore protection was also afforded by this system, when the cells were exposed to bile salt solutions (Guerin et al. 2003). Encapsulating *B. lactis* and *Lb. acidophilus* using calcium-induced-starch microencapsulates prevented their loss of viability due to oxygen sensitivity and effect of acid in yogurt (Kailasapathy and Champagne 2011). Synbiotic micro-encapsulation, where in, core consisting of probiotics cells and prebiotic carbohydrates forming a component of the encapsulation wall and coating materials and capsular matrix, was tried for improving viability. *B. infantis* Bb-O_2 was microencapsulated within a film-forming protein–carbohydrate–oil emulsion incorporating FOS. This protected the probiotic during non-refrigerated storage and in GIT transit (Crittenden et al. 2006).

10.7.2 Use of Prebiotics

Another approach to increasing the numbers of bifidobacteria in the GIT is the incorporation of prebiotics in the food. A prebiotic is a non-digestible dietary supplement that modifies the balance of the intestinal microflora stimulating the growth and/or activity of the beneficial microbes and suppressing potentially deleterious bacteria. Some examples are lactulose, galacto-, and fructooligosaccharides and resistant starch. Functional foods containing both probiotics and prebiotics are called synbiotics (Roberfroid 1998). The viability and activity of *B. pseudocatenulatum* G4, *B. longum* BB 536, and yogurt cultures were studied in yogurt containing 0.75% *Mangefira pajang* fibrous polysaccharides (MPFP) and inulin. This study showed better growth and activity of said strains in the presence of MPFP and inulin (Al-Sheraji et al. 2012a).

10.7.3 Exploiting Cellular Stress Responses

Ahn et al. (2001) examined the effect of oxygen stress on *B. longum* and found that a protein, Osp, was upregulated in an oxygen-tolerant *Bifidobacterium* strain. This study showed that exploitation of the oxidative stress response in bifidobacteria and the expression levels of Osp may provide suitable targets for enhancing the oxygen tolerance of bifidobacteria in functional food development and in the intestinal microbial ecosystem.

10.7.4 Genetic Characterization

Many researchers have focused their attention on the sequence information for intestinal lactobacilli and bifidobacteria (Turroni et al. 2011). Out of the 37 recognized bifidobacteria species, 10 bifidobacteria genomes have been entirely sequenced (Table 10.8). This genomics of health-promoting bacteria, called probiogenomics,

Table 10.8 Completed Bifidobacterial Genome Sequencing Projects

SPECIES	ACCESSION NUMBER	INSTITUTION
B. adolescentis ATCC15703	NC_008618.1	Gifu University, Life Science Research Center, Japan
B. animalis subsp. lactis AD011	NC_011835.1	Korea Research Inst. of Bioscience and Biotechnology
B. animalis subsp. lactis Bl-04	NC_012814	Danisco USA Inc
B. animalis subsp. lactis DSM 10140	NC_012815	Danisco USA Inc
B. bifidum PRL2010	CP001840	Laboratory of Probiogenomics, Department of genetics, University of Parma, Italy
B. bifidum S17	CP002220	University of Ulm, Germany
B. dentium Bd1	CP001750	University of Parma, Italy, University. College of Cork, Ireland
B. longum subsp. infantis ATCC15697	NC_011593.1	University of California Davis, USA
B. longum subsp. longum DJO10A	NC_010816.1	University of Minnesota
B. longum subsp. longum JDM301	NC_014169	Department of Medical Microbiology and Parasitology, Shanghai Jiao Tong University School of Medicine
B. longum subsp. longum NCC 2705	NC_004307.2	Nestle Research Center, Lausanne, Switzerland

Source: Turroni, F. et al., Int. J. Food Microbiol., 149, 37–44, 2011.

helps us in understanding the mechanisms of probiosis, probiotic adaptation to processing conditions as well as provides clues as to how probiotics interact with the host (Ventura et al. 2009; Turroni et al. 2011). Such genetic information opens up future possibilities for the development and design of more efficacious probiotic strains. Genetic manipulation approaches can be tried for increasing probiotic viability when exposed to stressful environments such as those encountered during functional food development or during gastric transit. Additionally, global gene, protein, and metabolite expression techniques are being used for providing evidence of probiotic adaptations in food products and survival and host–microbe interactions in the mammalian gut (Marco et al. 2006; Klaenhammer et al. 2007; Lebeer et al. 2008). The transcriptome profiling for *B. longum* during its growth in human milk showed that *B. longum* redirects sugar metabolism toward oligosaccharide consumption during growth in the milk matrix and that this activity might be relevant for the high abundance of this microorganism in the guts of breast-fed infants

(Gonzalez et al. 2008). Similarly, metabolomic and proteomic analyses of *B. longum* subsp. *infantis* ATCC 15697 during growth on human milk oligosaccharides resulted in the identification of oligosaccharide degradation pathways of this microorganism (Sela et al. 2008; Sanders and Marco 2010).

10.8 Therapeutic Aspects of Bifidus Products

The therapeutic value of bifidus products depends to a great extent on the bifidobacterial strains employed, other culture strains used, and the process of production. Tissier (1906) was the first to promote the therapeutic use of the bifidobacteria for the relief of digestive disorders (Kumar et al. 2012). He believed that the bifidobacteria displaced the putrefactive bacteria, which are responsible for gastric upsets, while, at the same time, re-establishing themselves as the dominant intestinal microbes (Samona et al. 1996). Bifidobacteria are naturally present in the dominant colonic microbiota of human, representing up to 25% of the cultivable fecal bacteria in adults and 80% in infants and play a prominent role in gut homeostasis and normal development (Kumar et al. 2012). Bifidobacteria are present in large numbers, particularly in the large intestine. In man, bifidobacteria begin to appear 3–4 days after birth and by day 5, bifidobacteria predominate the intestinal microflora. This predominance of bifidobacteria in the intestinal microflora is said to be an indication of good health in humans belonging to all age groups. Bifidobacteria are easily affected by host physiology, food, medication, age, stress, and so on and their absence in the GIT is an indication of abnormal health (Kanbe 1992).

The safety aspect of bifidobacteria as a probiotic is supported by its long history of consumption through fermented milks. Selected bifidobacteria strains used in these fermented milks show high survival in the GIT and exhibit probiotic properties in the colon (Berrada et al. 1991; Bouhnik et al. 1992; Pochart et al. 1992; Leahy et al. 2005; Picard et al. 2005). The ability of bifidobacteria to survive during gastrointestinal transit and subsequent colonization is highly strain specific (Berrada et al. 1991; Bouhnik et al. 1992; Marteau et al. 1992; Bezkorovainy 2001). Bifidobacteria may provide protection to the host from potentially harmful entities through production of inhibitory substances, blockade of adhesion sites, and immune stimulation

(Gibson and Wang 1994c). Clinical studies linking bifidobacteria with potential health benefits are summarized by Lee and O'Sullivan (2010).

To have an impact on the colonic flora, it is important that the probiotic strains exhibit antagonism against pathogenic bacteria (Saarela et al. 2000). Several mechanisms of protection have been suggested. These include the production of various acids, hydrogen peroxide, or bacteriocins, the competition for nutrients or adhesion receptors, antitoxin action, and stimulation of the immune system (Steer et al. 2000; Fooks and Gibson 2002; Rakoff-Nahoum et al. 2004). Many authors suggested that low molecular weight metabolites and secondary metabolites may be more important than bacteriocins, since these show a wide inhibitory spectrum against many harmful microbes (Saarela et al. 2000). The role of bacteriocins in the pathogen inhibition is limited, since bacteriocins have inhibitory effects only against closely related species (Holzapfel et al. 1998). A study by Slačanac et al. (2007) reported the inhibition of the *in vitro* growth of *Salmonella enteritidis* D by *B. longum* Bb-46. Bifidobacteria suppress the multiplication of putrefactive bacteria in GIT and thus prevent the formation of putrefactive products from amino acids and thereby prevent the losses of nutrients (Kanbe 1992). *In vitro* studies have shown the antibacterial activity of bifidobacteria against pathogenic *E. coli*, *Staphylococcus aureus*, *Shigella dysenteriae*, *Salmonella typhi*, *Candida albicans*, *Listeria monocytogenes* (Toure et al. 2003; Lučan et al. 2009), *Campylobacter jejuni* (Fooks and Gibson 2003), and so on. Acids produced by bifidobacteria, namely, lactic, acetic, and some amounts of formic acid, are thought to be responsible for this antibacterial action (Rasic and Kurmann, 1983). Also bifidobacteria can breakdown conjugated bile acids to free bile acids, which cause more effective inhibition than the conjugated forms and thus exert a beneficial effect on the intestinal microflora (Kanbe 1992). Imaoka et al. (2008) showed the anti-inflammatory activity of bifidobacteria fermented milk containing *B. bifidum* strain Yakult and *B. breve* strain Yakult against ulcerative colitis.

As probiotic agents, bifidobacteria have been studied for their efficacy in the prevention and treatment of a broad spectrum of animal and/or human GIT disorders, such as colonic transit disorders, intestinal infections, and colonic adenomas and cancer (Picard et al. 2005). The first proposed health benefits of bifidobacteria were for the prevention and treatment of diarrhea. Certain strains (e.g., *B. animalis* strain DN-173 010) have been reported to prevent or alleviate infectious diarrhea through effects on the immune system and resistance to colonization by pathogens (Picard et al. 2005). A fermented milk containing *B. lactis* FK 120 improved intestinal conditions (Shioya et al. 2000). Bifidobacteria have been shown to exhibit a protective effect against the devastating effects of acute diarrheal disease. Saavedra et al. (1994) reported a reduced incidence of rotaviral infection in infants when *B. bifidum* strain was given in conjunction with *Str. thermophilus* in a standard milk formula. Bae et al. 2002 reported the inhibitory effect of a *B. breve* strain toward the infectivity of a rotavirus in infants and young children. *B. lactis* strain Bb12 was found useful in prevention of acute diarrhea in infants (Chouraqui et al. 2004; Mohan et al. 2006). *B. lactis* strain has been reported to mediate a positive treatment against acute diarrhea in healthy children (Chouraqui et al. 2004). *B. longum* and *B. bifidum* were found to be effective in reducing the incidence and duration of antibiotic-associated diarrhea (Orrhage et al. 1994; Plummer et al. 2004). Kabeerdoss et al. (2011) tested the effect of a probiotic yogurt containing *B. lactis* Bb12 (10^9 in 200 ml) on fecal output of beta-defensin and immunoglobulin A in a group of young healthy women eating a defined diet. They found that *B. lactis* Bb12 increased secretory IgA output in feces and this property may explain the ability of probiotics to prevent GIT and lower respiratory tract infections. In a study to check the effects of a fermented milk product containing *B. lactis* DN-173 010 on abdominal distension and GIT transit in irritable bowel syndrome with constipation, Agrawal et al. (2008) found that the probiotic resulted in improvements in objectively measured abdominal girth and GIT transit, as well as reduced symptomatology. Bifidus preparations were used to treat constipation and irregular bowel movements. The metabolites produced by bifidobacteria, especially organic acids, are thought to stimulate the intestinal peristalsis and help with normal bowel movement. Administration of bifidobacteria leads to higher moisture content in feces and this may have contributed to the beneficial effect (Kanbe 1992). But in a randomized, double-blind, controlled trial, Tabbers et al. (2011) could not find any significant increase in stool frequency in constipated children compared to control group, when a fermented milk containing *B. lactis* DN-173010 was fed to them.

Bifidobacterial strains exhibit immunomodulatory properties. *Lb. rhamnosus* GG and *B. lactis* Bb-12 derived extracts have been shown to suppress lymphocyte proliferation *in vitro* (Mattila-Sandholm and Kauppila 1998). Further evidence for immunomodulation by these two strains was provided by a trial involving children with severe atopic eczema resulting from food allergy. Children fed *Lb. rhamnosus* GG and *B. lactis* Bb-12 showed a significant improvement in clinical symptoms compared to the placebo group (Mattila-Sandholm and Kauppila 1998; Alander and Mattila-Sandholm 2000). Studies with regard to *Bifidobacterium* strains revealed that a *B. lactis* strain can enhance the natural immune function by dietary consumption (Arunachalam et al. 2000; Chiang et al. 2000). *B. bifidum* is also being investigated as a strain that possibly exerts important immunomodulating effects in combination with other probiotics, when consumed in cheese (Medici et al. 2004). *B. infantis* may also play an immunoregulatory role in the suppression of Th2 cytokines during antigen sensitization (Lee et al. 2004).

Certain bifidobacteria are said to provide protection to the host from pro-carcinogenic activity of intestinal flora (Picard et al. 2005; O'Shea et al. 2012). Animal models have shown that *B. longum* and *B. breve* prevent DNA-induced damage by carcinogens and were effective against colorectal cancer (Pool-Zobel et al. 1996; Saikali et al. 2004). *Bifidobacterium* has also been investigated for the ability to produce conjugated linoleic acid, which has been associated with anticarcinogenicity (Coakley et al. 2003; Rosberg-Cody et al. 2004). Studies involving *B. animalis* have shown a 60%–90% decrease in aberrant crypt foci (ACF) incidence in rats (Tavan et al. 2002). *B. longum* has also been reported to play a role in the reduction of ACF occurrence in rats (Rowland et al. 1998). In addition, *B. longum* has been linked to antitumor activity (Reddy and Rivenson 1993; Singh et al. 1997).

Bifidobacteria have shown to increase the α- and β-galactosidase activities in the fecal samples after feeding with fermented milk containing these probiotics. This is an important probiotic quality as it supports lactose digestion in the intestine and compensates for lactase deficiency (De Vrese et al. 2001; Lourens-Hattingh and Viljoen 2001; Shah 2001). Jiang et al. (1996) investigated milk containing *B. longum* ATCC 15708 as a potential probiotic for treatment of lactose intolerance in certain individuals and was shown to have a positive effect on the reduction of lactose intolerance.

Bifidobacteria can breakdown α-casein present in human milk through its phosphoprotein phosphatase activity and thus improve absorption of human milk protein. Infants possessing bifidus microflora showed better nitrogen retention. Bifidobacteria are thought to promote the metabolism of amino acids.

The effect of probiotic *B. breve* for atopic dermatitis was assessed by Yoshida et al. (2010). The results suggested that *B. breve* may be beneficial for the treatment of atopic dermatitis. Hypocholesterolemic effect of yogurt containing *B. pseudocatenulatum* G4 or *B. longum* BB536 was studied (Al-Sheraji et al. 2012b). The efficacy of adding prebiotic oligosaccharide and probiotic *B. lactis* HN019 to milk, in preventing diarrhea, respiratory infections, and severe illnesses, in children aged 1–4 years in a peri-urban community-based setting in India was studied by Sazawal et al. (2010). They used a community-based, double-masked, randomized trial with four arms to evaluate the effect and concluded that delivery of prebiotic and probiotic through milk medium resulted in significant reduction of dysentery, respiratory morbidity, and febrile illness. Palaria et al (2012) investigated the effect of a synbiotic yogurt containing *B. animalis* subsp. *lactis* Bb-12 and inulin on the human intestinal bifidobacteria, clostridia, and enterobacteria. The results showed a significant increase in the total bifidobacterial numbers and significant decrease in clostridia, but not in case of enterobacteria. Synbiotic preparations are found to be more promising for delivering health benefits compared with probiotic only preparations (Steed et al. 2008). In an open label trial assessing the quality of life in ulcerative colitis patients receiving capsules containing prebiotic psyllium, *B. longum,* or both (synbiotic), questionnaire-based examinations have shown that the synbiotic group scored higher on social and systemic parameters, compared with the other two groups (Fujimori et al. 2009). A synbiotic preparation of *B. breve* Yakult and transgalactosylated oligosaccharides was found significantly more effective in reducing fecal excretion of *Salmonella enterica* serovar *typhimurium* in mice after pathogen challenge compared with either probiotic or prebiotic treatment (Asahara et al. 2001).

10.9 Requirements for Marketing Bifidus Products

A daily intake of probiotics is suggested to get the beneficial health effect on the host. With regard to product dose, the use of multiple probiotics, strain variations, variation in the consumer, and the health endpoint being tested result in making generalizations difficult. The doses between 10^6 and 10^9 colony forming units (cfu) daily are required and most clinical trials use doses within these ranges (Saavedra 2001). Therefore, a high probiotic count is recommended in the products at a minimum count of 10^6 cfu/g at the expiry date (Kurmann and Rasic 1991; Gomes and Malcata 1999). However, it has been suggested that the daily intake should be at least 10^8 cfu (Salminen et al. 1998b; Lourens-Hattingh and Viljoen 2001; Playne 2002). The International Dairy Federation (IDF) standard requires 10^6 cfu/g of bifidobacteria in fermented milks containing bifidobacteria at the time of sale (IDF 1992). In Japan, the Fermented Milks and Lactic Acid Beverages Association has established a standard that requires $\geq 10^7$ cfu/ml to be present in dairy products that claim to contain bifidobacteria (Ishibashi and Shimamura 1993; Martınez-Villaluenga et al. 2006). Likewise, The Swiss Food Regulation requires a minimum of 10^6 cfu/g of viable bifidobacteria.

10.10 Conclusions

The role of food as a tool for the health management has prompted the food industry to develop foods providing benefits and thus exploits the functional food market for health. Fermented milks containing probiotics have been extensively explored commercially and among those bifidus products are an important category worldwide. Adding bifidobacteria to milk and fermented milks improves the functionality of such products tremendously. But the therapeutic effects of such bifidus preparations on human GIT disorders demand more validation to be exclusively used as a therapy. In addition, probiotic effects are strain specific and must be demonstrated through appropriate clinical trials as per the requirement of regulatory agencies. Bifidus milk and products have potential in the infant food category, general health food, and personalized nutrition. This necessitates for a deeper understanding of *bifidobacteria* and human GIT. For realizing this, scientists from the fields of nutrition, microbiology and molecular biology should work together and exploit the advancing knowledge in nutrition, genetics of bifidobacteria, and the molecular methods. Also, scientists should have a close coordination with the food industry and legal regulatory agencies for bringing such products in the market. As a health-benefiting microbe, bifidobacteria have tremendous potential and the food and drug companies should find new and innovative ways to bring bifidobacterium as well as its products to the market shelves and finally to the consumer gut.

References

Agrawal, A., Houghton, L.A., Morris, J., Reilly, B., Guyonnet, D., Goupil Feuillerat, N., Schlumberger, A., Jakob, S., and Whorwell, P.J. 2008. Clinical trial: The effects of a fermented milk product containing *Bifidobacterium lactis* DN-173 010 on abdominal distension and gastrointestinal transit in irritable bowel syndrome with constipation. *Alimentary Pharmacology & Therapeutics* 29: 104–114.

Ahn, J.B., Hwang, H.J., and Park, J. 2001. Physiological responses of oxygen tolerant anaerobic *Bifidobacterium longum* under oxygen. *Journal of Microbiology and Biotechnology* 11: 443–451.

Alander, M. and Mattila-Sandholm, T. 2000. *Functional Foods for EU-Health in 2000*, 4th Workshop, FAIR CT96-1028, PROBDEMO, VTT Symposium 198, Rovaniemi, Finland.

Al-Sheraji, S.H., Ismail, A., Manap, M.Y., Mustafa, S., and Mohd Yusof, R.M. 2012a. Viability and activity of bifidobacteria during refrigerated storage of yoghurt containing *Mangifera pajang* fibrous polysaccharides. *Journal of Food Science* 77(11): M624–M630.

Al-Sheraji, S.H., Ismail, A., Manap, M.Y., Mustafa, S., Mohd Yusof, R.M., and Hassan, F.A. 2012b. Hypocholesterolaemic effect of yoghurt containing *Bifidobacterium pseudocatenulatum* G4 or *Bifidobacterium longum* BB536. *Food Chemistry* 135: 356–361.

Arunachalam, K., Gill, H.S., and Chandra, R.X. 2000. Enhancement of natural immune function by dietary consumption of *Bifidobacterium lactis* (HN019). *European Journal of Clinical Nutrition* 54: 263–267.

Asahara, T., Nomoto, K., Shimizu, K., Watanuki, M., and Tanaka, R. 2001. Increased resistance of mice to *Salmonella enteric* serovar *Typhimurium* infection by synbiotic administration of Bifidobacteria and transgalactosylated oligosaccharides. *Journal of Applied Microbiology* 91(6): 985–96.

Bae, E.A., Han, M.J., Song, M., and Kim, D.H. 2002. Purification of rotavirus infection-inhibitory protein from *Bifidobacterium breve* K-110. *Journal of Microbiology and Biotechnology* 12: 553–556.

Berrada, N., Lelemand, J.F., Laroche, G., Thouvenot, P., and Piaia, M. 1991. Bifidobacterium from fermented milks: Survival during gastric transit. *Journal of Dairy Science* 74: 409–413.

Bezkorovainy, A. 2001. Probiotics: Determinants of survival and growth in the gut. *American Journal of Clinical Nutrition* 73: 399S–405S.

Biavati, B., Vescovo, M., Torriani, S., and Bottazzi, V. 2000. Bifidobacteria: History, ecology, physiology and applications. *Annals of Microbiology* 50: 117–132.

Bouhnik, Y., Pochart, P., Marteau, P., Arlet, G., Goderel, I., and Rambaud, J.C. 1992. Fecal recovery in humans of viable *Bifidobacterium* sp. ingested in fermented milk. *Gastroenterology* 102: 875–878.

Charteris, W.P., Kelly, P.M., Morelli, L., and Collins, J.K. 2002. Edible table (bio)spread containing potentially probiotic *Lactobacillus* and *Bifidobacterium* species. *International Journal of Dairy Technology* 55: 44–56.

Chiang, B.L., Sheih, Y.H., Wang, L.H., Liao, C.K., and Gill, H.S. 2000. Enhancing immunity by dietary consumption of a probiotic lactic acid bacterium (*Bifidobacterium lactis* HN019): Optimization and definition of cellular immune responses. *European Journal of Clinical Nutrition* 54: 849–855.

Chouraqui, J.P., Van Egroo, L.D., and Fichot, M.C. 2004. Acidified milk formula supplemented with *Bifidobacterium lactis*: Impact on infant diarrhea in residential care settings. *Journal of Pediatric Gastroenterology and Nutrition* 38: 288–292.

Coakley, M., Ross, R.P., Nordgren, M., Fitzgerald, G., Devery, R., and Stanton, C. 2003. Conjugated linoleic acid biosynthesis by human-derived *Bifidobacterium* species. *Journal of Applied Microbiology* 94: 138–145.

Crittenden, R., Weerakkody, R., Sanguansri, L., and Augustin, M.A. 2006. Synbiotic microcapsules that enhance microbial viability during non-refrigerated storage and gastrointestinal transit. *Applied and Environmental Microbiology* 72(3): 2280–2282.

Crittenden, R.G., Morris, L.F., Harvey, M., Tran, L.T., Mitchell, H., and Playne, M.J. 2001. Selection of *Bifidobacterium* strain to complement resistant starch in a synbiotic yoghurt. *Journal of Applied Microbiology* 90(2): 268–278.

Dave, R.I., and Shah, N.P. 1997c. Effectiveness of ascorbic acid as an oxygen scavenger in improving viability of probiotic bacteria in yogurts made with commercial starter cultures. *International Dairy Journal* 7: 435–443.

De Vrese, A., Stegelmann, A., Richter, B., Fenselau, S., Laue, C., and Schrezenmeir, J. 2001. Probiotics-compensation for lactase insufficiency. *American Journal of Clinical Nutrition* 73: 421–429.

Dunne, C., Murphy, L., Flynn, S., O'Mahony, L., O'Halloran, S., Feeney, M., Morrissey, D. et al. 1999. Probiotics, from myth to reality. Demonstration of functionality in animal models of disease and in human clinical trials. *Antonie van Leeuwenhoek* 76: 279–292.

Farnworth, E.R. and Mainville, I. 2003. Kefir: A fermented milk product. In: *Handbook of Fermented Functional Foods*, ed. E.R. Farnworth, 77–111. Boca Raton, FL: CRC Press LLC.

Fooks, L.J. and Gibson, G.R. 2002. Probiotics as modulators of the gut flora. *British Journal of Nutrition* 88(S1): S39–S49.

Fooks, L.J. and Gibson, G.R. 2003. Mixed culture fermentation studies on the effects of synbiotics on the human intestinal pathogens *Campylobacter jejuni* and *Escherichia coli*. *Anaerobe* 9: 231–242.

Fujimori, S., Gudis, K., Mitsui, K., Seo, T., Yonezawa, M., Tanaka, S., Tatsuguchi, A., and Sakamoto, C. 2009. A randomized controlled trial on the efficacy of synbiotic versus probiotic or prebiotic treatment to improve the quality of life in patients with ulcerative colitis. *Nutrition* 25(5): 520–525.

Gibson, G.R. and Roberfroid, M.B. 1995. Dietary modulation of the human colonic microbiota: Introducing the concept of prebiotics. *Journal of Nutrition* 125: 1401–1412.

Gibson, G.R. and Wang, X. 1994a. Regulatory effects of bifidobacteria on the growth of other colonic bacteria. *Journal of Applied Bacteriology* 77: 412–420.

Gibson, G.R. and Wang, X. 1994b. Bifidogenic properties of different types of fructo-oligosaccharides. *Food Microbiology* 11: 491–498.

Gibson, G.R. and Wang, X. 1994c. Enrichment of bifidobacteria from human gut contents by oligofructose using continuous culture. *FEMS Microbiology Letters* 118: 121–128.

Gomes, A.M.P. and Malcata, F.X. 1999. *Bifidobacterium* spp. and *Lactobacillus acidophilus*: Biological, biochemical, technological and therapeutical properties relevant for use as probiotics. *Trends in Food Science & Technology* 10: 139–157.

Gonzalez, R., Klaassens, E.S., Malinen, E., de Vos, W.M., and Vaughan, E.E. 2008. Differential transcriptional response of *Bifidobacterium longum* to human milk, formula milk, and galactooligosaccharide. *Applied and Environmental Microbiology* 74(15): 4686–4694.

Guerin, D., Vuillemard, J.C., and Subirade, M. 2003. Protection of bifidobacteria encapsulated in polysaccharide-protein gel beads against gastric juice and bile. *Journal of Food Protection* 66: 2076–2084.

Gurakan, G.C., Cebeci, A., and Ozer, B. 2010. Probiotic dairy beverages: Microbiology and technology. In: *Development and Manufacture of Yoghurt and Other Functional Dairy Products*, ed. F. Yildiz, 181–183. Boca Raton, FL: CRC Press.

Helland, M.H., Wicklund, T., and Narvhus, J.A. 2004. Growth and metabolism of selected strains of probiotic bacteria in milk-and water-based cereal puddings. *International Dairy Journal* 14: 957–965.

Heller, K.J. 2001. Probiotic bacteria in fermented foods: Product characteristics and starter organisms. *American Journal of Clinical Nutrition* 73: 374S–379S.

Holzapfel, W.H., Haberer, P., Snel, J., Schillinger, U., and Huis Int Veld, J.H.J. 1998. Overview of gut flora and probiotics. *International Journal of Food Microbiology* 41: 85–101.

IDF. 1992. *General Standard of Identity for Fermented Milks*, 163. Brussels, Belgium: International Dairy Federation.

Imaoka, A., Shima, T., Kato, K., Mizuno, S., Uehara, T., Matsumoto, S., Setoyama, H., Hara, T., and Umesaki, Y. 2008. Anti-inflammatory activity of probiotic *Bifidobacterium*: Enhancement of IL-10 production in peripheral blood mononuclear cells from ulcerative colitis patients and inhibition of IL-8 secretion in HT-29 cells. *World Journal of Gastroenterology* 14(16): 2511–2516.

Ishibashi, N. and Shimamura, S. 1993. Bifidobacteria: Research and development in Japan. *Food Technology* 6: 126–135.

Ishibashi, N.R. and Yamazaki, S. 2001. Probiotics and safety. *American Journal of Clinical Nutrition.* 73: 465S–470S.

Isolauri, E., Arvola, T., Sutas, Y., Moilanen, E., and Salminen, P. 2000. Probiotics in the management of atopic eczema. *Clinical and Experimental Allergy* 30(11): 1605–1610.

Jiang, T., Mustapha, A., and Saviano, D.A. 1996. Improvement of lactose digestion in humans by ingestion of unfermented milk containing *Bifidobacterium longum. Journal of Dairy Science* 79: 750–757.

Kabeerdoss, J., Devi, R.S., Mary, R.R., Prabhavathi, D., Vidya, R., Mechenro, J., Mahendri, N.V., Pugazhendhi, S., and Ramakrishna, B.S. 2011. Effect of yoghurt containing *Bifidobacterium lactis* Bb12 on faecal excretion of secretory immunoglobulin A and human beta-defensin 2 in healthy adult volunteers. *Nutrition Journal* 10: 138.

Kailasapathy, K. 2002. Microencapsulation of probiotic bacteria: Technology and potential applications. *Current Issues in Intestinal Microbiology* 3: 39–48.

Kailasapathy, K. and Champagne, C. 2011. Fermentation and manufacture of probiotic enriched yoghurt. Chapter 3. In: *Synbiotic Yoghurt—A Smart Gut Food. Science, Technology and Applications*, eds. K. Kailasapaty, C. Champagne, and S. Moore, 44. New York: Nova Science Publishers, Inc.

Kailasapathy, K. and Chin, J. 2000. Potential survival and therapeutic of probiotic organisms with reference to *Lactobacillus acidophilus* and *Bifidobacterium* spp. *Immunology and Cell Biology* 78(1): 80–88.

Kanbe, M. 1992. Uses of intestinal lactic acid bacteria and health. In: *Functions of Fermented Milk*, eds. Y. Nakazawa and A. Hosono, 289–304. London: Elsevier Applied Science Publications.

Killer, J., Kopečný, J., Mrázek, J., Koppová, I., Havlík, J., Benada, O., and Kott, T. 2011. *Bifidobacterium actinocoloniiforme* sp. nov. and *Bifidobacterium bohemicum* sp. nov., from the bumblebee digestive tract. *International Journal of Systematic and Evolutionary Microbiology* 61: 1315–1321.

Killer, J., Kopecný, J., Mrázek, J., Rada, V., Benada, O., Koppová, I., Havlík, J., and Straka, J. 2009. *Bifidobacterium bombi* sp. nov., from the bumblebee digestive tract. *International Journal of Systematic and Evolutionary Microbiology* 59: 2020–2024.

Klaenhammer, T.R., Azcarate-Peril, M.A., Altermann, E., and Barrangon, R. 2007. Influence of the dairy environment on gene expression and substrate utilization in lactic acid bacteria. *Journal of Nutrition* 137(3 Suppl 2): 748S–750S.

Klaver, F.A.M., Kingma, F., and Weerkamp, A.H. 1993. Growth and survival of bifidobacteria in milk. *Netherlands Milk and Dairy Journal* 47: 151–164.

Kumar, C.L.P., Saroja, Y.S., Kumar, D.J.M., and Kalaichelvan, P.T. 2012. Bifidobacteria for life betterment. *World Applied Sciences Journal* 17 (11): 1454–1465.

Kurmann, J.A. and Rasic, J.L. 1991. The health potential of products containing bifidobacteria. In: *Therapeutic Properties of Fermented Milks*, ed. R. K. Robinson, 117–158. London: Elsevier.

Kurmann, J.A., Rasic, J.L., and Kroger, M. 1992. *Encyclopedia of Fermented Fresh Milk Products*, 200–281. New York: Van Nostrand Reinhold.

Lankaputhra, W.E.V. and Shah, N.P. 1996. A simple method for selective enumeration of *Lactobacillus acidophilus* in yogurt supplemented with *L. acidophilus* and *Bifidobacterium* spp. *Milchwissenschaft* 51: 446–451.

Leahy, S.C., Higgins, D.G., Fitzgerald, G.F., and van Sinderen, D. 2005. Getting better with bifidobacteria. A Review. *Journal of Applied Microbiology* 98: 1303–1315.

Lebeer, S., Vanderleyden, J., and De Keersmaecker, S.C.J. 2008. Genes and molecules of lactobacilli supporting probiotic action. *Microbiology and Molecular Biology Reviews* 72(4): 728–764.

Lee, H.-Y., Park, J.-H., Seok, S.-H., Cho, S.-A., Baek, M.-W., Kim, D.-J., Lee, Y.-H., and Park, J.-H. 2004. Dietary intake of various lactic acid bacteria suppresses type 2 helper T cell production in antigen-primed mice splenocyte. *Journal of Microbiology and Biotechnology* 14: 167–170.

Lee, J.H. and O'Sullivan, D.J. 2010. Genomic insights into bifidobacteria. *Microbiology and Molecular Biology Reviews* 74(3): 378–416.

Lievin, V., Peiffer, I., Hudault, S., Rochat, F., Brassart, D., Neeser, J., and Servin, A. 2000. *Bifidobacterium* strains from resident infant human gastrointestinal will microflora exert antimicrobial activity. *Gut* 47(5): 646–652.

Lin, D.C. 2003. Probiotics as functional foods. *Nutrition in Clinical Practice* 18(6): 497–506.

Lourens-Hattingh, A. and Viljoen, B.C. 2001. Yogurt as probiotic carrier food. *International Dairy Journal* 11: 1–17.

Lučan, M., Slačanac, V., Hardi, J., Mastanjević, K., Babić, J., Krstanović, V., and Jukić, M. 2009. Inhibitory effect of honey-sweetened goat and cow milk fermented with *Bifidobacterium lactis* Bb-12 on the growth of *Listeria monocytogenes*. *Mljekarstvo* 59(2): 96–106.

Manning, T.S. and Gibson, G.R. 2004. Prebiotics. *Best Practice & Research Clinical Gastroenterology* 18: 287–298.

Marco, M.L., Pavan, S., and Kleerebezem, M. 2006. Towards understanding molecular modes of probiotic action. *Current Opinion in Biotechnology* 17(2): 204–210.

Marteau, P., Pochart, P., Bouhnik, Y., Zidi, S., Goderel, I., and Rambaud, J.C. 1992. Survival of *Lactobacillus acidophilus* and Bifidobacterium sp. in the small intestine following ingestion in fermented milk. A rational basis for the use of probiotics in man. *Gastroenterol Clinical Biology* 16: 25–28.

Martınez-Villaluenga, C., Frıas, J., Gomez, R., and Vidal-Valverde, C. 2006. Influence of addition of raffinose family oligosaccharides on probiotics survival in fermented milk during refrigerated storage. *International Dairy Journal* 16: 768–774.

Matsumoto, M., Ohishi, H., and Benno, Y. 2004. H+− ATPase activity in *Bifidobacterium* with special reference to acid tolerance. *International Journal of Food Microbiology* 93: 109–113.

Mattila-Sandholm, T. and Kauppila, T. 1998. *Functional Food Research*. Presented at Europe 3rd Workshop, FAIR CT96-1028, PROBDEMO, VTT Symposium 187. Haikko, Finland.

Mättö, J., Alakomi, H.L., Vaari, A., Virkajarvi, I., and Saarela, M. 2006. Influence of processing conditions on *Bifidobacterium animalis* subsp. *lactis* functionality with a special focus on acid tolerance and factors affecting it. *International Dairy Journal* 16: 1029–1037.

Medici, M., Vinderola, C.G., and Perdigon, G. 2004. Gut mucosal immunomodulation by probiotic fresh cheese. *International Dairy Journal* 14: 611–618.

Misra, A. and Kulia, R. 1994. Use of *Bifidobacterium bifidum* for the manufacturing of bio-yoghurt and fruit bio-yoghurt. *Indian Journal of Dairy Science* 47: 192–197.

Miyake, T., Watanabe, K., Watanabe, T., and Oyaizu, H. 1998. Phylogenetic analysis of the genus Bifidobacterium and related genera based on 16S rDNA sequences. *Microbiology and Immunology* 42: 661–667.

Modler, H.W., Mckellar, R.C., and Yaguchi, M. 1990. Bifidobacteria and bifidogenic factors. *Canadian Institute of Food Science* 23: 29–41.

Mohan, R., Koebnick, C., Schildt, J. Schmidt, S., Mueller, M., Possner, M., Radke, M., and Blaut, M. 2006. Effects of *Bifidobacterium lactis* BB12 supplementation on intestinal microbiota of preterm infants: A double-blind placebo controlled randomized study. *Journal of Clinical Microbiology* 44: 4025–4031.

Muir, D.D., Tamime, A.Y., and Wszolek, M. 1999. Comparison of the sensory profiles of kefir, buttermilk and yogurt. *International Journal of Dairy Technology* 52: 129–134.

Oberman, H. and Libudzisz, Z. 1998. Fermented milks. In: *Microbiology of Fermented Food*, ed. B.J.B. Wood, 308–350. London: Blackie Academic & Professional.

Orrhage, K., Brismar, B., and Nord, C.E. 1994. Effects of supplements of *Bifidobacterium longum* and *Lactobacillus acidophilus* on intestinal microbiota during administration of clindamycin. *Microbial Ecology in Health and Disease* 7: 17–25.

O'Riordan, K.O., Andrews, D., Buckle, K., and Conway, P. 2001. Evaluation of microencapsulation of a Bifidobacterium strain with starch as an approach to prolonging viability during storage. *Journal of Applied Microbiology* 91: 1059–1066.

O'Shea, E.F., Cotter, P.D., Stanton, C., Ross, R.P., and Hill, C. 2012. Production of bioactive substances by intestinal bacteria as a basis for explaining probiotic mechanisms: Bacteriocins and CLA. *International Journal of Food Microbiology* 152(3): 189. doi:10.1016/j.ijfoodmicro.2011.05.025.

Palaria, A., Johnson-Kanda, I., and O'Sullivan, D.J. 2012. Effect of a synbiotic yogurt on levels of fecal bifidobacteria, clostridia and enterobacteria. *Applied and Environmental Microbiology* 78(4): 933–940.

Picard, C., Fioramonti, J., Francois, A., Robinson, T., Neant, F., and Matuchansky, C. 2005. Review article: Bifidobacteria as probiotic agents—Physiological effects and clinical benefits. *Alimentary Pharmacology & Therapeutics* 22(6): 495–512.

Playne, M.J. 2002. The health benefits of probiotics. *Food Australia* 54: 71–74.

Playne, M.J., Bennet, L.E., and Smithers, G.W. 2003. Functional dairy foods and ingredients. *Australian Journal of Dairy Technology* 58: 242–264.

Plummer, S., Weaver, M.A., Harris, J.C., Dee, P., and Hunter, J. 2004. *Clostridium difficile* pilot study: Effect of probiotic supplementation on the incidence of *C. difficile* diarrhoea. *International Microbiology* 7: 59–62.

Pochart, P., Marteau, P., Bouhnik, Y., Goderel, I., Bourlioux, P., and Rombard, J.C. 1992. Survival of bifidobacteria ingested via fermented milk during their passage through the human small intestine: An in vivo study using intestinal perfusion. *American Journal of Clinical Nutrition* 55: 78–80.

Pokusaeva, K., Fitzgerald, G.F., and van Sinderen, D. 2011. Carbohydrate metabolism in Bifidobacteria. *Genes and Nutrition* 6: 285–306.

Pool-Zobel, B.L., Neudecker, C., Domizlaff, I., Ji, S., Schillinger, U., Rumney, C., Moretti, M., Vilarini, I., Scassellati-Sforzolini, R., and Rowland, I. 1996. *Lactobacillus* and *Bifidobacterium*-mediated antigenotoxicity in the colon of rats. *Nutrition and Cancer* 26: 365–380.

Rakoff-Nahoum, S., Paglino, J., Eslami-Varzaneh, F., Edberg, S., and Medzhitov, R. 2004. Recognition of commensal microflora by toll-like receptors is required for intestinal homeostasis. *Cell* 118: 229–241.

Rasic, J.L. and Kurmann, J.A. 1983. Bifidobacteria and their Role. *Experientia Supplementum* 39: 1–295.

Reddy, B.S., and Rivenson, A. 1993. Inhibitory effect of *Bifidobacterium longum* on colon, mammary and liver carcinogenesis induced by 2-amino-3-methyl imidazol [4,5-f] quinoline, a food mutagen. *Cancer Research* 53: 3914–3918.

Reid, G., Sanders, M.E., Gaskins, H.R., Gibson, G.R., Mercenier, A., Rastall, R., Roberfroid, M., Rowland, I., Cherbut, C., and Klaenhammer, T.R. 2003. New scientific paradigms for probiotics and prebiotics. *Journal of Clinical Gastroenterology* 37: 105–118.

Reuter, G. 1990. Bifidobacteria cultures as components of yogurt-like products. *Bifidobacteria and Microflora* 9: 107–118.

Roberfroid, M.B. 1998. Prebiotics and synbiotics: Concepts and nutritional properties. *British Journal of Nutrition* 80: S197–S202.

Roberfroid, M., Gibson, G.R., and Delzenne, N. 1993. The biochemistry of oligofructose, a non-digestible fiber: An approach to calculate its caloric value. *Nutrition Reviews* 51: 137–146.

Rogosa, M. 1974. Genus III, *Bifidobacterium* Orla-Jensen. In: *Bergey's Manual of Determinative Bacteriology* (8th edition), eds. R.E. Buchanan and N.E. Gibbons, 669–676. Baltimore, MD: Williams & Wilkins.

Rosberg-Cody, E., Ross, R.P., Hussey, S., Ryan, C.A., Murphy, B.P., Fitzgerald, G.F., Devery, R. and Stanton, C. 2004. Mining the microbiota of the neonatal gastrointestinal tract for conjugated linoleic acid-producing bifidobacteria. *Applied and Environmental Microbiology* 70: 4635-4641.

Rothschild, P. 1995. Internal defenses. *Dairy Industries International* 60(2): 24–25.

Rowland, I.R., Rumney, C.J., Coutts, J.T., and Lievense, L.C. 1998. Effect of *Bifidobacterium longum* and inulin on gut bacterial metabolism and carcinogen-induced aberrant crypt foci in rats. *Carcinogenesis* 19: 281–285.

Saarela, M., Mogensen, G., Fonden, R., Matto, J., and Mattilla-Sandholm, T. 2000. Probiotic bacteria: Safety, functional and technological properties. *Journal of Biotechnology* 84: 197–215.

Saarela, M., Lahteenmaki, L., Crittended, R., Salminen, S., and Mattila-Sandholm, T. 2002. Gut bacteria and health foods-the European perspective. *International Journal of Food Microbiology* 78: 99–117.

Saavedra, J.M. 2001. Clinical applications of probiotic agents. *American Journal of Clinical Nutrition* 3(suppl): 1147S–1151S.

Saavedra, J.M., Bauman, N.A., Oung, I., Perman, J.A., and Yolken, R.H. 1994. Feeding of *Bifidobacterium bifidum* and *Streptococcus thermophilus* to infants' in-hospital for prevention of diarrhea and shedding of rotavirus. *Lancet* 344: 1046–1049.

Saikali, J., Picard, C., Freitas, M., and Holt, P.R. 2004. Fermented milks, probiotic cultures, and colon cancer. *Nutrition and Cancer* 49(1): 14–24.

Salminen, S. and Gueimonde, M. 2004. Human studies on probiotics: What is scientifically proven? *Journal of Food Science* 69: M137–M140.

Salminen, S., Vonwright, A., Morelli, L., Marteau, P., Brassart, D., de Vos, W.M. et al. 1998b. Demonstration of safety of probiotics—A Review. *International Journal of Food Microbiology* 44: 93–106.

Samona, A., Robinson, R.K., and Marakis, S. 1996. Acid production by bifidobacteria and yoghurt bacteria during fermentation and storage of milk. *Food Microbiology* 13: 275–280.

Sanders, M.E. and Marco, M.L. 2010. Food formats for effective delivery of probiotics. *Annual Review of Food Science Technology* 1: 65–85.

Sarkar, S. and Misra, A.K. 2010. Technological and dietetic characteristics of probiotic acidophilus milk. *British Food Journal* 112(3): 275–284.

Sazawal, S., Dhingra, U., Hiremath, G., Sarkar, A., Dhingra, P., Dutta, A., Verma, P., Menon, V.P., and Black, R.E. 2010. Prebiotic and probiotic fortified milk in prevention of morbidities among children: Community-based, randomized, double-blind, controlled trial. *PLoS One*. 5(8): e12164.

Schacht, E. and Syrazinski, A. 1975. Progurt, a new cultured product: Its manufacturing technology and dietetic value (in Spanish). *Industria Lechera* 646: 9–11.

Sela, D.A., Chapman, J., Adeuya, A., Kim, J.H., Chen, F., Whitehead, T.R., Lapidus, A. et al. 2008. The genome sequence of *Bifidobacterium longum* subsp. *infantis* reveals adaptations for milk utilization within the infant microbiome. *Proceedings of the National Academy of Sciences of the United States of America* 105(48): 18964–18969.

Shah, N.P. 2001. Functional foods from probiotics and prebiotics. *Food Technology* 55: 46–53.

Shah, N.P. 2011. *From Bulgarian Milks to Probiotics Fermented Milks*. Summilk, IDF World Dairy Summit.

Shah, N.P. and Lankaputhra, W.E.V. 1997. Improving the viability of *Lactobacillus acidophilus* and *Bifidobacterium spp*. in yogurt. *International Dairy Journal* 7: 349–356.

Shin, H.S., Lee, J.H., Pestka, J.J., and Ustunol, Z. 2000b. Viability of Bifidobacteria in commercial dairy products during refrigerated storage. *Journal of Food Protection* 63: 327–331.

Shioya, M., Nakaoka, K., Iizuka, N., and Benno, Y. 2000. Effect of fermented milk containing *Bifidobacterium lactis* FK 120 on the fecal flora, with special reference to Bifidum species, and fecal properties in healthy volunteers. *Food Health and Nutrition Research* 3: 19–32.

Singh, J., Rivenson, A., Tomita, M., Shimamura, S., Ishibashi, N., and Reddy, B.S. 1997. *Bifidobacterium longum*, a lactic acid producing intestinal microflora inhibit colon cancer and modulate the intermediate biomarkers of colon carcinogenesis. *Carcinogenesis* 18: 1371–1377.

Slačanac, V., Hardi, J., Čuržik, D., Pavlović, H., Lučan, M., and Vlainić, M. 2007. Inhibition of the in vitro growth of *Salmonella enteritidis* D by goat and cow milk fermented with probiotic bacteria *Bifidobacteriumlongum* Bb-46. *Czech Journal of Food Sciences* 25: 351–358.

Soccol, C.R., Vandenberghe, L.P., Spier, M.R., de Souza Vandenberghe, L.P., Spier, M.P., Medeiros, A.B.P., Yamaguishi, C.T., De Dea Lindner, J., Pandey, A., and Soccol, V.T. 2010. The potential of probiotics: A review. *Food Technology and Biotechnology* 48(4): 413–434.

Steed, H., Macfarlane, G.T., and Macfarlane, S. 2008. Prebiotics, synbiotics and inflammatory bowel disease. *Molecular Nutrition and Food Research* 52(8): 898–905.

Steer, T., Carpenter, H., Tuohy, K., and Gibson, G.R. 2000. Perspectives on the role of the human gut microbiota and its modulation by pro- and prebiotics. *Nutrition Research Reviews* 13: 229–254.

Tabbers, M.M., Chmielewska, A., Roseboom, M.G., Crastes, N., Perrin, C., Reitsma, J.B., Norbruis, O., Szajewska, H., and Benninga, M.A. 2011. Fermented milk containing *Bifidobacterium lactis* DN-173 010 in childhood constipation: A randomized, double-blind, controlled trial. *Pediatrics* 127(6): 1392–1399.

Talwalkar, A. and Kailasapathy, K. 2004. A Review of oxygen toxicity in probiotic yogurts: Influence on the survival of probiotic bacteria and protective techniques. *Comprehensive Reviews in Food Science and Food Safety* 3: 117–124.

Tamime, A.Y., Marshall, V.M.E., and Robinson, R.K. 1995. Microbiological and technological aspects of milks fermented by bifidobacteria. *Journal of Dairy Research* 62: 151–187.

Tannock, G.W. 1995. Internal renewal: The potential for modification of the normal microflora. In: *Normal Microflora: An Introduction to Microbes Inhabiting the Human Body*, 100. London: Chapman & Hall.

Tannock, G.W. 1999a. Analysis of the intestinal microflora: A renaissance. *Antonie van Leeuwenhoek* 76: 265–278.

Tannock, G.W. 1999b. Identification of Lactobacilli and Bifidobacteria. *Current Issues in Molecular Biology* 1(1): 53–64.

Tavan, E., Cayuela, C., Antonie, J.M., Trugnan, G., Chaugier, C., and Cassand, P. 2002. Effects of dairy products on heterocyclic aromatic amine-induced rat colon carcinogenesis. *Carcinogenesis* 23(3): 477–483.

Tissier, H. 1906. Traitement des infections intestinales par la méthode de la flore bactérienne de l'intestin. *C. R. Soc. Biol.* 60: 359–361.

Toure, R., Kheadr, E., Lacroix, C., Moroni, O., and Fliss, I. 2003. Production of antibacterial substances by bifidobacterial isolates from infant stool active against *Listeria monocytogenes*. *Journal of Applied Microbiology* 95: 1058–1069.

Turroni, F., van Sinderen, D., and Ventura, M. 2011. Genomics and ecological overview of the genus Bifidobacterium. *International Journal of Food Microbiology* 149: 37–44.

Ventura, M., Canchaya, C., Tauch, A., Chandra, G., Fitzgerald, G.F., Chater, K.F., and van Sinderen, D. 2007. Genomics of Actinobacteria: Tracing the evolutionary history of an ancient phylum. *Microbiology and Molecular Biology Reviews* 71: 495–548.

Ventura, M., O'Flaherty, S., Claesson, M.J., Turroni, F., Klaenhammer, T.R., van Sinderen, D., and O'Toole, P.W. 2009. Genome-scale analyses of health-promoting bacteria: Probiogenomics. *Nature Reviews Microbiology* 7: 61–71.

Wang, X. and Gibson, G.R. 1993. Effects of the in vitro fermentation of oligofructose and inulin by bacteria growing in the human large intestine. *Journal of Applied Bacteriology* 75: 373–380.

Yoshida, Y., Seki, T., Matsunaka, H., Watanabe, T., Shindo, M., Yamada, N., and Yamamoto, O. 2010. Clinical effects of probiotic *Bifidobacteriumbreve* supplementation in adult patients with atopic dermatitis. *Yonago Acta Medica* 53: 37–45.

11

Yogurt

Concepts and Developments

SARANG DILIP POPHALY, HITESH KUMAR, SUDHIR KUMAR TOMAR, AND RAMESHWAR SINGH

Contents

11.1	Introduction	312
11.2	Production	312
11.3	Yogurt Microbiology: Molecular Trading	313
	11.3.1 Peptides and Amino Acids	314
	11.3.2 Carbon Dioxide	315
	11.3.3 Formate	315
	11.3.4 Folate and p-Aminobenzoic Acid	316
	11.3.5 Long-Chain Fatty Acids	316
	11.3.6 Oxidative Stress	316
	11.3.7 Genetic Material Exchange	317
11.4	Yogurt Variants and Styles	317
	11.4.1 Type of Milk	317
	11.4.2 Consistency	317
	11.4.3 Heat-Treated Yogurt	318
	11.4.4 Fruit Yogurt	318
	11.4.5 Concentrated/Strained Yogurt	318
	11.4.6 Non-Dairy Yogurt	318
	11.4.7 Frozen Yogurt	318
	11.4.8 Dried Yogurt	318
	11.4.9 Enriched and Fortified Yogurts	319
11.5	Yogurt as Carrier to Probiotics	320
11.6	Conclusion	322
References		323

11.1 Introduction

Humans and microorganisms both have been closely associated through evolution process in one or other way to each other and have deeply impacted each other. With their domestication to form fermented food products, microbes are now also an indispensible part of the human food chain. It is a remarkable serendipity that different civilizations and societies in the different phases of development have shared a common tradition of consuming fermented products. These fermented products thus represent an important aspect of our sociocultural and scientific history.

Yogurt is one such popular fermented milk product consumed throughout the world and was first made by nomadic Turkish people living in Asia and this art was spread by these people in different parts of the world. Another school of thought suggests yogurt to be of Baltic origin, wherein the inhabitants of Thrace made a similar fermented product from sheep milk. Ancient sacred texts, such as the Vedas and the Bible, find the mention of yogurt-like products suggesting the deep-rooted religious and cultural association of this product to humans. Physicians in Middle East have been known to prescribe yogurt for curing stomach, liver, and intestinal disorders. Industrial revolution in Europe led to the commercialization and large-scale manufacturing of yogurt first undertaken by Danone in 1922 (Prajapati and Nair 2003). Although the fundamental steps for manufacture remained same, industrial adaptation resulted in standardized methodology and consistent quality of the product. It also leads to the organized scientific research on various technological, microbiological, sensory, and health-related aspects of yogurt.

11.2 Production

Yogurt is a classic example of how a microbial symbiotic process involving exchange of molecules is manifested and customized for human use and is one of the most complex fermented foods (German 2014). As necessary, while preparing any fermented product, the initial quality of milk is of paramount importance, which determines the quality, nutritional value, and market value of the finished product. Another important aspect that determines yogurt quality is culture purity and activity.

One should always have cognizance of the fact that in the preparation of fermented foods, the heat treatment step comes prior to fermentation, and thus, any slackening in culture addition or during incubation can not only compromise the quality of the product but may also put consumer's life on risk. The skill of yogurt making, as inherited from the ancestors, has been subjected to industrial standardization process devising an established protocol followed throughout the world.

For its preparation, the milk is first clarified and then standardized to achieve the desired fat content. Next, the total solid content of milk is increased to a level of 12%–13%, which ensures desired textural properties in the final product. The various ingredients are then blended together in a mix tank equipped with a powder funnel and an agitation system. The mix is then homogenized using high pressures of 2000–2500 psi. Besides, thoroughly mixing the stabilizers and other ingredients, homogenization also prevents creaming and wheying-off during incubation and storage. The homogenized mixture is then pasteurized using a continuous plate heat exchanger for 30 min at 85°C or 10 min at 95°C. These heat treatments, which are much more severe than fluid milk pasteurization, are necessary to create a relatively sterile environment and to denature whey proteins, which increases the viscosity and improves the texture of yogurt. The heat-treated mixture is then cooled to an optimum growth temperature (45°C) and yogurt starter is added. A ratio of 1:1, *Streptococcus thermophilus* to *Lactobacillus bulgaricus*, inoculation is then added to the milk base. Depending upon the final packaging of prepared product, the inoculated mix may be packed rapidly in cups or else fermented in a fermentation tank. A temperature of 43°C is maintained for 4–6 h under non-agitated conditions. The titratable acidity is carefully monitored until the acidity is 0.85%–0.90%. At this time, the jacket is replaced with cool water and agitation begins, both help in ceasing the fermentation process. The product is then cooled and stored at refrigeration temperatures (5°C) to slow down the physical, chemical, and microbiological degradation.

11.3 Yogurt Microbiology: Molecular Trading

Yogurt is a result of a unique and closely associated interaction of *Str. thermophilus* and *Lb. delbrueckii* subsp. *bulgaricus*. Both organisms have a symbiotic relationship in the micro-ecology of milk matrix, which brings about the characteristics of a typical yogurt. The primary driving force in such a mutualistic relationship is growth, which results in rapid acidification of milk, which is much slower if individual strains are allowed to bring about the fermentation processes. The chronicle of this microbial interaction could be traced to the evolutionary pressures faced by individual species and stochastic chances of their encounter with each other. Milk as a highly nutritious medium for microbial propagation, the microorganisms, which grew in medium, was subjected to a genomic decay over the course of evolution. This massive loss of genes leads to a situation wherein 10% of *Str. thermophilus* genes become psuedogenes (Bolotin et al. 2004) and *Lb. delbrueckii* subsp. *bulgaricus* also losing a huge set of active enzymes (Van de Guchte et al. 2006). The result of such a massive loss of functionality renders those depending for import of growth factors from the medium and open to proto-cooperation. The earlier known or proposed mechanisms of proto-cooperation are now validated and extended in light of genomic and proteomic evidences, obtained as a result of enormous data generated by whole genome sequencing of multiple strains of two organisms (Bolotin et al. 2004; Hao et al. 2011).

The co-culturing of these two species brings about significant changes and manifestations like higher acidification rates (Bautista et al. 1966), production of flavor compounds (Courtin and Rul 2004; Sieuwerts et al. 2008), increased polysaccharide production (Vaningelgem et al. 2004), and higher *Str. thermophilus* count (Herve-Jimenez et al. 2009). Co-culturing of *Str. thermophilus* with *Lb. bulgaricus* varied the expression of several (77) genes, involved in metabolism of nitrogen (24%), nucleotide base (21%), and iron (20%) (Herve-Jimenez et al. 2009). Some of the molecular exchange reactions and phenomena are placed below.

11.3.1 Peptides and Amino Acids

Lb. delbrueckii subsp. *bulgaricus* has a comparatively higher proteolytic activity than *Str. thermophilus*, but their mixed culture has enhanced proteolytic activity than monoculture (Rajagopal and Sandine 1990). The yogurt bacilli also have high auxotrophy for most amino acids and only are able to synthesize aspartate, asparagine, lysine, and threonine.

Thus, availability of several free amino acids is a growth-limiting factor for the yogurt microorganisms making it essential to degrade milk proteins into peptides and amino acids. *Lb. delbrueckii* subsp. *bulgaricus* harbors a battery of proteases and peptidases including a cell wall-associated protease PrtB and in some strains two extracellular peptidases (Hao et al. 2011). Together, these enzymes carry out the breakdown of casein resulting into peptides, which are then imported by various peptide transport systems. These peptides are further broken down intracellularly by endopeptidases and aminopeptidases ensuing free amino acids. The peptides released by the action of PrtB stimulate growth of *Str. thermophilus*, which lacks such an extensive proteolytic machinery (Beshkova et al. 1998; Courtin et al. 2002).

Besides providing *Str. thermophilus* with peptides, co-culturing with *Lb. delbrueckii* subsp. *bulgaricus* upregulates the synthesis of branched chain amino acids like arginine (Herve-Jimenez et al. 2009) and sulfur-containing amino acids (Herve-Jimenez et al. 2008). Interestingly, the gene cluster (*cbs-cblB(cglB)-cysE*) for metabolism of sulfur-containing amino acids in *Str. thermophilus* is putatively acquired from *Lb. delbrueckii* subsp. *bulgaricus* through a horizontal gene transfer (HGT) mechanism (Liu et al. 2009), suggesting intricately involved proto-cooperation mode in the two organisms.

11.3.2 Carbon Dioxide

Carbon dioxide is produced by *Str. thermophilus*, as a result of decarboxylation of urea by urease enzyme. Carbon dioxide is required as a precursor for biosynthesis of certain amino acids and nucleotides and promotes the growth of *Lb. delbrueckii* subsp. *bulgaricus*, but its concentration in milk is decreased as a result of heat treatment. This loss is compensated by carbon dioxide production by *Str. thermophilus* by degrading the urea present in milk, which promotes the growth of *Lb. delbrueckii* subsp. *bulgaricus* and thus is vital in proto-cooperation of the two organisms (Driessen et al. 1982).

11.3.3 Formate

Formate serves as a precursor for purine biosynthesis and is essential to be supplied, if the organism lacks the ability to synthesize it.

Lb. delbrueckii subsp. *bulgaricus* does not produce formate due to the absence of pyruvate formate lyase (*pfl*) (Hao et al. 2011). *Str. thermophilus*, on the other hand, produces high level of formate, when grown in milk through upregulation of *pfl* gene (Derzelle et al. 2005), suggestively providing *Lb. delbrueckii* subsp. *bulgaricus* with required formate (Sieuwerts et al. 2008).

11.3.4 Folate and p-Aminobenzoic Acid

Folate serves as an important cofactor for many enzymatic reactions and is an essential component of human diet. Although *in silico* analysis has revealed that *Lb. delbrueckii* subsp. *bulgaricus* possesses the molecular assembly for biosynthesis of folate, it does not have the ability to produce *p*-aminobenzoic acid (PABA), which serves as a key precursor for folate biosynthesis. *Str. thermophilus* is a prolific producer of folate and PABA, and thus complements *Lb. bulgaricus* for its requirement of these two molecules.

11.3.5 Long-Chain Fatty Acids

Long-chain fatty acids (LCFA) are required by *Lb. delbrueckii* subsp. *bulgaricus* for growth. However, it lacks the *de novo* synthesis of these fatty acids and hence collaborates with *Str. thermophilus*, which presumably releases LCFA.

11.3.6 Oxidative Stress

Oxidative stress has a unique role in symbiotic association of the two yogurt bacteria, and deoxygenated conditions could significantly reduce yogurt fermentation time (Horiuchi and Sasaki 2012). *Lb. delbrueckii* subsp. *bulgaricus* is involved in the production of hydrogen peroxide (Batdorj et al. 2007), which imparts oxidative stress to *Str. thermophilus* growing in co-culture. *Str. thermophilus* lacks enzymes like catalase and nicotinamide adenine dinucleotide (NADH) peroxidase, which are directly involved with H_2O_2 degradation; instead, it induces the gene encoding dpr (sequester iron) and also restricts iron import by downregulation of iron import assembly proteins like feoA and fatABC. Recently, a *Str. thermophilus* flavoprotein known as Nox (NADH oxidase) was reported as the major oxygen-consuming enzyme, which promoted the fermentation of milk. The unique feature of *Str. thermophilus* Nox is that it is able to reduce molecular oxygen to H_2O without producing hydrogen peroxide (Sasaki et al. 2014).

11.3.7 Genetic Material Exchange

Long symbiotic history of yogurt culture has not only ensured molecular complementation for growth but also impacted alterations in genomic content through HGT between the two species. *Str. thermophilus* harbors a 17-kb region having extensive similarity with *Lb. bulgaricus*, which imparted the methionine synthesis ability to the former (Pfeiler and Klaenhammer 2007). Symbiotic relationship has not only complemented their growth but also led to alterations in genomic content through HGT. *Lb. bulgaricus*, on the other hand, has obtained exopolysaccharides biosynthesis genes from *Str. thermophilus* (Hols et al. 2005; Liu et al. 2009).

11.4 Yogurt Variants and Styles

Yogurt is a product that can be made with multiple styles and formulations, as long as the basic culture remains unchanged. In fact, supplementation of adjunct starters is also practiced to impart additional functional features to yogurt, which may be categorized as one or other type of the product. The popular yogurt classification system can be summarized as follows.

11.4.1 Type of Milk

Cow milk has been traditionally used for making yogurt but availability of milk from different species in different regions of the world has led to the localized manufacturing of yogurt varieties from sheep, camel, goat, and buffalo milk.

11.4.2 Consistency

Plain yogurt may be manufactured according to the final consistency requirement of set type yogurt, stirred yogurt, and drinking yogurt.

11.4.3 Heat-Treated Yogurt

Such type of yogurt involves heat treatments like pasteurization, UHT, and so on post fermentation to inactivate the starter bacteria and enzymes in yogurt to impart a long shelf life to the product.

11.4.4 Fruit Yogurt

A vast selection of fruit and fruit pulps is available for making these types of yogurts. The sundae type yogurts have fruit puree layered at the bottom of the cups over which yogurt is set or else the puree and fruits could be blended throughout the yogurt body as in Swiss style yogurt.

11.4.5 Concentrated/Strained Yogurt

Strained yogurt popularly called as labneh is prepared by straining yogurt in a cloth or a paper bag to remove excess whey. It is low in lactose and whey proteins and rich in casein.

11.4.6 Non-Dairy Yogurt

Soy yogurt is an example of non-dairy yogurt, prepared by fermenting soymilk. Soy yogurt is a good alternative for vegetarians, who are disinclined to the products of animal origin. Adjunct cultures may be required for the preparation of soy yogurt as the yogurt starter may not be able to carry the fermentation of soymilk.

11.4.7 Frozen Yogurt

Frozen yogurt may be categorized as a type of ice cream or a frozen dessert. It is prepared by fermenting a yogurt mix and followed by freezing. It differs from ice cream by having a sharp acidic taste developed as a result of fermentation and a low fat content. Frozen yogurt may be classified into soft frozen yogurt, hard frozen yogurt, and mousse yogurt.

11.4.8 Dried Yogurt

Owing to its high nutritive value and water content, yogurt has a limited shelf life and contaminating spoilage organisms may cause rapid spoilage of the product. Dried yogurt forms are prepared primarily to extend the shelf life of yogurt, so that it can be used over a long period of time and also at regions where supply of fresh milk products is limited. Powdered yogurt can be used as an ingredient for the manufacturing of confectioneries, bakery foods, soup bases, dips, and sauces, and after reconstitution of yogurt powder, it can be used as a yogurt drink mixes with fruits or vegetables (Koc et al. 2010).

11.4.9 Enriched and Fortified Yogurts

Addition of various molecules or precursors to milk prior to fermentation gives a unique way of delivering bio-functional derivatives through yogurt. In this way, yogurt serves as a carrier of such molecules enhancing its nutritive and health-promoting features. These may include vitamins, trace minerals, herbs, and fibers (Table 11.1).

Table 11.1 Fortification of Yogurt with Different Molecules

CLASS OF MOLECULE FORTIFIED	FORTIFYING MOLECULE	TYPICAL CHARACTERISTICS	REFERENCES
Trace elements/ minerals	Calcium	Improved nutritive value, higher water holding capacity	Singh and Muthukumarappan (2008)
	Chromium	Improved nutritive value, more opaque white characteristics of yogurt	Achanta et al. (2007)
	Iron	Improved nutritional value	Hekmat and McMahon (1997), Achanta et al. (2007)
	Magnesium and Manganese	Fortification with Mg and Mn are beneficial for a healthy heart	Cueva and Aryana (2008), Achanta et al. (2007)
	Molybdenum	Improved nutritive value, more opaque white characteristics of yogurt	Achanta et al. (2007)
	Selenium	Biotransformation of inorganic selenium to safer and more bioavailable organic species	Alzate et al. (2007)
Vitamin	Vitamin C	Supplementing vitamin C to dairy products	Metzger (1962)
	Vitamin D	Improved glycemic status in type-II diabetic patients	Nikooyeh et al. (2011)

(*Continued*)

Table 11.1 (*Continued*) Fortification of Yogurt with Different Molecules

CLASS OF MOLECULE FORTIFIED	FORTIFYING MOLECULE	TYPICAL CHARACTERISTICS	REFERENCES
Fiber	Asparagus fiber	Improved nutritive value, modified rheological properties	Sanz et al. (2008)
	Barley β glucan	Post fermentation addition of barley glucan suggested for better preservation of its properties	Gee et al. (2007)
	Date fiber	Improved sensory characteristics and nutritive value	Hashim et al. (2009)
	Oat fiber	Reduced calories, improved body and texture	Fernández-Garía et al. (1998)
	Orange fiber	Modified rheological properties	Sendra et al. (2010)
	Soy fiber	Soy fiber supplementation requires either flavor or fruit pulp addition for better acceptability	Epstein et al. (1990), Kumar and Mishra (2003)
	Yam soluble fiber	Lowered syneresis and higher mouthfeel characteristics	Ramirez-Santiago et al. (2010)

11.5 Yogurt as Carrier to Probiotics

Probiotics need to be delivered to the host through a suitable matrix, which should support and preferably enhance their robustness and health-promoting properties. Various researchers have used yogurt as a carrier for probiotic bacteria resulting in products appropriately called probiotic yogurt or bio-yogurts (Lourens-Hattingh and Viljoen 2001). Formulation of such probiotic yogurt could be done in two broadways, either by using probiotic strains of yogurt starter or by supplementing yogurt with probiotic adjunct cultures. In general, *Str. thermophilus* and *Lb. delbrueckii* subsp. *bulgaricus* do not survive well in intestinal conditions and thus are not considered very potent probiotic candidates. However, few workers (Guarner et al. 2005; Mater et al. 2005) have reported strain-specific probiotic properties in yogurt starter cultures, which could be used for the preparation of probiotic yogurt. A more practical approach for making probiotic yogurt is using the established and well-documented probiotic cultures as adjunct cultures in yogurt. In this case, the yogurt starter carries the primary fermentation process and the adjunct culture exerts probiotic activities. The key functionality of such probiotic products like health-promoting attribute, sensory characteristics (Hekmat and Reid 2006) is largely dependent upon the adjunct strain (Table 11.2).

While preparing a probiotic yogurt, many factors should be considered to ensure that the product meets all the required legal and technological parameters. These factors include strain compatibility between yogurt starter and probiotic cultures, bacteriocin production by the cultures, sensory characteristics, and suitability of probiotics to yogurt processing conditions. Additives could be incorporated in the probiotic yogurt mix to enhance the survival or functionality of probiotics (Table 11.3). Yogurt may serve as an appropriate matrix for delivering the probiotic health benefits to the host (Saxelin 2008). The advantages of yogurt for delivery of probiotics can be summarized as follows:

- A fermented product, where incorporation of a probiotics seems technologically easy.
- Popularity of yogurt among people of different regions and age groups makes it an ideal choice for developing into a probiotic product.
- Milk matrix is suitable for growth of many probiotics.
- Storage temperature favors stabilization of probiotics in viable conditions for longer duration.
- Different presentations and styles of yogurt offer wide range of options for delivery of probiotics.

Table 11.2 Adjunct Probiotic Cultures Added to Yogurt

PROBIOTIC CULTURE/STRAIN	FUNCTIONALITY	REFERENCE
Bifidobacterium DN-173	Oral antibacterial activity	Caglar et al. (2005)
Lb. acidophilus LA14, *B. longum* BL 05	Immunoprotection	Lollo et al. (2013)
Lb. acidophilus LA-5 and *Bifidobacterium lactis* BB12	Anticholersteremic activity, Antioxidative activity	Ejtahed et al. (2011,2012)
Lb. rhamnosus GR-1	Anti-inflammatory activity	Lorea Baroja et al. (2007)
Lb. rhamnosus	Improved CD4 count in HIV positive individuals	Irvine et al. (2010)
Lb. rhamnosus	Management of acute diarrhea	Grandy et al. (2014)

Table 11.3 Effect of Different Ingredients on Functionality of Probiotic Yogurts

INGREDIENT(S)	FUNCTIONALITY	REFERENCE
Acid casein hydrolysate	Improved bifidobacteria count	Dave and Shah (1998)
Cysteine	Improved bifidobacteria count	Dave and Shah (1998)
Ginseng	Improved viability of probiotic bacteria	Cimo et al. (2013)
Glucose oxidase	Decreased oxidative stress, improved viability of probiotic bacteria	Cruz et al. (2012)
Inulin	Improved probiotic count	Akın et al. (2007)
Sodium calcium caseinate	Improved adhesiveness, firmness, and viscosity	Akalin et al. (2012)
Spice oleoresins (cardamom, cinnamon, nutmeg)	Better sensory quality	Illupapalayam et al. (2014)
Stevia, sucralose	Better sensory quality	Weber and Hekmat (2013)
Tryptone	Improved bifidobacteria count	Dave and Shah (1998)
Whey protein concentrate	Enhanced water holding capacity	Akalin et al. (2012)
β-Glucan	Improved probiotic count	Vasiljevic et al. (2007)

11.6 Conclusion

Yogurt is a classical fermented dairy product that has a long history of development and has impacted human life for centuries. The symbiotic association of *Str. thermophilus* and *Lb. delbrueckii* subsp. *bulgaricus* can be considered as a tradeoff using different molecules and enzymes for mutual existence in milk. Various technological and molecular interventions have led to high product quality and standardized methodology for industrial scale manufacturing of yogurt. An extensive diversity of variants has transformed the plain yogurt into products of exceptionally high market value and a wider consumer base worldwide. Yogurt, for its specific technological properties and consumer preference, is highly suited for delivering the probiotics to the host. A whole new range of probiotic yogurt products (bio-yogurts) are saturating the market, and the benefits and safety evaluation of such products need to be extensively considered. Finding of health-promoting characteristics, biochemical and molecular contribution of the involved species, and technological aspects are in phase of constant revelations that needs newly emerging tools and state-of-the-art techniques to explore the yogurts further.

References

Achanta, K., Aryana, K.J., and Boeneke, C.A. 2007. Fat free plain set yogurts fortified with various minerals. *LWT-Food Science and Technology* 40:424–429.

Akalin, A., Unal, G., Dinkci, N., and Hayaloglu, A. 2012. Microstructural, textural, and sensory characteristics of probiotic yogurts fortified with sodium calcium caseinate or whey protein concentrate. *Journal of Dairy Science* 95:3617–3628.

Akin, M., Akin, M., and Kirmaci, Z. 2007. Effects of inulin and sugar levels on the viability of yogurt and probiotic bacteria and the physical and sensory characteristics in probiotic ice-cream. *Food Chemistry* 104:93–99.

Alzate, A., Cañas, B., Pérez-Munguía, S., Hernández-Mendoza, H., Pérez-Conde, C., Gutiérrez, A.M., and Cámara, C. 2007. Evaluation of the in organic selenium biotransformation in selenium-enriched yogurt by HPLC-ICP-MS. *Journal of Agricultural and Food Chemistry* 55:9776–9783.

Batdorj, B., Trinetta, V., Dalgalarrondo, M., Prévost, H., Dousset, X., Ivanova, I., Haertlé, T., and Chobert, J.M. 2007. Isolation, taxonomic identification and hydrogen peroxide production by *Lactobacillus delbrueckii* subsp. *lactis* T31, isolated from Mongolian yoghurt: Inhibitory activity on food-borne pathogens. *Journal of Applied Microbiology* 103:584–593.

Bautista, E.S., Dahiya, R., and Speck, M. 1966. Identification of compounds causing symbiotic growth of *Streptococcus thermophilus* and *Lactobacillus bulgaricus* in milk. *Journal of Dairy Research* 33:299–307.

Beshkova, D.M., Simova, E.D., Frengova, G.I., Simov, Z.I., and Adilov, E.F. 1998. Production of amino acids by yogurt bacteria. *Biotechnology Progress* 14:963–965.

Bolotin, A., Quinquis, B., Renault, P., Sorokin, A., Ehrlich, S.D., Kulakauskas, S., Lapidus, A. et al. 2004. Complete sequence and comparative genome analysis of the dairy bacterium *Streptococcus thermophilus*. *Nature Biotechnology* 22:1554–1558.

Caglar, E., Sandalli, N., Twetman, S., Kavaloglu, S., Ergeneli, S., and Selvi, S. 2005. Effect of yogurt with Bifidobacterium DN-173 010 on salivary mutans streptococci and lactobacilli in young adults. *Acta Odontologica* 63:317–320.

Cimo, A., Soltani, M., Lui, E., and Hekmat, S. 2013. Fortification of probiotic yogurt with Ginseng (Panax quinquefolius) extract. *Food and Nutritional Disorders* 2:2. doi:10.4172/2324-9323.1000106.

Courtin, P., Monnet, V., and Rul, F. 2002. Cell-wall proteinases PrtS and PrtB have a different role in *Streptococcus thermophilus/Lactobacillus bulgaricus* mixed cultures in milk. *Microbiology* 148:3413–3421.

Courtin, P. and Rul, F. 2004. Interactions between microorganisms in a simple ecosystem: Yogurt bacteria as a study model. *Le Lait* 84:125–134.

Cruz, A., Castro, G.W.F., Faria, J.A., Bogusz, Jr., S., Granato, D., Celeguini, R.M.S., Lima-Pallone, J., and Godoy, T.H. 2012. Glucose oxidase: A potential option to decrease the oxidative stress in stirred probiotic yogurt. *LWT-Food Science and Technology* 47:512–515.

Cueva, O. and Aryana, K.J. 2008. Quality attributes of a heart healthy yogurt. *LWT-Food Science and Technology* 41:537–544.

Dave, R. and Shah, N. 1998. Ingredient supplementation effects on viability of probiotic bacteria in yogurt. *Journal of Dairy Science* 81: 2804–2816.

Derzelle, S., Bolotin, A., Mistou, M.-Y., and Rul, F.O. 2005. Proteome analysis of *Streptococcus thermophilus* grown in milk reveals pyruvate formate-lyase as the major upregulated protein. *Applied and Environmental Microbiology* 71:8597–8605.

Driessen, F., Kingma, F., and Stadhouders, J. 1982. *Evidence that* Lactobacillus bulgaricus *in Yogurt is Stimulated by Carbon Dioxide Produced by* Streptococcus thermophilus. Ede, the Netherlands: Netherlands Institute for Dairy Research.

Ejtahed, H.S., Mohtadi-Nia, J., Homayouni-Rad, A., Niafar, M., Asghari-Jafarabadi, M., and Mofid, V. 2012. Probiotic yogurt improves antioxidant status in type 2 diabetic patients. *Nutrition* 28:539–543.

Ejtahed, H.S., Mohtadi-Nia, J., Homayouni-Rad, A., Niafar, M., Asghari-Jafarabadi, M., Mofid, V., and Akbarian-Moghari, A. 2011. Effect of probiotic yogurt containing *Lactobacillus acidophilus* and *Bifidobacterium lactis* on lipid profile in individuals with type 2 diabetes mellitus. *Journal of Dairy Science* 94:3288–3294.

Epstein, E., Hoyda, D.L., and Streiff, P.J. 1990. Method of making fiber enriched yogurt. United States Patent US4971810 A.

Fernández-Garía, E., McGregor, J.U., and Traylor, S. 1998. The addition of oat fiber and natural alternative sweeteners in the manufacture of plain yogurt. *Journal of Dairy Science* 81:655–663.

Gee, V.L., Vasanthan, T., and Temelli, F. 2007. Viscosity of model yogurt systems enriched with barley β-glucan as influenced by starter cultures. *International Dairy Journal* 17:1083–1088.

German, J.B. 2014. The future of yogurt: Scientific and regulatory needs. *The American Journal of Clinical Nutrition*. doi:10.3945/ajcn.113.076844.

Grandy, G., Jose, Z., Soria, R., Castelú, J., Perez, A., Ribera, J.P., and Brunser, O. 2014. Use of Probiotic Yogurt in the management of acute diarrhoea in children. Randomized, double-blind, controlled study. *Open Journal of Pediatrics* 4:54.

Guarner, F., Perdigon, G., Corthier, G.R., Salminen, S., Koletzko, B., and Morelli, L. 2005. Should yoghurt cultures be considered probiotic? *British Journal of Nutrition* 93:783–786.

Hao, P., Zheng, H., Yu, Y., Ding, G., Gu, W., Chen, S., Yu, Z. et al. 2011. Complete sequencing and pan-genomic analysis of *Lactobacillus delbrueckii* subsp. *bulgaricus* reveal its genetic basis for industrial yogurt production. *PLoS One* 6:e15964.

Hashim, I., Khalil, A., and Afifi, H. 2009. Quality characteristics and consumer acceptance of yogurt fortified with date fiber. *Journal of Dairy Science* 92:5403–5407.

Hekmat, S. and McMahon, D.J. 1997. Manufacture and quality of iron-fortified yogurt. *Journal of Dairy Science* 80:3114–3122.

Hekmat, S. and Reid, G. 2006. Sensory properties of probiotic yogurt is comparable to standard yogurt. *Nutrition Research* 26:163–166.

Herve-Jimenez, L., Guillouard, I., Guedon, E., Boudebbouze, S., Hols, P., Monnet, V., Maguin, E., and Rul, F. 2009. Postgenomic analysis of *Streptococcus thermophilus* cocultivated in milk with *Lactobacillus delbrueckii* subsp. *bulgaricus*: Involvement of nitrogen, purine, and iron metabolism. *Applied and Environmental Microbiology* 75:2062–2073.

Herve-Jimenez, L., Guillouard, I., Guedon, E., Gautier, C., Boudebbouze, S., Hols, P., Monnet, V., Rul, F., and Maguin, E. 2008. Physiology of *Streptococcus thermophilus* during the late stage of milk fermentation with special regard to sulfur amino acid metabolism. *Proteomics* 8:4273–4286.

Hols, P., Hancy, F., Fontaine, L., Grossiord, B., Prozzi, D., Leblond-Bourget, N., Decaris, B. et al. 2005. New insights in the molecular biology and physiology of *Streptococcus thermophilus* revealed by comparative genomics. *FEMS Microbiology Reviews* 29:435–463.

Horiuchi, H. and Sasaki, Y. 2012. Short communication: Effect of oxygen on symbiosis between *Lactobacillus bulgaricus* and *Streptococcus thermophilus*. *Journal of Dairy Science* 95:2904–2909.

Illupapalayam, V.V., Smith, S.C., and Gamlath, S. 2014. Consumer acceptability and antioxidant potential of probiotic-yogurt with spices. *LWT-Food Science and Technology* 55:255–262.

Irvine, S.L., Hummelen, R., Hekmat, S., Looman, C.W., Habbema, J.D.F., and Reid, G. 2010. Probiotic yogurt consumption is associated with an increase of CD4 count among people living with HIV/AIDS. *Journal of Clinical Gastroenterology* 44:e201–e205.

Koc, B., Yilmazer, M.S., Balkır, P., and Ertekin, F.K. 2010. Spray drying of yogurt: Optimization of process conditions for improving viability and other quality attributes. *Drying Technology* 28:495–507.

Kumar, P. and Mishra, H. 2003. Optimization of mango soy fortified yogurt formulation using response surface methodology. *International Journal of Food Properties* 6:499–517.

Liu, M., Siezen, R.J., and Nauta, A. 2009. In silico prediction of horizontal gene transfer events in *Lactobacillus bulgaricus* and *Streptococcus thermophilus* reveals protocooperation in yogurt manufacturing. *Applied and Environmental Microbiology* 75:4120–4129.

Lollo, P.C.B., de Moura, C.S., Morato, P.N., Cruz, A.G., de Freitas Castro, G., Betim, C.B., Nisishima, L. et al. 2013. Probiotic yogurt offers higher immune-protection than probiotic whey beverage. *Food Research International* 54:118–124.

Lorea Baroja, M., Kirjavainen, P., Hekmat, S., and Reid, G. 2007. Anti-inflammatory effects of probiotic yogurt in inflammatory bowel disease patients. *Clinical and Experimental Immunology* 149:470–479.

Lourens-Hattingh, A. and Viljoen, B.C. 2001. Yogurt as probiotic carrier food. *International Dairy Journal* 11:1–17.

Mater, D.D., Bretigny, L., Firmesse, O., Flores, M.J., Mogenet, A., Bresson, J.L., and Corthier, G. 2005. *Streptococcus thermophilus* and *Lactobacillus delbrueckii* subsp. *bulgaricus* survive gastrointestinal transit of healthy volunteers consuming yogurt. *FEMS Microbiology Letters* 250:185–187.

Metzger, J. 1962. Vitamin c enriched yogurt. United States patent US3025164 A.

Nikooyeh, B., Neyestani, T.R., Farvid, M., Alavi-Majd, H., Houshiarrad, A., Kalayi, A., Shariatzadeh, N. et al. 2011. Daily consumption of vitamin D– or vitamin D+ calcium-fortified yogurt drink improved glycemic control in patients with type 2 diabetes: A randomized clinical trial. *The American Journal of Clinical Nutrition* 93:764–771.

Pfeiler, E.A. and Klaenhammer, T.R. 2007. The genomics of lactic acid bacteria. *Trends in Microbiology* 15:546–553.

Prajapati, J.B. and Nair, B.M. 2003. The history of fermented foods. In: *Fermented Functional Foods*, ed. E.D. Farnworth, 1–25. Boca Raton, FL: CRC Press.

Rajagopal, S. and Sandine, W. 1990. Associative growth and proteolysis of *Streptococcus thermophilus* and *Lactobacillus bulgaricus* in skim milk. *Journal of Dairy Science* 73:894–899.

Ramirez-Santiago, C., Ramos-Solis, L., Lobato-Calleros, C., Pena-Valdivia, C., Vernon-Carter, E., and Alvarez-Ramírezd, J. 2010. Enrichment of stirred yogurt with soluble dietary fiber from *Pachyrhizus erosus* L. Urban: Effect on syneresis, microstructure and rheological properties. *Journal of Food Engineering* 101:229–235.

Sanz, T., Salvador, A., Jimenez, A., and Fiszman, S. 2008. Yogurt enrichment with functional asparagus fibre. Effect of fibre extraction method on rheological properties, colour, and sensory acceptance. *European Food Research and Technology* 227:1515–1521.

Sasaki, Y., Horiuchi, H., Kawasahima, H., Mukai, T., and Yamamoto, Y. 2014. NADH Oxidase of *Streptococcus thermophilus* 1131 is required for the effective yogurt fermentation with *Lactobacillus delbrueckii* subsp. *bulgaricus* 2038. *Bioscience of Microbiota, Food and Health* 33:31–40.

Saxelin, M. 2008. Probiotic formulations and applications, the current probiotics market, and changes in the marketplace: A European perspective. *Clinical Infectious Diseases* 46:S76–S79.

Sendra, E., Kuri, V., Fernandez-Lopez, J., Sayas-Barbera, E., Navarro, C., and Perez-Alvarez, J. 2010. Viscoelastic properties of orange fiber enriched yogurt as a function of fiber dose, size and thermal treatment. *LWT-Food Science and Technology* 43:708–714.

Sieuwerts, S., de Bok, F.A., Hugenholtz, J., and van Hylckama Vlieg, J.E. 2008. Unraveling microbial interactions in food fermentations: From classical to genomics approaches. *Applied and Environmental Microbiology* 74:4997–5007.

Singh, G. and Muthukumarappan, K. 2008. Influence of calcium fortification on sensory, physical and rheological characteristics of fruit yogurt. *LWT-Food Science and Technology* 41:1145–1152.

Van de Guchte, M., Penaud, S., Grimaldi, C., Barbe, V., Bryson, K., Nicolas, P., Robert, C. et al. 2006. The complete genome sequence of *Lactobacillus bulgaricus* reveals extensive and ongoing reductive evolution. *Proceedings of the National Academy of Sciences* 103:9274–9279.

Vaningelgem, F., Zamfir, M., Adriany, T., and De Vuyst, L. 2004. Fermentation conditions affecting the bacterial growth and exopolysaccharide production by *Streptococcus thermophilus* ST 111 in milk based medium. *Journal of Applied Microbiology* 97:1257–1273.

Vasiljevic, T., Kealy, T., and Mishra, V. 2007. Effects of β-glucan addition to a probiotic containing yogurt. *Journal of Food Science* 72:C405–C411.

Weber, A., and Hekmat, S. 2013. The effect of Stevia rebaudiana on the growth and survival of *Lactobacillus rhamnosus* GR-1 and sensory properties of probiotic yogurt. *Journal of Food Research* 2:136–143.

12

Technology of Fresh Cheeses

YOGESH KHETRA, S.K. KANAWJIA, APURBA GIRI, AND RITIKA PURI

Contents

12.1	Introduction	330
12.2	Cottage Cheese	331
	12.2.1 Classification of Cottage Cheese	331
	12.2.2 Technology of Cottage Cheese	332
	12.2.2.1 Milk	332
	12.2.2.2 Milk Treatment	332
	12.2.2.3 Starter Culture	334
	12.2.2.4 Curd Formation	334
	12.2.2.5 Cutting and Cooking of Curd	335
	12.2.2.6 Washing and Dressing	335
	12.2.3 Shelf Life	336
12.3	Quarg Cheese	336
	12.3.1 Characteristics of Quarg Cheese	336
	12.3.2 Method of Production	336
12.4	Mozzarella Cheese	338
	12.4.1 Production Method	339
	12.4.2 Packaging, Preservation, and Storage	340
	12.4.3 Chemistry of "Stretch" of Mozzarella Cheese	342
12.5	Cream Cheese	342
12.6	Queso Blanco/Latin American White Cheese	343
12.7	Paneer—A Cooking Type Cheese	344
	12.7.1 Manufacturing Conditions	345
	12.7.1.1 Type of Milk	345
	12.7.1.2 Heat Treatment	345
	12.7.1.3 Coagulation of Milk	346
	12.7.1.4 Drainage of Whey and Curd Pressing	347
	12.7.1.5 Chilling of Paneer	347

		12.7.1.6 Packaging	347
		12.7.1.7 Yield of Paneer	347
	12.7.2	Paneer from Cow Milk	347
12.8	Whey Cheeses		348
	12.8.1	Principle of Whey Cheese Manufacture	348
	12.8.2	Whey Cheese Manufacture	349
		12.8.2.1 Ricotta Cheese	349
		12.8.2.2 Mysost, Gjetost, and Primost Cheese	349
	12.8.3	Shelf Life of Whey Cheese	350
	12.8.4	Use of Whey Cheese	350
References			351

12.1 Introduction

Cheese is a classical product of dairy, which has high nutritional significance owing to its richness in pre-digested proteins, fat, minerals, and vitamins. Cheese making involves a series of processes that convert milk into different varieties of cheeses known for their characteristic sensory attributes and nutritional value. Cheese is also considered as a functional food owing to its health benefits beyond basic nutrition (Renner 1993). Thus, cheese is a value-added dairy product that can be used in various forms as snacks, spread, and bread-mate. Cheese making involves a series of processes and even a small alteration in these processes can result in a different variety of cheese altogether. Thus, over 2000 varieties of cheese exist with different sensory properties, nutritive values, and uses.

Cheese has been classified on the basis of its composition, coagulating agent, and extent/manner of ripening. On the basis of ripening, cheese is broadly classified as ripened and unripened or fresh cheese.

Fresh cheeses include the products that are ready for consumption shortly after manufacture. These may be acid coagulated or rennet coagulated. On the basis of manner of coagulation, fresh cheese again may be classified as rennet curd cheese or fresh acid-curd cheese. For most of the varieties, starter culture and rennet are used in combination for development of desired quality characteristics. Some of the popular fresh cheese varieties consumed round the globe are discussed in this chapter. Ripened cheeses such as

Cheddar, Gouda, and Parmesan are not consumed just after manufacture and are kept at such a combination of time, temperature, and relative humidity that favors the desirable biochemical and textural changes (Upadhyay 2003).

12.2 Cottage Cheese

Cottage cheese is an unripened variety of cheese that is characterized as soft and mild acid cheese usually manufactured using skim milk. Its discrete and isolated curd particles of relatively uniform size differentiate it from other cheeses. The production of cottage cheese originated from Eastern and Central Europe. As the name suggests, this cheese was originally produced at small scale on family farms. Cream cottage cheese is another variant of cottage cheese that is prepared by blending *dry curd cottage cheese* and *cream dressing*. The cream cottage cheese contains at least 4% fat and not more than 80% moisture. Low fat cottage cheese contains 0.5%–2.0% fat and not more than 82.5% moisture (Guinee et al. 1993).

As per Food Safety and Standards Regulations (FSSR 2011), cottage cheese and creamed cottage cheese means soft unripen cheese obtained by coagulation of pasteurized skimmed milk of cow and/or buffalo or mixtures thereof with cultures of harmless lactic acid bacteria with or without the addition of other suitable coagulating enzymes. Creamed cottage cheese is a type of cottage cheese to which a pasteurized creaming mixture of cream, skimmed milk, condensed milk, nonfat dry milk, dry milk protein, and sodium/potassium/calcium/ammonium caseinate is added. It shall have a soft texture with a natural white color. It shall contain no more than 80% moisture. Milk fat in creamed cottage cheese should not be less than 4.0%.

12.2.1 Classification of Cottage Cheese

As per the method of coagulation, cottage cheese can be classified as acid curd or rennet curd.

Acid curd cottage cheese involves formation of coagulum through the action of starter culture while rennet curd cottage cheese primarily uses rennet for coagulation. The use of culture for coagulation results in acid and diacetyl flavor, while rennet curd is characterized by a pleasant and bland flavor (Kosikowski 1966).

12.2.2 Technology of Cottage Cheese

Cottage cheese can be manufactured by short, intermediate, or long set methods at varying time and temperature combination with starter culture. Alternatively, cottage cheese can also be manufactured employing direct acidification technique by using food-grade acid or acid whey (Figure 12.1). The method involves coagulation of milk caseins at or near isoelectric pH (4.6) to form a firm coagulum. The curd is then cut followed by whey drainage, washing, cooling, and optional creaming of the curd, and packaging.

Coagulation of casein may be carried out either by using acid or rennet. Acid coagulation results in less aggregation of casein due to high stability of curd particles against fusion and loss of moisture. Some essential requirements for the manufacture of good quality cottage cheese with recent developments are described hereunder.

12.2.2.1 Milk The skim milk should be pasteurized in holder method because higher heating temperatures result in a softer curd, which is easily broken while cutting and handling. Mainly protein in milk influences the curd firmness, quality, and yield of cheese. Caseins, especially β-casein and citrate content of milk, take vital role in the control of fines formation in whey. The proportion of α_{s1} casein is positively correlated to both fines and grit formation, whereas β-casein is negatively correlated (Dyurec and Zall 1985). Addition of sodium citrate (0.1%) and sodium caseinate (0.25%–0.55%) improves curd formation and reduces fines in the whey. In general, 8.5%–8.8% solid not fat (SNF) in skim milk is considered good to manufacture quality cottage cheese (Demott et al. 1984).

12.2.2.2 Milk Treatment To make good quality cottage cheese, the total and psychrophilic counts per ml of raw milk should be less than 50,000 and 10,000, respectively, and 1,000/ml in pasteurized milk. Pasteurization of milk is generally done at 63°C/30 min or at 71°C/15 s, though increasing pasteurization temperature improves the yield but at the cost of quality. Addition of increased amount

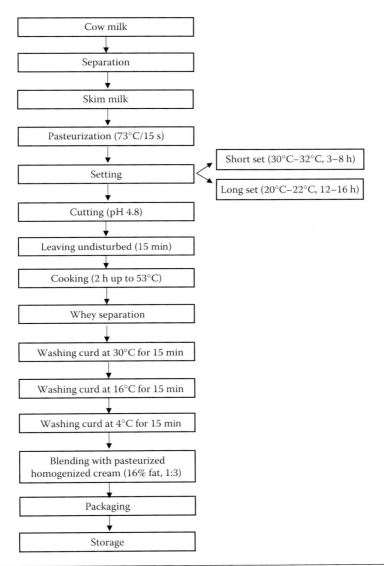

Figure 12.1 Manufacturing method for cottage cheese. (Data from Walstra, P. et al., Cheese varieties. In: *Dairy Technology: Principles of Milk Properties and Processes*, Marcel Dekker Inc, New York, 707–766, 1999.)

of calcium chloride and rennet and cutting the coagulum at high pH (pH 5.0) improve the quality and yield (Guinee et al. 1993; Makhal et al. 2015b). By ultrafiltration of milk, yield of cottage can be increased as compared to high heat treatment (Mattews et al. 1976).

Makhal et al. (2011) have suggested the use of κ-carragenan and tetrasodium pyrophosphate to increase the yield of direct acidified cottage cheese.

12.2.2.3 Starter Culture The type of starter culture influences the quality and yield of cottage cheese. In buffalo milk, single strain cultures have been found to improve the yield; while the multiple types mixed strain cultures improve the quality (Pandya et al. 1989). Freeze-dried lactic culture performs equally well as the conventional cultures. The major problem of agglutination in cottage cheese, which causes clumping and setting of starter bacteria as well as precipitation of casein at the bottom of the vat, is minimized by homogenization of skim milk or by the addition of defatted lecithin (0.5%) to the bulk culture. Homogenization of skim milk destroys milk agglutinins, whereas the addition of lecithin to starter culture or homogenization causes fragmentation of starter chains without affecting cell numbers or acid production.

12.2.2.4 Curd Formation Curd for cottage cheese is formed either by bacterial cultures or by direct acidification (Makhal et al. 2013) or by a combination of the two. Coagulation employing bacterial cultures can be accomplished either by long set method or short set method as mentioned in Table 12.1.

Direct acidification method involves addition of lactic acid/phosphoric acid to cold milk (2°C–12°C) to achieve a pH of 5.2 followed by addition of glucono-δ-lactone, which is slowly hydrolyzed to gluconic acid, resulting in a gradual reduction in pH to 4.6–4.8 (Makhal and Kanawjia 2008). Acid curd can also be made by mixing

Table 12.1 Coagulation Conditions for Short Set and Long Set Methods

PARTICULARS	SHORT SET	LONG SET
Coagulation time (h)	4–6	12–16
Setting temperature (°C)	30–32	21–22
Amount of starter added (%)	5–8	1–2

Source: Farkye, N.Y., Acid and acid/rennet—Curd cheeses, Part B: Cottage cheese. In *Cheese-Chemistry, Physics and Microbiology* (3rd Edition, Vol. 2), eds. P.F. Fox, P.L.H. McSweeney, T.M. Cogan, and T.P. Guinee, Elsevier Academic Press, London, 329–341, 2004.

33% sodium bisulfate solution with milk at 29°C, while maintaining the pH at 5.04–5.08, adjusting the pH to 5.2 without causing coagulation and adding a cold water slurry of glucono-δ-lactone and rennet (El-Batawy et al. 1993; Makhal et al. 2011).

12.2.2.5 Cutting and Cooking of Curd The curd is cut at pH 4.75–4.80. For buffalo milk, the curd is cut at titratable acidity of 0.72%. The desired cooking pH during cottage cheese making is 4.55–4.60 and the cooking temperature ranges from 50 to 55°C. However, cooking at 60°C for 20–30 min in a total cooking time of 90 min yields a good cottage cheese from cow milk. In case of buffalo milk, the final cooking temperature of 60°C in 75 min has been employed (Tiwari and Singh 1988).

12.2.2.6 Washing and Dressing Washing cools the curd, removes remaining whey, and adjusts the pH of the curd for optimum creaming. The volume of the wash water varies in the range of 30%–100% of the milk volume. Different treatments such as filtration, chlorination (5–8 ppm), and acidification (pH 5.5–6.0) of the wash water are practiced to avoid contamination and solubilization of curd.

Mechanized devices for washing and cooling of the curd have been developed, which involves pumping of cooked curd–whey mixture to the drainer where 90% of the whey is removed and the curd is mixed with wash water from the downstream washer. The curd in water is then pumped to the washer, the balance whey is removed, and the curd is washed with chilled water and cooled to 4°C–5°C. The cooled curd and water is then passed to the water curd drainer. The drained curd is then dropped into creamers, where cream dressing (12%–18% fat) is added to, and mixed with the dry curds after washing to obtain 4% fat in final product. The final pH of the cottage cheese should be 5.2. Use of 0.07%–0.3% stabilizer in cream helps it to adhere to the curd and reduces the amount of free cream. Fortification of milk solids followed by homogenization of the cream is recommended to prevent separation of the cream. Addition of hot cream at 88°C results in lower total and psychrophilic counts and higher flavor score than cottage cheese dressed with cream at 5°C (Ernstrom and Kale 1975).

12.2.3 Shelf Life

The shelf life of cottage cheese is about 2 weeks under refrigeration, which can be further improved by the following:

- Acidification of dressing mixture (Makhal et al. 2015a)
- Sterilization at 80°C–95°C after addition of stabilizer
- Addition of thymol MicroGARD™, pimaricin, and sorbic acid (Makhal et al. 2014)
- CO_2 or N_2 flushing
- Thermization at 65°C for 180 s

12.3 Quarg Cheese

12.3.1 Characteristics of Quarg Cheese

Quarg (German name—Speisequark) cheese is white in color or faintly yellowish. Body and texture is homogeneously soft, smooth, mildly elastic, and spreadable. The high moisture content (~82%) limits the shelf life of the product to 2–4 weeks at 8°C (Schulz-Collins and Senge 2004). The product neither should have free water or whey at surfaces nor should it be dry and grainy. Bacteriological deterioration, excess acidity, or bitter flavor development during storage are considered undesirable (Guinee et al. 1993; Mucchetti et al. 2000). As the product is high in moisture content, it is high in lactose or lactate and consequently relatively low in fat and protein. Quarg is low in Ca content because acid coagulation makes most of the calcium soluble resulting in its removal with whey.

12.3.2 Method of Production

Curd for quarg cheese is made by inoculating pasteurized skim milk with suitable mesophilic bacterial culture at the rate of 2% and kept at 23°C (Figure 12.2). After about 2 h of incubation (pH ~6.3), rennet (0.5–0.75 g Meito rennet/100 l of milk) is added and incubation at same temperature is continued for a total time of about 16 h. Whey separation from curd can be accomplished either by mechanical centrifugation separation or traditionally by hanging the curd in muslin cloth under cold conditions. Mechanical separation through

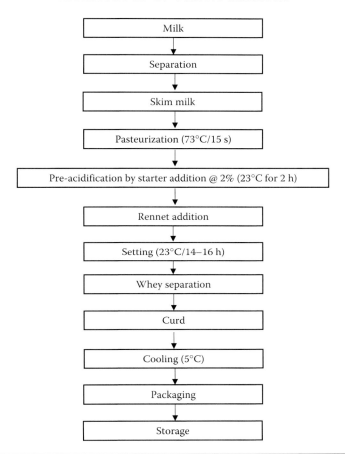

Figure 12.2 Manufacturing method for quarg cheese. (Data from Walstra et al., Cheese varieties. In: *Dairy Science and Technology*, CRC Press/Taylor & Francis group, 687–742, 1999.)

centrifugation paved the way for industrial manufacture of the product (Siggelkow 1984). To facilitate better separation of whey in traditional method of hanging, curd is cut and cooked in the same way as is done for other cheese varieties, but for quarg, size of cubes is usually kept bigger so as to prevent curd from excess firmness and results in softer curd (Gahane 2008; Kadiya 2009; Kumar 2012). For mechanical separation of whey, quarg separators are used. After coagulation, curd is stirred for about 15 min and this fermented gel is then fed to separators. In these separators, the clotted mass is separated from the whey by centrifugal force. Further, water content of curd can be controlled by varying flow rate. Alternatively, ultrafiltration of milk up to 3–3.5-fold concentration may be carried out and the retentate is then used for quarg making. This process results in higher yield as whey proteins and calcium phosphate are also incorporated in the curd (Jelen and Renz-Schauen 1989; Sharma and Reuter 1993; Pfalzer and Jelen 1994; Ottosen 1996; Walstra et al. 1999).

To increase the whey protein content and the yields of cheese, the cheese milk may be heated at higher temperatures (95°C–98°C) for about 2–5 min. This high heat treatment denatures whey proteins and results in their association with casein curd (Schulz-Collins and Senge 2004).

Quarg cheese is considered as an ideal vehicle for carrying functional ingredients such as dietary fibers, phytosterols, and prebiotics. Attempts have been made successfully for incorporation of inulin, oat fiber, soy fiber, and plant sterol esters (Gahane 2008). Synbiotic quarg cheese has been developed by Kadiya (2009) using probiotic *Lactobacillus casei* NCDC-298 and inulin as prebiotic. Kumar (2012) has developed sweetened quarg cheese wherein fructooligosaccharide, inulin, and sucralose were used.

The shelf life of quarg is limited to 2–4 weeks at refrigeration temperature. Major defects that occur upon storage are bitter and cheesy flavor (caused by proteolysis) and sharp acid taste caused by excessive acid production. Keeping quality may also be hampered by growth of yeast and molds. MicroGARD 100 (0.50%) or Nisin (250IU) have been used to extend the shelf life of quarg cheese up to 42 days without adversely affecting the quality characteristics (Kadiya 2009).

Quarg cheese is used as filling in pan cakes, pasta, dips or as stuffing in meat, chicken, fish, and so on. The product is best suited as an ingredient to cold dishes, soups, sauces, salad, and so on.

12.4 Mozzarella Cheese

Mozzarella cheese is a pasta-filata or stretched curd type of cheese that originated in Italy. Cheeses of this family are known for their unique stretching and melting properties imparted by plasticizing and kneading treatment of coagulated curd in hot water at appropriate pH. Mozzarella cheese is white in color with glossy surface and soft texture. It may be classified on the basis of its composition as mozzarella, low moisture mozzarella, and part skim mozzarella as given

Table 12.2 Classification of Mozzarella Cheese

TYPE	MOISTURE (%)	FDM (%)
Mozzarella	Minimum 52 and maximum 60	Minimum 45
Low-moisture mozzarella	Minimum 45 and maximum 52	Minimum 45
Low-moisture mozzarella	Minimum 45 and maximum 52	Minimum 30 and maximum 45
Part skim mozzarella	Minimum 52 and maximum 60	Minimum 30 and maximum 45

Source: Kindstedt, P., Pasta-filata cheeses. In: *Cheese-Chemistry, Physics and Microbiology*, eds. P.F. Fox, P.L.H. McSweeney, T.M. Cogan, and T.P. Guinee, 3rd Edition, Vol. 2, Elsevier Academic Press, London, 251–277, 2004.

in Table 12.2. As per FSSR, 2011, mozzarella cheese shall have maximum 60% moisture and minimum 35% fat on dry matter.

12.4.1 Production Method

The method of manufacturing mozzarella cheese is similar to that of Cheddar cheese with the primary difference that acidity of cheddared curd is brought up to 0.70–0.80 (pH 5.2–5.4), which is optimum for stretching or plasticizing of the curd. This plasticizing process provides the characteristic stretching property to the cheese.

Buffalo milk is preferred for making mozzarella cheese due to high yield, white color, and its characteristic aroma and texture (El-Koussy et al. 1995; Jana and Mandal 2011). However, it has been successfully manufactured using cow milk and mixed milk (Singh and Ladkani 1984; Bonassi et al. 1982). Cow or buffalo milk is standardized to 3%–6% fat for making mozzarella cheese of desired quality. Casein to fat ratio is adjusted to 0.7.

Acidification is a critical step in mozzarella cheese manufacturing, which can be done either by using starter culture or by direct acidification technique. In starter culture method, a mixed culture consisting of *Streptococcus salavarius* subsp. *thermophilus* and *Lb. delbrueckii* subsp. *bulgaricus* or *Lb. helveticus* is used for acidification, and the curd is cheddared up to pH 5.2 (Kindstedt 2004). Direct acidification method can be used in mozzarella cheese making as a rapid method through addition of an organic acid such as citric, lactic, or acetic acid (Najafi et al. 2011). This technique offers several advantages such as curtailed manufacturing time and expenses, elimination of propagation and maintenance of starter cultures, no possibility of starter failures due to bacteriophages and antimicrobial agents. The detailed method for manufacturing mozzarella cheese has been shown in Figures 12.3 and 12.4.

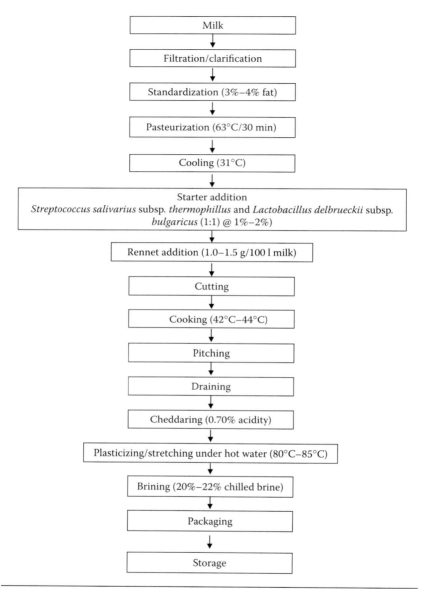

Figure 12.3 Traditional method for mozzarella cheese manufacturing.

12.4.2 Packaging, Preservation, and Storage

The high moisture content in mozzarella cheese makes it susceptible to spoilage due to microbial attack. The spoilage reported is mainly due to high initial count of mold, which can be reduced by applying

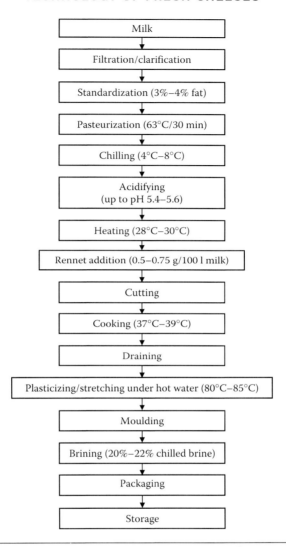

Figure 12.4 Direct acidification method for mozzarella cheese manufacturing.

vacuum packaging. Wrapping the cheese in parchment saran or vacuum packaging in cryopac, polyethylene, or cellophane pouches prior to refrigerated storage greatly reduces spoilage. Application of preservatives such as delvocid as well as employing sophisticated manufacturing processes such as microwave process and dielectric technique has also rendered mozzarella less prone to spoilage (Ghosh 1987).

12.4.3 Chemistry of "Stretch" of Mozzarella Cheese

Upon action of rennet on 105–106 bond of κ-casein, 1-105 part of the sequence of amino acids precipitates in the form of dicalcium paracaseinate with fat, minerals, and some lactose entrapped inside. When pH is reduced to 5.2–5.4 as is done in manufacturing mozzarella cheese, this dicalcium paracaseinate has the tendency to lose calcium ions resulting in the formation of monocalcium paracaseinate. This compound when heated to 54°C or higher becomes smooth, pliable, and stringy, which gives mozzarella cheese its desired characteristic of stretchability. However, if the acidification is excessive, which results in pH less than 5.2, monocalcium paracaseinate also loses calcium and forms paracaseinate, which is not desirable for optimum stretch of mozzarella cheese (Jana and Mandal 2011).

12.5 Cream Cheese

Cream cheese is a soft, unripened variety of cheese with slightly acidic and diacetyl flavor. It is most popular in North America. Cream cheese can be used as a spread, salad dressing and can be used to make dessert. There are two types of cream cheese depending upon fat content of initial mix and final composition. These are (1) double cream cheese—fat content 9%–11% in the initial mix; (2) single cream cheese—fat content 4.5%–5% in the initial mix.

During the manufacture of cream cheese, first milk is standardized as per the requirements of double cream cheese and single cream cheese by mixing the required amount of cream in milk. Then, the blend is heated at 50°C–55°C and homogenized (12–14 MPa). Pasteurization is carried out at 72°C/30–90 s. Then, starter culture is added and incubated. Depending on the starter culture and incubation, two conditions can be used: (1) short set incubation—milk is cooled to 31°C and after the addition of starter culture (@5%) milk is incubated for ~5 h; (2) long set incubation—milk is cooled to 22°C–23°C and after the addition of starter culture (@0.8–1.2%), incubated for 12–16 h. The mix is undisturbed until pH reaches to 4.5–4.8. The curd is cut, stirred, and heated to 50°C–70°C for whey separation from the curd. The hot curd is hanged overnight to remove whey from the curd or whey may be separated by centrifugal separator. Then the curd is cooled down to 10°C–20°C and mixed with salt (0.5%–1%)

TECHNOLOGY OF FRESH CHEESES

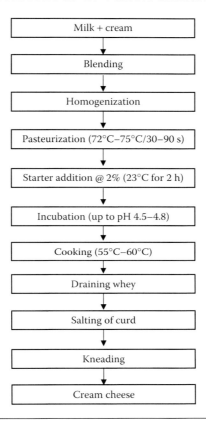

Figure 12.5 Manufacturing method for cream cheese. (Data from Kosikowski, F., *Cheese and Fermented Milk Foods*. Edwards Brothers, Inc, Ann Arbor, MI, 1966.)

and stabilizers (i.e., sodium alginate, xanthan gum, guar gum, carrageenan, etc.) and packaged as cold pack cream cheese. In the case of hot pack cream cheese, the curd is mixed with salt and stabilizer and heated to 70°C–85°C to improve its texture (Sanchez et al. 1996). Manufacturing steps of cream cheese are presented in Figure 12.5.

Both low and high pH in cream cheese are detrimental to its quality. Too high pH makes the product soft and lacks flavor, whereas too low pH makes the product grainy and acidic in flavor.

12.6 Queso Blanco/Latin American White Cheese

Queso Blanco is a product of acid and heat coagulation of milk. It a popular cheese of Central and South America. Unlike other cheeses, both acid and heat are used to coagulate curd. The method of manufacture resembles paneer and ricotta, popular cheeses of India and Italy, respectively, wherein heat and acid cause co-precipitation of milk proteins to get the desired product. For manufacturing Queso Blanco cheese with minimum fat loss in whey, high yield and good quality, protein: fat ratio of 1:1.2 (4.5% fat) has been suggested (Hill et al. 1982). Milk is heated at 85°C for 5 min to get most desirable qualities in the final product (Parnell-Clunies et al. 1985). Food-grade acids such as glacial acetic acid, citric acid, and lactic acid are used for coagulation of heat-treated milk without cooling. Acid is added in a quantity, which is sufficient to bring the pH to 5.2–5.3. After acidification, the coagulated mass is allowed to settle and the whey is drained off. Subsequent to whey removal, curd is salted (@2%–2.5% NaCl), hooped, and pressed. Then, the cheese block is cut into 250 or 500 g blocks and vacuum packaged and stored at 4°C (Kosikowski 1966; Farkye 2004).

In this method, some technical observations have been reported (Kosikowski 1966). First, replacing the organic acid with starter culture cannot be done because the starter may appear throughout the cheese as visible white particles. Second, pre-heating the milk to desired high temperature or plate heater is preferable. It is quicker, avoids accumulation of milk stone on the vat sides, and eliminates partial churning of the incoming milk. Moreover, the pressing is done at warm room temperature because the whey removal rate in cold room temperature will be less and hence will not result in the smooth compact body texture. Last, vacuum packaging at 29.5 in. using polyethylene cellophane pouches effectively prevents mold growth. Also, vacuum packaging makes the body of the cheese more cohesive and compact and improves its slicing property.

12.7 Paneer—A Cooking Type Cheese

Paneer, which is also known as Indian cheese, is a popular traditional Indian dairy product preferably made from buffalo milk. It is a south Asian variety of soft cheese that is used in a number of culinary preparations and snacks (Khan and Pal 2011). Coagulation is achieved by the combined action of acid and heat rather than rennet. Thus, no paracaseinate is formed in paneer. Due to the action of acid and heat, large structural protein aggregates are formed with entrained milk fat and other colloidal and soluble milk solids. Good quality paneer is characterized by a white color, sweet, mildly acidic, nutty flavor, spongy body and a closely knit texture. Paneer is highly nutritious since it retains fat, whey proteins along with casein, and minerals. Whey proteins are retained as the milk is heated to 90°C before coagulation and thus the denatured whey proteins interact with casein and remain in the precipitate part. Paneer contains approximately 54% moisture, 17.5% proteins, 25% fat, 2% lactose, and 1.5% minerals. According to FSSR (2011), paneer shall not contain more than 70% moisture and the fat content should not be less than 50% of dry matter.

12.7.1 Manufacturing Conditions

The process flow of manufacturing paneer is given in Figure 12.6 and the important variables/steps involved in the manufacture of paneer are as follows.

12.7.1.1 Type of Milk Buffalo milk produces best quality paneer. The paneer made from cow milk is too compact, fragile, and unsuitable for frying and cooking and low in yield (Singh and Kanawjia 1988). Buffalo milk produces good quality paneer due to the presence of higher amounts of casein, calcium, and phosphorus (Ghodekar 1989). Buffalo milk standardized to 6% fat has been recommended for preparing best quality paneer (Kanawjia and Singh 2000); however, buffalo milk standardized to 5%–6% fat has also been reported to produce good quality paneer (Singh and Kanawjia 1990).

12.7.1.2 Heat Treatment Heat treatment of milk significantly affects the quality characteristics and yield of paneer. Besides destroying all pathogenic microorganisms, heat treatment of milk also denatures whey proteins and co-precipitate colloidal calcium phosphate and whey proteins with casein upon acidification of milk and thus increases solid recovery (Khan and Pal 2011). Heat treatment at 90°C/no hold and 80°C/no hold has been recommended for buffalo milk and cow milk paneer, respectively (Vishweshwariaiah and Anantkrishna 1985; Sachdeva and Singh 1988).

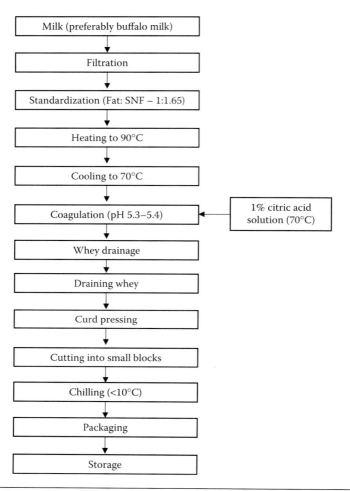

Figure 12.6 Manufacturing method for paneer.

12.7.1.3 Coagulation of Milk Paneer manufacturing involves coagulation of milk proteins using coagulants such as citric acid, tartaric acid, and lactic acid. Citric acid (1% solution) has been the most widely used coagulant for paneer making. About 2–2.5 g of citric acid is required for coagulating 1 kg of milk. Coagulants such as hydrochloric acid, phosphoric acid, and acidophilus whey can also be used for coagulation without affecting yield and quality of paneer. The optimum pH of coagulation of milk at 70°C is between pH 5.30 and 5.35 (Sachdeva and Singh 1987; Singh and Kanawjia 1988).

12.7.1.4 Drainage of Whey and Curd Pressing Curd is allowed to settle for about 5 min after proper coagulation, and then whey is allowed to drain through a strainer. It is desirable to maintain temperature of curd about 63°C during drainage of whey, which can be achieved by accelerating the process. Curd is then filled into hoops (38 × 28 × 10 cm) lined with cloth and pressure is applied on top of the hoop by placing a weight of about 45 kg (or 0.5 kg cm²) for 20–22 kg product. The hoop must have holes on all sides to facilitate expulsion of whey. Normally 15 min pressing is enough to expel the whey and obtain desired texture product.

12.7.1.5 Chilling of Paneer Paneer block is taken out of the press, cut into pieces of desired size (preferably 6″ × 6″), and transferred to chilled water tank for about 2–3 h. This not only helps in developing texture, but also speedily brings down the temperature of paneer. Utmost care should be taken to use chilled water of very good quality, particularly from bacteriological point of view.

12.7.1.6 Packaging Normally paneer blocks of required size/weight (weights 100, 250, and 500 g) are packaged in polyethylene pouches, heat sealed, and stored under refrigerated conditions (below 10°C). Some of the manufacturers have now started vacuum packaging of paneer in laminated or co-extruded films.

12.7.1.7 Yield of Paneer The yield of paneer depends on the total solids (fat and SNF) in the milk and the retention of milk solids and moisture in it. Generally, a yield of 20%–22% of paneer for buffalo milk and 16%–18% for cow milk is obtained under optimum conditions of manufacture.

12.7.2 Paneer from Cow Milk

To make paneer exclusively from cow milk, certain modifications in the conventional procedure have to be made. Addition of calcium chloride at the rate of 0.08%–0.1% to milk helps in getting a compact, sliceable, firm, and cohesive spongy body and closely knit texture (Sachdeva et al. 1991). The addition of 0.05% (w/v) calcium sulfate to cow milk has been recommended for better sensory scores in raw paneer (Mistry et al. 1992). In fried paneer, the sensory scores were best with 0.02% (w/v) added calcium sulfate. A higher temperature of coagulation (85°C) needs to be employed when making paneer from cow milk and the optimum pH of coagulation is in the range of 5.20–5.25.

12.8 Whey Cheeses

Whey is the largest by-product of the dairy industry. It is obtained during the manufacture of casein, cheese, paneer, chhana, and chakka. On an average, the manufacture of one ton of cheese results in the production of about 8 tons of liquid whey. The whey contains more than half the amount of solids present in the whole milk, including 20% of the protein (whey protein) and most of the lactose, minerals, and water-soluble vitamins. To exploit the nutritional significance of whey proteins and also to prevent valuable nutrients of whey to go in drain, it can be used for making cheese to recover whey solids mostly proteins. The manufacture of whey cheese is an age old practice and still manufactured all over the world by following traditional processes (Jelen and Buchheim 1976; Jelen 1992; Cabrera et al. 1995). There are several types of whey cheeses, which employ almost similar methods of manufacturing. Some of the whey cheeses are Gjetost, Fløtemysost, Mysosts or Primost in Norway; ricotta in Italy; and Manouri, Myzithra, Anthotyros in Greece. These whey cheeses mainly differ in the type of ingredients used for their manufacturing and the source of whey. As for example, Gjetost is made from goat's milk whey with added skim milk, Fløtemysost from cow's milk whey with fresh milk or cream, Blandet Gjetost from goat's milk whey with added cream, Gudbrandsdalost from cow's milk whey with goat's milk or cream, and Pultost Surost and Suprim from sour whey added with skim milk.

12.8.1 Principle of Whey Cheese Manufacture

The basic principle for making whey cheese involves thermal denaturation, sequential aggregation, and eventual precipitation of whey proteins from whey (Scott 1986). The whey cheese is made by concentration and/or heat and acid coagulation of whey, which may be supplemented with milk or milk fat. The addition of milk increases the yield of cheese as the co-precipitation of casein and whey protein occurs on heating and acidification. The addition of small amount of calcium salts also enhances the co-precipitation (Pitando et al. 2001). The other methods of increasing yield of whey cheese are condensation or ultrafiltration technique.

12.8.2 Whey Cheese Manufacture

12.8.2.1 Ricotta Cheese Ricotta is one of the most popular whey cheese, originally developed in Italy. Ricotta cheese is similar to cottage cheese except its moist, grainy surface. This cheese is manufactured from Cheddar cheese whey added with skim milk, whole milk, or buttermilk to raise the total milk solids as well as to improve the flavor. The fat content varies from 4% to 10% depending upon the type of whey used. Most commonly it is manufactured from sweet whey mixed with 5%–10% (v/v) milk that has been previously heated to 40°C–50°C. Salt is then added @0.1%(w/v), and the whey protein is coagulated by heating to 80°C–85°C with stirring followed by the addition of acidification agents like sour whey, citric acid, or white vinegar. Usually an aqueous solution of citric acid @0.11 kg/L is used. Heating and stirring are discontinued, when the precipitate (curd) rises to the surface of the hot whey. The precipitate is then dipped followed by draining for 4–6 h in a cool room (Pitando et al. 2001; Strieft et al. 1979). Usually, 1 kg of ricotta can be obtained from 15–20 L of whey (Mills 1986).

12.8.2.2 Mysost, Gjetost, and Primost Cheese Mysost and Gjetost are characterized by their darker brown color and coarse texture, whereas Primost is lighter in color, having caramelized flavor and smooth creamy body. Mysost is made from cow milk whey, whereas Gjetost is made from goat milk whey. Primost is manufactured from whey with the addition of milk fat. To make these cheeses, sweet whey is concentrated in a double effect evaporator to around 60% total solids followed by further concentration in an open pan up to 84% solids.

To promote Maillard browning, the concentrated mass is heated with continuous stirring till a plastic mass with brown color is formed. The plastic mass is transferred to a kneading box and stirred while cooling to prevent lactose crystallization (Whittier and Webb 1950).

12.8.3 Shelf Life of Whey Cheese

Fresh whey cheeses are very susceptible to microbial spoilage, by moulds, yeasts, and Enterobacteriaceae, especially under malice temperatures due to high pH (>6.0), a low salt content and high moisture content. Under aerobic conditions, these cheeses have a shelf life of less than 7 days. Papaioannou et al. (2007) packaged Greek whey cheese in modified atmosphere condition using two different gas compositions, that is, 30%/70% CO_2/N_2 and 70%/30% CO_2/N_2 and kept under storage temperature of 4°C and 12°C and compared with control cheese which was vacuum packed. Both gas compositions extended the shelf life of fresh Anthotyros cheese up to 20 days as compared to vacuum packaged samples. Pintado and Malcata (2000) investigated the effect of vacuum packaging on whey cheese kept at 4°C. Vacuum packaged cheese experienced extensive acidification, slight reduction in lactose, and moisture content and texture of whey cheese did not vary significantly.

12.8.4 Use of Whey Cheese

Whey cheese is now gaining popularity worldwide. Brown whey cheese is often consumed with sandwiches or Scandinavian type crisp bread. Manouri and Anthotyros are used as table cheese. Myzithra can also be used as table cheese but generally used in the preparation of certain foods and cheese pies. The fresh ricotta cheese is used in creamy desserts, cheese spreads, processed cheese foods, chip dips, cheesecakes, quiches, sour cream, cake filling, baked products, and Italian pasta products. In India, ricotta can be used in making traditional sweet delicacies. It can replace around 25%–50% of *khoa* (heat desiccated milk product) solids in *gulabjaman* (traditional Indian *khoa*-based sweet) manufacture and about 70%–100% chhana solids in manufacture of *sandesh*. Dried ricotta is suitable for grating and as a complement for other cheeses for more pronounced flavor. Ricotta cheese has been used to prepare a typical Piedmontese cheese product (Brus) via grating (or cutting up) of this cheese onto pasta (Pitando et al. 2001). Ricotta cheese can also be used as an ingredient in manufacturing processed cheese (Scott 1986).

References

Bonassi, I.A., de Camarco Carvalho, J.B., and Villares, J.B. 1982. Use of buffalo milk as raw material in the processing of Mozzarella cheese. *Archivos Latinoamericanos de Nutricion* 32(4): 903–912.

Cabrera, M.C.del, Menendez, T., Ootega, O., and Real, E. 1995. Manufacture of requeson from buffalo milk whey. *Alimentaria* 32: 275–278.

Demott, B.J., Hitchcock, J.P., and Sanders, O.G. 1984. Sodium concentration of selected dairy products and acceptability of sodium substitute in Cottage cheese. *Journal of Dairy Science* 67(7): 1539–1543.

Dyurec, Jr., D.J. and Zall, R.R. 1985. Effect of heating, cooling and storing milk on casein and whey proteins. *Journal of Dairy Science* 68(2): 273–280.

El-Batawy, M.A., Hewedi, H.M., Girgis, E.S., and Melry, W.A. 1993. Studies on cottage cheese. I. Effect of some technological parameters on making Cottage cheese by direct acidification. *Egyptian Journal of Food Science* 21(2): 113–125.

El-Koussy, L.A., Mustafa, M.B.M., Abdel-Kader, Y.I., and El-Zoghby, A.S. 1995. Properties of Mozzarella cheese as affected by milk type, yield-recovery of milk constituents and chemical composition of cheese. In: *Proceedings of the 6th Egyptian Conference for Dairy Science and Technology*. Egypt.

Ernstrom, C.A. and Kale, C.G. 1975. Continuous manufacture of Cottage cheese and other uncured cheese varieties. *Journal of Dairy Science* 58(7): 1008–1014.

Farkye, N.Y. 2004. Acid and acid/rennet—Curd cheeses, Part B: Cottage cheese. In: *Cheese-Chemistry, Physics and Microbiology* (3rd Edition, Vol. 2), eds. P.F. Fox, P.L.H. McSweeney, T.M. Cogan, and T.P. Guinee, 329–341. London: Elsevier Academic Press.

FSSR. 2011. *Food Safety and Standards Authority of India*. New Delhi, India: Ministry of Health and Family Welfare.

Gahane, H.B. 2008. *Development of Quarg Type Cheese with Enhanced Functional Attributes from Buffalo Milk* MTech dissertation. Karnal, India: National Dairy Research Institute.

Ghodekar, D.R. 1989. Factors affecting quality of paneer—A review. *Indian Dairyman* 41(3): 161–168.

Ghosh, B.C. 1987. *Production, Packaging and Preservation of Mozzarella Cheese from Buffalo Milk Using Microbial Rennet*. PhD thesis. Haryana, India: Kurukshetra University.

Guinee, T.P., Pudia, P.D., and Farkye, N.Y. 1993. Fresh acid curd cheese varieties. In: *Cheese—Chemistry, Physics and Microbiology. Vol. 2, Major Cheese Groups*, ed. P.F. Fox, 363–419. London: Chapman & Hall.

Hill, A.R., Bullock, D.H., and Irvine, D.M. 1982. Manufacturing parameters of Queso Blanco made from milk and recombined milk. *Canadian Institute of Food Science and Technology Journal* 15: 47–53.

Jana, A.H. and Mandal, P.K. 2011. Manufacturing and quality of Mozzarella cheese: A review. *International Journal of Dairy Science* 6(4): 199–226.

Jelen, P. 1992. Whey cheeses and beverages. In: *Whey and Lactose Processing*, ed. J.G. Zadow, 157–193. London: Elsevier Applied Science.

Jelen, P. and Buchheim, W. 1976. Norwagian whey cheese. *Food Technology* 30(11): 62–74.

Jelen, P. and Renz-Schauen, A. 1989. Quarg manufacturing innovations and their effects on quality, nutritive value, and consumer acceptance. *Food Technology* 43: 74–81.

Kadiya, K. 2009. *Development of Functional Quarg Cheese with Extended Shelf Life*. M Tech thesis. Karnal, India: National Dairy Research Institute.

Kanawjia, S.K. and Singh, S. 2000. Technological advances in paneer making. *Indian Dairyman* 52(10): 45–50.

Khan, S.U. and Pal, M.A. 2011. Paneer production: A review. *Journal of Food Science and Technology* 48(6): 645–660.

Kindstedt, P. 2004. Pasta-filata cheeses. In: *Cheese-Chemistry, Physics and Microbiology* (3rd Edition, Vol. 2), eds. P.F. Fox, P.L.H. McSweeney, T.M. Cogan, and T.P. Guinee, 251–277. London: Elsevier Academic Press.

Kosikowski, F. 1966. *Cheese and Fermented Milk Foods*. Ann Arbor, MI: Edwards Brothers, Inc.

Kumar, A. 2012. *Development of Sweetened Functional Soft Cheese*. M Tech dissertation. Karnal, India: National Dairy Research Institute.

Makhal, S., Giri, A., and Kanawjia, S.K. 2011. Effect of κ-carrageenan and tetrasodium pyrophosphate on the yield of direct acidified cottage cheese. *Journal of Food Science and Technology* 50(6): 1200–1205.

Makhal, S. and Kanawjia, S.K. 2008. Selection of pH at coagulation for manufacture of direct acidified Cottage cheese. *Indian Journal of Dairy Science* 61(4): 242–251.

Makhal, S., Kanawjia, S.K., and Giri, A. 2013. A dual-acidification process for the manufacture of direct-acidified Cottage cheese. *International Journal of Dairy Technology* 66(4): 552–561.

Makhal, S., Kanawjia, S. K., and Giri, A. 2014. Effectiveness of thymol in extending keeping quality of cottage cheese. *Journal of food science and technology* 51(9): 2022–2029.

Makhal, S., Kanawjia, S.K., and Giri, A. 2015a. Effect of microGARD on keeping quality of direct acidified Cottage cheese. *Journal of Food Science and Technology* 52(2): 936–943.

Makhal, S., Kanawjia, S. K., and Giri, A. 2015b. Role of calcium chloride and heat treatment singly and in combination on improvement of the yield of direct acidified Cottage cheese. *Journal of Food Science and Technology* 52(1): 535–541.

Mattews, M.E., So, S.E., Amundson, C.H., and Hill, C.G. 1976. Cottage cheese from ultrafiltered skim milk. *Journal of Food Science* 41(3): 619–623.

Mills, O. 1986. Sheep dairying in Britain-a future industry. *Journal of the Society of Dairy Technology* 39: 88–90.

Mistry, C.D., Singh, S., and Sharma, R.S. 1992. Physicochemical characteristics of paneer prepared from cow milk by altering its salt balance. *Australian Journal of Dairy Technology* 47(1): 23–27.

Mucchetti, G., Zardi, G., Orlandini, F., and Gostoli, C. 2000. The pre-concentration of milk by nanofiltration in the production of Quarg-type fresh cheeses. *Lait* 80: 43–50.

Najafi, M.B.H., Arianfar, A., and Ghaddosi, H.B. 2011. Study on physicochemical, rheological and sensory properties of Mozzarella cheese made by direct acidification. *American-Eurasian Journal of Agriculture and Environment Science* 1(3): 268–272.

Ottosen, N. 1996. *The Use of Membranes for the Production of Fermented Cheese.* Bulletin 311. Brussels, Belgium: International Dairy Federation.

Pandya, R.N., Tiwari, B.D., and Singh, S. 1989. Effect of processing variables on Cottage cheese prepared from buffalo milk. *Indian Journal of Dairy Science* 42: 568–571.

Papaioannou, G., Chouliara, I., Karatapanis, A.E., Kontominas, M.G., and Savvaidis, I.N. 2007. Shelf-life of a Greek whey cheese under modified atmosphere packaging. *International Dairy Journal* 17(4): 358–364.

Parnell-Clunies, E.M., Irvine, D.M., and Bullock, D.H. 1985. Heat treatment and homogenization of milk for Queso Blanco (Latin American white cheese) manufacture. *Canadian Institute of Food Science and Technology Journal* 18: 133–136.

Pfalzer, K. and Jelen, P. 1994. Manufacture of thermoquarg from mixtures of UF-retentate of sweet whey and skim milk. *Milchwissenschaft* 49: 490–494.

Pintado, M.E. and Malcata, F.X. 2000. Characterization of whey cheese packaged under vacuum. *Journal of Food Protection* 63(2): 216–221.

Pitando, M.E., Macedo, A.C., and Malcata, F.X. 2001. Review: Technology, chemistry and microbiology of whey cheeses. *Food Science and Technology International* 7(2): 105–116.

Renner, E. 1993. Nutritional aspects of Cheese. In: *Cheese: Chemistry Physics and Microbiology* (2nd Edition), ed. P.F. Fox, 557–580. London: Chapman & Hall.

Sachdeva, S., Prokopek, D., and Reuter, H. 1991. Technology of paneer from cow milk. *Japanese Journal of Dairy Food Science* 40(2): 85–90.

Sachdeva, S. and Singh, S. 1987. Use of non-conventional coagulants in the manufacture of paneer. *Journal of Food Science and Technology* 24(6): 317–319.

Sachdeva, S. and Singh, S. 1988. Optimisation of processing parameters in the manufacture of paneer. *Journal of Food Science and Technology* 25(3): 142–145.

Sanchez, C., Beauregard, J.L., Chassagne, M.H., Bimbenet, J.J., and Hardy, J. 1996. Effects of processing on rheology and structure of double Cream cheese. *Food Research International* 28: 547–552.

Schulz-Collins, D. and Senge, B. 2004. Acid- and acid/rennet curd cheeses. Part A: Quark, Cream cheese and related varieties. In: *Cheese-Chemistry, Physics and Microbiology* (3rd Edition, Vol. 2), eds. P.F. Fox, P.L.H. McSweeney, T.M. Cogan, and T.P. Guinee, 301–328. London: Elsevier Academic Press.

Scott, R. 1986. *Cheese Making Practice* (2nd Edition). London: Applied Science Publishers.

Sharma, D.K. and Reuter, H. 1993. Quarg-making ultrafiltration using polymeric and mineral membrane modules: A comparative performance study. *Lait* 73: 303–310.

Siggelkow, M.A. 1984. Modern methods in quarg production for consumer sale. *Dairy Industries International* 49(6): 17–21.

Singh, S. and Kanawjia, S.K. 1988. Development of manufacturing technique for paneer from cow milk. *Indian Journal of Dairy Science* 41(3): 322–325.

Singh, S. and Kanawjia, S.K. 1990. Effect of hydrogen peroxide and delvocid on enhancement of shelf life of recombined milk paneer. In: *Brief Communications of the XXIII International Dairy Congress* (Vol. II), 135. Montreal, Canada, October 9–12.

Singh, S. and Ladkani, B.C. 1984. *Standardization of Manufacturing Techniques of Mozzarella Cheeses*. Annual Report. Karnal, India: National Dairy Research Institute.

Strieft, P.J., Nilson, K.M., Duthie, A.H., and Atherton, H.V. 1979. Whey Ricotta cheese manufactured from fluid and condensed whey. *Journal of Food Protection* 42: 552–554.

Tiwari, B.D. and Singh, S. 1988. Innovations in technology of Cottage cheese. *Indian Dairyman* 40(4): 199–208.

Upadhyay, K.G. 2003. *Essentials of Cheese Making*. SMC College of Dairy Science. Gujarat, India: Agricultural University.

Vishweshwariaiah, L. and Anantakrishnan, C.P. 1985. Study on technological aspects of preparing paneer from cow's milk [Indigenous Milk Product]. *Asian Journal of Dairy Research (India)* 4(3): 171–176.

Walstra, P., Geurts, T.J., Noomen, A., Jellema, A., and Van Boekel, M.A.J.S. 1999. Cheese varieties. In: *Dairy Technology: Principles of Milk Properties and Processes*, 707–766. New York: Marcel Dekker Inc.

Walstra, P., Wouters, J.T.M., and Geurts, T.J. 1999. Cheese varieties. In: *Dairy Science and Technology*, 687–742. CRC Press/Taylor & Francis group.

Whittier, E.O. and Webb, B.H. 1950. *By Products of Milk*. New York: Reinhold.

13

Dahi, Lassi, and Shrikhand

S.V.N. VIJAYENDRA AND M.C. VARADARAJ

Contents

13.1	Introduction	355
13.2	Dahi	356
	13.2.1 Mistidoi	360
	13.2.2 Probiotic Dahi and Its Nutritional and Therapeutic Properties	361
13.3	Lassi	365
13.4	Shrikhand	367
13.5	Future Prospects	369
References		369

13.1 Introduction

Fermentation of foods is the conversion of raw materials by microorganisms and value addition to the end product, along with extending the shelf life of perishable raw food substrates. India is the largest milk producer in the world and home to popular fermented milk products such as *dahi*, *lassi*, and *shrikhand*. The basic details, such as the type of milk, method of preparation, and storage, have been well documented (Steinkraus 1995; Rati Rao et al. 2006). This chapter mainly focuses on the research carried out in recent years. The research has been focused mainly on technological innovations, nutritional and therapeutic properties of these products, and new advances that have been made to understand these aspects (Sarkar 2008; Khetra et al. 2011; Nagpal et al. 2012). Because of several nutritional and therapeutic functions it possesses, *dahi* could be considered as a functional food (Sarkar et al. 2011).

13.2 Dahi

Dahi is a well-known fermented milk product with a thick consistency and sweetish or slightly sour but refreshing taste. It is popular in India as well as in the Indian subcontinent. It is consumed as a part of a regular diet, along with cooked rice or wheat-based *chapati/roti*. It is delicious, easily digestible, and highly nutritious. It is also used as a base material for making *lassi* and *shrikhand* (Sarkar 2008), *raita*, *kadhi*, and several other ethnic fermented milk products such as *gheu*, *chhurpi*, and *mohi* (Tamang et al. 2012). Various steps involved in *dahi* preparation are indicated in Figure 13.1. The overall composition of *dahi* varies with type and quality of milk used. *Dahi* prepared from

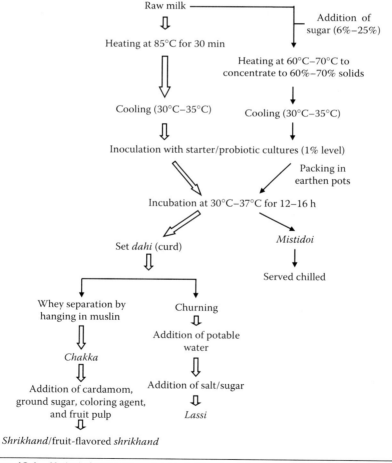

Figure 13.1 Method of preparation of *dahi*, *shrikhand*, and *lassi*.

whole milk contains 85%–88% water, 5%–8% fat, 3.2%–3.4% protein, 4.6%–5.2% lactose, 0.7%–0.75% ash, 0.5%–1.1% lactic acid, 0.12%–0.14% calcium, and 0.09%–0.11% phosphorus (Laxminarayana et al. 1952). Sarkar (2008) and Misra (1992) have reviewed various kinds of milk products; types of treatment; rate of inoculum used for the preparation of fermented products such as *dahi*, *lassi*, and *shrikhand*; packaging; and postproduction treatments of finished products to extend their shelf life. Technological innovations like thermization (mild heat treatment) and addition of biopreservatives are suggested to extend the shelf life of *dahi*, which is essential for its transportation to distant places for marketing purpose (Sarkar 2008).

In India, the estimated market size for *dahi* is 160 billion rupees (US$3.8 billion) and around 5.2 million tons (Anon 2008). The present status and scope of the *dahi* industry in India has been reviewed recently (Gawai and Prajapati 2012). Although *dahi* is produced commercially in India by cooperative dairies, its preparation is mainly confined to local *halwais* (sweets makers), shops, and restaurants. However, the use of starter cultures is restricted to only organized dairies. Many companies including multinationals are entering into the probiotic *dahi* market. It is gaining momentum day-by-day and is expected to reach $8 million by 2015 (Gawai and Prajapati 2012). In Nepal, *dahi*, along with other traditional fermented foods, is prepared on a large scale and consumed on the same day because of its perishable nature (Dahal et al. 2005).

The presence of various lactic cultures and their nutritional role in different fermented products has been reviewed very recently (Panesar 2011). The predominate genera of *lactis* isolated from *dahi* samples studied in Bangladesh included *Streptococcus* (50%), *Lactobacillus* (27%), *Enterococcus* (9%), *Leuconostoc* (5%), *Lactococcus* (5%), and *Pediococcus* (4%). Of these, *Str. bovis* dominated in 47% of samples. The other lactis present are *Lb. fermentum*, *Lb. delbrueckii* subsp. *bulgaricus*, *Lb. delbrueckii* subsp. *lactis*, *E. faecium*, *Str. thermophilus*, *Leu. mesenteroides* subsp. *mesenteroides*, *Leu. mesenteroides* subsp. *dextranicum*, *Lc. lactis* subsp. *lactis*, *Lc. raffinolactis*, and *P. pentosaceus*. Cocci (73%) were dominant over rods (27%) in 28 samples tested (Rashid et al. 2007). The predominant lactic acid bacteria (LAB) identified from ethnic fermented dairy products such as *dahi*, *mohi*, *chhurpi*, *somar*, *philu*, and *shyow* in the Himalayan regions of India, Bhutan, and

Nepal include *Lb. bifermentans*, *Lb. paracasei* subsp. *pseudoplantarum*, *Lb. kefir*, *Lb. hilgardii*, *Lb. alimentarius*, *Lb. paracasei* subsp. *paracasei*, *Lb. plantarum*, *Lc. lactis* subsp. *lactis*, *Lc. lactis* subsp. *cremoris*, and *E. faecium* (Dewan and Tamang 2007). *Lb. delbrueckii* subsp. *bulgaricus*, *Lb. paracasei*, and *Lb. rhamnosus* represented the dominant lactobacilli species present in *dahi* produced at different altitudes of Nepal (Koirala et al. 2014).

An exopolysaccharide (EPS)-producing non-ropy *Leuconostoc* sp. CFR2181 and *Lactobacillus* sp. CFR2182 were isolated from market *dahi* (Vijayendra et al. 2008, 2009). Use of such cultures can improve the texture of *dahi*, due to in situ production of EPS. Similarly, a nisin-producing *Lc. lactis* W8 was isolated from *dahi* (Mitra et al. 2010). Singh and Ramesh (2009) developed a facile method of template DNA preparation for polymerase chain reaction (PCR)–based detection and typing of LAB from *dahi* and other fermented foods such as *idli* batter and fermented vegetables. Native cultures of *Lb. plantarum*, *Lb. fermentum*, and *Lb. casei* subsp. *tolerans* that exhibit selected probiotic attributes were isolated from the samples of homemade *dahi*. As confirmatory evidence, PCR analysis based on 16S rRNA primers designed for the specificity of genus *Lactobacillus* resulted in positive amplification in these three conventionally characterized lactic cultures (Roopashri and Varadaraj 2009).

To increase the therapeutic and nutritional values of *dahi*, addition of *Lb. acidophilus* and *Bifidobacterium bifidum*, along with the technical lactic cultures for *dahi* preparation, was envisaged (Vijayendra and Gupta 1992). Associative growth behavior of *dahi* starter cultures with *B. bifidum* and *Lb. acidophilus* in buffalo skim milk was noticed, and nearly one log reduction in mesophilic lactis was observed. Associative growth increased titratable acidity marginally; however, it reduced volatile acidity from 36 to 15.8 ml, diacetyl from 4.05 to 2.80 ppm, and tyrosine from 0.46 to 0.36 µg (Vijayendra and Gupta 2013).

There is a great demand for milk fat, and its cost is also very high. Besides, when the fat is removed from large quantities of milk in dairy plants, huge quantities of skim milk is generated. To make use of this and to prevent environmental contamination, because it increases the BOD value, attempts have been made to prepare *dahi* using skim milk added with vegetable oil in place of milk fat. *Dahi* is prepared using different concentrations of reconstituted nonfat skim milk supplemented with vegetable fat (soybean oil, 4% v/v) and 0.5% gelatin, and the quality of *dahi* made with 20% of milk powder is similar to that of whole milk *dahi* (Begum et al. 2011).

Enriching buffalo milk with 50–100 mg calcium per 100 ml milk using calcium chloride, calcium lactate, or calcium gluconate caused a decrease in firmness of *dahi*. However, the sensory properties of *dahi* from milk enriched with calcium gluconate and calcium lactate were comparable to *dahi* made from control (without enrichment) milk (Ranjan et al. 2006). The effect of fermentation temperature (20.8°C–29.2°C), total solids (TS) content (13%–16% w/v), and the rate of inoculum (1%–4% v/v) on the acidification of buffalo milk fermented with *dahi* culture were explored simultaneously by following response surface methodology (Shiby and Mishra 2008). A good quality of gel was obtained with milk having TS of 14.5%–15%, a total inoculum level of 2.5%, and incubation at 27°C–29.2°C.

Trends in packaging of traditional fermented milk products and various materials used for packaging have been reviewed (Matkar and Khedkar 2008). Keeping the aspect of environmental pollution in view, use of plastics for food packaging is not advisable. Utilization of earthen pots for packing *dahi* is a well-known practice. However, some drawbacks are noticed with its use, such as the product can undergo shrinkage due to absorption of moisture by the pot, the heaviness of the pot, and the possibility of breakage during handling. To avoid the shrinkage, very recently, a process was standardized to coat the pots with shellac (60% solution in ethanol), a Food and Drug Administration (FDA)–approved coating material secreted by lac insects, and heating to 90°C for 1 h (Goyal and Goyal 2012).

The quality of *dahi* can vary based on the hygienic and sanitation conditions of where it is prepared. It is observed that the total plate count and fungal count in market samples of *dahi* collected in Bangladesh exceeded the acceptable limits of 10^7 CFU/g and 10^4 propagules/g, indicating poor hygienic conditions at either manufacturing or handling stages (Chowdhury et al. 2011). In fruit-based *dahi*, a native isolate of *Listeria monocytogenes* CFR 1302 was able to survive without any significant loss or death during 6 days of storage at 4°C (Jayanth 2010). Whey separation is one of the most common problem in *dahi* while storage. To reduce this problem, researchers are trying to use EPS-producing lactis as starter cultures. In situ production of EPS by *Lc. lactis* subsp. *lactis* B6 and KT 24 significantly

($p < .05$) reduced whey separation and improved rheological properties, besides increasing adhesiveness and stickiness of *dahi*, which contribute in giving a good mouthfeel (Behare et al. 2009a, 2009b). However, addition of EPS powder externally did not yield desirable quality of *dahi*.

TS content and inoculum level play a significant role on the hardness and syneresis of *dahi* (Shiby and Mishra 2012). Hardness and cohesiveness of buffalo milk *dahi* increased with the increase in inoculum level from 1% to 2.5%. *Dahi* with 14.5% TS (10% solids and 4.5% fat) had high scores for overall acceptability.

Presence of antibiotic-resistant lactis in *dahi* is a cause of concern. These cultures can transfer the resistant gene horizontally to other inherent flora in the intestine and can make the treatment difficult. Several lactic cultures that acquired erythromycin and tetracycline resistance genes from the environment were isolated from fermented foods such as *dahi*, buttermilk, *khoa*, and yogurt (Thumu and Halami 2012).

13.2.1 Mistidoi

Mistidoi, also known as *laldahi* or *payodhi*, is a sweetened *dahi* and is popular in the eastern part of India. It is characterized by firm light brown–colored body with cooked or caramelized flavor. There is increased demand for low fat and low sugar–containing food products, due to the health risks associated with these products. In this direction, research work was carried out to know the effect of reduced fat on the quality of *mistidoi* prepared from buffalo milk (Raju and Pal 2009). Excepting the increase in acidity of *mistidoi*, no other significant change in color, water activity, and firmness values was noticed with the decrease in fat from 5% to 3%. In another study, these authors reported that maltodextrin was the best bulking agent in the preparation of artificially sweetened (aspartame and acesulfame K) *mistidoi* using low-fat buffalo milk (Raju and Pal 2011). Addition of maltodextrin decreased syneresis; however, it increased water activity, viscosity, and acidity of the curd, when compared to other bulking agents (sorbitol and polydextrose). It also improved body and texture, besides recording the highest overall acceptability scores among all bulking agents tested. *Mistidoi* was prepared incorporating inulin as a texturizing and stabilizing agent to improve mouthfeel of the product (Chauhan and Abraham 2011), and as a dietary fiber (Raju and Pal 2014). Increasing the inulin concentration from 2.5% to 10% increased acidity, pH, firmness, and stickiness, whereas whey separation, flavor, body texture, and overall acceptability decreased. However, *mistidoi* prepared with 2.5% inulin had the best overall quality. Very recently, Raju et al. (2012) reviewed the preparation methods and composition of various variants of *mistidoi* such as low-, medium-, and high-fat *mistidoi*, dietetic *mistidoi* with dietary fiber, and low-sugar *mistidoi* with artificial sweeteners.

13.2.2 Probiotic Dahi and Its Nutritional and Therapeutic Properties

The nutritional and therapeutic values of *dahi* are high when compared to the milk used for its preparation. It is easily digestible when compared to milk. The importance and therapeutic characteristics of cow and buffalo milk *dahi* in diet are well documented in *Charaka Samhitā*, a famous document on Ayurveda (Prajapati and Nair 2003). Because of its unique texture, taste, and flavor, *juju dhou dahi*, a sweetened and flavored *dahi*, is in great demand in Bhaktapur and Kathmandu valley of Nepal (Karki 1986). However, use of goat milk, which is known for its better nutritional value, for the preparation of *dahi* was also studied (Palanidorai et al. 2009).

Fermented milks containing probiotic microorganisms are examples of best-known functional foods (Kurien et al. 2005). Various aspects of therapeutic nature of milk and fermented milk products have been reviewed recently (Nagpal et al. 2012). The Indian probiotic industry is currently valued at $2 million and is projected to grow further to around $8–$10 million by 2015 (Raja and Arunachalam 2011). The major brands one can see in the Indian probiotic industry are Amul, Mother Dairy, Nestle, Yakult Danone, and so on.

The nutritional quality of *dahi* prepared using cow milk and that of soymilk was compared (Ghosh et al. 2010). The folic acid, riboflavin, and thiamine contents (µg/g) were found to be 107, 13.41, and 358.88, respectively, in soymilk curd and 106.03, 2.30, and 110.32, respectively, in cow milk curd, indicating the nutraceutical quality of *dahi* prepared using either cow milk or soy milk, and it can be used as dietary supplement to improve the health.

With the current trend of early onset of diabetes in human population well below 40 years of age, there is an urgent need to prevent this trend or at least to delay it by few more years. In this direction, research is being carried out using *dahi* as a medium. The probiotic low-fat (2.5%) *dahi* containing *Lb. acidophilus* and *Lb. casei* delayed significantly the onset of dyslipidemia, glucose intolerance, hyperglycemia, hyperinsulinemia, and oxidative stress in high fructose–induced diabetic rats, thus reducing the risk of diabetes (Yadav et al. 2007a). Besides these findings, they have noticed low level of thiobarbituric acid-reactive substances and reduced glutathione in liver tissue, indicating a good antioxidant property of probiotic *dahi*. Using the same cultures, Yadav et al. (2007b) have also reported the production of conjugated linoleic acid and free fatty acids in probiotic *dahi*, thus increasing the therapeutic and nutritional values of this product. Yadav et al. (2006) earlier reported significantly low levels of fasting blood glucose levels, glycosylated hemoglobin, free fatty acid content, and insulin in high fructose–fed rats after feeding probiotic *dahi* containing *Lc. lactis* than plain *dahi*, indicating antidiabetic effect of probiotic *dahi*.

Feeding of probiotic *dahi* containing *Lb. acidophilus* NCDC14 and *Lb. casei* NCDC19 (73×10^8 CFU/g) to streptozotocin (STZ)-induced diabetic rats (15 g/day/rat) for 28 days significantly delayed reduction of insulin secretion during oral glucose tolerance test (Yadav et al. 2008a). In addition to this, STZ-induced oxidative damage in pancreatic tissues by inhibiting the lipid peroxidation was significantly suppressed by probiotic *dahi*, while formation of nitric oxide, leading to empowerment of antioxidant system of β-cells and reduction of insulin, was delayed.

Significant increase in counts of lactobacilli adherent to epithelial walls and reduction in the oxidative stress marker thiobarbituric acid-reactive species in intestinal tissues and glycosylation of hemoglobin was observed in STZ-induced diabetic rats fed with probiotic *dahi* containing *Lb. acidophilus* and *Lb. casei* (Yadav et al. 2008b). Preparation of acceptable quality of *dahi* incorporating artificial sweeteners like sac-sweet and sucrol for diabetic people was reported (Islam et al. 2010).

Elevation of total and ovalbumin-specific IgE in the serum of ovalbumin-injected mice was completely suppressed by feeding probiotic *dahi* prepared using *Lb. acidophilus* and *Lb. casei* (Jain et al. 2010). Besides this, increased production of T helper 1 (Th1) cell-specific cytokines, that is, interferon-γ and interleukin (IL)-2, and decreased content of Th2 cell-specific cytokines, that is, IL-4 and IL-6, were noticed in these mice. Feeding probiotic *dahi* also suppressed ovalbumin-stimulated lymphocyte proliferation.

Several species of *Lactobacillus* were isolated from market *dahi* samples, which are active against major foodborne pathogens due to the production of proteinaceous substance and hydrogen peroxide (Balasubramanyam and Varadaraj 1994, 1995). The culture filtrate of antilisterial LAB isolated from market *dahi* samples produced bacteriocin-like inhibitory substance, which was active against common Gram-positive pathogenic bacteria such as *Staphylococcus aureus*, *E. faecalis*, and *Bacillus cereus* (Singh et al. 2012). *Lc. lactis* W8 produced nisin concomitantly, while fermenting milk into dahi, and it reached maximum at 6 h of fermentation (Mitra et al. 2010). It showed antibacterial property against several spoilage and pathogenic bacteria including *L. monocytogenes*. When this *dahi* was inoculated with 5.2 log CFU/ml cells of *L. monocytogenes* and stored at 4°C, it became undetectable at 10 h.

DNA damage was significantly reduced (48.2%) in experimental rats fed with a diet containing wheat bran-*acidophilus casei dahi*-DMH (1,2-dimethylhydrazine) as compared to control group (87.8%) fed exclusively with DMH, indicating the protective potential of *acidophilus casei dahi* and wheat bran against DMH-induced molecular alteration in colonic cells during carcinogenesis (Kumar et al. 2010). Significant raise in antioxidant and level of detoxification enzymes (glutathione S-transferase and glutathione peroxidase) in *acidophilus casei dahi*–fed rats than the control group (DMH-fed rats) was observed along with reducing c-Myc protein level, thus reducing the risk of colon carcinogenesis (Arvind et al. 2009). This product also maintained low level of cancer-causing fecal enzymes (i.e., β-glucuronidase, β-glucosidase, and mucinase) in Wistar rats as compared to that of DMH-fed or *dahi* without probiotic cultures indicating the protective effect of probiotic cultures (Arvind et al. 2010). Very recently, acid-bile-resistant bacteriocin-producing *Lb. acidophilus* was isolated from a *dahi* sample (Mahmood et al. 2014). The bacteriocin, which has a molecular weight of ≤8.5 kDa, is stable between 60°C and 90°C and in a pH range of 4.5–6.5.

Consumption of probiotic *dahi* containing probiotic strains of *Lb. acidophilus* LaVK2 and *B. bifidum* BbVK3 significantly enhanced carcinogen detoxification activities of γ-glutamyltranspeptidase, UDP-glucuronosyltransferase, and quinone reductase in hepatic tissue of rats indicating chemoprotective effect of probiotic *dahi* (Shruti and Kansal 2011). Mohania et al. (2014) have recently reported that the probiotic *dahi* containing *Lb. acidophilus* and *B. bifidum* alone or an adjunct with piroxicam may have antineoplastic and antiproliferative activities.

The plasma cholesterol content in rats fed with hypercholesterolemic diet increased by only 39% upon the feeding of probiotic *dahi* containing *Lb. acidophilus* and *B. bifidum*, when compared to the rise with buffalo milk (68%) or *dahi* (62%), indicating the reduction of diet-induced hypercholesterolemia and atherogenic index by increasing high-density cholesterol (Rajpal and Kansal 2009a). Besides this, the magnitude of rise in plasma triacylglycerols by feeding on probiotic *dahi* was only 48% compared to 313% and 202% of feeding buffalo milk and *dahi*, respectively.

Fermented milk produced using *Lb. casei*, isolated from a commercial *dahi* sample, significantly reduced total cholesterol, LDL cholesterol, and HDL cholesterol levels in cholesterol-fed albino rats (Maity and Misra 2010). Increase in body weight and hypocholesterolemic effect of the freeze-dried probiotic cultures of *Lb. acidophilus* and *B. bifidum* in buffalo milk *dahi* and yogurt in experimental albino rats with no significant difference in sensory scores of the fermented products with the supplementation of probiotic cultures in technical starters has been reported (Vijayendra and Gupta 2012).

Probiotic *dahi* has shown its nutritional and therapeutic functionality not only in humans but also in canines (Kore et al. 2012). Labrador dogs that were fed with *dahi* made of *Lactobacillus* sp. (10^6–10^7/ml) for 6 weeks had significantly higher calcium absorption; reduced fecal pH and ammonia; increased quantities of short chain fatty acids, lactate, lactobacilli, and bifidobacteria; and reduced coliform count, thus improving the health of the hindgut.

Feeding of probiotic *dahi* containing probiotic *Lb. casei* to mice stimulated nonspecific immune response markers, increased phagocytic activity, and increased β-galactosidase activity (84%) in supernatant of cultured macrophages (Jain et al. 2008).

In experimental mice fed with probiotic *dahi* containing *Lb. casei*, remarkably low colonization of *Salmonella enteritidis* in liver and spleen was noticed than in mice fed with milk and plain *dahi*, and it enhanced innate and adaptive immunity (Jain et al. 2009). The potential benefiting effects of probiotic *dahi* containing *Lb. acidophilus* LaVK2 and *B. bifidum* BbVK3 on age-inflicted accumulation of oxidation products, antioxidant enzymes, and expression of biomarkers of aging were evaluated by feeding mice at levels of 5 g/day for 4 months. It increased catalase and glutathione peroxidase activity in tissues and red blood corpuscles, respectively, and decreased thiobarbituric acid-reactive substances in plasma (Kaushal and Kansal 2012). This probiotic *dahi* also reversed age-related decline in expression of biomarkers of aging, peroxisome proliferator–activated receptor-α, senescence marker protein-30, and klotho in hepatic and kidney tissues, indicating the potential of probiotic cultures to combat oxidative stress and molecular alterations associated with aging. Increased superoxide dismutase and catalase activity in red blood cells of rats fed with probiotic *dahi* containing *Lb. acidophilus* and *B. bifidum* was noticed (Rajpal and Kansal 2009c), thus increasing the antioxidant status in rats. With the stimulation in immune system, a substantial rise in phagocytic activity of macrophages was observed in experimental mice on feeding probiotic *dahi* containing *Lb. acidophilus* and *B. bifidum*, and it significantly reduced the colonization of *Shigella* in the intestine, liver, and spleen (Rajpal and Kansal 2009b). Improved peritoneal macrophage functions such as inciting nitric oxide and IL-6 and diminished production of PGE2, besides enhancing the lymphocyte stimulation and increased production of IL-2 in aging mice (16 weeks old) upon feeding with same probiotic *dahi*, thus reversed age-related reduction in immune functions in mice (Kaushal and Kansal 2011). It also reduced the myeloperoxidase activity, levels of IL-6, tumor necrosis factor alpha, and interferon significantly in ulcerative colitis mice, which were induced with dextran sodium sulfate (Jadhav et al. 2012).

13.3 Lassi

Lassi, otherwise known as buttermilk or *mahi* in Nepal (Dahal et al. 2005), is a popular drink in the Indian subcontinent. It is mostly consumed during summer months and is prepared by churning *dahi*, while separating ghee. It is consumed either as such or with addition of salt/sugar, spices, or flavoring agents for different taste and flavors. The steps involved in the preparation of *lassi* are indicated in Figure 13.1. It is produced by organized dairies in small quantities; however, it is mainly prepared by local vendors. It contains 96.2% water, 0.8% fat, 1.29% protein, 1.2% lactose, 0.44% lactic acid, 0.4% ash, 0.6% calcium, and 0.04% phosphorus (Rangappa and Achayya 1974). Probiotic culture of *Lb. acidophilus* along with *Str. thermophilus* were added to *lassi* to make it a therapeutic beverage and also to get a desirable flavor (Patidar and Prajapati 1998), and it retained a viable population of 15.86×10^6 CFU/g of *Str. thermophilus* and 20.4×10^7 CFU/g of *Lb. acidophilus*.

Extending the shelf life of *lassi* increases its marketability. Addition of nisaplin, a bacteriocin, at 200–500 IU/ml extended its shelf life up to 2 days at 30°C and up to 10 days under refrigerated condition (Naresh and Prasad 1996). A mild heat treatment between 50°C and 70°C for 5 min lowered its sensory score besides leading to wheying off (Ramana and Tiwari 1999). However, addition of carboxy methyl cellulose at 0.2% made it possible to heat treat at 55°C for 5 min without affecting its sensory properties. A ready-to-reconstitute spray-dried pineapple *lassi* powder blend having 25.04% protein, 9.97% fat, and 26.74% sugar, with a shelf life of 6 months at 37°C was developed (Ramesh Babu and Dasgupta 2005). Kadam et al. (2005) have prepared soymilk-blended *lassi* and reported that substitution of buffalo milk with soymilk up to 25% or addition of 10% skim milk powder to soymilk did not affect the acceptability of blended *lassi*. In another study on the preparation of *lassi* concentrate, Kadam et al. (2006) noticed that use of twofold concentrated milk with 6% sugar made it desirable even up to the ninth day. Carbonation (50 psi for 30 s) of cultured buttermilk (*lassi*) extended the shelf life of the product up to 12 weeks at 7°C, whereas only 5 weeks of shelf life was noticed for noncarbonated drink (Ravindra et al. 2014). In addition, carbonation arrested lipolysis and proteolysis, inhibiting the yeast and mold growth in the beverage.

Instant *lassi* was prepared by adding 50% v/v lactic acid to standardized cows' milk. Adjusting the pH of instant *lassi* to 3.8 and addition of 12% sugar had similar results in terms of color, appearance, and overall acceptability with that of control *lassi* (Hingmire et al. 2009). Preparation of fruit-flavored carbonated buttermilk by processing with ultrafiltration membrane was standardized; carbonation was found to increase sensory properties of the beverage and the acceptability was better than the product available in market (Shaikh and Rathi 2009). Use of EPS-producing *Str. thermophilus* IG16 for the preparation of lassi improved appearance, color, consistency, flavor, viscosity, and controlled whey separation in fat-free *lassi* (Behare et al. 2010). Addition of sweetener/sweetener blends (aspartame and acesulfame k) did not influence the quality of *lassi*, and these compounds were found to be stable during 5 days of storage and use of blends reduced the usage level by 38% (George et al. 2010, 2012). Fuzzy logic analysis revealed that market *lassi* was a better accepted product than the *lassi* made by reconstituting the *dahi* powder added with guar gum and locust bean gum (Routray and Mishra 2012).

13.4 Shrikhand

Shrikhand, a sweetened lactic fermented milk product, is mostly consumed in western and northern India. It has a pleasant aroma with refreshing taste, smooth and homogeneous texture, and firm consistency. The *shrikhand* preparation method is outlined in Figure 13.1. For its preparation, whole milk of cow (Kadam et al. 1984), buffalo (Reddy et al. 1984), and goat (Agnihotri and Pal 1996); ultrafiltered skim milk retentates (Ansari et al. 2006); and soymilk (Deshpande et al. 2008) are used. Although conventionally mesophilic lactic cultures are used for *shrikhand* preparation, typical yogurt cultures (*Str. thermophillus* and *Lb. delbruckii* subsp. *bulgaricus*) have also been used (Singh and Jha 2005). It reduced the curd setting time from 8–10 h to 4 h. The set *dahi* (curd) is drained by hanging in a muslin cloth for 6–8 h. To the solid mass thus obtained, known as *chakka*, ground sugar (45%) and flavoring substances such as cardamom and saffron are added and mixed uniformly. It has a composition of 35% moisture, 5% fat, 6% protein, 2% reducing sugar, and 55% sucrose. The content of soluble nitrogen, free fatty acids, acidity, and pH varied from manufacturer to manufacturer and was in the range of 0.12%–0.29%, 0.68%–0.78% oleic acid, 1.04%–1.54% lactic acid, and 3.9–4.3, respectively (Salunke et al. 2006). The synergy between dairy and nondairy products such as fruit pulps, vegetable oils, and soybean has been reviewed (Krupa et al. 2011). Preparation of fruit-flavored *shrikhand* using banana, guava, sapota (Dadarwal et al. 2005), papaya (Nigam et al. 2009; Mali et al. 2010), and apple pulps (Kumar et al. 2011) has been standardized. Rheological properties of *shrikhand* are well documented, and it exhibited a combination of several rheological properties such as thixotropy, weak gel-like viscoelasticity, long structural recovery timescales, and an apparent yield stress (Kulkarni et al. 2006). Addition of 5% sour whey concentrate to *chakka* increased the yield of *shrikhand* by 5% and did not vary in physicochemical and sensory properties from the *shrikhand* prepared by the conventional method (Giram et al. 2001).

Shrikhand is a rich source of minerals including calcium, magnesium, phosphorus, copper, iron, and zinc (Boghra and Mathur 2000). It has a shelf life of 35–40 days at 5°C or 2–3 days at room temperature (30°C–37°C). The shelf life is further enhanced up to 15 days at 35°C and up to 70 days at 8°C–10°C by subjecting *shrikhand* to post-production heat treatment at 70°C for 5 min (Prajapati et al. 1993).

Initially, *shrikhand* was prepared in households on a very small scale. However, currently it is being manufactured commercially on a large scale and is available all over India. The technological and microbiological aspects of *shrikhand* preparation are well reported in earlier studies (Mahajan et al. 1979; Patel and Abdel-Salam 1986; Patel and Chakraborty 1988; De and Patel 1990; Sharma 1998; Rati Rao et al. 2006). Ultrafiltration has been used to increase the yield in large-scale production of *shrikhand* (Sharma 1998). *Shrikhand* powder was prepared by either spray-drying the *shrikhand* (Mahajan et al. 1979) or colloidal milling of *chakka* followed by spray drying after adding ground sugar and flavoring agents (De and Patel 1990).

To reduce the high fat content for health conscious people, dietetic *shrikhand* was prepared by fermenting buffalo skim milk with *Lb. acidophilus* (NDRI-AH1) and *Str. salivarius* subsp. *thermophilus* (NDRI-YHS) (Subramanian et al. 1997). *Staph. aureus* could grow and produce thermostable deoxyribonuclease, which is closely related with enterotoxin production, even in the presence of *Lb. acidophilus* and *Lb. bulgaricus* in *shrikhand* (Varadaraj and Ranganathan, 1988). The behavior of *Escherichia coli* and *L. monocytogenes* in *shrikhand* added with pediocin K7 was predicted using response surface methodology (Jagannath et al. 2001a, 2001b). Feeding of *shrikhand* to albino mice led to induced production of IgG, indicating its therapeutic attribute in terms of an improvement in the immune system (Subramanian et al. 2005). Probiotic *shrikhand* was developed to enhance the probiotic quality, and *shrikhand* prepared using *Lb. acidophilus* and *Lb. sporogenes* had better sensory scores than that made with *Lb. rhamnosus* in terms of color, appearance, aroma, texture, taste, and overall acceptability of the product (Swapna et al. 2011).

13.5 Future Prospects

Trends of increasing demand for functional foods, especially of milk-based products, indicate a great potential for probiotic dairy products such as *dahi*, *lassi*, and *shrikhand*. Based on the knowledge generated on animal-feeding studies, research should be focused on human volunteers (clinical validation) to prove the efficacy of these probiotic strains. This can help in promoting probiotic products more effectively and on a large scale, so that general health of consumers can be maintained well to strengthen the concept of "health and wellness through foods" and their dependency on medicines can be reduced.

References

Agnihotri, M.K. and Pal, U.K. 1996. Production and quality of shrikhand from goat milk. *The Indian Journal of Small Ruminants* 2: 24–28.

Anon. 2008. Ministry of Food Processing Industries and Federation of Indian Chambers of Commerce and Industry. *Land of Opportunities: The Food Industry in India*. Gurgaon, India: Technopak Advisors Pvt. Ltd.

Ansari, M.I.A., Rai, P., Sahoo, P.K., and Datta, A.K. 2006. Manufacture of shrikhand from ultrafiltered skim milk retentates. *Journal of Food Science and Technology* 43: 49–52.

Arvind, Sinha, P.R., Singh, K.N., and Kumar, R. 2009. Alteration in oxidative stress and c-Myc protein by administration of acidophilus casei *dahi* on DMH induced colon carcinogenesis. *Current Topics in Nutraceutical Research* 7: 65–72.

Arvind, Sinha, P.R., Singh, K.N., and Kumar, R. 2010. Effect of acidophilus casei *dahi* (probiotic curd) on faecal enzyme activities in DMH induced colon carcinogenesis. *Milchwissenschaft* 65: 255–258.

Balasubramanyam, B.V. and Varadaraj, M.C. 1994. *Dahi* as a potential source of lactic acid bacteria active against foodborne pathogenic and spoilage bacteria. *Journal of Food Science and Technology* 31: 241–243.

Balasubramanyam, B.V. and Varadaraj, M.C. 1995. Antibacterial effect of *Lactobacillus* spp. on foodborne pathogenic bacteria in an Indian milk-based fermented culinary food item. *Cultured Dairy Products Journal* 30(1): 22–27.

Begum, J., Nahar, A., Islam, M.N., and Rahman, M.M. 2011. Qualitative characteristics of *dahi* prepared from non-fat dry milk fortified with vegetable oil. *Bangladesh Research Publications Journal* 6: 35–45.

Behare, P., Singh, R., and Singh, R.P. 2009a. Exopolysaccharide-producing mesophilic lactic cultures for preparation of fat-free *Dahi*—An Indian fermented milk. *Journal of Dairy Research* 76: 90–97.

Behare, P.V., Singh, R., Nagpal, R., Kumar, M., Tomar, S.K., and Prajapati, J.B. 2009b. Comparative effect of exopolysaccharides produced in situ or added as bioingredients on *dahi* properties. *Milchwissenschaft* 64: 396–400.

Behare, P.V., Singh, R., Tomar, S.K., Nagpal, R., Kumar, M., and Mohania, D. 2010. Effect of exopolysaccharide-producing strains of *Streptococcus thermophilus* on technological attributes of fat-free *lassi*. *Journal of Dairy Science* 93: 2874–2879.

Boghra, V.R. and Mathur, O.N. 2000. Physico-chemical status for major constituents and minerals at various stages of *shrikhand* production. *Journal of Food Science and Technology* 37: 111–115.

Chauhan, R. and Abraham, J. 2011. A comparative study of different inulin levels on quality parameters of symbiotic misti *dahi*. *Journal of Pure and Applied Microbiology* 5: 977–982.

Chowdhury, N.A., Paramanik, K., and Zaman, W. 2011. Study on the quality assessment of curd (*dahi*) locally available in Bangladesh market. *World Journal of Dairy and Food Sciences* 6: 15–20.

Dadarwal, R., Beniwal, B.S., and Singh, R. 2005. Process standardization for preparation of fruit flavoured *shrikhand*. *Journal of Food Science and Technology* 42: 22–26.

Dahal, N.R., Karki, T.B., Swamylingappa, B., Li, Q., and Gul, G. 2005. Traditional foods and beverages of Nepal—A review. *Food Reviews International* 21: 1–25.

De, A. and Patel, R.S. 1990. Technology of *shrikhand* powder. *Cultured Dairy Products Journal* 25(2): 21, 23–24, 26, 28.

Deshpande, S., Bargale, P.C., and Jha, K. 2008. Suitability of soy milk for development of *shrikhand*. *Journal of Food Science and Technology* 45: 284–286.

Dewan, S. and Tamang, J.P. 2007. Dominant lactic acid bacteria and their technological properties isolated from the Himalayan ethnic fermented milk products. *Antonie von Leeuwenhoek* 92: 343–352.

Gawai, K. and Prajapati, J.B. 2012. Status and scope of *dahi* industry in India. *Indian Dairyman* 64(7): 46–50.

George, V., Aurora, S., Sharma, V., Wadhwa, B.K., and Singh, A.K. 2012. Stability, physico-chemical, microbial and sensory properties of sweetener/sweetener blends in *lassi* during storage. *Food and Bioprocess Technology* 5: 323–330.

George, V., Aurora, S., Wadhwa, B.K., Singh, A.K., and Sharma, G.S. 2010. Optimization of sweetener blends for the preparation of *lassi*. *International Journal of Dairy Technology* 63: 256–261.

Ghosh, D., Ray, L., and Chattopadhyay, P. 2010. Nutraceutical potential of cow milk and soy milk curd (*dahi*). *Indian Chemical Engineer* 52: 336–346.

Giram, S.D., Barbind, R.P., Pawar, V.S., Sakhale, B.K., and Agarkar, B.S. 2001. Studies of fortification of sour whey concentrate in chakka for preparation of shrikhand. *Journal of Food Science and Technology* 38: 294–295.

Goyal, G.K. and Goyal, S. 2012. Packaging of *dahi* in eco-friendly shellac coated earthen pots. *Indian Dairyman* 64(9): 60–62.

Hingmire, S.R., Lembhe, A.F., Zanjad, P.N., Pawar, V.D., and Machewad, G.M. (2009). Production and quality evaluation of instant *lassi*. *International Journal of Dairy Technology* 62: 80–84.

Islam, M.N., Akhter, F., Masum, A.K.M., Khan, M.A.S., and Asaduzzaman, M. 2010. Preparation of *dahi* for diabetic patient. *Bangladesh Journal of Animal Sciences* 39: 144–150.

Jadhav, S.R., Shandilya, U.K., and Kansal, V.K. 2012. Immunoprotective effect of probiotic *dahi* containing *Lactobacillus acidophilus* and *Bifidobacteriumbifidum* on dextran sodium sulfate-induced ulcerative colitis in mice. *Probiotics and Antimicrobial Proteins* 4: 21–26.

Jagannath, A., Ramesh, A., Ramesh, M.N., Chandrashekar, A.C., and Varadaraj, M.C. 2001a. Predictive model for the behavior of *Listeria monocytogenes* Scott A in *shrikhand*, prepared with a biopreservative, pediocin K7. *Food Microbiology* 18: 335–343.

Jagannath, A., Ramesh, M.N., and Varadaraj, M.C. 2001b. Predicting the behaviour of *Escherichia coli* introduced as a post processing contaminant in s*hrikhand*, a traditional sweetened lactic fermented milk product. *Journal of Food Protection* 64: 462–469.

Jain, S., Yadav, H., and Sinha, P.R. 2008. Stimulation of innate immunity by oral administration of *dahi* containing probiotic *Lactobacillus casei* in mice. *Journal of Medicinal Food* 11: 652–656.

Jain, S., Yadav, H., and Sinha, P.R. 2009. Probiotic *dahi* containing *Lactobacillus casei* protectsagainst *Salmonella enteritidis* infection and modulates immune response in mice. *Journal of Medicinal Food* 12: 576–583.

Jain, S., Yadav, H., Sinha, P.R., Kapila, S., Naito, Y., and Marotta, F. 2010. Anti-allergic effects of probiotic *dahi* through modulation of the gut immune system. *Turkish Journal of Gastroenterology* 21: 244–250.

Jayanth, H.S. 2010. *Studies on the Characterization and DNA-Based Detection of Listeria Species in Dairy Foods*. PhD thesis, Mysore, India: University of Mysore.

Kadam, P.S., Kalepatil, R.K., Avhad, V.B., Sharma, S.K., and Motey, R.H. 2006. Some studies on preparation of *lassi* concentrate. *Journal of Soils and Crops* 16: 121–126.

Kadam, P.S., Kedare, S.D., Ghatge, P.U., and Ingole, A.J. 2005. Studies on preparation of soya milk blended *lassi*. *Journal of Soils and Crops* 15: 366–369.

Kadam, S.J., Bhosle, D.N., and Chavan, I.G. 1984. Studies on preparation of *chakka* from cow milk. *Journal of Food Science and Technology* 21: 180–182.

Karki, T. 1986. Some Nepalese fermented foods and beverages. In: *Traditional Foods, Some Products and Technologies*. Mysore, India: Central Food Technological Research Institute.

Kaushal, D. and Kansal, V.K. 2011. Age-related decline in macrophage and lymphocyte functions in mice and its alleviation by treatment with probiotic *dahi* containing *Lactobacillus acidophilus* and *Bifidobacterium bifidum*. *Journal of Dairy Research* 78: 404–411.

Kaushal, D. and Kansal, V.K. 2012. Probiotic *Dahi* containing *Lactobacillus acidophilus* and *Bifidobacterium bifidum* alleviates age-inflicted oxidative stress and improves expression of biomarkers of ageing in mice. *Molecular Biology Reports* 39: 1791–1799.

Khetra, Y., Raju, P.N., Hati, S., and Kanawjia, S.K. 2011. Health benefits of traditional fermented milk products. *Indian Dairyman* 63(9): 54–60.

Koirala, R., Ricci, G., Taverniti, V., Ferrario, C., Malla, R., Shrestha, S., Fortina, M.G., and Guglielmetti, S. 2014. Isolation and molecular characterization of lactobacilli from traditional fermented *dahi* produced at different altitudes in Nepal. *Dairy Science and Technology* 94: 397–408.

Kore, K.B., Pattanaik, A.K., Sharma, K., and Mirajkar, P.P. 2012. Effect of feeding traditionally prepared fermented milk *dahi* (curd) as a probiotics on nutritional status, hindgut health and haematology in dogs. *Indian Journal of Traditional Knowledge* 11: 35–39.

Krupa, H., Jana Atanu, H., and Patel, H.G. 2011. Synergy of dairy with non-dairy ingredients or product: A review. *African Journal of Food Science* 5: 817–832.

Kulkarni, C., Belsare, N., and Lele, A. 2006. Studies on shrikhand rheology. *Journal of Food Engineering* 74: 169–177.

Kumar, A., Singh, N.K., Sinha, P.R., and Kumar, R. 2010. Intervention of *acidophilus-casei dahi* and wheat bran against molecular alteration in colon carcinogenesis. *Molecular Biology Reports* 37: 621–627.

Kumar, S., Bhat, Z.F., and Kumar, P. 2011. Effect of apple pulp and *Celosia argentea* on the quality characteristics of shrikhand. *American Journal of Food Technology* 6: 817–826. doi:10.3923/ajft.2011.

Kurien, A., Puniya, A.K., and Singh, K. 2005. Selection of prebiotic and *Lactobacillus acidophilus* for synbiotic yogurt preparation. *Indian Journal of Microbiology* 45: 45–50.

Laxminarayana, H., Nambudripad, V.K.N., Lakshmi, N., Anantaramaiah, S.N., and Sreenivasamurthy, V. 1952. Studies on *dahi*, II: General survey of the quality of market *dahi*. *Indian Journal of Veterinary Science and Animal Husbandry* 22: 13–25.

Mahajan, B.M., Mathur, O.N., Bhattacharya, D.C., and Srinivasan, M.R. 1979. Production and shelf life of spray dried *shrikhand* powder. *Journal of Food Science and Technology* 16: 9–10.

Mahmood, T., Masud, T., and Sohail, A. 2014. Some probiotic and antibacterial properties of *Lactobacillus acidophilus* cultured from *dahi* a native milk product. *International Journal of Food Sciences and Nutrition* 65: 582–588. doi:10.3109/09637486.2014.880666.

Maity, T.K. and Misra, A.K. 2010. Hypocholesterolemic effect of *Lactobacillus casei* isolated from *dahi* (Indian yoghurt) in albino rats. *Milchwissenschaft* 65: 140–143.

Mali, R.S., Dhapke, D.H., and Zinjarde, R.M. 2010. Effect of papaya pulp on the quality and cost structure of *shrikhand*. *Journal of Soils and Crops* 20: 290–294.

Matkar, S.P. and Khedkar, J.N. 2008. Trends in packaging of traditional fermented milk products. *Indian Dairyman* 60(7): 47–51.

Misra, A.K. 1992. Commercial production of *dahi* by the dairy industry. *Indian Dairyman* 44: 501–503.

Mitra, S., Chakrabartty, P.K., and Biswas, S.R. 2010. Potential production and preservation of *dahi* by *Lactococcus lactis* W8, a nisin-producing strain. *LWT—Food Science and Technology* 43: 337–342.

Mohania, D., Kansal, V.K., Kruzliak, P., and Kumari, A. 2014. Probiotic *dahi* containing *Lactobacillus acidophilus* and *Bifidobacterium bifidum* modulates the formation of aberrant crypt foci, mucin depleted foci and cell proliferation on 1, 2-dimethylhydrazine induced colorectal carcinogenesis in Wistar rats. *Rejuvenation Research* 17: 325–333. doi:10.1089/rej.2013.1537.

Nagpal, R., Behare, P.V., Kumar, M., Mohania, D., Yadav, M., Jain, S., Menon, S. et al. 2012. Milk, milk products and disease free health: An updated overview. *Critical Reviews in Food Science and Nutrition* 52: 321–333.

Naresh, K. and Prasad, D.N. 1996. Preservative action of nisin in *lassi* under different storage temperatures. *Indian Journal of Animal Science* 66: 525–528.

Nigam, N., Singh, R., and Upadhayay, P.K. 2009. Incorporation of *chekka* by papaya pulp in the manufacture of *shrikhand*. *Journal of Dairying, Foods & Home Science* 28: 115–118.

Palanidorai, R., Akila, N., and Siva Kumar, T. 2009. Goat milk *dahi*. *Tamilnadu Journal of Veterinary and Animal Science* 5: 114–115.

Panesar, P.S. 2011. Fermented dairy products: Starter cultures and potential nutritional benefits. *Food and Nutrition Sciences* 2: 47–51.

Patel, R.S. and Abdel-Salam, M.H. 1986. *Shrikhand*: An Indian analogue of western quarg. *Cultured Dairy Products Journal* 21(1): 6–7.

Patel, R.S. and Chakraborty, B.K. 1988. *Shrikhand*: A review. *Indian Journal of Dairy Science* 41: 109–115.

Patidar, S.K. and Prajapati, J.B. 1998. Standardization and evaluation of *lassi* prepared using *Lb. acidophilus* and *S. thermophilus*. *Journal of Food Science and Technology* 35: 428–431.

Prajapati, J.B. and Nair, B.M. 2003. The history of fermented foods. In: *Handbook of Fermented Functional Foods*, ed. E.R. Farnworth, 1–25. Boca Raton, FL: CRC Press.

Prajapati, J.P., Upadhyay, K.G., and Desai, H.K. 1993. Comparative quality appraisal of heated *shrikhand* stored at ambient temperature. *Australian Journal of Dairy Technology* 47: 18–22.

Raja, B.R. and Arunachalam, K.D. 2011. Market potential for probiotic nutritional supplements in India. *African Journal of Business Management* 5: 5418–5423.

Rajpal, S. and Kansal, V.K. 2009a. Probiotic *Dahi* containing *Lactobacillus acidophilus* and *Bifidobacterium bifidum* attenuates diet induced hypercholesterolemia in rats. *Milchwissenschaft* 64: 21–25.

Rajpal, S. and Kansal, V.K. 2009b. Probiotic *Dahi* containing *Lactobacillus acidophilus* and *Bifidobacterium bifidum* stimulates immune system in mice. *Milchwissenschaft* 64: 147–150.

Rajpal, S. and Kansal, V.K. 2009c. Probiotic *dahi* containing *Lactobacillus acidophilus* and *Bifidobacterium bifidum* stimulates antioxidant enzyme pathways in rats. *Milchwissenschaft* 64: 287–290.

Raju, P.N., Khetra, Y., Pal, D., Kanawjia, S.K., and Singh, A.K. 2012. Misti*dahi* variants for health conscious consumers. *Indian Dairyman* 64(9): 58–64.

Raju, P.N. and Pal, D. 2014. Effect of dietary fibers on physico-chemical, sensory and textural properties of *misti dahi*. *Journal of Food Science and Technology* 51: 3124–3133. doi:10.1007/s13197-012-0849-y.

Raju, P.N. and Pal, D. 2009. The physico-chemical, sensory, and textural properties of *mistidahi* prepared from reduced fat buffalo milk. *Food Bioprocess Technology* 2: 101–108.

Raju, P.N. and Pal, D. 2011. Effect of bulking agents on the quality of artificially sweetened *mistidahi* (caramel colored sweetened yoghurt) prepared from reduced fat buffalo milk. *LWT—Food Science and Technology* 44: 1835–1843.

Ramana, B.L.V. and Tiwari, B.D. 1999. Effect of processing variables on sensory and physico-chemical qualities of heat treated *lassi*. *Indian Journal of Dairy Science* 42: 272–278.

Ramesh Babu, D. and Dasgupta, D.K. 2005. Development of pineapple juice-milk/*lassi* powders. *Journal of Food Science and Technology* 42: 241–245.

Rangappa, K.S. and Achayya, K.T. 1974. *Indian Dairy Products*, 119–123. New Delhi, India: Asia Publishing House.

Ranjan, P., Arora, S., Sharma, G.S., Sindhu, J.S., and Singh, G. 2006. Sensory and textural profile of curd (*dahi*) from calcium enriched buffalo milk. *Journal of Food Science and Technology* 43: 38–40.

Rashid, M.H., Togo, K.K., Ueda, M., and Miyamoto, T. 2007. Identification and characterization of dominant lactic acid bacteria isolated from traditional fermented milk *dahi* in Bangladesh. *World Journal of Microbiology and Biotechnology* 23: 125–133.

Rati Rao, E., Vijayendra, S.V.N., and Varadaraj, M.C. 2006. Fermentation biotechnology of traditional foods of the Indian subcontinent. In: *Food Biotechnology* (2nd Edition), eds. K. Shetty, G. Paliyath, A. Pometto, and R.E. Levin, 1759–1794. Boca Raton, FL: CRC Press.

Ravindra, M.R., Jayaraj Rao, K., Nath, B.S., and Ram, C. 2014. Carbonated fermented dairy drink—Effect on quality and shelf life. *Journal of Food Science and Technology* 51: 3397–3403. doi:10.1007/s13197-012-0854-1.

Reddy, K.K., Ali, M.P., Rao, B.V., and Rao, T.J. 1984. Studies on production and quality of *shrikhand* from buffalo milk. *Indian Journal of Dairy Science* 37: 293–297.

Roopashri, A.N. and Varadaraj, M.C. 2009. Molecular characterization of native isolates of lactic acid bacteria, bifidobacteria and yeasts for beneficial attributes. *Applied Microbiology and Biotechnology* 83: 1115–1126.

Routray, W. and Mishra, H.N. 2012. Sensory evaluation of different drinks formulated from *dahi* (Indian yogurt) powder using fuzzy logic. *Journal of Food Processing and Preservation* 36: 1–10.

Salunke, P., Patel, H.A., and Thakar, P.N. 2006. Physico-chemical properties of *shrikhand* sold in Maharashtra state. *Journal of Food Science and Technology* 43: 276–281.

Sarkar, S. 2008. Innovations in Indian fermented milk products—A review. *Food Biotechnology* 22: 78–97.

Sarkar, S., Sur, A., Pal, R., Sarkar, K., Majhi, R., Biswas, T., and Banerjee, S. 2011. Potential of *dahi* as a functional food. *Indian Food Industry* 30(1): 27–32.

Shaikh, M.F.B. and Rathi, S.D. 2009. Utilisation of buttermilk for the preparation of carbonated fruit-flavoured beverages. *International Journal of Dairy Technology* 62: 564–570.

Sharma, D.K. 1998. Ultrafiltration for manufacture of indigenous milk products: *Chhana* and *shrikhand*. *Indian Dairyman* 50(8): 33–38.

Shiby, V.K. and Mishra, H.N. 2008. Modeling of acidification kinetics and textural properties in *dahi* (Indian yogurt) made from buffalo milk using response surface methodology. *International Journal of Dairy Technology* 61: 284–289.

Shiby, V.K. and Mishra, H.N. 2012. Effect of starter culture level on textural and sensory properties of buffalo milk *dahi* (curd). *Egyptian Journal of Dairy Science* 40: 15–23.

Shruti, S. and Kansal, V.K. 2011. Effect of feeding probiotic *dahi* containing *Lactobacillus acidophilus* and *Bifidobacterium bifidum* on enzymes that catalyze carcinogen activation and detoxification in rats. *Milchwissenschaft* 66: 244–247.

Singh, A.K., Mukherjee, S., Adhikari, M.D., and Ramesh, A. 2012. Fluorescence-based comparative evaluation of bactericidal potency and food application potential of anti-listerial bacteriocin produced by lactic acid bacteria isolated from indigenous samples. *Probiotics and Antimicrobial Proteins* 4: 122–132.

Singh, A.K. and Ramesh, A. 2009. Evaluation of a facile method of template DNA preparation for PCR-based detection and typing of lactic acid bacteria. *Food Microbiology* 26: 504–513.

Singh, R. and Jha, Y.K. 2005. Effect of sugar replacers on sensory attributes, biochemical changes and shelf-life of *shrikhand*. *Journal of Food Science and Technology* 42: 199–202.

Steinkraus, K.H. 1995. *Handbook of Indigenous Fermented Foods*. New York: Marcel Dekker.

Subramanian, B.S., Kumar, C.N., and Venkateshaiah, B.V. 2005. Therapeutic properties of dietetic-*shrikhand* prepared using LAB. *Mysore Journal of Agricultural Science* 39: 399–403.

Subramanian, B.S., Naresh Kumar, C., Narasimhan, R., Shanmugam, A.M., and Khan, M.M.H. 1997. Selection of level and type of LAB starter in the preparation of dietetic *shrikhand*. *Journal of Food Science and Technology* 34: 340–342.

Swapna, G., Brahmaprakash, G.P., and Chavannavar, S.V. 2011. Challenges associated with the development of lactic acid bacteria and probiotic containing fermented *shrikhand*. *Environment and Ecology* 29: 1018–1022.

Tamang, J.P., Tamang, N., Thapa, S., Dewan, S., Tamang, B., Yonzan, H., Rai, A.K., Chettri, R., Chakrabarty, J., and Kharel, N. 2012. Microorganisms and nutritional value of ethnic fermented foods and alcoholic beverages of North East India. *Indian Journal of Traditional Knowledge* 11: 7–25.

Thumu, S.C.R. and Halami, P.M. 2012. Presence of erythromycin and tetracycline resistance genes in lactic acid bacteria from fermented foods of Indian origin. *Antonie van Leeuwenhoek* 102: 541–551. doi:10.1007/s10482-012-9749-4.

Varadaraj, M.C. and Ranganathan, B. 1988. Fate of *Staphylococcus aureus* in *shrikhand* prepared with *Lactobacillus acidophilus* and *Lb. bulgaricus*. *Indian Journal of Dairy Science* 41: 363–366.

Vijayendra, S.V.N. and Gupta, R.C. 1992. Therapeutic importance of bifidobacteria and *Lactobacillus acidophilus* in fermented milks. *Indian Dairyman* 44: 595–599.

Vijayendra, S.V.N. and Gupta, R.C. 2013. Associative growth behavior of *dahi* and yoghurt starter cultures with *Bifidobacterium bifidum* and *Lactobacillus acidophilus* in buffalo skim milk. *Annals of Microbiology* 63: 461–469. doi:10.1007/s13213-012-0490-z.

Vijayendra, S.V.N. and Gupta, R.C. 2012. Assessment of probiotic and sensory properties of *dahi* and yoghurt prepared using bulk freeze-dried cultures in buffalo milk. *Annals of Microbiology* 62: 939–947.

Vijayendra, S.V.N., Palanivel, G., Mahadevamma, S., and Tharanathan, R.N. 2008. Physico-chemical characterization of an exopolysaccharide produced by a non-ropy strain of *Leuconostoc* sp. CFR 2181 isolated from *dahi*–An Indian traditional lactic fermented milk product. *Carbohydrate Polymers* 72: 300–307.

Vijayendra, S.V.N., Palanivel, G., Mahadevamma, S., and Tharanathan, R.N. 2009. Physico-chemical characterization of a new heteropolysaccharide produced by a native isolate of heterofermentative *Lactobacillus* sp. CFR-2182. *Archives of Microbiology* 191: 303–310.

Yadav, H., Jain, S., and Sinha, P.R. 2006. Effect of *dahi* containing *Lactobacillus lactis* on the progression of diabetes induced by a high fructose diet in rats. *Bioscience, Biotechnology and Biochemistry* 70: 1255–1258.

Yadav, H., Jain, S., and Sinha, P.R. 2007a. Antidiabetic effect of probiotic *dahi* containing *Lactobacillus acidophilus* and *Lactobacillus casei* in high fructose fed rats. *Nutrition* 23: 62–68.

Yadav, H., Jain, S., and Sinha, P.R. 2007b. Production of free fatty acids and conjugated linoleic acid in probiotic *dahi* containing *Lactobacillus acidophilus* and *Lactobacillus casei* during fermentation and storage. *International Dairy Journal* 17: 1006–1010.

Yadav, H., Jain, S., and Sinha, P.R. 2008a. Oral administration of *dahi* containing probiotic *Lactobacillus acidophilus* and *Lactobacillus casei* delayed the progression of streptozotocin-induced diabetes in rats. *Journal of Dairy Research* 75: 189–195.

Yadav, H., Jain, S., and Sinha, P.R. 2008b. The effect of probiotic *dahi* containing *Lactobacillus acidophilus* and *Lactobacillus casei* on gastropathic consequences in diabetic rats. *Journal of Medicinal Food* 11: 62–68.

14

Indonesian Dadih

INGRID SURYANTI SURONO

Contents

14.1	Introduction	377
14.2	Product Description of Dadih	379
14.3	Nutrition Profile of Dadih	381
14.4	Important Factors and Physicochemical Changes during Dadih Fermentation	382
	14.4.1 Fresh Raw Buffalo Milk	383
	14.4.2 Natural LAB	383
	14.4.3 Indigenous Enzymes	385
	14.4.4 Bamboo Tubes as Container of Dadih Product	386
	14.4.5 Biochemical Changes during Buffalo Milk Fermentation	387
	14.4.5.1 Carbohydrate	387
	14.4.5.2 Protein	388
	14.4.5.3 Lipids	388
	14.4.5.4 Flavor Compounds	388
	14.4.5.5 Cell Mass	388
14.6	Probiotic Properties of LAB Isolated from Dadih	389
	14.6.1 Probiotic Properties of LAB Strains from Dadih	389
	14.6.2 Novel Probiotics	392
14.7	Dadih Probiotic for Human Health Promotion	393
	14.7.1 Immune System and Nutritional Status	393
14.8	Concluding Remarks	395
References		395

14.1 Introduction

Indonesia is the largest archipelago blessed with one of the richest mega-biodiversities on the globe, rich in ethnic and cultural diversities, and also home to one of the most diverse cuisines and traditional fermented foods. Dadih is an Indonesian traditional fermented milk produced and consumed by the West Sumatran Minangkabau ethnic group of Indonesia as one of the characteristic traditional foods of Minangkabau culture. Native people named it as dadiah, and it is a very popular dairy product in Bukittinggi, Padang Panjang, Solok, Lima Puluh Kota, and Tanah Datar (Surono and Hosono 1996). It serves as a significant dairy product in the diet of the Minangkabau resembling yogurt and is similar to dahi of India.

Dadih is one of the most distinct products in microbial quality and an important fermented milk. It is a yogurt-like product with a distinctive thicker consistency, smooth texture, and pleasant flavor and has provided safety, portability, and novelty to milk nutrients for the indigenous people in West Sumatra. Currently, dadih is consumed as a traditional food, served at weddings and while giving an honorable title "Datuk" in West Sumatra during the ethnic tradition or "adat" ceremony. Generally, dadih is consumed during breakfast with rice after adding sliced shallot and chili ("sambal"), or it is mixed with palm sugar and coconut milk, being served as a topping of steamed traditional glutinous rice flakes, a corn flake-like product, called "ampiang dadih."

The manufacturing method of dadih is quite similar to the dahi of India, except for the heat treatment of raw milk and the starter cultures being incorporated. In dahi making, the raw cow or buffalo milk, or a combination of both, is pasteurized and then fermented using leftover dahi from the previous lot as starter cultures (Indian Standard Institution 1980). In Indonesia, dadih is still a homemade product by the traditional way, involving the milk of water buffaloes. No heat is applied to buffalo milk while manufacturing dadih. The milk is neither boiled nor inoculated with any starter culture. The fresh unheated buffalo milk is placed in bamboo tubes covered with banana leaves, incubated at the ambient temperature ($28°C–30°C$) overnight, and allowed to ferment naturally until it acquires a thick consistency (Akuzawa and Surono 2002). Various indigenous lactic acid bacteria (LAB) involved in the dadih fermentation may vary from time to time as well as from one place to another due to the natural fermentation without any starter culture involved (Surono and Hosono 2000; Akuzawa and Surono 2002).

Interestingly, even though the process did not implement the hygiene practice, there was no product failure and no food poisoning was reported among people who consumed dadih. Instead, the older generation believes that consuming dadih may provide a beneficial effect to their health. Due to this fact, it is very interesting to explore the indigenous LAB involved during dadih fermentation. Some dadih LAB have antimutagenicity, hypocholesterolemic properties, antipathogenic properties, and immunomodulatory properties (Surono and Hosono 1996; Pato et al. 2004; Surono et al. 2011).

Microbial ecology of dadih is little known. However, recent development in molecular biology techniques and sequencing may allow us to gain more insight on the complexity, beauty, and potential advanced applications of these understudied, invaluable resources for microbial bioprospecting as well as functional foods. Scientific literature and data on production is also limited, and marketing and standard specifications for production and quality control of dadih are not well established. Still, it remains challenging to study dadih preparation and supposed to be well documented on a scientific basis. The technological parameters, the biochemical changes, and the keeping quality of dadih should be further studied, in order to scale up production. Nevertheless, this chapter covers the available aspects of production of dadih.

14.2 Product Description of Dadih

Yogurt-like products are made widely in various regions in the world, including the Mediterranean area, Asia, Africa, and central Europe. Dadih and dahi are Indonesian and Indian yogurts, respectively, made out of buffalo or cow milk that seems to share the same root word. The body and texture of yogurt depends largely on the composition of milk employed in its manufacture, whereas the manufacture of dadih and dahi is simpler than western-type yogurt (Surono and Hosono 2011).

As dadih is made from buffalo milk, it produces thick bodied product due to its higher total solids content as compared to cow milk, as a consequence of higher fat and casein content in buffalo milk (Table 14.1). The higher protein content in buffalo milk results in custard-like consistency at the end of fermentation. In addition, higher fat content enrich the flavor developed in the dadih products. Dadih is

Table 14.1 Composition of Mammals Milks

MAMMALS	PERCENT COMPOSITION OF MAMMALS MILKS					
	FAT	CASEINS	WHEY PROTEINS	LACTOSE	ASH	TOTAL SOLIDS
Cow	3.7	2.8	0.6	4.8	0.7	12.7
Water buffalo	7.4	3.2	0.6	4.8	0.8	17.2
Goat	4.5	2.5	0.4	4.1	0.8	13.2
Sheep	7.4	4.6	0.9	4.8	1.0	19.3
Mare	1.9	1.3	1.2	6.2	0.5	11.2
Sow	6.8	2.8	2.0	5.5	–	18.8

Sources: Fernandes, C.F. et al., Fermented dairy products and health. In: *The Lactic Acid Bacteria: The Lactic Acid Bacteria in Health and Disease*, ed. B.J.B. Wood, 297–339. Elsevier Applied Science, London, 1992; Chandan, R.C. and Shahani, K.M., Other fermented dairy products. In: *Biotechnology, Biotechnology: Enzymes, Biomass, Food and Feed*, Vol. 9., eds. H.-J. Rehm and G. Reed. Wiley-VCH Verlag GmbH, Weinheim, Germany, 1995.

manufactured in bamboo tubes, which is hygroscopic aided in keeping the product from wheying off (Figure 14.1).

The milk composition of various mammals used in yogurt and yogurt-like manufacture is shown in Table 14.1. The nutritive value of water buffalo milk products is higher than cow milk products because of the higher concentrations of protein, fat, lactose, minerals, and vitamins in buffalo milk (Walstra et al. 1999).

Dadih making is carried out by natural fermentation. The buffalo milk was poured in bamboo tubes and kept overnight at room temperature, which stimulate the mesophilic indigenous LAB derived from the fresh raw milk to dominate and grow. Consequently, the fermentation of dadih is much longer than yogurt, 24 and 4 h, respectively, due to different types of LAB involved in the fermentation process at the incubation temperature, 28°C–30°C and 45°C, respectively, besides thicker consistency of dadih.

Hosono et al. (1989) reported that *Leuconostoc paramesenteroides* was the dominant strain of LAB encountered in dadih. Surono and Nurani (2001) found that *Lactobacillus* sp., *Lactococcus* sp., and *Leuconostoc* sp. were dominant in dadih. Surono (2003) reported that among 20 colonies of dadih LAB isolated from Bukittinggi, West Sumatra: five strains were of as *Lactococcus lactis* subsp. *lactis*; three strains of *Lb. brevis*; and three each of *Lb. plantarum*, *Lb. casei*, *Lb. paracasei*, and *Leu. mesenteroides*.

Figure 14.1 Traditional way of dadih making in Padang Panjang West Sumatra using fresh raw buffalo milk in bamboo tube.

The milk is fermented by indigenous LAB of the buffalo milk. Its natural fermentation provides different strains of indigenous lactic bacteria involved in each fermentation (Akuzawa and Surono 2002). The natural indigenous LAB observed in dadih could be derived from the bamboo tubes, buffalo milk, or banana leaves involved in milk fermentation, and buffalo milk has been observed to contribute the most, while bamboo tubes and banana leaves as well as personal hygiene practice may also contribute. Hence, diverse indigenous LAB were reported involved in dadih fermentation due to traditional way of dadih manufacture.

Dadih does not meet any standards such as the national standard for yogurt, Standard National Indonesia 2981:2009, and international standard of yogurt and fermented milk available such as the U.S. Federal Standards of Identity, International Dairy Federation, or Codex Standard for Fermented Milks, which requires pasteurizing the milk, since there is no heat application to the buffalo milk as a raw material of dadih in home industry.

14.3 Nutrition Profile of Dadih

Chemical composition/nutrition profile of dadih may vary from time to time and from place to place of production due to its traditional way of manufacture. However, as dadih is made out of buffalo milk, both protein and fat content are much higher than yogurt made from cow milk, as shown in Table 14.2.

Table 14.2 Nutritional Profile and Microbial Counts of Buffalo Milk Dadih, Dahi, and Cow Milk Yogurt

SAMPLE	PH	TA	PROTEIN (%)	FAT (%)	CARBOHYDRATE (%)	MOISTURE (%)	YEAST COUNTS (CFU/G)
Dadih A[a]	4.1	1.278	5.93	5.42	3.34	84.35	2.0×10^5
Dadih B[a]	4.0	1.32	7.57	6.48	3.79	81.03	2.7×10^9
Dahi[b]	Not detected	0.5–1.1	3.2–3.4	5.0–8.0	4.6–5.2	85–88	1.32×10^8 to 2.46×10^8
Yogurt[c]	3.85	1.49	3.47	3.25	4.66	87.90	1×10^6 to 2×10^6

Sources: [a] Yudoamijoyo, M. et al., *Japan J. Dairy Food Sci.*, 32, A7, 1983.
[b] Gandhi, D.N. and Natrajan, A.M., *Preparation of a Good Quality Dahi (Curd) and Probiotic Milk Products*. http://s3-ap-southeast-.amazonaws.com/jigsydney/general/PDF/49212~Dahi-Making.pdf, 2006.
[c] USDA Agricultural Research Service National Nutrient Data Base for Standard Reference Release 27. http://ndb.nal.usda.gov/ndb/foods/show/105?fgcd=&manu=&lfacet=&format=&count=&max=35& offset=&sort=&qlookup=Yogurt%2C+plain%2C+whole+milk%2C+8+grams+protein+per+8+ounce.
Note: Dadih A: collected from a village near Bukittinggi, West Sumatra. Dadih B: purchased from local market in Padang Panjang, West Sumatra.
TA, titratable acidity (as lactic acid).

Buffalo milk yogurt is reported to have higher contents of protein, total solid, carbohydrate, and ash than those reported for cow milk yogurt (Fundora et al. 2001; Lindmark-Månsson et al. 2003), indicating higher nutrient density in buffalo milk dadih as well as dahi. During the dadih fermentation, bacteria convert milk into curd and predigest milk protein. As a result, the hydrolysis of protein in buffalo milk supports the growth of LAB and provides more nutritious dadih to consumers, which is easily digested as compared to the buffalo milk.

14.4 Important Factors and Physicochemical Changes during Dadih Fermentation

Fresh raw buffalo milk, diverse natural LAB cultures, fermentation temperature, and time are the important factors during spontaneous dadih fermentation. Traditional way of dadih fermentation gives many variations in the characteristics, quality, and acceptability of traditional dadih, inevitable due to the unregulated condition of the spontaneous natural fermentation, while in fermentation technology, the milk, microbes, and the environment are supposed to be controlled properly. Moreover, information about detailed characteristics of bacteria or other microorganisms contributing to these fermentation processes is scanty. A good quality dadih is firm with uniform consistency, creamy-white color, pleasant aroma, and acidic taste. The surface is smooth and glossy, and a cut surface is trim and free from cracks and air bubbles.

14.4.1 Fresh Raw Buffalo Milk

Most of the buffalo milk in West Sumatra and nearby areas is produced in the villages by farmers with small land holdings; by landless agricultural laborers, mostly produced in small quantities of 2–4 l; and by small and marginal farmers in numerous and widely scattered villages. Conditions under which milk is produced in the villages are far from satisfactory, mainly due to the economic backwardness of the producers. Hence, significant portion of the milk is being fermented immediately into traditional dadih due to lack of refrigeration and transportation facilities.

The water buffalo, as a domesticated cattle animal of the bovine subfamily, is fed with natural feed grass, which is free from antibiotics. Hence, no antibiotic residue is in the buffalo milk, which may give inhibitory impact on the growth of the natural starter and cause product failures of dadih, as well as no possible antibiotic resistance of indigenous LAB is in the buffalo milk.

The buffalo milk is not pasteurized, and hygiene practice is not implemented properly during dadih making. Buffalo milk contains higher total solids and has 100% more fat content than cow milk, which makes it creamier and thicker, makes it suitable for the manufacture of traditional dadih, and contributes to the aroma development. The nutritional value of yogurt is similar to the milk from which it is made except for the partial loss of lactose due to fermentation.

14.4.2 Natural LAB

The raw milk of different domestic animals is a natural niche and habitat of lactic bacteria (Surono 1996). The diverse indigenous LAB cultures involved in the spontaneous fermentation play important roles in curd formation. A mixture of natural mesophilic starter cultures derived from fresh unheated buffalo milk, containing specific genera, species, and strains of LAB is involved in the manufacture of dadih developing flavor, body, and texture characteristics of the dadih product. As a result of LAB culture growth, transformation of chemical, physical, microbiological, sensory, nutritional, and physiological attributes in basic milk medium is observed. Traditional dadih might be used for strain hunt so that bacterial repository of wider biological diversity to be used as starter cultures as well as beneficial effect to human health can be established.

Diverse microflora has been seen in dadih. Surono et al. (1983) reported the finding of yeast-like fungi at 1.1×10^7 CFU/g, identified as *Endomyces lactis*, which is commonly found in dairy products. Imai et al. (1987) reported that the major bacterial species responsible for dadih fermentation were *Lb. casei* subsp. *casei* and *Lb. plantarum*. LAB in spontaneous fermented dadih from various parts of West Sumatra have been reported. *Leu. paramesenteroides* was the predominant strain of LAB encountered, responsible for producing aromatic compounds such as diacetyl, acetic acid, and other volatile compounds (Hosono et al. 1989). The lactic acid bacterial count of 4.3×10^8 CFU/g was detected in fresh dadih, dominated by LAB, which was 4.0×10^8 CFU/g. *Leu. paramesenteroides* was the dominant strain of the LAB encountered. *Lc. cremoris*, *Lc. lactis*, *Lb. casei* subsp. *casei*, and *Lb. casei* subsp. *rhamnosus* were also found. Several strains belonging to *E. faecalis* subsp. *liquefaciens* were also found in dadih. This fact indicates that the way of manufacturing dadih did not implement hygiene practices, because those microbes belong to the *Enterococci* group.

Total viable LAB were found in the range of 1.42×10^8 to 3.80×10^8 CFU/g in dadih originated from Bukittinggi and Padang Panjang area of West Sumatra (Surono and Nurani 2001). Table 14.3 shows various LAB cultures involved in the manufacture of commercial yogurt and dadih. Dadih involves mesophilic cultures grown at 25°C–28°C and takes 6–18 h to ferment, whereas yogurt production uses thermophilic cultures, which grow faster at 45°C, and requires only 3–4 h fermentation time. Apart from the huge milk microflora, microbial isolates of dadih have also been reported to exhibit probiotic attributes.

Table 14.3 Lactic Acid Bacteria Involved in the Manufacture of Yogurt and Dadih with Major Function as Lowering Acidity, Curd and Texture Formation, Aroma, and Flavor

PRODUCT	PRIMARY MICROBE(S)	SECONDARY MICROBE(S)	INCUBATION TEMPERATURE AND TIME
Dadih	Lb. plantarum, Lc. lactis subsp. lactis, Lc. lactis subsp. cremoris, Lc. lactis subsp. lactis var. diacetylactis, E. faecium, Lb. brevis	Leu. lactis, Leu. mesenteroides subsp. cremoris	25°C–28°C/12–18 h
Yogurt	Lb. delbrueckii subsp. bulgaricus, Str. thermophilus	Lb. acidophilus, B. longum/bifidum/infantis, Lb. casei/lactis/jugurti/helveticus	43°C–45°C/2.5 h

14.4.3 Indigenous Enzymes

Milk contains a large number of indigenous enzymes, with differing functions, stability to processing, impact on dairy products, and significance for consumer safety (e.g., antimicrobial enzymes). Some enzymes are of interest for their beneficial activity (e.g., lactoperoxidase), some for use as indices of processing (e.g., alkaline phosphatase), and some for effects on the quality of dairy products (e.g., plasmin, lipoprotein lipase), which may be either positive or negative for different products. The indigenous proteolytic enzyme in milk is desirable for cheese ripening, but not for the stability of gel formation in yogurt.

Bovine milk contains several proteases including plasmin, plasminogen, plasminogen activators, thrombin, cathepsin D, acid milk proteases, and amino peptidase. Indigenous enzymes in milk can potentially influence the quality of milk and dairy products. The two sets of enzymes, bacterial enzymes mainly derived from LAB and indigenous native protease, may interact during dadih fermentation. The system is presumably optimized to function properly at body temperature.

Neither thermal breakdown of fresh unheated buffalo milk protein nor denaturation and coagulation of milk albumins and globulins occurred as a consequence of no heat application to buffalo milk prior to dadih making, but the custard-like consistency of dadih is formed. In dadih manufacture, fresh unheated buffalo milk is kept overnight at room temperature, which permits protein degradation catalyzed by bacterial or native proteases.

Thickening, gelation, and coagulation of buffalo milk occurs during spontaneous fermentation of fresh unheated buffalo milk attributed to the proteolytic activity from either milk proteases such as plasmin or proteases of bacterial origin. This fact might be due to strong indigenous proteolytic enzymes derived from the buffalo milk itself or from the indigenous natural LAB in the buffalo milk. The presence of natural indigenous proteolytic enzymes from milk showed more rapid fermentation (Gassem and Frank 1991; Kelly and Fox 2006). Lactic bacteria cultures coincidentally may interfere with the milk enzyme system to their advantage.

The presence of alpha-caseins in freshly drawn milk clearly indicates the presence of active plasmin, and hence the activation of plasminogen, in the udder. During incubation of milk, concomitant processes that either increase (i.e., plasminogen activation) or reduce (autolysis) the activity of plasmin occur. Gassem and Frank (1991) reported that yogurt made from milk pretreated with microbial protease had higher firmness, syneresis, and apparent viscosity than the untreated product. Yogurt made from milk treated with plasmin had significantly lower firmness and apparent viscosity and, after 8 days, lower syneresis as compared to the untreated. Proteolysis of milk accelerates fermentation of yogurt.

The proteolytic system is composed of proteinases, which initially cleaves the milk protein to peptides, cleaved by peptidases into small peptides and amino acids. Apart from lactic streptococcal proteinases, there are several other proteinases from non-lacto-streptococcal origin such as serine type of proteinases, for example, proteinases from *Lb. acidophilus*, *Lb. plantarum*, *Lb. delbrueckii* ssp. *bulgaricus*, *Lb. lactis*, and *Lb. helveticus*. Aminopeptidases are important for the development of flavor in fermented milk products (Law and Haandrikman 1997). Production of volatile compounds (e.g., diacetyl and acetaldehyde) contributes to aroma of dairy products.

14.4.4 Bamboo Tubes as Container of Dadih Product

Bamboo tube is used for fermenting dadih, because of its hygroscopic properties, and its bitter taste gives protection against ants. There are two kinds of bamboo tubes, namely, *bamboo gombong* (*Gigantochloa verticillata*) and *bamboo ater* (*Gigantochloa atter*). The native Minangkabau prefers to use *bamboo gombong*. Azria (1986) reported that microbial load of inner part of bamboo tube was 2.5×10^2 to 1.0×10^3 CFU/cm^2, dominated by acid-producing proteolytic bacteria, which may play an important role in dadih fermentation.

14.4.5 Biochemical Changes during Buffalo Milk Fermentation

Understanding of the transformation of buffalo milk into dadih is necessary to appreciate its nutritional and health properties. The major changes brought about in dadih fermentation process due to LAB resulted in specific health benefits documented in literature.

The changes in the milk constituents during dadih manufacture are related to various steps in the fermentation process. Fermentation of lactose may produce different metabolite products depending on the bacteria involved in the fermentation, but they generally include energy for the bacterial cell to grow. Adenosine triphosphate, either lactic acid and/or carbon dioxide, and some volatile compounds will be produced. When LAB get into milk, they convert the lactose into lactic acid, precipitate proteins in the milk, and form curd. The formation of lactic acid changed the taste to sour in dadih products compared with the buffalo milk. Dadih, like yogurt and other fermented milk products, in particular, is low in lactose because of having been fermented, which means many lactose-intolerant individuals can consume it. Fermentation of lactose to form lactic acid is an important means of preventing, or limiting, milk spoilage due to the growth of contaminating bacteria and their enzyme activity.

14.4.5.1 Carbohydrate Lactose is the major carbohydrate in the buffalo milk. A consortium of LAB, which could be homofermentative and heterofermentative natural starter cultures producing lactic acid, with the involvement of beta-galactosidase from lactic starter cultures, results in coagulation of buffalo milk beginning at pH below 5.0 and completing at 4.6. Texture, body, and acid flavor of dadih owe their origin to lactic acid produced during fermentation.

Small quantities of flavor compounds are generated through carbohydrate catabolism, via volatile fatty acids, ethanol, acetoin, acetic acid, butanone, diacetyl, and acetaldehyde. Homolactic starter cultures in dadih such as *Lactobacilli*, *Lactococci*, *Pediococci*, and *Streptococci* yield lactic acid as 95% of the fermentation output, and lactic acid acts as a preservative. Heterolactic starter cultures, such as *Lb. brevis*, *Lb. fermentum*, and *Leuconostoc* sp., contribute to flavor compounds. There are two important roles of lactic acid in dadih manufacture: (1) helps to destabilize the casein micelles and (2) gives the dadih its distinctive and characteristic sharp acidic taste.

14.4.5.2 Protein The proteolytic system in dadih fermentation is composed of proteinases that initially cleave milk protein into peptides and peptidases that cleave peptides to small peptides and amino acids, and the transport system is responsible for cellular uptake of small peptides and amino acids. LAB have a complex proteolytic system capable of converting milk casein into free amino acids and peptides necessary for their growth. LAB are nutritionally fastidious in nature and require several amino acids and vitamins for their growth. The overall proteolytic system of LAB is very weak, but it is sufficient to permit exponential growth in milk. The indigenous enzymes derived from raw buffalo milk may contribute to make amino acids and peptides available supporting the growth of natural LAB starter cultures.

14.4.5.3 Lipids A weak lipase activity originated from microbial contaminants of milk results is the liberation of minor amounts of free fatty acids and volatile fatty acids. In addition, natural lipases from the raw buffalo milk may contribute to the lipid metabolism. Buffalo milk contains about twice as much butterfat as cow milk and higher amounts of total solids and casein (Table 14.1), resulting in creamy textures and rich flavor profiles of dadih. Unfortunately, research on the lipid metabolism during dadih manufacture is scant.

14.4.5.4 Flavor Compounds Lactic acid, acetaldehyde, acetone, diacetyl, and other carbonyl compounds produced by fermentation constitute key flavor compounds of dadih as a result of carbohydrate fermentation, proteolytic enzyme activities, and lipid metabolisms.

14.4.5.5 Cell Mass During fermentation, natural LAB starter multiply to a count of 10^5 to 10^9 CFU/g (Yudoamijoyo et al. 1983; Hosono et al. 1989; Surono and Nurani 2001) and occupy about 1% volume of dadih product. These LAB cells contain cell walls, enzymes, nucleic acids, cellular proteins, lipids, and carbohydrates. Beta-galactosidase activity contributes a major conversion of lactose into LAB in dadih, which is beneficial for lactase deficient people.

14.6 Probiotic Properties of LAB Isolated from Dadih

The latest discoveries on gut microbiota open doors to the new mechanisms of immunity, intestinal permeability, insulin sensitivity, and weight management. Moreover, prevention and treatment of obesity reduce colon cancer, oral health such as bad breath and tooth decay, skin care such as antiaging and biomoisturizer, and manage psychological stress and the gut/brain relation could be managed by gut microbiome. Including dadih, several dairy products have been reported to consist of probiotic bacteria, which when consumed alive and in an adequate amount confer health benefit to the host (FAO/WHO 2002).

14.6.1 Probiotic Properties of LAB Strains from Dadih

Even though no heat is applied to buffalo milk while manufacturing dadih, and no good hygiene practice is implemented, there is neither product failure nor foodborne disease incidents reported among people who consumed dadih. Instead, the older generation believes that consuming dadih may provide a beneficial effect to their health. This fact has inspired more exploration of the powerful indigenous LAB involved during dadih fermentation, excluding the contaminants and the pathogens from the milk itself as well as environmental surroundings. The selection of new natural probiotics that inhibit or displace a specific pathogen can be used for further assessment, product development, and human clinical interventions on prevention or treatment of infection caused by the pathogen. This would promote human health and have a positive economic impact, especially in developing countries such as Indonesia.

Many criteria have been suggested for the selection of probiotics, including safety, tolerance to gastrointestinal conditions, ability to adhere to the gastrointestinal mucosa, and competitive exclusion of pathogens (Collins et al. 1998; Ouwehand et al. 2002; Collado et al. 2005).

Surono and Nurani (2001) found that *Lactobacillus* sp., *Lactococcus* sp., and *Leuconostoc* sp. were dominant in dadih. Further, to explore the probable probiotic isolates, Surono (2003) screened the 20 colonies of LAB isolated from dadih originated from Bukittinggi, West Sumatra, and 10 strains of dadih lactic bacteria, *Lc. lactis* subsp. *lactis*. *Leu. mesenteroides* and *Lb. casei* each had a moderate survival rate (in the range of 4.83–5.49 log CFU/ml) for 2 h at pH 2.0, while commercial starters such as *Lb. casei* Rolly C, *Str. thermophillus* Rolly T, and *B. breve* Rolly B also had the same range of survival, which was 5.34, 5.41, and 5.33 log CFU/ml, respectively, and reduced two to three log cycles from the initial concentration of viable counts.

Similarly, bile-salt tolerance of testing of dadih lactic bacteria strains *Lc. lactis* subsp. *lactis* exhibited good tolerance. Additionally, one strain of *Lc. lactis* subsp. *lactis* had good survival rate (8.11 log CFU/ml) in the presence of lysozyme after 60-min incubation.

Adhesion to the intestinal mucosa would allow colonization, although transient, of the human intestinal tract (Alander et al. 1999) and has been related to the ability to modulate the immune system, especially during its development (Schiffrin et al. 1997). Hence, adhesion is one of the main selection criteria for new probiotic strains (Havenaar et al. 1992; Salminen et al. 1999).

Collado et al. (2007a) reported that all the five strains of dadih origin showed good adhesion property, and the most adhesive was *Lb. plantarum* strain IS-10506. All LAB strains isolated from dadih fermented milk were able to significantly reduce the adhesion levels of all the pathogens tested. *Lb. plantarum* IS-10506 and *E. faecium* IS-27526 had the highest inhibition abilities. Taken together, *Lb. plantarum* IS-10506 and *E. faecium* IS-27526 have the best adhesion and inhibition properties (Collado et al. 2007a) and show inhibitory, competitive, and displacing properties against pathogens. Hence, they are promising candidates for future probiotics.

Pathogen inhibition by LAB plays an important role to protect against infection through a natural competitive barrier against pathogens in the gastrointestinal tract. Dharmawan et al. (2006) reported that among 10 LAB strains isolated from dadih on human intestinal mucosal surface, there were autoaggregation between *E. faecium* IS-27526 and *E. coli* as well as *S. typhimurium* and *H. pylori*, which might be another mechanism by which probiotics prevent the attachment of pathogens directly on the intestinal surface (Bibiloni et al. 1999; Ouwehand et al. 1999; Tuomola et al. 1999; Canzi et al. 2005). The competence of both strains IS-16183 and IS-7257 (renamed as IS-10506) in competing with *E. coli* was also reported (Collado et al. 2007a).

Aggregation ability is related to cell adherence properties (Vandevoorde et al. 1992; Boris et al. 1997; Del Re et al. 2000). Collado et al. (2007b) reported that all five strains isolated from Indonesian fermented milk product tested show higher percentages of autoaggregation. *Lb. plantarum* IS-10506 and *E. faecium* IS-23427 and IS-27526 presented the highest adhesion to hydrocarbons and also the highest autoaggregation abilities. Further, Surono et al. (2010) reported a significant increase of viable fecal LAB of rats after 3 days of administration with *Lb. plantarum* IS-10506 and *Lb. plantarum* IS-20506 at 1.2×10^{10} to 1.6×10^{10} CFU/day each, by 3.25–3.5 and 0.35–0.65 log cycles, respectively, and they continued the increment after 7 days, by 1.8–2.0 and 2.1–2.3 log cycles, respectively. The abilities of dadih LAB isolates in detoxifying mutagens and cyanobacterial toxins have been reported. The mutagen absorbed and bound to the cell wall, while the cyanobacterial toxin was being metabolized (Hosono et al. 1990; Surono and Hosono 1996; Surono et al. 2008; Surono et al. 2009).

Hosono et al. (1990) reported the ability of 36 LAB strains isolated from dadih to bind mutagenic compounds such as amino acid pyrolysates, namely, 3-amino-1,4-dimethyl-5*H*-pyrido[4,3-b] indole (Trp-P1), 3-amino-1-methyl-5*H*-pyrido[4,3-b] indole (Trp-P2), and 2-amino-6-methyldipyrido[1,2-a:3',2'-d]imidazole (Glu-P1). *Leu. paramesenteroides* R-62 showed the highest binding ability toward Trp-P1, Trp-P2, and Glu-P1 at 99.74%, 99.65%, and 45.2%, respectively. Surono and Hosono (1996) reported the antimutagenic properties of milk cultured with LAB from dadih against mutagenic properties of terasi, an Indonesian condiment, and *Lb. casei* subsp. *casei* R-52 found to have the most. Surono et al. (2009) also reported that *E. faecium* IS-27526 exhibited in vivo antimutagenic property toward Trp-P1.

Microcystins are the main toxins produced by cyanobacteria, cyclic peptides classified as hepatotoxins, and tumor promoters. A provisional guideline level with a limit of 1 μg of microcystin-LR (MC-LR) per liter in drinking water has been established for the protection of human health protection (WHO 1998). Microcystins are chemically stable compounds (Harada 1996; Lahti et al. 1997). Conventional drinking water treatment has only limited efficacy in removing dissolved MC-LR (Svrcek and Smith 2004).

Surono et al. (2008) reported high MC-LR removal by viable *Lb. plantarum* from dadih. *Lb. plantarum* strains IS-10506 and IS-20506 at 8.6×10^{10}–1.2×10^{11} CFU and 7.6×10^{10}–1.6×10^{11} CFU viable cells, respectively, showed MC-LR removal performances at both 22°C and 37°C, but MC-LR removal at 22°C was higher after 30-h incubation, with 75% and 81% of 100 μg/l MC-LR, respectively. Another study (Nybom et al. 2008) reported the highest removal of 95% by *Lb. plantarum* strain IS-20506 (37°C, 10^{11} CFU/ml) with 1%–2% glucose supplementation and 75% in PBS as compared to other probiotic commercial strains tested.

Many other researchers reported hypocholesterolemic activity of dadih. Hosono and Tonooka (1995) reported that *Lc. lactis* subsp. *lactis* biovar. diacetylactis R-43 and R-22 of dadih origin showed high-cholesterol-binding abilities, 33.91% and 29.73%, respectively. Surono (2003) reported that *Lc. lactis* subsp. *lactis* strain IS-10285 and IS-29862 possess taurocholate-deconjugating abilities. Pato et al. (2004) found that rats fed with fermented milk made from *Lc. lactis* subsp. *lactis* strain IS-10285 showed significant ($p < .05$) lower total bile acids in serum.

All these attributes show dadih as potential health-benefitting product.

14.6.2 Novel Probiotics

The aforementioned probiotic properties of several strains isolated from dadih may provide the evidence that how strong the indigenous LAB derived from the fresh raw buffalo milk are in combating the contaminants, both spoil bacteria and pathogens, during the spontaneous fermentation of dadih. The results of research are confirming each other and demonstrating how important molecular identification for probiotic candidate. Moreover, several teams from different labs in Indonesia, Singapore, Japan, and Finland had been involved in dadih research.

Lb. plantarum strain IS-10506 and *E. faecium* had been proved to have in vitro and in vivo probiotic properties.

14.7 Dadih Probiotic for Human Health Promotion

Presently, many dairy products are popular for their therapeutic attributes. Every geographical region have at least one significant dairy product that impart health benefits. Dadih, Indonesian traditional fermented buffalo milk, is believed by the natives to be beneficial for consumers' health. This might be contributed by the probiotic properties exerted by the indigenous LAB present in dadih. Human trials have been conducted to Indonesian subjects based on in vitro and in vivo probiotic properties on two probiotic strains from dadih for their health-promotion properties.

14.7.1 Immune System and Nutritional Status

Several studies on health promotion had been conducted on *Lb. plantarum* IS-10506 originated from dadih as listed in Tables 14.4 and 14.5. Surono et al. (2011) reported a pilot randomized controlled trial study on *E. faecium* IS-27526 isolated from dadih. The supplementation of lyophilized *E. faecium* IS-27526 (2.31×10^8 CFU/day) in 125 ml ultra high temperature low-fat milk for a period of 90 days significantly increased total salivary secretory IgA (sIgA) level and bodyweight of the children ($p < .05$) compared to the placebo. Changes of total salivary sIgA levels were significantly higher in underweight children supplemented with probiotic. Weight gain was observed significantly in children with normal bodyweight supplemented with probiotic. A 90-day randomized doubleblind placebo-controlled pre–post trial has been conducted to four groups of Indonesian children aged 12–24 months, namely, placebo, probiotic, zinc, and combination of probiotic and zinc. Groups of 12 children each were supplemented with microencapsulated *Lb. plantarum* IS-10506 of dadih origin, at 10^{10} CFU/day as probiotic and 20 mg zinc sulfate monohydrate (8 mg zinc elemental) was supplemented as zinc. Blood and stool were collected at baseline and end line. Fecal sIgA was assessed by enzyme-linked immuno assay and serum zinc concentrations by ICP-MS. Fecal sIgA increased significantly in probiotic group, 30.33 ± 3.32 μg/g ($p < .01$), and in probiotic and zinc group, 27.55 ± 2.28 μg/g ($p < .027$), as compared to placebo group, 13.58 ± 2.26 μg/g. Changes of serum zinc concentration in the combination of probiotic and zinc group

Table 14.4 List of In Vitro, In Vivo, and Human Clinical Trials of Scientific-Based Evidences on Probiotic *Lb. plantarum* IS-10506 from Dadih

PROPERTIES	TYPES OF RESEARCH	REFERENCE(S)
Adhesion, colonization of strain IS-10506, increased viable fecal LAB of Balb/c mice, suppression of the allergic reaction by establishing new balance of Th1/Th2	In vivo	Endaryanto (2007)
Adhesion, colonization of strain IS-10506, increased viable fecal LAB of Sprague–Dawley rats	In vivo	Surono et al. (2010)
Adhesion, colonization of strain IS-10506, increased viable fecal LAB of apparently healthy adults	Human clinical trial	Surono et al. (2009)
Adhesion, colonization, co-aggregation, pathogen exclusion of strain IS-10506	In vivo	Dharmawan et al. (2006), Collado et al. (2007a, 2007b)
Combination of strain IS-10506 with zinc, increased fecal sIgA, zinc, and selenium level for 12–24 months of apparently healthy children	Human clinical trial	Surono et al. (2014)
Increased fecal sIgA of children living with HIV received antiretroviral therapy	Human clinical trial	Brahmantya (2013)
Strains IS-10506 and IS-20506 repaired brush border damage induced by LPS *E. coli* by repairing the brush border protein of Wistar rats	In vivo	Ranuh Gunadi Reza (2008)
Strain IS-10506 increased CD4+ cell counts of adult living with HIV strain IS-10506	Human clinical trial	Surono et al. (2010)
Strain IS-10506 modulate immune response of model infection of elderly Wistar rats induced by LPS *E. coli*	In vivo	Kadir (2013)
Strain IS-10506 significantly increased TGF-β and IL-4 of children living with HIV received antiretroviral therapy	Human clinical trial	Radhiah (2013)
Strain IS-10506 significantly reduced blood lipopolysaccharide level in HIV-infected children with antiretroviral therapy		Dwiastuti (2013)

showed the highest elevation at end line. A combination of probiotic *Lb. plantarum* IS-10506 at 10^{10} CFU/day and 8 mg zinc elemental supplementation showed potential ability in improving zinc status of preschool children. Supplementing probiotic *Lb. plantarum* IS-10506 and zinc for 90 days resulted in a significant increase of humoral immune response as well as improved zinc status of the young children (Surono et al. 2014).

Table 14.5 List of In Vitro, In Vivo, and Human Clinical Trials of Scientific-Based Evidences on Probiotic *E. faecium* IS-27526 from Dadih

PROPERTIES	TRIAL	REFERENCE(S)
Adhesion, colonization of strain IS-27526, increased salivary sIgA and bodyweight gain of apparently healthy young children	Human clinical trial	Surono (2011)
Adhesion, colonization of strain IS-27526, increased viable fecal LAB of apparently healthy young children, increased serum IgA	Human clinical trial	Riewpassa (2004)
Adhesion, colonization of strain IS-27526, increased viable fecal LAB of young children, increased fecal sIgA of young children	Human study	Catur Adi (2011)
Adhesion, colonization, co-aggregation, pathogen exclusion of dadih strains	In vitro	Dharmawan et al. (2006), Collado et al. (2007a, 2007b)

14.8 Concluding Remarks

Recent advancement led to investigate the attributes of traditional foods. Dadih is an important traditional food of Indonesia. It consists of many natural microflora that make it nutritious and a healthy product. Dadih contains various LAB microflora with probiotic attributes. However, if the flora of dadih is changed, similarly the composition of dadih also changes. Standard protocol and microflora are yet to be formulated to make this product more significant. Further investigation is needed to explore the complete characteristics of this Indonesian product.

References

Akuzawa, R. and Surono, I.S. 2002. Fermented milks of Asia. In: *Encyclopedia of Dairy Sciences*, eds. H. Roginski, J.W. Fuquay, and P.F. Fox, 1045–1048. London: Academic Press Ltd.

Alander, M., Satokari, R., Korpela, R., Saxelin, M., Vilpponen-Salmela, T., Mattila-Sandholm, T., and von Wright, A. 1999. Persistence of colonization of human colonic mucosa by a probiotic strain, *Lactobacillus rhamnosus* GG, after oral consumption. *Applied and Environmental Microbiology* 65: 351–354.

Azria, D. 1986. *Mikrobiologi Dalam Pembuatan Dadih Susu Sapi* (*Microbiology of Cow Milk Dadih*). Graduate manu. Bogor, Indonesia: Bogor Agricultural University.

Bibiloni, R., Pérez, P.F., and De Antoni, G.L. 1999. Will a high adhering capacity in a probiotic strain guarantee exclusion of pathogens from intestinal epithelia? *Anaerobe* 5: 519–524.

Boris, S., Suarez, J.E., and Barbes, C. 1997. Characterization of the aggregation promoting factor from *Lactobacillus gasseri*, a vaginal isolate. *Journal of Applied Microbiology* 83(4): 413–420.

Brahmantya, H. 2013. *Effect of* Lactobacillus plantarum *IS-10506 Supplementation on Changes of Serum T Cell CD4+ Cells, Rasio T Cells CD4+/CD8+ and secretory IgA Feses Children Treated with Antiretroviral (ARV)*. Pediatric dissertation. Surabaya, Indonesia: Airlangga University.

Canzi, E., Guglielmetti, S., Mora, D., Tamagnini, I., and Parini, C. 2005. Conditions affecting cell surface properties of human intestinal bifidobacteria. *Antonie van Leeuwenhoek* 88: 207–219.

Catur Adi, A. 2011. *Efficacy of Enriched Fish Isolate Protein and Powder of Cat Fish* (Clarias garlepinus), *Biscuit Complementary Feeding and Microencapsulated Probiotic* Enterococcus faecium *IS-27526 on Low Bodyweight Young Children (2–5 Years)*. PhD dissertation. Bogor, Indonesia: Bogor Agricultural University.

Chandan, R.C. and Shahani, K.M. 1995. Other fermented dairy products. In: *Biotechnology: Enzymes, Biomass, Food and Feed*, Vol. 9., eds. H.-J. Rehm and G. Reed. Weinheim, Germany: Wiley-VCH Verlag GmbH.

Collado, M.C., Gueimonde, M., Hernandez, M., Sanz, Y., and Salminen, S. 2005. Adhesion of selected *Bifidobacterium* strains to human intestinal mucus and its role in enteropathogen exclusion. *Journal of Food Protection* 68: 2672–2678.

Collado, M.C., Surono, I.S., Meriluoto, J., and Salminen, S. 2007a. Cell surface properties of indigenous dadih lactic acid bacteria and their interactions with pathogens. *Journal of Food Science* 72: M89–M93.

Collado, M.C., Surono, I.S., Meriluoto, J., and Salminen, S. 2007b. Potential probiotic characteristics of *Lactobacillus* and *Enterococcus* strains isolated from traditional dadih fermented milk against pathogen intestinal colonization. *Journal of Food Protection* 70: 700–705.

Collins, J.K., Thornton, G., and Sullivan, G.O. 1998. Selection of probiotic strains for human application. *International Dairy Journal* 8: 487–490.

Del Re, B., Sgorbati, B., Miglioli, M., and Palenzona, D. 2000. Adhesion, autoaggregation and hydrophobicity of 13 strains of *Bifidobacterium longum*. *Letters in Applied Microbiology* 31: 438–442.

Dharmawan, J., Surono, I.S., and Lee, Y.K. 2006. Adhesion properties of indigenous dadih lactic acid bacteria on human intestinal mucosal surface. *Asian–Australasian Journal of Animal Sciences* 19: 751–755.

Dwiastuti, S. 2013. *Changes of Blood LPS Level of Children Living with HIV and Treated with Antiretroviral (ARV) Therapy After Supplementation with Probiotic* Lactobacillus Plantarum *IS-10506*. Pediatric dissertation. Surabaya, Indonesia: Airlangga University.

Endaryanto, A. 2007. *Immunoegulatory of Th1, Treg and Th2 Through TLR2 and TLR4 by LGG and* L. plantarum *IS-10506 in Suppressing Allergic Reaction*. PhD dissertation. Surabaya, Indonesia: Airlangga University.

Fernandes, C.F., Chandan, R.C., and Shahani, K.M. 1992. Fermented dairy products and health. In: *The Lactic Acid Bacteria in Health and Disease*, ed. B.J.B. Wood, 297–339. London: Elsevier Applied Science.

Food and Health Agricultural Organization of the United Nations and World Health Organization. 2002. *Guidelines for the Evaluation of Probiotics in Food*. Working group report. Washington, DC: Food and Health Agricultural Organization of the United Nations and World Health Organization.

Fundora, O., González, M.E., Lezcano, O., Montejo, A., Pompa, N., and Enriquez, A.V. 2001. A comparative study of milk composition and stability of Murrah river buffaloes and Holstein cows grazing star grass. *Cuban Journal of Agricultural Science* 35: 219–222.

Gandhi, D.N. and Natrajan, A.M. *Preparation of a Good Quality Dahi (Curd) and Probiotic Milk Products*. http://s3-ap-southeast-.amazonaws.com/jigsydney/general/PDF/49212~Dahi-Making.pdf. Accessed on May 29, 2014.

Gassem, M.A. and Frank, J.F. 1991. Physical properties of yogurt made from milk treated with proteolytic enzymes. *Journal of Dairy Science* 74: 1503–1511.

Harada, K.-I. 1996. Chemistry and detection of microcystins. In: *Toxic Microcystis*, eds. M.F. Watanabe, K.-I. Harada, W.W. Carmichael, and H. Fujiki, 103–148. Boca Raton, FL: CRC Press.

Havenaar, R., ten Brink, B., and Huis in't Veld, J.H.J. 1992. Selection of strains for probiotic use. In: *Probiotics: The Scientific Basis*, ed. R. Fuller, 209–224. London: Chapman & Hall.

Hosono, A. and Tono-oka, T. 1995. Binding of cholesterol with lactic acid bacteria cells. *Milchwissenschaft* 50: 556–560.

Hosono, A., Wardoyo, R., and Otani, H. 1989. Microbial flora in "dadih," a traditional fermented milk in Indonesia. *Lebensmittel-Wissenschaft & Technologie* 22: 20–24.

Hosono, A., Wardoyo, R., and Otani, H. 1990. Binding of amino acid pyrolyzates by lactic acid bacteria isolated from dadih. *Lebensmittel-Wissenschaft & Technologie* 23: 149–153.

Imai, K., Tekeuchi, M., Sakane, T., and Ganjar, I. 1987. Bacterial flora in Dadih. *IFO Research Communications* 13: 13–16.

Indian Standard Institution. 1980. *Specification of Dahi: IS: 9617*. New Delhi, India: Bureau of Indian Standards.

Kadir, S. 2013. *The Effect of Probiotic Lactobacillus plantarum IS-10506 Administration on Immune Response (Total CD4+, IL-2, IL-6, IgA) of Elderly in Rodent Model*. PhD dissertation. Surabaya, Indonesia: Post Graduate Faculty of Public Health, Airlangga Univesity.

Kelly, A.L. and Fox, P.F. 2006. Indigenous enzymes in milk: A synopsis of future research requirements. *International Dairy Journal* 16: 707–715.

Lahti, K., Rapala, J., Fardag, M., Maija, N., and Sivonen, K. 1997. Persistence of cyanobacterial hepatoxin, microcystin-LR, in particulate material and dissolved in lake water. *Water Research* 31: 1005–1012.

Law, J. and Haandrikman, A. 1997. Proteolytic enzymes of lactic acid bacteria. *International Dairy Journal* 7: 1–11.

Lindmark-Månsson, H., Fondén, R., and Pettersson, H.-E. 2003. Composition of Swedish dairy milk. *International Dairy Journal* 13: 409–425.

Nybom, S.M., Collado, M.C., Surono, I.S., Salminen, S.J., and Meriluoto, J.A. 2008. Effect of glucose in removal of microcystin-LR by viable commercial probiotic strains and strains isolated from dadih fermented milk. *Journal of Agriculture and Food Chemistry* 56: 3714–20.

Ouwehand, A.C., Isolauri, E., Kirjavainen, P.V., and Salminen, S. 1999. Adhesion of four *Bifidobacterium* strains to human intestinal mucus from subjects in different age groups. *FEMS Microbiology Letter* 172: 61–64.

Ouwehand, A.C., Salminen, S., Tolkko, S., Roberts, P., Ovaska, J., and Salminen, E. 2002. Resected human colonic tissue: New model for characterizing adhesion of lactic acid bacteria. *Clinical and Diagnostic Laboratory Immunology* 9: 184–186.

Pato, U., Surono, I.S., Koesnandar, and Hosono, A. 2004. Hypocholesterolemic effect of indigenous dadih lactic acid bacteria by deconjugation of bile salts. *Asian–Australasian Journal of Animal Sciences* 17: 1741–1745.

Radhiah, S. 2013. *Changes on Immune Response TH1, TH2, TREG DAN TH17 of Children Living with HIV Supplemented with Probiotic* Lactobacillus plantarum *IS-10506*. Pediatric dissertation. Surabaya, Indonesia: Airlangga University.

Ranuh Gunadi Reza, I.G.M. 2008. *Effect of* Lactobacillus plantarum *IS-10506 and IS-20506 on Repairment of Brush Border Damage Induced by LPS* E. coli *by Protein Expression of Brush, Border Galectin-4, Myosin-1a, Occluding Dan ZO-1*. PhD dissertation. Surabaya, Indonesia: Airlangga University.

Riewpassa, F. 2004. *Effect of Fish Protein Concentrate Biscuit and Probiotic Complementary Feeding on Serum IgA and Nutritional Status of Children Younger than Five*. PhD dissertation. Bogor, Indonesia: Bogor Agricultural University.

Salminen, S., Ouwehand, A.C., Benno, Y., and Lee, Y.K. 1999. Probiotics: How should they be defined? *Trends in Food Science and Technology* 10: 107–110.

Schiffrin, E.J., Brassard, D., Servin, A.L., Rochat, F., and Donnet-Hughes, A. 1997. Immune modulation of blood leukocytes in humans by lactic acid bacteria: Criteria for strain selection. *American Journal of Clinical Nutrition* 66: 515–520.

Surono, I.S. 2003. In vitro probiotic properties of indigenous dadih lactic acid bacteria. *Asian–Australasian Journal of Animal Sciences* 16: 726–731.

Surono, I.S., Collado, M.C., Salminen, S., and Meriluoto, J. 2008. Effect of glucose and incubation temperature on metabolically active *Lactobacillus plantarum* from dadih in removing microcystin-LR. *Food and Chemical Toxicology* 46: 502–507.

Surono, I.S. and Hosono, A. 1996. Antimutagenicity of milk cultured with lactic acid bacteria from dadih against mutagenic terasi. *Milchwissenschaft* 51: 493–497.

Surono, I.S. and Hosono, A. 2000. Performance of dadih lactic cultures at low temperature milk application. *Asian–Australasian Journal of Animal Sciences* 13: 495–498.

Surono, I.S. and Hosono, A. 2011. Starter cultures. In: *Encyclopedia of Dairy Sciences*, ed. H. Roginski, J.W. Fuquay, and P.F. Fox, 477–482. London: Academic Press Ltd.

Surono, I.S., Khomsan, A., Sobariah, E., and Nurani, D. 2010. Effect of oxygenated water and probiotic administration on fecal microbiota of rats. *Microbiology Indonesia* 4: 17–21.

Surono, I.S., Martono, P.D., Kameo, S., Suradji, E.W., Koyama, H. 2011. Novel probiotic *Enterococcus faecium* IS-27526 supplementation increased total salivary sIgA level and bodyweight of pre-school children: A pilot study. *Anaerobe* 17 (6): 496–500.

Surono, I.S., Martono, P.D, Kameo, S., Suradji, E.W., and Koyama, H. 2014. Effect of probiotic *L. plantarum* IS-10506 and zinc supplementation on humoral immune response and zinc status of Indonesian pre-school children. *Journal of Trace Elements in Biology and Medicine* 28: 465–469.

Surono, I.S. and Nurani, D. 2001. *Exploration of Indigenous Dadih Lactic Bacteria for Probiotic and Starter Cultures*. Domestic research collaborative grant-URGE-IBRD World Bank Project 2000-2001. Republic of Indonesia: Research Report. Directorate General of Higher Education, Ministry of Education and Culture.

Surono, I.S., Pato, U., Koesnandar, and Hosono, A. 2009. In vivo antimutagenicity of dadih probiotic bacteria towards Trp-P1. *Asian–Australasian Journal of Animal Sciences* 22: 119–123.

Surono, I.S., Saono, J.K.D., Tomomatsu, A., Matsuyama, A., and Hosono, A. 1983. Traditional milk products made from buffalo milk by use of higher plants as coagulants in Indonesia. *Japanese Journal of Dairy and Food Science* 32: A103–A110.

Svrcek, C. and Smith, D.W. 2004. Cyanobacteria toxin and the current state of knowledge on water treatment options: A review. *Journal of Environmental Engineering and Science* 3: 155–185.

Tuomola, E.M., Ouwehand, A.C., and Salminen, S. 1999. The effect of probiotic bacteria on the adhesion of pathogens to human intestinal mucus. *FEMS Immunology and Medical Microbiology* 26: 137–142.

USDA Agricultural Research Service National Nutrient Data Base for Standard Reference Release 27. http://ndb.nal.usda.gov/ndb/foods/show/105?fgcd=&manu=&lfacet=&format=&count=&max=35&offset=&sort=&qlookup=Yogurt%2C+plain%2C+whole+milk%2C+8+grams+protein+per+8+ounce. Accessed on May 7, 2015.

Vandevoorde, L., Christiaens, H., and Verstraete, W. 1992. Prevalence of coaggregation reactions among chicken lactobacilli. *Journal of Applied Bacteriology* 72: 214–219.

Walstra, P., Geurts, T., Noomen, A., Jellema, A., and Van Boekel, M. 1999. *Dairy Technology—Principles of Milk, Properties and Processes*. New York: Marcel Dekker.

WHO. 1998. *Guidelines for Drinking-Water Quality. Second Edition, Addendum to Volume 2, Health Criteria and Other Supporting Information*. Geneva, Switzerland: World Health Organization.

Yudoamijoyo, M., Tirza, Z., Herastuti, S.R., Tomomatsu, A., Matsuyama, A., and Hosono, A. 1983. Chemical composition and microbiological properties of yogurt. *Japanese Journal of Dairy and Food Science* 32: A7.

15
Traditional Fermented Dairy Products of Turkey

OZLEM ISTANBULLU AND BULENT KABAK

Contents

15.1	Introduction	401
15.2	Turkish Yogurts	402
15.3	*Torba* Yogurt	403
15.4	*Kurut*	404
15.5	*Ayran*	406
15.6	*Kefir*	408
15.7	*Koumiss*	408
15.8	Turkish Cheeses	409
	15.8.1 Turkish White Cheese	411
	15.8.2 *Kashar* Cheese	414
	15.8.3 *Tulum* Cheese	416
	15.8.4 *Çökelek* Cheese	418
	15.8.5 *Herby* Cheese	419
References		420

15.1 Introduction

Fermented milk products are produced following the fermentation of milk by specific groups of microorganisms, resulting in a decrease in pH and in the subsequent coagulation of milk proteins, with the microorganisms remaining active as long as they do not undergo heat treatment (Turkish Food Codex 2001). Fermented dairy products have originated in various parts of the world. These traditional products have been made with cow, buffalo, goat, sheep, mare, and camel milks. In the world, there are around 400 fermented milk products with traditional and cultural value. The diversity of such fermented products derives from the heterogeneity of traditions found in the world, cultural preferences, different geographical areas where they are produced, the staple and/or by-products used for fermentation, and the processing methods that are related to the microorganisms involved (Marshall and Mejia 2011). Examples of well-known dairy products include yogurt originating from the Balkan and Middle Eastern countries, *kefir* originating from Caucasus Mountains, *koumiss* originating from Mongolia, and *dahi* and *lassi* originating from India.

Fermented milks constitute an important part of the national diet in Turkey. Yogurt, *ayran*, *cacık*, and traditional cheeses have been made in individual households throughout Anatolia for centuries. The most common groups of microorganisms involved in the traditional dairy fermentations are lactic acid bacteria (LAB), such as the genera *Lactobacillus*, *Lactococcus*, *Streptococcus*, *Leuconostoc*, and *Pediococcus*, and yeasts. Most of these fermented products are produced by either naturally occurring microflora or the "backslopping method," which involves reintroducing part of the previous fermentation as an inoculum into the new fermentation. Thus, the overall microbial profiles and their precise contribution to the fermentation process are not well known. This is the case for many yogurts that are produced by the backslopping method under household conditions, to meet the nutritional needs of villagers in some regions of Turkey (Kabak and Dobson 2011). This review aims to describe the production processes of the most common Turkish traditional fermented milks and to highlight some of the microbiological and biochemical properties of the fermented products.

15.2 Turkish Yogurts

Yogurt has been produced traditionally by the backslopping method in individual households throughout the Middle Eastern countries, Turkey, and Balkan countries for centuries. The raw material for yogurt production is usually cows' milk, although the milk from other mammals, such as buffalo, sheep, and goat, is equally suitable for fermentation (Robinson 2003). The product is obtained through the lactic acid fermentation of thermophilic yogurt culture, which consists of *Str. thermophilus* and *Lb. delbrueckii* subsp. *bulgaricus* in different ratios (Surono and Hosono 2003). These yogurt bacteria metabolize lactose available in the milk to lactic acid and aroma compounds such as acetaldehyde and diacetyl (Beshkova et al. 1998).

Since 1960s, there has been a significant increase in the popularity of yogurt worldwide. The volume of yogurt production is about 1 million kg in 2012 in Turkey, accounting for 13% of the total industrialized cows' milk. In Turkey, there is a legal requirement for both *Str. thermophilus* and *Lb. delbrueckii* subsp. *bulgaricus* to be included in the starter culture of any fermented dairy product that is subsequently labeled as "yogurt." Many different types of yogurt are available in Turkey, for example, *plain* yogurt, *low-fat* yogurt, *nonfat* yogurt, *Turkish-type creamy* yogurt, *filtered* yogurt, *sac* yogurt, *Silivri* yogurt, *Silifke* yogurt, *winter* yogurt, and *salted* yogurt, and these will therefore not be discussed any further at this point.

Yogurt is also used for the manufacture of other products such as *cacık* and *haydari* in Anatolia. *Cacık* is made of yogurt mixed with chopped cucumbers, crushed garlic, and salt; diluted with water to a low consistency; and garnished with mint and/or sumac. Similar products exist in Greece (*tzatziki*) and Balkan countries (*tarator*). In winter months, lettuce is used instead of cucumber for the production of *cacık*, called as *winter cacık*. Haydari is obtained by mixing yogurt or *torba* yogurt with crushed garlic, dill, and salt, and it is served with *rakı*, a traditional high alcoholic beverage.

15.3 *Torba* Yogurt

Torba yogurt (*filtered* yogurt) is a concentrated fermented product, which has been made in the Anatolia region for centuries. Several types of concentrated yogurt including *tulum* yogurt, *winter* yogurt, *kese* yogurt, *süzme*, and *pesküten* have also been produced traditionally. The product is also known as *labneh* (22%–26% total solids) in various Middle Eastern and Balkan countries (Kaaki et al. 2012), *laban zeer* in Egypt, *besa* in Bulgaria, *skyr* in Iceland, and *chakka* and *shrikhand* in India (Nsabimana et al. 2005).

Torba yogurt is made from cow, goat, or sheep milks. Traditionally, yogurt is strained in a special cloth bag for 10–14 h to remove the whey and is then packed and stored at a cool temperature (4°C) (Figure 15.1). Additionally, salt can be added to *torba* yogurt in order to enhance the shelf life of the product (Yaygın 1999). *Torba* yogurt is also used for the manufacture of another product called *kurut* in Anatolia.

Torba yogurt contains about 70%–82% moisture, 22%–30% dry matter, 4.5%–9.2% protein, 6%–16% fat, 4.3%–9.2% lactose, 1%–1.3% ash,

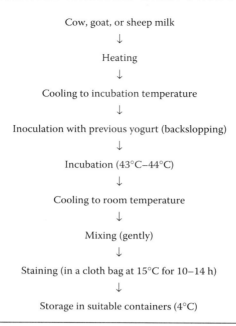

Figure 15.1 Traditional *torba* yogurt manufacture.

1.5%–2.3% lactic acid, and 4.1–4.2 pH (Kırdar and Gün 2001; Tekinşen et al. 2008; Güler and Şanal 2009). The mineral composition of *torba* yogurt is as follows: 1060–1807 mg kg^{-1} calcium, 373–785 mg kg^{-1} sodium, 1527–1894 mg kg^{-1} potassium, 1639–2180 mg kg^{-1} phosphorus, and 425–478 mg kg^{-1} magnesium (Güler and Şanal 2009). About 65% of the normal yogurt is separated as whey. The overall nutrient content is reduced significantly during the preparation of *torba* yogurt, with average losses of sodium, potassium, calcium, and phosphorus of 70.2%, 68.2%, 65.6%, and 50.2%, respectively, being reported. The mean loss of protein during production of *torba* yogurt has been determined to be 7.3% (Nergis and Seçkin 1998). While limited data is available on the microbiology of *torba* yogurt, LAB count in the range of 1.1×10^5–8.5×10^6 CFU g^{-1} has been reported (Hocalar et al. 2004).

15.4 Kurut

Kurut is a sun-dried fermented milk product, traditionally consumed in the East Anatolia of Turkey. *Kurut* is derived from the Turkish word *kurutmak*, which means "dry" (Patır and Ateş 2002). The product has also been known as *keş*, *geşk*, *keşk*, *kask*, *çörten*, *çortan*, *tarak*, and *terne* throughout Anatolia (Atasever 2007; Soltani and Güzeler 2013). There are also similar products with different names such as *labneh* and *kishk* in Middle Eastern countries, *jub-jub* in Syria, and products such as *madeer* in Saudi Arabia (Abd El-Salam 2003; Soltani and Güzeler 2013) and *oggt* in the Arabian Peninsula (Alabdulkarım et al. 2012). The product is usually consumed during the winter months.

Although the method of preparing *kurut* differs slightly in different regions of Turkey, the product is prepared traditionally from low-fat yogurt or *ayran*. The yogurt is poured onto a cloth bag and filtered for 1 day to remove the water. The concentrated yogurt is then poured into a large pot and cut into small pieces with spoons into round, oval, or conical shapes of approximately 30–60 g in weight and around 4–5 cm in diameter. Salt (1%–3%) and cream (5%–10%) are typically added before the shaping process. These shaped pieces are then dried in the sun for 1–2 weeks. The pH of the final product ranges from 4.0 to 4.3. Before serving, *kurut* can be softened by placing it in lukewarm water to give the required consistency, or it can be served after grating and granulating (Kamber 2008a). Typically, around 1 kg of *kurut* can be obtained from 16 to 17 kg of low-fat yogurt. The production of *kurut* has not industrialized yet; however, it is stated that thin-layer dryers or convective-type dryers could be used for the industrial production (Karabulut et al. 2007).

The composition of *kurut* can vary greatly, but typical ranges are 12.1%–13.4% moisture, 25.5%–49.7% protein, 15.5%–45.9% fat, 6.7%–12.9% salt, 9.9%–11.9% ash, and 1.83%–2.91% titratable acidity (Patır and Ateş 2002; Atasever 2007; Kamber 2008a). However, traditional liquid *kashks* contain about 74.5% moisture, 25.5% dry matter, 13.7% protein, 2% fat, 2.5% salt, 2.8% ash, and 4.2 pH (Soltani and Güzeler 2013). *Kurut* has a high nutritional value, with high protein content, and contains minerals such as calcium, potassium, and phosphorus (Patır and Ateş 2002). *Kurut* made from yak milk contains large amounts of calcium (1400 mg kg^{-1}), phosphorus (1460 mg kg^{-1}), magnesium (154 mg kg^{-1}), potassium (1380 mg kg^{-1}), zinc (5.7 mg kg^{-1}), vitamin B_1 (0.3 mg kg^{-1}), vitamin B_2 (6.4 mg kg^{-1}), vitamin B_6 (0.3 mg kg^{-1}), and vitamin C (17.4 mg kg^{-1}) (Zhang et al. 2008). The product can be stored for up to a year at ambient temperature without any deterioration.

LAB levels in the range of 2×10^2–1.9×10^6 CFU g^{-1} have been reported in *kurut* samples, with more than half of the samples analyzed containing large numbers of yeasts/molds (about 10^4 CFU g^{-1}), which can be attributed to post-contamination of *kurut* surfaces (Kamber 2008a). However, *kurut* samples from yak milk in China contain large numbers of LAB (1.5×10^8 CFU g^{-1}) and yeasts (2.1×10^7 CFU g^{-1}) (Zhang et al. 2008).

15.5 Ayran

Ayran is a very popular yogurt-based beverage and a traditional Turkish fermented milk drink, which is consumed mainly during the summer months in Anatolia. The history of *ayran* is based on the Middle East from AD 552 to AD 745, to Göktürks who are the first Turkish people (Codex Alimentarius Commission 2011). *Ayran* has several names and varying composition throughout the Middle Eastern and Balkan countries, *doogh* in Iran (Kiani et al. 2008), *lassi* in southern Asia, and *tahn* in Armenia (Kabak and Dobson 2011). The product is obtained by mixing yogurt with water or obtained from milk with modification of dry matter content by the metabolic activity of symbiotic cultures of *Str. thermophilus* and *Lb. delbrueckii* subsp. *bulgaricus* with or without the addition of salt (maximum concentration of 1%). A flow diagram for the production of *ayran* is represented in Figure 15.2. The pH value of finished product is 4.3–4.4, and the shelf life is between 10 and 15 days (Köksoy and Kılıç 2003).

The total production of *ayran* was 508,444 tons in 2012, accounting for 6.4% of the total industrialized cows' milk (TUIK 2012). Taking into consideration homemade *ayran*, an estimated 1,250,000 tons/year of *ayran* are consumed in Turkey.

The composition of *ayran* varies and is dependent on the type of milk used, the efficiency of fat removal, and the dilution rate. The typical average composition of *ayran* is as follows: 94% water, 5% total solids, 1.2% fat, 1.7% protein, 0.7% salt, 1% ash, and 0.5%–1% titratable acidity (Yaygın 1999). The product can be also divided into three categories called "whole-fat *ayran*" (milk fat $\geq$ 1.8%), "low-fat *ayran*" (1.2% > milk fat $\geq$ 1.8%), and "nonfat *ayran*" (milk fat $\leq$ 0.5%).

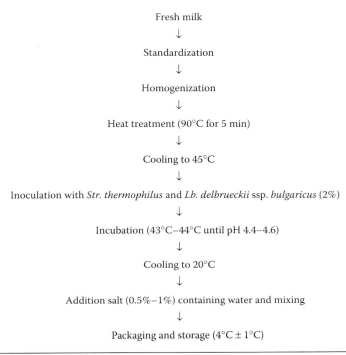

Figure 15.2 Manufacturing steps for *ayran* production.

The product has viable yogurt bacteria count of $>10^6$ CFU ml^{-1} during the storage period (Özünlü 2005), which is normally sufficient to be effective as a human dietary adjunct.

While the main flavor compound of *ayran* is acetaldehyde, yogurt bacteria produce lactic acid, together with other flavor components such as acetone, acetoin, diacetyl, pyruvic acid, formic acid, and CO_2 (Tamime and Robinson 1999). Like other acidic milk beverages, *ayran* is prone to textural instability during storage due to its low pH. The main textural defects, which have been reported, are low viscosity and serum separation (up to 30%). Serum separation can be prevented, however, with the addition of hydrocolloid stabilizers such as pectin, guar gum, and gelatin (Köksoy and Kılıç 2004).

Because of the growing trend of *ayran* production and trade, the Codex Alimentarius Commission decided to elaborate a standard for *ayran* production in order to ensure food safety, essential quality, hygiene, and labeling requirements for the purposes of protecting the health of the consumers and ensuring fair practices in food trade. The regional standard for *ayran* will be adopted in the Codex Alimentarius Commission on July 2015.

15.6 *Kefir*

Kefir is an acidic and mildly alcoholic fermented milk beverage from the Caucasus Mountains (Liu and Lin 2000; Fontán et al. 2006). The word *kefir* originates from the Turkish word *keyif*, which means "good feeling." The Caucasian people discovered that the fresh milk carried in leather pouches would occasionally ferment to produce an effervescent beverage (Irigoyen et al. 2005). While it has been widely consumed in Soviet countries for centuries, it has become increasingly popular in European countries, Japan, and the United States due to its health-promoting effects. Although less well known than yogurt, compositional analysis of *kefir* indicates that it may contain bioactive ingredients that give it unique health benefits, establishing it as a potentially important probiotic product (Farnworth 1999).

Kefir differs from other fermented milk products in that the milk is produced as a result of a fermentation involving a mixed microflora confined to a matrix of discrete *kefir grains*, which are recovered for subsequent fermentation (Marshall and Cole 1985). Several homofermantative lactobacilli including *Lb. kefiranofaciens* and *Lb. kefir* are known to produce this polysaccharide (Toba et al. 1987). The microbiological and chemical properties of *kefir* are well described by Kabak and Dobson (2011).

15.7 *Koumiss*

Koumiss is also known as *kımız*, *airag*, *kumys*, and *kumis* in Turkey, Mongolia, Kazakhstan, Kyrgyzstan, and in some regions of Russia (Danova et al. 2005). *Koumiss* is milky-gray in color, lightly and naturally carbonated, and traditionally made from mare's milk. The product has a sharp alcoholic and acidic taste. The fermentation process and the type of product obtained is similar to *kefir* but is produced from a liquid starter culture in contrast to the solid *kefir grains*. The main fermentation metabolites are lactic acid

(0.7%–1.8%), ethanol (0.6%–2.5%), and CO_2 (0.5%–0.9%) (Tamime et al. 1999). The alcohol content in *koumiss* is slightly higher than in *kefir*, because the lactose content in mares' milk is higher than in cows' milk, which is fermented into *kefir* (Kabak and Dobson 2011).

Three main groups of microorganisms have been identified in *koumiss* fermentation, lactobacilli such as *Lb. delbrueckii* subsp. *bulgaricus* (Akuzawa and Surono 2003), *Lb. salivarus*, *Lb. buchneri* (Danova et al. 2005), *Lb. plantarum* (Danova et al. 2005; Wang et al. 2008), *Lb. casei* (Wang et al. 2008; Ya et al. 2008), *Lb. helveticus*, and *Lb. fermentum* (Wang et al. 2008); lactose-fermenting yeasts such as *Saccharomyces lactis*, *Torula koumiss* (Tamime et al. 1999), and *Kluyveromyces lactis* (Akuzawa and Surono 2003); and non-lactose-fermenting yeast such as *S. unisporus* (Montanari and Grazia 1997).

15.8 Turkish Cheeses

Cheese plays an important role in the diet of people in Anatolia. It is believed that cheese evolved in the Fertile Crescent between the Tigris and Euphrates rivers, in Mesopotamia, some 8000 years ago (Fox 1993). This area now forms a part of Iraq, Iran, Turkey, and Syria. The word *peynir* or *penir*, which means cheese, is encountered for the first time in Mamluk culture, and Turks learned this word from Persian at the time of the migration of the Turks from Middle Asia to Anatolia (Kamber 2008b). The Turkish Statistical Institute reported that cheese production in Turkey was 564,191 tons in 2012 (TUIK 2012).

Cheeses are not only consumed at breakfast but also used as an ingredient for dishes such as *pide*, *börek*, *kuymak*, *pasta*, and *salads* or in desserts such as *künefe*, *kemalpaşa*, and *hoşmerim*. Cheese is manufactured mainly from cows' milk. Sheep, goat, and buffalo milks are used only for the production of some pickled cheeses. There are more than 100 types of cheeses in Turkey. However, when they are grouped according to their similarities, there are around 30 different kinds of cheese, which are manufactured in different parts of Turkey. Table 15.1 lists different kinds of cheeses produced in seven geographic regions of Turkey.

Table 15.1 Traditional Cheese Varieties Made in Different Regions of Turkey

REGIONS OF TURKEY	NAME OF TRADITIONAL CHEESES
Aegean	*Afyon tulum* cheese, *Armola* cheese, *Aydın cıvık* cheese, *Aydın çiğ kesik* cheese, *Aydın tulum* cheese, *Izmir tulum* cheese, *Kopanisti* cheese, *Milas Kırk tokmak* cheese, *Kuru çökelek* cheese, *Posa* cheese, *Karaburun lor* cheese, *Kuru ezme* cheese, *Sepet* cheese, and *Tire çamur* cheese
Black Sea	*Abaza* cheese, *Aho* cheese, *Artvin anzlat* cheese, *Artvin eridik* cheese, *Artvin imansız* cheese, *Artvin yümme* cheese, *Cabaaltı çökelek* cheese, *Civil* cheese, *Giresun uzayan* cheese, *Golot* cheese, *Kadina* cheese, *Kargı tulum* cheese, *Karin kaymağı* cheese, *Kars lor* cheese, *Kes* cheese, *Koleta* cheese, *Külek* cheese, *Küp çökelek* cheese, *Kurçi* cheese, *Minzi* cheese, *Ogma* cheese, *Samsun çiğ kesik* cheese, *Tel* cheese, *Tonya kashar* cheese, *Yayla* cheese, and *Yusufeli molded* cheese
Central Anatolia	*Cihanbeyli molded* cheese, *Divle tulum* cheese, *Ereğli cloth tulum* cheese, *Gölbaşı tulum* cheese, *Kayseri küp* cheese, *Kesmük* cheese, *Kızılcıhamam çömlek* cheese, *Konya küp* cheese, *Konya molded* cheese, *Küpecik* cheese, *Niğde gödelek* cheese, *Örgü* cheese, and *Yozgat çanak* cheese
Eastern Anatolia	*Bitlis küp* cheese, *Civil* cheese, *Dövme* cheese, *Erzincan şavak (tulum)* cheese, *Erzurum tel* cheese, *Göçer (nomad)* cheese, *Karin kaymağı* cheese, *Kars çökelek* cheese, *Kars gravyer* cheese, *Kars kashar* cheese, *Kerti lor* cheese, *Malatya çökelek* cheese, *Motal* cheese, *Tomas (serto)* cheese, and *Van Herby* cheese
Marmara	*Abaza* cheese, *Ayvalık Kirli hanım* cheese, *Bursa dil* cheese, *Çerkez* cheese, *Edremit sepet* cheese, *Ezine* cheese, *Mengen* cheese, *Mihaliç (Kelle)* cheese, *Sirvatka lor* cheese, and *Trakya kashar* cheese
Mediterranean	*Antalya lor* cheese, *Çimi tulum* cheese, *Çoban tulum* cheese, *Dolaz (tort)* cheese, *Hatay künefe* cheese, *Hatay sünme* cheese, *Hatay testi (cara)* cheese, *Hatay tulum çökelek*, *Hellim* cheese, *Maraş parmak (finger)* cheese, *Sürk* cheese, *Sütçüler tortu (ekşimik)* cheese, and *Yalvaç küp* cheese
Southeastern Anatolia	*Ergani* cheese, *Herby* cheese, *Örgü* cheese, *Sıkma* cheese, *Urfa topak* cheese, and *Urfa white* cheese

The quality and characteristics of cheese vary according to the raw milk even for the same type of cheese. The cow's milk is used in every region; however, goat's milk or sheep's milk is also used either alone or in a mixture depending on the season and availability. Moreover, cheese-making process may vary in each region, which has come along traditionally throughout time. For instance, some traditional cheeses such as *yörük* cheese can be manufactured both brined or *tulum* type (Kamber 2008b). On the other hand, specific herbs are added to cheese in some provinces of Turkey. As the seasonal and geographical conditions affect the composition of milk obtained, the nutritious values are highly correlated with the feeding options of the animals. Therefore, the cheeses produced in every region have a specific characteristic and aroma together with the manufacturing methods, especially the technique used, makes the cheese unique in addition to its flavor, for example, *tulum* cheese, which has a specific ripening method and packing material.

While a few cheese varieties are produced in large-scale cheese factories, most of the types are still produced traditionally by women at home in villages or in small-scale cheese units. The most popular types of cheeses manufactured and consumed commonly in Turkey are Turkish white cheese (white pickled cheese), *kashar* cheese, *tulum* cheese, *çökelek* cheese, and herby cheese.

15.8.1 Turkish White Cheese

Turkish white cheese is a semihard cheese ripened in brine (~12–14 g/100 g NaCl solution). It has a smooth, wet coat, close texture (without gas holes) with a salty acidic taste. The cheeses are cubical or rectangular, typically $7 \times 7 \times 7$ or $7 \times 7 \times 10$ cm^3, weighing about 350 or 500 g. In general, Turkish white cheese is allowed to age for at least 1 month before its consumption. Even though Turkish white cheese has a soft texture when fresh, after ripening for 3 months in brine, it can be classified as a semihard or semisoft variety (Hayaloglu et al. 2002).

Cheeses similar to Turkish white cheese are widespread throughout the Mediterranean, European, American, Middle Eastern, and Balkan countries and are known as *sirene* in Bulgaria, *telamme* in Romania, *feta* in Greece, *primonski sir* and *grobnicki sir* in former Yugoslavia, *brinsa* in Israel, *domiati* in Egypt, and *queso blanco* in the United States (Kamber 2008b).

According to the Turkish Standard Institute, the product is divided into four categories called as "full-fat cheese," "semifat cheese," "low-fat cheese," and "nonfat cheese," depending on the percent fat-in-dry matter (Turkish Standard Institute 1995):

- Full-fat cheese, containing at least 45% fat-in-dry matter
- Semifat cheese, containing 30%–44% fat-in-dry matter
- Low-fat cheese, containing at least 20%–29% fat-in-dry matter
- Nonfat cheese, containing <20% fat-in-dry matter

Turkish white cheese can be made from any kind of milk, including cow, goat, and ewe milk, and from their combinations (Kocak et al. 2005). Figure 15.3 is a flow diagram for the production of Turkish white cheese. The raw milk is clarified, standardized for the adjustment of casein-to-fat ratio, and then heated at 60°C–75°C for 5–10 min. However, raw milk can be used without any heat treatment in farms. After that, the milk is cooled to 32°C, transferred from the milking vessel to cheese vats, and inoculated with

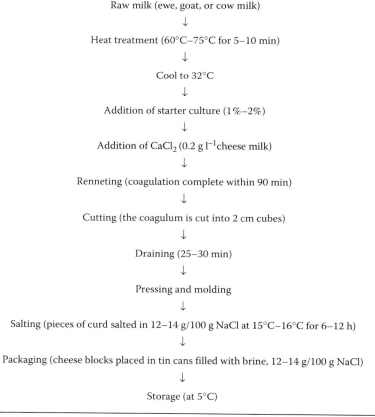

Figure 15.3 Turkish white cheese production.

1%–2% starter culture and $CaCl_2$ at a level of 0.2 g l^{-1} cheese milk. In the nonindustrial production process, starter cultures are generally not used, and therefore, acid production by naturally occurring microorganisms is likely to be rather variable. The inoculated milk is held for 30 min, and liquid rennet is added (~10 g kg^{-1} cheese milk) to obtain a firm coagulum in 90 min. The milk starts to form a solid jelly-like mass after 30–45 min, and the gel has formed after 75–90 min. The coagulum (curd) is cut into cubes (~2 cm) with special curd-cutting knives and allowed to rest in the whey for 5–10 min. The curds are transferred to square-shaped stainless steel or wooden molds with cheese clothes. Pressure is then applied at room temperature for about 4 h or until whey drainage has stopped or decreased to a low level by placing weights on top of the molds. After that, the weights are removed, the cheese cloth is opened, and the cheese mass is divided with cutting knives into $7 \times 7 \times 7$ or $7 \times 7 \times 10$ cm^3 pieces, weighing 350–500 g. The curd pieces are placed in brine (12–14 g/100 g NaCl) at 15°C–16°C for 6–12 h. The brined cheese pieces are placed in cans of 1, 2, 3, 5, or 18 kg and filled with brine, and the containers are closed. The cheese is ripened in the cans at 12°C–15°C for 60–120 days and is then ready for consumption (Hayaloglu et al. 2002).

The chemical composition of Turkish white cheese varies widely as follows: 35%–48% total solids, 20%–50% fat-in-dry matter, 6%–22% fat, 13%–38% protein, 2%–7.2% salt, 3.6%–6.6% ash, and 0.55%–3.8% lactic acid (Hayaloglu et al. 2002; Kamber 2008b). These differences are due to several factors, namely, the differences in methods of manufacture, chemical composition of raw material, and ripening period. In Turkish white cheese, the mineral compositions vary from 637 to 908, 430 to 807, 290 to 1511, and 25 to 40 mg per 100 g for calcium, phosphorus, sodium, and magnesium, respectively (Demirci 1988; Ayar et al. 2006). The yield of fresh Turkish white cheese (kg cheese/100 l raw material) made from sheep milk is 26%–28% while that from goat or cow milk is 15%–16% (Ucuncu 1999).

While starter cultures are generally not used in the traditional manufacturing of Turkish white cheese, it has been recently commercially introduced in the market. Cheese starters are used primarily to convert lactose to lactic acid, which would affect several aspects of cheese manufacture such as coagulant activity, retention of coagulant in the curds, rate of proteolysis during storage, cheese yield, cheese moisture, and rate of pH decline in the cheese (Pappas et al. 1996). Several thermophilic and/or mesophilic bacteria are used individually or in combination in the production of Turkish white cheese, including *Lactococcus lactis* subsp. *lactis*, *Lc. lactis* subsp. *cremoris* (Uysal 1996; Gursoy et al. 2001; Kayagil and Candan 2009), *Lb. casei* (Yildiz et al. 1989), *Lb. brevis*, *Lb. paracasei*, *Lb. helveticus*, and *Lb. delbrueckii* subsp. *bulgaricus* (Gursoy et al. 2001). The combinations of four coccus-shaped strains (*Lc. lactis* subsp. *lactis* and *Lc. lactis* subsp. *cremoris* as acid producer and *Lc. lactis* subsp. *diacetylactis* and *Leu. mesenteroides* subsp. *cremoris* as aroma producer) have been suggested by European firms producing commercial starter cultures (Kayagil and Candan 2009).

Salt is not only a source of dietary sodium but also an integral part of Turkish white cheese process. Salt plays a major role in suppressing the proliferation of unwanted microorganisms, including pathogens and regulating the growth of desirable organisms, including the LAB. In addition, salt is a vital part of Turkish white cheese process as it controls moisture, chemical changes, physical changes in cheese proteins that influence cheese texture, protein solubility and protein conformation, and flavor development (Pappas et al. 1996; Sutherland 2003).

15.8.2 Kashar *Cheese*

The word *kashar* is derived from the Hebrew word *kaşerde*, which means "approved for consumption by the rabbi, permissible, able to eaten freely." According to legend, *kashar* cheese was made first by a Jewish girl in Thessaloniki (Inal 1990; Kamber 2008b).

Kashar cheese is the second most consumed cheese in Turkey, with annual production of around 80,000 tons after Turkish white cheese. The cheese is semihard or hard, depending on maturation period. It is also known as *kashkaval*, *kasseri*, and *caciocavallo* in Bulgaria, Greece, and Italy, respectively (Güler et al. 2004; Kurultay et al. 2004). The most famous *kashar* cheeses are made traditionally in Kars, Kırklareli, Edirne, Kocaeli, Muş, Trabzon, Kadırga, Bayburt, and Tonya provinces of Turkey (Kamber 2008b). The product is divided into two categories as "fresh *kashar* cheese" and "old or matured *kashar* cheese" after ripening according to the Turkish Standards (Turkish Standard Institute 1999). The weight of cheese varies from 250 g to 2–3 kg and from 5 to 25 kg for fresh *kashar* and matured *kashar* cheeses, respectively. The ripening period could be of 1–2 weeks at 12°C–16°C and 3–10 months at 2°C–3°C to reach its characteristic structure and flavor (Kinik et al. 2005).

Traditionally, *kashar* cheese is made from ewe milk. However, cow milk is commonly used in its production in recent years (Kamber 2008b). For the commercial production of *kashar* cheese, first the fat content of milk is standardized to 2.5%–3% for full-fat and 0.6%–0.7% for low-fat cheese production. Milk is pasteurized at 65–68 for 10 min and then cooled to 33°C. Thermophilic starter cultures are added at a rate of 0.7%–1% (w/v) as well as 0.02% $CaCl_2$ (w/v) and held for 30 min until the pH reached 6.3 for starter maturation. Then the liquid calf rennet is added in order to coagulate the milk within 45 min. After that, the coagulum is cut into cubes of 1 cm^3 with a curd knife and left to settle for around 10 min. The temperature of curd is slowly raised to 40°C. At the end of the cooking, one-third of the whey is drained from each batch. The cheese curd is fermented until its pH reaches 5.2–5.3, and the remaining whey is drained. After fermentation, the cheese portions are boiled in 5%–6% brine at 75°C–80°C for 2 min and dried at 15°C for 3 days. The blocks of cheeses are vacuum and shrink packaged and stored at 4°C–6°C for 3 months (Var et al. 2004). Figure 15.4 shows a schematic representation of *kashar* cheese manufacturing process. The yield of matured *kashar* cheese is 18%.

The chemical composition of *kashar* cheese can vary greatly, but typical ranges are 35%–55% moisture, 21%–34% protein, 7%–34% fat, 3%–5% NaCl, 3%–5% ash, and 1%–1.6% titratable acidity (Inal 1990; Var et al. 2004; Kamber 2008b). The mineral composition of the product is as follows: 965–2254 mg kg^{-1} calcium, 450–2855 mg kg^{-1} potassium, 1164–9122 mg kg^{-1} sodium, 65–210 mg kg^{-1} magnesium, 0.17–3.75 mg kg^{-1} iron, 8–21 mg kg^{-1} zinc, and 0.04–0.39 mg kg^{-1} manganese.

LAB play an important role in the development of required characteristics of *kashar* cheese. A mixture of *Lc. lactis* and *Str. cremoris* (Çağlar and Çakmakçı 1998; Tunçtürk et al. 2010) has been used as a starter for the production of *kashar* cheese. Also, different combinations of *Str. thermophilus* and *Lb. bulgaricus* (Şengül et al. 2010) can be added to the milk. LAB counts in *kashar* cheese fall in the range of 3×10^5–7×10^7 CFU g^{-1} during the 90 days of maturation (Var et al. 2004).

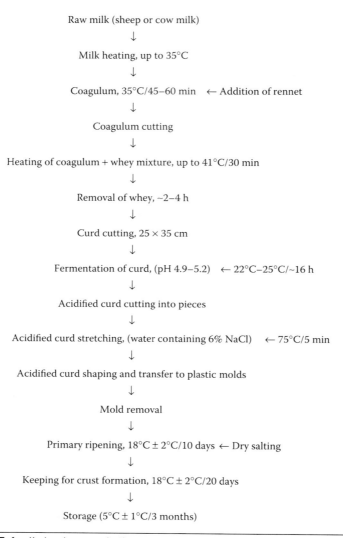

Figure 15.4 *Kashar* cheese production.

15.8.3 Tulum *Cheese*

Tulum cheese, the third largest consumed cheese in Turkey after Turkish white cheese and *kashar* cheese, is a traditional Turkish goat milk cheese. The cheese is manufactured with different production methods throughout the country, except Thrace region, and is called by various names. The most famous *tulum* cheeses are *Erzincan tulum* in the Eastern Anatolian region, *Izmir tulum* in the Aegean region, *Divle tulum* in the central

Anatolian region, *Kargı tulum* in the northern Anatolian/central Black Sea region, and *Çimi tulum* in the Mediterranean region.

The product is a hard cheese and ripened up to 90–100 days in traditional goatskin (rarely sheepskin) casing called *tulum* (Kinik et al. 2005). However, the growing difficulty and expenses involved in procuring skills led to an increasing commercial use of plastic and can containers (Kinik et al. 2005; Bayar and Ozrenk 2011). It has a dry rind with a homogeneous body without eyeholes (FAO 1990).

Traditionally, the product is made from either goat or ewe milk, but now cow milk can also be used for the commercial production of *tulum* cheese. The production of *tulum* cheese is as follows (Inal 1990; Kocak et al. 2005; Kamber 2008b):

- Standardized milk is heated to 35°C–40°C for inoculation of rennet.
- The milk is inoculated with 1%–2% liquid rennet (1% of yogurt can also be added as starter) and coagulation is obtained within 60–150 min.
- After cutting the coagulum (0.5–3 cm size pieces), it is heated to 50°C–60°C for 10–15 min, and the curd is transferred to cheese bags with a ladle and hung up for draining for 30 min at room temperature.
- The coagulum is pressed until the curd reaches the desired moisture.
- Subsequently, it is broken into small pieces (5 cm diameter) with hand to be dry-salted with 2.5% salt.
- Salted curd is tightly stuffed in a goatskin/sheepskin casing and pressed.
- The casing opening is sewn, and the needle is inserted into the skin for 40–50 times.
- *Tulum* cheese is ripened under anaerobic conditions at 6°C–8°C for 90–120 days.
- The weights of *tulum* cheese vary from 40 to 70 kg.

As there is no standard manufacturing method for the product, differences can be expected in the composition and appearance of *tulum* cheese depending on the processing methods used. The composition of *tulum* cheese varies within the following ranges: 30%–50% moisture, 50%–70% dry matter, 6%–55% fat, 18%–26% protein, 4%–8%

ash, 3%–5% salt, and 1%–3% titratable acidity (Akyüz 1981; Inal 1990). *Kargı tulum* cheese is produced mainly in Kargı, Corum province, and contains about 65% dry matter, 20% fat, 21% protein, 3.7% salt, 0.6% lactic acid, and 2.7% total free fatty acids (Dinkçi et al. 2012).

While differences can be found in the flora of *tulum* cheeses throughout the season, LAB, *Lc. lactis* subsp. *lactis* (Tuncer et al. 2008), *Lc. lactis* subsp. *cremoris*, *Lb. casei* subsp. *casei*, *Lb. plantarum*, *Lb. curvatus*, *Leu. mesenteroides* subsp. *cremoris*, *Leu. mesenteroides* subsp. *dextranicum*, *Leu. lactis* (Ateş and Patır 2001; Öksüztepe et al. 2005), *E. faecium*, *E. faecalis*, *Str. lactis*, *Str. cremoris* (Bostan et al. 1992), and *Pediococcus* ssp. (Öksüztepe et al. 2005) are found in *tulum* cheeses. Viable counts of 2×10^7 CFU g^{-1} and 2.5×10^7 CFU g^1 for streptococci and lactobacilli have been found in *Kargı tulum* cheese, respectively (Dinkçi et al. 2012).

15.8.4 Çökelek *Cheese*

Çökelek cheese or shortened *çökelek*, which resembles *lor* cheese, is a traditional popular acidic cheese made of yogurt or *ayran*. It is manufactured mainly by boiling of set yogurt made from cow's milk and then straining in a special cloth bag (Simsek and Sagdıc 2010). The method of making *çökelek* differs slightly in the different provinces of Anatolia (Yerlikaya et al. 2011). A list of major *çökelek* types is shown in Table 15.2. *Çökelek* is usually made from cow's milk, and its production involves the following (Kırdar 2005):

- Standardization of fat content (1.5%)
- Heat treatment at 90°C for 10–15 min
- Cooling to 42°C–45°C
- Incubation with 1.5%–3% yogurt bacteria
- Fermentation for 2–3 days until souring occurs
- Dilution with drinkable water, stirring and churning for separation of butter granules
- Heat treatment at 85°C–90°C for 5 min, cooling and pouring into a cheese cloth bag
- Addition of 4%–5% salt and draining at room temperature for 24 h

Table 15.2 Major Classes of *Çökelek* Cheese

NAME AND STYLES	PRODUCTION PROVINCES	PRINCIPLE OF MANUFACTURING
Cobaltı or *cabaaltı* çökelek cheese	Kastamonu/Inebolu	It is made from first milk (colostrum) that ewe produces after lambing
Tortu or *ekşimik*	Isparta/Sütçüler	It is prepared by adding milk to the whey of yogurt made from ewe's or goat's milk
Kurçi or *kurçta*	Bayburt, Rize, Erzurum	It is prepared by coagulating the *ayran*
Giresun çökelek cheese	Giresun	It is made from *ayran*
Kars çökelek cheese	Kars	It is prepared by using buttermilk
Milas kırk tokmak cheese	Muğla/Milas	It is made from buttermilk left over from the churning process
Kelle çökelek cheese	Denizli/Çal, Çivril and Bekilli	It is made from pasteurized skim milk. Curd is heated and filtered. Obtained cheese is mixed with salted water prepared with eggs. It is consumed after rinsing
Kuru çökelek cheese	Izmir, Aydın	It is made of salted whey and/or cheese mixed with black cumin seeds. It is ripened in *tulum*
Minzi or *Minci*	Trabzon, Rize, Artvin	It is made from yogurt

Çökelek has the following composition: 60%–80% moisture, 20%–40% dry matter, 1%–5% fat, 8% protein, 1%–6% salt, 1%–4% ash, 1%–2% lactic acid, and 3.5–4 pH (Kırdar 2005; Öksüztepe et al. 2007; Kamber 2008b; Simsek and Sagdic 2010). Yogurt bacteria are predominant microorganisms in *çökelek* as it is mostly made from yogurt or *ayran*. The count of heterofermentative LAB in *çökelek* varies from 8.8×10^6 to 1×10^7 CFU g^{-1} (Önganer and Kırbağ 2009).

15.8.5 Herby Cheese

Herby cheese, also known as Van Herby cheese, is another popular semihard Turkish cheese mainly produced in eastern and southeastern parts of Turkey and characterized by a typical flavor and aroma. It is a dry-salted cheese, traditionally made from sheep milk with the addition of aromatic herbs. It can be also made from cow, goat, or buffalo milks (Tarakçi et al. 2004).

More than 20 types of herbs, including *Allium*, *Thymus*, *Silene*, *Ferula*, *Anethum*, and *Mentha* species, have been used individually or as appropriate mixtures. Herbs are harvested in their vegetative period of the springtime and usually added into the cheeses in two ways. One way is the addition of the fresh herbs after washing and slicing. The other way is to make pickle by adding brine (16% salt) to the washed and sliced herbs. About 20 days later, the herbs are ready to be added into the cheese. They can also be stored for a long time and used when it is needed (Kavaz et al. 2013).

Traditionally, the milk is coagulated with calf rennet at the milking temperature. After cutting the coagulum, the whey is removed, and the previously prepared herbs (0.5–2 kg per curd obtained from 100 l of milk) are added into the curd. Subsequently, the curd is pressed for about 3 h to remove the remaining whey and cut into blocks. After that, cheese blocks are ripened in brine or dry-salted and placed in plastic containers. All containers filled with cheese are turned upside down and kept in a cool place or placed in the soil to allow the moisture to come out. In this position, the bottom of the container is always left open to enable the moisture loss. This cheese is ripened for 3 months to get the desired taste and flavor (Tarakçi et al. 2004).

Herby cheese contains 37%–58% dry matter, 17%–24% fat, 19%–21% protein, 3.7%–7.3% salt, 0.48%–1.84% titratable acidity, and 4.3–5.6 pH (Çoşkun and Özturk 2002; Tarakçi et al. 2004; Işleyici and Akyüz 2009). LAB isolated from Van Herby cheese include *Lb. plantarum*, *Lb. pseudoplantarum*, *Lb. casei*, *Lb. tolerans*, *Lb. delbrueckii* subsp. *bulgaricus*, *Lb. delbrueckii* subsp. *lactis*, *Lb. delbrueckii* subsp. *delbrueckii*, *Lb. helveticus*, *Lb. fermentum*, *Lb. büchnerii*, *Lb. brevis*, *Lb. rhamnosus*, *Leuconostoc* ssp., *Pediococcus* ssp., *E. faecalis*, *E. faecium*, *Lc. lactis* subsp. *lactis*, *Lc. lactis* subsp. *lactis* biovar. *diacetylactis*, and *Lc. lactis* subsp. *cremoris* (Işleyici and Akyüz 2009).

References

Abd El-Salam, M. H. 2003. Fermented milks: Middle East. In: *Encyclopedia of Dairy Sciences*, eds. H. Roginski, J. W. Fuquay, and P. F. Fox, 1041–1045. London: Academic Press.

Akuzawa, R. and Surono, I. S. 2003. Fermented milks: Asia. In: *Encyclopedia of Dairy Sciences*, eds. H. Roginski, J. W. Fuquay, and P. F. Fox, 1045–1049. London: Academic Press.

Akyüz, N. 1981. Erzincan (Şavak) tulum peynirinin yapılışı ve bileşimi. *Atatürk Üniversitesi Ziraat Fakültesi Dergisi* 12: 85–112.

Alabdulkarim, B., Arzoo, S., and Osman, M. S. E. 2012. Effect of packaging materials on the physico-chemical, microbiological and sensory quality of cooked oggtt. *World Applied Sciences Journal* 17: 951–957.

Atasever, M. A. 2007. *Erzurum ve Bingöl yöresinden toplanan kurut örneklerinin mikrobiyolojik ve kimyasal nitelikleri*. Yüksek Lisans Tezi. Erzurum, Turkey: Atatürk Üniv. Sağlık Bil. Enstitüsü. 69.

Ateş, G. and Patır, B. 2001. Starter kültürlü tulum peynirinin olgunlaşması sırasında duyusal, kimyasal ve mikrobiyolojik niteliklerinde meydana gelen değişimler üzerine araştırmalar. *Fırat Üniversitesi Sağlık Bilimleri Dergisi* 15: 45–56.

Ayar, A., Akın, N., and Sert, D. 2006. Bazı peynir çeşitlerinin mineral kompozisyonu ve beslenme yönünden önemi. *Türkiye 9. Gıda Kongresi*, 319–326. Bolu, Turkey: Mayıs.

Bayar, N. and Ozrenk, E. 2011. The effect of quality properties on tulum cheese using different package materials. *African Journal of Biotechnology* 10: 1393–1399.

Beshkova, D., Simova, E., Frengova, G., and Simov, Z. 1998. Production of flavour compounds by yoghurt starter cultures. *Journal of Industrial Microbiology & Biotechnology* 20: 180–186.

Bostan, K., Uğur, M., and Çiftcioğlu, G. 1992. Tulum peynirlerinde laktik asit bakterileri ve küf florası. *Istanbul Üniveristesi Veteriner Fakültesi Dergisi* 17: 111–118.

Çağlar, A. and Çakmakçı, S. 1998. Use of protease and lipase enzymes by different methods to accelerate kaşar cheese ripening. *Gıda* 23: 291–301.

Codex Alimentarius Commission. 2011. *Joint FAO/WHO Food Standards Programme*, REP 11. Thirty-fourth season, Geneva, Switzerland, July 4–9.

Çoşkun, H. and Öztürk, B. 2002. Otlu peynir adı altında üretilen peynirlerin bazı mikrobiyolojik ve kimyasal özellikleri. *Gıda Teknolojisi* 6: 44–48.

Danova, S., Petrov, K., Pavlov, P., and Petrova, P. 2005. Isolation and characterization of *Lactobacillus* strains involved in koumiss fermentation. *International Journal of Dairy Technology* 58: 100–105.

Demirci, M. 1988. Ülkemizin önemli peynir çeşitlerinin mineral madde düzeyi ve kalori değeri. *Gıda* 13: 17–21.

Dinkçi, N., Ünal, G., Akalın, A. S., Varol, S., and Gönç, S. 2012. Kargı tulum peynirinin kimyasal ve mikrobiyolojik özellikleri. *Ege Üniversitesi Ziraat Fakültesi Dergisi* 49: 287–292.

FAO (Food and Agricultural Organization). 1990. The technology of traditional milk products in developing countries. *FAO Animal Production and Health Paper 85*, Rome, Italy.

Farnworth, E. R. 1999. Kefir: From folklore to regulatory approval. *Journal of Nutraceuticals, Functional & Medical Foods* 1: 57–68.

Fontán, M. C. G., Martínez, S., Franco, I., and Carballo, J. 2006. Microbiological and chemical changes during the manufacture of kefir made from cows' milk, using a commercial starter culture. *International Dairy Journal* 16: 762–767.

Fox, P. F. 1993. *Cheese: Chemistry, Physics and Microbiology*. London: Chapman & Hall.

Güler, Z., Gün, I., and Uraz, T. 2004. The use of fungal lipase to accelerate the ripening of Kashar cheese. *Milchwissenschaft* 59: 277–279.

Güler, Z. and Şanal, H. 2009. The essential mineral concentration of torba yoghurts and their wheys compared with yoghurt made with cows', ewes' and goats' milks. *International Journal of Food Science and Nutrition* 60: 153–164.

Gursoy, A., Gursel, A., Senel, E., Deveci, O., and Karademir, E. 2001. Yağ içeriği azaltılmış Beyaz peynir üretiminde ısıl işlem uygulanan *Lactobacillus helveticus* ve *Lactobacillus delbrueckii* subsp. *bulgaricus* kültürlerinin kullanımı. *GAP II. Tarım Kongresi, Şanlıurfa Ekim* 24–26, 269–278.

Hayaloglu, A. A., Guven, M., and Fox, P. F. 2002. Microbiological, biochemical and technological properties of Turkish white cheese "Beyaz Peynir." *International Dairy Journal* 12: 635–648.

Hocalar, B., Kemağlıoğlu, K., and Dokuzoğuz, F. 2004. Geleneksel bir süt ürünü: Torba yoğurdu. *Paper presented at Geleneksel Gıdalar Sempozyumu*, Van, Turkey: Yüzüncü Yıl Üniversitesi.

Inal, T. 1990. *Süt ve süt ürünleri hijyen ve teknolojisi*. Istanbul, Turkey: Final Ofset.

Irigoyen, A., Arana, I., Castiella, M., Torre, P., and Ibáñez, F. C. 2005. Microbiological, physicochemical, and sensory characteristics of kefir during storage. *Food Chemistry* 90: 613–620.

Işleyici, Ö. and Akyüz, N. 2009. Van ilinde satışa sunulan otlu peynirlerde mikrofloranın ve laktik asit bakterilerinin belirlenmesi. Yüzüncü Yıl Üniversitesi Veteriner Fakültesi Dergisi 20: 59–64.

Kaaki, D., Kebbe Baghdadi, O., Najim, N. E., and Olabi, A. 2012. Preference mapping of commercial Labneh (strained yogurt) products in the Lebanese market. *Journal of Dairy Science* 95: 521–532.

Kabak, B. and Dobson, A. D. W. 2011. An introduction to the traditional fermented foods and beverages of Turkey. *Critical Reviews in Food Science and Nutrition* 51: 248–260.

Kamber, U. 2008a. The manufacture and some quality characteristics of kurut, a dried dairy product. *International Journal of Dairy Technology* 61: 146–150.

Kamber, U. 2008b. The traditional cheeses of Turkey: Cheeses common to all regions. *Food Reviews International* 24: 1–38.

Karabulut, I., Hayaloglu, A.A., and Yildirim, H. 2007. Thin-layer drying characteristics of kurut, a Turkish dried dairy by-product. *International Journal Food Science and Technology* 42: 1080–1086.

Kavaz, A., Bakırcı, I., and Kaban, G. 2013. Some physico-chemical properties and organic acid profiles of herby cheeses. *Kafkas Üniversitesi Veteriner Fakültesi Dergisi* 19: 89–95.

Kayagil, F. and Candan, G. 2009. Effects of starter culture combinations using isolates from traditional cheese on the quality of Turkish white cheese. *International Journal of Dairy Technology* 62: 387–396.

Kiani, H., Mousavi, S. M. A., and Emam-Djorneh, Z. 2008. Rheological properties of Iranian yoghurt drink, doogh. *International Journal of Dairy Science* 3: 71–78.

Kinik, O., Gursoy, O., and Seckin, A. K. 2005. Cholesterol content and fatty acid composition of most consumed Turkish hard and soft cheeses. *Czech Journal of Food Sciences* 23: 66–172.

Kırdar, S. and Gün, I. 2001. Burdur'da süzme yoğurt üretimi teknolojisi üzerine bir araştırma. *Gıda* 26: 99–107.

Kırdar, S. S. 2005. Burdur yöresi geleneksel peynirleri: çökelek peyniri. *Presented at I. Burdur sempozyumu*, Kasım, Burdur, Turkey.

Kocak, C., Aydemir, S., and Seydim, Z. B. 2005. Levels of proteolysis in important types of Turkish cheeses. *Gıda* 30: 395–398.

Köksoy, A. and Kılıç, M. 2003. Effects of water and salt level on rheological properties of ayran, a Turkish yoghurt drink. *International Dairy Journal* 13: 835–839.

Köksoy, A. and Kılıç, M. 2004. Use of hydrocolloids in textural stabilization of a yoghurt drink, ayran. *Food Hydrocolloids* 18: 593–600.

Kurultay, S., Yasar, K., and Oksuz, O. 2004. The effect of different curd pH and stretching temperatures on the chemical properties of Kashar cheese. *Milchwissenschaft* 59: 386–388.

Liu, J.-R. and Lin, C.-W. 2000. Production of kefir from soymilk with or without added glucose, lactose or sucrose. *Journal of Food Science* 65: 716–719.

Marshall, E. and Mejia, D. 2011. *Traditional Fermented Food and Beverages for Improved Livelihoods*. FAO diversification booklet number 21. Rome, Italy: Food and Agriculture Organization of the United Nations, 819.

Marshall, V. M. and Cole, W. M. 1985. Methods for making kefir and fermented milk based on kefir. *Journal of Dairy Research* 52: 451–456.

Montanari, G. and Grazia, L. 1997. Galactose-fermenting yeasts as fermentation microorganisms in traditional koumiss. *Food Technology and Biotechnology* 35: 305–308.

Nergis, C. and Seçkin, A. K. 1998. The losses of nutrients during the production of strained (Torba) yoghurt. *Food Chemistry* 61: 13–16.

Nsabimana, C., Jiang, B., and Kossah, R. 2005. Manufacturing, properties and shelf life of labneh: A review. *International Journal of Dairy Technology* 58: 129–137.

Öksüztepe, G., Patır, B., and Çalıcıoğlu, M. 2005. Identification and distribution of lactic acid bacteria during the ripening of Şavak tulum cheese. *Turkish Journal* of *Veterinary and Animal Sciences* 29: 873–879.

Öksüztepe, G. A., Patır, B., Dikici, A., Bozkurt, Ö. P., and Çalıcıoğlu, M. 2007. Microbiological and chemical quality of cokelek marketed in Elazığ. *Fýrat Üniversitesi Saðlýk Bilimleri Dergisi* 21: 27–31.

Önganer, A. N. and Kırbağ, S. 2009. Diyarbakır'da taze olarak tüketilen çökelek peynirlerinin mikrobiyolojik kalitesi. *Erciyes Üniversitesi Fen Bilimleri Enstitüsü Dergisi* 25: 24–33.

Özünlü, B. M. 2005. *Ayran Kalitesinde Etkili Bazı Parametreler Üzerine Araştırmalar*. PhD Thesis. Ankara, Turkey: Ankara Üniversitesi Fen Bilimleri Enstitüsü

Pappas, C. P., Kondyli, E., Voutsinas, L. P., and Malatou, H. 1996. Effects of starter level, draining time and aging on the physicochemical, organoleptic and rheological properties of Feta cheese. *International Journal of Dairy Technology* 49: 73–78.

Patır, B. and Ateş, G. 2002. Kurutun mikrobiyolojik ve kimyasal bazı nitelikleri üzerine araştırmalar. *Turkish Journal* of *Veterinary and Animal Sciences* 26: 785–792.

Robinson, R. K. 2003. Yoghurt: Role of starter cultures. In: *Encyclopedia of Dairy Sciences*, eds. H. Roginski, J. W. Fuquay, and P. F. Fox, 1059–1063. London: Academic Press.

Şengül, M., Erkaya, T., and Fırat, N. 2010. Çiğ ve pastörize sütten üretilen kaşar peynirlerinin olgunlaşma süresince bazı mikrobiyolojik özelliklerinin karşılaştırılması. *Atatürk Üniversitesi Ziraat Fakültesi Dergisi* 41: 149–156.

Simsek, B. and Sagdic, O. 2010. Determination of fatty acids and chemical characteristics of çökelek cheese from cows milk using of *L. helveticus* and/ or yoghurt bacteria. *Food Science and Technology Research* 16: 179–184.

Soltani, M. and Güzeler, N., 2013. The production and quality properties of liquid kashks. *Gıda* 38: 1–7.

Surono, I. S. and Hosono, A. 2003. Fermented milks. In: *Encyclopedia of Dairy Sciences*, eds. H. Roginski, J. W. Fuquay, and P. F. Fox, 1018–1023. London: Academic Press.

Sutherland, B. J. 2003. Salting of cheese. In: *Encyclopedia of Dairy Sciences*, eds. H. Roginski, J. W. Fuquay, and P. F. Fox, 293–300. London: Academic Press.

Tamime, A. Y., Muir, D. D., and Wszolek, M. 1999. Kefir, koumiss and kishk. *Dairy Industries International* 64: 32–33.

Tamime, A. Y. and Robinson, R. K. 1999. *Yoghurt Science and Technology* (2nd edition). Boca Raton, FL: CRC Press.

Tarakçi, Z., Coşkun, H., and Tunçtürk, Y. 2004. Some properties of fresh and ripened herby cheese, a traditional variety produced in Turkey. *Food Technology and Biotechnology* 42: 47–50.

Tekinşen, K. K., Nizamlıoğlu, M., Bayar, N., Telli, N., and Köseoğlu, I. E. 2008. Konya'da üretilen süzme (torba) yoğurtların bazı mikrobiyolojik ve kimyasal özellikleri. *Veterinary Bilimleri Dergisi* 24: 69–75.

Toba, T., Arihara, K., and Adachi, S. 1987. Comparative study of polysaccharides from kefir grains, an encapsulated homofermentative *Lactobacillus* species and *Lactobacillus kefir*. *Milchwissenschaft* 42: 565–568.

TUIK (Turkish Statistical Institute). 2012. Change ratios of production of milk and milk products, 2012–2013. http://www.turkstat.gov.tr.

Tuncer, Y., Ozden, B., and Akçelik, M. 2008. Tulum peynirlerinden izole edilen *Lactococcus lactis* subsp. *lactis* YBML9 ve YBML21 suşları tarafından üretilen bakteriyosinlerin kısmi karakterizasyonları. *Süleyman Demirel Üniversitesi Fen Bilimleri Dergisi* 12: 141–148.

Tunçtürk, Y., Ocak, E., and Zorba, Ö. 2010. The effect of different homogenization pressures on chemical, biochemical, microbiological, and sensorial properties of Kashar cheese. *Yüzüncü Yıl Üniversitesi Tarım Bilimleri Dergisi* 20: 88–99.

Turkish Food Codex. 2001. *Communique on Fermented Milk*. Official Gazette.

Turkish Standard Institute. 1995. *White cheese standards (TS-591)*. Ankara, Turkey: The Institute of Turkish Standards.

Turkish Standard Institute. 1999. *Kashar Cheese Standards (TS-3272)*. Ankara, Turkey: The Institute of Turkish Standards.

Ucuncu, M. 1999. *Süt teknolojisi*, Ege Üniversitesi Mühendislik Fakültesi Yayınları, yayın no: 32, Izmir, Turkey: Ege Üniversitesi Basımevi.

Uysal, H. R. 1996. Değişik miktarlarda kültür kullanılarak üretilen beyaz peynirlerde proteoliz düzeyi üzerine araştırmalar. *Ege Üniversitesi Ziraat Fakültesi Dergisi* 33: 107–114.

Var, I., Sahan, N., Kabak, B., Yasar, K., and Karaca, O. B. 2004. Effect of some fat replacers on microflora of low-fat Kashar cheese during ripening. *Archiv für lebensmittelhygiene* 56: 97–120.

Wang, J., Chen, X., Liu, W., Yang, M., and Zhang, H. 2008. Identification of *Lactobacillus* from koumiss by conventional and molecular methods. *European Food Research and Technology* 227: 1555–1561.

Ya, T., Zhang, Q., Chu, F., Merritt, J., Bilige, M., Sun, T., Du, R., and Zhang, H. 2008. Immunological evaluation of *Lactobacillus casei* Zhang: A newly isolated strain from koumiss in Inner Mongolia, China. *BMC Immunology* 9: 1–9.

Yaygın, H. 1999. *Yoğurt teknolojisi*. Akdeniz Üniversitesi Yayın no: 75. Antalya, Turkey.

Yerlikaya, O., Torunoğlu, F. A., Kınık, Ö., and Akbulut, N. 2011. Ülkemizde tüketilen çökelek peynirleri ve üretim şekillerine bir bakış. *Dünya Gıda*, 2: 48–52.

Yildiz, F., Kocak, C., Karacabey, A., and Gursel, A. 1989. Türkiye'de kaliteli salamura beyaz peynir üretim teknolojisinin belirlenmesi. *Doga Türk Veterinerlik ve Hayvancılık Dergisi* 13: 384–392.

Zhang, H., Xu, J., Wang, J., Menghebilige, Sun, T., Li, H. and Guo, M. 2008. A survey on chemical and microbiological composition of kurut, naturally fermented yak milk from Qinghai in China. *Food Control* 19: 578–586.

16
Cheese Whey Fermentation

CARLA OLIVEIRA, GIULIANO DRAGONE, LUCÍLIA DOMINGUES, AND JOSÉ A. TEIXEIRA

Contents

16.1	Introduction	427
16.2	Cheese Whey Composition	428
16.3	Fermented Whey-Based Beverages	430
	16.3.1 Functional Beverages	431
	16.3.2 Alcoholic Whey Beverages	434
	16.3.3 Bioethanol	436
16.4	Whey Fermentation	437
	16.4.1 Whey Fermentation by *S. cerevisiae*	437
	16.4.2 Whey Fermentation by *Kluyveromyces* ssp.	439
	16.4.3 Whey Fermentation by Recombinant *S. cerevisiae* Strains	446
References		450

16.1 Introduction

Cheese whey (or milk whey), the green-yellowish liquid resulting from the separation of milk casein during cheese production and casein manufacture, is the main by-product of dairy industries due to the large volumes produced and their nutritional composition (Siso 1996). The worldwide production of this by-product was estimated at around 184 million tons in 2012 (Bulatović et al. 2012b), and future projections (Figure 16.1) estimate that this amount will rise by more than 2% per year until 2019.

Although almost half of the global whey production is currently processed into several value-added food products (Mandal et al. 2012), the other half is still not exploited but often discharged into water bodies (e.g., rivers and lakes) or disposed into land (Aktaş et al. 2006).

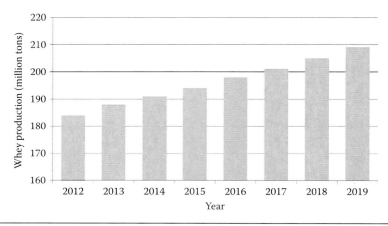

Figure 16.1 Long-term projections of worldwide cheese whey production. (Adapted from Bulatović, M.L. et al., *Hem. Ind.*, 66, 565–577, 2012b.)

Whey disposal represents a significant loss of resources and a serious environmental hazard because of its high biological oxygen demand (BOD) (40–60 g/l) and chemical oxygen demand (COD) (50–80 g/l) values (Athanasiadis et al. 2004). Continuous dumping of whey into land can contribute to serious groundwater contamination and affect the properties of soil, reducing crop yields (Panesar et al. 2007). The high values of BOD and COD are mainly due to the presence of lactose mass fraction (Domingues et al. 1999a).

16.2 Cheese Whey Composition

Cheese whey constitutes about 85%–90% of the milk volume used for cheese production, and it contains more than half of the nutrients present in the original milk (Panesar et al. 2007). The most abundant nutrients are lactose (4.5%–5% w/v), proteins (0.6%–0.8% w/v), lipids (0.4%–0.5% w/v), and mineral salts (8%–10% of dried extract), which are mainly composed of NaCl, KCl, and calcium salts (Siso 1996). Whey proteins have a higher biological value in comparison with casein of milk or egg proteins due to the high content of branched-chain essential amino acids (Smithers 2008). These proteins consist of β-lactoglobulin, α-lactalbumin, lactoferrin, lactoperoxidase, bovine serum albumin, thermostable fractions of proteose peptones, immunoglobulins, and bioactive peptides. The presence of essential amino acids such as lysine, cysteine, methionine, and cystin imparts anticarcinogenic properties to whey proteins (Singh et al. 2011). Whey also contains considerable quantities of citric and lactic (0.05% w/v) acids, nonprotein nitrogen compounds (urea and uric acid), and B group vitamins (Siso 1996).

The composition of whey varies according to its origin (e.g., goat, sheep, or cow) and the procedure employed for casein removal from liquid milk. As an example, sheep whey has a total nitrogen/dry matter ratio much higher than the one existing in bovine whey, doubling the content in soluble proteins (Carvalho et al. 2013).

Considering the cheese-making technology employed, cheese whey can be defined as acid whey or sweet whey. Acid whey (pH $\leq$ 5.1) is mainly obtained from nonfat milk used in the production of cottage, ricotta, or similar fresh cheeses through the coagulation of casein by lactic fermentation or addition of mineral/organic acids. Sweet whey (pH $\geq$ 5.6) is mainly obtained from milk used in the manufacture of Cheddar, mozzarella, and Swiss cheese through the coagulation of casein by rennet, an industrial casein-clotting preparation containing chymosin or other casein-coagulating enzymes (Jarvis et al. 2007). Table 16.1 shows the average composition of acid and sweet dry whey.

It is noteworthy that acid whey has higher ash and lower protein contents than sweet whey; therefore, the use of acid whey in food applications has been more limited due to its high saline content and acidic flavor (Siso 1996).

As lactose is the major component of cheese whey solids in addition to proteins, minerals, and water-soluble vitamins, numerous biotechnological processes have been developed to utilize whey to make useful products of industrial importance (Panesar et al. 2007). Recent research attempts have tried to develop technologies that use cheese whey as feedstock for the production of value-added products such as single-cell proteins, organic acids (i.e., lactic, succinic, and propionic), enzymes, methane, oligosaccharides, and ethanol. The production of beverages by bioconversion of cheese whey has also been recognized as an alternative of great interest for the utilization of this by-product (Dragone et al. 2009). Among the different possibilities for cheese whey utilization presented in Figure 16.2, the use of fermentation processes in beverage and ethanol production from this by-product will be described further.

Table 16.1 Average Composition of Sweet and Acid Dry Whey[a]

CONSTITUENT	SWEET	ACID
• Proteins, $N \times 6.38$ (%)	12.9	11.7
• Fat (%)	1.1	0.5
• Lactose (%)	74.5	73.4
• Total ash (%)	8.3	10.8
• Vitamins		
• Vitamin A (IU)	30	59
• Vitamin C (mg)	1.5	0.9
• Vitamin E (mg)	0.03	0.00
• Thiamine (B_1) (mg)	0.5	0.62
• Riboflavin (B_2) (mg)	2.2	2.06
• Pyridoxine (B_6) (mg)	0.6	0.62
• Vitamin B_{12} (µg)	2.4	2.5
• Pantothenic acid (mg)	5.6	5.6
• Niacin (mg)	1.3	1.2
• Folate (µg)	12.0	0.33
• Minerals		
• Calcium (mg)	796	2050
• Phosphorus (mg)	932	1349
• Sodium (mg)	1079	968
• Potassium (mg)	2080	2289
• Magnesium (mg)	176	199
• Zinc (mg)	1.97	6.3
• Iron (mg)	0.9	1.2
• Copper (mg)	0.07	0.05
• Selenium (µg)	27.2	27.3

Source: Jarvis, J.K. et al., *Handbook of Dairy Foods and Nutrition* (3rd edition), CRC Press, Boca Raton, FL, 2007.

[a] Values per 100 g of dry whey.

16.3 Fermented Whey-Based Beverages

The manufacture of beverages through lactic or alcoholic fermentations that can provide desirable sensory properties has been considered an interesting approach for whey valorization. Lactic fermentations of cheese whey typically use conventional starter microorganisms or probiotic strains, whereas alcoholic

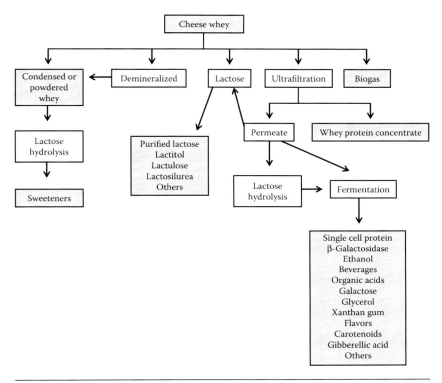

Figure 16.2 Scheme of different utilizations of cheese whey. (Adapted from Mollea, C. et al., Valorisation of cheese whey, a by-product from the dairy industry. In: *Food Industry*, ed. I. Muzzalupo, InTech, Rijeka, Croatia, 549–588, 2013.)

fermentations commonly use *Kluyveromyces* yeast strains (Koutinas et al. 2007; Skudra et al. 1998).

16.3.1 Functional Beverages

The production of functional whey-based beverages by fermentation with lactic acid bacteria (LAB) has been recognized, in economic terms, as the most favorable way of whey processing (Bulatović et al. 2012a).

LAB have been widely used as starter cultures in the fermented food industry due to their metabolic activity on sugars, proteins, and lipids, thus promoting food digestibility and preservation as well as improving the texture and sensory profile of the end product (Pescuma et al. 2008). Cheese whey fermentation by LAB could decrease the high lactose content in whey, producing mainly lactic acid and other metabolites such as aroma compounds that contribute to the flavor and increasing carbohydrate solubility and sweetness of the beverage (Mauriello et al. 2001).

Species of LAB commonly used in the production of functional whey-based beverages are *Lactobacillus acidophilus*, *Lb. delbrueckii* subsp. *bulgaricus*, *Lb. reuteri*, *Lb. rhamnosus*, *Lb. casei*, *Lb. plantarum*, *Lb. helveticus*, *Lb. paracasei*, *Streptococcus thermophilus*, *Propionibacterium freudenreichii* subsp. *shermanii*, *Enterococcus faecium*, *Bifidobacterium animalis* subsp. *lactis*, and *Bifido. bifidum* (Bulatović et al. 2012a).

It has been demonstrated that cheese whey fermentation by *Lb. delbrueckii* subsp. *bulgaricus* and *Str. thermophilus* (lactic yogurt culture) produces a more intense yogurt flavor (especially when threonine, a possible precursor of acetaldehyde, is added to the whey) than that obtained when skim milk is fermented. This suggests the possibility of producing functional whey-based beverages with similar sensory profiles to those of fermented milk drinks or with some flavor attributes of drinking yogurt, following manufacturing procedures conventionally used for milk (Gallardo-Escamilla et al. 2005).

Lb. johnsonii can be used successfully in the manufacture of a yogurt-like beverage from reconstituted sweet whey (Bulatović et al. 2012a). The strain *Lb. johnsonii* NRRL B-2178 had probiotic characteristics and showed high viability in the presence of whey proteins during storage of the beverage.

A strawberry-flavored probiotic beverage with acceptable flavor has also been produced by incorporating up to 65% acid cheese whey in its formulation (Castro et al. 2013). The beverage was prepared by inoculating the milk–cheese whey mixture with 1% (w/v) of starter culture (*Str. salivarius* subsp. *thermophilus* and *Lb. delbrueckii* subsp. *bulgaricus*) and 2% (w/v) of probiotic culture (*Lb. acidophilus*). After incubation at 45°C until the pH value reached 4.7, the beverage was cooled down to 8°C and stored under refrigeration.

Maity et al. (2008) developed a healthy whey-based beverage by using *Lb. rhamnosus*, *Bifido. bifidum*, and *P. freudenreichii* subsp. *shermanii*. The fermented probiotic drink was produced by adding 4% of mixed culture (1:1:1) into deproteinized whey (4.6% lactose, 0.62% ash, 0.48% fat, and 0.5% protein) adjusted to pH 6.4 and incubated at 37°C during 8 h. The authors demonstrated that the final product met the probiotic criteria by maintaining bacterial population ($>10^8$ CFU/ml), therapeutic properties, and optimum sensory qualities with a shelf life of 5 days. The fermented whey beverage preserved an acceptable flavor.

Additionally, *Lb. acidophilus*, *Lb. plantarum*, and *E. faecium* has been used in the manufacture of fermented probiotic beverage from milk permeate enriched with 10% cheese whey (Pavunc et al. 2009). At the end of the controlled fermentation, probiotic strains produced 7.4 g/l lactic acid, pH was decreased to 4.7, and the number of live cells was around 10^8 CFU/ml. The number of viable count of probiotic bacteria was maintained at around 10^7 CFU/ml during 28 days of the preservation at 4°C.

Hernandez-Mendoza et al. (2007) developed a whey-based probiotic beverage by inoculating reconstituted whey containing pectin and sucrose with *Lb. reuteri* and *Bifido. bifidum*. After 30 days of storage, the beverage retained an acceptable flavor as well as probiotic bacteria counts higher than 10^6 CFU/ml.

Kefir grains were also used as starter culture in the production of functional cheese whey-based beverages. These grains are irregular granules that vary in size (3–35 mm in diameter) and contain LAB (*Lactobacillus*, *Lactococcus*, and *Leuconostoc*), acetic acid bacteria, and yeasts coupled together with casein and complex sugars by a matrix of polysaccharides (Guzel-Seydim et al. 2005). The polysaccharide matrix, designated as kefiran, is produced by LAB and commonly associated to the therapeutic properties of kefir (Tada et al. 2007).

Magalhães et al. (2010) reported that the main characteristics of kefir grains were maintained, when using whole or deproteinized cheese whey instead of milk during the production of kefir-like beverages. They also proved that the microbial community present in the fermented beverages obtained from milk, whole cheese whey, and deproteinized cheese whey did not show significant differences. In addition, a recent study (Magalhães et al. 2011) demonstrated that kefir grains were able to utilize lactose from whole cheese whey and deproteinized cheese whey and produce similar amounts of ethanol, lactic acid, and acetic acid to those obtained during milk fermentation. Moreover, the concentration of higher alcohols (2-methyl-1-butanol, 3-methyl-1-butanol, 1-hexanol, 2-methyl-1-propanol, and 1-propanol), ester (ethyl acetate), and aldehyde (acetaldehyde) in cheese whey-based kefir and milk kefir beverages was also produced in similar amounts.

16.3.2 Alcoholic Whey Beverages

Cheese whey can be used as raw material for alcoholic beverage production once its main constituent (about 70% of dry matter) is the disaccharide lactose, a carbohydrate composed of D-glucose and D-galactose. Alcoholic whey beverages may be classified as beverages containing less than 1.5% alcohol (beverages with low alcohol content), whey beer, whey wine, whey champagne, and whey spirit (distillate).

Whey-based beverages with low alcohol content are typically produced from whey permeate by fermenting the lactose with yeast strains *K. fragilis* or *K. lactis* to the desired alcohol content (0.5%–1%), adding flavoring, sweetener, and bottling. During fermentation, a certain amount of lactose is converted into lactic acid, which gives a refreshing sour taste to the beverage, while the rest is metabolized into ethanol (Jelicic et al. 2008). One product belonging to this category is a fermented whey beverage from Poland containing 0.6%–0.7% lactic acid and 0.8%–0.95% alcohol and is produced by the inoculation of acid whey permeate with kefir grains using 5 h incubation at 25°C (Sienkiewicz and Riedel 1990).

Whey beer can be produced with or without the addition of malt, and it can be fortified with minerals or can contain starch hydrolysates and vitamins (Jelicic et al. 2008). It has been seen that cheese whey can be considered as a suitable raw material for beer production due to several reasons: lactose in whey is only slightly sweet and does not alter the taste of the finished beverage; whey protein content and quantity of minerals in the colloidal state form the basis for a high degree of CO_2 binding; and a caramel-like flavor, similar to that of kilned malt, can be developed, mainly as a consequence of the browning reaction of lactose. However, some problems may also occur, including the loss of beer foam owing to the presence of fat, as well as undesirable odor and taste due to low solubility of whey proteins (Sienkiewicz and Riedel 1990).

Interestingly, sweet cheese whey has been successfully employed as raw material for the production of whey wines of acceptable quality (Larson and Yang 1976). The method used involved the deproteinization of Cheddar cheese whey by heat (82°C during 5 min), the addition of sulfur dioxide (100 ppm) to stabilize the whey before fermentation, and the fermentation of lactose normally present in whey by *K. fragilis*. As lactose yielded only 2% or 3% alcohol, it was necessary to add dextrose (about 22%) and to ferment it with *Saccharomyces cerevisiae* to increase the alcohol content of whey wine. Yeast nutrients such as nitrogen and addition of B vitamins were found to be unnecessary for whey wine fermentation, as cheese whey itself contained sufficient nutrients for yeast growth. In addition, sweet whey wine must be pasteurized or handled aseptically to prevent secondary fermentation. Whey wines blended with fruit and berry wines were found acceptable by taste panels.

Yoo and Mattick (1969) also evaluated the feasibility of producing whey wine from both acid and sweet cheese whey through lactose fermentation by *K. fragilis*. These authors found that the maximum ethanol production from cheese whey was obtained with a lactose concentration of 12%. Lactose in acid cheese whey (pH 4.2) fermented more rapidly and produced more alcohol than sweet whey (pH 5.7). They produced an acceptable whey wine with an ethanol concentration of 10% when 16% sucrose was added to a 10% acid whey solution.

The first production of a sparkling beverage from acid cheese whey (whey champagne) was described by Anatovskiy and Yaroshenko (1950). According to that study, the deproteinized whey was pasteurized, cooled, inoculated with *Lb. acidophilus* or *Lactococcus lactis*, and fermented for 2 h at 42°C. After the addition of the yeast culture and 8%–10% of sugar, the whey was filled into champagne-type bottles, sealed, and kept at 7°C–9°C for 3–4 days to produce ethanol and CO_2.

The production of distilled beverages from cheese whey has not been extensively explored. However, a recent study (Dragone et al. 2009) has demonstrated the possibility of producing whey spirit (35.4% v/v ethanol) of acceptable organoleptic characteristics by distillation of the broth obtained from continuous whey permeate fermentation by *K. marxianus*. Higher alcohols, which greatly contribute to the aroma and essential character of a distillate, were the compounds with the highest concentration in the beverage. Total concentration of higher alcohols [541 g/(l h) absolute alcohol] was above the minimum concentration [140 g/(l h)] required by the European legislation for that group of compounds in distillates. Isoamyl (3-methyl-1-butanol), isobutyl (2-methyl-1-propanol), 1-propanol, and isopentyl (2-methyl-1-butanol) alcohols were the higher alcohols found in highest quantities (Table 16.2).

Table 16.2 Concentration of Major Volatile Compounds in Whey Spirit Obtained from Cheese Whey Permeate

	CONCENTRATION (MG/L)
ALDEHYDES	
Acetaldehyde	36.7
ESTERS	
Ethyl acetate	138
ALCOHOLS	
2-Methyl-1-butanol (isopentyl alcohol) + 3-methyl-1-butanol (isoamyl alcohol)	176 + 887
2-Methyl-1-propanol (isobutyl alcohol)	542
1-Propanol	266
2-Butanol	34.7
2-Phenylethanol	7.8
ACIDS	
Acetic acid	79.9

Source: Dragone, G. et al., *Food Chem.*, 112(4), 929–935, 2009.

Besides the higher alcohols, four other compounds (1 ester, 1 aldehyde, and 2 acids) were identified as major components in whey spirit (concentrations higher than 6.0 mg/l) and 30 compounds were identified as minor components, including another 7 higher alcohols, 11 esters, 3 aldehydes, 5 acids, and 4 terpenes, with concentrations varying from 26.5 to 1740 µg/l. Despite the low concentration of these minor compounds, their presence is very important because they show a synergistic effect that determines the characteristic aroma of the beverage. It should also be emphasized that no methanol was found in whey spirit (probably due to the lack of pectin in cheese whey), which is of great significance, because the presence of this compound in concentrations higher than the maximum legal limit [1000 g/(l h) of 100% vol. ethanol] can be detrimental for the quality of the distillate.

16.3.3 Bioethanol

Whey bioethanol is the principal product obtained through the fermentation of the lactose contained in cheese whey into ethanol by some microorganisms, mainly the yeasts *K. fragilis*, *K. marxianus*, and *S. cerevisiae*. Bioethanol is potable and thus can be used in the food and beverage industries, in pharmaceutical and cosmetics industries, and also in the chemical industries or as an alternative environmental fuel (Prazeres et al. 2012). The production of ethanol by cheese whey fermentation has long been considered a possible solution for whey bioremediation, as it allows for the reduction of the whey polluting load and ethanol production simultaneously (Guimarães et al. 2010). The first studies on this topic date from the 1940s (see, e.g., Browne 1941; Rogosa et al. 1947). Since then, three main strategies have been employed for the fermentation of cheese whey (or simply whey) to ethanol, as detailed later: the fermentation of hydrolyzed whey by wild *S. cerevisiae* strains, the fermentation of whey by different *Kluyveromyces* species, and by recombinant lactose-consuming *S. cerevisiae* strains.

16.4 Whey Fermentation

Whey fermentation is explained in Subsections 16.4.1 through 16.4.3.

16.4.1 Whey Fermentation by S. cerevisiae

The first applications of the non-lactose-utilizing yeast *S. cerevisiae* in whey fermentations involved the use of hydrolyzed whey. Soon after, recombinant DNA technology was used to create lactose-fermenting strains (see Subsection 16.4.3). The fermentation of hydrolyzed whey by *S. cerevisiae* can be developed in two steps, consisting of lactose pre-hydrolyzation with an appropriate enzyme (β-galactosidase) followed by fermentation of the resulting mixture of glucose and galactose, or can be developed in a single step by co-immobilizing the enzyme and the yeast or with mixed cultures (using both free and co-immobilized cells).

S. cerevisiae cannot utilize lactose but can utilize galactose, which is taken up by a permease encoded by the gene *GAL2* (Nehlin et al. 1989). Once inside the cell, catabolism of galactose proceeds through the Leloir pathway. Nevertheless, the ethanol yield and productivity from galactose are significantly lower than those from glucose. In addition, when *S. cerevisiae* uses the mixture of glucose and galactose as a carbon source, it manifests diauxic growth and lower yields of ethanol production (Mehaia and Cheryan 1990; O'Leary et al. 1977), because the presence of glucose inhibits the uptake and metabolism of galactose through a process known as catabolite repression (Gancedo 1998). To overcome that problem, Bailey et al. (1982) isolated catabolite repression-resistant mutants of an industrial *S. cerevisiae* strain capable of utilizing glucose and galactose simultaneously, using 2-deoxyglucose as a selection agent. One of these mutants fermented completely a mixture of 10% glucose plus 10% galactose in less than 37 h, producing approximately 90 g/l of ethanol (Bailey et al. 1982). In a continuous fermentation system with cell recycling, an ethanol productivity of 13.6 g/(l h) was attained from feed medium containing an equimolar mixture of glucose and galactose (15% total sugar), resulting in a residual sugar concentration below 1% (Terrell et al. 1984). Nevertheless, the fermentative capacity of this strain in whey media was not studied.

Besides catabolite repression, other disadvantages of using prehydrolyzed lactose solutions is the high price of the enzyme and its failure to hydrolyze lactose completely (Panesar and Kennedy 2012). Nevertheless, the fermentation of whey by *S. cerevisiae* cells co-immobilized with the β-galactosidase enzyme has been evaluated by some authors (Hahn-Hägerdal 1985; Lewandowska and Kujawski 2007; Roukas and Lazarides 1991; Staniszewski et al. 2009). Roukas and Lazarides (1991) compared the performance of a co-immobilized enzyme treatment (*S. cerevisiae* co-immobilized with β-galactosidase) in shake-flask fermentations of deproteinized cheese whey to that of a treatment using acid prehydrolyzed whey lactose. Enzyme co-immobilization resulted in a slower rate and a lower extent of substrate utilization, thus giving a lower maximum ethanol concentration (13.5 vs. 16.7 g/l) but resulting in a better ethanol yield (95% vs. 89% theoretical) (Roukas and Lazarides 1991). In a continuous concentrated whey fermentation (15% lactose), with *S. cerevisiae* cells co-immobilized with β-galactosidase in calcium alginate, four times more ethanol titer (52 g/l) and productivity [4.5 g/(l h)] were obtained, compared to immobilized *K. fragilis* cells (Hahn-Hägerdal 1985). Lewandowska and Kujawski (2007) carried out a study to improve the effectiveness of a semicontinuous ethanol fermentation of lactose mash (a lactose concentrate [12%]—dried permeate from milk ultrafiltration) combined with a pervaporation module. The fermentation was conducted with a biocatalyst immobilized in calcium alginate, consisting of *S. cerevisiae* cells co-immobilized with β-galactosidase cross-linked with glutaraldehyde. In this work, a 5-l fermenter with a water jacket operated in circulation with mash feeding through the biocatalyst layer packed in a perforated cylinder was used. The productivity of ethanol with an average concentration of 15.6% m/v amounted to about 530 g/day. Afterward, mathematical models were developed for a similar fermentation–pervaporation system using a whey medium as substrate (Staniszewski et al. 2009). In what concerns co-culturing, a mixed culture of *S. cerevisiae* and permeabilized *K. marxianus* cells (as the source of β-galactosidase) resulted in higher ethanol productivity [1.0 g/(l h)] in concentrated whey-based medium (6.5% lactose), as compared to direct fermentation using *K. marxianus* [0.45 g/(l h)] (Rosenberg et al. 1995). A mixed culture of *S. cerevisiae* and *K. marxianus* cells immobilized in calcium alginate also led to improved whey fermentation [ethanol productivity of 0.88 g/(l h)] (Guo et al. 2010).

16.4.2 Whey Fermentation by Kluyveromyces *ssp.*

As the main strategy for bioethanol production, the lactose in whey is directly fermented using lactose-utilizing yeasts that are efficient in fermenting lactose. This approach seems more practical and economical than the previous one (Panesar and Kennedy 2012). Although the yeasts that assimilate lactose aerobically are widespread, those that ferment lactose are rather rare, yet including, for example, special strains of *Kluyveromyces* ssp. (Guimarães et al. 2010). Among the several microorganisms evaluated for producing ethanol directly from lactose/whey, *K. fragilis* is the yeast of choice for most of the commercial plants (Panesar and Kennedy 2012). In the research reports, besides *K. fragilis*, *K. marxianus* has been largely used, as we will see ahead in this chapter. The method for the production of ethanol was developed initially using the yeast *Candida pseudotropicalis* under anaerobic conditions (Rogosa et al. 1947). Nevertheless, this yeast has been less used in whey fermentations than *Kluyveromyces* ssp. (e.g., Ghaly and El-Taweel 1995, 1997a, 1997b; Koushki et al. 2012; Szczodrak et al. 1997). Besides the current distinction between the *Candida* and *Kluyveromyces* species, these are all reported as synonyms of *K. marxianus* since 1998: *K. fragilis* is now included in the *K. marxianus* species and *C. pseudotropicalis* (synonym of *Candida kefyr*) is the anamorph (asexual) form of *K. marxianus* (Guimarães et al. 2010).

Recently, *K. marxianus* demonstrated greater ethanol-producing ability from cheese whey permeate than *C. kefyr* (Koushki et al. 2012).

The reaction describing the bioconversion of lactose to ethanol reveals a theoretical maximum value of 0.538 g of ethanol per gram of lactose consumed (i.e., 4 mol of ethanol produced per mol of lactose consumed) (Domingues et al. 2010). The direct fermentation of the lactose contained in cheese whey (~50 g/l) results in low ethanol titer (2%–3% v/v). In order to make the process economically feasible, fermentation should begin with concentrated whey (usually with a lactose concentration as high as 150–200 g/l) to attain high ethanol yield (corresponding to the maximum theoretical value of 80.7–107.6 g/l). Therefore, whey is usually concentrated by ultrafiltration (whey permeate), whereas dried whey is resuspended to the desired final lactose concentration. However, the use of concentrated whey results in several inhibitory effects, specifically, osmotic sensitivity (due to high lactose concentration), low ethanol tolerance, and inhibition by high salt concentrations, leading to low fermentations with high residual sugar (e.g., Grubb and Mawson 1993; Janssens et al. 1983; Vienne and von Stockar 1985; Zafar et al. 2005). The extent of such negative effects seems to be strain-dependent, although the fermentation conditions, especially oxygen and nutrients availability, may as well play a key role in this regard (Guimarães et al. 2010). Optimization of specific experimental conditions seems to circumvent those problems. Several studies report a positive effect of nutritional supplementation (e.g., Janssens et al. 1983; Kargi and Ozmihci 2006; Mahmoud and Kosikowski 1982; Vienne and von Stockar 1985), oxygen control (e.g., Silveira et al. 2005; Varela et al. 1992), and fed-batch operation (Ozmihci and Kargi 2007d) on the fermentative performance of *K. marxianus* and *K. fragilis* in concentrated whey media. Alternatively, robust yeast strains with enhanced tolerance toward osmotic and/or ethanol stresses may be naturally selected or developed by using metabolic engineering and applied for efficient fermentation of concentrated whey (Guimarães et al. 2010).

The production of ethanol from whey by *Kluyveromyces* ssp. was done using raw cheese whey, cheese whey powder solution, cheese whey permeate from ultrafiltration, and also deproteinized cheese whey. Most studies have been performed in batch systems (shake-flasks and bioreactors), but fed-batch (semicontinuous) fermentation and continuous fermentation, using different bioreactor designs, have also been employed using both nonconcentrated and concentrated whey. Table 16.3 lists the compilation of the ethanol volumetric productivities and maximum ethanol titers calculated from the data shown in some of the studies on whey fermentation by *Kluyveromyces* yeasts.

Several authors have analyzed the fermentation of raw cheese whey to ethanol (e.g., Ariyanti and Hadiyanto 2013; Christensen et al. 2011; Ghaly and El-Taweel 1995, 1997b; Kourkoutas et al. 2002; Zafar and Owais 2006). As already mentioned, direct fermentation of whey results in low ethanol titers. For instance, it was verified that crude whey (non-deproteinized, nondiluted, and nonsterilized whey) could be used to obtain bioethanol through lactose fermentation by *K. marxianus*, but only 11% of ethanol yield (2.1 g/l) was achieved in 22 h (Zafar and Owais 2006). In order to increase ethanol yield, different yeast immobilization strategies have been conducted. For example, Kourkoutas et al. (2002) developed a novel system for high temperature alcoholic fermentation of liquid whey (5% lactose), consisting of *K. marxianus* IMB3 cells (a thermotolerant strain) immobilized on delignified cellulosic material in batch system at 45°C. Under these conditions, 7.3 g/l ethanol was produced with 90% of lactose consumed. Fermenting at high temperatures enables cost savings regarding the refrigeration of the process but leads to heating and ethanol evaporation concerns. Recently, in continuous fermentation with nonsterilized whey, using *K. marxianus* cells immobilized in calcium alginate, a maximum yield and productivity of 20 g/l and 2.5–4.5 g/(l h) ethanol, respectively, were obtained at the dilution rate of 0.2/h (Christensen et al. 2011). The experiments conducted in this work showed that *K. marxianus* was able to take over live LAB present in the nonpasteurized whey and produce ethanol efficiently. This is a great advantage from an industrial point of view because the pasteurization/sterilization of whey adds expense to the process.

A series of studies was conducted to validate the production of ethanol from different concentrations of cheese whey powder solution using the yeast *K. marxianus* in batch (Kargi and Ozmihci 2006; Ozmihci and Kargi 2007d), fed-batch (Ozmihci and Kargi 2007c), and continuous fermentative processes, the last ones with and without yeast cells immobilization (Ozmihci and Kargi 2007a, 2007b, 2008)

Table 16.3 Selected Studies on Ethanol Production from Cheese Whey by *K. fragilis* and *K. marxianus*

ORGANISM	WHEY PREPARATION	LACTOSE CONTENT	BIOREACTOR/OPERATION TYPE	ETHANOL PRODUCTIVITY [G/(L H)]	ETHANOL TITER (G/L)	REFERENCE
K. fragilis	Concentrated whey permeate	24%	3-l bottles/batch	0.2	80	Gawel and Kosikowski (1978)
			14-l bioreactor/batch	0.6	72	Mahmoud and Kosikowski (1982)
	Deproteinized whey powder + 0.5% peptone (a) plus ergosterol, linoleic acid, and Tween 80	15% 20%[a]	1-l flasks/batch	2.0 1.4	71 86	Janssens et al. (1983)
		10%	6-l bioreactor/continuous (with cell recycling)	7.1	47	Janssens et al. (1984)
	Concentrated whey permeate	15%	Packed-bed column/continuous (with immobilized cells)	1.1	13	Hahn-Hägerdal (1985)
	Concentrated whey solution	15%	Tubular reactor/continuous (with immobilized cells)	17.2	18	Gianetto et al. (1986)
	Concentrated whey permeate	10%	Fed-batch	3.3	64	Ferrari et al. (1994)
	Deproteinized whey	5.5%	Fluidized-bed reactor/continuous (with immobilized cells)	14.5	20	Kleine et al. (1995)
	Deproteinized whey powder	20%	Shake-flasks	1.8	81	Dragone et al. (2011)

(*Continued*)

Table 16.3 (Continued) Selected Studies on Ethanol Production from Cheese Whey by *K. fragilis* and *K. marxianus*

ORGANISM	WHEY PREPARATION	LACTOSE CONTENT	BIOREACTOR/OPERATION TYPE	ETHANOL PRODUCTIVITY [G/(L H)]	ETHANOL TITER (G/L)	REFERENCE
K. marxianus	Deproteinized whey + yeast extract and salts	6.5%	5-l bioreactor/batch	0.52	26	Rosenberg et al. (1995)
		10%	2-l bioreactor/batch	3.1	43	Grba et al. (2002)
			2-l bioreactor/fed-batch	4.9	59	
	Whey permeate solution	17%	1-l flasks/batch	1.0–1.5	76–80	Silveira et al. (2005)
	Concentrated whey powder solution	15%	Shake-flasks	0.4	80	Kargi and Ozmihci (2006)
		10%	5-l bioreactor/continuous	0.74	32	Ozmihci and Kargi (2007a)
		10%–12.5%	5-l bioreactor/continuous	0.54	29	Ozmihci and Kargi (2007b)
		12.5%	5-l bioreactor/fed-batch	5.3	63	Ozmihci and Kargi (2007c)
		7.5%	Shake-flasks	0.55	40	Ozmihci and Kargi (2007d)
	Whey powder solution	5%	Packed-bed column/continuous (with immobilized cells)	0.4	20	Ozmihci and Kargi (2008)
	Whey permeate solution	5%	5-l UASB reactor/continuous	0.36	16	Jędrzejewska and Kozak (2011)

[a] Guimarães et al. (2010), Dragone et al. (2011), and Jędrzejewska and Kozak (2011).

(Table 16.3). High ethanol titers were obtained in shake-flasks with 150 g/l of initial lactose concentration (80 g/l; Kargi and Ozmihci 2006). In a continuous fermenter, the highest ethanol concentration (3.7% v/v, 29 g/l), productivity [0.54 g/(l h)], and ethanol yield coefficient ($Y_{P/S}$) were obtained with the feed lactose content of 100 or 125 g/l (Ozmıhci and Kargi 2007c). In a packed-bed column bioreactor operating in continuous, using olive pits as support particles for yeast cells attachment, 0.54 g of ethanol per gram of lactose was obtained from cheese whey powder solution with 5% lactose, at an hydraulic residence time of 50 h, reaching a productivity of 0.4 g/(l h) (Ozmıhci and Kargi 2008). Higher ethanol productivities were attained in fed-batch fermentation (Ozmihci and Kargi 2007d). For optimization, the concentration of cheese whey powder in the feed was varied between 51 and 408 g/l. The ethanol yield coefficient was almost constant (0.54 ± 0.02 g ethanol per gram of substrate) for lactose concentrations between 25 and 150 g/l but decreased at 200 g/l due to high osmotic pressure at high sugar concentrations. The highest ethanol titer (63 g/l) and productivity [5.3 g/(l h)] were also achieved for a lactose concentration in the feed of 125 g/l.

The highest ethanol productivities were reported for continuous fermentations with immobilized *K. fragilis* cells, either using concentrated or non-concentrated whey (Table 16.3). For example, a productivity of 17.2 g/(l h) ethanol was obtained by concentrated whey fermentation (15% lactose) in a tubular continuous bioreactor with *K. fragilis* cells immobilized in charcoal pellets (Gianetto et al. 1986). Also, *K. fragilis* cells immobilized in plant material produced 14.5 g/(l h) ethanol by deproteinized whey fermentation (5.5% lactose) in a continuous fluidized-bed reactor (Kleine et al. 1995).

Recently, deproteinized cheese whey powder has been used for ethanol production by *K. fragilis* in Erlenmeyer flasks (Dragone et al. 2011, Table 16.3). According to the statistical analysis, only the initial lactose concentration had a significant effect on the ethanol production, among all the variables studied (i.e., fermentation temperature, initial lactose concentration, and inoculum concentration). Initial lactose concentrations up to 200 g/l favored the bioconversion to ethanol, whereas lactose concentrations between 200 and 250 g/l drastically affected this reaction. The best conditions for maximizing ethanol production were 200 g/l of initial lactose concentration in whey, inoculum concentration of 1 g/l, and temperature of 35°C, resulting in 80.95 g/l of ethanol after 44 h of fermentation (Dragone et al. 2011).

Scotta is also an effluent of the dairy industry and is a by-product obtained after ricotta cheese production, which is an Italian whey cheese made from sheep milk (or cow, goat, or Italian buffalo milk whey leftover from the production of cheese). Scotta is mainly produced in Italy but also in other countries located in the Mediterranean region (Sansonetti et al. 2009). It is estimated that Italian cheese production amounts to about 1.0 Mt per year, thus determining significant environmental problems related to its disposal (Sansonetti et al. 2009). Due to severe thermal treatment and the addition of acid salts during ricotta processing, scotta has different characteristics with respect to raw cheese whey. Scotta still contains lactose but does not contain proteins, and nowadays it is considered as a waste and is not reused in any way (Zoppellari and Bardi 2012). The feasibility of using scotta for ethanol production has been recently evaluated in batch (aerobic and anaerobic) and semicontinuous fermentations involving the yeast *K. marxianus* (Sansonetti et al. 2009, 2011; Zoppellari and Bardi 2012). Results demonstrated that this waste is an excellent substrate for fermentation, and under some experimental conditions, it exhibits identical or even better performance, comparing to both raw cheese whey and deproteinized whey (Sansonetti et al. 2009; Zoppellari and Bardi 2012). Response surface methodology was applied to optimize the scotta-ethanol bioprocess (Sansonetti et al. 2010). The effect of temperature, pH, agitation rate, and initial lactose concentration was evaluated by a central composite design. The best operating conditions were temperature (32°C–35°C), pH (5.4), agitation (195 rpm), and initial lactose concentration (40 g/l), and the model ensured a good fitting of the observed data. Scotta fermentation was also simulated by using pure neural network and multiple hybrid neural models (Saraceno et al. 2010, 2011). The development of kinetic models to describe the ethanol production from whey and scotta using *Kluyveromyces* ssp. has received increasing attention in the last decade (e.g., Longhi et al. 2004; Ozmihci and Kargi 2007a; Sansonetti et al. 2010; Zafar et al. 2005). Such mathematical models are essential tools for the optimization and industrial implementation of fermentation systems. Very recently, Sansonetti and collaborators (2013) developed a biochemically structured model to describe the continuous fermentation of lactose to ethanol by *K. marxianus*, which allowed the determination of the metabolic coefficients. It was suggested that the model developed provides a solid basis for the rational design of optimized fermentation of cheese whey (Sansonetti et al. 2013).

16.4.3 Whey Fermentation by Recombinant S. cerevisiae *Strains*

The third strategy for bioethanol production involves the direct fermentation of the lactose in whey by using genetically engineered *S. cerevisiae* strains capable of utilizing lactose.

S. cerevisiae is usually the first choice for industrial processes involving alcoholic fermentation for the following reasons: (1) its good fermentative capacity and ethanol tolerance, allowing to produce up to 20% (v/v) ethanol; (2) its GRAS (generally regarded as safe) status; (3) its capacity to grow rapidly under anaerobic conditions, which helps circumventing the oxygenation problems inherent to large-volume industrial fermentations; (4) the extensive industrial and scientific knowledge accumulated that makes it one of the best-studied organisms; and (5) the possibility to use its biomass as animal feed (co-product), which is important for industrial process economics (Guimarães et al. 2010). However, wild *S. cerevisiae* strains are unable to metabolize lactose. Therefore, different recombinant approaches have been employed to obtain lactose-consuming *S. cerevisiae* strains, such as the generation of hybrids of *S. cerevisiae* and *Kluyveromyces* ssp. by protoplasts fusion, the expression of heterologous β-galactosidases, and the simultaneous expression of genes that code for the β-galactosidase and lactose permease system of *K. lactis* (Domingues et al. 2010). The first approach has resulted in hybrid strains of *S. cerevisiae* and *K. lactis* which are able to ferment lactose in sweet and salted whey, respectively (Tahoun et al. 1999, 2002). More recently, Guo et al. (2012) reported the construction of a *K. marxianus* and *S. cerevisiae* hybrid with efficient ethanol production from cheese whey powder solution (80 g/l initial lactose). The hybrid strain produced 3.8% (v/v) ethanol in 72 h, while the parental *K. marxianus* strain only produced 3.1% (v/v) ethanol in 84 h under the same conditions. The fermentative capacity of the fusant was maintained for at least 20 generations, suggesting its genetic stability (Guo et al. 2012).

In the second approach, several recombinant *S. cerevisiae* strains expressing the β-galactosidase from the filamentous fungi *A. niger*, the yeast *K. lactis*, and the bacterium *E. coli* have been constructed (Oliveira et al. 2011). Although some recombinant strains growing well in lactose have been obtained, their application for whey fermentation was poorly explored as in most cases the main goal was the production of heterologous β-galactosidase. As the β-galactosidase from *E. coli* and *K. lactis* is cytosolic, lactose has to be transported to the cytoplasm, or these enzymes have to be directed to the extracellular medium to obtain recombinants that are able to utilize lactose. The *K. lactis* transport system for lactose functions in *S. cerevisiae* (see third recombinant approach) but not that of *E. coli* (Casadaban et al. 1983). To produce β-galactosidase from *E. coli* and *K. lactis* in recombinant *S. cerevisiae* into the extracellular medium, two different strategies can be devised: (1) their secretion, using different yeast secretion signal sequences and fusion partners, and (2) their release to the culture medium, using spontaneous lysis strategies or the chemical permeabilization of the cells (Oliveira et al. 2011). For example, autolytic *S. cerevisiae* cells expressing the *lacZ* gene (coding for *E. coli* β-galactosidase) (Porro et al. 1992) produced up to 9 g/l of ethanol (60%–70% of the theoretical conversion yield) from lactose-rich medium (6% lactose) supplemented with whey, reaching a productivity of 1 g/(l h), with complete lactose consumption (Compagno et al. 1995). However, it should be noted that cell lysis affects negatively the downstream processing, which is not industrially attractive.

The β-galactosidase from *A. niger* has the advantage of being extracellular. Flocculent recombinant *S. cerevisiae* strains secreting *A. niger* β-galactosidase from episomal plasmids (Domingues et al. 2000, 2002), or in which the β-galactosidase gene was integrated in the genomic δ-sequences in multicopies (Oliveira et al. 2007), were constructed. Besides secreting high-levels of recombinant β-galactosidase, the recombinant strains produced ethanol from lactose/whey with close to theoretical yields in batch and in high-cell-density continuous fermentations with complete lactose utilization (Domingues et al. 2002, 2005; Oliveira et al. 2007). The use of these strains in the dairy industry is very attractive for the simultaneous production of ethanol and β-galactosidase and whey treatment. Furthermore, the recombinant enzyme can be easily recovered by ultrafiltration and used to obtain other products within the dairy industry, as for instance, lactose-free products or hydrolyzed whey syrups (Domingues et al. 2010).

The simultaneous expression of the lactose permease and intracellular β-galactosidase of *K. lactis* (encoded by *lac12* and *lac4* genes, respectively) in *S. cerevisiae* has resulted in recombinants able to ferment whey lactose with high efficiency. In the first works concerning this strategy, however, limited success was obtained for either lactose consumption or ethanol production rates (Rubio-Texeira et al. 1998; Sreekrishna and Dickson 1985). Conversely, a flocculent *S. cerevisiae* strain transformed with the same plasmid used in a previous work (Sreekrishna and Dickson 1985), but with a different selection procedure (Domingues et al. 1999b), presented a good lactose metabolization ability and ethanol production after an adapting period in lactose medium (Domingues et al. 1999a). The advantage of using flocculent cells is that their sedimentation characteristics allow for a simplest recovery of the extracellular product and high productivities in high-cell-density continuous fermentations performed in bioreactors designed to retain these cells, as the air-lift bioreactor (Figure 16.3).

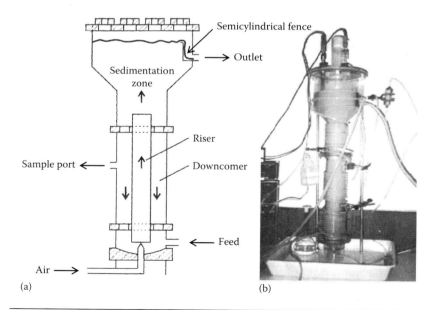

Figure 16.3 (a) Schematic representation and (b) photography of the air-lift bioreactor. (Adapted from Domingues, L. et al., *Bioeng. Bugs.*, 1, 164–171, 2010.)

The constructed flocculent strain was successfully used in long-term continuous lactose fermentations in air-lift bioreactor (Domingues et al. 1999a, 2001), resulting in high ethanol productivities from lactose. Specifically, this strain produced about 10 g/(l h) of ethanol in such system using cheese whey permeate as substrate (Domingues et al. 2001). While producing ethanol, the recombinant *S. cerevisiae* strain cleared the cheese whey permeate of most organic substances, allowing for a significant reduction in its pollutant load. When using twofold concentrated cheese whey permeate, a fermentation product with 5% (w/v) ethanol was obtained (Domingues et al. 2001).

The same recombinant flocculent strain, after a long-term evolutionary engineering process, showed improved performance in terms of lactose fermentation and flocculation capacity (Guimarães et al. 2008a, 2008b, 2008c). The evolved strain produced 7% (v/v) ethanol from concentrated cheese whey (150 g/l lactose) in shake-flask fermentations (Guimarães et al. 2008a). After supplementation of whey with 10 g/l of corn steep liquor the production of ethanol was 7.4% (v/v), with a superior productivity of 1.2 g/(l h), as compared to the productivity obtained with nonsupplemented whey [0.46 g/(l h)] (Silva et al. 2010). During five consecutive batches in a 5.5-l air-lift bioreactor, the average ethanol productivity was 0.65 g/(l h) and ethanol accumulated up to 8% (v/v) with lactose-to-ethanol conversion yields over 80% of theoretical (Silva et al. 2010). Unfortunately, it is not possible to operate continuously with flocculent cells in concentrated cheese whey due to the deflocculating effect attributed to the high-salt concentration (Domingues et al. 2001). Anyway, these flocculent strains present important advantages for industrial application due to the easy and inexpensive cell recycling for repeated batch operation and simplification of the downstream processing, as well as improved ethanol tolerance and cell viability (Silva et al. 2010; Zhao and Bai 2009).

Recently, Zou et al. (2013) have constructed *S. cerevisiae* lactose-consuming strains by the genomic integration of the *K. marxianus* *lac12* and *lac4* genes and simultaneous disruption of specific gene-encoding regions. The target genes were *ATH1* and *NTH1*, to abolish the activity of acid/neutral trehalase, aiming to improve osmotic stress tolerance, and *MIG1*, to relieve glucose repression. In concentrated cheese whey powder solutions, the recombinant strain with the *lac4* and *lac12* genes integrated into the *MIG1* and *NTH1* gene-encoding regions, respectively, produced 63.3 g/l ethanol from approximately 150 g/l initial lactose in 120 h. On the other hand, the strain with the *lac4* and *lac12* genes integrated into the *ATH1* and *NTH1* gene-encoding regions, respectively, consumed only 63.7% of the initial lactose and thus produced less ethanol (35.9 g/l). The authors claimed that relieving glucose repression is an effective strategy for constructing lactose-consuming *S. cerevisiae* strains (Zou et al. 2013).

The fermentation of whey to ethanol using the most effective recombinant-engineered *S. cerevisiae* strains developed so far needs now to be scaled up to provide evidence of industrial significance. These recombinant strains, as well as the best lactose-fermenting *Kluyveromyces* wild strains, may also be further improved by using metabolic engineering and systems biology toolboxes to attain higher fermentative performance. In spite of some examples of industrial implementation using wild lactose-fermenting yeasts, namely, in Ireland, New Zealand, the United Sates, and Denmark, the whey-ethanol fermentation technology should be upgraded in order to enhance the economic attractiveness of this bioprocess and thus to improve its competitiveness with the currently established processes, which use cane sugar and cornstarch as substrates, or with emerging second-generation technologies that use lignocelluloses biomass as raw material (Guimarães et al. 2010).

References

Aktaş, N., Boyacı, İ.H., Mutlu, M., and Tanyolaç, A. 2006. Optimization of lactose utilization in deproteinated whey by *Kluyveromyces marxianus* using response surface methodology (RSM). *Bioresource Technology* 97(18): 2252–9.

Anatovskiy, A. and Yaroshenko, V. 1950. Preparation of sparkling whey. *Molochnaya Promyshlennost* 11(9): 31.

Ariyanti, D. and Hadiyanto, H. 2013. Ethanol production from whey by *Kluyveromyces marxianus* in batch fermentation system: Kinetics parameters estimation. *Bulletin of Chemical Reaction Engineering & Catalysis* 7: 179–84.

Athanasiadis, I., Paraskevopoulou, A., Blekas, G., and Kiosseoglou, V. 2004. Development of a novel whey beverage by fermentation with kefir granules. Effect of various treatments. *Biotechnology Progress* 20(4): 1091–5.

Bailey, R.B., Benitez, T., and Woodward, A. 1982. *Saccharomyces cerevisiae* mutants resistant to catabolite repression: Use in cheese whey hydrolysate fermentation. *Applied and Environmental Microbiology* 44(3): 631–9.

Browne, H.H. 1941. Ethyl alcohol from fermentation of lactose in whey. *Chemical Engineering News* 19: 1272–3.

Bulatović, M.L., Rakin, M.B., Mojović, L.V., Nikolić, S.B., Sekulić, M.S.V., and Vuković, A.J.D. 2012a. Selection of *Lactobacillus* strains for functional whey-based beverage production. *Journal of Food Science and Engineering* 2: 705–11.

Bulatović, M.L., Rakin, M.B., Mojović, L.V., Nikolić, S.B., Sekulić, M.S.V., and Vuković, A.J.D. 2012b. Whey as a raw material for the production of functional beverages. *Hemijska Industrija* 66(4): 565–77.

Carvalho, F., Prazeres, A.R., and Rivas, J. 2013. Cheese whey wastewater: Characterization and treatment. *Science of the Total Environment* 445–446: 385–96.

Casadaban, M.J., Martinezarias, A., Shapira, S.K., and Chou, J. 1983. β-Galactosidase gene fusions for analyzing gene expression in *Escherichia coli* and yeast. *Methods Enzymol* 100: 293–308.

Castro, W.F., Cruz, A.G., Bisinotto, M.S., Guerreiro, L.M.R., Faria, J.A.F., Bolini, H.M.A., Cunha, R.L., and Deliza, R. 2013. Development of probiotic dairy beverages: Rheological properties and application of mathematical models in sensory evaluation. *Journal of Dairy Science* 96(1): 16–25.

Christensen, A.D., Kadar, Z., Oleskowicz-Popiel, P., and Thomsen, M.H. 2011. Production of bioethanol from organic whey using *Kluyveromyces marxianus*. *Journal of Industrial Microbiology & Biotechnology* 38(2): 283–9.

Compagno, C., Porro, D., Smeraldi, C., and Ranzi, B.M. 1995. Fermentation of whey and starch by transformed *Saccharomyces cerevisiae* cells. *Applied Microbiology and Biotechnology* 43(5): 822–5.

Domingues, L., Dantas, M.M., Lima, N., and Teixeira, J.A. 1999a. Continuous ethanol fermentation of lactose by a recombinant flocculating *Saccharomyces cerevisiae* strain. *Biotechnology and Bioengineering* 64(6): 692–7.

Domingues, L., Guimarães, P.M.R., and Oliveira, C. 2010. Metabolic engineering of yeast strains for lactose/whey metabolisation. *Bioengineering Bugs* 1: 164–71.

Domingues, L., Lima, N., and Teixeira, J.A. 2001. Alcohol production from cheese whey permeate using genetically modified flocculent yeast cells. *Biotechnology and Bioengineering* 72(5): 507–14.

Domingues, L., Lima, N., and Teixeira, J.A. 2005. *Aspergillus niger* β-galactosidase production by yeast in a continuous high cell density reactor. *Process Biochemistry* 40(3–4): 1151–4.

Domingues, L., Onnela, M.L., Teixeira, J.A., Lima, N., and Penttila, M. 2000. Construction of a flocculent brewer's yeast strain secreting *Aspergillus niger* β-galactosidase. *Applied Microbiology and Biotechnology* 54(1): 97–103.

Domingues, L., Teixeira, J.A., and Lima, N. 1999b. Construction of a flocculent *Saccharomyces cerevisiae* fermenting lactose. *Applied Microbiology and Biotechnology* 51(5): 621–6.

Domingues, L., Teixeira, J.A., Penttila, M., and Lima, N. 2002. Construction of a flocculent *Saccharomyces cerevisiae* strain secreting high levels of *Aspergillus niger* β-galactosidase. *Applied Microbiology and Biotechnology* 58(5): 645–50.

Dragone, G., Mussatto, S.I., Almeida e Silva, J.B., and Teixeira, J.A. 2011. Optimal fermentation conditions for maximizing the ethanol production by *Kluyveromyces fragilis* from cheese whey powder. *Biomass & Bioenergy* 35(5): 1977–82.

Dragone, G., Mussatto, S.I., Oliveira, J.M., and Teixeira, J.A. 2009. Characterisation of volatile compounds in an alcoholic beverage produced by whey fermentation. *Food Chemistry* 112(4): 929–35.

Ferrari, M.D., Loperena, L., and Varela, H. 1994. Ethanol production from concentrated whey permeate using a fed-batch culture of *Kluyveromyces fragilis*. *Biotechnology Letters* 16(2): 205–10.

Gallardo-Escamilla, F.J., Kelly, A.L., and Delahunty, C.M. 2005. Influence of starter culture on flavor and headspace volatile profiles of fermented whey and whey produced from fermented milk. *Journal of Dairy Science* 88(11): 3745–53.

Gancedo, J.M. 1998. Yeast carbon catabolite repression. *Microbiology and Molecular Biology Reviews* 62(2): 334–61.

Gawel, J. and Kosikowski, F.V. 1978. Improving alcohol fermentation in concentrated ultrafiltration permeates of cottage cheese whey. *Journal of Food Science* 43(6): 1717–9.

Ghaly, A.E. and El-Taweel, A.A. 1995. Effect of micro-aeration on the growth of *Candida pseudotropicalis* and production of ethanol during batch fermentation of cheese whey. *Bioresource Technology* 52(3): 203–17.

Ghaly, A.E. and El-Taweel, A.A. 1997a. Continuous ethanol production from cheese whey fermentation by *Candida pseudotropicalis*. *Energy Sources* 19(10): 1043–63.

Ghaly, A.E. and El-Taweel, A.A. 1997b. Kinetic modelling of continuous production of ethanol from cheese whey. *Biomass & Bioenergy* 12(6): 461–72.

Gianetto, A., Berruti, F., Glick, B.R., and Kempton, A.G. 1986. The production of ethanol from lactose in a tubular reactor by immobilized cells of *Kluyveromyces fragilis*. *Applied Microbiology and Biotechnology* 24(4): 277–81.

Grba, S., Stehlik-Tomas, V., Stanzer, D., Vahcic, N., and Skrlin, A. 2002. Selection of yeast strain *Kluyveromyces marxianus* for alcohol and biomass production on whey. *Chemical and Biochemical Engineering Quarterly* 16(1): 13–6.

Grubb, C.F. and Mawson, A.J. 1993. Effects of elevated solute concentrations on the fermentation of lactose by *Kluyveromyces marxianus* Y-113. *Biotechnology Letters* 15(6): 621–6.

Guimarães, P.M.R., François, J., Parrou, J.L., Teixeira, J.A., and Domingues, L. 2008a. Adaptive evolution of a lactose-consuming *Saccharomyces cerevisiae* recombinant. *Applied and Environmental Microbiology* 74(6): 1748–56.

Guimarães, P.M.R., Le Berre, V., Sokol, S., François, J., Teixeira, J.A., and Domingues, L. 2008b. Comparative transcriptome analysis between original and evolved recombinant lactose consuming *Saccharomyces cerevisiae* strains. *Biotechnology Journal* 3: 1591–7.

Guimarães, P.M.R., Teixeira, J.A., and Domingues, L. 2008c. Fermentation of high concentrations of lactose to ethanol by engineered flocculent *Saccharomyces cerevisiae*. *Biotechnology Letters* 30(11): 1953–8.

Guimarães, P.M.R., Teixeira, J.A., and Domingues, L. 2010. Fermentation of lactose to bio-ethanol by yeasts as part of integrated solutions for the valorisation of cheese whey. *Biotechnology Advances* 28(3): 375–84.

Guo, X.W., Wang, R.S., Chen, Y.F., and Xiao, D.G. 2012. Intergeneric yeast fusants with efficient ethanol production from cheese whey powder solution: Construction of a *Kluyveromyces marxianus* and *Saccharomyces cerevisiae* AY-5 hybrid. *Engineering in Life Sciences* 12(6): 656–61.

Guo, X.W., Zhou, J., and Xiao, D. 2010. Improved ethanol production by mixed immobilized cells of *Kluyveromyces marxianus* and *Saccharomyces cerevisiae* from cheese whey powder solution fermentation. *Applied Biochemistry and Biotechnology* 160(2): 532–8.

Guzel-Seydim, Z., Wyffels, J.T., Seydim, A.C., and Greene, A.K. 2005. Turkish kefir and kefir grains: Microbial enumeration and electron microscobic observation. *International Journal of Dairy Technology* 58(1): 25–9.

Hahn-Hägerdal, B. 1985. Comparison between immobilized *Kluyveromyces fragilis* and *Saccharomyces cerevisiae* coimmobilized with β-galactosidase, with respect to continuous ethanol production from concentrated whey permeate. *Biotechnology and Bioengineering* 27(6): 914–6.

Hernandez-Mendoza, A., Robles, V.J., Angulo, J.O., De La Cruz, J., and Garcia, H.S. 2007. Preparation of a whey-based probiotic product with *Lactobacillus reuteri* and *Bifidobacterium bifidum*. *Food Technology and Biotechnlogy* 45(1): 27–31.

Janssens, J.H., Bernard, A., and Bailey, R.B. 1984. Ethanol from whey— Continuous fermentation with cell recycle. *Biotechnology and Bioengineering* 26(1): 1–5.

Janssens, J.H., Burris, N., Woodward, A., and Bailey, R.B. 1983. Lipid-enhanced ethanol production by *Kluyveromyces fragilis*. *Applied and Environmental Microbiology* 45(2): 598–602.

Jarvis, J.K., McBean, L.D., and Miller, G.D. 2007. *Handbook of Dairy Foods and Nutrition* (3rd edition). Boca Raton, FL: CRC Press.

Jędrzejewska, M. and Kozak, K. 2011. Ethanol production from whey permeate in a continuous anaerobic bioreactor by *Kluyveromyces marxianus*. *Environmental Technology* 32(1–2): 37–42.

Jelicic, I., Botanic, R., and Tratnik, L. 2008. Whey based beverages—New generation of dairy products. *Mljekarstvo* 58(3): 257–74.

Kargi, F. and Ozmihci, S. 2006. Utilization of cheese whey powder (CWP) for ethanol fermentations: Effects of operating parameters. *Enzyme and Microbial Technology* 38(5): 711–18.

Kleine, R., Achenbach, S., and Thoss, S. 1995. Whey disposal by deproteinization and fermentation. *Acta Biotechnologica* 15(2): 139–48.

Kourkoutas, Y., Dimitropoulou, S., Kanellaki, M., Marchant, R., Nigam, P., Banat, I.M., and Koutinas, A.A. 2002. High-temperature alcoholic fermentation of whey using *Kluyveromyces marxianus* IMB3 yeast immobilized on delignified cellulosic material. *Bioresource Technology* 82(2): 177–81.

Koushki, M., Jafari, M., and Azizi, M. 2012. Comparison of ethanol production from cheese whey permeate by two yeast strains. *Journal of Food Science and Technology-Mysore* 49(5): 614–9.

Koutinas, A.A., Athanasiadis, I., Bekatorou, A., Psarianos, C., Kanellaki, M., Agouridis, N., and Blekas, G. 2007. Kefir-yeast technology: Industrial scale-up of alcoholic fermentation of whey, promoted by raisin extracts, using kefir-yeast granular biomass. *Enzyme and Microbial Technology* 41(5): 576–82.

Larson, P.K. and Yang, H.Y. 1976. Some factors involved in the clarification of whey wine. *Journal of Milk and Food Technology* 39: 614–8.

Lewandowska, M. and Kujawski, W. 2007. Ethanol production from lactose in a fermentation/pervaporation system. *Journal of Food Engineering* 79(2): 430–7.

Longhi, L.G.S., Luvizetto, D.J., Ferreira, L.S., Rech, R., Ayub, M.A.Z., and Secchi, A.R. 2004. A growth kinetic model of *Kluyveromyces marxianus* cultures on cheese whey as substrate. *Journal of Industrial Microbiology & Biotechnology* 31(1): 35–40.

Magalhães, K.T., Dragone, G., de Melo Pereira, G.V., Oliveira, J.M., Domingues, L., Teixeira, J.A., Almeida e Silva, J.B., and Schwan, R.F. 2010. Production of fermented cheese whey-based beverage using kefir grains as starter culture: Evaluation of morphological and microbial variations. *Bioresource Technolology* 101(22): 8843–50.

Magalhães, K.T., Dragone, G., de Melo Pereira, G.V., Oliveira, J.M., Domingues, L., Teixeira, J.A., Almeida e Silva, J.B., and Schwan, R.F. 2011. Comparative study of the biochemical changes and volatile compound formations during the production of novel whey-based kefir beverages and traditional milk kefir. *Food Chemistry* 126(1): 249–53.

Mahmoud, M.M. and Kosikowski, F.V. 1982. Alcohol and single cell protein production by *Kluyveromyces* in concentrated whey permeates with reduced ash. *Journal of Dairy Science* 65(11): 2082–7.

Maity, T.K., Kumar, R., and Misra, A.K. 2008. Development of healthy whey drink with *Lactobacillus rhamnosus*, *Bifidobacterium bifidum* and *Propionibacterium freudenreichii* subsp. *shermanii*. *Mljekarstvo* 58(4): 315–25.

Mandal, S., Puniya, M., Sangu, K.P.S., Dagar, S.S., Singh, R., and Puniya, A.K. 2012. Whey, a waste or a resource: The shifting perception after valorization. In: *Valorization of Food Processing By-Products*, ed. M. Chandrasekaran, 617–45. Boca Raton, FL: CRC Press/Taylor & Francis Group.

Mauriello, G., Moio, L., Moschetti, G., Piombino, P., Addeo, F., and Coppola, S. 2001. Characterization of lactic acid bacteria strains on the basis of neutral volatile compounds produced in whey. *Journal of Applied Microbiology* 90(6): 928–42.

Mehaia, M.A. and Cheryan, M. 1990. Ethanol from hydrolyzed whey permeate using *Saccharomyces cerevisiae* in a membrane recycle bioreactor. *Bioprocess Engineering* 5(2): 57–61.

Mollea, C., Marmo, L., and Bosco, F. 2013. Valorisation of cheese whey, a by-product from the dairy industry. In: *Food Industry*, ed. I. Muzzalupo, 549–88. Rijeka, Croatia: InTech.

Nehlin, J.O., Carlberg, M., and Ronne, H. 1989. Yeast galactose permease is related to yeast and mammalian glucose transporters. *Gene* 85(2): 313–9.

O'Leary, V.S., Green, R., Sullivan, B.C., and Holsinger, V.H. 1977. Alcohol production by selected yeast strains in lactase-hydrolyzed acid whey. *Biotechnology and Bioengineering* 19(7): 1019–35.

Oliveira, C., Guimarães, P.M., and Domingues, L. 2011. Recombinant microbial systems for improved β-galactosidase production and biotechnological applications. *Biotechnology Advances* 29(6): 600–9.

Oliveira, C., Teixeira, J.A., Lima, N., Da Silva, N.A., and Domingues, L. 2007. Development of stable flocculent *Saccharomyces cerevisiae* strain for continuous *Aspergillus niger* β-galactosidase production. *Journal of Bioscience and Bioengineering* 103(4): 318–24.

Ozmihci, S. and Kargi, F. 2007a. Continuous ethanol fermentation of cheese whey powder solution: Effects of hydraulic residence time. *Bioprocess and Biosystems Engineering* 30(2): 79–86.

Ozmihci, S. and Kargi, F. 2007b. Effects of feed sugar concentration on continuous ethanol fermentation of cheese whey powder solution (CWP). *Enzyme and Microbial Technology* 41(6–7): 876–80.

Ozmihci, S. and Kargi, F. 2007c. Ethanol fermentation of cheese whey powder solution by repeated fed-batch operation. *Enzyme and Microbial Technology* 41(1–2): 169–74.

Ozmihci, S. and Kargi, F. 2007d. Kinetics of batch ethanol fermentation of cheese-whey powder (CWP) solution as function of substrate and yeast concentrations. *Bioresource Technology* 98(16): 2978–84.

Ozmihci, S. and Kargi, F. 2008. Ethanol production from cheese whey powder solution in a packed column bioreactor at different hydraulic residence times. *Biochemical Engineering Journal* 42(2): 180–5.

Panesar, P.S. and Kennedy, J.F. 2012. Biotechnological approaches for the value addition of whey. *Critical Reviews in Biotechnology* 32(4): 327–48.

Panesar, P.S., Kennedy, J.F., Gandhi, D.N., and Bunko, K. 2007. Bioutilisation of whey for lactic acid production. *Food Chemistry* 105(1): 1–14.

Pavunc, A.L., Turk, J., Beganović, J., Frece, J., Mahnet, S., Kirin, S., and Šušković, J. 2009. Production of fermented probiotic beverages from milk permeate enriched with whey retentate and identification of present lactic acid bacteria. *Mljekarstvo* 59(1): 11–9.

Pescuma, M., Hebert, E.M., Mozzi, F., and de Valdez, G.F. 2008. Whey fermentation by thermophilic lactic acid bacteria: Evolution of carbohydrates and protein content. *Food Microbiology* 25(3): 442–51.

Porro, D., Martegani, E., Ranzi, B.M., and Alberghina, L. 1992. Lactose/whey utilization and ethanol production by transformed *Saccharomyces cerevisiae* cells. *Biotechnology and Bioengineering* 39(8): 799–805.

Prazeres, A.R., Carvalho, F., and Rivas, J. 2012. Cheese whey management: A review. *Journal of Environmental Management* 110: 48–68.

Rogosa, M., Browne, H.H., and Whittier, E.O. 1947. Ethyl alcohol from whey. *Journal of Dairy Science* 30(4): 263–9.

Rosenberg, M., Tomaska, M., Kanuch, J., and Sturdik, E. 1995. Improved ethanol production from whey with *Saccharomyces cerevisiae* using permeabilized cells of *Kluyveromyces marxianus*. *Acta Biotechnologica* 15(4): 387–90.

Roukas, T. and Lazarides, H.N. 1991. Ethanol production from deproteinized whey by β-galactosidase coimmobilized cells of *Saccharomyces cerevisiae*. *Journal of Industrial Microbiology* 7(1): 15–8.

Rubio-Texeira, M., Castrillo, J.I., Adam, A.C., Ugalde, U.O., and Polaina, J. 1998. Highly efficient assimilation of lactose by a metabolically engineered strain of *Saccharomyces cerevisiae*. *Yeast* 14(9): 827–37.

Sansonetti, S., Curcio, S., Calabrò, V., and Iorio, G. 2009. Bio-ethanol production by fermentation of ricotta cheese whey as an effective alternative non-vegetable source. *Biomass & Bioenergy* 33(12): 1687–92.

Sansonetti, S., Curcio, S., Calabrò, V., and Iorio, G. 2010. Optimization of ricotta cheese whey (RCW) fermentation by response surface methodology. *Bioresource Technology* 101(23): 9156–62.

Sansonetti, S., Hobley, T.J., Calabrò, V., Villadsen, J., and Sin, G. 2011. A biochemically structured model for ethanol fermentation by *Kluyveromyces marxianus*: A batch fermentation and kinetic study. *Bioresource Technology* 102(16): 7513–20.

Sansonetti, S., Hobley, T.J., Curcio, S., Villadsen, J., and Sin, G. 2013. Use of continuous lactose fermentation for ethanol production by *Kluveromyces marxianus* for verification and extension of a biochemically structured model. *Bioresource Technology* 130: 703–9.

Saraceno, A., Curcio, S., Calabrò, V., and Iorio, G. 2010. A hybrid neural approach to model batch fermentation of "ricotta cheese whey" to ethanol. *Computers and Chemical Engineering* 34(10): 1590–96.

Saraceno, A., Sansonetti, S., Calabro, V., Iorio, G., and Curcio, S. 2011. A comparison between different modeling techniques for the production of bio-ethanol from dairy industry wastes. *Chemical and Biochemical Engineering Quarterly* 25(4): 461–9.

Sienkiewicz, T. and Riedel, C.L. 1990. *Whey and Whey Utilization: Possibilities for Utilization in Agriculture and Foodstuffs Production*. Gelsenkirchen, Germany: Verlag Th. Mann.

Silva, A.C., Guimarães, P.M., Teixeira, J.A., and Domingues, L. 2010. Fermentation of deproteinized cheese whey powder solutions to ethanol by engineered *Saccharomyces cerevisiae*: Effect of supplementation with corn steep liquor and repeated-batch operation with biomass recycling by flocculation. *Journal of Industrial Microbiology & Biotechnology* 37(9): 973–82.

Silveira, W.B., Passos, F., Mantovani, H.C., and Passos, F.M.L. 2005. Ethanol production from cheese whey permeate by *Kluyveromyces marxianus* UFV-3: A flux analysis of oxido-reductive metabolism as a function of lactose concentration and oxygen levels. *Enzyme and Microbial Technology* 36(7): 930–6.

Singh, S., Khemariya, P., and Rai, A. 2011. Process optimization for the manufacture of lemon based beverage from hydrolyzed whey. *Journal of Food Science and Technology*. doi:10.1007/s13197-011-0563-1.

Siso, M.I.G. 1996. The biotechnological utilization of cheese whey: A review. *Bioresource Technology* 57(1): 1–11.

Skudra, L., Blija, A., Sturmović, E., Dukaļska, L., Áboltiņš, A., and Krkliņa, D. 1998. Studies on whey fermentation using lactic acid bacteria *L. acidophilus* and *L. bulgaricus*. *Acta Biotechnologica* 18(3): 277–86.

Smithers, G.W. 2008. Whey and whey proteins—From "gutter-to-gold." *International Dairy Journal* 18(7): 695–704.

Sreekrishna, K. and Dickson, R.C. 1985. Construction of strains of *Saccharomyces cerevisiae* that grow on lactose. *Proceedings of the National Academy of Sciences of the United States of America* 82(23): 7909–13.

Staniszewski, M., Kujawski, W., and Lewandowska, M. 2009. Semi-continuous ethanol production in bioreactor from whey with co-immobilized enzyme and yeast cells followed by pervaporative recovery of product—Kinetic model predictions considering glucose repression. *Journal of Food Engineering* 91(2): 240–9.

Szczodrak, J., Szewczuk, D., Rogalski, J., and Fiedurek, J. 1997. Selection of yeast strain and fermentation conditions for high-yield ethanol production from lactose and concentrated whey. *Acta Biotechnologica* 17(1): 51–61.

Tada, S., Katakura, Y., Ninomiya, K., and Shioya, S. 2007. Fed-batch coculture of *Lactobacillus kefiranofaciens* with *Saccharomyces cerevisiae* for effective production of kefiran. *Journal of Bioscience and Bioengineering* 103(6): 557–62.

Tahoun, M.K., El-Nemr, T.M., and Shata, O.H. 1999. Ethanol from lactose in salted cheese whey by recombinant *Saccharomyces cerevisiae*. *Zeitschrift Fur Lebensmittel-Untersuchung Und-Forschung a-Food Research and Technology* 208(1): 60–4.

Tahoun, M.K., El-Nemr, T.M., and Shata, O.H. 2002. A recombinant *Saccharomyces cerevisiae* strain for efficient conversion of lactose in salted and unsalted cheese whey into ethanol. *Nahrung* 46(5): 321–6.

Terrell, S.L., Bernard, A., and Bailey, R.B. 1984. Ethanol from whey: Continuous fermentation with a catabolite repression-resistant *Saccharomyces cerevisiae* mutant. *Applied and Environmental Microbiology* 48(3): 577–80.

Varela, H., Ferrari, M.D., Loperena, L., and Lareo, C. 1992. Effect of aeration rate on the alcoholic fermentation of whey by *Kluyveromyces fragilis*. *Microbiologia* 8(1): 14–20.

Vienne, P. and von Stockar, U. 1985. An investigation of ethanol inhibition and other limitations occurring during the fermentation of concentrated whey permeate by *Kluyveromyces fragilis*. *Biotechnology Letters* 7(7): 521–6.

Yoo, B.W. and Mattick, J.F. 1969. Utilization of acid and sweet cheese whey in wine production. *Journal of Dairy Science* 52(6): 900–6.

Zafar, S. and Owais, M. 2006. Ethanol production from crude whey by *Kluyveromyces marxianus*. *Biochemical Engineering Journal* 27(3): 295–8.

Zafar, S., Owais, M., Salleemuddin, M., and Husain, S. 2005. Batch kinetics and modelling of ethanolic fermentation of whey. *International Journal of Food Science and Technology* 40(6): 597–604.

Zhao, X.Q. and Bai, F.W. 2009. Yeast flocculation: New story in fuel ethanol production. *Biotechnology Advances* 27(6): 849–56.

Zoppellari, F. and Bardi, L. 2012. Production of bioethanol from effluents of the dairy industry by *Kluyveromyces marxianus*. *New Biotechnology*. doi:10.1016/j.nbt.2012.11.017.

Zou, J., Guo, X., Shen, T., Dong, J., Zhang, C., and Xiao, D. 2013. Construction of lactose-consuming *Saccharomyces cerevisiae* for lactose fermentation into ethanol fuel. *Journal of Industrial Microbiology & Biotechnology* 40(3–4): 353–63.

PART IV
DAIRY PRODUCTS BASED ON THE TYPE OF FERMENTATION
Fungal Lactic Fermentations

17

Milk Kefir

Structure and Microbiological and Chemical Composition

ROSANE FREITAS SCHWAN, KARINA TEIXEIRA MAGALHÃES-GUEDES, AND DISNEY RIBEIRO DIAS

Contents

17.1	Introduction	462
17.2	Structure of Kefir Grains	463
	17.2.1 Fluorescence Staining and Confocal Laser Scanning Microscopy Imaging	463
	17.2.2 Scanning Electron Microscopy	464
17.3	Microbial Composition	466
	17.3.1 Kefir Grains	466
	17.3.1.1 Lactic Acid Bacteria	466
	17.3.1.2 Acetic Acid Bacteria	466
	17.3.1.3 Yeasts	467
	17.3.2 Kefir Beverages	467
	17.3.2.1 Lactic Acid Bacteria	467
	17.3.2.2 Acetic Acid Bacteria	468
	17.3.2.3 Yeasts	468
17.4	Chemical Composition	469
	17.4.1 Kefir Flavor	469
	17.4.2 Impact of Environmental Factors on Flavor	473
	17.4.2.1 Grain Activity	473
	17.4.2.2 Grain-to-Milk Ratio	474
	17.4.2.3 Starter	474
	17.4.2.4 Incubation Temperature	474
	17.4.2.5 Effect of pH	475
	17.4.2.6 Storage Conditions	475

17.5   Production and Uses of Kefir   476
17.6   Final Considerations   477
References   477

17.1 Introduction

Kefir, a fermented milk beverage, has its origin in the Caucasus region, but it is consumed throughout the world. Kefir grains, a cluster of microorganisms held together by a polysaccharide matrix referred to as kefiran, serve as the fermentation inoculum. These grains are small, irregularly shaped, yellowish-white, hard granules that resemble miniature cauliflower blossoms. Kefir is a probiotic food (Otles and Cagindi 2003; Guedes et al. 2014), and it is well known that probiotic foods are associated with health benefits that are of significant current interest of food industry (Ahmed et al. 2013).

Kefir grains are an example of symbiosis between yeast and bacteria. A variety of yeast and bacteria species that form the kefir grains have been isolated and identified (Angulo et al. 1993; Garrote et al. 2001; Simova et al. 2002; Loretan et al. 2003; Latorre-Garcia et al. 2007; Jianzhong et al. 2009; Bergmann et al. 2010; Magalhães et al. 2010; Miguel et al. 2010; Magalhães et al. 2011c; Puerari et al. 2012; Gao et al. 2013). The *Lactobacillus* genus is the most common genus during the fermentation period. Other lactic bacteria such as species from the *Leuconostoc*, *Lactococcus*, and *Streptococcus* genera are also commonly detected (Angulo et al. 1993; Garrote et al. 2001; Loretan et al. 2003; Jianzhong et al. 2009; Magalhães et al. 2010; Miguel et al. 2010; Magalhães et al. 2011c; Puerari et al. 2012; Gao et al. 2013), as well as Gram-negative bacteria in the *Acetobacter* genera (Puerari et al. 2012; Gao et al. 2013). The yeast isolates include species from the following genera: *Kluyveromyces*, *Candida*, *Torulaspora*, *Pichia*, *Cryptococcus*, *Debaromyces*, *Kazachstania*, *Lachancea*, *Torulaspora*, *Zygosaccharomyces*, *Trichosporon*, and *Saccharomyces* (Angulo et al. 1993; Simova et al. 2002; Loretan et al. 2003; Latorre-Garcia et al. 2007; Bergmann et al. 2010; Magalhães et al. 2010; Magalhães et al. 2011c; Puerari et al. 2012; Garofalo et al. 2015).

The microorganisms present in kefir perform three types of fermentation: lactic, alcoholic, and acetic. The flavor compounds responsible for the typical kefir taste can be divided into two groups: major and minor compounds (secondary metabolites). The only major compound is lactic acid; the minor compound group includes flavor compounds produced during the microbial stationary growth phase. Carbonyl compounds (acetaldehyde, ethanol, diacetyl, acetoin, 2-butanone, and ethyl acetate), volatile organic acids (formic, acetic, propionic, and butyric), and nonvolatile acids (lactic, pyruvic, oxalic, and succinic) are secondary metabolites and are classified as kefir flavor-forming compounds (Guzel-Seydim et al. 2000; Ostlie et al. 2003; Magalhães et al. 2011a, 2011b).

In this chapter, a comprehensive and detailed discussion of the structure, microbiological and chemical composition, production, and utilization of kefir will be presented.

17.2 Structure of Kefir Grains

17.2.1 Fluorescence Staining and Confocal Laser Scanning Microscopy Imaging

Microscale examination of the structure of kefir grains was performed by fluorescently probing the distribution of cells (bacteria and yeast) and polysaccharides by CLSM examination (Figure 17.1) (Magalhães et al. 2010). The microbial biomass visualized in Figure 17.1b covers a great portion of the external surface localized both within and between the inner regions (Figure 17.1c and 17.1d), that is, the polysaccharide matrix. This polysaccharide matrix, called kefiran, is produced by lactic acid bacteria and usually associated with the therapeutic properties of kefir (Tada et al. 2007). Kefiran has frequently been claimed to be effective against a variety of complaints and diseases. Several studies have investigated the antitumor activity, antibacterial and antifungal activities (Otles and Cagindi 2003; Silva et al. 2009). Recently, the potential of kefiran to modulate key steps in the virulence of *Bacillus cereus* in the context of intestinal infections has been reported (Medrano et al. 2009). *Lb. kefiranofaciens* and several other unidentified species of *Lactobacillus* have been pointed by several authors as the major producers of the kefiran polymer in kefir grains (Tada et al. 2007). A similar distribution pattern was observed in the internal surface of the grains, with macroclusters of

464 FERMENTED MILK AND DAIRY PRODUCTS

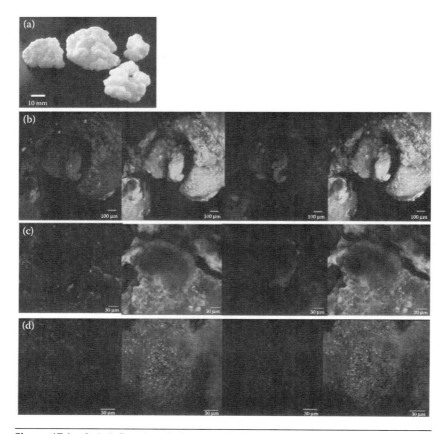

Figure 17.1 Grain kefir under naked-eye examination (a). Outer portion of the kefir grains with distribution of cells (bacteria and yeast) and polysaccharides (b). Inner portion of the kefir grains with distribution of cells (bacteria and yeast) and polysaccharides (c and d). (Data from Magalhães, K.T. et al., *Bioresour. Technol.*, 101, 8843–8850, 2010.)

yeasts distributed within the grain's matrix, essentially composed of polysaccharides and bacteria.

17.2.2 Scanning Electron Microscopy

A complex and tightly packed biofilm has been observed around the grains, whereas the interior is composed primarily of unstructured material. Figures 17.2 and 17.3 show the association of the kefir microbiota using scanning electron microscopy (SEM) (Magalhães et al. 2011c). The kefir grains have a smooth exterior surface (Figure 17.2a) covered

MILK KEFIR 465

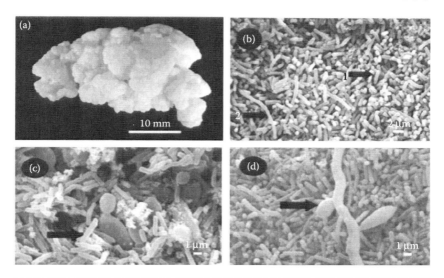

Figure 17.2 (a) Kefir grain viewed with the naked eye. (b–d) External surface of kefir grain. Arrow 1 (b)—short bacteria. Arrow 2 (b)—long bacteria. Arrow (c)—bacteria. Arrow (d)—yeasts. (Data from Magalhães, K.T. et al., *Brazilian J. Microbiol.* 42, 693–702, 2011c.)

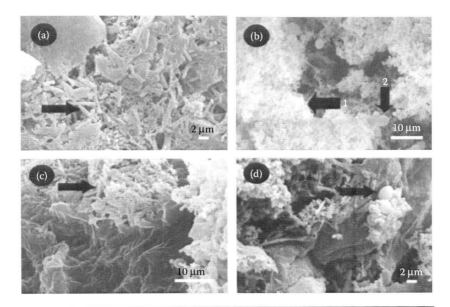

Figure 17.3 Internal surface of kefir grain. Arrows (a and c)—bacteria. Arrow 1 (b)—polysaccharide. Arrow 2 (b)—yeasts. Arrow (d)—yeasts. (Data from Magalhães, K.T. et al., *Brazilian J. Microbiol.* 42, 693–702, 2011c.)

by an agglomerate of microorganisms (Figure 17.2b through 17.2d). The microbiota on the exterior of the grain is dominated by bacilli cells (short and curved long) growing in association with lemon-shaped yeast cells (Figure 17.2b through 17.2d). Fewer microbial cells were observed on the inner portion than on the outer portion (Figure 17.3). Fibrillar material (polysaccharide kefiran) was observed on the outer as well as the inner portion of the grains (Figures 17.2c and 17.3b).

Together, CLSM and SEM revealed the structure and relative proportion of microbiota and polysaccharides on the kefir grains.

17.3 Microbial Composition

17.3.1 Kefir Grains

Kefir grains are microbial rich, cauliflower-like structures typically consisting of three groups of microorganisms living in symbiotic association. These include LAB, acetic acid bacteria (AAB), and yeasts (Loretan et al. 2003; Chen et al. 2008; Magalhães et al. 2010, 2011c; Puerari et al. 2012; Garofalo et al. 2015).

17.3.1.1 Lactic Acid Bacteria The major LAB population may be either homofermentative or heterofermentative (Lin et al. 1999) and comprises 65%–80% of the total microbial population (Wouters et al. 2002). Of the lactobacilli, 74.5% are heterofermentative and 25.7% are homofermentative (Angulo et al. 1993). The same distribution pattern was reported by Garrote et al. (2001): 20 isolates of heterofermentative lactobacilli were detected compared with 16 homofermentative isolates. In general, Kefir grains contain lactobacilli, lactococci (Garrote et al. 2001; Witthuhn et al. 2005; Chen et al. 2008; Miguel et al. 2010; Magalhães et al. 2011c; Gao et al. 2013), and *Leuconostoc* sp. (Garrote et al. 2001; Jianzhong et al. 2009; Gao et al. 2013).

17.3.1.2 Acetic Acid Bacteria AAB represent only 20% of the total microbial population and are usually present in lower counts (<10^5 CFU g^{-1}) (Garrote et al. 2001; Magalhães et al. 2011c; Miguel et al. 2010). However, counts as high as 10^8 CFU g^{-1} were reported by Abraham and De Antoni (1999). In other studies, AAB were not detected (Witthuhn et al. 2005), and sometimes they were considered to be contaminants (Angulo et al. 1993). According to Rea et al. (1996), AAB may stimulate the growth of other organisms because they are vitamin B_{12} producers. Koroleva (1988) reported that the consistency of kefir can be improved by using a starter containing at least 10^6 CFU g^{-1} AAB. Thus, the presence of AAB may be important for good kefir consistency and therefore a quality product.

17.3.1.3 Yeasts The yeasts present in kefir grains have been reported to be either lactose fermenting and/or non-lactose fermenting (Simova et al. 2002; Loretan et al. 2003; Latorre-Garcia et al. 2007). There are usually fewer yeast than LAB (approximately 10^4–10^5 CFU g^{-1}) (Latorre-Garcia et al. 2007), but in some grains, the yeast accounted for a larger percentage of the total microbes than LAB (50% and 31.2%, respectively) (Angulo et al. 1993; Zajsek and Gorsek 2010).

17.3.2 Kefir Beverages

Kefir beverages owe their microbial composition to the presence of kefir grains. Once in milk, kefir grain microorganisms are released and continue to multiply (Kroger 1993) using the available nutrients in the milk, especially lactose, which serves as the carbon and energy source. It is therefore expected that both kefir grains and beverages should have a similar composition. Although the microbial profiles of kefir grains and kefir beverages are very similar, the use of kefir beverage as an inoculum to make a new batch of kefir is not recommended and the grains are essential to produce traditional kefir (Simova et al. 2002). It is also preferable to use kefir grains rather than a mixture of pure cultures as a starter (Assadi et al. 2000). According to Marshall (1984), the integrity of the grains is essential to achieve the effervescent character, typical yeast flavor, and creamy texture of kefir (Simova et al. 2002).

17.3.2.1 Lactic Acid Bacteria LAB population of the beverage has been reported to be higher than the yeast population by some researchers (Witthuhn et al. 2005; Ertekin and Guzel-Seydim 2010), especially the lactobacilli population (10^8–10^9 CFU ml^{-1}). In contrast, others (Rea et al. 1996; Beshkova et al. 2002) have found more lactococci (10^9 CFU ml^{-1}) than lactobacilli (10^6 CFU ml^{-1}). The *Leuconostoc* species was the second major group of microorganisms isolated from an Irish kefir beverage, with a count of 10^8 CFU ml^{-1}. *Leuconostoc* sp. grow poorly in milk and are usually found in association with lactococci (Rea et al. 1996).

17.3.2.2 Acetic Acid Bacteria AAB are typically found in kefir beverage at levels between 10^8 and 10^9 cfu ml^{-1} (Rea et al. 1996; Loretan et al. 2003; Magalhães et al. 2011c); however, AAB are not always present in kefir beverage and are sometimes considered contaminants (Angulo et al. 1993).

17.3.2.3 Yeasts Lactose- or non-lactose-fermenting yeast have been reported to exist in kefir beverage at levels of 10^4–10^5 CFU ml^{-1} (Loretan et al. 2003; Jianzhong et al. 2009; Miguel et al. 2010). Yeasts are generally present in kefir beverage (Simova et al. 2002; Zajsek and Gorsek 2010; Magalhães et al. 2010, 2011c; Puerari et al. 2012).

The overall microbial composition of kefir is complex and varies among geographic regions. The environment (i.e., cultivation, preservation, and storage conditions) is the principal factor leading to the microbial diversity of kefir grains (Latorre-Garcia et al. 2007; Miguel et al. 2010; Magalhães et al. 2011c). Tables 17.1 and 17.2 show the different microorganisms isolated from kefir grains and beverages.

Table 17.1 Species of Bacteria Present in Kefir (Grains and Beverages)

MICROORGANISMS	REFERENCES
Lb. brevis, Str. salivarium subsp. *thermophillus, Lb. casei* subsp. *rhamnosus, Lb. casei* subsp. *tolerans, Lb. fermentum, Lb. gasseri, Lb. viridescens, Lc.* subsp. *lactis, Str. durans, Str. salivarium* subsp. *thermophillus, Lb. helveticus, Lb. bulgaricus, Str. durans*	Angulo et al. (1993), Frengova et al. (2002), Loretan et al. (2003), Jianzhong et al. (2009), Kesmen and Kacmaz (2011)
Lb. casei subsp. *pseudoplantarum, Lb. delbrueckii* subsp. *bulgaricus*	Simova et al. (2002)
Lb. delbrueckii subsp. *lactis, Acetobacter acetic*	Marshall (1984)
Lb. fermentum, Leu. mesenteroides subsp. *dextranicum*	Witthuhn et al. (2005), Puerari et al. (2012)
Lb. kefir (also known as *kefiri*), *Lb. lactis paracasei* subsp. *paracasei, Lb. lactis parabuchneri, Lb. lactis paracasei, Lb. lactis paracasei* subsp. *tolerans, Lb. kefiranogrum, Lb. lactis* subsp. *lactis, Lb. kefiranofaciens, Lb. kefiranofaciens* subsp. *kefirgranum, Lb. plantarum*	Garrote et al. (2001), Chen et al. (2008), Jianzhong et al. (2009), Miguel et al. (2010), Magalhães et al. (2011c), Puerari et al. (2012)
Acetobacter sp., *Acetobacter lovaniensis, Acetobacter syzgii*	Miguel et al. (2010), Magalhães et al. (2011c), Puerari et al. (2012)
Lc. lactis subsp. *cremoris, Lc. lactis* subsp. *cremoris, Leu. dextranicum, Leu. Kefir*	Kuo and Lin (1999), Garofalo et al. (2015)
Lb. filant, Lc. Lactis	Koroleva et al. (1988)
Shewanella sp., *Pseudomonas* sp., *Acinetobacter* sp., *Pelomonas* sp., *Dysgonomonas* sp., *Weissella* sp.	Gao et al. (2013)

Table 17.2 Species of Yeasts Present in Kefir (Grains and Beverages)

MICROORGANISMS	REFERENCES
Candida famata	Bergmann et al. (2010)
Candida holmii	Witthuhn et al. (2005)
Candida inconspicua	Simova et al. (2002), Bergmann et al. (2010)
Candida kefyr, Saccharomyces unisporus	Angulo et al. (1993), Witthuhn et al. (2005)
Candida krusei	Latorre-Garcia et al. (2007), Garofalo et al. (2015)
Candida mari, Kluyveromyces marxianus var. *lactis, Pichia* sp., *Pichia fermentans*	Simova et al. (2002)
Candida sake, Saccharomyces humaticus	Latorre-Garcia et al. (2007)
Cryptococcus humicolus, Zygosaccharomyces sp.	Witthuhn et al. (2005)
Debaromyces hansenii, Zygosaccharomyces florentinus	Loretan et al. (2003)
Kazachstania aerobia	Magalhães et al. (2011c)
Kazachstania exigua	Jianzhong et al. (2009)
Kazachstania unispora	Jianzhong et al. (2009), Magalhães et al. (2010), Puerari et al. (2012)
Kluyveromyces fragilis	Jianzhong et al. (2009)
Kluyveromyces lactis, Lachancea meyersii	Angulo et al. (1993), Loretan et al. (2003), Magalhães et al. (2011c)
Kluyveromyces marxianus	Loretan et al. (2003), Jianzhong et al. (2009)
Saccharomyces lipolytic, Saccharomyces turicensis	Lin et al. (1999)
Saccharomyces cerevisiae	Loretan et al. (2003), Latorre-Garcia et al. (2007), Jianzhong et al. (2009), Bergmann et al. (2010), Magalhães et al. (2010, 2011c), Puerari et al. (2012), Garofalo et al. (2015)

17.4 Chemical Composition

17.4.1 Kefir Flavor

Lactic acid is the major acid produced during kefir fermentation (Belin et al. 1992). Carbonyl compounds (acetaldehyde, ethanol, diacetyl, acetoin, 2-butanone, and ethyl acetate), volatile organic acids (formic, acetic, propionic, and butyric) and nonvolatile acids (lactic, pyruvic, oxalic, and succinic) are the secondary metabolites and are classified as flavor-forming compounds (Magalhães et al. 2011a, 2011b; Puerari et al. 2012).

Table 17.3 shows the microbiological and chemical characteristics of some LAB involved in kefir fermentation. The ratio and type of flavor compounds produced by these microorganisms differ according to the species and/or strains present. This variation in the LAB composition can greatly affect the final product quality (Maurellio et al. 2001).

The major/secondary end products formed by LAB include lactic acid, acetaldehyde, diacetyl, acetoin, acetone, ethanol, CO_2, and acetic acid (Table 17.3). Lactic acid, whose concentration in kefir varies from 0.80% to 1.15% (m v^{-1}), is a nonvolatile, odorless compound responsible for the characteristic acidity of fermented products. Lactic acid results from the degradation of lactose by the homofermentative and heterofermentative LAB present in kefir grains. Homofermentation produces only two moles of lactic acid and two ATPs per mole of glucose consumed, whereas heterofermentation produces one mole each of lactic acid, ethanol, and CO_2 and 1 ATP per mole of glucose. In the presence of oxygen, nicotinamide adenine dinucleotide (NAD+) can be regenerated by NADH oxidases and peroxidases, leaving acetyl-P available for conversion to acetic acid (Garrote et al. 2001). LAB responsible for lactic acid synthesis are either homofermentative or heterofermentative. The homofermentative bacteria are better acid producers than the heterofermentative bacteria (Rea et al. 1996). LAB belonging to the genus *Lactococcus* are homofermentative. Species of this genus are generally used in dairy fermentation for their ability to lower the pH to approximately 4.5. *Lc. lactis* subsp. *lactis* and *Lc. lactis* subsp. *cremoris* belong to this genus and are the principal species used as dairy starter cultures (Chen et al. 2008). Lactobacilli such as *Lb. delbrueckii* subsp. *bulgaricus*, *Lb. delbrueckii* subsp. *lactis*, *Lb. helveticus*, and *Lb. acidophilus* are primarily added to dairy foods for their probiotic benefits (Otles and Cagindi 2003).

Acetaldehyde is the compound responsible for the characteristic "fresh fruit" aroma of yogurt (Ott et al. 2002). This compound is one of the principal aroma compounds found in kefir, with concentrations ranging from 0.5 to 10 mg l^{-1} (Guzel-Seydim et al. 2000).

Diacetyl is also a desirable constituent of many dairy products, and its presence at very low concentrations up to 5 mg l^{-1} is responsible for the buttery aroma of milk products (Guzel-Seydim et al. 2000). In contrast, diacetyl is considered undesirable in brewery and wine industries, because it causes undesirable flavors (Belin et al. 1992). Diacetyl is

Table 17.3 Characteristics of LAB Involved in Milk Kefir Fermentation

GENUS	MORPHOLOGY	OPTIMUM TEMPERATURE	SPECIES	MAJOR END-PRODUCTS	SECONDARY END-PRODUCTS
Streptococcus	Cocci	40°C–44°C	Str. thermophillus	L(+) lactic acid	Acetaldehyde, acetone, acetoin, diacetyl, ethanol
Lactobacillus	Rod	40°C–44°C	Lb. delbrueckii subsp. bulgaricus	D(−) lactic acid	Acetaldehyde, acetone, acetoin, diacetyl, ethanol
		40°C–44°C	Lb. helveticus	DL lactic acid	Acetaldehyde, acetic acid, diacetyl, ethanol
		40°C–44°C	Lb. delbrueckii subsp. lactis	D(−) lactic acid	Acetaldehyde, acetone, diacetyl, ethanol
		40°C–44°C	Lb. acidophillus	DL lactic acid	Acetaldehyde, ethanol
		25°C–30°C	Lb. casei subsp. casei	L(+) lactic acid	Acetic acid, ethanol
		25°C–30°C	Lb. kefir	DL lactic acid	Acetic acid, acetaldehyde, etanol, CO_2
Lactococcus	Cocci	25°C–30°C	Lc. Lactis subsp. lactis	L(+) lactic acid	Acetaldehyde, acetone, diacetyl, ethanol
		25°C–30°C	Lc. Lactis subsp. cremoris	L(+) lactic acid	Acetaldehyde, acetone, diacetyl, ethanol
		25°C–30°C	Lc. Lactis subsp. diacetylactis	L(+) lactic acid, acetaldehyde, diacetyl, acetoin, CO_2	Acetone, ethanol
Leuconostoc	Oval	25°C–30°C	Leu. mesenteroides subsp. cremoris	D(−)lactic acid, acetic acid, diacetyl, acetoin, CO_2	Ethanol
		25°C–30°C	Leu. mesenteroides subsp. dextranicum	D(−)lactic acid, acetic acid, diacetyl, acetoin, CO_2	Ethanol
		25°C–30°C	Leu. lactis	D(−)lactic acid, acetic acid, diacetyl, acetoin, CO_2	Ethanol

Sources: Marshall, V.M.E., *Biochem. Soc. Trans.*, 12, 1150–1152, 1984; Sloff-Coste, C.J., *Danone World Newsl.*, 5, 1–5, 1994.

considered to be an important aroma compound of kefir (Beshkova et al. 2003) and has been found at varying concentrations (0.30–1.85 mg l^{-1}) in kefir beverages (Beshkova et al. 2003; Aghlara et al. 2009).

Acetoin has been reported to be present in good-quality kefir beverages at a concentration of 9 mg l^{-1} (Guzel-Seydim et al. 2000). Acetoin is formed during citrate metabolism by the decarboxylation of α-AL via acetolactate decarboxylase or the reduction of diacetyl by diacetyl reductase. At concentrations encountered in cultured products, acetoin is usually flavorless and odorless and, therefore, would be of little flavor value (Guzel-Seydim et al. 2000).

Acetone is a normal constituent of milk and cheese and has been found in kefir beverage at concentrations ranging from 0.6 to 4.91 mg l^{-1} (Beshkova et al. 2003; Aghlara et al. 2009). Acetone plays only a minor role in kefir's organoleptic characteristics, and it is believed that acetone concentrations below 1 mg l^{-1} are unlikely to have a significant effect on flavor (Aghlara et al. 2009). Acetone originates from citrate and lactose metabolism, and its production appears to be strain-dependent. Some lactobacilli strains such as *Lb. bulgaricus* and *Lb. helveticus*, as well as streptococci cultures such as *S. lactis*, *S cremoris*, and *Str. diacetylactis*, are able to synthesize acetone in small quantities (Beshkova et al. 2003).

The ethanol concentration of kefir has been reported to vary from 0.01% to 2.5% (m v^{-1}), depending on the starter and the method used to prepare the kefir (Beshkova et al. 2003; Magalhães et al. 2011c). Ethanol is formed during the conversion of acetaldehyde to ethanol by alcohol dehydrogenase, an enzyme present in both yeasts and LAB (Guzel-Seydim et al. 2000). Yeasts and *Leuconostoc* are considered the principal producers of ethanol. However, because no ethanol is produced during co-metabolism of lactose and citrate by *Leuconostoc*, yeasts are effectively the main ethanol producers (Guzel-Seydim et al. 2000). Two types of yeasts may be present in kefir: non-lactose- and lactose-fermenting yeasts. Lactose-fermenting yeasts do not have sufficient alcohol dehydrogenase activity, and the final beverage obtained has a weaker yeast flavor than beverages prepared with non-lactose-fermenting yeasts (Simova et al. 2002; Beshkova et al. 2003). Carbon dioxide originating from alcoholic fermentation and heterofermentation gives kefir its subtle effervescence (Liu et al. 2002).

Acetic acid is a short-chain volatile fatty acid that has been identified in kefir at concentrations between 200 and 850 mg l^{-1}

(Garrote et al. 2001). However, Guzel-Seydim et al. (2000) did not detect any acetic acid in kefir, and Magalhães et al. (2011c) found low acetic acid concentrations (0.273 mg l^{-1}) in milk kefir beverages but noted that this did not affect the organoleptic nature of the beverage. Acetic acid gives a vinegar-like flavor, but in kefir, this flavor is not predominant. Acetic acid is unlikely to be a product of lipolysis because natural lipase in milk is destroyed during pasteurization (Kondyli et al. 2002). Biosynthesis of acetic acid may use various amino acids, for example, *Str. diacetylactis* is able to form acetic acid from glycine, alanine, and leucine (Liu et al. 2002). Acetate can also be formed from pyruvate in the absence or presence of oxygen. In the former case and under substrate limitation, pyruvate is cleaved by pyruvate-formate lyase to form formate and acetyl-CoA. Acetyl-CoA is phosphorylated to yield acetyl-P, which is then converted to acetic acid by acetate kinase. In presence of oxygen, NAD$^+$ can be regenerated by NADH oxidases and peroxidases, resulting in acetyl-P available for conversion to acetic acid (Garrote et al. 2001).

17.4.2 Impact of Environmental Factors on Flavor

Variations in the storage conditions, growth medium, and environment can affect the development of a community of microorganisms that contribute to the characteristics of the grains (Otles and Cagindi 2003; Schoever and Britz 2003). Therefore, the organoleptic features of kefir are directly linked to the microorganisms present in the grains and can be influenced by changes in the intrinsic factors of the grains (grain activity, grain-to-milk ratio, and starter) or by one or several environment factors such as the incubation temperature, effect of pH, and storage conditions.

17.4.2.1 Grain Activity Kefir grains kept dried, freeze-dried, or frozen must be considered inactive grain forms because the microorganisms are in lag phase. Therefore, they must be activated to their exponential growth phase before use. Physically, this can be seen when the grains float to the surface of milk or when the milk clots. No standard method of activation exists; however, the recommended activation process is incubation in pasteurized full cream or even skimmed milk at room temperature (20°C–25°C) for 18–24 h (Schoevers and Britz 2003).

The activation process can last up to 1 week if frozen grains are used (Witthuhn et al. 2005) and up to 1 month for lyophilized grains (Simova et al. 2002). The grains are transferred daily (Schoevers and Britz 2003; Magalhães et al. 2011c) into a new batch of milk or thrice weekly (Angulo et al. 1993; Guzel-Seydim et al. 2000).

17.4.2.2 Grain-to-Milk Ratio The impact of the inoculum size on the characteristics of kefir beverage, especially the pH, lactococci concentration, apparent viscosity, and CO_2 content, has been shown. Significant differences in the characteristics of the kefir were obtained with an inoculum size of 1% versus 10%. The inoculum size (1%) resulted in a high viscosity, low acidity beverage, whereas the 10% inoculum size produced a low viscosity, highly acidic, effervescent product (Garrote et al. 2001). Some authors (Kuo and Lin 1999) agree that an inoculum size of 5% (m v^{-1}) is suitable for making a traditional high-quality, refreshing kefir beverage with a smooth texture, pleasant flavor, and a prickling, slight yeasty taste associated with a non-bitter clear acid (Assadi et al. 2000).

17.4.2.3 Starter The strains present in the starter, as either pure cultures or kefir grains, can affect the quality of kefir. Indeed, it has been shown that the amount of aroma compounds is dependent on the initial strains present (Liu et al. 2002). Burrow et al. (1970) demonstrated that the amount of diacetyl produced by *Str. diacetylactis* strains varies from 0.07 to 3.72 mg l^{-1}, whereas none of the other lactococci strains isolated from kefir produced diacetyl (Yuksekdag et al. 2004). The irreversible conversion of diacetyl to acetoin, which is further reduced to 2.3 butanediol, and volatilization are responsible for the low level of diacetyl and acetoin in cultured products, especially if long incubation periods are used (Ostlie et al. 2003). Acetaldehyde, which is toxic to the human organism, may be reduced to ethanol by alcohol dehydrogenase rather than being excreted. Thus, accumulation of acetaldehyde in the growth medium will depend on the level of alcohol dehydrogenase activity (Ostlie et al. 2003).

17.4.2.4 Incubation Temperature The incubation temperature is another important parameter in the manufacture of the final kefir because it can enhance or inhibit the activity of a specific group of microorganisms

(Zajsel and Gorsek 2010), resulting in specific desirable or even undesirable flavors. A good kefir is obtained with an inoculum size of 5% (m v^{-1}) and an incubation temperature of 25°C (Zajsel and Gorsek 2010). However, according to Koroleva (1988), fermentation at 25°C–27°C leads to an atypical product, whereas fermentation at a lower temperature (20°C–22°C) permits the development of all the characteristic microorganisms. Consequently, the production cycles should last 24 h and consist of two steps: fermentation at 20°C–22°C for 10–12 h and maturation at 8°C–10°C for the remaining 12 h.

17.4.2.5 Effect of pH Citrate permease (Cit-P) is the key enzyme in the citrate metabolic pathway because it is the means by which citrate is transported into the cell (Samarzija et al. 2001). Thus, possibly due to pH constraints, citrate uptake may limit the rate of citrate utilization and may therefore directly affect the yield of aroma compounds. Studies have demonstrated that Cit-P optimum activity lies between pH 4.5 and 5.5 in *Lc. lactis* subsp. *diacetylactis* and between pH 5.0 and 6.0 in some other species. Under these conditions, citrate is metabolized and converted to flavor compounds. Thus, a lack of flavor in kefir may be attributed to inadequate pH due to a short fermentation time or the absence of a diacetyl producer such as *Lc. lactis* subsp. *diacetylactis* among kefir grains microbiota (Samarzija et al. 2001).

17.4.2.6 Storage Conditions The absence of diacetyl in dairy products such as cultured buttermilk and sour cream is mainly due to the irreversible conversion of diacetyl to acetoin by diacetyl reductase. Diacetyl reductase is common in LAB; however, its activity varies among species and among strains within species. This enzyme has been found in several species, including strains of *Str. diacetylactis*, *Lactobacillus cremoris*, and *Lactobacillus dextranicum*. The reduction of diacetyl proceeds rapidly at high temperatures and decreases with decreasing temperatures. Therefore, to stabilize and to even increase the diacetyl content of cultured products such as kefir, refrigeration at 4°C–5°C is recommended. However, it is interesting to note that at 16°C–18°C, *Str. diacetylactis* has been reported to possess 100 units of diacetyl reductase per milligram of enzyme protein and was able to reduce 9 mg l^{-1} of diacetyl in 10 min. This finding highlights the importance of choosing the right combination of species in a mixed culture. Diacetyl reductase has also been found in coliforms (*E. coli*) and psychrophilic bacteria (*Pseudomonas putrefaciens, P. fragi*). Although diacetyl reductase activity is generally low in *Leuconostoc* and *Lactococcus* species, the opposite is true for coliforms and psychrophilic species, which exhibit activities ranging from 3 to 345 units mg^{-1} of enzyme protein. Thus, defects in refrigerated cultured products where diacetyl is the main flavor compound may be attributed to contamination by spoilage psychrophilic bacteria (Bassit et al. 1995).

17.5 Production and Uses of Kefir

The "traditional" method of producing kefir uses raw unpasteurized, pasteurized, or UHT-treated milk (Figure 17.4). The milk is poured into a clean suitable container, and kefir grains are added. The contents are left to stand at room temperature for approximately 24 h. The cultured milk is filtered to separate and retrieve the kefir grains from the liquid kefir. This fermented milk is appropriate for consumption. The grains are mixed to additional fresh milk, and the process is repeated. This simple process can be performed on an indefinite basis because kefir grains are a living ecosystem complex that can be

Figure 17.4 Milk kefir beverage production. Kefir grains (1) are added to milk and are left to stand at room temperature for fermentation for 18–24 h (2), the milk is then fermented forming the kefir beverage (3), after which they are filtered (4) and are ready to start another cycle. The fermented milk that results from step 3 is appropriate for consumption.

preserved forever, as long as there is feeder. Because active kefir grains are continually cultured in fresh milk to prepare kefir, the grains increase in volume or in biological mass.

Milk kefir beverage production is not the only industrial application being pursued. The conversion of agricultural and industrial waste materials to commercially valuable products is of great interest, especially if the final product is a nutritive food source. Studies have investigated new nutritive kefir-based beverage technologies, including cheese whey-based kefir beverages (Magalhães et al. 2011a, 2011b), cocoa pulp-based kefir beverages (Puerari et al. 2012), and a walnut milk kefir beverage (Cui et al. 2013).

17.6 Final Considerations

Kefir is an example of the beneficial coexistence of bacteria and yeast in equilibrium. The importance of the symbiotic relationship between yeast and bacteria in kefir seems clear because they are both required to produce the components that are needed to provide a health benefit. The importance of probiotics in the food industry is currently growing, and the symbiotic relationships between different microorganisms and how these interactions result in nutritional and therapeutic benefits that can cure and/or prevent human diseases and other disorders should be investigated further.

References

Abraham, A.G. and De Antoni, L.G. 1999. Characterization of kefir grown in cow's milk and soya milk. *Journal of Dairy Research* 66: 327–333.

Aghlara, A., Mustafa, S., Manap, Y.A., and Mohamad, R. 2009. Characterization of headspace volatile flavour compounds formed during kefir production: Application of solid phase microextraction. *International Journal of Food Properties* 12: 808–818.

Ahmed, Z., Wang, Y., Ahmad, A., Tariq, K., Nisa, M., Ahmad, H., and Afreen, A. 2013. Kefir and health: A contemporary perspective. *Critical Reviews in Food Science and Nutrition* 53: 422–434.

Angulo, L., Lopez, E., and Lema, C. 1993. Microflora present in kefir grains of the Galicia region (North-West of Spain). *Journal of Dairy Research* 60: 263–267.

Assadi, M.M., Pourahmad, R., and Moazami, N. 2000. Use of isolated kefir starter cultures in kefir production. *World Journal of Microbiology and Biotechnology* 16: 541–543.

Bassit, N., Boquien, C.-Y., Picque, D., and Corrieu, G. 1995. Effect of temperature on diacetyl and acetoin production by *Lactococcus lactis* subsp. *lactis* biovar *diacetylactis* CNRZ 483. *Journal of Dairy Research* 62: 123–129.

Belin, J.M. 1992. Microbial biosynthesis for the production of food flavours. *Trends in Food Science and Technology* 3: 11–14.

Bergmann, R.S.O., Pereira, M.A., Veiga, S.M.O.M., Schneedorf, J.M., de Mello Silva Oliveira, N., and Fiorini, J.E. 2010. Microbial profile of a kefir sample preparation—Grains in natura and lyophilized and fermented suspension. *Ciência e Tecnologia de Alimentos* 30: 1022–1026.

Beshkova, D.M., Simova, E.D., Frengova, G.I., Simov, Z.I., and Dimitrov, Zh.P. 2003. Production of volatile aroma compounds by kefir starter cultures. *International Dairy Journal* 13: 529–535.

Beshkova, D.M., Simova, E.D., Simov, Z.I., Frengova, G.I., and Spasov, Z.N. 2002. Pure cultures for making kefir. *Food Microbiology* 19: 537–544.

Burrow, C.D., Sandine, W.E., Elliker, P.R., and Speckman, C. 1970. Characterization of diacetyl negative mutants of *Streptococcus diacetylactis*. *Journal of Dairy Science* 53: 121–125.

Chen, H.-S., Wang, S.-Y., and Chen, M.-J. 2008. Microbiological study of lactic acid bacteria in kefir grains by cultured-dependent and cultured-independent methods. *Food Microbiology* 25: 492–501.

Cui, X.-H., Chen, S.-J., Wang, Y., and Han, J.-R. 2013. Fermentation conditions of walnut milk beverage inoculated with kefir grains. *LWT—Food Science and Technology* 50: 349–352.

Ertekin, B. and Guzel-Seydim, Z.B. 2010. Effect of fat replacers on kefir quality. *Journal of the Science of Food and Agriculture* 90: 543–548.

Frengova, G.I., Simova, E.D., Beshkova, D.M., and Simov, Z.I. 2002. Expolysaccharides produced by lactic acid bactéria of kefir grains. *Zeitschrift für Naturforsch* 57: 805–810.

Gao, J., Gu, F., He, J., Xiao, J., Chen, Q., Ruan, H., and He, G. 2013. Metagenome analysis of bacterial diversity in Tibetan kefir grains. *Europe Food Research Technology* 236: 549–556.

Garofalo, C., Osimani, A., Milanovic, V., Aquilanti, L., De Filippis, F., Stellato, G., Di Mauro, S. et al. 2015. Bacteria and yeast microbiota in milk kefir grains from different Italian regions. *Food Microbiology* 49: 123–133.

Garrote, G.L., Abraham, A.G., and De Antoni, G.L. 2001. Chemical and microbiological characterization of kefir grains. *Journal of Dairy Research* 68: 639–652.

Guedes, J.D.S., Magalhães-Guedes, K.T., Dias, D.R., Schwan, R.F., and Braga-Junior, R.A. 2014. Assessment of biological activity of kefir grains by laser biospeckle technique. *African Journal of Microbiology Research* 28: 2639–2642.

Guzel-Seydim, Z.B., Seydin, A.C., Greene, A.K., and Bodine, A.B. 2000. Determination of organic acids and volatile flavor substances in kefir during fermentation. *Journal of Food Composition and Analysis* 13: 35–43.

Jianzhong, Z., Xiaoli, L., Hanhu, J., and Mingheng, D. 2009. Analysis of the microflora on Tibetan kefir grains using denaturating gradient gel electrophoresis. *Food Microbiology* 26: 770–775.

Kesmen, Z. and Kacmaz, N. 2011. Determination of lactic microflora of kefir grains and kefir beverages by using cultured-dependent and cultured-independent methods. *Journal of Food Science* 76: 276–283.

Kondyli, E., Katsiari, M.C., Masouras, T., and Voutsinas, L.P. 2002. Free fatty acids and volatile compounds of low-fat feta-type cheese made with a commercial adjunct culture. *Food Chemistry* 79: 199–205.

Koroleva, N.S. 1988. Technology of kefir and kumys. *International Dairy Federation* 227: 96–100.

Kroger, M. 1993. Kefir. *Cultured Dairy Products Journal* 28: 26–29.

Kuo, C.Y. and Lin, C.W. 1999. Taiwanese kefir grains: Their growth, microbial and chemical composition of fermented milk. *Australian Journal of Dairy Technology* 54: 19–23.

Latorre-Garcia, L., Castilho-Agudo, L., and Del Polaina, J. 2007. Taxonomical classification of yeast isolated from kefir based on the sequence of their ribossomal RNA genes. *World Journal of Microbiology and Biotechnology* 23: 758–791.

Lin, C.W., Chen, H.L., and Liu, J.R. 1999. Identification and characterization of lactic acid bacteria and yeast isolated from kefir grains in Taiwan. *Australian Journal of Dairy Technology* 54: 14–18.

Liu, J.R., Chen, M.-J., and Lin, C.-W. 2002. Characterization of polysaccharide and volatile compounds produced by kefir grains grown in soymilk. *Food Chemistry and Toxicology* 67: 104–108.

Loretan, T., Mostert, J.F., and Viljoen, B.C. 2003. Microbial floral associated with South African household kefir. *South African Journal of Science* 99: 92–95.

Magalhães, K.T., Dias, D.R., Pereira, G.V.M., Oliveira, J.M., Domingues, L., Teixeira, J.A., de Almeida e Silva, J.B., and Schwan, R.F. 2011a. Chemical composition and sensory analysis of cheese whey-based beverages using kefir grains as starter culture. *International Journal of Food Science and Technology* 46: 871–878.

Magalhães, K.T., Dias, D.R., Pereira, G.V.M., Oliveira, J.M., Domingues, L., Teixeira, J.A., de Almeida e Silva, J.B., and Schwan, R.F. 2011b. Comparative study of the biochemical changes and volatile compounds during the production of novel whey-based kefir beverages and traditional milk kefir. *Food Chemistry* 126: 249–253.

Magalhães, K.T., Pereira, G.V.M., Campos, C.R., Dragone, G., and Schwan, R.F. 2011c. Brazilian kefir: Structure, microbial communities and chemical composition. *Brazilian Journal of Microbiology* 42: 693–702.

Magalhães, K.T., Pereira, M.A., Nicolau, A., Dragone, G., Domingues, L., Teixeira, J.A., de Almeida e Silva, J.B., and Schwan, R.F. 2010. Production of fermented cheese whey-based beverage using kefir grains as starter culture: Evaluation of morphological and microbial variations. *Bioresource Technology* 101: 8843–8850.

Marshall, V.M.E. 1984. Fermented milks: New developments in the biochemistry of the starter cultures. *Biochemical Society Transactions* 12: 1150–1152.

Maurellio, G., Moio, L., Moschetti, G., Piombino, P., Addeo, F., and Coppola, S. 2001. Characterization of lactic acid bacteria strains on the basis of neutral volatile compounds produced in whey. *Journal of Applied Microbiology* 90: 928–942.

Medrano, M., Hamet, M.F., Abraham, A.G., and Pérez, P.F. 2009. Kefiran protects Caco-2 cells from cytopathic effects induced by *Bacillus cereus* infection. *Antonie van Leeuwenhoek* 96: 505–513.

Miguel, M.G.C.P., Cardoso, P.G., Lago, L.A., and Schwan, R.F. 2010. Diversity of bacteria present in milk kefir grains using culture-dependent and culture-independent methods. *Food Research International* 43: 1523–1528.

Ostlie, H.M., Helland, M.H., and Narvhus, J.A. 2003. Growth and metabolism of selected strains of probiotics in milk. *International Journal of Food Microbiology* 87: 17–27.

Otles, S. and Cagindi, O. 2003. Kefir: A probiotic dairy-composition, nutritional and therapeutic aspects. *Pakistan Journal of Nutrition* 2: 54–59.

Ott, A., Germond, J.E., and Chaintreau, A. 2002. Origin of acetaldehyde production by *Streptococcus thermophilus*. *Applied Environment Microbiology* 68: 5656–5662.

Puerari, C., Magalhães, K.T., and Schwan, R.F. 2012. New cocoa pulp-based kefir beverages: Microbiological, chemical composition and sensory analysis. *Food Research International* 48: 634–640.

Rea, M.C., Lennartsson, T., Dillon, P., Drinan, F.D., Reville, W.J., Heapes, M., and Cogan, T.M. 1996. Irish kefir-like grains: Their structure, microbial composition and fermentation kinetics. *Journal of Applied Bacteriology* 81: 83–94.

Samarzija, D., Antunac, N., and Lukac-Havranek, J. 2001. Taxonomy, physiology and growth of *Lactococcus lactis*: A review. *Mljekarstvo* 51: 35–48.

Schoevers, A. and Britz, T.J. 2003. Influence of different culturing conditions on kefir grain increase. *International Journal of Dairy Technology* 56: 183–187.

Silva, K.R., Rodrigues, S.A., Filho, L.X., and Lima, A.S. 2009. Antimicrobial activity of broth fermented with kefir grains. *Applied Biochemistry and Biotechnology* 152: 316–325.

Simova, E., Beshkova, D., Angelov, A., Hristozova, Ts., Frengova, G., and Spasov, Z. 2002. Lactic acid bactéria and yeast in kefir grains and kefir made from them. *Journal of Industrial Microbiology & Technology* 28: 1–6.

Sloff-Coste, C.J. 1994. Lactic acid bacteria. *Danone World Newsletter* 5: 1–5.

Tada, S., Katakura, Y., Ninomiya, K., and Shioya, S. 2007. Fed-batch coculture of *Lactobacillus kefiranofaciens* with *Saccharomyces cerevisiae* for effective production of kefiran. *Journal of Bioscience and Bioengineering* 103: 557–562.

Witthuhn, R.C., Schoeman, T., and Britz, T.J. 2005. Characterization of the microbial population at different stages of kefir production and kefir mass cultivation. *International Dairy Journal* 15: 383–389.

Wouters, J.T.M., Ayad, E.H.E., Hugenholtz, J., and Smit, G. 2002. Microbes from raw milk for fermented dairy products. *International Dairy Journal* 12: 91–109.

Yuksekdag, Z.N., Beyatli, Y., and Aslim, B. 2004. Determination of some characteristic coccoid forms of lactic acid bacteria isolated from Turkish kefirs with natural probiotic. *Lebensmittel-Wissenshaft und-Technologie* 37: 663–667.

Zajsek, K. and Gorsek, A. 2010. Effect of natural starter culture activity on ethanol content in fermented dairy products. *International Journal of Dairy Technology* 63: 113–118.

18

Koumiss

Nutritional and Therapeutic Values

TEJPAL DHEWA, VIJENDRA MISHRA, NIKHIL KUMAR, AND K.P.S. SANGU

Contents

18.1	Koumiss: An Introduction	483
18.2	Historical Background	484
18.3	Microflora of Koumiss	485
18.4	Methods of Production	486
	18.4.1 Strong Koumiss	487
	18.4.2 Moderate Koumiss	487
	18.4.3 Light Koumiss	489
18.5	Nutritional Properties of Koumiss	489
	18.5.1 Amino Acid	489
	18.5.2 Minerals	489
	18.5.3 Vitamins	489
	18.5.4 Lactose	490
	18.5.5 Fat	490
	18.5.6 Protein	490
18.6	Therapeutic Values of Koumiss	490
18.7	Conclusions	494
References		494

18.1 Koumiss: An Introduction

Fermented milk products have long been well documented as the important components of our nutritional diet, and their medicinal properties have been experienced since the ancient days of civilization. Evidences show that fermented milk products have been produced since around 10,000 BC. Fermented milk products are popularly known with the names of cultured dairy foods, cultured milk, or cultured dairy products. Mainly, such kind of dairy foods are produced by the process of fermentation using lactic acid bacteria (LAB) (i.e., *Lactobacillus*, *Lactococcus*, and *Leuconostoc*). The fermentation enhances the shelf life of products along with the improvement in the taste and digestibility of milk. Several lactobacilli strains are involved in the production of a wide range of cultured milk products with different tastes. The consumer's interest in fermented milk products is accelerated due to the newer food-processing techniques, changing social attitudes, and evidences of health benefits. Some of the cultured dairy foods (i.e., koumiss, bioghurt, yakult, actimel) are already available to the consumers in the market worldwide, as therapeutic and dietetic products.

Koumiss (also signified as *kumiss* or *coomys*), a fermented dairy product traditionally made from mare milk by fermentation, originated from the nomadic tribes of Central Asia. Although it is an important drink for the peoples of the Central Asian steppes, Bulgar, Turkic, and Mongol origin, it is equally important to the Bashkirs, Kalmyks, Kazakhs, Kyrgyz, Mongols, Uyghurs, and Yakuts. Koumiss is mildly alcoholic and has a sour taste. Koumiss is similar to kefir but is produced from the liquid starter, in contrast to solid *kefir* "grains." Even in the areas of the world where koumiss is a popular product, mare milk remains a limited commodity. Hence, for commercial production of koumiss, cow milk is used, which is richer in fat and protein but lower in lactose than mare milk. Prior to fermentation, the cow milk is fortified to allow a comparable fermentation.

18.2 Historical Background

Koumiss is known as the ancient fermented beverage of dairy origin. The name *koumiss* was derived from a tribe. Historical facts reveal that Greeks and Romans used to prepare such types of beverages from milk. More than 2500 years ago, the nomads in Southeast Russia and the Scythia tribes in Central Asia used to make koumiss from mare milk. In fifth century BC, Herodotus (a famous Greek historian) documented the use of koumiss. William Lubuluqi, a French missionary, came to China in the thirteenth century and gave a detailed methodology of koumiss to the Mongolians. This is recorded in William Lubuluqi's work *Lubuluqi Journey to the East*. Preparation of koumiss in China is also reported in history. Around 1500 BC, domestic horses were used to produce fermented milk products. Famous persons of Chinese origin attained popularity in koumiss preparation during the Han Yuan Dynasty (202 BC to AD 202) and Yuan Dynasty (AD 1271–1368). Presently, fermented equine milk is widely produced with various names and consumed in Russia, Mongolia, Kazakhstan, and Eastern Europe, mainly for its therapeutic value. In Mongolia, koumiss is the national drink (*Airag*), and a high-alcoholic drink made by distilling koumiss (*Arkhi*) is also produced (Kanbe 1992), with *per capita* consumption of koumiss about 50 l per year. Koumiss is still made in remote areas of Mongolia by traditional methods but with increased demand, elsewhere it is nowadays produced on industrial scale under controlled conditions.

18.3 Microflora of Koumiss

Due to microbial activities in fermented milk products, there is a decrease in pH and consequently there is coagulation of milk proteins. The lactic cultures play a major role in the making of traditional dairy products. The LAB are a diverse group of microorganisms (*Lactococcus*, *Lactobacillus*, *Leuconostoc*, *Pediococcus*, and *Streptococcus*) that are Gram-positive and catalase-negative, reductase-negative, oxidase-negative, as well as non-motile and non-spore forming. In yogurt and some other functional dairy products, *Bifidobacterium* is added. In addition, yeasts are also present in significant numbers and contribute to the flavor, texture, and nutritional value of some of the traditional dairy products. Most of these fermented products are produced either by naturally occurring microflora or by reintroducing part of the previous fermentation (as an inoculum into the new fermentation).

There are studies that describe the involvement of different microorganisms in koumiss fermentation, and mainly three groups have been reported to be involved (Table 18.1), but the microbial composition in original koumiss is quite variable. The major microflora is *Lactobacillus delbrueckii*, *Lb. bulgaricus*, *Lb. acidophilus*, and lactose-fermenting yeasts (*Torula koumis* and *Saccharomyces lactis*). Lactic streptococci, coliforms, and some spore-forming bacilli are also found. In order to produce good-quality koumiss, some new starters such as *Streptococcus lactis*, *Lb. bulgaricus*, and *Saccharomyces lactis* are also used. To improve the specific aroma, *Acetobacter aceti* (0.2%) can be added. Hao et al. (2010)

Table 18.1 Microbial Diversity in Koumiss

GROUP	MICROORGANISM(S)	REFERENCE(S)
Lactobacilli	Lb. casei	Wang et al. (2008), Ya et al. (2008)
	Lb. delbrueckii subsp. bulgaricus	Akuzawa and Surono (2003)
	Lb. helveticus, Lb. fermentum	Wang et al. (2008)
	Lb. salivarus, Lb. buchneri	Danova et al. (2000), Wang et al. (2008)
Lactose-fermenting yeasts	Klu. lactis	Akuzawa and Surono (2003)
	S. lactis, Torula koumiss	Taminme and Robinson (1999)
Nonlactose-fermenting yeast	S. unisporus	Montanari and Grazia (1997)

studied bacterial biodiversity in traditionally fermented koumiss by denaturing gradient-gel electrophoresis. The results revealed a novel microbial profile and extensive biodiversity in koumiss. The dominant LAB were *Lb. acidophilus*, *Lb. helveticus*, *Lb. fermentum*, and *Lb. kefiranofaciens*. Commonly encountered bacterial species, such as *Enterococcus faecalis*, *Lactococcus lactis*, *Lb. paracasei*, *Lb. kitasatonis*, *Lb. kefiri*, *Leuconostoc mesenteroides*, *Str. thermophilus*, *Lb. buchneri*, and *Lb. jensenii*, were sometimes found in this product. In addition, *Lb. buchneri*, *Lb. jensenii*, and *Lb. kitasatonis* that were never isolated earlier by culture-dependent methods were identified for the first time in the Xinjiang koumiss.

18.4 Methods of Production

In general, koumiss is made by fermenting raw unpasteurized mare milk and recognized as an important nutritional resource that is beneficial to the elderly, to convalescents, and to infants (Marconi and Panfili 1998). The energy value of mare milk (480 kcal kg/l) is, however, slightly lower than that of human and cow milk, due to lower fat content. The average composition of mare milk is 2.1% protein, 1.2% fat, 6.4% lactose, and 0.4% ash (Malacarne et al. 2002). As the cost and availability is a major factor in its use, studies were also conducted using bovine milk for koumiss production. Conventionally, this fermentation took place in horsehide containers that might be left on the top of a yurt and turned over occasionally. Nowadays, a wooden vat, plastic barrel, or stainless steel tanks are used in place of the leather containers.

As a traditional practice to prepare koumiss in northern or western China, skin that is partially filled with mares' milk is hung at the door during the season for making such types of beverages, and the passersby, who are familiar with the practice, give such skin a good punch as they walk by, agitating the contents, so the milk would turn into koumiss rather than coagulate and spoil.

In controlled production, to prepare koumiss, mare/cow milk is heated at 90°C–92°C for 5–10 min and cooled to 26°C–28°C, and the starter is added (one part of *Lb. bulgaricus* grown at 37°C in cow milk for 7 h and two parts of *Torula* ssp. at 30°C for 15 h) at an inoculum rate of 10%–30% to give an initial acidity of 0.45% lactic acid and is then incubated at 28°C till an acidity of 0.7%–0.8% is achieved. It is then cooled to 20°C, while stirring for 1–2 h. The product is bottled, capped, and allowed to ripen at 6°C–8°C for 1–3 days (Figure 18.1). The finished product contains alcohol between 0.7% and 2.5% (Table 18.2). Koumiss itself has a low level of alcohol, like small beer, the common drink of medieval Europe. Though koumiss can be strengthened through freeze distillation, it may also be distilled into the spirit known as *araka* or *arkhi*. Three types of koumiss are produced depending on the lactic acid content.

18.4.1 Strong Koumiss

It is produced by the LAB (*Lb. bulgaricus* and *Lb. rhamnosus*) that acidify the milk to pH 3.3–3.6 and whose conversion ratio of lactose into lactic acid is about 80%–90%. Due to its low pH and higher lactic acid content, it falls in the category of strong koumiss.

18.4.2 Moderate Koumiss

Moderate koumiss contains *Lactobacillus* bacteria (*Lb. acidophilus*, *Lb. plantarum*, *Lb. casei*, and *Lb. fermentum*) with restricted acidification properties that lower the pH to 3.9–4.5 at the end of fermentation, and the conversion ratio averages 50%.

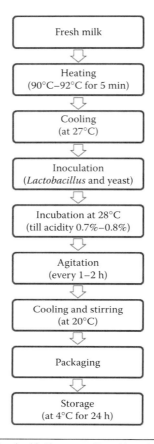

Figure 18.1 General steps involved in the preparation of koumiss

Table 18.2 Physiochemical and Biological Characteristics of Finished Koumiss

CONTENT	AMOUNT (%)
Ash	0.4–0.5
Carbon dioxide	0.5–0.9
Ethanol	0.6–2.5
Lactic acid	0.7–1.8
Lactose	3.5–4.3
Lipid	0.6–1.3
Protein	1.7–1.9
Total solids	10.6–11.3

Note: *Organoleptic properties*: milky-gray in color, lightly and naturally carbonated, sharp alcoholic and acidic taste; *specific gravity*: 1.03; pH: 3.8–4.0.

18.4.3 Light Koumiss

This type of Koumiss is a slightly acidified product (pH 4.5–5.0) and is formed by *Str. thermophilus* and *Str. cremoris*.

18.5 Nutritional Properties of Koumiss

Koumiss contains several components such as protein, fat, lactose, minerals, enzymes, vitamins, immune cells, and pigments. The composition of koumiss depends on the source of milk (mare or cow milk, fat content: 1.1% and 3.9%, protein: 3.5% and 3.3%, sugar: 6.1% and 4.7%, calcium: 90 and 120 mg, respectively), and some bioactive components may be formed through the microbial fermentation. Hence, in finished koumiss, there may be slight differences because of these varying contents. For example, mare milk is rich in lactalbumin, peptone, amino acids, essential fatty acids, vitamins, and minerals as compared to cow and sheep milk. The general, nutritional characteristics of koumiss are listed in Sections 18.5.1 through 18.5.6 (Lv and Wang 2009).

18.5.1 Amino Acid

Koumiss contains almost all essential amino acids that are required for humans. Hence, koumiss fulfills all the nutritional requirements of the consumer related to amino acids' requirement.

18.5.2 Minerals

Koumiss contains different types of minerals beneficial to humans, for example, phosphorus, calcium, magnesium, zinc, iron, manganese, and copper. In addition, the ratio of calcium to phosphorus is 2:1, which is similar to human milk (Ha et al. 2003).

18.5.3 Vitamins

Koumiss has a quite high vitamin C content that contributes certain pharmacological actions. In addition, it is also rich in vitamins A, B, E, B_1, B_2, B_{12}, pantothenic acid, and bacteriocins.

18.5.4 Lactose

High concentration of lactose in milk (6%–7%) favors microbial fermentation, as it is decomposed by the starter cultures into lactic acid, alcohol, and other small molecules. The content of lactose in koumiss is about 1.4%–4.5%.

18.5.5 Fat

Fat is the most important component of koumiss that contains essential fatty acids (such as linoleic and linolenic acids). In general, mare milk contains more essential fatty acids than cow milk (Huo et al. 2003). Such types of fatty acids are health promoting and required by humans.

18.5.6 Protein

Protein content in koumiss is 1.7%–2.2% and depends on the milk source (e.g., in mare milk, protein is lower than that of cow milk). The ratio of casein to whey protein is 1:1, which is close to human milk (Ha et al. 2003).

18.6 Therapeutic Values of Koumiss

Some of the valuable properties of fermented milks such as therapeutic and prophylactic properties make these to cure many lifestyle diseases. There are ways in which regular intake of naturally fermented foods benefit our health by improving the digestive system. The regular intake of fermented milk products may lead to the following major advantages:

- Easily absorbed and show high rate of assimilation than sweet whole milk [assimilation of milk is 32% to that of fermented milks (91%) in an hour].
- Health-promoting bacteria present in naturally fermented foods produce enzymes that can metabolize foods in the intestines and, hence, making the nutrients easier to absorb. Besides, these bacteria produce certain vitamins, making the fermented foods richer and healthier in nutrients.

- Better assimilation of fermented milk products is because of the partial peptonization and intensity of secretion of the ferment by digestive tract glands.
- Essential amino acid, methionine, plays a significant role in the removal of excessive fat from the liver and improves the condition of patients suffering from arteriosclerosis, hypertension, heart diseases, and chronic inflammation of liver.
- Stimulate appetite because of their pleasant, refreshing, and pungent taste and also improve the functioning of the central nervous and respiratory systems.
- Increases the secretion of gastric juices, which maintains the desirable ratio of calcium and phosphorus leading to a high digestive capability.
- Daily intake improves the immune response, as around 80% of the immune tissue of the body is concentrated around the intestines.

It has been documented that if intestinal bacteria are removed, then the total immune defense is greatly reduced. It is confirmed that gut bacteria–free animals had lower levels of important white blood cells and other defense chemicals in their blood. When naturally occurring bacteria were reintroduced into the animal's intestinal system, the white blood cells were activated and the immune system became resilient. The good bacteria from naturally fermented foods make chemicals that pass through the wall of the intestine and stimulate the formation of immune cells in the body system.

Among the fermented milks, koumiss is more valuable because of its dietary nature. In fermentation, the lactose in mare milk gets converted into lactic acid, ethanol, and CO_2, and the milk becomes a source of nutrition for people who are unable to digest lactose. During milk fermentation, the composition of the minerals remains unchanged, while those of proteins, carbohydrates, and vitamins and to some extent fat constituents change. These changes produce some special physiological effects as microorganisms and their metabolites play a very important role in the improvement of the dietary and therapeutic values of these types of milk products by producing lactic acid, alcohol, carbon dioxide, antibiotics, and vitamins.

As raw mare milk contains almost 40% more lactose than cow milk, the unfermented mare milk is generally not consumed because it is laxative in nature. It is claimed that drinking 190 ml of koumiss in a day would give a lactose-intolerant person severe intestinal symptoms. In addition, there are reports that support the following health benefits derived from the consumption of koumiss.

High blood pressure is considered as a risk for the development of cardiovascular and end stage of renal diseases. Zha et al. (1994) studied that the regular consumption of koumiss will reduce the blood cholesterol levels and control the formation of blood lipids. Thus, it can increase blood flow and decrease blood pressure, preventing the breakdown of vessels and formation of clots. An interest in the peptides that can lower the blood pressure of hypertensive patients has been developed, yet the research on the bioactive peptides derived from liquid milk is limited. Peptides from the hydrolysis of equine β-casein may have a positive action on human health (Doreau and Martin-Rosset 2002). Chen et al. (2010) reported the presence of four novel angiotensin I-converting enzyme-inhibitory peptides in koumiss, which may enhance the beneficial effects of it on cardiovascular health. Peptides with trophic or protective activity have been identified in asinine milk (Salimei 2011).

The effects of koumiss on the immune system and its promotion of antibacterial activities have been studied. The results exhibited that it could significantly increase the immune system of experimental animals. Fresh mare milk can enhance the thymus and spleen index, as well as strengthen the functions of macrophages and increase the ratio of hemolysin in blood serum. In addition, fresh mare milk increases the weight of the immune organs of rats and enhances the normal immune functions, regulates cell immune abilities, and controls abnormal body fluid immune functions. Ya et al. (2008) studied in four separate experiments via oral administration of live and heat-killed LcZhang (a novel LAB, *Lb. casei* Zhang from koumiss) to BALB/c mice for several consecutive days and investigated the immuno-modulating capacity of LcZhang in vivo by analyzing the profile of T cell subpopulations, cytokines, and immunoglobulin concentrations induced in blood serum and intestinal fluid in BALB/c mice. In this experiment, only live bacteria produced a wide range of immune responses that include the increased production of interferon-γ and depression of tumor necrosis factor-α level. In addition, interleukin-2 and its receptor gene transcription increased considerably, but the proportion of T cell subsets appeared to be unaffected. LcZhang was also capable of inducing gut mucosal responses by enhancing the production of secretory immunoglobulin A as well as influencing the systemic immunity via cytokines released to the circulating blood.

Koumiss may also be used for beautification of skin as it can act as a skin moisturizer. As a beautifying agent, it can make the skin white, soft, and smooth after applying. Bao et al. (1995) prepared a face cream and masks using koumiss to cure chloasma (which is a patchy brown or dark brown skin discoloration that usually occurs on a woman's face and may result from hormonal changes, as in case of pregnancy), with a total effectiveness of more than 90%. Mu and Bai (2003) reported that koumiss can adjust the sugar metabolism by reducing the sugar content in a patient's blood and enhance insulin secretion.

The mare milk is of minor importance in terms of milk production as compared to bovine, sheep, and goat. In many respects, equine milk resembles human milk and is claimed to have special therapeutic properties (Lozovich 1995); it is becoming increasingly important in Europe, particularly in France, Italy, Hungary, and the Netherlands. In Russia and Mongolia, koumiss is used for the management of digestive and cardiovascular diseases (Lozovich 1995; Levine 1998). The microflora of koumiss develops a biological barrier on the walls of the stomach and intestines and also secretes antimicrobial metabolites, which prevents the growth of pathogens.

Koumiss contains some essential and rare compounds required for the smooth functioning of neurological systems. Ha et al. (2003) studied that koumiss can improve blood circulation in brain and blood supply functions. Hence, it can cure various disorders of the neurological systems as well as the digestive system.

Initially, Mongolian doctors used koumiss to cure tuberculosis and included it in their clinical practices. Zha (1987) mentioned that koumiss was used to cure tuberculosis effectively during every summer and autumn at the Ximeng Mongolian Medical Research Institute. It was proved that the causative agent of tuberculosis (*Mycobacterium tuberculosis*) cannot survive in the mare milk because of the anti-tuberculosis element generated by the microflora of koumiss. When koumiss is used in clinical practice to treat tuberculosis, a 60%–91% rate of recovery was reported, which was confirmed by lab techniques such as X-rays and tuberculosis test, and the disappearance of the symptoms are an indication of effective treatment.

18.7 Conclusions

Fermented dairy products have had a pivotal role in the human diet for ages. Koumiss, a fermented equine milk product, is widely consumed in Russia and Western Asia mainly for its therapeutic value. In Mongolia, koumiss, called *airag*, is the national drink and distilled koumiss, *arkhi*, is also produced (Kanbe 1992; Ørskov 1995). Both koumiss and kefir belong to the yeast-lactic fermentation group, where alcoholic fermentation using yeasts is used in combination with lactic acid (Tamine and Marshall 1984). In traditional koumiss, a part of the previous day's batch is used to inoculate the fresh mare milk and the fermentation takes place over 3–8 hours with the indigenous microbial population that include *Lb. delbrueckii* subsp. *bulgaricus*, *Lb. casei*, *Lc. lactis* subsp. *lactis*, *Kluyveromyces fragilis*, and *S. unisporus* (Litopoulou-Tzanetaki and Tzanetakis 2000). Koumiss is still made in Mongolia by this method, but with increased demand elsewhere, it is now produced under carefully controlled conditions. It also has established therapeutic values such as lowering of blood pressure, dissolving blood clots, strengthening the vital organs, enhancing digestibility, and improved immunity.

References

Akuzawa, R. and Surono, I.S. 2003. Fermented milks: Asia. In: *Encyclopaedia of Dairy Sciences*, eds. H. Roginski, J.W. Fuquay, and P.F. Fox, 1045–1049. London: Academic Press.

Bao, D., Wu, J., Ma, F., Wang, S., and Hao, B. 1995. The 144 samples of chloasma treatment with Koumiss facial mask. *The Chinese Journal of Dermatovenerology* 12: 110–110.

Chen, Y., Wang, Z., Chen, X., Liu, Y., Zhang, H., and Sun, T. 2010. Identification of angio 1-converting enzyme inhibitory peptides from koumiss, a traditional fermented mare's milk. *Journal of Dairy Science* 93: 884–892.

Danova, S., Petrov, K., Pavlov, P., and Petrova, P. 2000. Isolation and characterization of *Lactobacillus* strains involved in koumiss fermentation. *International Journal of Dairy Technology* 58: 100–105.

Doreau, M. and Martin-Rosset, W. 2002. Dairy animals: Horse. In: *Encyclopaedia of Dairy Sciences*, eds. H. Roginski, J.A. Fuquay, and P.F. Fox, 630–637. London: Academic Press.

Ha, S., Leng, A.M.G., and Mang, L. 2003. Koumiss and its medicinal values. *China Journal of Chinese Material Medica* 28(1): 11–14.

Hao, Y., Zhao, L., Zhang, H., Zhai, Z., Huang, Y., Liu, X., and Zhang, L. 2010. Identification of the bacterial biodiversity in koumiss by denaturing gradient gel electrophoresis and species-specific polymerase chain reaction. *Journal of Dairy Science* 93(5): 1926–1933.

Huo, Y., Rong, H.S.S., and Leng, A.M.G. 2003. A study of koumiss nutritional composition and molecules. *Inner Mongolia Husbandry* 6: 22–23.

Kanbe, M. 1992. Traditional fermented milk of the world. In: *Functions of Fermented Milk: Challenges for the Health Sciences*, eds. Y. Nakazawa and A. Hosono. London: Elsevier Applied Science.

Levine, M.A. 1998. Eating horses: The evolutionary significance of hippophagy. *Antiquity* 72: 90–100.

Litopoulou-Tzanetaki, E. and Tzanetakis, N. 2000. Fermented milk. In: *Encyclopaedia of Food Microbiology*, eds. R. Robinson, C. Batt, and P. Patel, 774–805. London: Elsevier.

Lozovich, S. 1995. Medical uses of whole and fermented mare milk in Russia. *Cultured Dairy Products Journal* 30: 18–21.

Lv, J. and Wang, L. 2009. Bioactive components in kefir and koumiss. In: *Bioactive Components in Milk and Dairy Products*, ed. Y.W. Park, 251–260. Ames, IA: Wiley-Blackwell.

Malacarne, M., Martuzzi, F., Summer, A., and Mariani, P. 2002. Protein and fat composition of mare's milk: Some nutritional remarks with reference to human and cow's milk. *International Dairy Journal* 12: 869–877.

Marconi, E. and Panfili, G. 1998. Chemical composition and nutritional properties of commercial products of mare milk powder. *Journal of Food Composition and Analysis* 11: 178–187.

Montanari, G. and Grazia, L. 1997. Galactose-fermenting yeasts as fermentation microorganisms in traditional koumiss. *Food Technology and Biotechnology* 35: 305–308.

Mu, Z. and Bai, Y. 2003. Mare milk. *Journal of Inner Mongolia Agricultural University* 24(1): 116–120.

Ørskov, E.R. 1995. A traveler's view of outer Mongolia. *Outlook on Agriculture* 24: 127–129.

Salimei, E. 2011. Animals that produce dairy foods: Donkey. In: *Encyclopaedia of Dairy Sciences*, eds. J.W. Fuquay, P.F. Fox, and P.L.H. McSweeney, Vol. 1, 365–373. London: Elsevier.

Tamine, A.Y. and Marshall, V.M.E. 1984. Microbiology and technology of fermented milk. In: *Advances in the Microbiology and Biochemistry of Cheese and Fermented Milk*, eds. F.L. Davies, and B.A. Law, 118–122. London: Elsevier Applied Science.

Tamine, A.Y. and Robinson, R.K. 1999. *Yoghurt: Science and Technology*. Cambridge: Woodhead Publishing Ltd.

Wang, J., Chen, X., Liu, W., Yang, M., Airidengcaicike, and Zhang, H. 2008. Identification of lactobacillus from koumiss by conventional and molecular methods. *European Food Research and Technology* 227(5): 1555–1561.

Ya, T., Zhang, Q., Chu, F., Merritt, J., Bilige, M., Sun, T., Du, R., and Zhang, H. 2008. Immunological evaluation of *Lactobacillus casei* Zhang: A newly isolated strain from koumiss in inner Mongolia, China. *BMC Immunology* 9: 68.

Zha, M. 1987. *Kumiss Therapy*. Huhehaote, China: Inner Mongolian People's Publishing House.

Zha, M., Liu, Y., and Jin, Z. 1994. 50 clinical samples of koumiss fat reduction and anti-clotting. *Inner Mongol Journal of Traditional Chinese Medicine* 13: 16.

19
Viili

MING-JU CHEN AND SHENG-YAO WANG

Contents

19.1	Introduction	497
19.2	Microorganisms in Viili Starter Cultures	498
	19.2.1 Lactic Acid Bacteria	499
	19.2.2 Fungi and Yeasts	500
	19.2.3 Characteristics of the Viili Microorganisms during Culture	500
19.3	Production of Viili and Viili Products	501
	19.3.1 Manufacture of Viili	502
	19.3.2 Other Viili Products	504
	19.3.3 Sources of Milk	505
19.4	Slime Properties of Viili	507
	19.4.1 Composition and Sugar Components of the EPS	507
	19.4.2 EPS Biosynthesis	508
19.5	Functionality of Viili	509
	19.5.1 Antioxidative Effect	509
	19.5.2 ACE Inhibitory Activity	510
	19.5.3 Immunomodulatory Effects	510
	19.5.4 Antitumor Activity	511
References		511

19.1 Introduction

Different cultural origins, sources, and processing methods have diversified soured milk into a variety of products such as dahi, dadih, kefir, koumiss, langfil, and viili (Mistry 2004; Chen et al. 2006; Dharmawan et al. 2006). Viili is a ropy fermented milk that originated in Scandinavia. The name *viili* is Swedish and describes mesophilic fermented milk that is claimed to have various functional benefits and the potential to improve health (Kitazawa et al. 1991, 1993, 1996; Nakajima et al. 1992; Ruas-Madiedo et al. 2006; Chiang et al. 2011).

Viili has a pleasant sharp taste and a good diacetyl aroma linked to a stringy texture and can be cut easily with a spoon. Diacetyl is an important flavor impact compound in viili. This domestic fermented milk is typically consumed at breakfast and is also a popular snack food among children. The product is eaten either on its own or together with cereals, muesli, or fruit. A very traditional way of eating viili is by mixing it with cinnamon and sugar (Leporanta 2003).

Viili was historically made on farms in large wooden buckets, and at a later stage, families also started to make it at home. The industrial manufacture of viili began in Finland in the 1950s. Today, this product has grown to be an important fermented milk product in Finland. Annual consumption stands at more than 4.5 kg/capita (Leporanta 2003). The most traditional viili culture consists of mesophilic lactic acid bacteria (LAB) forming a culture in the body of the milk together with a surface-growing yeast-like fungus (Leporanta 2003; Boutrou and Guéguen 2005; Kahala et al. 2008; Wang et al. 2008, 2010), thus all viili cultures also contain yeasts. The effect of the yeasts on the viili is not clear, but it is believed that yeasts in viili may provide the product's unique flavor and induce the LAB to produce more exopolysaccharides (EPS) (Wang et al. 2008). The following sections describe viili manufacture and the microorganisms that have been found to be present in viili cultures. The health benefits of viili product and the microorganisms isolated from viili culture are also discussed.

19.2 Microorganisms in Viili Starter Cultures

During the traditional home manufacture of viili, the propagation of the microflora was initiated by adding a small quantity of a previously prepared product. Nowadays, the production of viili has evolved to industrial-scale manufacture; nevertheless, traditional starter cultures containing mixed populations of LAB and fungi are still used. A stable and constant starter culture, which is necessary for manufacturing a quality fermented beverage, can be difficult to sustain due to the complex microbiological composition of viili culture. Therefore, knowledge of microbiological profiles of viili cultures is important to the product's characteristics because, first, safety and quality control are crucial to viili products and, second, their probiotic effects need to be evaluated to determine any health benefits. Several studies have identified a number of different various microorganisms in viili starter cultures using selective growth medium, morphological characterization, biochemical characterization, and molecular methods (Tsutsui et al. 1998; Lin et al. 1999; Leporanta 2003; Wang et al. 2008, 2011; Chen et al. 2011).

19.2.1 Lactic Acid Bacteria

The numbers of LAB present in ropy fermented milk are around 10^8–10^9 CFU/g (Wang et al. 2008; Uchida et al. 2009). A number of studies have isolated and identified various LAB from viili products. Kahala et al. (2008) identified that by phenotypic methods, polymerase chain reaction (PCR), and pulsed-field gel electrophoresis, the makeup of three industrial dairy starters is used for the manufacture of viili. Four LAB strains, namely, the acid-producing strains *Lactococcus lactis* subsp. *cremoris* and *Lc. lactis* subsp. *lactis*, together with the aroma-producing strains *Lactobacillus lactis* subsp. *lactis* biovar. *diacetylactis* and *Leuconostoc mesenteroides* subsp. *cremoris*, were found. Wang et al. (2008) identified two LAB strains, *Lc. lactis* subsp. *cremoris* and *Leu. mesenteroides* subsp. *mesenteroides*, in Taiwan viili culture by PCR-denaturing gradient gel electrophoresis (PCR-DGGE) and 16S rDNA sequencing. Interestingly, the ropy fermented milk products from various different countries all seem to contain similar LAB strains, mainly *Lc. lactis* and *Leu. mesenteroides* (Tsutsui et al. 1998; Leporanta 2003; Boutrou and Guéguen 2005; Kahala et al. 2008; Wang et al. 2008; Uchida et al. 2009). Wang et al. (2008) concluded that the number of distinct LAB species in ropy fermented milk products may not be very lengthy, even when these products are produced in different parts of the world. However, Chen et al. (2011) identified a very different microbial profile in a viili culture, including *Lb. plantarum*, *Streptococcus thermophilus*, *Lb. paracasei*, and *Bacillus cereus*. The strains other than *Lc. lactis* and *Leu. mesenteroides* that are found in viili products have entered the culture from the local environment.

19.2.2 Fungi and Yeasts

Traditionally, viili has been made from unhomogenized milk in which a layer of cream has formed on the surface. It is on this layer of cream that fungi grow, specifically *Geotrichum candidum*, which produces a velvet-like surface on the viili. *G. candidum* provides the viili with a typical musty aroma. As viili cups are tightly sealed, this limits the amount of oxygen available and thus also limits the growth of this mold.

Most traditional viili cultures also contain yeast species (Kontusaari et al. 1985; Wang et al. 2008), but most authors do not specify what these species consist of. Only Wang et al. (2008), who used a combination of conventional microbiological cultivation and PCR-DGGE, was able to do this and identified three yeast species from viili cultures in Taiwan. *Kluyveromyces marxianus*, *Saccharomyces unisporus*, and *Pichiafermentans* were identified in viili starter cultures at levels of 58%, 11%, and 31%, respectively. Viili yeasts produce carbon dioxide and alcohol, which together contribute to the typical yeast flavor.

19.2.3 Characteristics of the Viili Microorganisms during Culture

The fermentation of viili starters is characterized by a rapid increase in the number of LAB from an initial value of 10^6 CFU/g to 10^9 CFU/g after the first 20 h of fermentation, and this level then remains stable for the rest of process (Wang et al. 2008). A study of the microbial dynamics at different fermentation stages demonstrated that *Lc. lactis* subsp. *cremoris* was the dominant bacterial species in the samples, followed by *Leu. mesenteroides* subsp. *mesenteroides*; furthermore, there was no change in this proportion as the fermentation progressed. Uchida et al. (2009) also reported that *Lc. lactis* subsp. *cremoris* was the dominant strain in all samples of Japanese domestic ropy fermented milk. A high population of *Lc. lactis* subsp. *cremoris*, which results in the production of a high level of EPS, is necessary to produce the ropiness of the fermented milk product.

The viili fungus, *G. candidum*, contributes to the typical musty aroma of this produce. Lipases and proteases produced by *G. candidum* release fatty acids and peptides that are metabolized by the microbial populations, and this contributes to the development of viili's distinctive flavors and other qualities (Bertolini et al. 1995; Holmquist 1998). *G. candidum* also neutralizes the curd by catabolizing the lactic acid produced by the LAB and by releasing ammonia during the metabolism of the amino acids (Greenberg and Ledford 1979).

The viili yeast, *K. marxianus*, is able to utilize both lactose and galactose as carbon sources, which explains why this strain is the primary yeast in the viili samples. The participation of *K. marxianus* in the starter culture ensures the metabolism of lactose via alcohol fermentation and the formation of the product's typical yeasty flavor. In addition to *K. marxianus*, various lactose-negative yeasts seem to play very important roles in the formation of the flavor of viili. Simova et al. (2002) reported that the product's typical yeasty flavor and aroma were absent, when ropy fermented products were produced by *K. marxianus* only.

Additionally, it is worthwhile mentioning that there are remarkable differences in the cell surface properties, auto-aggregation, co-aggregation, and biofilm formation between the kefir and viili strains (Wang et al. 2012). The LAB and yeasts in viili do not show any significant auto-aggregation and biofilm formation, both of which are necessary for forming starter grains. Grain formation begins with the self-aggregation of LAB strains to form small granules. Biofilm-producing strains then begin to attach to the surface of the granules and co-aggregate with other organisms and components in the milk to form the grains (Wang et al. 2012). The surface properties of viili microorganism explain why there is no starter grain formation in viili starter cultures.

19.3 Production of Viili and Viili Products

Viili was historically made on farms, and later families also started to make it at home. Viili culture began to be sold by families at the Helsinki market square in the 1920s, which was before dairy companies began selling it in their shops (Leporanta 2003). The industrial production of viili only began during the late 1950s.

19.3.1 Manufacture of Viili

To make homemade viili (Figure 19.1), raw milk is pasteurized. Active starter (5%, w/v), usually from the previous day viili, is then inoculated into the milk, and the mixture is incubated at 20°C for 20 h. After fermentation, the viili is cooled to 4°C and is ready to consume. However, the ropy nature of viili products seems to become instable if it is produced at higher fermentation temperatures and, when they are frequent, transfers of starter cultures.

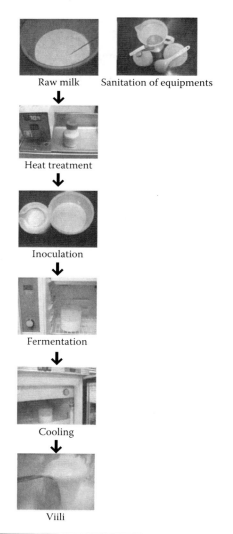

Figure 19.1 Homemade viili.

For the industrial production of viili (Figure 19.2), the milk is separated and standardized to a fat content in the range of 1.0%–3.5%. Traditionally, viili has been made from unhomogenized milk, which is the reason why a layer of cream forms on its surface. After standardization, milk is then pasteurized and cooled to 20°C. A mixed culture of *Lc. lactis* subsp. *cremoris*, *Lc. lactis* subsp. *lactis*, and *G. candidum* is then inoculated and mixed in well. The mixture is then packed in a cup and removed to a storage area for ripening, where the viili fermentation is conducted for 20 h at 20°C. After fermentation, the viili is cooled to below 6°C. The shelf life of viili is around 3 weeks. There are a wide range of different types of viili on the market, including products with different fat contents, ones with reduced lactose and

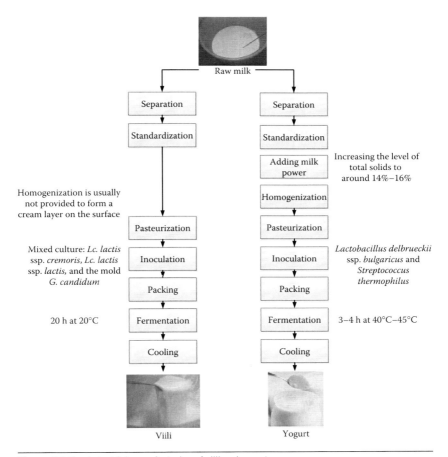

Figure 19.2 Industrial manufacturing of viili and yogurt.

flavored varieties. Viili is also made commercially from homogenized milk without mold growing on the surface and from milk other than cow's, but traditional viili is still the most popular in the Finnish region (Leporanta 2003).

When viili is fermented in a cup, it can be compared to set-type yogurt. However, the starter cultures and the manufacturing processing are quite different between these two fermented dairy products. To produce set-type yogurt (Figure 19.2), raw milk has its level of total solids increased to around 14%–16%. The milk is then homogenized and pasteurized. The heat treatment allows the milk to be held at high temperature for a period for 5–30 min, which provides better texture and prevent whey separation. The milk is then inoculated with a bacterial culture in which *Lactobacillus delbrueckii* subsp. *bulgaricus* and *S. thermophilus* are the dominant microorganisms. The mixtures are packed in a cup and removed to fermentation storage, where the yogurt fermentation is conducted for 3–6 h at 40°C–45°C. The set yogurt has a firm body and smooth texture, which is very different from viili with its stringy/ropy texture.

19.3.2 Other Viili Products

The particular property that defines viili is its slime, which prevents graininess and synergesis, thickening the viili product (Kahala et al. 2008). Thus, viili and its EPS-producing bacterial cultures might be an alternative that may improve the quality of products, and even to provide additional functionality, one possibility is low-fat cheese. Recently, a low-fat stretching-type cheese was made using viili bacterium *Lc. lactis* subsp. *cremoris* and has been commercialized in Taiwan. The cheese processing includes cream separation, pasteurization, inoculation with viili *Lc. lactis* subsp. *cremoris*, coagulation with rennet, cutting and cooking, whey separation, kneading, and cooling (Figures 19.3 and 19.4). This cheese not only shows a good overall acceptability but also has demonstrated antioxidative and angiotensin converting enzyme (ACE) inhibitory activities (Chiang et al. 2011).

Several studies (Awad et al. 2005; Hassan and Awad 2005) have indicated that *Lc. lactis* subsp. *cremoris*, a ropy strain, is able to amplify moisture reservation and amend the melting, viscoelastic, and textural characteristics of reduced-fat Cheddar cheese. Dabour et al. (2005)

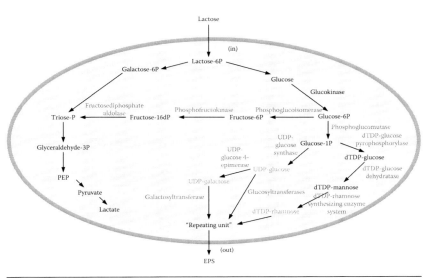

Figure 19.3 EPS biosynthesis in *Lc. lactis* for EPS secretion. (Based on Looijesteijn, P.J. et al., *Appl. Environ. Microbiol.*, 65, 5003–5008, 1999; Hugenholtz, J. et al., *J. Biotechnol.*, 77, 17–23, 2000.)

also showed that EPS-producing *Lc. lactis* subsp. *cremoris* might be an alternative to commercial stabilizers because this strain is able to effectively improve the moisture content of reduced-fat Cheddar cheese.

In addition to low-fat cheese, viili is also used to make soft cream cheese at home. To make soft viili cheese, viili is gently heated and but not allowed to burn. It changes first from thick viili to liquid, and then small curds are formed after continuously heating to just below boiling temperature. The curds then are separated from the whey by a double layer of fine cheese cloth. A very soft cheese can be produced in only 1 h, but a more spreadable cheese needs several hours.

19.3.3 Sources of Milk

Viili products can be made using milk from a variety of sources. However, the source of the milk affects the microbial distribution in the viili products, which then further affects the physical and chemical properties of the viili. Using bovine and reconstituted milk samples, the slime producer, *Lc. lactis* subsp. *cremoris* TL1, is the most dominant species found in samples. In contrast, a non-slime strain, *Lc. lactis* subsp. *cremoris* TL4, was found to be the most prevalent LAB

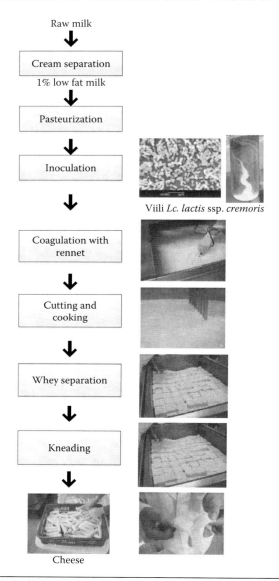

Figure 19.4 Industrial manufacturing of stretching style cheese with viili *Lc. lactis* subsp. *cremoris* in Taiwan.

species in goat's milk viili (Wang et al. 2008). Thus, the goat's milk viili show a lower viscosity than all other samples ($p < .05$). Furthermore, goat's milk, which has a very low proportion of α_{s1}-casein and a higher degree of casein micelle dispersion, also influences the viili's rheological properties (Wszolek et al. 2001).

19.4 Slime Properties of Viili

The characteristic ropy consistency of viili is caused by the slime-forming variants of *Lc. lactis* subsp. *lactis* and *Lc. lactis* subsp. *cremoris* found in the starter culture (Nakajima et al. 1992). Both strains are able to produce slime as capsule or exocellular EPS. Scanning electron micrographs of milk gels prepared using a ropy strain, *Lc. lactis* subsp. *cremoris*, showed that EPS was present in the form of a network that attached the bacterial cells to the protein matrix. A thick network of slime also attached the casein micelle-clusters to each other producing casein conglomerates, which seems to be the origin of the characteristic consistency of viili (Toba et al. 1990).

19.4.1 Composition and Sugar Components of the EPS

The composition of the EPS excreted by *Lc. lactis* subsp. *cremoris* consists of 3%–47% protein and 29%–85% carbohydrate (Macura and Twonsley 1984; Nakajima et al. 1990). The sugar components are most frequently galactose, glucose, and often rhamnose (Nakajima et al. 1990), with repeating units of "→4-β-glucopyranosyl-(1→4)-β-D-galactopyranosyl-(1→4)-β-D-glucopyranosyl—(1→," and groups of α-L-rhamnopyranosyl and α-D-galactopyranosyl-1-p attached to each side of galactopyranosyl (Nakajima et al. 1990, 1992; Sletmoen et al. 2003). The composition and sugar components of EPS are strain and medium dependent. The EPSs secreted by *Lc. lactis* subsp. *cremoris* SBT 0495, ARH53, ARH74, ARH 84, ARH 87, and B30 are composed of repeating units of galactose, glucose, and rhamnose with a phosphodiester structural element (Nakajima et al. 1990; Yang et al. 1999). In contrast, the EPS produced by *Lc. lactis* subsp. *cremoris* H414 is a homopolymer consisting of galactose with a branched-pentasaccharide repeating unit (Gruter et al. 1992). Marshall et al. (1995) reported that *Lc. lactis* subsp. *cremoris* strain LC33 was able to generate two different EPSs. One contains glucose, galactose, rhamnose, glucosamine, and phosphate. The other was composed of galactose, glucose, and glucosamine with branched terminal galactose moieties. The slime material obtained from *Lc. lactis* subsp. *cremoris* SBT0495 supernatant in whey permeate medium consisted of 42% carbohydrate and 21% protein with large amount of mannans present in the cultured medium (Yang et al. 1999). Additionally, *Lc. lactis* produced more EPS on glucose than on fructose as the sugar substrate, although the transcription level of the *eps* gene cluster was independent of the sugar source (Looijesteijn et al. 1999).

19.4.2 EPS Biosynthesis

Recent studies have demonstrated that the ropy phenotype of *Lc. lactis* subsp. *cremoris* is associated with an 18.5–30 MDa plasmid (Looijesteijn et al. 1999; Knoshaug et al. 2007). Several enzymes under positive or negative control are involved in polysaccharide synthesis and excretion. Therefore, the plasmid may code for certain proteins that are involved in slime production (Kontusaari and Forsén 1988). The metabolic pathways that are involved in the biosynthesis of the EPS molecules from the milk sugar lactose are shown in Figure 19.3. The biosynthesis can be roughly divided into three steps: the intracellular formation of EPS precursors, the sugar nucleotides, and the formation of a repeating unit on a lipid carrier that is located at the cytoplasmic membrane (Looijesteijn et al. 1999; Hugenholtz et al. 2000). The repeating unit of EPS produced by most of *Lc. lactis* subsp. *cremoris* strains is composed of glucose, galactose, rhamnose, and phosphate (Nakajima et al. 1990; Yang et al. 1999). The sugar nucleotides UDP-glucose, UDP-galactose, and dTDP-rhamnose are the donors of the monomers used for the biosynthesis of the pentasaccharide unit. The last step of EPS formation involves transportation of the repeating units across the membrane to the outer layer of the bacteria and polymerization of several hundred to several thousand repeating units to form the final EPS (Looijesteijn et al. 1999; Hugenholtz et al. 2000).

The enzymes involved in EPS formation by *Lc. lactis* subsp. *cremoris* include the enzymes responsible for the original metabolism of the carbohydrate, enzymes leading to sugar nucleotide synthesis and interconversion, glycosyl transferases that form the repeating unit attached to the glycosyl carrier lipid, and translocases and polymerases that form the polymer. The genes encoding the enzymes involved in the biosynthesis of EPS by *Lc. lactis* subsp. *cremoris* are borne by an EPS plasmid. The gene products *EpsD*, *EpsE*, *EpsF*, and *EpsG* are glycosyl transferases and are required for the synthesis of the EPS backbone

(Van Kranenburg et al. 1997, 1999). The presence of the EPS genes on a plasmid has been suggested to be the cause of EPS expression instability at higher temperatures and when there are frequent transfers of starter culture (Vedamuthu and Neville 1986; Cerning et al. 1992). Forsén et al. (1973) showed that the slime-forming capacity is stable when the bacteria are grown at 17°C but is lost when they are grown at 30°C.

19.5 Functionality of Viili

Viili and its isolated bacteria have been reported to possess several health benefits including antioxidative effect, ACE inhibitory activity, anti-tumor activity, and immunomodulatory effects.

19.5.1 Antioxidative Effect

Endogenous metabolic processes and exogenous chemicals in the human body or in a food system are able to in some circumstances produce highly active reactive oxygen species that are able to could oxidize biomolecules, which results in tissue damage and cell death; this may lead to inflammation, diabetes, genotoxicity, cancer, and accelerated aging (Biswas et al. 2010). Viili and cheese made with viili bacteria have been demonstrated to contain α,α-diphenyl-β-picylhydrazl, which has a free radical-scavenging effect and Fe^{2+}-chelating ability (Chiang et al. 2011). Viili EPS also possesses a strong antioxidant activity in the gastric intestine in vivo (Liu et al. 2012). The antioxidative nature of viili and its products may help the human body to reduce oxidative damage. Several antioxidative effects peptides with free radical-scavenging activities have been identified in fermented dairy products (Kudoh et al. 2001; Hernández-Ledesma et al. 2005; Pritchard et al. 2010). Moreover, some studies have shown that certain *Lc. lactis* subsp. *cremoris* strains possess proteolytic abilities that aid the digestion of the milk protein into peptides and free amino acids (Tan and Konings 1990; Alting et al. 1995). The change in antioxidant activity noted in low-fat cheeses made with viili is probably associated with the viable populations of LAB present in the cheese and the levels of antioxidative peptides.

19.5.2 ACE Inhibitory Activity

ACE is known to associate with hypertension and congestive heart failure. Viili and cheese made with viili have been demonstrated to have a strong inhibitory effect on ACE activity (Chiang et al. 2011). The proteolytic activity of the starters and the rate of proteolysis both seem to play an important role in the inhibitory activity of these dairy products. The proteolytic system of *Lc. lactis* has been studied, and the system consists of a cell wall–bound proteinase and several intracellular peptidases (Tan and Konings 1990; Alting et al. 1995). *Lc. lactis*, because this organism seems to possess high proteolytic activity, will produce a viili that should demonstrate a better ACE inhibitory effect. Moreover, many studies have reported that the type and concentration of bioactive peptides present in a product are able to significantly affect the functional properties of the dairy product (Ong and Shah 2008a, 2008b). ACE inhibitory peptides are released from α_{s1}-casein and β-caseins and also from α-lactalbumin and β-lactoglobulin during fermentation of milk by various microbiological cultures.

19.5.3 Immunomodulatory Effects

Viili and its LAB have been demonstrated to have immunoregulatory effects in vitro and in vivo, including antiallergic effects and anticolitis effects (Huang et al. 2010). Viili and the viili bacterium *Lc. lactis* subsp. *cremoris* TL1 have been shown to induce the production of the helper cell type I (Th1) cytokine tumor necrosis factor-α, the proinflammatory cytokine interleukin (IL)-6, and T regulatory cell (Treg) cytokine IL-10 in vitro, which suggests that viili may be beneficial and improve the Th1/Th2 balance. Oral feeding of the viili bacterium *Lc. lactis* subsp. *cremoris* TL1 has been shown to suppress total immunoglobulin IgE and ovalbumin (OVA)-specific IgE levels in the serum of OVA-sensitized mice. Suppression of IgE production is an important target when treating allergies.

Additionally, in vivo effects of the viili bacterium *Lc. lactis* subsp. *cremoris* TL1 on the regulation of dextran sulfate sodium (DSS) intestinal physiology have been demonstrated. This strain is able to ameliorate DSS-induced colitis as exemplified by a significant attenuation of the bleeding score and a reduction in colon shortening. Histological analysis also showed crypt regeneration and epithelial restitution in the colon among the animals in *Lc. lactis* subsp. *cremoris* TL1 treated group. These findings suggest that the viili isolated strain, *Lc. lactis* subsp. *cremoris* TL1, has a potential direct anti-inflammatory activity with respect to epithelial cells and that this may lead to inhibition of neutrophil accumulation in the mucosal region of the DSS-colitis mice.

19.5.4 Antitumor Activity

A few studies have identified that viili possesses antitumor activity. Kitazawa et al. (1992) reported that viili and one of the starter bacteria, *Lc. lactis* subsp. *cremoris* KVS20, exhibited antitumor activity inhibiting the metastasis of Lewis lung carcinoma, as well as acting to reduce the growth of solid and ascetic forms of sarcoma-180 in vivo. The antitumor effect of viili might be due to increase in cytotoxic activity of the macrophages stimulated by *Lc. lactis* subsp. *cremoris* KVS20. Liu et al. (2012) observed the senescence of HepG2 cancer cells after treatment with viili EPS, which supports the idea that viili EPS might have anti-tumor activity.

References

Alting, R.C., Engels, W.J.M., Schalkwijk, S., and Exterkate, F.A. 1995. Purification and characterization of cystathionine β-lyase from *Lactococcus lactis* subsp. *cremoris* B78 and its possible role in flavor development in cheese. *Applied and Environmental Microbiology* 61: 4037–4042.

Awad, S., Hassan, A.N., and Halaweish, F. 2005. Application of exopolysaccharide-producing cultures in reduced-fat Cheddar cheese: Composition and proteolysis. *Journal of Dairy Science* 88: 4195–4203.

Bertolini, M.C., Schrag, J.D., Cygler, M., Ziomek, E., Thomas, D.Y., and Vernet, T. 1995. Expression and characterization of *Geotrichum candidum* lipase I gene, comparison of specificity profile with lipase II. *European Journal of Biochemistry* 228: 863–869.

Biswas, M., Haldar, P.K., and Ghosh, A.K. 2010. Antioxidant and free-radical-scavenging effects of fruits of *Dregea volubilis*. *Journal of Natural Science, Biology and Medicine* 1(1): 29–34.

Boutrou, R. and Guéguen, M. 2005. Interest in *Geotrichum candidum* for cheese technology. *International Journal of Food Microbiology* 102: 1–20.

Cerning, J., Bouillanne, C., Landon, M., and Desmazeaud, M. 1992. Isolation and characterization of exopolysaccharides from slime-forming mesophilic lactic acid bacteria. *Journal of Dairy Science* 75: 692–699.

Chen, H.C., Lin, C.W., and Chen, M.J. 2006. The effects of freeze drying and rehydration on survival of microorganisms in kefir. *Asian-Australasian Journal of Animal Sciences* 19: 126–130.

Chen, T., Tan, Q., Wang, M., Xiong, S., Jiang, S., Wu, Q., Li, S., Luo, C., and Wei, H. 2011. Identification of bacterial strains in viili by molecular taxonomy and their synergistic effects on milk curd and exopolysaccharides production. *African Journal of Biotechnology* 10: 16969–16975.

Chiang, M.L., Chen, H.C., Wang, S.Y., Hsieh, Y.L., and Chen, M.J. 2011. Use of Taiwanese ropy fermented milk (TRFM) and *Lactococcus lactis* subsp. *cremoris* isolated from TRFM in manufacturing of functional low-fat cheeses. *Journal of Food Science* 76: M504–M510.

Dabour, N., Kheadr, E., Fliss, I., and LaPointe, G. 2005. Impact of ropy and capsular exopolysaccharide-producing strains of *Lactococcus lactis* subsp. *cremoris* on reduced-fat Cheddar cheese production and whey composition. *International Dairy Journal* 15: 459–471.

Dharmawan, J., Surono, I.S., and Kunm, L.Y. 2006. Adhesion properties of indigenous dadih lactic acid bacteria on human intestinal mucosal surface. *Asian-Australasian Journal of Animal Sciences* 19: 751–755.

Forsén, R., Raunio, V., Myllymaa, R., Nousiainen, R., and Päkkilä, M. 1973. Studies on the slime forming group N *Streptococcus* strains. I. Differentiation between some lactic streptococcus strains by polyacrylamide gel electrophoresisof soluble cell proteins. *Acta Universitatis Ouluensis: Scientiae rerum naturalium* 3: 1.

Greenberg, R.S. and Ledford, R.A. 1979. Deamination of glutamic and aspartic acids by *Geotrichum candidum*. *Journal of Dairy Science* 62: 368–372.

Gruter, M., Leeflang, B.R., Kuiper, J., Kamerling, J.P., and Vliegenthart, J.F.G. 1992. Structure of the exopolysaccharide produced by *Lactococcus lactis* subsp. *cremoris* H414 grown in a defined mediumor skimmed milk. *Carbohydrates Research* 231: 273–291.

Hassan, A.N. and Awad, S. 2005. Application of exopolysaccharide-producing cultures in reduced-fat Cheddar cheese: Cryo-scanning electron microscopy observations. *Journal of Dairy Science* 88: 4214–4220.

Hernández-Ledesma, B., Miralles, B., Amigo, L., Ramos, M., and Recio, I. 2005. Identification of antioxidant and ACE-inhibitory peptides in fermented milk. *Journal of the Science of Food and Agriculture* 85: 1041–1048.

Holmquist, M. 1998. Insights into the molecular basis for fatty acyl specificities of lipases from *Geotrichum candidum* and *Candida rugosa*. *Chemistry and Physics of Lipids* 93: 57–65.

Huang, I.N., Dai, T.Y., Wang, S.Y., and Chen, M.J. 2010. Inhibitory effect of Taiwanese ropy fermented milk in an ovalbumin-induced allergy mouse model. *Journal of Dairy Science* 93 S(1): 807.

Hugenholtz, J., Looijesteijn, E., Starrenburg, M., and Dijkema, C. 2000. Analysis of sugar metabolism in an EPS producing *Lactococcus lactis* by ^{31}P NMR. *Journal of Biotechnology* 77: 17–23.

Kahala, M., Mäki, M., Lehtovaara, A., Tapanainen, J.-M., Katiska, R., Juuruskorpi, M., Juhola, J., and Joutsjoki, V. 2008. Characterization of starter lactic acid bacteria from the Finnish fermented milk product viili. *Journal of Applied Microbiology* 105: 1929–1938.

Kitazawa, H., Itoh, T., Tomioka, Y., Mizugaki, M., and Yamaguchi, T. 1996. Induction of IFN-γ and IL-1α production in macrophages stimulated with phosphopolysaccharide produced by *Lactococcus lactis* ssp. *cremoris*. *International Journal of Food Microbiology* 31: 99–106.

Kitazawa, H., Toba, T., Itoh, T., Kumano, N., Adachi, S., and Yamaguchi, T. 1991. Antitumoral activity of slime-forming, encapsulated *Lactococcus lactis* ssp. *cremoris* isolated from Scandinavian ropy sour milk "viili." *Animal Science and Technology* 62: 277–283.

Kitazawa, H., Yamaguchi, T., and Itoh, T. 1992. B-Cell Mitogenic activity of slime products produced from slime-forming, encapsulated *Lactococcus lactis* ssp. *cremoris*. *Journal of Dairy Science* 75: 2946–2951.

Kitazawa, H., Yamaguchi, T., Miura, M., Saito, T., and Itoh, T. 1993. B-cell mitogen produced by slime-forming encapsulated *Lactococcus lactis* ssp. *cremoris* isolated from ropy sour milk, viili. *Journal of Dairy Science* 76: 1514–1519.

Knoshaug, E.P., Ahlgren, J.A., and Trempy, J.E. 2007. EPS in *Lactococcus lactis* ssp. *cremoris* ropy 352: Evidence for novel gene organization. *Applied and Environmental Microbiology* 73: 897–905.

Kontusaari, S.I. and Forsén, R.I. 1988. Finnish fermented milk "Villi": Involvement of two cell surface proteins in production of slime by *Streptococcus lactis* ssp. *cremoris*. *Journal of Dairy Science* 71: 3197–3202.

Kontusaari, S.I., Vuokila, P.T., and Forsén, R.I. 1985. Immunochemical study of triton x-100-soluble surface components if slime-forming, encapsulated *Streptococcus cremoris* from the fermented milk product viili. *Applied and Environmental Microbiology* 50: 174–176.

Kudoh, Y., Matsuda, S., Igoshi, K., and Oki, T. 2001. Antioxidative peptide from milk fermented with *Lactobacillus delbrueckii* subsp. *bulgaricus* IFO13953. *Nippon Shokuhin Kagaku Kaishi* 48: 44–50.

Leporanta, K. 2003. Viili and Långfil—Exotic fermented products from Scandinavia. *Valio Foods & Functionals*. Online Available: http://www.valio.fi/portal/page/portal/valiocom/Valio_Today/Publications/valio_foods___functionals05102006130335/2003.pdf.

Lin, C.W., Chen, H.L., and Liu, J.R. 1999. Identification and characterisation of lactic acid bacteria and yeasts isolated from kefir grains in Taiwan. *Australian Journal of Dairy Technology* 54: 14–18.

Liu, L., Wu, J., Zhang, J., Li, Z., Wang, C., Chen, M., Wang, Y., Sun, Y., Wang, L., and Luo, C. 2012. A compatibility assay of ursolic acid and foodborne microbial exopolysaccharides by antioxidant power and anti-proliferative properties in hepatocarcinoma cells. *Journal of Food, Agriculture and Environment* 10: 111–114.

Looijesteijn, P.J., Boels, I.C., Kleerebezen, M., and Hugenholtz, J. 1999. Regulation of exopolysaccharide production by *Lactococcus lactis* subsp. *cremoris* by the sugar source. *Applied and Environmental Microbiology* 65: 5003–5008.

Macura, D. and Townsley, P.M. 1984. Scandinavian ropy milk-identification and characterization of endogenous ropy lactic streptococci and their extracellular excretion. *Journal of Dairy Science* 67: 735.

Marshall, V.M., Cowie, E.N., and Moreton, R.S. 1995. Analysis and production of two exopolysaccharides from *Lactococcuslactis* subsp. *cremoris* LC330. *Journal of Dairy Research* 62: 621–628.

Mistry, V.V. 2004. Fermented liquid milk products. In: *Handbook of Food and Beverage Fermentation Technology*, eds. Y.H. Hui, L. Meunier-Goddik, Å.S. Hansen, J. Josephsen, W. Nip, P.S. Stanfield, and F. Toldrá, 1–16. New York: Marcel Dekker, Inc.

Nakajima, H., Suzuki, Y., Kaizu, H., and Hirota, T. 1992. Cholesterol lowering activity of ropy fermented milk. *Journal of Food Science* 57: 1327–1329.

Nakajima, H., Toyoda, S., Toba, T., Itoh, T., Mukai, T., Kitazwa, H., and Adachi, S. 1990. A novel phosphophlysaccharide from slime-forming *Lactococcus lactis* subspecies *cremoris* SBT 0495. *Journal of Dairy Science* 73: 1472–1477.

Ong, L. and Shah, N.P. 2008a. Influence of probiotic *Lactobacillus acidophilus* and *L. helveticus* on proteolysis, organic acid profiles, and ACE-inhibitory activity of Cheddar cheeses ripened at 4, 8, and 12°C. *Journal of Food Science* 73: M111–M120.

Ong, L. and Shah, N.P. 2008b. Release and identification of angiotensin-converting enzyme inhibitory peptides as influenced by ripening temperatures and probiotic adjuncts in Cheddar cheeses. *LWT Food Science and Technology* 41: 1555–1566.

Pritchard, S.R., Phillips, M., and Kailasapathy, K. 2010. Identification of bioactive peptides in commercial Cheddar cheese. *Food Research International* 43: 545–548.

Ruas-Madiedo, P., Gueimonde, M., de los Reyes-Gavilán, G.C., and Salminen, S. 2006. Short communication: Effect of exopolysaccharide isolated from "viili" on the adhesion of probiotics and pathogens to intestinal mucus. *Journal of Dairy Science* 89: 2355–2358.

Simova, E., Beshkova, D., Angelov, A., Hristozova, Ts., Frengova, G., and Spasov, Z. 2002. Lactic acid bacteria and yeasts in kefir grains and kefir made from them. *Journal of Industrial Microbiology and Biotechnology* 28: 1–6.

Sletmoen, M., Maurstad, G., Sikorski, P., Paulsen, B.S., and Stokke, B.T. 2003. Characterization of bacterial polysaccharides: Steps towards single-molecular studies. *Carbohydrate Research* 338: 2459–2475.

Tan, P.S.T. and Konings, W.N. 1990. Purification and characterization of an aminopeptidase from *Lactococcus lactis* subsp. *cremoris* Wg2. *Applied and Environmental Microbiology* 56: 526–532.

Toba, T., Nakajima, H., Tobitani, A., and Adachi, S. 1990. Scanning electron microscopic and texture studies on characteristic consistency of Nordic ropy sour milk. *International Journal of Food Microbiology* 11: 313–320.

Tsutsui, S., Kikuchi, M., Takahashi, S., and Ariga, H. 1998. Finland ropy sour milk "viili," analyzed for its acidity, viscosity and constitution. *Journal of Rakuno Gakuen University* 22: 231–237.

Uchida, K., Akashi, K., Motoshima, H., Urishima, T., Arai, I., and Saito, T. 2009. Microbiota analysis of Caspian sea yogurt, a ropy fermented milk circulated in Japan. *Animal Science Journal* 80: 187–192.

Van Kranenburg, R., Marugg, J.D., Van Swam, I.I., Willem, J., and DeVos, W.M. 1997. Molecular characterization of the plasmid-encoded *eps* gene cluster essential for exopolysaccharide biosynthesis in *Lactococcus lactis*. *Molecular Microbiology* 24: 387–397.

Van Kranenburg, R., Van Swam, I.I., Marugg, J.D., Kleerebezem, M., and De Vos, W.M. 1999. Exopolysaccharide biosynthesis in *Lactococcus lactis* NIZO B40: Functional analysis of the glycosyl transferase genes involved in synthesis of the polysaccharide backbone. *Journal of Bacteriology* 181: 338–340.

Vedamuthu, E.R. and Neville, J.M. 1986. Involvement of a plasmid in production of ropiness (mucoidness) in milk cultures by *Streprococcus cremoris* MS. *Applied and Environmental Microbiology* 51: 677.

Wang, S.Y., Chen, H.C., Liu, J.R., Lin, Y.C., and Chen, M.J. 2008. Identification of yeasts and evaluation of their distribution in Taiwanese kefir and viili starters. *Journal of Dairy Science* 91: 3798–3805.

Wang, S.Y., Chen, H.C., Dai, T.Y., Huang, I.-N., Liu, J.R., and Chen, M.J. 2011. Identification of lactic acid bacteria in Taiwanese ropy fermented milk and evaluation of their microbial ecology in bovine and caprine milk. *Journal of Dairy Science* 94: 623–635.

Wang, S.Y., Chen, K.N., Lo, Y.M., Chiang, M.L., Chen, H.C., Liu, J.L., and Chen, M.J. 2012. Investigation of microorganisms involved in biosynthesis of the kefir grain. *Food Microbiology* 32: 274–285.

Wang, Y., Liu, Y., Luo, C., Ding, L.C., Sun, X.N., and Liu, X.C. 2010. Initial isolation and identification of lactic acid bacteria in viili. *Journal of Anhui Agricultural Sciences* 38: 2831–2832.

Wszolek, M., Tamime, A.Y., Muir, D.D., and Barclay, M.N.I. 2001. Properties of kefir made in Scotland and Poland using bovine, caprine and ovine milk with different starter cultures. *LWT Food Science and Technology* 34: 251–261.

Yang, Z., Huttunen, E., Staaf, M., Widmalm, G., and Tenhu, H. 1999. Separation, purification and characterisation of extracellular polysaccharides produced by slime-forming *Lactococcus lactis* ssp. *cremoris* strains. *International Dairy Journal* 9: 631–638.

PART V
HEALTH BENEFITS

20
Probiotic Dairy Products

SURAJIT MANDAL, SUBROTA HATI, AND CHAND RAM

Contents

20.1	Introduction	520
20.2	Probiotics—Functional Food Ingredients	521
20.3	Selection of Probiotic Strains for Functional Dairy Products	522
20.4	Currently Used Probiotics	524
	20.4.1 Human Probiotics	525
20.5	Dairy Products as Delivery Vehicle of Live Probiotics	525
20.6	Probiotics in Dairy Products: An Overview	526
20.7	Market Status of Probiotics	527
20.8	Probiotics in Dairy Products	528
	20.8.1 Probiotics in Cheese	528
	20.8.2 Probiotics in Fresh Fermented Milks	530
	20.8.3 Probiotics in Non-Fermented Dairy Products	531
20.9	Manufacturing of Probiotic Functional Dairy Products	533
	20.9.1 Probiotic Yogurt	533
	20.9.2 Probiotic Yogurt Drink	534
	20.9.3 Probiotic Dahi	535
	20.9.4 Probiotic Lassi	536
	20.9.5 Probiotic Shrikhand	537
	20.9.6 Probiotic Ice Cream	538
	20.9.7 Probiotic Kulfi	539
	20.9.8 Probiotic Milk Chocolate	540
20.10	Challenges for Probiotic Dairy Products	541
20.11	Criteria for Effective Probiotic Foods	541
	20.11.1 Probiotic Effective Dose	541
	20.11.2 Hurdles Affecting Probiotics' Survival during Processing and Storage	542
	20.11.3 Improving Probiotics' Survival in Dairy Products	543

20.11.4	Selection of Appropriate Probiotic Strains	543
20.11.5	Protection against Oxygen	544
	20.11.5.1 Ascorbic Acid	544
	20.11.5.2 Cysteine	544
	20.11.5.3 Use of Air-Tight Containers	545
20.11.6	Two-Stage Fermentation	545
20.11.7	Addition of Micronutrients	546
20.11.8	Stress Adaptation	546
20.11.9	Technological Interventions	547
20.12 Summary		548
References		548

20.1 Introduction

The market of functional foods, "foods that promote health beyond providing basic nutrition," is flourishing worldwide nowadays. The functional foods contain probiotics—"the live microbial feed supplements that beneficially affect the individual by modulating the intestinal microbial balance positively" (Shah 2000)—and are growing in a big leap. The major probiotic functional foods belong to dairy, and there is a synergistic effect between components of milk and probiotics. The natural buffering of stomach acid by milk proteins also enhances the stability of consumed probiotics. Besides, milk products provide a number of high-quality nutrients including calcium, protein, bioactive peptides, and conjugated linoleic acids. In addition to the valuable nutrients, higher pH, solid-not-fat content, and fat content of dairy products compose a very favorable milieu of probiotics. Therefore, consumptions of foods containing probiotics and prebiotics are on the rise due to their positive effects on human health. A number of foods including yogurt, kefir, ice cream, frozen fermented dairy desserts, freeze-dried yogurt, cheese, spray-dried milk powder, fruit juices, cheese-based dip, lassi, dahi, beverages, infant formula, bars, and so on are employed as delivery vehicles for the probiotics. The significant challenges are stability and survival of probiotics during processing, during storage, as well as during their passage via gastrointestinal tract (GIT). As per International Dairy Federation recommendations, probiotics should be active and present $\geq 10^7$ CFU/g of product and 10^8–10^9 viable cells are needed for daily intake for beneficial effects.

20.2 Probiotics—Functional Food Ingredients

Probiotics were first conceptualized by the Nobel Laureate Elie Metchnikoff at the turn of twentieth century in Russia. He believed that the fermenting bacillus (now known as *Lactobacillus*) contained in the fermented milk products consumed by Bulgarian peasants positively influenced the microflora of colon and decreased toxemia. Positive effects associated with probiotics include cholesterol and/or triglyceride reduction, antitumor properties, improved lactose tolerance and stimulation of immune system via nonpathogenic means, and increase in natural resistance to infectious disease of GIT (Collins and Gibson 1999). The term *probiotic* means "for life" was defined by Food and Agriculture Organization of the United Nations as "microorganisms administered in adequate amounts which confer a beneficial health effect on the host" (FAO/WHO 2001, pp. 1–34).

Probiotics, along with a large and diverse range of other bacteria, colonize the human body. These bacteria play a significant role in human physiology, that is, probiotics lower the pH of intestinal tract, so that pathogens cannot survive. Further, growing beneficial bacteria compete with pathogens for nutrients and space in GIT. Probiotics are also associated with preventing diarrhea, relieving constipation, building immunity, preventing skin allergy, and improving women urogenital health (Shah 2000). The two most commonly used probiotics, that is, *Lactobacillus acidophilus* and *Bifidobacterium bifidum* and other lactic acid producers have exhibited positive health effect.

Although most of the researches are directed toward the use of intestinal probiotics, however, over the period of time, microbes consisting of not only lactic acid bacteria (i.e., lactobacilli, streptococci, enterococci, lactococci, bifidobacteria, etc.) but also *Bacillus* ssp. and fungi (i.e., *Saccharomyces* ssp. and *Aspergillus* ssp.) have also been used. Selection of suitable strains is a vital factor in determining the success of the probiotic foods. The desirable factors related to health promotion or sustenance serve as important criteria for the strain selection (Holzapfel et al. 1998). Key criteria for the selection of an appropriate strain comprised related origin, identity, safety and resistance (e.g., to mutations, environmental stress, and antimicrobial factors prevailing in upper GIT), technical aspects (i.e., growth properties in vitro and during processing, survival, and viability during transport, storage, etc.), and a number of functional aspects. The primary probiotics associated with dairy are *Lactobacillus* and bifidobacteria. *Lactobacillus* ssp. include *Lb. acidophilus*, *Lb. amylovorus*, *Lb. plantarum*, *Lb. crispatus*, *Lb. gasseri*, *Lb. casei*, *Lb. paracasei*, *Lb. rhamnosus*, *Lb. reuteri*, *Lb. johnsonni*, while *Bifidobacterium* species include *B. longum*, *B. adolescentis*, *B. animalis*, *B. bifidum*, *B. breve*, *B. infantis*, and *B. lactis* (Reid 1999). Some of these and closely related microbes have been used in making cultured dairy products and are generally regarded as safe (GRAS). This means there is a minimal concern to their use as dietary adjuncts in dairy products.

Health beneficial effects of probiotic foods include recovery from diarrhea, alleviation of lactose intolerance and malabsorption, relieved constipation, treating colitis, enhanced immune response, inhibition of pathogens and translocation, stimulation of gastrointestinal immunity, reduction of infection from common pathogens, reduction of risk of certain cancers by detoxification of carcinogens, lowered serum cholesterol, reduced blood pressure in hypertensive, treatment of food allergies, synthesize nutrients (i.e., folic acid, niacin, riboflavin, vitamins B_6, B_{12}), increased nutrient bioavailability, improved reproductive health, enhanced efficacy of vaccines, and so on.

20.3 Selection of Probiotic Strains for Functional Dairy Products

Probiotics should exhibit desirable characteristics such as maintenance of viability during processing and storage, ease of application in products, and resistance to the physicochemical processing of foods (Prado et al. 2008). On the other hand, traditional starter cultures are in use as these have the ability to impart desirable organoleptic qualities of cultured dairy products. The probiotics should be chosen to elicit health or nutritional benefits after their uptake. But in some dairy products, probiotics are mixed to make the market value rather to impart the specific function and health benefit. Therefore, manufacturers can claim higher prices for the probiotic product. Thus, proper selection of strain to be included as a probiotic is of utmost importance. The different probiotics exhibit distinctive properties that can affect their survival in foods, fermentation pattern, and other probiotic attributes. The strain selection therefore becomes a critical parameter to ensure the fermentation or probiotic performance (Klaenhammer 2003). The key points for selection of probiotics comprises safety, functional (i.e., survival in host, appropriate adherence, colonization, anitimicrobial production, antigenotixic activity, immune-modulation, exclusion of pathogens), and technological characteristics including growth in milk and other food base, sensory properties, phage resistance, and viability (Mishra and Prasad 2000; Sabikhi and Mathur 2001). Safety concern of probiotics includes many stipulations like origin from healthy GIT of human, non-virulence, and antibiotic resistance characteristics. Strains used should preferably be of human origin, as these can function better in an environment similar to that from where these are isolated. The human origin strains can be adhesive and better colonize the human GIT, which is the first step in promoting colonization resistance. Probiotics should possess GRAS status. To survive and grow in vivo conditions at desired site of administration, the probiotics must be able to tolerate low pH of stomach and high concentration of both conjugated and deconjugated bile acids of small intestine (Collins et al. 1998).

The following are desirable selection criteria for probiotic strains:

1. *Suitability*
    - Identification, that is, taxonomy by phylogenetic analysis and rRNA sequencing
    - Nonpathogenic and nontoxic having GRAS status
    - Normal inhabitant of the species selection and obtained from a healthy individual
2. *Technological Appropriateness*
    - Able to grow and acid production in milk
    - Able to survive in high biomass density
    - Genetically stable to maintain phenotypic and physiological functionality

- Impart desirable organoleptic characteristics when introduced in foods or fermentation processes
- Mass production and storage: sufficient growth, recovery, concentration, freezing, dehydration, storage, distribution, and so on
- Metabolically stable to maintain desired attributes during product preparation, storage, and delivery

3. *Competitiveness*
   - Ability of survival, proliferation, and execution of metabolic activity in vivo
   - Ability to compete with the normal microflora, and resistant to bacteriocins, acids, and other antimicrobial agents produced by residing microflora and other inhibitory agents present in milk
   - Adherence, colonization, and retention capability
   - Resistant to bile salts and acid

4. *Manifestation and Efficiency*
   - Able to elicit one or more clinically validated health benefits
   - Able to produce primary and secondary antimicrobial metabolites (i.e., bacteriocins, hydrogen peroxide, organic acids, or other inhibitors)
   - Anticarcinogenic and anticarcinogenic
   - Anti-inflammatory
   - Antimutagenic
   - Having antagonistic ability toward pathogenic microflora
   - Immunostimulatory
   - Synthesis and release of bioactive compounds (i.e., enzymes, vaccines, peptides) (Klaenhammer 2003)

20.4 Currently Used Probiotics

The basic concept behind probiotics is pretty straightforward: reestablishment of lost microflora of intestine with new beneficial one. Many different microbes are added to dairy products for their probiotic potential. Probiotics have been designed for delivery in food or dairy products, via supplementation or fermentation. These include *Lactobacillus* ssp., such as *Lb. acidophilus*, *Lb. casei*, *Lb. delbrueckii* subsp. *bulgaricus*, *Lb. reuteri*, *Lb. brevis*, *Lb. cellobiosus*, *Lb. carvatus*,

Lb. fermentum, and *Lb. plantarum*; Gram-positive cocci, such as *Lactococcus lactis* subsp. *cremoris*, *Str. thermophilus*, *Enterococcus faecium*, *Str. diacetylactis*, and *Str. intermedius*; and *Bifidobacterium* ssp., such as *B. bifidum*, *B. adolescentis*, *B. animalis*, *B. infantis*, *B. longum*, and *B. thermophilum*. Nonpathogenic microbes that occupy important niches in the host gut or tissues, such as yeasts, enterococci, and *Enterobacteriaceae*, are also being used as human and animal probiotics although *Lactobacillus* and *Bifidobacterium* are the most common species of probiotics for the production of fermented milks and other dairy products (Fuller 1992).

20.4.1 Human Probiotics

- *B. longum* (SBT2928, BB536)
- *B. breve* (Yakult)
- *B. lactis* (BB12)
- *Lb. acidophilus* (NCFM, SBT2062)
- *Lb. casei* (Shirota, CRL431, DN014001, immunits)
- *Lb. delbrueckii* subsp. *bulgaricus* (2038)
- *Lb. fermentum*
- *Lb. johnsonii* (La1, Lj 1)
- *Lb. paracasei* (CRL431, F19)
- *Lb. plantarum* (299V)
- *Lb. reuteri* (SD 2112)
- *Lb. rhamnosus* (GG, 271, GR1)
- *Lb. salivarius* (UCC118)
- *Str. thermophilus* (1131) (Sanders and Huisin't Veld 1999)

20.5 Dairy Products as Delivery Vehicle of Live Probiotics

Milk or its products provide an excellent carrier for probiotics, as most of these can readily utilize lactose as an energy source for their growth. An important requirement for the growth of these microbes in GIT is provided by the milk constituents. Milk proteins also provide important protection to the probiotics during passage through stomach (Charteris et al. 1998). Yogurts and fermented dairy drinks have long been an ideal vehicle for delivering probiotics to the human GIT, which explains the widespread use of probiotic cultures in dairy products. Probiotics are also increasingly introduced into nondairy beverages (i.e., fruit juices or energy drinks). Cheese is another promising delivery vehicle of some of the probiotics to humans.

20.6 Probiotics in Dairy Products: An Overview

The benefits of dairy products to consumers are a part of the stories until the perceptions of probiotics aroused. Now, the investigation on fermented milks and yogurt-containing probiotics is more systematic. Functional foods are prepared as a whole food or food ingredients with positive effects on health of host apart from the nutritional aspects. Fermented milk with probiotics is the emergent class of functional foods (Mattila-Sandholm et al. 2002). Fermented milk supplemented with synbiotics (i.e., probiotics and prebiotics together) meets all the standards of a functional food. These foods promote health to avert diseases and in general are used to indicate a food that contains some health-promoting components beyond the basic nutrients (Berner and O'Dannell 1998). Functional foods have been variously termed as *neutraceuticals, designer foods, medicinal foods, therapeutic foods, super foods, foodiceuticals, medifoods*, and so on (Finley 1996; ADA 2009).

The concept of synbiotics is still in its infancy, but combining a probiotic with a prebiotic in a single food product, the expected benefits are improved survival during the passage of probiotics through the upper GIT (Collins and Gibson 1999; Fooks et al. 1999). A significant effect on the composition of fecal flora was observed on feeding volunteers, a daily supplement of synbiotic composed of 125 ml of *Lactobacillus* fermented milk containing 2.75 g oligofructose for 7 weeks (Gibson et al. 1995). Martin (1996) studied the effects of bifidogenic growth factors on survival of *B. longum, B. infantis*, and *B. adolescentis* in different dairy products by supplementing 10% solid skim milk containing *B. longum* or *B. infantis* with 0.5% fructo-oligosaccharides (FOS) and 0.5% lactulose. FOS and lactulose used (0.5%) does not significantly affect *B. longum* or *B. infantis* in skim milk during incubation at 37°C for 48 h. Dubey and Mistry (1996) studied the effect of bifidogenic factors on growth characteristics of bifidobacteria in infant food formulae. They incorporated a bifidogenic factor, lactulose, or FOS (0.5%) into infant formulae, which were then inoculated (2.5%) with *B. bifidum, B. breve, B. infantis*, or *B. longum* or their mixture. Lactulose did not influence maximal counts or generation time in either formula for any species except *B. infantis*, which had lower counts. Change in pH and production of biochemical metabolites and maximal bacterial counts were not influenced by the presence of FOS in infant formulae. Kunz et al. (2001) developed a synbiotic low-fat soft curd cheese and investigated the effect of probiotic dietary fibers on sensory quality. They reported that the combination of oligofructose and wheat fiber increased the sensory and the nutritional quality of the low-fat soft curd cheese. Use of inulin and oligofructose in fruit yogurts, milk-based drinks, milk, spreads, cheese, and ice cream has also been reported (Coussement 1996). Number of symbiotic dairy products containing bifidobacteria and lactulose are already available in Japanese markets: Hounyu milk powder for adults (lactulose—8.3 g/100 g and bifidobacteria $>3 \times 10^7$), Sawayaka sour milk (lactulose 4 g/100 g and bifidobacteria $>10^8$), and so on. (Mizota et al. 1987). Some synbiotic dairy products of Europe include Sym balance, mixture of *Lb. reuterii*, *Lb. acidophilus*, and *Lb. casei* along with inulin and John après Jour (Young 1998).

20.7 Market Status of Probiotics

With increasing awareness among consumers, there has been a boom in the growth and production of probiotic foods world over. There have been long-term interests in the use of cultured milks products with various strains of probiotics to improve health of humans (Salminen et al. 1998). Traditionally, probiotics have been added to yogurt and other fermented foods. A number of commercial probiotic dairy foods are available in market. There are over 70 bifidus and acidophilus products produced including sour cream, buttermilk, yogurt, powdered milk, frozen desserts, and so on. More than 53 different types of dairy products containing probiotics are marketed in Japan alone (Hilliam 2000). The largest markets for functional foods and supplements are the United States, Europe. and Japan, accounting for 33.6%, 28.2%, and 20.9%, respectively (Blandon et al. 2007). Functional dairy products market has increased by 43% during 2004–2009 with an amount of $4,840 million in 2009 (Mintel 2009). This growth is heavily supported by the segment of probiotic yogurts, where Danonn Activia is the leader (Anon 2008). Probiotics and dairy products contribute the largest functional food market accounting for

Table 20.1 Commercial Probiotic Products and Their Manufacturers Available on the Market

PRODUCT CATEGORY	PRODUCT MANUFACTURER	PROBIOTIC STRAINS
Actimel	Danone	*Lb. caseidefensis*
Activia	Danone	*B. animalis* DN173010
Chamyto	Nestle	*Lb. johnsonii, Lb. helveticus*
Danito	Danone	*Lb. casei*
Fermented milk Yakult	Yakult	*Lb. casei* Shirota
Nesvita (stirred and drinkable)	Nestle	*B. animalis* subsp. *lactis*
Yogurt-like beverages	Danone	*B. animalis* DN173010

nearly 33% of the broad market, while cereal products have just over 22% (Leatherhead Food International 2006). Consumption of yogurt has increased because many consumers associate yogurt with health benefits (Hekmat and Reid 2006). A new market report from Transparency Market Research titled "Probiotics Market (Dietary Supplements, Animal Feed, Foods & Beverages)—Global Industry Analysis, Market Size, Share, Trends, Analysis, Growth and Forecast, 2012–2018" reported that global probiotics demand was worth $27.9 billion in 2011 and is expected to reach $44.9 billion in 2018, growing at a compound annual growth rate of 6.8% from 2013 to 2018. Demand from Asia-Pacific and Europe dominated the global market, with Asia-Pacific expected to be the most prominent market in the future. Increasing scientific evidence and health awareness among consumers are the driving factors for the rapid growth of probiotic market (Table 20.1).

20.8 Probiotics in Dairy Products

20.8.1 Probiotics in Cheese

Probiotics are dominated by yogurts and its drinks in European markets. Cheeses can also be an alternative food vehicle for the delivery of viable probiotics (Ong and Shah 2009). Cheese (especially cheddar) may offer certain advantages over other probiotic products such as yogurt or milk because of reduced acidity and high fat content, and texture of cheddar cheese may protect probiotics during passage through the GIT (Stanton et al. 1998). Taking advantage of the growing market for functional dairy products and the opportunity to develop a commercially viable probiotic cheese, several cheese brands

Table 20.2 Probiotic Cheeses

PRODUCT	REFERENCE
Cheddar cheese	Ong and Shah (2009)
Cottage cheese	Blanchette et al. (1996)
Crescenza cheese	Burns et al. (2008)
Feta cheese	Kailasapathy and Masondole (2005)
Synbiotic petit-suisse cheeses	Cardarelli et al. (2008)
Turkish Beyaz probiotic cheese	Kilic et al. (2009)
Turkish white cheese	Kasımoglu et al. (2004)
White-brined cheese	Yilmaztekin et al. (2004)

and dairy entrepreneurs have tried to penetrate the functional dairy products market with probiotic cheese (Table 20.2).

Probiotic cheeses (i.e., cheddar-like) were produced with microfiltered milk standardized with cream enriched with native phosphocaseinateretentate and fermented by *B. infantis* (Daigle et al. 1999) and found that it exhibited good viability during storage for 12 weeks. *Bifidobacterium* ssp. survived well in cheddar and gouda cheeses (Stanton et al. 1998), and probiotic cheddar cheeses can be manufactured containing high levels of *Lb. paracasei* (10^8 CFU/g) at relatively low cost by conventional manufacturing procedures. The possibility of improving nutritive value of goat milk by adding probiotics was assessed (Gomes and Malcata 1998). They observed the growth of *B. lactis* up to 3×10^8 CFU/g that was dependent on the physicochemical characteristics of cheese but *Lb. acidophilus* did not grow substantially in any of the experimental cheeses, and maximum numbers were less than 6×10^7 CFU/g. Song et al. (2001) used *Lb. helveticus* CU 631 that has antagonistic effects against *Helicobacter pyroli* to make cream cheese. An extensive proteolysis occurred during ripening by probiotics was also observed. A synbiotic low-fat soft curd cheese was developed and the effect of dietary fibers, one of which was prebiotic, on sensory quality was studied (Kunz et al. 2001), and they reported that combination of oligofructose and wheat fiber increased the sensory and the nutritional quality of the low-fat soft curd cheese. Cruz et al. (2009a) emphasized that the making of probiotic cheese should have minimum changes, when compared to traditional products that make the production of functional cheeses favorable. The growth capacity of probiotic *Lb. paracasei* A13, *B. bifidum* A1, and *Lb. acidophilus* A3 in a probiotic fresco cheese commercialized in Argentina was studied (Vinderola et al. 2009) during manufacturing and refrigerated storage at 5°C and 12°C for 60 days, and a negative impact on sensory quality was observed.

20.8.2 Probiotics in Fresh Fermented Milks

Increases in sales and turnover are attributed to the growth in the probiotic yogurts. The healthy image of yogurt is well known (O'-Carroll 1996). Five starters (*B. longum* and *Lb. delbrueckii* subsp. *bulgaricus*; *B. longum*, *Lb. delbrueckii* subsp. *bulgaricus*, and *Str. salivarius* subsp. *thermophilus*; *B. longum*, *Str. salivarius* subsp. *Thermophilus*, and *Lb. acidophilus*; *B. longum*, *Lb. acidophilus*, and *Lb. casei* subsp. *casei*; and *B. longum* and *Lb. acidophilus*) were used to produce set-type of probiotic yogurt. Starters were added to recombined milk at 3 g/100 g in which bifidobacteria accounted for 50% of the starter blend. Initial counts of *Bifidobacterium* were 10^7–10^8 CFU/g. No spoilage microbes were detected over 21 days of storage, and yogurt was of acceptable organoleptic quality. With some combinations of probiotic starter resulting in yogurt of better quality than controls made using conventional starters were noticed. Sarkar and Misra (1998) studied the incorporation of *Propionibacterium freudenreichii* subsp. *shermanii* MTCC 1371 and *B. bifidum* NCDC into cultures for the making of probiotic yogurt. An incubation temperature of 42°C ± 1°C instead of 37°C ± 1°C was suggested for yogurt manufacture in 4 h. They showed that use of *P. freudenreichii* subsp. *shermanii* NTCC 1371 + *B. bifidum* NCDC with yogurt YH-3 culture produced probiotic yogurt with good nutritional and therapeutic qualities. Fermented beverages were produced from goat milk using 2% of inoculum of yogurt culture DVS-YC 180 (*S. thermophilus* and *Lb. delbrueckii* subsp. *bulgaricus*) and mixed ABT 4 (*Lb. acidophilus*, *Str. thermophilus*, and *Bifidobacterium* ssp.) cultures (Bozanic et al. 2001). They observed that probiotic yogurt displayed slightly better sensory properties. Samples of acidified milk and orange-flavored probiotic yogurt, manufactured in Cuba from 70% buffalo milk and 30% cow milk, were stored at a temperature of 4°C in 0.5-l plastic containers (Iniguez et al. 2001). The average shelf life was 10 days for acidified milk and 11 days for orange-flavored yogurt. Study also showed that the main characters affected by prolonged storage were odor, flavor, and development of surface fungal growth. Survival of yogurt and probiotics and their antimicrobial effects were assessed during making and storage for 21 days (El-Rahman 2000). *B. bifidum* and *Lb. acidophilus* survived in the yogurt, and all samples had higher antimicrobial effect against *S. typhimurium*, *E. coli*, *B. cereus*, and *Candida albicans* compared with *S. aureus*. They observed that the antimicrobial effect was increased during storage, and yogurt made with probiotics had higher flavor, texture, and overall acceptability than did samples without probiotics (Table 20.3).

20.8.3 Probiotics in Non-Fermented Dairy Products

Various non-fermented foods have been studied as delivery vehicle of probiotics, and the buffering capacity of these foods, which is the major factor affecting pH, and the rate of gastric emptying, also the quantity and physiological state of the bacteria, have an important influence on the survival of the GIT transit (Bertazzoni-Minelli et al. 2004). Ice cream being the most preference of consumer, probiotic ice cream is a suitable vehicle for delivering beneficial microbes such as *Lactobacillus* and *B. bifidum* to consumers (Hekmat and McMohan 1997). Ice creams are food products that show great potential for use

Table 20.3 Probiotic Fresh Fermented Dairy Products

PRODUCT	REFERENCE
Acidophilus "sweet" drink	Speck (1978)
Acidophilus butter and progurt	Gomes and Malcata (1999)
Acidophilus milk drink	Itsaranuwat et al. (2003)
Biogarde, mil-mil, and acidophilus milk with yeasts	Gomes and Malcata (1999)
Dahi	Yadav et al. (2007)
Fermented lactic beverages supplemented with oligofructose and cheese whey	Castro et al. (2009)
Frozen yogurt	Davidson et al. (2000)
Low-fat yogurt	Penna et al. (2007)
Mango soy fortified probiotic yogurt	Kaur et al. (2009)
Regular full-fat yogurt	Aryana and Mcgrew (2007)
Stirred fruit yogurt	Kailasapathy et al. (2008)
Synbiotic acidophilus milk	Amiri et al. (2008)
Traditional Greek yogurt	Maragkoudakisa et al. (2006)

as vehicles for probiotics, with the advantage of foods being consumed by all age groups (Cruz et al. 2009b).

Probiotic ice cream mix was made by adding 25% and 50% of commercial cultured milks fermented with *Lb. acidophilus* and *B. bifidum* to a control ice cream mix (Christiansen et al. 1996). The number of viable *Lb. acidophilus* and *B. bifidum* was 0.5×10^7–1.0×10^7 CFU/ml. Younis et al. (1998) prepared ice cream containing viable *Lb. acidophilus* and *B. bifidum* by direct acidification or by culturing with starter. The most acceptable probiotic product was produced by direct acidification with lactic acid or culturing at pH 6.0. Culture viability over a 12-month period was compared between commercial frozen concentrates of *Lb. acidophilus*, *B. lactis*, and *Lb. paracasei* subsp. *paracasei* when incorporated into a low-fat ice cream and stored at −25°C (Haynes and Playne 2002). The observations of study revealed that full-fat ice cream offered no extra protection for cultures over the low-fat product during storage, with the low-fat formulation showing improved survival of all three cultures during the freezing process. Additionally, direct culture addition was a suitable method for producing a low-fat probiotic ice cream that is able to retain acceptable levels of viable microorganisms over an extended shelf life. Effects of inulin and sugar levels on sensory properties of probiotic ice cream added with *Lb. acidophilus* and *B. lactis* were studied by Akin et al. (2007). Overall, probiotic ice cream gave a good total impression and did not have any marked-off flavor during the storage period. Hekmat and McMahon (1992) studied the survival of *Lb. acidophilus* and *B. bifidum* in ice cream and evaluated its sensory properties and concluded that ice cream could be used as a good source for delivering probiotics to consumers. Probiotic ice cream was prepared by incorporating the freeze-dried cultures of *Lb. acidophilus*, *Lb. delbrueckii* subsp. *bulgaricus*, and *B. bifidum* in ice cream mix at 1% and 2% using batch freezer. *Lb. acidophilus* and *B. bifidum* at 1% level yielded the superior results in acceptable form of ice cream and survival of cultures (Vijayageetha et al. 2011).

Milk chocolates were prepared by incorporating with free or encapsulated *Lb. casei* NCDC 298 and inulin (Mandal et al. 2012), and they found that *Lactobacillus* counts were remained above 8.0 log CFU/g till 60 days under refrigerated condition. Yeasts and molds and coliforms

Table 20.4 Probiotic Non-Fermented Dairy Products

PRODUCT	REFERENCES
Beverages, infant formula, bars	Katz (2001)
Chocolate mousse	Aragon-Alegro et al. (2007)
Mayonnaise	Khalil and Mansour (1998)
Probiotic ice cream	Kailasapathy and Sultana (2003), Akin et al. (2007)
Probiotics frozen ice milk	Sheu et al. (1993)
Synbiotic ice cream	Homayouni et al. (2008)
Synbiotic milk chocolate	Mandal et al. (2012)

were absent in the products during the storage. Sensory panelists liked the milk chocolate with encapsulated lactobacilli (Table 20.4).

20.9 Manufacturing of Probiotic Functional Dairy Products

Presently, numerous dairy food claims probiotic attributes and are present in market worldwide. The probiotic cultures used are mainly belonging to genus *Lactobacillus* and *Bifidobacterium*. Bacteria widely studied for probiotic dairy products manufacturing are *Lb. acidophilus*, *Lb. casei*, *Lb. rhamnosus*, *Lb. plantarum*, *B. bifidum*, *B. longum*, and so on. For preparation of probiotic fermented dairy products such as yogurt and dahi, probiotic bacteria are used as sole starter cultures to carry over the fermentation or in combination of compatible dairy starter cultures such as yogurt cultures, dahi cultures, and so on. Sometimes, probiotic fermented or non-fermented dairy products such as drinkable yogurt, lassi, shrikhand, kulfi, and so on can be prepared by incorporating probiotic bacterial cell biomass prepared separately. Nowadays, researches for development of non-fermented dairy products with added probiotic cell biomass are being carried out for diversification of probiotic application beyond the traditional fermented probiotic dairy products. Now, general outlines for preparation of some probiotic fermented and non-fermented dairy products will be discussed in the following section.

20.9.1 Probiotic Yogurt

Yogurt remains a milk-based fermented milk that is found in set solid and stirred fluid form. Additionally, fruits and flavors may also be supplemented in yogurt. The majority of yogurts consumed worldwide are manufactured with cultures with probiotic growth optima of 37°C–45°C, namely, *Lb. delbrueckii* subsp. *bulgaricus* and *Str. thermophilus*. Milk of good quality is clarified followed by standardization to set the desired fat content. Afterward, ingredients are then blended together in a mix tank with agitation system. The mixture is then heated for 30 min at 85°C or 10 min at 95°C for making a relatively sterile and conducive environment for the growth of starter culture as well as this time–temperature combination will denature and coagulate the whey proteins to enhance the viscosity and texture. Subsequently, the mixture is homogenized at pressures of 2000–2500 psi before final heat treatment. Besides thoroughly mixing the ingredients, homogenization prevents creaming and wheying off during incubation and storage. Stability, consistency, body, and texture are enhanced by homogenization. After the final heat treatment, the mix is cooled to 37°C. Probiotic yogurt is prepared by incorporating compatible probiotic cultures along with yogurt cultures and incubated at 37°C (Figure 20.1).

20.9.2 Probiotic Yogurt Drink

Yogurt drink is prepared from yogurt/probiotic yogurt by breaking the curd and mixing with sugar syrup/salt and flavor followed by packing and stored under refrigerated conditions. During mixing, stable probiotic biomass can also be supplemented for preparing the probiotic yogurt drink (Figure 20.2).

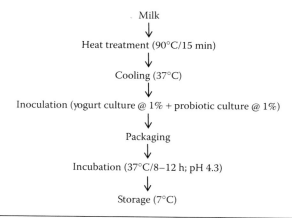

Figure 20.1 Preparation of probiotic yogurt.

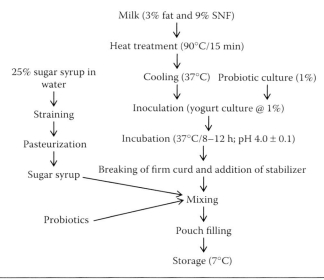

Figure 20.2 Preparation of probiotic yogurt drink.

20.9.3 Probiotic Dahi

Dahi is one of the most popular Indian traditional fermented milk products that is similar to yogurt and is consumed by people throughout the south Asian subcontinent including Bangladesh, Pakistan, Nepal, and Bhutan. In contrast to starter cultures of yogurt (a mixed culture of *Lb. delbrueckii* subsp. *bulgaricus* and *Str. thermophilus*), the starter culture of dahi is not well defined. There is no strict definition of dahi as for yogurt. Dahi contains various strains of lactic acid bacteria. Traditionally, at household level and halwai's, dahi is prepared by the backslopping method. However, production of dahi with an individual culture of *Lc. lactis* or a combination of cultures containing lactobacilli and lactococci has been reported. The major problem of dahi is the rapid deterioration and short shelf life due to frequent microbial contamination, a major concern for the health of consumers. The traditional method for preparation of dahi invariably involves a small scale, either in consumers' household or in the sweet makers' shop in urban areas. In the household, milk is boiled, cooled to about 37°C and inoculated with 0.5%–1.0% of starter (i.e., previous day's dahi or buttermilk), and allowed to set overnight. It is then stored under refrigeration and consumed. Dahi containing probiotic *Lb. casei* is prepared to modulate the immune response against *Salmonella enteritidis* infection. Probiotic dahi is more efficacious in protecting against *S. enteritidis* infection by enhancing innate and adaptive immunity than fermented milk and plain dahi. Prefeeding of probiotic dahi may strengthen the consumer's immune system and may protect infectious agents like *S. enteritidis*. Like yogurt, probiotic dahi can be prepared with compatible probiotic cultures along with dahi culture by incubating suitably (Figure 20.3).

20.9.4 Probiotic Lassi

Lassi is a refreshing summer beverage, popular in North India. Glasses of fresh, frothy lassi are invariably pressed upon a visitor. Variants of lassi include buttermilk, *Chhach*, and *Mattha* that are consumed with relish in other regions too. Lassi is a white to creamy white, viscous liquid, with a sweetish, rich aroma and mild to high acidic taste. It is flavored with either salt or sugar and other condiments, depending on regional preferences. Lassi is obtained from pasteurized milk or part of skim milk, cultured with lactic and aroma/flavor producing microorganisms. The term *lassi* is also used for a phospholipid-rich fluid fraction obtained as a by-product during the churning of dahi while making Makkhan (i.e., desi fresh butter). Lassi making was earlier confined to the cottage sector or homes. It was mainly a rural product. Now it is commercially prepared in several parts of North India. Commercial lassi is viscous, cultured fluid milk, containing a characteristic pleasing aroma and flavor. It is packaged in traditional

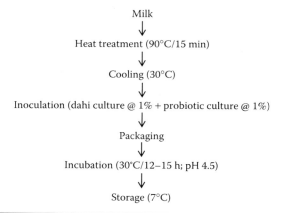

Figure 20.3 Preparation of probiotic dahi.

milk cartons/sachets/UHT boxes. Salted lassi is marketed in a number of cities in the southern region of India. Aseptically packaged long-life lassi is also getting popular in the market. Lassi is prepared from dahi/probiotic dahi by breaking the curd and mixing with sugar syrup/salt and flavor followed by packing and stored under refrigerated conditions.

Like probiotic yogurt drink, probiotic lassi is prepared from dahi/probiotic dahi by breaking the curd and mixing with sugar syrup/salt and flavor followed by packing and stored under refrigerated conditions. During mixing, stable probiotic biomass can also be supplemented for preparing the probiotic lassi (Figure 20.4).

20.9.5 Probiotic Shrikhand

Shrikhand is a semisolid, sweetish-sour fermented milk product prepared from dahi (curd). The technology involves (1) coagulation of milk by fermentation with culture to obtain acidic curd or dahi, (2) preparation of chhaka by draining of whey from the curd, (3) blending of additives such as sugar, color, flavor, and spices to reach a desired level of composition and consistency. From probioitic dahi, chakka can be prepared after draining of whey, and shrikhand can be prepared by mixing the chakka with ground sugar, color, flavor, spices,

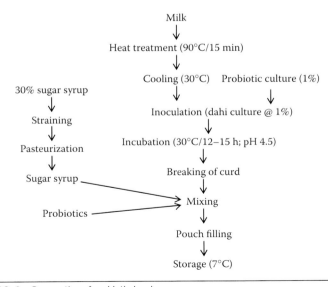

Figure 20.4 Preparation of probiotic lassi.

and so on. During the mixing stage, probiotic biomass can be supplemented at desired level of viable cells for preparing the functional probiotic shrikhand from traditionally prepared chakka from plain dahi (Figure 20.5).

20.9.6 Probiotic Ice Cream

First, fresh cow milk is standardized to 10% fat, 11% nonfat milk solids, 15% sugar, and 0.3% stabilizer followed by mixing of liquid ingredients and warming to 50°C in a water-filled double-jacketed steam vat. Sugar and skim milk powder are added to the container with constant stirring after the temperature of the content was raised to 60°C. Later the stabilizer is added to the container at 65°C. The ingredients are all well dispersed, and the mix is filtered through a muslin cloth to remove the undissolved material in the mix. The mix is homogenized at 60°C by using a two-stage homogenizer maintaining 2500 psi during first stage and 500 psi during second stage of homogenization. The mix is pasteurized at 68°C for 30 min in a double-jacketed vat in order to safeguard the health of consumer. Then probiotics are inoculated, when the temperature of the pasteurized mix reached to the desired temperature of inoculation, that is,

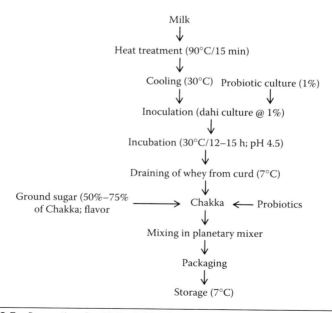

Figure 20.5 Preparation of probiotic shrikhand.

37°C depending on the type of culture. After the incubation period, fermented ice cream mix is frozen in a freezer. The ice cream is packed in polystyrene cups and kept at −18°C. Alternately, in pasteurized and cooled ice cream mix, stable and active probiotic cell biomass can be incorporated at desired probiotic level followed by freezing in preparation of probiotic ice cream.

20.9.7 Probiotic Kulfi

Kulfi is a popular South Asian dessert made with boiled milk. It comes in many flavors, including pistachio, malai, mango, cardamom (*elaichi*), saffron (*kesar*), the more traditional flavors, as well as newer variations such as apple, orange, peanut, and avocado. Where western ice creams are whipped with air or overrun, kulfi contains no air; it is solid-dense frozen milk. Therefore, kulfi is not ice cream but a related and distinct product in the frozen dairy-based dessert category. Kulfi is a natural dessert made with pure milk and contains no eggs. Kulfi is prepared by boiling milk until it is reduced to half. Then sugar is added and the mixture is boiled for another 10 min, and a teaspoon of corn flour is added to it after making it into a paste using water. On adding the paste, the mixture thickens and is boiled for some more time. Then flavorings, dried fruits, cardamom, and so on are added. The mixture is then cooled, put in molds, and frozen.

As per the PFA standards, kulfi is the frozen product obtained from cow or buffalo milk or a combination thereof or obtained from cream and/or other milk products, with or without the addition of cane sugar, dextrose, liquid glucose, eggs, fruits, fruit juices, preserved fruits, nuts, chocolate, edible flavor, and permitted food colors. It may contain permitted stabilizers and emulsifiers, not exceeding 0.5% by weight. The mixture shall be suitably heated before freezing. The product shall contain not less than 10% milk fat, 3.5% proteins, and 36% total solids except that when any of the aforesaid preparation contains fruits or nuts or both, the content of milk fat shall not be less than 8.0% by weight. In pasteurized and cooled kulfi mix, stable and active probiotic cell biomass can be incorporated at desired probiotic level followed by filling in kulfi molds and freezing in ice salt mix in preparation of probiotic kulfi (Figure 20.6).

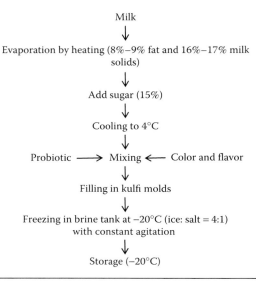

Figure 20.6 Preparation of probiotic kulfi.

20.9.8 Probiotic Milk Chocolate

Milk chocolate is prepared at laboratory scale with following ingredients: milk cream (55% fat)—300 g, skim milk powder—400 g, sugar—800 g, liquid glucose—300 g, cocoa powder—75 g, cocoa butter—100 g, lecithin—10 g, carboxymethyl cellulose—10 g, sodium benzoate—6 g, salt—0.6 g, and chocolate flavor—6 ml. Almost 50% sugar syrup is prepared and heated to boil. The dry mixture of carboxymethyl cellulose and salt is mixed to it. Separately, half of the total skim milk powder and milk cream is reconstituted in required amount of water, and it is added to the boiling sugar syrup. The content is boiled up to a semithick consistency and after that cocoa butter is added. After thorough mixing, liquid glucose is mixed and heated up to the pat forming stage. Dry-mix remaining half of skim milk powder, cocoa powder, and sodium benzoate, and then lactobacilli (probiotic) (free or encapsulated—10^8 CFU/g final product), inulin (5% of final product), and flavor are mixed uniformly. The whole mass is allowed to set and cool on stainless steel tray overnight at refrigerated conditions. Chocolate mass is cut in small rectangular pieces and wrapped with aluminum foil. The product is stored at $7°C \pm 1°C$ (Figure 20.7) (Mandal et al. 2012).

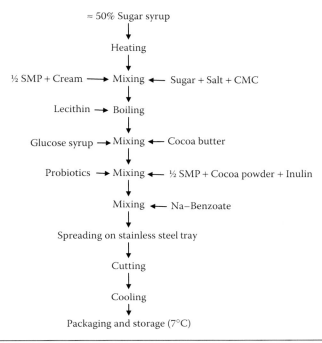

Figure 20.7 Preparation of probiotic milk chocolate.

20.10 Challenges for Probiotic Dairy Products

The dairy products claiming probiotic need to mention shelf life and species of probiotics used on product labels. However, name of specific strain and number of viable cells are not mentioned in product level. Hence, the basis challenge for any probiotic food is survival and concentration of probiotics. The survival of probiotic in any food faces many problems such as adaptability, oxygen stress, and storage conditions. The success rate of probiotic claim and their validation is very low presently.

20.11 Criteria for Effective Probiotic Foods

20.11.1 Probiotic Effective Dose

Currently, there are no legal recommendations for consumption of probiotics through foods. Adequate numbers of viable cells, that is, "therapeutic minimum," need to be consumed regularly to exert its probiotic effect. The IDF (1997) proposed that in probiotic foods, "the specific microbes shall be viable, active and abundant at the level of at least 10^7 CFU/g in the product to the date of minimum durability" (Ouwehand and Salminen 1998, pp. 749–756). It has been suggested that approximately 10^9 CFU/day of probiotics is necessary to elicit health effects. Based on daily consumption of 100 g of a probiotic food, it has been suggested that a product should contain at least 10^7 CFU/g, a level paralleling current Japanese recommendations (Ishibashi and Shimamura 1993). The ingestion of 10^6–10^9 viable cells per day is necessary for humans in order to develop beneficial effects (Lee and Salminen 1995). Kurmann and Rasic (1991) suggested achieving optimal potential therapeutic effects, and the number of probiotics in a probiotic food should meet a suggested minimum ($\geq 10^6$ CFU/ml) at satisfactory level. One should aim to consume 10^8 live probiotic cells per day. Regular consumption of 400–500 g/week of bioyogurt, containing 10^6 viable cells/ml, would provide these numbers (Tamime et al. 1995). The food industry considers 10^6 CFU/g at the time of consumption as the minimum level of probiotics necessary to perform their nutritional benefits (Boylston et al. 2004). It is important to report that these bacteria should be present in a dairy food to a minimum level of 10^6 CFU/g, or the daily intake should be about 10^8 CFU/g, with the aim to compensate for the possible reduction in the number of the probiotics during passage through the gut (Shah 2000).

20.11.2 Hurdles Affecting Probiotics' Survival during Processing and Storage

Probiotic strains should be selected on their technological suitability. Additionally, bacteria should not lose functional heath properties due to technological limitations. Hence, the viability of probiotics has been both marketing and technological concern for many industrial products. Probiotics are extremely susceptible to environmental conditions such as oxygen, processing and preservation treatments, acidity, and salt concentration, which collectively affect the overall viability of probiotics (Lankaputhra et al. 1996). For instance, high concentration of oxygen in yogurts can show adverse effect on extended survival of probiotics, particularly the oxygen-sensitive bifidobacteria (Talwalkar and Kailasapathy 2003).

Other inhibitory substances such as lactic acid and cold storage can affect the survival of probiotics (Shah 2000). The survival of bifidobacteria in fermented dairy products depends on the strain of bacteria used, fermentation conditions, and storage temperature (Shin et al. 1996). The viability of bifidobacteria significantly decline during storage of frozen yogurt (Laroia and Martin 1991). The survival of bifidobacteria in yogurt is quite low because the pH of yogurt ranges from 4.2 to 4.6 (Adhikari et al. 2000). Lankaputhra et al. (1996) reported that only three out of nine bifidobacterial strains were viable in the pH range of 3.7–4.3. Klaver et al. (1993) found that 14 out of 17 strains lost their viability in fermented milks in the first week of storage. Hughes and Hoover (1995) reported the similar observation. *Lb. acidophilus* tolerates acidity, but a rapid decrease in their number has been observed under acidic conditions (Lankaputhra and Shah 1995). Bifidobacteria are not as acid tolerant as *Lb. acidophilus*, the growth of the latter microorganism ceases below pH 4.0, while the growth of *Bifidobacterium* ssp. is retarded below pH 5.0 (Shah 1997). Ravula and Shah (1998) also reported that very high levels of probiotics do not survive in fermented frozen dairy desserts.

20.11.3 Improving Probiotics' Survival in Dairy Products

Viability of probiotics can be improved by appropriate selection of acid- and bile-resistant strains, use of oxygen impermeable containers, two-step fermentation, stress adaptation, incorporation of micronutrients such as peptides and amino acids, sonication of yogurt bacteria, and microencapsulation (Shah 2000).

20.11.4 Selection of Appropriate Probiotic Strains

The basis for selection of probiotics include safety, functional aspects (i.e., survival, adherence, colonization, antimicrobial production, immune stimulation, antigenotoxic activity, prevention of pathogens), and technological details such as growth in milk and other food base, sensory properties, phage resistance, and viability (Sabikhi and Mathur 2001). One of the most important characteristics of probiotics is their ability to survive through the acid in the human stomach and bile in the intestine. Clark et al. (1993) studied the survival of

B. infantis, B. adolescentis, B. longum, and *B. bifidum* in acidic conditions and reported that *B. longum* survived the best. Clark and Martin (1994) reported that *B. longum* tolerated bile concentration of as high as 4.0%, whereas Ibrahim and Bezkorovainy (1993) found *B. longum* to be the least resistant to bile. Lankaputhra and Shah (1995) have shown that among six strains of lactobacilli, three *Lb. acidophilus* survived best under acidic conditions and two *Lb. acidophilus* showed the best tolerance to bile. Among nine strains of *Bifidobactereium* ssp., *B. longum* and *B. pseudolongum* survived best under acidic conditions and *B. longum, B. pseudolongum,* and *B. infntis* showed the best tolerance to bile. Thus, selection of appropriate strains on the basis of acid and bile tolerance would help to improve viability of these probiotics (Shah 2000).

20.11.5 Protection against Oxygen

20.11.5.1 Ascorbic Acid Ascorbic acid (vitamin C), an oxygen scavenger, is permitted in fruit juices and other food products as an additive. Milk and milk products supply only 10%–15% of the daily requirements of vitamin C (Rasic and Kurmann 1978). The viability of yogurt and probiotics was evaluated during manufacture and storage of yogurt supplemented with ascorbic acid using four commercial starter cultures (Dave and Shah 1997). The viable counts of *Str. thermophilus* were lower, whereas *Lb. delbrueckii* subsp. *bulgaricus* were higher, with increased ascorbic acid levels. Therefore, fortification with ascorbic acid may increase the nutritive value as well as viability of probiotics.

20.11.5.2 Cysteine Cysteine, a sulfur-containing amino acid, could provide amino nitrogen as a growth factor while reducing the redox potential, both of which might favor the growth of anaerobic bifidobacteria. Collins and Hall (1984) reported improved viability of some *Bifidobacterium* species in reconstituted milk supplemented with 0.05% cysteine. Dave and Shah (1997) reported that cysteine at 50 mg/l promoted the growth of *Str. thermophiles* and decreased incubation time to reach pH 4.5 and also affected redox potential, and a slight decrease in redox potential is beneficial for survival, however, when the concentration of cysteine is increased beyond 50 mg/l affected the growth of *S. thermophilus*. Viability of bifidobacteria was improved by incorporating cysteine into yogurt.

20.11.5.3 Use of Air-Tight Containers *Lb. acidophilus* is microaerophilic, and bifidobacteria are anaerobic. As bifidobacteria are anaerobic, oxygen toxicity is an important consideration, as it can easily dissolve in milk. Dave and Shah (1997) studied the survival of yogurt and probiotics in yogurt made in plastic containers and glass bottles. The increase in numbers and survival of *Lb. acidophilus* during storage was directly affected by the dissolved oxygen content, which was shown to be higher in yogurts made in plastic containers than glass. For the samples stored in glass bottles, the counts remained higher than those stored in plastic cups. Bifidobacteria multiplied better in glass bottles than in plastic cups. The initial counts of the bifidobacteria population were 1.6-fold higher in yogurt prepared in glass bottles than in plastic cups. Although the acid contents were similar in products stored in glass bottles and plastic cups at 4°C, the survival rate was 30%–70% higher in products fermented and stored in glass bottles than in plastic cups. Better survival and viability of bifidobacteria in deaerated milk has been observed (Klaver et al. 1993). Shah (2000) suggested to store AB or ABE products in glass containers or to increase the thickness of the packaging materials. Improved viability of *Lb. acidophilus* in product made in glass bottles in which oxygen permeation was minimal and in yogurt supplemented with vitamin C and oxygen scavenging agent (Dave and Shah 1997). To exclude oxygen during the production of bifidus milk products, special equipment is required to provide an anaerobic environment (Shah 2000).

20.11.6 Two-Stage Fermentation

Poor survival of *Lb. acidophilus* and bifidobacteria is due to the production of inhibitory substances such as acid and hydrogen peroxide by yogurt bacteria, which are essential for the typical yogurt flavor. Generally, yogurt bacteria grow faster as compared to probiotics during fermentation and produce acids that could reduce the viability. Lankaputhra and Shah (1997) studied the effect of two-stage fermentation on viability of probiotics in yogurt. Initial fermentation with probiotics was carried out for 2 h, followed by fermentation with yogurt culture. This allowed the probiotics to be in their final stage of lag phase and thus could dominate the flora, resulting in higher counts. They reported that the initial counts of probiotics have increased by four to five times in the product manufactured by two-step fermentation. The counts after 6 weeks of storage were more than 10^7 CFU/g, and *Lb. acidophilus* 2309 and *B. longum* 1941 after 6 weeks of storage were 6.85 and 7.93 log CFU/g and 7.60 and 8.84 log CFU/g for single- and two-step processes, respectively.

20.11.7 Addition of Micronutrients

Probiotics grow slowly in milk due to lack of proteolytic activity, and thus yogurt bacteria are added to reduce the fermentation time. *Lb. delbrueckii* subsp. *bulgaricus* produces essential amino acids because of its proteolytic nature, and the symbiotic relationship of *Lb. delbrueckii* subsp. *bulgaricus* and *Str. thermophilus* is well established, where the latter also produces growth factors for the former microorganism. However, *Lb. delbrueckii* subsp. *bulgaricus* produces lactic acid during refrigerated storage. Post-acidification causes loss of viability of probiotics. To overcome the loss of viability of probiotics due to post-acidification, use of starter cultures with no *Lb. delbrueckii* subsp. *bulgaricus* and incorporation of micronutrients (peptides and amino acids) through casein hydrolysate are essential to reduce the fermentation time and to improve the viability of probiotics (Dave and Shah 1998).

20.11.8 Stress Adaptation

The viability of *Lb. acidophilus* can be improved by adapting the microorganisms to harsh environments, which cause the loss in viability. Adaptation increases the survival rate of microorganisms in harsh conditions more than those that had been shifted directly to lethal acidic conditions. Mechanisms contributing to the adaptive acid tolerance response include the induction of a new pH homeostasis system and the synthesis of a new set of proteins known as stress proteins (Foster 1993). Stress proteins can be transiently or constitutively produced and contribute to the adaptive acid tolerance response (Foster 1993). Transiently produced proteins confer only short-term protection to microorganisms, while constitutive production of proteins is associated with extended survival periods. The stress responses to acid also fully or partially protect the cell against other stress responses with the overlap of production of proteins (Foster and Hall 1990).

20.11.9 Technological Interventions

Probiotic suppliers have developed a variety of proprietary techniques to preserve and protect the integrity of these tiny living microorganisms. Freeze drying is the most commonly used method to protect probiotics from stomach acid and to deliver a high number of live microorganisms to the small and large intestines. Bacterial cultures are coated with polysaccharides to ensure stability throughout the manufacturing process and to protect the microorganisms from stomach acid in "Probio-Tec" process. The main purpose is to protect the microorganisms from heat, light, air, and moisture. LiveBac is a new-patented process, based on microencapsulation, which protects microorganisms from air and moisture, thus greatly extending product shelf life even without refrigeration. Using the controlled delivery technology, tablets or capsules can be either "programmed" to release active ingredients into the GIT at a constant rate or pulsed at precisely time intervals. These can also be used in products that contain several active ingredients, which can be distributed to specific locations in the digestive tract, thus optimizing uptake. Packaging is also an important consideration to improve probiotics survivability. Probiotics and probiotics-containing products are packaged in air-tight and impermeable containers. One can preserve probiotics for a long time, if these are kept frozen without any damage to the product, as frozen condition slows down the die-off process considerably.

Enteric coating technology protects probiotics during the passage through the gastric barrier. The enteric coating prevents the solubilization of capsules in the stomach and protects probiotics against acid shock. The process can be applied to any type of capsule and is approved as GRAS. Probiotics are available that are stable at ambient temperatures by packaging, so as to protect them from the adverse conditions of oxygen, moisture, light, and heat, for example, in gel capsules, similar to vitamin capsules. The important aspect of this technology is its ability to deliver high dose of viable probiotics in the jejunum and the ileum, which can escape the harsh processing and storage condition and harsh acidity of the stomach and bile in the intestine. However, the application of such large capsules is not feasible for incorporation in food systems. Microencapsulation is the most suitable alternative technology to offer the best protection to the probiotic cells resulting from the freeze drying and milling, and thus microencapsulated probiotics can be used in numerous food systems (Mandal et al. 2006, 2012).

20.12 Summary

Dairy foods may well turn out to be an ideal vehicle for reintroducing probiotics to human GIT. Problems associated with probiotics incorporation into foods are survival and stability during processing, preservation, storage, and GIT transition, and probiotic foods should include probiotics at suitable level until consumption. Approach to improve probiotics survivability is the selection of suitable strains for food application, stress adaptation, physical separation by microencapsulation, and so on.

References

ADA (American Dietetic Association). 2009. Position of the American Dietetic Association: Functional foods. *Journal of the American Dietetic Association* 109: 735–746.

Adhikari, K., Mustapha, A., Grun, I.V., and Fernando, L. 2000. Viability of microencapsulated bifidobacteria in set yoghurt during refrigerated storage. *Journal of Dairy Science* 83: 1946–1951.

Akin, M.B., Akin, M.S., and Kirmaci, Z. 2007. Effects of inulin and sugar levels on the viability of yogurt and probiotic bacteria and the physical and sensory characteristics in probiotic ice cream. *Food Chemistry* 104: 93–99.

Amiri, Z.R., Khandelwal, P., Aruna, B.R., and Sahebjamnia, N. 2008. Optimization of process parameters for preparation of synbiotic acidophilus milk via selected probiotics and prebiotics using artificial neural network. *Journal of Biotechnology* 136: 460.

Anon, 2008. More cheese trends. *Dairy Foods* 109(3): 76–77.

Aragon-Alegro, L.C., Alarcon-Alegro, J.H., Cardarelli, H.R., Chiu, M.C., and Saad, S.M.I. 2007. Potentially probiotic and synbiotic chocolate mousse. *LWT—Food Science and Technology* 40: 669–675.

Aryana, K.J. and Mcgrew, P. 2007. Quality attributes of yogurt with *Lactobacillus casei* and various prebiotics. *LWT—Food Science and Technology* 40: 1808–1814.

Berner, L.A. and O'Dannell, J.A. 1998. Functional foods and health claims legislation: Application to dairy foods. *International Dairy Journal* 8: 355–362.

Bertazzoni-Minelli, E., Benini, A., Marzotto, M., Sbarbati, A., Ruzzenente, O., Ferrario, R., Hendricks, H., and Dellaglio, F. 2004. Assessment of novel probiotic *Lactobacillus casei* strains for the productions of functional dairy foods. *International Dairy Journal* 14: 723–736.

Blanchette, L., Roy, D., Belanger, G., and Gauthier, S.F. 1996. Production of cottage cheese using dressing fermented by bifidobacteria. *Journal of Dairy Science* 79: 8.

Blandon, J., Cranfield, J., and Henson, S. 2007. *International Food Economy Research Group: Department of Food, Agricultural and Resource Economics.* http://www4.agr.gc.ca/resources/prod/doc/misb/fbba/nutra/pdf/u_of_guelph_functional_foods_review_final_25jan2008_en.pdf. Accessed March 30, 2010.

Boylston, T.D., Vinderola, C.G., Ghoddusi, H.B., and Reinheimer, J.A. 2004. Incorporation of bifidobacteria into cheese: Challenges and rewards. *International Dairy Journal* 14: 375–387.

Bozanic, R., Tratnik, L., and Parat, M. 2001. Acceptability of yoghurt and probiotic yoghurt from goat's milk. *Mljekarstvo* 51: 317–326.

Burns, P., Patrignani, F., Serrazanetti, D., Vinderola, G.C., Reinheimer, J.A., Lanciotti, R., and Guerzoni, M.E. 2008. Probiotic Crescenza cheese containing *Lactobacillus casei* and *Lactobacillus acidophilus* manufactured with high-pressure homogenized milk. *Journal of Dairy Science* 91: 500–512.

Cardarelli, H.R., Buriti, F.C.A., Castro, I.A., and Saad, S.M.I. 2008. Inulin and oligofructose improve sensory quality and increase the probiotic viable count in potentially synbiotic petit-suisse cheese. *LWT—Food Science and Technology* 41: 1037–1046.

Castro, F.P., Cunha, T.M., Ogliari, P.J., Teófilo, R.F., Ferreira, M.M.C., and Prudencio, E.S. 2009. Influence of different content of cheese whey and oligofructose on the properties of fermented lactic beverages: Study using response surface methodology. *LWT—Food Science and Technology* 42: 993–997.

Charteris, W.P., Kelly, P.M., Morelli, L., and Collins, J.K. 1998. Development and application of an in-vitro methodology to determine the transit tolerance of potentially probiotic *Lactobaillus* and *Bifidobacterium* species in the upper gastrointestinal tract. *Journal of Applied Microbiology* 84: 759–768.

Christiansen, P.S., Edelsten, D., Kristiansen, J.R., and Nielsen, E.W. 1996. Some properties of ice cream containing *Bifidobacterium bifidum* and *Lactobacillus acidophilus*. *Milchwissenschaft* 51: 502–504.

Clark, P.A., Cotton, L.N., and Martin, J.H. 1993. Selection of bifidobacteria for use as dietary adjuncts in cultured dairy foods: II-tolerance to simulated pH of human stomachs. *Cultured Dairy Products Journal* 28: 11–14.

Clark, P.A. and Martin, J.H. 1994. Selection of bifidobacteria for use as dietary adjuncts in cultured dairy foods: III-tolerance to simulated bile of human stomachs. *Cultured Dairy Products Journal* 29: 18–21.

Collins, E.B. and Hall, B.J. 1984. Growth of bifidobacteria in milk and preparation of *Bifidobacterium infantis* for a dietary adjunct. *Journal of Dairy Science* 67: 1376–1380.

Collins, J.K., Thornton, G., and Sullivan, G.O. 1998. Selection of probiotic strains for human applications. *International Dairy Journal* 8: 487–490.

Collins, M.D. and Gibson, G.R. 1999. Probiotics, prebiotics and synbiotics: Approaches for modulating the microbial ecology of the gut. *American Journal of Clinical Nutrition* 67: 1052S–1057S.

Coussement, P. 1996. Innovations using inulin and oligofructose. *Deutsche-Milchwirtschaft* 47: 697–698.

Cruz, A.G., Antunes, A.E.C., Sousa, A.L.O.P., Faria, J.A.F., and Saad, S.M.I. 2009b. Ice cream as a probiotic food carrier. *Food Research International* 42: 1233–1253.

Cruz, A.G., Buriti, F.C.A., Souza, C.H.B., Faria, J.A.F., and Saad, S.M.I. 2009a. Probiotic cheese: Health benefits, technological and stability aspects. *Trends in Food Science & Technology* 20: 344–354.

Daigle, A., Roy, D., Belanger, G., and Vuillemard, J.C. 1999. Production of probiotic cheese (Cheddar-like cheese) using enriched cream fermented by *Bifidobacterium infantis*. *Journal of Dairy Science* 82: 1081–1091.

Dave, R.I. and Shah, N.P. 1997. Characteristics of bacteriocin produced by *Lb. acidophilus* LA-1. *International Dairy Journal* 7: 707–715.

Dave, R.I. and Shah, N.P. 1998. Ingredients supplementation effects on viability of probiotic bacteria in yoghurt. *Journal of Dairy Science* 81: 2804–2816.

Davidson, R.H., Duncan, S.E., Hackney, C.R., Eigel, W.N., and Boling, J.W. 2000. Probiotic culture survival and implications in fermented frozen yogurt characteristics. *Journal of Dairy Science* 83: 666–673.

Dubey, U.K. and Mistry, V.V. 1996. Effects of bifidogenic factors on growth characteristics of bifidobacteria in infant formulas. *Journal of Dairy Science* 79: 1156–1163.

El-Rahman, A.E.R.M.A. 2000. Survival of probiotic bacteria in yoghurt and their antimicrobial effect. *Alexandria Journal of Agricultural Research* 45: 63–80.

FAO/WHO. 2001. Evaluation of health and nutritional properties of probiotics in food including powder milk with live lactic acid bacteria. Córdoba, Spain: Food and Agriculture Organization of the United Nations, World Health Organization.

Finley, J.W. 1996. Designer foods: Is there a role for supplementation/fortification? *Advances in Experimental Medicine and Biology* 401: 213–220.

Fooks, L.J., Fuller, R., and Gibson, G.R. 1999. Prebiotics, probiotics and human gut microbiology. *International Dairy Journal* 9: 53–61.

Foster, J.W. 1993. The acid tolerance response of *Salmonella typhimurium* involves transient synthesis of key acid shock proteins. *Journal of Bacteriology* 175: 1981–1987.

Foster, J.W. and Hall, H.K. 1990. Adaptive acidification tolerance response of *Salmonella typhimurium*. *Journal of Bacteriology* 172: 771–778.
Fuller, R. 1992. *Probiotics—The Scientific Basis*. London: Chapman & Hall.
Gibson, G.R., Beatty, E.R., Wang, X., and Cumming, J.H. 1995. Selective stimulation of bifidobacteria in the human colon by oligofructose and inulin. *Gastroenterology* 108: 975–982.
Gomes, A.M.P. and Malcata, F.X. 1998. Development of probiotic cheese manufactured from goat milk: Response surface analysis via technological manipulation. *Journal of Dairy Science* 81: 1492–1507.
Gomes, A.M.P. and Malcata, F.X. 1999. *Bifidobacterium* spp. and *Lactobacillus acidophilus*: Biological, biochemical, technological and therapeutic properties relevant for use as probiotics. *Trends in Food Science & Technology* 10: 139–157.
Haynes, I.N. and Playne, M.J. 2002. Survival of probiotic cultures in low-fat ice-cream. *Australian Journal of Dairy Technology* 57: 10–14.
Hekmat, S. and McMahon, D.J. 1992. Survival of *Lactobacillus acidophilus* and *Bifidobacterium bifidum* in ice cream for use as a probiotic food. *Journal of Dairy Science* 75: 1415–1422.
Hekmat, S. and Mcmahon, D.J. 1997. Manufacture and quality of iron fortified yohgurt. *Journal of Dairy Science* 80: 3114–3122.
Hekmat, S. and Reid, G. 2006. Sensory properties of probiotic yogurt is comparable to standard yogurt. *Nutrition Research* 26: 163–166.
Hilliam, M. 2000. Functional food: How big is the market? *World of Food Ingredients* 12: 50–53.
Holzapfel, W.H., Haberer, P., Snel, J., Schillinger, U., and Huis in't Veld, J.H.J. 1998. Overview of gut flora and probiotics. *International Journal of Food Microbiology* 41: 85–101.
Homayouni, A., Azizi, A., Ehsani, M.R., Yarmand, M.S., and Razavi, S.H. 2008. Effect of microencapsulation and resistant starch on the probiotic survival and sensory properties of synbiotic ice cream. *Food Chemistry* 111: 50–55.
Hughes, D.B. and Hoover, D.G. 1995. Viability and enzymatic activity of bifidobacteria in milk. *Journal of Dairy Science* 78: 268–276.
Huis in't Veld, J.H., Havennar, R., and Marteau, P. 1994. Establishing a scientific basis for probiotic R and D. *Trends in Biotechnology* 12: 6–8.
Ibrahim, S.A. and Bezkorovainy, A. 1993. Inhibition of *Escherichia coli* by bifidobacteria. *Journal of Food Protection* 56: 713–715.
Iniguez, C., Cardoso, F., de-Villavicencio, M.N., and Rodriguez, M. 2001. Storage life of fermented, probiotic milk made from a mixture of buffalo and cow's milk. *Alimentaria* 38: 69–72.
Ishibashi, N. and Shimamura, S. 1993. Bifidobacteria: Research and development in Japan. *Food Technology* 47: 126–134.
Itsaranuwat, P., Al-Haddad, K.S.H., and Robinson, R.K. 2003. The potential therapeutic benefits of consuming 'health-promoting' fermented dairy products: a brief update. *International Journal Dairy Technology* 56: 203–210.
Kailasapathy, K., Harmstorf, I., and Phillips, M. 2008. Survival of *Lactobacillus acidophilus* and *Bifidobacterium animalis* spp. *lactis* in stirred fruit yogurts. *LWT—Food Science and Technology* 41: 1317–1322.

Kailasapathy, K. and Masondole, L. 2005. Survival of free and microencapsulated *Lactobacillus acidophilus* and *Bifidobacterium lactis* and their effect on texture of feta cheese. *Australian Journal of Dairy Technology* 60: 252–258.

Kailasapathy, K. and Sultana, K. 2003. Survival and β-D-galactosidase activity of encapsulated and free *Lactobacillus acidophilus* and *Bifidobacterium lactis* in ice cream. *Australian Journal of Dairy Technology* 58: 223–227.

Kasımoglu, A., Goncuoglu, M., and Akgun, S. 2004. Probiotic white cheese with *Lactobacillus acidophilus*. *International Dairy Journal* 14: 1067–1073.

Katz, F. 2001. Active cultures add function to yoghurt and other foods. *Food Technology* 55: 46–50.

Kaur, H., Mishra, H.N., and Umar, P. 2009. Textural properties of mango soy fortified probiotic yogurt: Optimization of inoculum level of yogurt and probiotic culture. *International Journal of Food Science and Technology* 44: 415–424.

Khalil, A.H. and Mansour, E.H. 1998. Alginate encapsulated bifidobacteria survival in mayonnaise. *Journal of Food Science* 63: 702–705.

Kilic, G.B., Kuleansan, H., Eralp, I., and Karahan, A.G. 2009. Manufacture of Turkish Beyaz cheese added with probiotic strains. *LWT—Food Science and Technology* 42: 1003–1008.

Klaenhammer, T.R. 2003. Probiotics and prebiotics. In: *Food Microbiology: Fundamentals and Frontires*, eds. M.P. Doyle, L.R. Beuchat, and J.M. Thomas, 797–811. Washington, DC: ASM.

Klaver, F.A.M., Kingma, F., and Weerkamp, A.H. 1993. Growth and survival of bifidobacteria in milk. *Netherlands Milk and Dairy Journal* 47: 151–164.

Kunz, B., Schuth, S., Stefer, B., and Strater, S. 2001. Product development of a synbiotic low-fat soft curd cheese focussing on the sensory effects of dietary fibre additives. *Ernahrungs-Umschau* 48: 195–199.

Kurmann, J.A. and Rasic, J.L. 1991. The health potential of products containing *bifidobacteria*. In: *Therapeutic Properties of Fermented Milk*, ed. R.K. Robinson, 117–158. London: Elsevier Applied Food Science.

Lankaputhra, W.E.V. and Shah, N.P. 1995. Survival of *Lactobacillus acidophilus* and *Bifidobacterium* spp. in the presence of acid and bile salts. *Cultured Dairy Product Journal* 30: 2–7.

Lankaputhra, W.E.V. and Shah, N.P. 1997. Improving viability of *Lactobacillus acidophilus* and bifidobacteria in yoghurt using two-step fermentation and neutralized mix. *Food Australia* 49: 363–366.

Lankaputhra, W.E.V., Shah, N.P., and Britz, M.L. 1996. Survival of bifidobacteria during refrigerated storage in the presence of acid and hydrogen peroxide. *Milchwissenschaft* 51: 65–70.

Laroia, S. and Martin, J.H. 1991. Effect of pH on survival of *Bifidobacterium bifidum* and *Lactobacillus acidophilus* in frozen fermented dairy desserts. *Cultured Dairy Product Journal* 26: 3–21.

Leatherhead Food International. 2006. The international market for functional foods. In: *Functional Food Market Report*. London: Leatherhead Food International Publication.

Lee, Y.K. and Salminen, S. 1995. The coming of age of probiotics. *Trends in Food Science & Technology* 6: 241–245.

Mandal, S., Hati, S., Puniya, A.K., Singh, R., and Singh, K. 2012. Development of synbiotic milk chocolate using encapsulated *Lactobacillus casei* NCDC 298. *Journal of Food Processing and Preservation* 37: 1031–1037.

Mandal, S., Puniya, A.K., and Singh, K. 2006. Effect of alginate concentrations on survival of microencapsulated *Lactobacillus casei* NCDC 298. *International Dairy Journal* 16: 1190–1195.

Maragkoudakisa, P.A., Miarisa, C., Rojeza, P., Manalisb, N., Magkanarib, F., Kalantzopoulosa, G., and Tsakalidou, E. 2006. Production of traditional Greek yogurt using Lactobacillus strains with probiotic potential as starter adjuncts. *International Dairy Journal* 16: 52–60.

Martin, J.H. 1996. Technological considerations for incorporating bifidobacteria and bifidogenic facrors into dairy products. *International Dairy Federation Bulltein* 313: 49–51.

Mattila-Sandholm, T., Myllärinen, P., Crittenden, R., Mogensen, G., Fondén, R., and Saarela, M. 2002. Technological challenges for future probiotic foods. *International Dairy Journal* 12: 173–182.

Mintel, 2009. *Study on the Functional Food in the US*. Chicago, IL: Mintel.

Mishra, V. and Prasad, D.N. 2000. Probiotics and their potential health benefits. *Indian Dairyman* 52: 7–13.

Mizota, T., Tamura, Y., Tomita, M., and Okanogi, S. 1987. Lactulose as a sugar with physilogical significance. *International Dairy Federation Bulltein* 212: 69–76.

Nutraceuticals World. 2013. Global probiotics market expected to reach $44.9 billion by 2018. http://www.nutraceuticalsworld.com/contents/view_breaking-news/2013-01-11/global-probiotics-market-expected-to-reach-449-billion-by-2018. Accessed April 28, 2015.

O'-Carroll, P. 1997. Formulating Yoghurt. *The World of Food Ingredients*.

Ong, I. and Shah, N.P. 2009. Probiotic Cheddar cheese: Influence of ripening temperatures on survival of probiotic microorganisms, cheese composition and organic acid profiles. *Food Science and Technology* 42: 1260–1268.

Ouwehand, A.C. and Salminen, S. 1998. The health effects of cultured milk products with viable and nonviable bacteria. *International Dairy Journal* 8: 749–756.

Penna, A.L.B., Gurram, S., and Cánovas, G.V.B. 2007. High hydrostatic pressure processing on microstructure of probiotic low-fat yogurt. *Food Research International* 40: 510–519.

Prado, F.C., Parada, J.L., Pandey, A., and Soccol, C.R. 2008. Trends in non-dairy probiotic beverages. *Food Research International* 41: 111–123.

Rasic, J.L. and Kurmann, J.A. 1978. *Yogurt: Scientific Grounds, Technology, Manufacture and Preparations*. Copenhagen, Denmark: Technical Dairy Publishing House.

Ravula, R.R. and Shah, N.P. 1998. Viability of probiotic bacteria in ferment dairy desserts. *Food Australia* 50: 136–139.

Reid, G. 1999. The scientific basis for probiotic strains of *Lactobacillus*. *Applied and Environmental Microbiology* 65: 3763–3766.

Sabikhi, L. and Mathur, B.N. 2001. Probiotic cultures in dairy foods—A review. *Indian Journal of Dairy Science* 54: 178–185.

Salminen, S., Deighton, M.A., Benno, Y., and Gorbach, S.L. 1998. Lactic acid bacteria in health and disease. In: *Lactic Acid Bacteria*, eds. S. Salminen and A. Von Wright, 211–253. Hong Kong, China: Marcel Dekker, Inc.

Sanders, M.E. and Huisin't Veld, J. 1999. Bringing a probiotic-containing functional food to market: Microbiological, product, regulatory and labeling issues. *Antonie van Leeuwenhoek* 79: 293–315.

Sarkar, S. and Misra, A.K. 1998. Selection of starter cultures for the manufacture of probiotic yoghurt. *Egyptian Journal of Dairy Science* 26: 295–307.

Shah, N.P. 1997. Isolation and enumeration of bifidobacteria in fermented milk products: A review. *Milchwissenschaft* 52: 71–76.

Shah, N.P. 2000. Probiotic bacteria: Selective enumeration and survival in dairy foods. *Journal of Dairy Science* 83: 894–910.

Sheu, T.Y., Marshall, R.T., and Heymann, H. 1993. Improving survival of culture bacteria in frozen desserts by microentapment. *Journal of Dairy Science* 76: 1902–1907.

Shin, H.S., Lee, J.H., Pestka, J.J., and Ustunol, Z. 1996. Viability of bifidobacteria in commercial dairy products during refrigerated storage. *Journal of Dairy Science* 79: 124S–129S.

Song, E.H., Won, B.R., and Yoon, Y.H. 2001. Production of probiotic cream cheese by utilizing *Lactobacillus helveticus* CU 631. *Journal of Animal Science and Technology* 43: 919–930.

Speck, M.L. 1978. Acidophilus food products. *Developments in Industrial Microbiology* 19: 95–101.

Stanton, C., Gardiner, G., Lynch, P.B., Collins, J.K., Fitzgerald, G., Ross, R., Klaenhammer, T.R. et al. 1998. Probiotic cheese. *International Dairy Journal* 8: 491–496.

Talwalkar, A. and Kailasapathy, K. 2003. Effect of microencapsulation on oxygen toxicity in probiotic bacteria. *Australian Journal of Dairy Technology* 58: 36–39.

Tamime, A.Y., Marshall, V.M.E., and Robinson, R.K. 1995. Microbiological and technical aspects of milks fermented by bifidobacteria. *Journal of Dairy Research* 62: 151–187.

Vijayageetha, V., Khursheed Begum, S.K., and Reddy, Y.K. 2011. Technology and quality attributes of probiotic ice cream. *Tamilnadu Journal of Veterinary Animal Sciences* 7: 299–302.

Vinderola, G., Prosello, W., Molinari, F., Ghilberto, D., and Reinheimer, J. 2009. Growth of *Lactobacillus paracasei* A13 in Argentinian probiotic cheese and its impact on the characteristics of the product. *International Journal of Food Microbiology* 135: 171–174.

Yadav, H., Jain, S., and Sinha, P.R. 2007. Production of free fatty acids and conjugated linoleic acid in probiotic dahi containing *Lactobacillus acidophilus* and *Lactobacillus casei* during fermentation and storage. *International Dairy Journal* 17: 1006–1010.

Yilmaztekin, M., Ozer, B.H., and Atasoy, F. 2004. Survival of *Lactobacillus acidophilus* La-5 and *Bifidobacterium bifidum* BB-02 in white-brined cheese. *International Journal of Nutrition* and *Food Sciences* 55: 53–60.

Young, J. 1998. European market development in prebiotic and probiotic contaning food stuffs. *British Journal of Nutrition* 80: S231–S233.

Younis, M.F., Dawood, A.H., Hefny, A.A., and El-Sayed, R.M. 1998. Manufacture of probiotic ice-cream. *Presented at Proceedings of the 7th Egyptian Conference for Dairy Science and Technology*, 373–380, Cairo, Egypt.

21

HEALTH BENEFITS OF FERMENTED PROBIOTIC DAIRY PRODUCTS

ARTHUR C. OUWEHAND AND JULIA TENNILÄ

Contents

21.1	Introduction	558
21.2	Probiotics versus Starter Cultures	559
21.3	Influence of the Food Matrix on Probiotic Functionality	560
	21.3.1 Influence of Other Cultures	560
	21.3.2 Influence of Other Ingredients/Components	560
	21.3.3 Influence of Storage Conditions	561
	21.3.4 Influence of the Matrix on Probiotic Health Efficacy	561
21.4	Health Benefits of Probiotic Dairy Products	562
	21.4.1 Yogurt	562
	21.4.2 Gastrointestinal Health	562
	21.4.2.1 Infectious Diarrhea	563
	21.4.2.2 Antibiotic-Associated Diarrhea	563
	21.4.2.3 Necrotizing Enterocolitis	563
	21.4.2.4 *Helicobacter pylori* Infection	563
	21.4.2.5 Intestinal Transit	564
	21.4.2.6 Irritable Bowel Syndrome	564
	21.4.2.7 Inflammatory Bowel Disease	564
	21.4.3 Immune Health	565
	21.4.3.1 Allergy/Autoimmune Disease	565
	21.4.3.2 Respiratory Tract Infections	565
	21.4.4 Other Health Benefits	566
	21.4.4.1 Weight Management	566
	21.4.4.2 Metabolic Syndrome	566
	21.4.4.3 Reduction of Serum Cholesterol	566
21.5	Conclusions	567
References		567

21.1 Introduction

Dairy products in general have been documented to show various health benefits. Many of these benefits relate to the fact that dairy foods are good source of several essential nutrients such as protein, calcium, and vitamin D. These nutritional benefits can be enhanced by fermentation. That is, however, not the topic of this chapter, as it will focus on the benefits provided by probiotics contained in the fermented dairy products. The most widely accepted definition of probiotics is the one proposed by a joint work group of the Food and Agriculture Organization (FAO) and World Health Organization (WHO): "Live microorganisms which when administered in adequate amounts confer a health benefit on the host" (FAO/WHO 2002). Although the definition does not state it, as a rule of thumb, the minimum dose for probiotic efficacy is assumed to be 10^9 CFU/dose. It is, however, likely that the dose depends on the health target, the particular strain used, and possibly the matrix in which it is administered. Strains from many species and genera have been used as probiotics. For fermented dairy products, probiotics tend to be limited to species from the genera *Bifidobacterium* and *Lactobacillus*. Strains from the genera will not be discussed here. Nor will strains from other genera, such as *Saccharomyces* or *Bacillus*, be discussed even though they are used as probiotics in supplements. Examples of probiotic species used fermented dairy products are given in Table 21.1.

Traditionally, probiotic benefits are considered to be strain specific and should not be extrapolated from one strain to another, not even when from the same species. However, as mentioned earlier and as can be seen in Table 21.1, probiotics tend to be selected from a limited number of species. Also, the source from which they were isolated is usually limited to dairy products and intestinal/fecal samples. This makes that probiotic strains have often comparable properties; thus, although they may differ in some properties, there will be many properties that are shared between strains or even whole species and genera (by virtue of their taxonomic relatedness) as illustrated in Figure 21.1. In particular, *Bifidobacterium animalis* subsp. *lactis* strains exhibit small differences between them, which whole genome sequence comparison indicates (Briczinski et al. 2009).

Table 21.1 Examples of Probiotic Strains Used in Fermented Dairy Products

GENUS	SPECIES	STRAIN(S)
Bifidobacterium	animalis subsp. lactis	Bb-12, 420, BI-04, Bi-07, HN019, DN-173 010
	longum	BB536, BI-05, SBT 2928
Lactobacillus	acidophilus	NCFM, LA-5, 74-2
	casei/paracasei	Shirota, Lc-11, CRL-431, Lpc-37, F19, DN-114 001
	fermentum	ME-3, PCC
	gasseri	SBT2055
	johnsonii	La1
	plantarum	299v (used in fermented oat product)
	reuteri	DSM 17938
	rhamnosus	GG, HN001, Lc705, LB21, IMC 501
Lactococcus	lactis	L1A
Propionibacterium	freudenreichii subsp. shermanii	JS

Note: While strains from other genus and species are used as probiotics, these are usually not used as fermented dairy products.

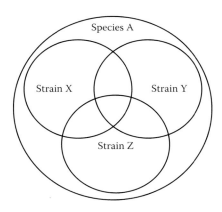

Figure 21.1 Graphical representation of commonalities and differences between strains from the same species. (Modified after ILSI Concise Monograph: "Probiotics, Prebiotics and the Gut Microbiota.")

21.2 Probiotics versus Starter Cultures

Besides probiotics, fermented dairy products will often contain other organisms usually from group of lactic acid bacteria (LAB) such as lactococci, *Streptococcus thermophilus*, and propionibacteria. These, so-called starter cultures, are added for their technological properties and are described in detail elsewhere in this book. For some fermented probiotic products, these are, however, of importance as they aid in the growth of probiotic strains. As probiotics used in fermented dairy products belong to the same group of LAB or bifidobacteria, these also grow in the product and may contribute to the flavor of the product through symbiotic interactions. This should be taken into account when formulating a probiotic fermented dairy product; for example, products that rely (partially) on fermentation by bifidobacteria tend to have a milder taste.

21.3 Influence of the Food Matrix on Probiotic Functionality

21.3.1 Influence of Other Cultures

There are two ways by which probiotics can be incorporated into fermented dairy products. These can be added after the fermentation process has been completed or, more common, can be added together with the starter culture. The latter strategy has the obvious advantage that the strain will multiply in the milk and less need to be added, thus reducing cost. As mentioned earlier, it requires that the strain contributes favorably to the flavor of the product or is neutral in this respect. Furthermore, the probiotic and the starter need to match so the probiotic does indeed get the opportunity to grow and its survival is not negatively affected.

By definition, probiotics need to be viable at consumption and also need to be present in sufficient amounts. The viable counts of the probiotics need, therefore, to be guaranteed at the end of shelf life. The viable counts should preferably be at least the same as the dose administered in an intervention trial used to document a particular health benefit. As a rule of thumb, this is usually at least 10^9 colony-forming units/portion.

21.3.2 Influence of Other Ingredients/Components

Besides other cultures, various ingredients and dairy components may influence the survival of probiotics. Of the ingredients, in particular food preparations may influence the survival of probiotics. Fruits that contain high levels of benzoic acid have a negative influence on the survival of probiotics. Understandably, fruit preparations that cause a substantial reduction in pH have also a negative influence on probiotic survival. Vitamin C, on the other hand, has a protective effect, possibly by reducing oxidate stress (Shah et al. 2010). Prebiotics, however, do not seem to have any influence on probiotic survival, which may relate to the fact that the probiotics are not supposed to grow on the prebiotic during manufacture or storage; there would be no or less prebiotic left for a health effect. Additional sugar may have a protective effect on probiotic viability. Obviously, preservatives of any kind are to be avoided.

21.3.3 Influence of Storage Conditions

Because the viability of probiotics needs to be maintained, a probiotic dairy product cannot be heat-treated, or as mentioned earlier, use preservatives to extend shelf-life. The products, therefore, need to be stored, refrigerated, or frozen in the case of ice cream. Also during storage, the pH is a major controlling factor and should not be too low (i.e., below 4). Oxygen tension should be kept low as well. Exposure to light should be avoided; this possibly creates reactive oxygen species as total anaerobiosis cannot always be guaranteed (Kiviharju et al. 2004).

21.3.4 Influence of the Matrix on Probiotic Health Efficacy

Probiotic health efficacy studies are often performed with a dietary supplement format instead of a fermented dairy product. There are various reasons for this. The logistics of a study with fermented dairy products is challenging; getting sufficient good quality product (with adequate counts) to the volunteers requires careful planning. Also preparing an indistinguishable placebo is difficult, especially in products where the probiotic contributes to the fermentation.

To what extent the results obtained with a probiotic supplement can be translated to a probiotic dairy product is not entirely clear.

A limited number of studies have compared fecal recovery of probiotics after administration in various formats (i.e., cheese, yogurt, skim milk, water, and supplements), suggesting some difference for some, *P. freudenreichii* subsp. *shermanii* JS and *B. animalis* subsp. *lactis* Bb-12, but not all strains, *Lb. acidophilus* NCFM, *Lb. rhamnosus* GG, and *Lb. rhamnosus* Lc705 (Saxelin et al. 2010; Varcoe et al. 2002).

Of course, fecal recovery is only part of the story, but what really counts are the health benefits. Even less studies have assessed this; *Lb. rhamnosus* HN001 appears to elicit the same increase in phagocytic and cytotoxic activity in elderly subjects when consumed in reconstituted fat-free milk or cheese (Gill 2001a; Ibrahim et al. 2010). There are, however, too few studies on the influence of the matrix on probiotic health efficacy to form any definite opinion on this. The influence should, however, not be overestimated as it is likely that the diet is a bigger source of variability than the probiotic matrix.

21.4 Health Benefits of Probiotic Dairy Products

The suggested health benefits of probiotics are summarized here, regardless of the matrix in which these were tested.

21.4.1 Yogurt

It is a matter of taste whether yogurt can be considered a probiotic or not. As described elsewhere, yogurt consists of milk fermented by *Lb. delbrueckii* subsp. *bulgaricus* and *S. thermophilus*. These species produce β-galactosidase. Upon ingestion, the bacteria are lysed by bile acids in the small intestine and release the β-galactosidase hydrolyzing the lactose from the yogurt in the digesta. This contributes to a reduction in the symptoms of lactose intolerance in lactose maldigesting subjects, a health benefit that has received a positive opinion from the European Food Safety Authority (European Food Safety Authority 2010). Although heat-treated yogurt tends to have a similar effect, it is more pronounced with yogurt containing viable bacteria. There is, therefore, an argument to consider yogurt a probiotic food (Guarner et al. 2005). Relief of lactose intolerance has also been reported for other LAB such as *Lb. acidophilus* NCFM (Montes et al. 1995). However, the case for yogurt is exceptionally strong.

21.4.2 Gastrointestinal Health

Being consumed, it is not surprising that the prime objective for probiotics has traditionally been in intestinal health with various diarrheas as main target.

21.4.2.1 Infectious Diarrhea Infectious diarrheas can have many different etiologies: viral, bacterial, protozoal, or chemical. The various causes make it also a challenging target. In children, the effect of probiotics on the duration of rotavirus diarrhea has been well documented for various strains (Van Niel et al. 2002). In adults, travelers' diarrhea is relatively common, but probiotics have shown inconsistent results (McFarland 2007; Ritchie and Romanuk 2012). This is most likely due to the heterogeneity of its causes.

21.4.2.2 Antibiotic-Associated Diarrhea Antibiotics have provided great medical progress in the treatment of infectious diseases. However, because of their antimicrobial nature, one of their side effects is a disturbance of the intestinal microbiota with antibiotic-associated diarrhea (AAD) as a consequence. Many probiotic strains and combinations of probiotic strains have been documented to reduce the incidence and/or duration of AAD. It is among one of the best documented health benefits of probiotics (Hempel et al. 2012). Probiotics do not replace the microbes that are reduced by antibiotic therapy but support the microbiota in maintaining its composition (Engelbrektson et al. 2009).

21.4.2.3 Necrotizing Enterocolitis Premature infants and in particular very low birth weight infants are at risk of developing necrotizing enterocolitis (NEC). The mortality of NEC is substantial. Probiotics provide one of the few preventive treatment options. While many probiotics on their own do not provide a significant reduction in NEC incidence, meta-analyses clearly indicate a reduced incidence and reduced mortality (Wang et al. 2012). Although the etiology of NEC is poorly understood, it is likely to have an infectious component. By competitive exclusion, probiotics may prevent the establishment of potentially pathogenic microbes.

21.4.2.4 Helicobacter pylori *Infection* *H. pylori* can be a resident of the gastric mucosa. It is the causative agent of gastric ulcer and is thought to cause gastric cancer. Most probiotics have been selected to survive gastric conditions, and many produce antimicrobial components; probiotics could therefore be effective against *H. pylori* infection. This does, however, not appear to be the case. Nevertheless, probiotics can play a valuable role in *H. pylori* eradication, which has serious side effects. Probiotics have been observed to reduce the side effects of *H. pylori* eradication therapy, thereby improving compliance and also otherwise improving the success rate (Wang et al. 2013).

21.4.2.5 Intestinal Transit Slow colonic transit is associated with constipation though not all constipation is caused by slow transit. It is thought to increase the risk of, among others, colorectal cancer and diverticulitis. Sufficient fluid and fiber intake are recommended but not always sufficient. Selected probiotic strains have been documented to shorten intestinal transit (Ibrahim and Ouwehand 2011; Waller et al. 2011). The mechanism by which probiotics are able to shorten colonic transit remains to be established; short chain fatty acid production may be one of the contributing factors.

21.4.2.6 Irritable Bowel Syndrome Irritable bowel syndrome (IBS) is a common functional gastrointestinal disorder, estimated to affect 2%–20% of people in industrialized countries (Rey and Talley 2009). Nowadays IBS is usually diagnosed according to the Rome III criteria, which comprises prolonged and recurrent abdominal pain or discomfort along with disturbances in bowel movements. In addition, patients often experience other gastrointestinal symptoms such as bloating, distension, and flatulence. Although the exact mechanisms behind IBS are still unclear, visceral sensory and motor disturbance, alterations in intestinal microbiota, in addition to increased gut permeability and inflammation may have an important role. Probiotics, such as *Bifidobacterium infantis*, have shown beneficial effects on symptoms of IBS (Brenner et al. 2009); however, the beneficial effects are often symptom specific and seem to depend on the specific strain, dose, and duration (Clarke et al. 2012). The proposed mechanisms behind the efficacy of probiotics include decreasing pain radiation and colonization of pathogenic/gas-producing bacteria in the intestine and modulating the host immunity (Clarke et al. 2012).

21.4.2.7 Inflammatory Bowel Disease Inflammatory bowel disease (IBD) is a collection of chronic inflammatory conditions of the intestine; ulcerative colitis; inflammation of restricted to the colon, Crohn's disease, an inflammation that can be anywhere from mouth to anus and pouchitis; and inflammation of an ileo-anal pouch that has been constructed after severe ulcerative colitis. The etiology of the disease is poorly understood, but the microbiota is clearly involved; among others a lack or low level of *Faecalibacterium prausnitzii* (Sokol et al. 2009). The disease can be brought into remission with anti-inflammatory drugs but tends to flare up after some time (Hedin et al. 2007). Probiotics have not been successful in inducing remission but have been found to prolong remission.

21.4.3 Immune Health

The second most common target for probiotics is immune health. The intestine is the body's main immune organ where 80% of the antigen-producing cells reside. These antigen-producing cells, however, can migrate outside the intestine and exert health affects beyond the intestine (Brandtzaeg 2011). Probiotics not so much boost the immune system but rather modulate its activity, increasing immune activity in case of infections but reducing it in the case of inflammation and allergy/autoimmune disease.

21.4.3.1 Allergy/Autoimmune Disease The potential of probiotics in autoimmune diseases such as IBD has been described earlier.

The etiology of atopic eczema is thought to be related to a low microbial exposure of the neonate/infant and possibly already the expecting mother. Selected probiotic strains appear to be able to provide a safe exposure that is required for the immune system to develop correctly and not lead to oversensitivity. This may result in a reduced prevalence of atopic eczema in at-risk children (Pelucchi et al. 2012). Treatment of allergies, in particular after childhood, has been less successful with probiotics (Lee et al. 2008) though sporadic positive results have been reported for selected probiotic strains (Roessler et al. 2008).

21.4.3.2 Respiratory Tract Infections Although they may be not the most obvious target for probiotics, several studies have documented reduced duration and/or incidence of respiratory tract infections (Hao et al. 2011). It is thought that modulation of the immune system is the mechanism by which probiotics elicit this effect, most likely innate immunity (Gill 2001b) as the adaptive immune system takes substantial time to develop. It can, however, not be excluded that a change in the oropharyngeal microbiota is a contributing factor as well.

21.4.4 Other Health Benefits

A number of other potential targets for probiotics have been identified, such as metabolic syndrome, weight management, and cholesterol reduction.

21.4.4.1 Weight Management The observation that the intestinal microbiota in obese subjects is different from lean subjects (Lyra et al. 2010) has stimulated the thought that probiotics might be able to play a role here as well. In most cases, this would involve probiotics other than the traditional bifidobacteria and lactobacilli. Human studies have not been reported, but animal studies indicate some potential (Aronsson et al. 2010). It will, however, take time before this is a feasible health target for probiotics.

21.4.4.2 Metabolic Syndrome Influencing the effects of obesity with probiotics may be a target for the distant future; however, managing the consenquences of metabolic syndrome may be a more feasible short-term goal. Metabolic syndrome is characterized by a chronic subclinical inflammation combined with insulin resistance and eventually type II diabetes. The inflammation is thought to be caused by leakage of lipopolysaccharide (LPS) from intestinal microbiota. This provides multiple targets for probiotics, reducing inflammation, suppressing Gram-negatives, and improving intestinal barrier function to further reduce LPS leakage. Selected probiotics have indeed been reported to improve the different subconditions of metabolic syndrome (Andreasen et al. 2010).

21.4.4.3 Reduction of Serum Cholesterol Already in the 1970s there have been reports that fermented milks would be able to reduce serum cholesterol (Mann 1977). Animal studies with selected probiotic strains often confirmed these anecdotic reports. Human studies indicate a moderate benefit from the consumption of probiotics in people's serum cholesterol levels (Guo et al. 2011). With the widespread introduction of plant stanols/sterols, however, the option of reducing serum cholesterol with probiotics appears to have moved to the background.

21.5 Conclusions

Numerous strains of probiotics have been documented to exert various health benefits. Their inclusion in various fermented dairy products, however, may require some practical considerations in order to guarantee sufficient numbers of viable organisms/consumption up to end of shelf life.

References

Andreasen, A. S., Larsen, N., Pedersen-Skovsgaard, T., Berg, R. M. G., Moller, K., Svendsen, K. D., Jakobsen, M., and Pedersen, B. K. 2010. Effects of *Lactobacillus acidophilus* NCFM on insulin sensitivity and the systemic inflammatory response in human subjects. *British Journal of Nutrition* 104: 1831–1838.

Aronsson, L., Huang, Y., Parini, P., Korach-Andre, M., Hakansson, J., Gustafsson, J. A., Pettersson, S., Arulampalam, V., and Rafter, J. 2010. Decreased fat storage by *Lactobacillus paracasei* is associated with increased levels of angiopoietin-like 4 protein (ANGPTL4). *PLoS One* 5: e13087.

Brandtzaeg, P. 2011. The gut as communicator between environment and host: Immunological consequences. *European Journal of Pharmacology* 668(Suppl 1): S16–S32.

Brenner, D. M., Moeller, M. J., Chey, W. D., and Schoenfeld, P. S. 2009. The utility of probiotics in the treatment of irritable bowel syndrome: A systematic review. *The American Journal of Gastroenterology* 104: 1033–1050.

Briczinski, E. P., Loquasto, J. R., Barrangou, R., Dudley, E. G., Roberts, A. M., and Roberts, R. F. 2009. Strain-specific genotyping of *Bifidobacterium animalis* subsp. lactis by using single-nucleotide polymorphisms, insertions, and deletions. *Applied and Environmental Microbiology* 75: 7501–7508.

Clarke, G., Cryan, J. F., Dinan, T. G., and Quigley, E. M. 2012. Review article: probiotics for the treatment of irritable bowel syndrome—focus on lactic acid bacteria. *Alimentary Pharmacology & Therapeutics* 35: 403–413.

Engelbrektson, A., Korzenik, J. R., Pittler, A., Sanders, M. E., Klaenhammer, T. R., Leyer, G., and Kitts, C. L. 2009. Probiotics to minimize the disruption of faecal microbiota in healthy subjects undergoing antibiotic therapy. *Journal of Medical Microbiology* 58: 663–670.

European Food Safety Authority. 2010. Scientific opinion on the substantiation of health claims related to live yoghurt cultures and improved lactose digestion (ID 1143, 2976) pursuant to Article 13(1) of Regulation (EC) No 1924/2006. *EFSA Journal* 8: 1763.

FAO/WHO. 2002. *Guidelines for the Evaluation of Probiotics in Food.* http://www.who.int/foodsafety/publications/fs_management/probiotics2/en/.

Gill, H. S. 2001a. Probiotic supplementation to enhance natural immunity in the elderly: Effects of a newly characterized immunostimulatory strain *Lactobacillus rhamnosus* HN001 (DR20™) on leucocyte phagocytosis. *Nutrition Research* 21: 183.

Gill, H. S. 2001b. Enhancement of immunity in the elderly by dietary supplementation with the probiotic *Bifidobacterium lactis* HN019. *American Journal of Clinical Nutrition* 74: 833.

Guarner, F., Perdigon, G., Corthier, G., Salminen, S., Koletzko, B., and Morelli, L. 2005. Should yoghurt cultures be considered probiotic? *British Journal of Nutrition* 93: 783–786.

Guo, Z., Liu, X. M., Zhang, Q. X., Shen, Z., Tian, F. W., Zhang, H., Sun, Z. H., Zhang, H. P., and Chen, W. 2011. Influence of consumption of probiotics on the plasma lipid profile: A meta-analysis of randomised controlled trials. *Nutrition, Metabolism, and Cardiovascular Diseases* 21: 844–850.

Hao, Q., Lu, Z., Dong, B. R., Huang, C. Q., and Wu, T. 2011. Probiotics for preventing acute upper respiratory tract infections. *Cochrane Database of Systematic Reviews* CD006895. doi: 10.1002/14651858.CD006895.pub3.

Hedin, C., Whelan, K., and Lindsay, J.O. 2007. Evidence for the use of probiotics and prebiotics in inflammatory bowel disease: A review of clinical trials. *The Proceedings of the Nutrition Society* 66: 307–315.

Hempel, S., Newberry, S. J., Maher, A. R., Wang, Z., Miles, J. N. V., Shanman, R., Johnsen, B., and Shekelle, P. G. 2012. Probiotics for the prevention and treatment of antibiotic-associated diarrhea: A systematic review and meta-analysis. *Journal of the American Medical Association* 307: 1959–1969.

Ibrahim, F. and Ouwehand, A. C. 2011. Potential of probiotics in reducing intestinal transit time. http://www.touchgastroenterology.com.

Ibrahim, F., Ruvio, S., Granlund, L., Salminen, S., Viitanen, M., and Ouwehand, A. C. 2010. Probiotics and immunosenescence: cheese as a carrier. *FEMS Immunology and Medical Microbiology* 59: 53–59.

Kiviharju, K., Leisola, M., and von Weymarn, N. 2004. Light sensitivity of *Bifidobacterium longum* in bioreactor cultivations. *Biotechnology Letters* 26: 539–542.

Lee, J., Seto, D., and Bielory, L. 2008. Meta-analysis of clinical trials of probiotics for prevention and treatment of pediatric atopic dermatitis. *The Journal of Allergy and Clinical Immunology* 121: 116–121.e111.

Lyra, A., Lahtinen, S., Tiihonen, K., and Ouwehand, A. C. 2010. Intestinal microbiota and overweight. *Beneficial Microbes* 1: 407–421.

Mann, G. V. 1977. A factor in yogurt which lowers cholesteremia in man. *Atherosclerosis* 26: 335–340.

McFarland, L. V. 2007. Meta-analysis of probiotics for the prevention of traveler's diarrhea. *Travel Medicine and Infectious Disease* 5: 97–105.

Montes, R. G., Bayless, T. M., Saavedra, J. M., and Perman, J. A. 1995. Effect of milks inoculated with *Lactobacillus acidophilus* or a yogurt starter culture in lactose-maldigesting children. *Journal of Dairy Science* 78: 1657–1664.

Pelucchi, C., Chatenoud, L., Turati, F., Galeone, C., Moja, L., Bach, J.-F., and La Vecchia, C. 2012. Probiotics supplementation during pregnancy or infancy for the prevention of atopic dermatitis: A meta-analysis. *Epidemiology* 23: 402–414.

Rey, E. and Talley, N. J. 2009. Irritable bowel syndrome: Novel views on the epidemiology and potential risk factors. *Digestive and Liver Disease* 41: 772–780.

Ritchie, M. L. and Romanuk, T. N. 2012. A meta-analysis of probiotic efficacy for gastrointestinal diseases. *PLoS One* 7: e34938.

Roessler, A., Friedrich, U., Vogelsang, H., Bauer, A., Kaatz, M., Hipler, U. C., Schmidt, I., and Jahreis, G. 2008. The immune system in healthy adults and patients with atopic dermatitis seems to be affected differently by a probiotic intervention. *Clinical and Experimental Allergy* 38: 93–102.

Saxelin, M., Lassig, A., Karjalainen, H., Tynkkynen, S., Surakka, A., Vapaatalo, H., Jarvenpaa, S., Korpela, R., Mutanen, M., and Hatakka, K. 2010. Persistence of probiotic strains in the gastrointestinal tract when administered as capsules, yoghurt, or cheese. *International Journal of Food Microbiology* 144: 293–300.

Shah, N. P., Ding, W. K., Fallourd, M. J., and Leyer, G. 2010. Improving the stability of probiotic bacteria in model fruit juices using vitamins and antioxidants. *Journal of Food Science* 75: M278–282.

Sokol, H., Seksik, P., Furet, J. P., Firmesse, O., Nion-Larmurier, I., Beaugerie, L., Cosnes, J., Corthier, G., Marteau, P., and Dore, J. 2009. Low counts of *Faecalibacterium prausnitzii* in colitis microbiota. *Inflammatory Bowel Diseases* 15: 1183–1189.

Van Niel, C. W., Feudtner, C., Garrison, M. M., and Christakis, D. A. 2002. Lactobacillus therapy for acute infectious diarrhea in children: A meta-analysis. *Pediatrics* 109: 678–684.

Varcoe, J., Zook, C., Sui, J., Leighton, S., Busta, F., and Brady, L. 2002. Variable response to exogenous *Lactobacillus acidophilus* NCFM consumed in different delivery vehicles. *Journal of Applied Microbiology* 93: 900–906.

Waller, P. A., Gopal, P. K., Leyer, G. J., Ouwehand, A. C., Reifer, C., Stewart, M. E., and Miller, L. E. 2011. Dose-response effect of *Bifidobacterium lactis* HN019 on whole gut transit time and functional gastrointestinal symptoms in adults. *Scandinavian Journal of Gastroenterology* 46: 1057–1064.

Wang, Q., Dong, J., and Zhu, Y. 2012. Probiotic supplement reduces risk of necrotizing enterocolitis and mortality in preterm very low-birth-weight infants: an updated meta-analysis of 20 randomized, controlled trials. *Journal of Pediatric Surgery* 47: 241–248.

Wang, Z.-H., Gao, Q.-Y., and Fang, J.-Y. 2013. Meta-analysis of the efficacy and safety of *Lactobacillus*-containing and *Bifidobacterium*-containing probiotic compound preparation in *Helicobacter pylori* eradication therapy. *Journal of Clinical Gastroenterology* 47: 25–32.

22

Validation of Health Claims for Fermented Milks

EDWARD FARNWORTH AND FARAH HOSSEINIAN

Contents

22.1	Introduction	571
22.2	Characterizing the Fermented Milk Product	574
	22.2.1 Characterizing the Microorganism(s)	574
	22.2.2 Microorganisms Nomenclature	575
	22.2.3 Methodology to Characterize the Microorganism(s)	576
	22.2.3.1 Defining the Dose for the Fermented Milk Product	577
	22.2.3.2 Dose Response Effects	578
	22.2.3.3 Food Matrix	578
22.3	Experiments Designed to Validate Fermented Milk Product Efficacy	579
	22.3.1 Learning from Past Experience	579
	22.3.1.1 Reasons for Rejected Claims	580
	22.3.1.2 Demonstrating Cause and Effect	582
	22.3.1.3 Food versus Drug Effects	582
	22.3.2 Yogurt and Lactose Digestion	583
22.4	Possible Health Claims for Fermented Milk Products	584
22.5	Conclusions	589
References		589

22.1 Introduction

The production and consumption of fermented milks have a long history (Prajapati and Nair 2003), and the health aspects of eating fermented milk products are accepted worldwide (Tamime 2002;

Farnworth, 2004). Before the advent of pasteurization and refrigeration, the quick spoilage of raw milk limited its uses (Ross et al. 2002). The initial fermented milks were probably discovered by accident, when unprotected raw milk was inoculated by environmental bacteria and yeasts. Although many of these contaminating microorganisms caused spoilage, others lowered the pH of milk, changed its viscosity, and protected the fermented milk from spoilage organisms through their production of antimicrobial peptides. Thus, initially, fermenting milk was seen as a way to prolong the shelf life of milk (Ross et al. 2002).

As the science of microbiology advanced, the bacteria that were most efficient at fermenting foods including milk were identified (Bourdichon et al. 2012). Through trial and error, it was shown that, in the case of yogurt, only two bacteria—*Streptococcus thermophilus* (formerly known as *Str. salivarius* subsp. *thermophilus*) and *Lactobacillus delbrueckii* subsp. *bulgaricus*—were needed to produce a consistent product acceptable to consumers. In some jurisdictions, these two bacteria are named in official yogurt specifications (Tamime 2002).

The studies carried out by Tissier, Moro, and Metchnikoff in the early 1900s advanced the idea that some bacteria could be beneficial to human health (Tamime 2002). It was Metchnikoff who suggested that the consumption of yogurt was linked to longevity (Metchnikoff 1908).

Subsequent research has shown that a wide range of bacteria have potential health benefits, and this has led to the manufacture of many fermented milk products that contain additional bacteria besides those necessary to produce the fermentation (Tamime 2002).

It is generally believed that the beneficial effects of consuming fermented milk products depend on the microorganisms arriving at their site of action in the gastrointestinal tract (GIT) alive and in large enough numbers to produce an effect. This requirement has been addressed by the use of encapsulation techniques that protect sensitive bacteria during the fermentation and processing steps and during transit through the upper GIT (Sultana et al. 2000; Kailasapathy 2002; Urbanska et al. 2007). It was reasoned that perhaps the bacteria that were used to produce a fermented milk product were also producing health-promoting bioactives that are produced as the fermenting bacteria hydrolyze milk protein (Soomro et al. 2002). Today, in addition to the two bacteria used to make yogurt, some fermented milk products contain other bacteria that have been added because of their possible ability to produce health-promoting metabolites. The potential health benefits of these fermented milk products have encouraged a worldwide demand (Tamime 2002).

Human-feeding trials have been carried out to study the effects of consuming fermented milk products that contained live bacteria that could be beneficial to health (probiotics). During these experiments, researchers measured the effects on the intestinal microbiota and effects on metabolism and health. Data showed that the live bacteria that were consumed persisted in the GIT, only as long as the fermented product was eaten, and the bacteria that were being consumed passed through the GIT and did not become part of the intestinal microbiota (Alander et al. 1999). As the characterization of the human intestinal microbiota advanced, it was apparent that the population of bacteria in the human gut was large in size and diverse in its makeup causing some to question whether feeding exogenous bacteria could have an impact on the metabolism and health of the host (O'Sullivan et al. 1992). Data showed that bacteria fed in clinical trials could not be found in fecal material 1 week after the end of the feeding trial (Goossens et al. 2003). This, in part, could explain why the feeding of exogenous bacteria, in relatively low numbers, did not result in the implantation in the GIT. It was proposed that a better way of altering the gut microbiota would be to stimulate the growth in situ of beneficial bacteria by feeding nutrients that specifically encouraged the growth of certain target bacteria that had been linked to health promotion. The emphasis has been on the study of particular sugars (prebiotics) that stimulate the growth of bifidobacteria and other potential beneficial bacteria (Gibson and Rastall 2006).

Fermented milk products present the unique opportunity to provide beneficial bacteria, namely, probiotics, and ingredients that promote the growth of beneficial bacteria that are already resident in the GIT, namely, prebiotics, in one product. Today, many fermented milk products can be found in the marketplace, which advertise that they contain probiotic bacteria and prebiotics.

Health regulatory officials in many jurisdictions have followed the growth of fermented milk products that claim directly or indirectly that they are healthy because they contain probiotics and prebiotics. Several jurisdictions now have detailed guidelines about label health claims for all foods and the data that are needed to substantiate such claims (Sanders et al. 2005; Van Loveren et al. 2012). Fermented milk products present some particular challenges in terms of the type of data required to support a health claim (Farnworth 2008). As more and more research teams plan and carry out experiments to generate data to support an eventual health claim, it is important that they understand the criteria the health regulatory officials will use to judge their data. Most early petitions to health regulatory officials for a food health claim for fermented milk products and other probiotic products have been unsuccessful because experiments designed to advance knowledge of which bacteria are beneficial to health and how these exert their effects often did not provide details that would satisfy regulatory officials. Therefore, publishing in a high-impact journal is not necessarily a guarantee that health regulatory officials will find in that paper the information they need to judge whether a health claim is supported or not.

This chapter will give details about the design and procedures that health regulatory officials are requiring for clinical trials that are being carried out to test the health-promoting properties of fermented milk products. By doing this, it may be helpful in explaining why, to date, so few health claims have been approved for fermented milk products and other probiotic products in any of the major jurisdictions. By giving details of some of the judgments that have been handed down to date, it is hoped that petitioners of future petitions will understand what data and information need to be supplied to validate and to support a successful health claim. This will allow researchers to design and carry out clinical trials that clearly support and validate health claims for fermented milk products.

22.2 Characterizing the Fermented Milk Product

22.2.1 Characterizing the Microorganism(s)

The validation of a health claim for a fermented food begins with a complete description of the food including a detailed characterization of the microorganisms it contains. Failure to adequately characterize the organism contained in the fermented food can result in a rejection of a proposed food health claim (EFSA 2011a). With very few exceptions, petitions for a health claim in the major jurisdictions have been for products that contain bacteria. However, some petitions for products containing yeasts can be found.

The list of potential bacteria that are beneficial to health continues to grow (Collins et al. 1998; Lewis and Freedman 1998; Reid and Burton 2002; Soomro et al. 2002; Sanders 2003; Niers et al. 2007). At the same time, new methods for the identification and characterization of microorganisms are being developed (Farnworth 2003; Vaughan et al. 2005; Ammor et al. 2007; Mathys et al. 2008). Today, health regulatory officials are demanding better characterization of the microorganisms contained in foods that carry a health claim as a way to ensure that the proposed benefits are characteristic of the microorganisms(s) present and as a way for consumers to distinguish between products.

22.2.2 Microorganisms Nomenclature

The use of scientific names for microorganisms presents a dilemma. The scientific names given to microorganisms are often long and can contain a mixture of words, letters, and numbers. This is necessary because many microorganisms are closely related, and the naming system for microorganisms was established for scientific purposes, not for food labeling purposes. Consumers are both confused and intimidated by the strange sounding names of microorganisms. Also, stating that a product contains "bacterial cultures" is too nonspecific. Some manufacturers have given proprietary microorganisms simpler names. Often the name that is used has not been approved or accepted by the larger scientific community. Health regulatory officials have expressed concern when microorganisms are given names that imply the health benefit the microorganism is purported to have, because this could represent an implied health claim.

The task of naming of a microorganism in a petition for a health claim is made easier if the microorganism has been obtained from an internationally recognized culture depository and carries the same name used by that depository. The Budapest Treaty set out regulations for the "International Recognition of the Deposit of Microorganisms for the Purposes of Patent Procedure" (WIPO 1977). To date, depositories have been set up in 23 countries for a wide variety of microorganisms. A large number of the submissions to European Food Safety

Authority (EFSA) have referenced the organism under consideration as being obtained from, or having been compared to, an organism in one of these recognized depositories. The Belgian Co-ordinated Collections of Microorganisms (BCCM/LMG) and the French National Collection of Cultures of Microorganisms are most often referenced in EFSA petitions. A review of the opinions made by the EFSA review panel for fermented foods would indicate that using this type of information to identify a microorganism is acceptable to the EFSA review panel.

22.2.3 Methodology to Characterize the Microorganism(s)

EFSA has concluded that the microorganism has been fully characterized at the species and strain level, when both phenotypic (i.e., cell morphology, colonial morphology, carbohydrate fermentation pattern, enzymatic activity profile, antimicrobial resistance pattern, PAGE) and genotypic (i.e., DNA-DNA hybridization, 16S rRNA gene sequence analysis, 16S/23S intergenic spacer region sequence analysis, elongation factor *tuf* gene sequence analysis, species-specific PCR, AFLP, MLST, RAPD, PFGE) methods have been used, where appropriate. In addition, petitioners are expected to supply references to the methods of identification and characterization they have used.

Only when these two criteria (i.e., genotypic, phenotypic) are fulfilled is the microorganism to be sufficiently characterized. EFSA has stated that in the case of combinations of several bacteria, their review panel considers that if one microorganism used in the combination is not sufficiently characterized, the combination proposed is not sufficiently characterized (EFSA 2012a). It is perhaps noteworthy then that very few submissions to EFSA for approval of a health claim for a fermented milk product or probiotic bacteria have been criticized for inadequate characterization. Because of the different properties of even closely related microorganisms, only data for the particular organism in the food that is being submitted for approval are acceptable. It can be assumed that as more sensitive methods are developed that can characterize microorganisms and more accurately distinguish between closely related organisms, the level of scientific proof required to characterize a microorganism submitted for a food health claim will increase.

Health Canada (HC) has taken a more proactive approach to the problem of characterizing probiotic bacteria used in fermented foods that have potential health benefits. They have published a list of bacteria that they have judged to be sufficiently characterized and have sufficient experimental data to be called probiotic (Table 22.1, Health Canada 2009a). According to HC, probiotic (bacteria) include bacteria belonging to a group broadly defined as lactic acid bacteria and to the genus *Bifidobacterium*. Presumably, any petition for a food product using a listed microorganism would not have to have a detailed characterization or validation of its effects. It is noteworthy that the list of acceptable health claims acceptable to HC is so short, compared to the purported effects of products submitted to and rejected by EFSA.

22.2.3.1 Defining the Dose for the Fermented Milk Product Accurate estimates of the number of bacteria that inhabit the GIT vary, and the number of bacteria differ in the various parts of the GIT. The colon, which is the area of the GIT targeted by most probiotic products, can contain as many as 10^9–10^{12} cfu per ml contents (Blaut and Clavel 2007). The question of how many exogenous bacteria need to be included in a fermented milk product to exert an effect on so large an established bacterial population is an important one. It would be reasonable to assume that the numbers of bacteria required would

Table 22.1 Probiotic Bacteria and Acceptable Non-Strain-Specific Health Claims

BACTERIAL SPECIES ACCEPTED AS HAVING PROBIOTIC PROPERTIES
- *B. adolescentis; B. animalis* subsp. *animalis; B. animalis* subsp. *lactis* synonym; *B. lactis; B. bifidum; B. breve; B. longum* subsp. *infantis* comb. nov.; *B. longum* subsp. *longum* subsp. nov.
- *Lb. acidophilus; Lb. casei; Lb. fermentum; Lb. gasseri; Lb. johnsonii; Lb. paracasei; Lb. plantarum; Lb. rhamnosus; Lb. salivarius*

ACCEPTABLE PROBIOTIC HEALTH CLAIM
- Naturally forms part of the gut flora
- Provides live microorganisms that naturally form part of the gut flora
- Contributes to healthy gut flora
- Provides live microorganisms that contribute to healthy gut flora

Source: Health Canada, *Accepted Claims about the Nature of Probiotic Microorganisms in Food.* http://www.hc-sc.gc.ca/fn-an/label-etiquet/claims-reclam/probiotics_claims-allegations_probiotiques-eng.php, 2009a.

depend on the bacteria being added to the food and the targeted metabolic/disease process.

In order to be effective, sufficient numbers of bacteria must reach the site of action. The Fermented Milks and Lactic Acid Bacteria Beverages Association of Japan set a minimum of 10^7 bifidobacteria/g or ml for fermented milk products (Ishibashi and Shimamura 1993). CODEX (2003) has set a minimum of 10^6/g of microorganisms added to fermented milk and yogurt, in addition to those added to produce the product. HC has stated that "A serving of stated size of a product should contain a minimum level of 1.0×10^9 cfu of one of the eligible microorganism(s) that is the subject of the claim" (Health Canada 2009a).

22.2.3.2 Dose Response Effects It is important that measuring and reporting the viability and the number of microorganisms that are fed in a clinical trial to be able to establish a clear cause and effect be done. Because of the complexity of designing and carrying out clinical trials to test fermented foods, dose response–type experiments testing with humans have been limited to the testing of probiotic products given in capsules (Christensen et al. 2006; Larsen et al. 2006; Gao et al. 2010). Fortunately, it does not appear that demonstrating a dose response is necessary for obtaining regulatory approval. In the case of EFSA, demonstrating cause and effect appears to be sufficient. Dose response data are requested by HC as a way of showing causality (Health Canada 2009b).

From a practical point of view, it is important that an effective dose of the responsible microorganism can be contained in a realistic portion (dose).

22.2.3.3 Food Matrix The properties of microorganisms are often defined under lab conditions, and therefore, when these microorganism are added to milk to produce a fermented milk product, interactions with the milk protein, fat, sugars, and other nutrients need to be taken into consideration. In the case of milk, food matrix effects can be both positive and negative. Some microorganisms possess enzymes that allow them to grow and reproduce in milk, while others do not. The fermentation process and subsequent processing and packaging procedures may also affect the numbers and properties of viable microorganisms in the fermented milk product. It is imperative, therefore, that as many details about the production and processing of the fermented milk product fed in a clinical trial are included in any reports of results and that any conclusions from the trial apply only to the fermented milk product as fed.

22.3 Experiments Designed to Validate Fermented Milk Product Efficacy

The design and procedures for an experiment that is destined to be published in a peer-reviewed scientific journal may not be the same as that would be used to generate data to validate and substantiate a food label health claim. This becomes obvious when the quality assessment tool of HC is applied to a published paper being submitted to substantiate a food health claim (Health Canada 2009b). Even if an article is acceptable to be published in an international refereed journal with a high impact factor, important details required by HC may be missing from the original experimental plan. Also, the way the experiment is reported in the literature may not provide sufficient information to contribute to the substantiation of the proposed food health claim. This is particularly true for early publications that described experiments to test the health benefits of fermented milk products and were never intended to be used to substantiate a food health claim.

The use of the CONSORT checklist when planning, conducting, and reporting (publishing) clinical trials may help ensure that details are not omitted that could result in a published article not being considered as being detailed enough to help substantiate a food health claim (Moher et al. 2010; Schulz et al. 2010). Successful petitions to obtain a food health claim for any type of food have been few in number (Farnworth 2008).

22.3.1 Learning from Past Experience

To date, it appears that only one petition for a health claim for a fermented milk product or a probiotic microorganism has received approval by EFSA (2010a). This poor success rate has caused much criticism of the EFSA review system. In 2009, EFSA did publish a guidance document that contained a short section titled "What are pertinent studies for substantiation of a claim?" (EFSA 2009a).

In general, the advice from EFSA has been to use the best science to substantiate a claim that their expert panel will then judge as being adequate or not. Tables 22.2 through 22.4 are summaries of some of the comments that have been received back from the EFSA review panel to explain their reasons for rejection of claims for fermented milk products and probiotics. Even though the opinions are given on a case-by-case basis, they give much insight into the type and level of data needed to support a successful food health claim submission. Knowing what has not been acceptable in the past could be used in the planning of future experiments.

22.3.1.1 Reasons for Rejected Claims The reasons for rejection by the EFSA review panel fall into three broad categories: errors/weaknesses in the experimental design and methodology used to test the efficacy of the food products, errors and omissions in the statistical analysis used in clinical trials, and the failure to establish a clear cause-and-effect

Table 22.2 Reasons for Rejection by EFSA—Experimental Design and Methodology

- No information was given about validation of the questionnaire used for subjective measurements (EFSA 2012b)
- That inclusion criteria (at least two symptoms perceived as induced by stress) were not sufficiently defined (EFSA 2012b)
- Insufficient information on the randomization and blinding (EFSA 2012c)
- Studies provided on the effects of individual constituents cannot be used for substantiation of a claim on the combination (EFSA 2012a)
- Studies were uncontrolled and no conclusions can be drawn from these pilot studies (EFSA 2012b)
- Pilot study was underpowered (EFSA 2012b)
- Procedure used following discharge could have resulted in un-blinding of patients due to the different size of the bottles for the two products (EFSA 2010b)
- Comparison was not preplanned in the study protocol and that no scientific justification was provided for it (EFSA 2010c)
- Food products tested were of quite different nature (EFSA 2010d)
- Number of subjects ingesting each product was not reported (EFSA 2010d)
- Possible effects of the matrices carrying the strains were not evaluated (EFSA 2010d)
- Small sample size (EFSA 2009b, 2010d)
- Short intervention periods (EFSA 2009c)
- No intestinal microbiota analysis studies have been performed during or after the administration (EFSA 2008c)
- Post-randomization, the study was not sufficiently controlled for confounders that could potentially have affected the outcome (EFSA 2008d)

Table 22.3 Reasons for Rejection by EFSA—Statistics

- That the process of randomization was insufficiently described (EFSA 2012b)
- No correction for multiple testing was performed in the statistical analysis of the results
- Weaknesses in the statistical analyses (EFSA 2012c)
- No correction for multiple testing was performed in the statistical analysis of the results (EFSA 2012d)
- No scenario has been applied for the missing data imputation for more conservative (EFSA 2010b)
- High dropout rate was not taken into account appropriately in the analysis of the data (EFSA 2010c)
- Lack of power calculation (EFSA 2010d)
- Inconsistent exclusions of subjects for the statistical analyses (EFSA 2009b)
- Statistical analysis did not exploit the data available for different timepoint (EFSA 2009b)

Table 22.4 Reasons for Rejection by EFSA—Cause and Effect

- Animal and in vitro studies cannot predict the occurrence of an effect of a combination of *Lactobacillus fermentum* 57A, *Lactobacillus plantarum* 57B, and *Lactobacillus gasseri* 57C on defense against vaginal pathogens in vivo in humans (EFSA 2012e)
- No evidence has been provided by the applicant to establish that the claimed effect, "beneficial modulation of the intestinal microflora," is a beneficial physiological effect (EFSA 2011a)
- Increasing the number of any group of microorganisms, including lactobacilli and/or bifidobacteria, is not considered in itself a beneficial physiological effect (EFSA 2011a)
- Stimulation of immunological responses is not a beneficial physiological effect *per se* but needs to be linked to a beneficial physiological or clinical outcome (EFSA 2012f)
- Panel considers that the claimed effect is related to the treatment of a disease and does not comply with the criteria laid down in Regulation (EC) No. 1924/2006 (EFSA 2012f)
- Status of the gastrointestinal tract in subjects with diarrhea due to a gastrointestinal infection may not be comparable to the status of the gastrointestinal tract in subjects without gastrointestinal infection (EFSA 2012h)
- Results obtained in young children provide information about the adult population (EFSA 2011a)
- Symptoms used to define the occurrence of URTI (i.e., running nose, sore throat, fever, and cough) are non-specific, and that the presence of one or more of these symptoms is not an appropriate measure of the occurrence of URTI in the study population (EFSA 2010c)
- Evidence obtained for one strain cannot be extrapolated to another, given the strain specificity of the effects (EFSA 2009d)
- Results obtained in studies with seriously ill patients or from studies in animals or conducted in vitro and ex vivo can be extrapolated to the general human population (EFSA 2009e)
- Frequent symptoms of cold and/or influenza were common among the subjects of both groups (with about 50% reporting such symptoms during the trial), which are considered to limit the significance of the results in a study measuring immune parameters (EFSA 2009b)
- In vitro studies are not sufficient to predict in vivo efficacy in humans (EFSA 2008e)

relationship for the food product and its purported beneficial effect. A review of the many opinions made by the EFSA review panel show that over the years, both the submissions and the opinions have increased in quality and detail.

22.3.1.2 Demonstrating Cause and Effect Of particular note is the observation that most of the submissions to EFSA have been rejected because of the failure to demonstrate a cause-and-effect relationship between the fermented food eaten and changes to health and/or metabolism. In several cases, this was due to the fact that the effect claimed for the fermented food was considered too vague and ill defined. The EFSA review panel reported that terms such as oral flora, digestive health, intestinal flora, digestive system, natural defenses/immune system, and immune system were too general and nonspecific descriptors of an effect. Similarly, it appears that terms such as *normal* or *stable* intestinal microbial populations are difficult to define and therefore should be avoided. This contrasts with the HC-proposed probiotic food health claims that refer to gut flora and healthy gut flora, without defining them (Table 22.1). For the EFSA review panel, effects shown in animal or in vitro models are not accepted as proof of effect in humans. HC also emphasizes the use of human data to validate a proposed claim (Health Canada 2009b).

22.3.1.3 Food versus Drug Effects In many jurisdictions (Canada, Europe, the United States, Australia), there is a distinction between foods and drugs. Drugs are to be used to cure, prevent, or reduce/eliminate the effect(s) of diseases, not so for foods. In some cases, the EFSA review panel has rejected the submission, because the purported effect was on a disease, and therefore beyond the allowed purpose of a food health claim. Establishing the line between a food and a drug has been difficult and frustrating for petitioners, and many still do not understand why this distinction is necessary.

Contributing to the challenge of demonstrating cause and effect is the fact that very few biomarkers or end points (either real or surrogate) have been defined that could be used to show the beneficial effect of consuming a fermented milk product (Sanders 2003). The ideal end point would be easy to measure (technically, from the participants' perspective), be sensitive to the desired effect of the intervention, measure an important change or benefit to the target population, and be an objective measurement (Shane et al. 2010). It is evident from the decisions that have come back from the EFSA review panel that showing that consuming a food product containing live microorganisms and results of finding those microorganisms in fecal/digesta samples is by itself not sufficient to claim a cause and effect. It is the consequences of the presence of these exogenous microorganisms in the GIT that demonstrate the effect. This requirement can also be seen in the claims for prebiotic products, where both changes to the intestinal microbiota, and changes to measurable health/metabolism parameters are necessary to establish a food health cause-and-effect relationship for a proposed prebiotic.

22.3.2 Yogurt and Lactose Digestion

The sole exception to the inability to demonstrate a cause and effect due to the use of an inappropriate biomarker/end point is demonstrated in the petition to claim that the consumption of live yogurt cultures improves lactose digestion (EFSA 2010a). The petitioners used the breath hydrogen concentration method to measure lactose digestion. This method is based on the measurement of the concentration of hydrogen released into the breath from lactose hydrolyzed after ingestion of a certain amount of lactose (usually 18 g or higher). The method is not specific, and it is generally used in clinical practice for the diagnosis of lactose maldigestion. However, the EFSA review panel was satisfied that this biomarker of lactose malabsorption was acceptable, and the experiments that used this method demonstrated an acceptable cause and effect.

The challenge in the future will be to find other acceptable biomarkers that can be used in clinical trials to measure parameters that are appropriate. This is a particularly difficult activity, in as much as most health regulatory officials wish to have evidence that any biomarker that is being used to demonstrate a treatment effect has been endorsed by the scientific community at large.

Designing a clinical trial to validate the efficacy of any food product as to its health-promoting properties presents many challenges. When that food product is a fermented food that contains live microorganism, even more challenges are evident (Farnworth 2000, 2008). The International Scientific Association for Probiotics and Prebiotics

Table 22.5 ISAPP Suggestions to Ensure a High Quality Experiment to Test Prebiotics and Probiotics

STUDY DESIGN AND METHODOLOGY
- Use methods (such as intent-to-treat) to prevent bias between comparison groups, randomization and allocation of participants, delivery of care, assessment of outcomes, and differences in completion rates
- Choose outcome measurements based on their clinical relevance
- Use sufficient power to detect clinically significant effects, depending on the anticipated magnitude of effect
- Recruit a variety of participants for trials
- Use a placebo as similar to the active intervention as possible
- Use several different placebos if appropriate
- Length of the follow-up period depends on the study question, outcome anticipated, and resources
- Use the assessment of baseline data in treatment groups to allow for appropriate statistical analysis
- Use an appropriate method to measure adherence to the protocol

INTERPRETATION OF RESULTS
- It must be scientifically appropriate to extrapolate from the study sample to the population that is the subject of the health claim
- Small clinical effects (not necessarily statistically significant) may be important

Source: Shane, A.L. et al., *Gut Microbes*, 1, 243–253, 2010.

(ISAPP) acknowledged the increasing number of studies being published that were designed to test the health-promoting properties of prebiotic and probiotic foods. The lack of standards that should be used in designing and conducting such experiments prompted ISAPP to publish a review article after their 2009 meeting to give guidance to researchers on study design, target populations, selection of placebo and probiotic microorganisms, duration of follow-up, outcome and end point measurements, safety assessments, and regulatory considerations (Shane et al. 2010). Table 22.5 is a summary of aspects of study design and interpretation suggested by ISAPP.

22.4 Possible Health Claims for Fermented Milk Products

The number of potential microorganisms that can be used to ferment milk or that can be added to fermented milk to increase its health-promoting properties is large and continues to grow. Worldwide, it appears that it is EFSA that has received the most petitions for regulatory approval of a health claim for fermented milk products and/or probiotic bacteria. A summary of the judgments of the EFSA Panel on Dietetic Products, Nutrition and Allergies reveals the large diversity of the claims that have been sought (Table 22.6). Many claims are targeted at gastrointestinal conditions and events, while others target other parts of the body. Some of the claims are very precise in their wording, for example, "to maintaining individual intestinal microbiota in subjects receiving antibiotic treatment" or "reduction of flatulence and bloating," but others are more general, for example, "beneficial modulation of intestinal microflora" or "treatment of disease" or "bowel motor function." Stimulation of the immune system and protection against pathogenic organisms are reported to be the beneficial effects in several petitions. In some cases, the beneficial effects are attributed to one microorganism, but claims for a combination of microorganisms have also been made.

Table 22.6 List of Bacteria Included in Petitions to EFSA for Fermented Milk Products

BACTERIA/PRODUCT	PROPOSED EFFECT	REFERENCE
Lb. casei DN-114001 and yogurt bacteria (Actimel®)	Reduction of Clostridium difficile toxins in the gut of patients receiving antibiotics and reduced risk of acute diarrhea in patients receiving antibiotic	EFSA (2010b)
Danacol®	Enriched with plant sterols/stanols and lowering/reducing blood cholesterol and reduced risk of (coronary) heart disease	EFSA (2009f)
Yogurt	Rich in fiber and protein, and reduction of the sense of hunger	EFSA (2008e)
Lb. helveticus	Reduction of arterial stiffness	EFSA (2008f)
Lb. paracasei LPC 01	Relieve symptoms typically associated with irritable bowel syndrome	EFSA (2012g)
Lb. casei DG CNCM I-1572	Decreasing potentially pathogenic gastrointestinal microorganisms	EFSA (2012i)
Lb. gasseri PA 16/8, B. bifidum M 20/5, and B. longum SP 07/3	Maintenance of upper respiratory tract defense against pathogens	EFSA (2012j)
Lb. fermentum 57A, Lb. plantarum 57B, and Lb. gasseri 57C	Defense against vaginal pathogens	EFSA (2012e)
Lb. rhamnosus CNCM I-1720 and Lb. helveticus CNCM I-1722	Defense against pathogenic gastrointestinal microorganisms	EFSA (2012h)

(Continued)

Table 22.6 (*Continued*)   List of Bacteria Included in Petitions to EFSA for Fermented Milk Products

BACTERIA/PRODUCT	PROPOSED EFFECT	REFERENCE
Lb. delbrueckii subsp. *bulgaricus* AY/CSL (LMG P-17224) and S. *thermophilus* 9Y/CSL (LMG P-17225)	Beneficial modulation of intestinal microflora	EFSA (2011a)
Lb. *rhamnosus* GR-1 (ATCC 55826) and Lb. *reuteri* RC-14 (ATCC 55845)	Defense against vaginal pathogens	EFSA (2011b)
Lb. *johnsonii* NCC 533 (La1) (CNCM I-1225)	Improving immune defense against pathogenic gastrointestinal microorganisms	EFSA (2011c)
Lb. *rhamnosus* ATCC 53103 (LGG)	Gastrointestinal health and maintenance of tooth mineralization	EFSA (2011d)
Lb. *rhamnosus* GG	Maintenance of defense against pathogenic gastrointestinal microorganisms	EFSA (2011c)
Lb. *paracasei* LMG P-22043	Decreasing potentially pathogenic gastrointestinal microorganisms	EFSA (2011e)
Lb. *johnsonii* BFE 6128	Natural defenses/immune system and skin health	EFSA (2011f)
Lb. *fermentum* ME-3	Decreasing potentially pathogenic gastrointestinal microorganisms	EFSA (2011g)
Lb. *rhamnosus* LB21 NCIMB 40564	Maintaining individual intestinal microbiota in subjects receiving antibiotic treatment	EFSA (2010e)
Lb. *plantarum* BFE 1685	Natural defenses/immune system	EFSA (2011h)
Lb. *plantarum* 299v	Reduction of flatulence and bloating and protection of DNA, proteins, and lipids from oxidative damage	EFSA (2011i)
Lb. *casei* DN-114 001 and yogurt (Actimel®)	Reduction of *Clostridium difficile* toxins in the gut of patients receiving antibiotics and reduced risk of acute diarrhea in patients receiving antibiotics	EFSA (2010b)
Lb. *gasseri* CECT5714 and Lb. *coryniformis*	Natural defense/immune system	EFSA (2010f)
Lb. *fermentum*	Maintenance of the upper respiratory tract defense against pathogens by maintaining immune defenses	EFSA (2010g)
Lb. *paracasei* B21060	Decreasing potentially pathogenic gastrointestinal microorganisms and maintenance of a normal intestinal transit time reduction of gastrointestinal discomfort	EFSA (2010h)

(*Continued*)

Table 22.6 (Continued) List of Bacteria Included in Petitions to EFSA for Fermented Milk Products

BACTERIA/PRODUCT	PROPOSED EFFECT	REFERENCE
Yogurt cultures	Improved lactose digestion	EFSA (2010a)
Lb. plantarum 299 (DSM 6595, 67B)	Decreasing potentially pathogenic intestinal microorganisms	EFSA (2010i)
Lb. reuteri ATCC 55730	Natural defense	EFSA (2010j)
Lb. casei strain shirota	Maintenance of the upper respiratory tract defense against pathogens by maintaining immune defenses	EFSA (2010c)
Lb. rhamnosus IMC 501® and Lb. paracasei IMC 502® (Synbio)	Maintenance and improvement of intestinal well-being	EFSA (2010d)
Lb. rhamnosus LB21 NCIMB 40564	Decreasing potentially pathogenic intestinal microorganisms digestive health reduction of mutans streptococci in the oral cavity	EFSA (2010k)
Lb. plantarum 299v (DSM 9843)	Immune system	EFSA (2010l)
Lb. plantarum BFE 1685	Decreasing potentially pathogenic intestinal microorganisms	EFSA (2010m)
Lb. gasseri CECT5714 and Lb. coryniformis CECT5711	Decreasing potentially pathogenic intestinal microorganisms and improvement of intestinal transit	EFSA (2009g)
Lb. johnsonii BFE 6128	Decreasing potentially pathogenic intestinal microorganisms	EFSA (2009d)
Lb. paracasei 8700:2 (DSM 13434, 240HI)	Decreasing potentially pathogenic intestinal microorganisms	EFSA (2009h)
Lb. reuteri ATCC 55730	Decreasing potentially pathogenic intestinal microorganisms	EFSA (2009i)
Lb. plantarum 299v (DSM 9843)	Decreasing potentially pathogenic intestinal microorganisms	EFSA (2009j)
Lb. rhamnosus HN001 (AGAL NM97/09514)	Decreasing potentially pathogenic intestinal microorganisms	EFSA (2009k)
Lb. casei F19 (LMG P-17806)	Bowel motor function	EFSA (2009l)
Lb. plantarum 299 (DSM 6595, 67B)	Immune system	EFSA (2009e)
Regulat®	Enhancement/modulation/improvement/regulation of the activity of the immune system	EFSA (2009b)
Lb. plantarum 299v (DSM 9843)	Improve iron absorption	EFSA (2009c)
Lb. plantarum, Lb. rhamnosus, and B. longum (LACTORAL)	Improvement of the general immunity	EFSA (2008e)

(Continued)

Table 22.6 (Continued) List of Bacteria Included in Petitions to EFSA for Fermented Milk Products

BACTERIA/PRODUCT	PROPOSED EFFECT	REFERENCE
Lb. plantarum, Lb. rhamnosus, and B. longum (LACTORAL)	Building of the natural intestinal barrier	EFSA (2008g)
Lb. plantarum, Lb. rhamnosus, and B. longum (LACTORAL)	Maintenance of natural intestinal microflora during travel	EFSA (2008a)
Lb. plantarum, Lb. rhamnosus, and B. longum (LACTORAL)	Normal functioning of the alimentary tract	EFSA (2008g)
Lb. plantarum, Lb. rhamnosus, and B. longum (LACTORAL)	Intestinal tract colonization	EFSA (2008d)
Lb. rhamnosus GG (LGG®, ATCC 53103), Lb. rhamnosus Lc705 (DSM 7061), P. freudenreichii subsp. shermanii JS (DSM 7067), and B. animalis subsp. lactis Bb12 (DSM 15954) or B. breve 99 (DSM13692) (LGG® MAX)	Gastrointestinal discomfort	EFSA (2008b)
Lb. helveticus (Evolus®)	Reduce arterial stiffness	EFSA (2008d)
Lb. rhamnosus GG and Lb. reuteri (Regulat®. pro kid BRAIN)	Mental and cognitive development of children	EFSA (2008h)
Lb. rhamnosus GG and Lb. reuteri (Regulat®.pro.kid IMMUN)	Immune system of children during growth	EFSA (2008i)
B. animalis subsp. lactis Bb-12	Immune defense against pathogens and decreasing potentially pathogenic gastrointestinal microorganisms and natural immune function and reduction of symptoms of inflammatory bowel conditions and maintenance of normal blood LDL cholesterol concentrations	EFSA (2011j)
B. longum BB536	Improvement of bowel regularity and normal resistance to cedar pollen allergens and decreasing potentially pathogenic gastrointestinal microorganisms	EFSA (2011k)
B. bifidum, B. breve, B. infantis, and B. longum	Decreasing potentially pathogenic intestinal microorganisms	EFSA (2009m)
B. animalis Lafti B94 (CBS118.529)	Decreasing potentially pathogenic intestinal microorganisms	EFSA (2009n)
B. animalis subsp. lactis Bb-12	Decreasing potentially pathogenic intestinal microorganisms	EFSA (2009o)

As has been pointed out earlier, only one submission has been approved by EFSA, and this appears to be the only successful petition in any jurisdiction related to a fermented milk product. It would appear that there is a long list of microorganisms, with a variety of potential health-promoting properties that could be used in fermented milk products. However, at this time, there is insufficient experimental evidence to validate these claims.

22.5 Conclusions

To date, petitions to health regulatory agencies to obtain a food health claim for fermented foods, including fermented milk products, have not be very successful. The main reason for rejections has been that the data submitted have not demonstrated a convincing cause-and-effect relationship. It is agreed by both petitioners and regulatory officials that any claim must be based on good scientific evidence. However, because of problems particularly related to clinical trials involving fermented foods containing live microorganisms when consumed, and a misunderstanding about what details and aspects of an experiment that are considered essential by regulatory officials, many clinical trials in the past were poorly planned, executed, and reported. Poorly planned, executed, and reported if the objective was to use the data to validate and support a petition for a food health claim. The result has been rejected petitions. Now that we are able to read opinions of the panels judging these petitions, we can start to understand what deficiencies led to rejection. We are now in a better position to give guidance to researchers who wish to conduct experiments to produce data to validate and support a petition for a food health claim for a fermented milk product.

References

Alander, M., Satokari, R., Korpela, R., Saxelin, M., Vilpponen-Salmela, T., Mattila-Sandholm, T., and von Wright, A. 1999. Persistence of colonization of human colonic mucosa by a probiotic strain, *Lactobacillus rhamnosus* GG, after oral consumption. *Applied and Environmental Microbiology* 65: 351–354.

Ammor, M., Delgado, S., Alvarez-Martin, P., Margolles, A., and Mayo, B. 2007. Reagenless identification of human bifidobacteria by intrinsic fluorescence. *Journal of Microbiological Methods* 69: 100–106.

Blaut, M. and Clavel, T. 2007. Metabolic diversity of the intestinal microbiota: Implications for health and disease. *Journal of Nutrition* 137: 751S–755S.

Bourdichon, F., Casaregola, S., Farrokh, C., Frisvad, J.C., Gerds, M.L., Hammes, W.P., Harnett, J. et al. 2012. Food fermentations: Microorganisms with technological beneficial use. *International Journal of Food Microbiology* 154: 87–97.

Christensen, H.R., Larsen, C.N., Kaestel, P., Rosholm, L.B., Sternberg, C., Michaelsen, K.F., and Frokiaer, H. 2006. Immunomodulating potential of supplementation with probiotics: A dose-response study in healthy young adults. *FEMS Immunology and Medical Microbiology* 47: 380–390.

CODEX. 2003. *CODEX Standard for Fermented Milks*. STAN 243-2003. Rome, Italy.

Collins, J., Thornton, G., and Sullivan, G. 1998. Selection of probiotic strains for human applications. *International Dairy Journal* 8: 487–490.

EFSA. 2008a. LACTORAL and maintenance of natural intestinal microflora during travel—Scientific substantiation of a health claim related to LACTORAL (a combination of three probiotic strains: *Lactobacillus plantarum*, *Lactobacillus rhamnosus*, *Bifidobacterium longum*) and maintenance of natural intestinal microflora during travel, pursuant to Article 14 of Regulation (EC) No 1924/2006[1]—Scientific opinion of the panel on dietetic products, nutrition and allergies. *EFSA Journal* 863: 861–868. doi:810.2903/j.efsa.2008.2863.

EFSA. 2008b. LGG® MAX and gastro-intestinal discomfort—Scientific substantiation of a health claim related to LGG® MAX and reduction of gastro-intestinal discomfort pursuant to Article 13(5) of Regulation (EC) No 1924/2006[1]—Scientific opinion of the panel on dietetic products, nutrition and allergies. *EFSA Journal* 853: 851–815. doi:810.2903/j.efsa.2008.2853.

EFSA. 2008c. Milk product, rich in fibre and protein, and reduction of the sense of hunger—Scientific opinion of the panel on dietetic products, nutrition and allergies. *EFSA Journal* 894: 891–899. doi:810.2903/j.efsa.2008.2894.

EFSA. 2008d. Evolus® and reduce arterial stiffness—Scientific substantiation of a health claim related to *Lactobacillus helveticus* fermented Evolus® low-fat milk products and reduction of arterial stiffness pursuant to Article 14 of the Regulation (EC) No 1924/2006[1]—Scientific opinion of the panel on dietetic products, nutrition and allergies. *EFSA Journal* 824: 821–812. doi:810.2903/j.efsa.2008.2824.

EFSA. 2008e. LACTORAL and improvement of the general immunity—Scientific substantiation of a health claim related to LACTORAL (a combination of three probiotic strains: *Lactobacillus plantarum*, *Lactobacillus rhamnosus*, *Bifidobacterium longum*) and improvement of the general immunity pursuant to Article 14 of Regulation (EC) No 1924/2006[1]—Scientific opinion of the panel on dietetic products, nutrition and allergies. *EFSA Journal* 860: 861–868. doi:810.2903/j.efsa.2008.2860.

EFSA. 2008f. LACTORAL and building of the natural intestinal barrier—Scientific substantiation of a health claim related to LACTORAL (a combination of three probiotic strains: *Lactobacillus plantarum*, *Lactobacillus rhamnosus*, *Bifidobacterium longum*) and building of the natural intestinal barrier pursuant to Article 14 of Regulation (EC) No 1924/2006[1]—Scientific opinion of the panel on dietetic products, nutrition and allergies. *EFSA Journal* 859: 851–859. doi:810.2903/j.efsa.2008.2859.

EFSA. 2008g. LACTORAL and normal functioning of the alimentary tract—Scientific substantiation of a health claim related to LACTORAL (a combination of three probiotic strains: *Lactobacillus plantarum*, *Lactobacillus rhamnosus*, *Bifidobacterium longum*) and normal functioning of the alimentary tract pursuant to Article 14 of Regulation (EC) No 1924/2006[1]—Scientific opinion of the panel on dietetic products, nutrition and allergies. *EFSA Journal* 861: 861–869. doi:810.2903/j.efsa.2008.2861.

EFSA. 2008h. Regulat®.pro.kid BRAIN and mental and cognitive development of children—Scientific substantiation of a health claim related to regulat®.pro.kid BRAIN and mental and cognitive developments of children pursuant to Article 14 of Regulation (EC) No 1924/2006[1]—Scientific opinion of the panel on dietetic products, nutrition and allergies. *EFSA Journal* 829: 821–810. doi:810.2903/j.efsa.2008.2829.

EFSA. 2008i. Regulat®.pro.kid IMMUN and immune system of children—Scientific substantiation of a health claim related to regulat®.pro.kid IMMUN and immune system of children during growth pursuant to Article 14 of Regulation (EC) No 1924/2006[1]—Scientific opinion of the panel on dietetic products, nutrition and allergies. *EFSA Journal* 782: 781–789. doi:710.2903/j.efsa.2008.2782.

EFSA. 2009a. EFSA panel on dietetic products, nutrition and allergies (NDA); Frequently Asked Questions (FAQ) related to the assessment of Article 14 and 13.5 health claims applications on request of EFSA. *EFSA Journal* 7(9): 1339. doi:10.2903/j.efsa.2009.1339.

EFSA. 2009b. Regulat® and "the immune system"—Scientific substantiation of a health claim related to Regulat® and "enhancement/modulation/improvement/regulation of the activity of the immune system" pursuant to Article 13(5) of Regulation (EC) No 1924/2006 [1]. *EFSA Journal* 1182: 1181–1189. doi:1110.2903/j.efsa.2009.1182.

EFSA. 2009c. *Lactobacillus plantarum* 299v (DSM 9843) and improve iron absorption—Scientific substantiation of a health claim related to *Lactobacillus plantarum* 299v (DSM 9843) and improve iron absorption pursuant to Article 13(5) of Regulation (EC) No 1924/2006. *EFSA Journal* 999: 991–999. doi:910.2903/j.efsa.2009.2999.

EFSA. 2009d. Scientific opinion on the substantiation of health claims related to *Lactobacillus johnsonii* BFE 6128 and decreasing potentially pathogenic intestinal microorganisms (ID 989) pursuant to Article 13(1) of Regulation (EC) No 1924/2006. *EFSA Journal* 7: 1239. doi:1210.2903/j.efsa.2009.1239.

EFSA. 2009e. Scientific opinion on the substantiation of health claims related to *Lactobacillus plantarum* 299 (DSM 6595, 67B) and immune system (ID 1077) pursuant to Article 13(1) of Regulation (EC) No 1924/2006. *EFSA Journal* 7: 1241. doi:1210.2903/j.efsa.2009.1241.

EFSA. 2009f. Danacol® and blood cholesterol Scientific substantiation of a health claim related to a low fat fermented product (Danacol®) enriched with plant sterols/stanols and lowering/reducing blood cholesterol and reduced risk of (coronary) heart disease pursuant to Article 14 of Regulation (EC) No 1924/2006. *EFSA Journal* 1177: 1–12. doi:1110.2903/j.efsa.2009.1177.

EFSA. 2009g Scientific opinion on the substantiation of health claims related to "*Lactobacillus gasseri* CECT5714 and *Lactobacillus coryniformis* CECT5711" and decreasing potentially pathogenic intestinal microorganisms and improvement of intestinal transit (ID 937) pursuant to Article 13 of Regulation (EC) No 1924/2006. *EFSA Journal* 7: 1238. doi:1210.2903/j.efsa.2009.1238.

EFSA. 2009h. Scientific opinion on the substantiation of health claims related to *Lactobacillus paracasei* 8700:2 (DSM 13434, 240HI) and decreasing potentially pathogenic intestinal microorganisms (ID 1074) pursuant to Article 13(1) of Regulation (EC) No 1924/2006. *EFSA Journal* 7: 1240. doi:1210.2903/j.efsa.2009.1240.

EFSA. 2009i. Scientific opinion on the substantiation of health claims related to *Lactobacillus reuteri* ATCC 55730 and decreasing potentially pathogenic intestinal microorganisms (ID 904) pursuant to Article 13(1) of Regulation (EC) No 1924/2006. *EFSA Journal* 7: 1243. doi:1210.2903/j.efsa.2009.1243.

EFSA. 2009j. Scientific opinion on the substantiation of health claims related to *Lactobacillus plantarum* 299v (DSM 9843) and decreasing potentially pathogenic intestinal microorganisms (ID 1084) pursuant to Article 13(1) of Regulation (EC) No 1924/2006. *EFSA Journal* 7: 1242. doi:1210.2903/j.efsa.2009.1242.

EFSA. 2009k. Scientific opinion on the substantiation of health claims related to *Lactobacillus rhamnosus* HN001 (AGAL NM97/09514) and decreasing potentially pathogenic intestinal microorganisms (ID 908) pursuant to Article 13(1) of Regulation (EC) No 1924/2006. *EFSA Journal* 7: 1244. doi:1210.2903/j.efsa.2009.1244.

EFSA. 2009l. Scientific opinion on the substantiation of health claims related to *Lactobacillus casei* F19 (LMG P-17806) and bowel motor function (ID 893) pursuant to Article 13(1) of Regulation (EC) No 1924/2006. *EFSA Journal* 7: 1237. doi:1210.2903/j.efsa.2009.1237.

EFSA. 2009m. Scientific opinion on the substantiation of a health claim related to a combination of bifidobacteria (*Bifidobacterium bifidum*, *Bifidobacterium breve*, *Bifidobacterium infantis*, *Bifidobacterium longum*) and decreasing potentially pathogenic intestinal microorganisms pursuant to Article 14 of Regulation (EC) No 1924/2006. *EFSA Journal* 7: 1420. doi:1410.2903/j.efsa.2009.1420.

EFSA. 2009n. Scientific opinion on the substantiation of health claims related to *Bifidobacterium animalis Lafti* B94 (CBS118.529) and decreasing potentially pathogenic intestinal microorganisms (ID 867) pursuant to Article 13(1) of Regulation (EC) No 1924/2006. *EFSA Journal* 7: 1232. doi:1210.2903/j.efsa.2009.1232.

EFSA. 2009o. Scientific opinion on the substantiation of health claims related to fermented dairy products and decreasing potentially pathogenic intestinal microorganisms (ID 1376) pursuant to Article 13(1) of Regulation (EC) No 1924/2006. *EFSA Journal* 7: 1233. doi:1210.2903/j.efsa.2009.1233.

EFSA. 2010a. Scientific opinion on the substantiation of health claims related to live yoghurt cultures and improved lactose digestion (ID 1143, 2976) pursuant to Article 13(1) of Regulation (EC) No 1924/2006. *EFSA Journal* 8: 1763. doi:1710.2903/j.efsa.2010.1763.

EFSA. 2010b. Scientific opinion on the substantiation of a health claim related to fermented milk containing *Lactobacillus casei* DN-114 001 plus yoghurt symbiosis (Actimel®), and reduction of Clostridium difficile toxins in the gut of patients receiving antibiotics and reduced risk of acute diarrhoea in patients receiving antibiotics pursuant to Article 14 of Regulation (EC) No 1924/2006. *EFSA Journal* 8: 1903. doi:1910.2903/j.efsa.2010.1903.

EFSA. 2010c. Scientific opinion on the substantiation of a health claim related to *Lactobacillus casei* strain Shirota and maintenance of the upper respiratory tract defence against pathogens by maintaining immune defences pursuant to Article 13(5) of Regulation (EC) No 1924/2006. *EFSA Journal* 8: 1860. doi:1810.2903/j.efsa.2010.1860.

EFSA. 2010d. Scientific opinion on the substantiation of a health claim related to Synbio, a combination of *Lactobacillus rhamnosus* IMC 501® and *Lactobacillus paracasei* IMC 502®, and maintenance and improvement of intestinal well being pursuant to Article 13(5) of Regulation (EC) No 1924/2006. *EFSA Journal* 8: 1773. doi:1710.2903/j.efsa.2010.1773.

EFSA. 2010e. Scientific opinion on the substantiation of health claims related to *Lactobacillus rhamnosus* LB21 NCIMB 40564 and contribution to maintaining individual intestinal microbiota in subjects receiving antibiotic treatment (ID 1061) pursuant to Article 13(1) of Regulation (EC) No 1924/2006. *EFSA Journal* 9: 2029. doi:2010.2903/j.efsa.2010.2029.

EFSA. 2010f. Scientific opinion on the substantiation of health claims related to *Lactobacillus gasseri* CECT5714 and *Lactobacillus coryniformis* CECT5711 and "natural defence/immune system" (ID 930) pursuant to Article 13(1) of Regulation (EC) No 1924/2006. *EFSA Journal* 8: 1803. doi:1810.2903/j.efsa.2010.1803.

EFSA. 2010g. Scientific opinion on the substantiation of health claims related to *Lactobacillus fermentum* CECT5716 and maintenance of the upper respiratory tract defence against pathogens by maintaining immune defences (ID 916) pursuant to Article 13(1) of Regulation (EC) No 1924/2006. *EFSA Journal* 8: 1802. doi:1810.2903/j.efsa.2010.1802.

EFSA. 2010h. Scientific opinion on the substantiation of health claims related to *Lactobacillus paracasei* B21060 and decreasing potentially pathogenic gastro-intestinal microorganisms (ID 2959), maintenance of a normal intestinal transit time (ID 2959) and reduction of gastro-intestinal discomfort (ID 2959) pursuant to Article 13(1) of Regulation (EC) No 1924/2006. *EFSA Journal* 8: 1804. doi:1810.2903/j.efsa.2010.1804.

EFSA. 2010i. Scientific opinion on the substantiation of health claims related to *Lactobacillus plantarum* 299 (DSM 6595, 67B) (ID 1078) and decreasing potentially pathogenic intestinal microorganisms pursuant to Article 13(1) of Regulation (EC) No 1924/2006. *EFSA Journal* 8: 1726. doi:1710.2903/j.efsa.2010.1726.

EFSA. 2010j. Scientific opinion on the substantiation of health claims related to *Lactobacillus reuteri* ATCC 55730 and "natural defence" (ID 905) pursuant to Article 13(1) of Regulation (EC) No 1924/2006. *EFSA Journal* 8: 1805. doi:1810.2903/j.efsa.2010.1805.

EFSA. 2010k. Scientific opinion on the substantiation of health claims related to *Lactobacillus rhamnosus* LB21 NCIMB 40564 and decreasing potentially pathogenic intestinal microorganisms (ID 1064), digestive health (ID 1064), and reduction of *mutans streptococci* in the oral cavity (ID 1064) pursuant to Article 13(1) of Regulation (EC) No 1924/2006. *EFSA Journal* 8: 1487. doi:1410.2903/j.efsa.2010.1487.

EFSA. 2010l. Scientific opinion on the substantiation of health claims related to *Lactobacillus plantarum* 299v (DSM 9843) and "immune system" (ID 1081), pursuant to Article 13(1) of Regulation (EC) No 1924/2006. *EFSA Journal* 8: 1488. doi:1410.2903/j.efsa.2010.1488.

EFSA. 2010m. Scientific opinion on the substantiation of health claims related to *Lactobacillus plantarum* BFE 1685 and decreasing potentially pathogenic intestinal microorganisms (ID 992) pursuant to Article 13(1) of Regulation (EC) No 1924/2006. *EFSA Journal* 8: 1471. doi:1410.2903/j.efsa.2010.1471.

EFSA. 2011a. Scientific opinion on the substantiation of a health claim related to a combination of *Lactobacillus delbrueckii* subsp. *bulgaricus* AY/CSL (LMG P-17224) and *Streptococcus thermophilus* 9Y/CSL (LMG P-17225) and "beneficial modulation of intestinal microflora" pursuant to Article 14 of Regulation (EC) No 1924/2006. *EFSA Journal* 9: 2288. doi:2210.2903/j.efsa.2011.2288.

EFSA. 2011b. Scientific opinion on the substantiation of a health claim related to *Lactobacillus rhamnosus* GG and maintenance of defence against pathogenic gastrointestinal microorganisms pursuant to Article 13(5) of Regulation (EC) No 1924/2006. *EFSA Journal* 9: 2167. doi:2110.2903/j.efsa.2011.2167.

EFSA. 2011c. Scientific opinion on the substantiation of health claims related to *Lactobacillus johnsonii* NCC 533 (La1) (CNCM I-1225) and improving immune defence against pathogenic gastro-intestinal microorganisms (ID 896), and protection of the skin from UV-induced damage (ID 900) pursuant to Article 13(1) of Regulation (EC) No 1924/2006. *EFSA Journal* 9(6): 2231. doi:10.2903/j.efsa.2011.2231.

EFSA. 2011d. Scientific opinion on the substantiation of health claims related to *Lactobacillus rhamnosus* ATCC 53103 (LGG) and "gastro-intestinal health" (ID 906) and maintenance of tooth mineralisation (ID 3018) pursuant to Article 13(1) of Regulation (EC) No 1924/2006. *EFSA Journal* 9(6): 2233. doi:10.2903/j.efsa.2011.2233.

EFSA. 2011e. Scientific opinion on the substantiation of health claims related to *Lactobacillus paracasei* LMG P-22043 and decreasing potentially pathogenic gastro-intestinal microorganisms (ID 2964) and reduction of gastro-intestinal discomfort (ID 2964) pursuant to Article 13(1) of Regulation (EC) No 1924/2006. *EFSA Journal* 9(4): 2027. doi:10.2903/j.efsa.2011.2027.

EFSA. 2011f. Scientific opinion on the substantiation of health claims related to *Lactobacillus johnsonii* BFE 6128 and "natural defences/immune system" (ID 990) and "skin health" (ID 991) pursuant to Article 13(1) of Regulation (EC) No 1924/2006. *EFSA Journal* 9(4): 2026. doi:10.2903/j.efsa.2011.2026.

EFSA. 2011g. Scientific opinion on the substantiation of health claims related to *Lactobacillus fermentum* ME-3 and decreasing potentially pathogenic gastro-intestinal microorganisms (ID 3025) pursuant to Article 13(1) of Regulation (EC) No 1924/2006. *EFSA Journal* 9(4): 2025. doi:10.2903/j.efsa.2011.2025.

EFSA. 2011h. Scientific opinion on the substantiation of health claims related to *Lactobacillus plantarum* BFE 1685 and "natural defences/immune system" (ID 993) pursuant to Article 13(1) of Regulation (EC) No 1924/2006. *EFSA Journal* 9(4): 2028. doi:10.2903/j.efsa.2011.2028.

EFSA. 2011i. Scientific opinion on the substantiation of health claims related to *Lactobacillus plantarum* 299v and reduction of flatulence and bloating (ID 902), and protection of DNA, proteins and lipids from oxidative damage (ID 1083) pursuant to Article 13(1) of Regulation (EC) No 1924/2006. *EFSA Journal* 9(4): 2037. doi:10.2903/j.efsa.2011.2037.

EFSA. 2011j. Scientific opinion on the substantiation of health claims related to *Bifidobacterium animalis* ssp. *lactis* Bb-12 and immune defence against pathogens (ID 863), decreasing potentially pathogenic gastro-intestinal microorganisms (ID 866), "natural immune function" (ID 924), reduction of symptoms of inflammatory bowel conditions (ID 1469) and maintenance of normal blood LDL-cholesterol concentrations (ID 3089) pursuant to Article 13(1) of Regulation (EC) No 1924/2006. *EFSA Journal* 9(4): 2047. doi:10.2903/j.efsa.2011.2047.

EFSA. 2011k. Scientific opinion on the substantiation of health claims related to *Bifidobacterium longum* BB536 and improvement of bowel regularity (ID 3004), normal resistance to cedar pollen allergens (ID 3006), and decreasing potentially pathogenic gastro-intestinal microorganisms (ID 3005) pursuant to Article 13(1) of Regulation (EC) No 1924/2006. *EFSA Journal* 9(4): 2041. doi:10.2903/j.efsa.2011.2041.

EFSA. 2012a. Scientific opinion on the substantiation of health claims related to a combination of *Lactobacillus helveticus* CNCM I-1722 and *Bifidobacterium longum* subsp. *longum* CNCM I-3470 and alleviation of psychological stress (ID 938) and "maintains the balance of healthy microbiota that helps to strengthen the natural defence" (ID 2942) (further assessment) pursuant to Article 13(1) of Regulation (EC) No 1924/2006. *EFSA Journal* 10: 2849. doi:2810.2903/j.efsa.2012.2849.

EFSA. 2012b. Scientific opinion on the substantiation of a health claim related to *Bifidobacterium animalis* subsp. *lactis* LMG P-21384 and changes in bowel function (ID 2940, further assessment) pursuant to Article 13(1) of Regulation (EC) No 1924/2006. *EFSA Journal* 10: 2851. doi:2810.2903/j.efsa.2012.2851.

EFSA. 2012c. Scientific opinion on the substantiation of health claims related to a combination of *Lactobacillus rhamnosus* CNCM I-1720, *Lactobacillus helveticus* CNCM I-1722, *Bifidobacterium longum* subsp. *longum* CNCM I-3470 and *Saccharomyces cerevisiae var. boulardii* CNCM I-1079 and defence against pathogenic gastro-intestinal microorganisms (ID 3017, further assessment) pursuant to Article 13(1) of Regulation (EC) No 1924/2006. *EFSA Journal* 10: 2853. doi:2810.2903/j.efsa.2012.2853.

EFSA. 2012d. Scientific opinion on the substantiation of health claims related to a combination of *Lactobacillus fermentum* 57A, *Lactobacillus plantarum* 57B and *Lactobacillus gasseri* 57C and defence against vaginal pathogens (ID 934, further assessment) pursuant to Article 13(1) of Regulation (EC) No 1924/2006. *EFSA Journal* 10: 2719. doi:2710.2903/j.efsa.2012.2719.

EFSA. 2012e. Scientific opinion on the substantiation of health claims related to various microorganisms and changes in bowel function, and digestion and absorption of nutrients (ID 960, 961, 967, 969, 971, 975, 983, 985, 994, 996, 998, 1006, 1014), decreasing potentially pathogenic gastro-intestinal microorganisms (ID 960, 967, 969, 971, 975, 983, 985, 994, 996, 998, 1006, 1014), and stimulation of immunological responses (ID 962, 968, 970, 972, 976, 984, 986, 995, 997, 999, 1007, 1015) (further assessment) pursuant to Article 13(1) of Regulation (EC) No 1924/2006. *EFSA Journal* 10: 2857. doi:2810.2903/j.efsa.2012.2857.

EFSA. 2012f. Scientific opinion on the substantiation of health claims related to *Lactobacillus paracasei* LPC 01 (CNCM I-1390) and treatment of disease (ID 3055, further assessment) pursuant to Article 13(1) of Regulation (EC) No 1924/2006. *EFSA Journal* 10: 2850. doi:2810.2903/j.efsa.2012.2850.

EFSA. 2012g. Scientific opinion on the substantiation of health claims related to a combination of *Lactobacillus rhamnosus* CNCM I-1720 and *Lactobacillus helveticus* CNCM I-1722 and defence against pathogenic gastro-intestinal microorganisms (ID 939, further assessment) pursuant to Article 13(1) of Regulation (EC) No 1924/2006. *EFSA Journal* 10: 2720. doi:2710.2903/j.efsa.2012.2720.

EFSA. 2012h. Scientific opinion on the substantiation of health claims related to non-characterised micro-organisms (ID 2936, 2937, 2938, 2941, 2944, 2965, 2968, 2969, 3035, 3047, 3056, 3059, further assessment) pursuant to Article 13(1) of Regulation (EC) No 1924/2006. *EFSA Journal* 10: 2854[2836 pp.]. doi:2810.2903/j.efsa.2012.2854.

EFSA. 2012i. Scientific opinion on the substantiation of health claims related to *Lactobacillus casei* DG CNCM I-1572 and decreasing potentially pathogenic gastro intestinal microorganisms (ID 2949, 3061, further assessment) pursuant to Article 13(1) of Regulation (EC) No 1924/2006. *EFSA Journal* 10: 2723. doi:2710.2903/j.efsa.2012.2723.

EFSA. 2012j. Scientific opinion on the substantiation of health claims related to a combination of *Lactobacillus gasseri* PA 16/8, *Bifidobacterium bifidum* M 20/5 and *Bifidobacterium longum* SP 07/3 and maintenance of upper respiratory tract defence against pathogens (ID 931, further assessment) pursuant to Article 13(1) of Regulation (EC) No 1924/2006. *EFSA Journal* 10: 2718. doi:2710.2903/j.efsa.2012.2718.

Farnworth, E.R. 2000. Designing a proper control for testing the efficacy of a probiotic product. *Journal of Nutraceuticals, Functional and Medical Foods* 2: 55–63.

Farnworth, E.R. 2003. *The Future of Fermented Foods*, 361–378. Boca Raton, FL: CRC Press.

Farnworth, E.R. 2004. The beneficial health effects of fermented foods: Potential probiotics around the world. *Journal of Nutraceuticals, Functional and Medical Foods* 4: 93–117.

Farnworth, E.R. 2008. The evidence to support health claims for probiotics. *Journal of Nutrition* 138: 1250S–1254S.

Gao, X.W., Mubasher, M., Fang, C.Y., Reifer, C., and Miller, L.E. 2010. Dose-response efficacy of a proprietary probiotic formula of *Lactobacillus acidophilus* CL1285 and *Lactobacillus casei* LBC80R for antibiotic-associated diarrhea and *Clostridium difficile*-associated diarrhea prophylaxis in adult patients. *American Journal of Gastroenterology* 105: 1636–1641.

Gibson, G.R. and Rastall, R.A. 2006. Prebiotics: Development and application. West Sussex: John Wiley & Sons Ltd.

Goossens, D., Jonkers, D., Russel, M., Stobberingh, E., Van den Bogaard, A., and Stockbrugger, R. 2003. The effect of *Lactobacillus plantarum* 288v on the bacterial composlition andmetabolic activity in faeces of healthy volunteers: A placebo-controlled study on the onset and duration of effects. *Alimentary Pharmacology & Therapeutics* 18: 495–505.

Health Canada. 2009a. *Accepted Claims about the Nature of Probiotic Microorganisms in Food*. http://www.hc-sc.gc.ca/fn-an/label-etiquet/claims-reclam/probiotics_claims-allegations_probiotiques-eng.php.

Health Canada. 2009b. *Guidance Document for Preparing a Submission for Food Health Claims*. Section 5.1.7 Step 7. Evaluate Study Quality. http://www.hc-sc.gc.ca/fn-an/legislation/guide-ld/health-claims_guidance-orientation_allegations-sante-eng.php#a5-1-7.

Ishibashi, N. and Shimamura, S. 1993. Bifidobacteria: Research and development in Japan. *Food Technology* 47(6): 126–135.

Kailasapathy, K. 2002. Microencapsulation of probiotic bacteria: Technology and potential applications. *Current Issues in Intestinal Microbiology* 3: 39–48.

Larsen, C.N., Nielsen, S., Kaestel, P., Brockmann, E., Bennedsen, M., Christensen, H.R., Eskesen, D.C., Jacobsen, B.L., and Michaelsen, K.F.

2006. Dose-response study of probiotic bacteria *Bifidobacterium animalis* subsp *lactis* BB-12 and *Lactobacillus paracasei* subsp *paracasei* CRL-341 in healthy young adults. *European Journal of Clinical Nutrition* 60: 1284–1293.

Lewis, S.J. and Freedman, A.R. 1998. Review article: The use of biotherapeutic agents in the prevention and treatment of gastrointestinal disease. *Alimentary Pharmacology & Therapeutics* 12: 807–822.

Mathys, S., Lacroix, C., Mini, R., and Meile, L. 2008. PCR and real-time PCR primers developed for detection and identification of *Bifidobacterium thermophilum* in faeces. *BMC Microbiology* 8: 179.

Metchnikoff, E. 1908. *The Prolongation of Life: Optimistic Studies*. New York: G.P. Putnam's Sons.

Moher, D., Hopewell, S., Schulz, K.F., Montori, V., Gotzsche, P.C., Devereaux, P.J., Elbourne, D., Egger, M., and Altman, D.G. 2010. CONSORT 2010 explanation and elaboration: Updated guidelines for reporting parallel group randomised trials. *Journal of Clinical Epidemiology* 63: e1–37.

Niers, L.E., Hoekstra, M.O., Timmerman, H.M., van Uden, N.O., de Graaf, P.M., Smits, H.H., Kimpen, J.L., and Rijkers, G.T. 2007. Selection of probiotic bacteria for prevention of allergic diseases: Immunomodulation of neonatal dendritic cells. *Clinical & Experimental Immunology* 149: 344–352.

O'Sullivan, M.G., Thorton, G., O'Sullivan, G.C., and Collins, J.K. 1992. Probiotic bacteria: Myth or reality? *Trends Food Science and Technology* 3: 309–314.

Prajapati, J.B., and Nair, B.M. 2003. The history of fermented foods. In: *Handbook of Fermented Functional Foods*, ed. E.R. Farnworth, 1–26. Boca Raton, FL: CRC Press.

Reid, G. and Burton, J. 2002. Use of *Lactobacillus* to prevent infection by pathogenic bacteria. *Microbes and Infection* 4: 319–324.

Ross, R.P., Morgan, S., and Hill, C. 2002. Preservation and fermentation: Past, present and future. *The International Journal of Food Microbiology* 79: 3–16.

Sanders, M.E. 2003. Probiotics: Considerations for human health. *Nutrition Reviews* 61: 91–99.

Sanders, M.E., Tompkins, T., Heimbach, J.T., and Kolida, S. 2005. Weight of evidence needed to substantiate a health effect for probiotics and prebiotics: Regulatory considerations in Canada, E.U., and U.S. *European Journal of Nutrition* 44: 303–310.

Schulz, K.F., Altman, D.G., and Moher, D. 2010. CONSORT 2010 statement: Updated guidelines for reporting parallel group randomized trials. *Annals of Internal Medicine* 152: 726–732.

Shane, A.L., Cabana, M.D., Vidry, S., Merenstein, D., Hummelen, R., Ellis, C.L., Heimbach, J.T. et al. 2010. Guide to designing, conducting, publishing and communicating results of clinical studies involving probiotic applications in human participants. *Gut Microbes* 1: 243–253.

Soomro, A.H., Masud, T., and Anwaar, K. 2002. Role of Lactic Acid Bacteria (LAB) in food preservation and human health—A review. *Pakistan Journal of Nutrition* 1: 20–24.

Sultana, K., Godward, G., Reynolds, N., Arumugaswamy, R., Peiris, P., and Kailasapathy, K. 2000. Encapsulation of probiotic bacteria with alginate-starch and evaluation of survival in simulated gastrointestinal conditions and in yoghurt. *International Journal of Food Microbiology* 62: 47–55.

Tamime, A.Y. 2002. Fermented milks: A historical food with modern applications—A review. *European Journal of Clinical Nutrition* 56(Suppl 4): S2–S15.

Urbanska, A.M., Bhathena, J., and Prakash, S. 2007. Live encapsulated *Lactobacillus acidophilus* cells in yogurt for therapeutic oral delivery: Preparation and in vitro analysis of alginate-chitosan microcapsules. *Canadian Journal of Physiology and Pharmacology* 85: 884–893.

Van Loveren, H., Sanz, Y., and Salminen, S. 2012. Health claims in Europe: Probiotics and prebiotics as case examples. *Annual Review of Food Science Technology* 3: 247–261.

Vaughan, E.E., Heilig, H.G., Ben-Amor, K., and de Vos, W.M. 2005. Diversity, vitality and activities of intestinal lactic acid bacteria and bifidobacteria assessed by molecular approaches. *FEMS Microbiology Reviews* 29: 477–490.

WIPO. 1977. *Depository Institutions Having Acquired the Status of International Depositary Authority Under the Budapest Treaty*. Geneva, Switzerland: WIPO. http://www.wipo.int/treaties/en/registration/budapest/pdf/ida.pdf.

PART VI
Quality Assurance, Packaging, and Other Prospects

23

Quality Assurance of Fermented Dairy Products

H.V. RAGHU, NIMISHA TEHRI, LAXMAN NAIK, AND NARESH KUMAR

Contents

23.1	Introduction	604
23.2	Microbiological Quality and Safety Issues	605
	23.2.1 Quality Issues	605
	23.2.2 Safety Issues	606
23.3	Microbiological Risk Assessment	608
	23.3.1 Risk Assessment	608
	23.3.1.1 Hazard Identification	609
	23.3.1.2 Hazard Characterization	610
	23.3.1.3 Exposure Assessment	611
	23.3.1.4 Risk Characterization	612
	23.3.2 Risk Management	613
	23.3.3 Risk Communication	613
23.4	Quality Assurance of Fermented Milk Products	613
	23.4.1 Hygiene Measures	613
	23.4.2 Good Manufacturing Practices	615
	23.4.3 GMP at Processing Condition	615
	23.4.3.1 Plant Hygiene	616
	23.4.3.2 Personnel Hygiene	617
	23.4.3.3 Equipment Hygiene	617
	23.4.3.4 Production and Process Controls	618
	23.4.3.5 Air Hygiene	619
	23.4.3.6 Water Hygiene	619
	23.4.4 Monitoring the Process Plant	619

23.5 HACCP-Based Food Safety Management System   620
    23.5.1 HACCP   620
    23.5.2 Food Safety Management System   626
    23.5.3 Verification Procedure   626
23.6 Regulatory Aspects of Fermented Milk Products   628
23.7 Conclusions   629
References   630

23.1 Introduction

Fermented milk products are prepared by the fermentation of milk with the help of starter cultures, resulting in compositional changes and low pH of the product with or without coagulation. These cultures have been viable, active, and ample in the product to the date of minimum stability (Codex 2010). Selective growth of specific starter culture in milk leads to the production of these fermented milk products such as yogurt, dahi, shrikhand, chakka, and different varieties of cheeses. These products are produced in order to fulfill the need of extending the keeping quality of milk in the absence of cooling (Kosikowski and Mistry 1997). Diversity of these products may vary with composition, texture, taste, aroma, and flavor. Starter cultures or lactic acid bacteria (LAB) are used in the preparation of fermented milk products; are known to produce specific metabolites during fermentation, such as peptides, fatty acids, and simple sugars, all these contribute toward better digestibility of the fermented foods; and offer nutritional as well as therapeutic qualities (Abeer et al. 2009).

Quality problems associated with the fermented milk products such as poor technology, hygiene, sanitation, and shelf life are well documented (Nout 2001). Major sources of contamination of fermented milk products are processing equipment, human handlers, air, starter cultures, packaging materials, and so on. Microorganisms such as *L. monocytogenes*, *Salmonella* ssp., *Escherichia coli* O157:H7, *Shigella* ssp., *Yersinia enterocolitica*, and *Staphylococcus aureus* were also reported (Tatini and Kauppi 2003), as the most frequently associated foodborne pathogens that are able to persist in fermented milk products (Alm 1983; Ahmed et al. 1986; Schaak and Marth 1988; Pazakova et al. 1997; Canganella et al. 1998; Dineen et al. 1998; Tekinşen and Özdemir 2006). In this chapter, we discuss the quality and safety aspects, regulatory aspects, microbiological criteria, and different approaches for controlling the entry and growth of microorganisms in fermented milk products through quality assurance and science-based microbiological risk assessment (MRA).

23.2 Microbiological Quality and Safety Issues

23.2.1 Quality Issues

All the fermented milk products are prepared by heat processing of raw milk, which leads to the destruction of initial microflora. The antimicrobial products such as organic acid (lactic acid, acetic acid, etc.), hydrogen peroxide (H_2O_2), and bacteriocins (nisin) are produced during the preparation of the fermented milk products, which inhibit the growth of undesirable microorganisms, namely, spoilage and pathogenic microorganisms in the final products (Adams and Mitchell 2002), which results in longer shelf life of fermented milk as compared to non-fermented milks. There are several factors, which are related with the production of poor-quality fermented milk products, such as the use of poor-quality raw milk and unhygienic practices during preparation, handling, and storage. Loose packing during transportation is further known to deteriorate the keeping quality of fermented milk products. Mixed cultures used to produce fermented milk products under unhygienic conditions are also found to contain both desirable and undesirable bacteria.

The coliforms, anaerobic spores, yeasts, and molds may cause spoilages in the final fermented milk products (Sperber and Doyle 2009). Fox et al. (2000) described the occurrence of yeasts in different varieties of cheeses. The dominant yeast is *Debaryomyces hansenii* followed by *Kluyveromyces lactis*, *Yarrowia lipolytica*, and *Trichosporon beigelii*, which occur in virtually all varieties of cheeses including Limburger, Tilsit, Roquefort, Weinkase, Camembert, Romadour, Cabrales, and St. Nectaire. The most common spoilage of cheese is caused by *Candida* ssp., *Kluyveromyces marxianus*, *Geotrichum candidum*, *D. hansenii*, and *Pichia* ssp. (Johnson 2001). Proteolysis takes place during the ripening of cheeses and results in the production of amino acids associated with the decrease in acidity, which helps in the development of *Clostridia*, especially *Clostridium tyrobutyricum*, which in turn leads to the release of butyric acid and gas (Klijn et al. 1995). Cheeses most often affected are Swiss, Emmental, Gouda, and Edam, which have high pH and moisture content and low salt content (Fox et al. 2000). As spores are accumulated in cheese curd, as few as 1 spore/ml of milk can cause gassiness defects in cheese (Sperber and Doyle 2009). The defect in large wheels of rindless Swiss cheese may require more than 25 CFU/ml of spores (Dasgupta and Hull 1989).

The primary cause of late blowing cheese is attributed to the growth and survival of an anaerobe Gram-positive spore-forming bacterium such as *Cl. tyrobutyricum*, and other *Clostridia* species such as *Clostridium sporogenes*, *Clostridium beijerinckii* (Garde et al. 2011), and *Clostridium butyricum* also significantly cause this defect in cheese (Klijn et al. 1995; Cocolin et al. 2004; Le Bourhis et al. 2005). Spores of *Clostridium* species are known to cause gassiness in processed cheese unless their numbers are low, the pH is not higher than 5.8, the salt concentration is at least 6% of the serum, and the cheese is held at 20°C or lower (Kosikowski and Mistry 1997). *Eubacterium* sp., facultative anaerobic bacteria, is known to cause gassiness in Cheddar cheese, and is able to grow at pH 5.0–5.5 in the presence of 9.5% salt (Myhr et al. 1982). In Swiss cheese, an unusual white-spot defect is caused by a thermoduric *Enterococcus faecalis* subsp. *liquefaciens*. *E. faecalis* contamination in cheese always results in undesirable eye formation and loss of flavor as this bacterium is inhibitory to *Propionibacteria* and *Lactobacillus fermentum* (Nath and Kostak 1985). Yogurt and fermented milks are having high acidity, and they are selective for their growth and cause yeasty, off-flavors and gassy appearance (Fleet 1990; Rohm et al. 1992), fruity flavor, and egg odor. The second most important spoilage-causing group of microorganisms in fermented milk products include mold-like *Penicillium* ssp., *Cladosporium* ssp. (Hocking and Faedo 1992), and *Byssochlamysnivea nivea* (Pitt and Hocking 1999) and cause off-flavor defects.

23.2.2 Safety Issues

Fermented milk products are safe for consumption. During the fermentation process, the starter culture aids in the development of antimicrobial peptides, and some of the organic acids also produced, which decreases the pH of the product, creating an unfavorable environment for the growth of the pathogens. This is the major reason why the fermented products are rarely associated with foodborne diseases (Rubin et al. 1982; Giraffa et al. 1994; Varnam and Sutherland 1994). However, some pathogens have an ability to survive and grow at low pH of fermented milk products (Caro and Garcia-Armesto 2007). *L. monocytogenes* is a lactic-acid producer that can survive numerous stresses including higher acidity (Cole et al. 1990). Cormac et al. (1996) established this finding by demonstrating the enhanced survival of acid-adapted bacteria in food having lactic acid such as yogurt. *Salmonella typhimurium* and *S. aureus* are also well-known to acclimatize to acidic environs, and this encourages their persistence in dairy products (Leyer and Johnson 1993; Leyer et al. 1995; Pazakova et al. 1997). *Staphylococcus* ssp., *E. coli* O157:H7, *L. monocytogenes*, and *Salmonella* ssp. have been shown to survive under stressful environments, including low pH (Gahan et al. 1996; Lou and Yousef 1997; Beales 2004; Benkerroum 2013).

Another pathogen of concern is coagulase-positive *S. aureus*, which is highly resistant to salt (concentration of 10%) and relatively to a low water activity (Aw—0.86); this produces preformed toxins in the food, which cause food intoxication. The growth and production of *Cl. botulinum* in some cheese varieties may lead to the outbreak of botulism. During 1974, in France and Switzerland, the outbreak of botulism was attributed to the consumption of Brie cheese containing botulinum toxins (Johnson et al. 1990). Singh and Prakash (2008) reported the presence of higher number of *L. monocytogenes* in curd compared to cottage cheese. Another survey conducted by Yabaya and Idris (2012) on bacteriological quality of commercial yogurt at different markets within Kaduna Metropolis revealed predominant growth of *S. aureus* and *E. coli*.

Because of higher moisture content in soft variety cheese, it provides a favorable environment for the growth of pathogens. Generally, soft cheeses are made from pasteurized milk, unless the cheese is handled carefully; there is not much danger of heavy pathogenic contamination. However, there have been more than a few reports of cheese-borne food infections, and food poisoning cases are reported. Many such outbreaks have been associated with *L. monocytogenes*. In a few cases, there have been fatalities associated with cheese-borne listeriosis. Mexican-style white soft cheese has been implicated in some cases of outbreaks. In France, listeriosis was associated with Brie cheese (Simon et al. 2002).

E. coli serotype O157:H7 has been found to exhibit unusual heat and acid tolerance and has high life-threatening potential to cause serious cheese-related food infection and fatalities. *E. coli* being implicated in several cases of food poisoning results from the consumption of French Brie and Camembert in the United States and Scandinavia (D'Aoust 1989). Raw milk soft cheeses may potentially carry different *Salmonella* strains. *Salmonella*, implicated in foodborne illness, result from the consumption of Vacherin in Switzerland and French Brie and Camembert in Scandinavia (D'Aoust 1989). *E. coli* O157:H7, *L. monocytogenes*, and *Y. enterocolitica* are three of the most important pathogens that can lead to foodborne diseases through consumption of contaminated fermented milk products (including yogurt, kefir) (Morgan et al. 1993; Mead et al. 1999; Gulmez and Guven 2003). Pathogens that are found in fermented foods come from the respective raw materials or from handlers. Some fermented foods may be contaminated by molds, which produce mycotoxins. Olasupo et al. (2002) reported that among microorganisms of public health concern, *S. aureus* and *Klebsiella* ssp. were isolated from wara, while *E. coli*, *Salmonella* ssp., and *Klebsiella* ssp. were isolated from nono. Ogi and kunu-zaki contained *B. subtilis*, *E. coli*, *S. aureus*, *Klebsiella* ssp., and *E. faecalis*. Some of the major foodborne outbreaks occurred in fermented milk products are enlisted in Table 23.1.

23.3 Microbiological Risk Assessment

23.3.1 Risk Assessment

Risk assessment is a component of the risk analysis framework along with risk management and risk communication as given by the Codex Alimentarius Commission (CAC). The CAC has an extensive array of applications in food safety-like sanitary measures that achieve specific food-safety goals, develop broad food-safety policies, and elaborate standards for food (Rocourt et al. 2003).

Risk assessment of microbiological hazards in foods has been recognized as a significant area of work by the CAC. Risk assessment for

Table 23.1 Outbreaks of Illness Associated with Fermented Milk Products

YEAR	COUNTRY	CASES (DEATH)	PRODUCT	CAUSATIVE AGENT	REFERENCE
1985	United States	142 (48)	Soft cheeses	L. monocytogenes	Linnan et al. (1988)
1989–90	Denmark	26 (6)	Hard and blue cheese	L. monocytogenes	Jensen et al. (1994)
1991	United Kingdom	16	Yogurt	E. coli O157	Morgan et al. (1993)
2001	France	45	Brie cheese	Salmonella serotype infantis	Simon et al. (2002)
2001	Italy, Germany	4.4 (45)	Hard cheese	L. monocytogenes	Rudolf and Scherer (2001)
2004	Portugal	1.6 (4.0)	Cheese	L. monocytogenes	Mena et al. (2004)
2006	Germany	6 (1)	Hard cheese	L. monocytogenes	EFSA (2007)
2007	Turkey	4.8 (250)	Semi-hard cheese	L. monocytogenes	Colak et al. (2006)
2007	Czech Republic	78 (13) in 3 outbreaks	Soft cheese	L. monocytogenes	EFSA (2007)
2007	Norway	21 (5)	Raw milk soft cheese	L. monocytogenes	EFSA (2009)
2008	Greece	40 (10)	Soft cheese	L. monocytogenes	Filiousis et al. (2009)

microbiological hazards in foods is defined by the CAC as a scientific-based process consisting of four components (CAC 1999): hazard identification, exposure assessment, hazard characterization, and risk characterization (Figure 23.1).

23.3.1.1 Hazard Identification "The identification of biological, chemical and physical agents capable of causing adverse health effects and which may be present in a particular food or group of foods" (CAC 2004). In MRA, the microorganisms are usually identified as a hazard based on the epidemiological data on illnesses associated with the infectious diseases caused by the microorganisms. Classification and identification of pathogen, which have the ability to transmit through fermented milk products, is an important step in the overall risk assessment.

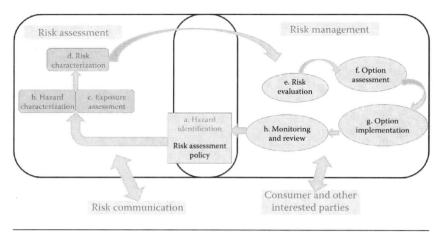

Figure 23.1 Microbiological risk assessment by Codex Alimentarius Commission.

Different pathogens gain entry into a product as a result of postprocessing contamination; during processing and storage, raw materials are often the principal source of hazards. The range of fermented milk foods are huge (Wood 1998), made using raw milk, and there is considerable deviation in the unit operations involved in the processes, storage, and consumption (Nout 2001). As a result, most of the microbiological hazards potentially associated with fermented milk foods are correspondingly large. It is possible to edit the extensive lists obtained by examining epidemiological data from outbreaks, where fermented milk foods have been involved. For example, a list of pathogens associated with fermented milks like cheese is substantial and includes most common and some less common foodborne bacterial pathogens such as *L. monocytogenes, Bacillus cereus, Cl. botulinum, Cl. perfringens, Mycobacterium avium* complex, *S. aureus, Streptococcus pyogenes, Brucella melitensis, Campylobacter jejuni/coli, E. coli* (O27:H20; O124:B17; O157), *Salmonella,* and *Shigella sonnei* (Nichols et al. 1996).

23.3.1.2 Hazard Characterization Hazard characterization is defined in the CAC *Procedural Manual, Fourteenth Edition* (CAC 2004) as "The qualitative and/or quantitative evaluation of the nature of the adverse health effects associated with biological, chemical and physical agents, which may be present in foods." In MRA, this step provides a qualitative or quantitative description of the severity and duration of adverse effects that may result from the uptake of microorganism or its toxin in food (CAC 1999). A microbiological dose–response model describes the probability of a specified response from exposure to a specific pathogen in a specified population as a function of the ingested dose. The biological basis for microbiological dose–response models derives from the major steps in the disease process: exposure, infection, illness, and consequences. The issue of response derives from the interactions between the pathogen, the host, and the food matrix.

23.3.1.3 Exposure Assessment   It is defined in the Codex (CAC 2004) as "The qualitative and/or quantitative evaluation of the likely intake of biological, chemical and physical agents via food as well as exposures from other sources if relevant." Exposure assessment in MRA includes an assessment of the extent of actual or anticipated human exposure to microbiological pathogens or microbiological toxins, that is, an estimate of the likelihood of their occurrence in foods at the time of consumption and their level, within various levels of uncertainty (CAC 1999). Predictive microbiology models can be useful to assess the growth, survival, or death of microorganisms as a function of the food and environmental conditions encountered from raw materials to the food consumed and are particularly important, when making quantitative estimates.

Fermented foods are prime examples of the hurdle or multiple barrier approach to food preservation, where the overall antimicrobial effect seen is the aggregate effect of a number of different factors. For example, when an acid stress preceded a low water activity stress, there was a greater lethal effect on *E. coli* when the stresses were applied in the reverse order (Shadbolt et al. 2001).

Considerable efforts have been devoted to the isolation and study of LAB antimicrobials such as bacteriocins, and this has tended to obscure the fact that the principal inhibitory contribution of LAB during lactic fermentations is the production of organic acid at levels up to and exceeding 100 mm and the consequent decrease in pH (Yusof et al. 1993). The antimicrobial activity of fermented milk products due to fermentation greatly focuses on the inhibition of bacterial pathogens (Dethmers et al. 1975; Kantor and Potter 1975; Nout et al. 1989; Wood and Adams 1992). With toxigenic organisms, such as *Cl. botulinum* or *S. aureus*, preventing growth can effectively ensure the safety, assuming that initial numbers are well below those necessary to produce a harmful level of toxin (Warnekulasuriya et al. 1998; Perez et al. 2001). Any inactivation of bacterial hazards may not be sufficient to eliminate risk entirely.

Predictive modeling is normally used in surveillance data analysis, which describes how the conditions predominant during production, processing, and storage will affect the growth and survival of pathogenic microorganisms in fermented milk products (Van Gerwen and Zwietering 1998). When *E. coli* O157:H7 is present in the milk, its numbers have been shown to increase during the early stages of production of several cheese types, including hard cheeses, Camembert, cottage cheese, and smear-ripened cheese (Arocha et al. 1992; Reitsma and Henning 1996; Ramsaran et al. 1998; Maher et al. 2001; Saad et al. 2001). Although the numbers of *E. coli* O157:H7 in cheeses generally fall later in the production process (Reitsma and Henning 1996; Maher et al. 2001), after the final heating stage in cottage cheese production, *E. coli* initially present in the fresh cheese at 10^7 CFU g^{-1} was undetectable (Arocha et al. 1992). Most cheeses do not experience the high temperature scalding used in cottage cheese production, but pasteurization of the milk used in cheese production is an extremely valuable critical control point (CCP). Surveillance data from England and Wales, which showed that, from 1437 soft cheeses sampled, significantly more of those produced from pasteurized milk conformed to the Public Health Laboratory Service guidelines for ready-to-eat foods (94%) than did raw (unpasteurized) milk soft cheeses (71%) (Nichols et al. 1996).

23.3.1.4 Risk Characterization It is defined as "The qualitative and/or quantitative estimation, including attendant uncertainties, of the probability of occurrence and severity of known or potential adverse health effects in a given population based on hazard identification, hazard characterization and exposure assessment" (CAC 2004). Due to their vast diversity, quantitative risk assessment methodology cannot be applied to fermented foods. A more firmly defined announcement of purpose does simplify this, and a quantitative risk assessment has been described in covering the risk of human listeriosis from consumption of raw milk soft cheeses (Bemrah et al. 1998).

Overall, the epidemiological data direct that fermented foods have a good safety record, particularly in view of the large quantities consumed worldwide. Knorr (1998) has suggested that fermented foods of various types can comprise 30% of our food supply. Further, development of risk assessments in this capacity offers a powerful and valuable tool in successfully managing the food safety hazards, allowing these stimulating and attractive foods to be consumed with confidence.

23.3.2 Risk Management

The process, distinct from risk assessment, of weighing policy alternatives in consultation with all interested parties, considered risk assessment and other factors relevant for the health protection of consumers and for the promotion of fair trade practices and, if needed, selected appropriate prevention and control options (FAO 2005).

23.3.3 Risk Communication

The interactive exchange of information and opinions throughout the risk analysis process concerning risk, risk-related factors, and risk perceptions, among risk assessors, risk managers, consumers, industry, the academic community, and other interested parties include the explanation of risk assessment findings and the basis of risk management decisions (FAO 2005).

23.4 Quality Assurance of Fermented Milk Products

23.4.1 Hygiene Measures

Foodborne pathogens associated with fermented milk products are capable of forming biofilm on contact surfaces, and contamination of product often occurs during processing rather than by survivors from the raw milk.

In order to prevent the contamination, of fermented milk products with foodborne contaminants, it is essential to focus and direct the attention on hygiene in dairy plant production facilities. Sanitation measures including washing with detergents and disinfection of clean surfaces must be carried out properly. Preparations based on iodoform are recommended for use in the dairy industry because their residues do not inactivate the starter cultures. Quaternary ammonium compounds are not recommended for direct use on surfaces, which are in contact with food, because even the smallest residues of these compounds inactivate the starter cultures, but they are very efficient for washing of floors, drains, walls, and cold stores.

Steam can be used as a substitute for chemical disinfection. Use of steam should be limited to disinfection of dairy equipment that is tough for washing and cleaning and closed systems due to forming of aerosol and condensation on equipment. Combinations of detergents and disinfectants are not recommended because disinfectants require precise time of contact with the equipment or surface in order to be efficient. Accessories/utensils used for washing, after use, must be washed and disinfected using quaternary ammonium compounds (600–1000 ppm). Also, as much as possible, the equipment, floors, accessories/utensils, and all other work surfaces need to be kept dry.

By introduction of HACCP as a new way of control in the process of production and processing, the risk of contamination of products with these pathogenic microorganisms is reduced. All food business operators have an authorized responsibility to produce safe food (Regulation 178 2002). Control of *pathogens* is required at all stages in the production chain, and a joint approach is necessary to prevent the proliferation of the pathogen in the final product. The challenges faced by industries for controlling pathogens are considerable given its ubiquitous nature, high resistance to heat, salt and acidic pH, and arguably, most importantly, its ability to grow and survive at or below commercial refrigeration temperatures.

The following general recommendations apply to all sectors of the food chain for its implementation:

- Good hygiene practices (GHP) and good manufacturing practices (GMP)
- Food safety management system (FSMS) based on the principles of HACCP
- Microbiological criteria, as appropriate, when validating and verifying the correct functioning of the HACCP-based principles and hygiene control measures

23.4.2 Good Manufacturing Practices

GMPs as defined by the Food and Drug Administration, in 21 CFR part 110, are the minimum sanitary and processing requirements for food companies. GMPs are fairly broad and general and can be used to help guide the development of standard operating procedures that are very specific (Mucklow 1998).

During the production of fermented milk products, raw material, culture propagation, addition of culture, processing, post-processing contamination through air, equipment, human handler, storage, packaging materials, and distribution system are mainly affected on the quality and safety. Therefore, the following guidelines are for developing GMP for fermented milk product operations. These GMPs are not designed to control specific hazards but are intended to provide guidelines to help processors produce safe and wholesome products (Harris and Blackwell 1999).

Like any other products, for fermented milk products, strict compliance to the hygiene recommendations of GMPs must be followed in order to achieve full control of the entire process and detection of the risk areas where faults may occur and cause loss. In case of fermented milks, faults are due to incorrect fermentation, gel/curd treatment, and recontamination during handling.

In this respect, the following precautions need to be strictly followed to assure (Yadav et al. 1993; Robinson 2002; Marth and Steele 2005) the product quality and safety.

1. Correct cleaning and sanitization of room, plant, and equipment
2. Use superior quality (microbiological) raw materials
3. Use of pure and active selected cultures under aseptic conditions
4. Adequate heat treatment of milk and ingredients
5. Fermentation with controlled temperature and time
6. Gel breaking under constant conditions
7. Aseptic filling
8. Hygienic packaging and storage

23.4.3 GMP at Processing Condition

GMPs have been divided, as they usually are, into several categories: personnel, buildings and facilities, equipment, and production and process controls (Yadav et al. 1993; Marth and Steele 2005). GMPs are necessary at the following points during processing (STTA Report 2005):

1. Pasteurizer's gaskets are worn out, and milk spills on the floor from the free spaces between adjacent plates.
2. pH meter is not calibrated, as often as it should, there are no calibrating solutions in the laboratory, and the operating and calibration manual is missing.
3. Plant does not have a moisture balance, to determine the moisture content of the cheeses.
4. There is an extensive use of wooden utensils, which enter into contact with the milk, after pasteurization.
5. Lamps in the ceiling have no protective covers.
6. Hoses and various other materials are usually lying on the floor.
7. Production personnel do not wear appropriate garments and there is no enforcement or supervision of basic hygienic practices.
8. Production personnel lack basic scientific and technological training regarding the meaning of what they do, from measuring titratable acidity to predicting cheese yield based on milk composition.

23.4.3.1 Plant Hygiene Dairy plants hygiene can be achieved by adopting adequate program in place to monitor and control all elements and maintain appropriate record.

1. *Location of plant*: The plant must be located in an area that should not be surrounded by pollution sources such as dust, allergens, and contaminants.
2. *Building and facilities*: Building should be well constructed in such a way that it should favor the accurate performance of all operations.
3. *Floors, walls, and ceiling materials*: These should be durable, cleanable, and sloped for the liquids to drain. Preferably walls should be light colored and well joined.
4. *Washrooms, lunch halls, and change rooms*: These should not be directly linked to food production and processing units.
5. *Drainage and sewage system*: There should be a separate facility for drainage of inedible waste, and its storage in plant before disposal

should be in a hygienic way. There should be no connection between human waste effluent and plant effluent.
6. *Sanitation program*: Plants must have an adequate sanitation program in place and maintain appropriate records.
7. *Chemicals and cleaning*: Chemicals must be used in accordance with the manufacturer's recommendations (CAC 2004).
8. *Pest control program*: Dairy plants should have an adequate, effective, safe, and written pest control program in place. Written pest control program should include the name of a contact person at the establishment of pest control program, the list of chemicals and methods used, the frequency of treatment and inspection, pest survey, and control reports.

23.4.3.2 Personnel Hygiene Personnel hygiene is very important during production and processing of fermented dairy products (Ramírez Vela and Martín Fernández 2003). For personnel hygiene in food production area, the following points should be taken care of:

1. *Cleanliness*: Including hygienic practices, proper outer garments, proper footwear, and personal cleanliness, washing of hands thoroughly by all persons at the entrance of food production unit of the plant, after handling contaminated materials, and after using toilet facilities.
2. Persons should not be diseased or carrier of any pathogen that can transmit through food to the consumers. Tobacco, gums, and other eatables should not be permitted in the production area. Jewelry must be removed in food-handling area.
3. *Education and training*: Everyone should have high level of proficiency essential for production of clean and safe food (CAC 2004).
4. *Supervision*: Responsibility for the production of food should be given to skilled managerial staffs.

23.4.3.3 Equipment Hygiene Dairy plants must use hygienically designed equipment for the manufacture of different fermented milk products. The installation and maintenance of the equipment should be done in such a way that it should not proliferate microorganisms, which lead to contamination of product during its processing in the production area (CAC 2004).

- *Control measures during designing of equipment*: Contamination of fermented and other dairy products can be prevented by taking care of the following points:
  - Construction of equipment using corrosion resistant material
  - Food contact surfaces must be nonabsorbent, nontoxic, smooth, and free from pitting
  - Chemicals, lubricants, coatings, and paints used on equipment surfaces should be non-reactive type
- *Equipment and utensils* should be designed and of such materials and workmanship as to be adequately cleanable and properly maintained.
- Instruments for measuring and recording acidity, temperature, pH, and so on should be perfect (calibration), specific, and well preserved.
- *Control measures during installation of equipment*: Equipment must be installed in proper way that makes them easily accessible for cleaning, sanitizing, maintenance, and inspection.
- *Calibration of equipment*: Appropriate protocols must be established for calibration of equipment.
- *Preventive maintenance program*: This program specifies necessary servicing of equipment and frequency, including replacement of parts, responsible person, methods of monitoring, verification activities, and records to be kept.

23.4.3.4 Production and Process Controls

1. *Processes and controls*: All operations (receiving, inspecting, transporting, etc.) should be conducted according to adequate sanitation principles. This includes inspection and storage of raw materials.
2. *Equipment, utensils, and food containers* should be maintained through appropriate cleaning and sanitizing procedures.
3. *Storage and transportation of food* should be done under conditions that will protect the food against physical, chemical, and microbial contamination.

4. *Buildings and facilities*:
    a. *Plant and grounds*, including plant construction and design, equipment storage, waste disposal, and pest control, should enhance food safety.
    b. *Sanitary facilities and controls*, including cleaning and sanitizing chemicals, water supply, toilet facilities, hand washing facilities, plumbing, floor drainage, sanitary traps, and sewer disposal, should be such that opportunities for producing unsafe food are minimized.

23.4.3.5 Air Hygiene Plants can be equipped with air filtration facility as it can reduce the risk of airborne contamination of food products. Installing ozone generators is another option and, if these generators are allowed to operate for 5–6 h overnight, most bacterial and fungal spores will be killed.

23.4.3.6 Water Hygiene Water should be evaluated for microbiological, chemical, and physical quality. Water should be purified by chlorination, ozonization, or UV treatment. Potable and non-potable water supply should not come in contact with each other.

23.4.4 Monitoring the Process Plant

The acidity of fermented product means that spoilage is often associated with yeasts and molds, and the latter in particular often have their origin in the microbial population of the air. The control of the atmosphere within the factory environment will depend on the level of air cleanliness that is essential for completion of a particular operation (Hoolasi 2005). It is important, however, that plants designed to induce air flow through a filling room or production area can also act as a source of contamination (Tamime and Robinson 1985). Packaging materials stored adjacent to the filling line can also cause problems, as can the unnecessary movement of personnel, and these aspects of plant operation deserve constant attention (Tamime and Robinson 1999).

Different methods have been devised to monitor the hygiene of dairy equipment surfaces, thus contributing to maintaining production

Table 23.2 Suggested Standards for Dairy Equipment Surfaces Prior to Pasteurization/Heat Treatment

COLONY FORMING UNITS/100 CM²	CONCLUSION
500 (coliforms <10)	Satisfactory
500–2500	Dubious
>2500 (coliforms >100)	Unsatisfactory

Source: Mostert, J.F. and Peter, J.J., Quality control in the dairy industry. *Dairy Microbiology Handbook: The Microbiology of Milk and Milk Products* (3rd Edition, pp. 655–736), John Wiley, London, 2002.

of high-quality products and, at the same time, ensuring compliance with legal requirements (Mostert and Peter 2002; Hoolasi 2005).

Enumeration of aerobic plate count, coliforms, yeasts, and mold counts is the most frequent parameter carried out to assess the microbiological status of surfaces (Tamime and Robinson 1999). The most commonly used methods for surface assessment are outlined by Mostert and Peter (2002), which include surface agar contact method (swab, rinse, adhesive tape, and vacuum methods), agar contact method (replicate organisms direct agar contact, agar slice, and direct surface agar plating methods), and the ATP bioluminescence method. Some suggested standards for prior to pasteurization/heat treatment, according to Mostert and Peter (2002), are shown in Table 23.2.

23.5 HACCP-Based Food Safety Management System

23.5.1 HACCP

HACCP is a methodology and a management system used to identify, prevent, and control food safety hazards. In the development of an HACCP plan, five preliminary tasks need to be accomplished before the application of seven principles of HACCP, which are as follows (Joint FAO/WHO CAC, Joint FAO/WHO Food Standards Programme, and World Health Organization 2003; Mortimore and Wallace 2013):

1. *Step 1*: Assemble HACCP team—assembling a multidisciplinary team.
2. *Step 2*: Describe the final product—describe the full characteristics of the product, including relevant safety data such as

composition; physical/chemical structure including Aw, pH, and so on; packaging; stability and storage conditions; and distribution method.

3. *Step 3*: Identify the products intended use—likely uses of the product by the consumer or end user.
4. *Step 4*: Construct the process flow diagram—it should be constructed by the HACCP team covering all the steps in the operation.
5. *Step 5*: Verify the flow diagram—HACCP team should confirm the processing task against the flow diagram during all steps and hours of operation and adjust the flow diagram where appropriate.
6. *Implement principle 1*: Conduct a hazard analysis—list all biological, physical, and chemical hazard related to incoming material, processing, ingredients product flow, and so on.
7. *Apply principle 2*: Identify CCPs—the Codex guidelines define a CCP (Joint FAO/WHO CAC, Joint FAO/WHO Food Standards Programme, and World Health Organization 2003) as "a step at which control can be applied and is essential to prevent or eliminate a food safety hazard or reduce it to an acceptable level." The determination of CCP in an HACCP system can be facilitated by the application of a decision tree. Hazards that are not fully controlled by GMPs should be analyzed to determine whether they are CCPs or not.
8. *Employ principle 3*: Establish critical limits—each CCP/critical limit is established and specified. Critical limits are defined as criteria that separate acceptability from unacceptability (Mortimore and Wallace 2013).
9. *Implement principle 4*: Establish CCP monitoring procedures—monitoring is the scheduled measurement or observation of a CCP relative to its critical limits. The monitoring procedure must be able to detect loss of control at the CCP. Therefore, it is important to specify fully how, when, and by whom monitoring is to be performed.
10. *Organize principle 5*: Establish corrective action—intended to prevent deviations from occurring, perfections are rarely, if ever, achievable. Because of variations in CCPs for different

food operations and the diversity of possible deviations, specific corrective action plans must be developed for each CCP.
11. *Institute principle 6*: Establish recordkeeping procedures—efficient and accurate record keeping is essential to the application of an HACCP system. HACCP procedures should be documented. Documentation and record keeping should be appropriate to the nature and size of the operation.
12. *Principle 7*: Establish verification procedures—establishing method for verification. Verification and auditing methods, procedures, and tests including random sampling and analysis can be used to determine if the HACCP system is working correctly.

These methodologies are used to develop HACCP plan. A HACCP plan is a document that describes how an organization plans to manage and control its food safety hazards. A HACCP plan for fermented milk products is shown in Figure 23.2. The HACCP procedure is targeted at food safety management (pathogens and their toxins), but as an approach in the context of broader quality management, it can be effectively applied to microbiological spoilage, foreign-body contaminations, or pesticide contamination. Hazard description and CCP for fermented milk products during different stages of their manufacturing is summarized in Table 23.3.

The only way of ensuring that every package of product from a given production line is safe, from a chemical or microbiological standpoint, is to test every package. Clearly, such a suggestion is totally impractical, so that, instead, a representative group of packages is withdrawn against a sampling plant appropriate for the product and the history of the plant. However, while this approach is essential to confirm that present standards of hygiene are being met and that potential contaminants are at a low level or absent, the procedure can never prevent some spoiled packages from reaching the consumer. Consequently, the emphasis within quality assurance has turned to the avoidance of problems, a concept that forms the basis of HACCP. In particular, the system identifies seven aspects of production that merit constant attention, and these aspects are enshrined in seven principles (Tamime and Robinson 1999). In any HACCP system, it is vital that the different stages, within each principle, be considered in order and that the

QUALITY ASSURANCE

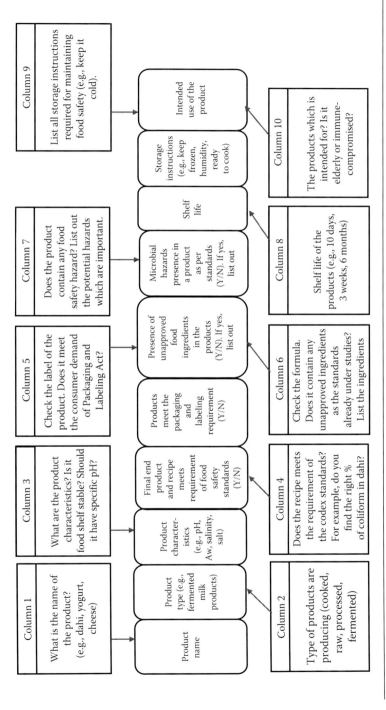

Figure 23.2 Generic HACCP plan for pathogen control in fermented milk products.

624 FERMENTED MILK AND DAIRY PRODUCTS

Table 23.3 Hazard Description and Critical Control Point for Fermented Milk Products

Incoming Material/Process Step		List All Biological, Chemical, and Physical Hazards Related to Ingredients, Incoming Materials, Processing, Product Flow, and so on	Q1. Could a Control Measure(s) Be Used at Any Process Step?	Q2. Is It Likely That Contamination with the Identified Hazard Could Occur in Excess of the Acceptable Level or Could Increase to an Unacceptable Level?	Q3. Is This Control Measure Specifically Designed to Eliminate or Reduce the Likely Occurrence of the Identified Hazard to an Acceptable Level?	Q4. Will a Subsequent Step Eliminate the Identified Hazard or Reduce Its Likely Occurrence to an Acceptable Level?	CCP Number GMP: Proceed to Next Identified Hazard
Receiving of raw milk	B	Microorganisms	No	Yes	No	Yes	CCP1
	C	Antibiotics, aflatoxin M_1	Yes	—	—	—	GMP
	P	Dirt, flies, hair, and so on	Yes	—	—	—	GMP
Filtration/clarification	C	Detergents residues	yes	—	—	—	GMP
Standardization	B	Microbial	Yes				GMP
Preheating at 60°C/few minutes	B	Microbial	Yes				GMP
Homogenization at 76 kg/cm²	C	Detergent residues	Yes				GMP
Pasteurization at 80°C–90°C/15–20 min	B	Microbial	No	Yes	No	Yes	CCP2

(*Continued*)

Table 23.3 (Continued) Hazard Description and Critical Control Point for Fermented Milk Products

Cooling at 20°C–25°C	B	Microbial	Yes	—	GMP
Starter culture inoculation	B	Microbial	No	Yes	CCP3
Packaging	B	Microbial	Yes	—	GMP
	C	Detergents and preservatives	Yes	—	GMP
	P	Pieces of packing material, insects, flies, and so on	Yes	—	GMP
Incubation at 20°C–25°C	B	Microbial	Yes	—	GMP
Cooling to 5°C	B	Microbial	Yes	—	GMP
Curd is ready and stored under refrigeration (5°C)	B	Microbial	Yes	—	CP

B, biological; C, chemical; P, physical.

required information and conclusion be completed for each stage before moving on to the next. HACCP is designed as a structured approach, and the proper sequencing of activities is crucial to obtain an effective output.

23.5.2 Food Safety Management System

ISO 22000:2005 specifies requirements for an FSMS, where an organization in the food chain needs to demonstrate its ability to control food safety hazards in order to ensure that the food is safe at the time of human consumption. It does not matter how complex the organization is or what size it is, ISO 22000 can help ensure the safety of its food products. The food chain consists of the entire sequence of stages and operations involved in the creation and consumption of food products. This includes every step from initial production to final consumption. More precisely, it includes the production, processing, distribution, storage, and handling of all food and food ingredients. It also includes organizations that produce materials that will eventually come into contact with food or food ingredients (ISO 22000:2005).

23.5.3 Verification Procedure

1. Every establishment shall validate the HACCP plan's adequacy in controlling food safety hazards identified during the hazard analysis and shall verify that the plan is being effectively implemented.
    a. Initial validation: Upon completion of the hazard analysis and development of the HACCP plan, the establishment shall conduct activities designed to determine that the HACCP plan is functioning as intended. During this HACCP plan validation period, the establishment shall repeatedly test the adequacy CCPs, critical limits, monitoring and recordkeeping procedures, and corrective actions set forth in the HACCP plan. Validation also encompasses reviews of the records themselves, routinely generated by the HACCP system, in the context of other validation activities.

b. Ongoing verification activities: Ongoing verification activities include but are not limited to
   i. The calibration of process-monitoring instruments.
   ii. Direct observations of monitoring activities and corrective actions.
   iii. The review of records generated and maintained in accordance with the verification procedure.
c. Reassessment of the HACCP plan: Every establishment shall reassess the adequacy of the HACCP plan at least annually and whenever any changes occur that could affect the hazard analysis or alter the HACCP plan. Such changes may include, but are not limited to, changes in raw materials or source of raw materials; product formulation; processing methods or systems; production volume; personnel; packaging; finished product distribution systems; or the intended use or consumers of the finished product. The reassessment shall be performed by an individual trained in accordance with HACCP system of this part. The HACCP plan shall be modified immediately, whenever a reassessment reveals that the plan no longer meets the requirements.
2. Reassessment of the hazard analysis: Any establishment that does not have an HACCP plan because a hazard analysis has revealed no food safety hazards that are reasonably likely to occur shall reassess the adequacy of the hazard analysis whenever a change occurs that could reasonably affect whether a food safety hazard exists. Such changes may include, but are not limited to, changes in raw materials or source of raw materials, product formulation, processing methods or systems, production volume, packaging, finished product distribution systems, or the intended use or consumers of the finished product (USDA 1999).

Verification applies methods, procedures, product tests, and other evaluations, other than monitoring, to determine compliance with the HACCP plan; that is, it demonstrates that the HACCP plan and its application is consistently controlling the process, so that the product meets the food safety requirements (Robinson 2002). The HACCP team should specify methods and frequency of verification procedures, which might include (Hoolasi 2005; Pacheco and Galindo 2010)

1. The examination of raw milk, cream, skimmed milk powder, and starter culture for microbial pathogens such as *Salmonella*, *L. monocytogenes*, *E. coli*, *Bacillus cereus*, and *S. aureus* and the examination of dahi/curd for acidity/pH, total viable count, coliform, *E. coli*, yeast and molds, *Salmonella*, coagulase-positive *S. aureus*, and *L. monocytogenes*.
2. Review of complaints from consumers or regulatory agencies and outcome of these investigations into these complaints, if these were substantiated, indicating that the HACCP plan did not completely control the process.
3. Auditing all monitoring and corrective actions records to establish whether the HACCP plan is being fully implemented and demonstrates control.
4. A review of validation records and, if appropriate, the application of more searching tests at selected CCPs to confirm the efficacy of control measures.

23.6 Regulatory Aspects of Fermented Milk Products

With a growing world economy, there will be liberalization of food trade, developments in food science and technology, growing consumer demand, improvements in transport and communication, and increased international trade in fermented food. Food protection measures are also essential in view of agricultural production being the central point of the economies of most underdeveloping countries (FAO 2003). CAC (2004) is an intergovernmental body that harmonizes food standards at the global level. It has proved to be most successful in achieving the harmonization of international food quality and safety standards. It has framed international standards for a wide variety of food products and specific requirements covering hygiene, food contaminants, labeling pesticide residues, food additives, veterinary drug residues, and so on.

The conclusion of the Uruguay Round of Multilateral Trade Negotiations in Marrakesh led to the establishment of the WTO on January 1, 1995, and to the coming into force of the agreement on the application of sanitary and phytosanitary (SPS) measures and the agreement on technical barriers to trade. Both these agreements are relevant in understanding the requirements for food protection measures at the national level and the rules under which food is traded internationally (Silverglade 2000). The SPS agreement confirms the right of WTO member countries to apply measures to protect human, animal, and plant life and health. The agreement covers all relevant laws, decrees, and regulations; testing, inspection, certification, and approval procedures; and packaging and labeling requirements directly related to safety of milk products including fermented milk products. The agreement encourages use of international standards, guidelines, or recommendations where they exist and identifies those from Codex relating to food additives, veterinary drugs and pesticide residues, contaminants, methods of analysis and sampling, and codes and guidelines of hygienic practices to be consistent with provisions of SPS (FAO 2003). Thus, the Codex standards serve as a benchmark for comparison of national SPS measures (Gruszczynski 2008). The regulatory issues related with probiotic fermented products cover manufacturing as well as shelf life and labeling. The regulatory aspects concerned with this are safety and ensuring the proper labeling of the products and communicating the health attributes in a clear manner. Under ideal situation for dairy products, where the products are sold, the manufacturer should have a GMP facility and its procedures approved by the health protection branch (Canada), federal drug administration, or equivalent agency (Reid 2001).

23.7 Conclusions

Worldwide there is a great consumer demand for fermented milk products. Milk, being a highly nutritious food, serves as an excellent growth medium for a wide range of microorganisms. Processing methodology for fermented milk preparation is specific, and it is impossible to avoid contamination of milk with microorganisms; therefore, the microbial content of milk is a major feature in determining its quality. Hence, these products should be produced under a good hygienic condition and formally executed through quality/safety management systems such as ISO 9000, total quality management, and HACCP. All samples should formally undergo testing procedures as prescribed in the production protocol given by authorized bodies, and it should conform to the national and international standards. Finally, this can be an effective and rational means of assuring food safety at each step starting from farm to fork.

References

Abeer, A.A., Ali, A., and Dardir, H.A. 2009. Hygienic quality of local traditional fermented skimmed milk (Laban Rayb) sold in Egypt. *World Journal of Dairy and Food Science* 4(2): 205–209.

Adams, M. and Mitchell, R. 2002. Fermentation and pathogen control: A risk assessment approach. *International Journal of Food Microbiology* 79: 75–83.

Ahmed, A.H., Moustafa, M., and El-Bassiony, T.A. 1986. Growth and survival of *Y. enterocolitica* in yoghurt. *Journal of Food Protection* 4: 983–985.

Alm, L. 1983. Survival rate of *Salmonella* and *Shigella* in fermented milk products with and without added gastric juice: An in vitro study. *Progress in Food and Nutrition Science* 7: 19–28.

Arocha, M.M., Mc Vey, M., Loder, S.D., Rupnow, J.H., and Bullerman, L. 1992. Behavior of enterohaemorrhagic *E. coli* O157:H7 during manufacture of cottage cheese. *Journal of Food Protection* 55: 379–381.

Beales, N. 2004. Adaptation of microorganisms to cold temperatures, weak acid preservatives, low pH, and osmotic stress: A review. *Comprehensive Reviews in Food Science and Food Safety* 3: 1–20.

Bemrah, N., Sanaa, M., Cassin, M.H., Griffiths, M.W., and Cerf, O. 1998. Quantitative risk assessment of human listeriosis from consumption of soft cheese made from raw milk. *Preventive Veterinary Medicine* 37: 129–145.

Benkerroum, N. 2013. Traditional fermented foods of North African countries: Technology and food safety challenges with regard to microbiological risks. *Comprehensive Reviews in Food Science and Food Safety* 12: 54–89.

CAC. 1999. *Report of the 31st Session of the Codex Committee on Food Additives and Contaminants*. The Hauge, the Netherlands, March 17–21, 1997, ALINORM 99/12. Rome, Italy: Codex Alimentarius Commission.

CAC. 2004. *Code of Hygienic Practice for Milk and Milk Products*. CAC/RCP 57-2004 (2nd Edition). Procedural manual, 14th Edition. Rome, Italy: FAO.

Canganella, F., Ovidi, M., Paganini, S., Vettraino, A.M., Bevilacqua, L., and Trovatelli, L. D. 1998. Survival of undesirable microorganisms in fruit yoghurts during storage at different temperatures. *Food Microbiology* 15: 71–77.

Caro, I. and Garcia-Armesto, M.R. 2007. Occurrence of shiga toxin-producing *E. coli* in Spanish raw ewe's milk cheese. *International Journal of Food Microbiology* 116: 410–413.

Cocolin, L., Innocente, N., Biasutti, M., and Comi, G. 2004. The late blowing in cheese: A new molecular approach based on PCR and DGGE to study the microbial ecology of the alteration process. *International Journal of Food Microbiology* 90: 83–91.

Codex. 2010. *Codex Standard for Fermented Milks.* CODEX STAN 243-2003. http://www.codexalimentarius.org/download/standards/400/. CXS_243e.pdf.

Colak, H., Hampikyan, H., Bingol, E.B., and Ulusoy, B. 2006. Prevalence of *L. monocytogenes* and *Salmonella* spp. in Tulum cheese. *Food Control* 18(5): 576–579.

Cole, M., Jones, M., and Holyoak, C., 1990. The effect of pH, salt concentration and temperature on the survival and growth of *L. monocytogenes*. *Journal of Applied Bacteriology* 69: 63–72.

Cormac, G.M., Gahan, B., and Colin, H. 1996. Acid adaptation of *L. monocytogenes* can enhance survival in acidic foods and during milk fermentation. *Applied Environmental Microbiology* 62: 3128–3132.

D'Aoust, J.Y. 1989. Manufacture of dairy products from unpasteurized milk: A safety assessment. *Journal of Food Protection* 52: 906–912.

Dasgupta, A.R. and Hull, R.R. 1989. Late blowing of Swiss cheese. Incidence of *Cl. tyrobutyricum* in manufacturing milk. *Australian Journal of Dairy Technology* 44: 82–87.

Dethmers, A.E., Rock, H., Fazio, T., and Johnston, R.W. 1975. Effect of added sodium nitrate on sensory quality and nitrosamine formation in thuringer sausage. *Journal of Food Science* 40: 491–495.

Dineen, S.S., Takeuchi, K., Soudah, J.E., and Boor, K.J. 1998. Persistence of *E. coli* O157:H7 in dairy fermentation systems. *Journal of Food Protection* 61: 1602–1608.

EFSA. 2007. Scientific opinion of the panel on biological hazards on a request from the European Commission on request for updating the former SCVPH opinion on *Listeria monocytogenes* risk related to ready-to-eat foods and scientific advice on different levels of *Listeria monocytogenes* in ready-to-eat foods and the related risk for human illness. *The EFSA Journal* 599: 1–42.

EFSA. 2009. Report of task force on zoonoses data collection on proposed technical specifications for a survey on *Listeria monocytogenes* in selected categories of ready-to-eat food at retail in the EU. *The EFSA Journal* 300: 1–66.

FAO. 2003. *Assuring Food Safety and Quality: Guidelines for Strengthening National Food Control Systems.* Rome, Italy: Food and Agriculture Organization of the United Nations World Health Organization.

FAO. 2005. *Food Safety Risk Analysis. PART I: An Overview and Framework Manual.* (Provisional Edition). Rome, Italy: FAO.

Filiousis, G., Johansson, A., Frey, J., and Perreten, V. 2009. Prevalence, genetic diversity and antimicrobial susceptibility of *Listeria monocytogenes* isolated from open-air food markets in Greece. *Food Control* 20: 314–317.

Fleet, G.H. 1990. Yeasts in dairy products. *Journal of Applied Bacteriology* 68: 99–211.

Fox, P.F., Guinee, T.P., Cogan, T.M., and McSweeney, P.L.H. 2000. Fundamentals of cheese science. Gaithersburg, MD: Aspen Publishers, Inc.

Gahan, C.G., O'Driscoll, B., and Hill, C. 1996. Acid adaptation of *L. monocytogenes* can enhance survival in acidic foods and during milk fermentation. *Applications of Environmental Microbiology* 62: 3128–3132.

Garde, S., Arias, R., Gaya, P., and Nuñez, M. 2011. Occurrence of *Clostridium* spp. in bovine milk and Manchego cheese with late blowing defect: Identification and characterization of isolates. *International Dairy Journal* 21: 272–278.

Giraffa, G., Carmanti, D., and Tarelli, G.T. 1994. Inhibition of *L. innocua* in milk by bacteriocin-producing *Enterococcus faecium*. *Journal of Food Protection* 58: 621–623.

Gruszczynski, L. 2008. *The SPS Agreement within the Framework of WTO Law. The Rough Guide to the Agreement's Applicability*. http://ssrn.com/abstract=1152749.

Gulmez, M. and Guven, A. 2003. Survival of *E. coli* O157:H7, *L. monocytogenes* 4b and *Y. enterocolitica* O3 in different yogurt and kefir combinations as pre-fermentation contaminant. *Journal of Applied Microbiology* 95: 631–636.

Harris, K. and Blackwell, J. 1999. Guidelines for developing good manufacturing practices (GMPs), standard operating procedures (SOPs) and environmental sampling/testing recommendations (ESTRs). Ready-to eat (RTE) products. Desarrollado por los Representantes de las Industrias Productoras de Alimentos Listos para el Consumo. EE.UU.

Hocking, S.L. and Faedo, M. 1992. Fungi causing thread mould spoilage of vacuum packaged Cheddar cheese during maturation. *International Journal of Food Microbiology* 16: 123–130.

Hoolasi, K. 2005. *A HACCP Study on Yoghurt Manufacture*. MSc thesis, Department of Operations & Quality Management, Durban Institute of Technology.

ISO 22000:2005. *Food Safety Management Systems—Requirements for Any Organization in the Food Chain*. ISO/TC 34/SC 17. Geneva, Switzerland.

Jensen, A., Frederiksen, W., and Gerner-Smidt, P. 1994. Risk factors for listeriosis in Denmark, 1989-1990. *Scandinavian Journal of Infectious Diseases* 26: 171–178.

Johnson, M.E. 2001. Cheese products. In: E.H. Marth and J.L. Steele (Eds.), *Applied Dairy Microbiology* (2nd Edition, pp. 345–384). New York: Marcel Dekker.

Johnson, M.E., Riesterer, B.S., and Olson, N.F. 1990. Influence of nonstarter bacteria on calcium lactate crystallization on the surface of Cheddar cheese. *Journal of Dairy Science* 73: 1145–1149.

Joint FAO/WHO Codex Alimentarius Commission, Joint FAO/WHO Food Standards Programme, and World Health Organization. 2003. *Codex Alimentarius: Food Hygiene, Basic Texts*. Food & Agriculture Organization. Rome, Italy.

Kantor, M.A. and Potter, N.N. 1975. Persistence of echovirus and poliovirus in fermented sausage. *Journal of Food Science* 40: 491–495.

Klijn, N., Nieuwendorf, F.F.J., Hoolwerf, J.D., Vander Waals, C.B., and Weerkamp, A.H. 1995. Identification of *Cl. butyricum* as the causative agent of late blowing in cheese by species–species PCR amplification. *Applied and Environmental Microbiology* 61: 2919–2924.

Knorr, D. 1998. Technology aspects related to microorganisms in functional foods. *Trends in Food Science and Technology* 9: 295–306.

Kosikowski, F.V. and Mistry, V.V. 1997. *Cheese and Fermented Milk Foods. Vol. I. Origins and Principles* (pp. 260, 344). Westport, CT: F. V. Kosikowski, L.L.C.

Le Bourhis, A.G., Saunier, K., Doré, J., Carlier, J.P., Chamba, J.F., Popoff, M.R., and Tholozan, J.L. 2005. Development and validation of PCR primers to assess the diversity of *Clostridium* spp. in cheese by temporal temperature gradient gel electrophoresis. *Applied and Environmental Microbiology* 71: 29–38.

Leyer, G.J. and Johnson, E.A. 1993. Acid adaptation promotes survival of *Salmonella* spp. in cheese. *Applied and Environmental Microbiology* 58: 2075–2080.

Leyer, G.J., Wang, L., and Johnson, E.A. 1995. Acid adaptation of E. coli O157:H7 increases survival in acidic foods. *Applied and Environmental Microbiology* 61: 3752–3755.

Linnan, M.J., Mascola, L., Lou, X.D., Goulet, V., May, S., Salminen, C., and Hird, D.W. et al. 1988. Epidemic listeriosis associated with Mexican-style cheese. *New England Journal of Medicine* 319(13): 823–828.

Lou, Y. and Yousef, A.E. 1997. Adaptation to sub lethal environmental stresses protects *L. monocytogenes* against lethal preservation factors. *Applied and Environmental Microbiology* 63: 1252–1255.

Maher, M.M., Jordan, K.N., Upton, M.E., and Coffey, A. 2001. Growth and survival of *E. coli* O157:H7 during the manufacture and ripening of a smear-ripened cheese produced from raw milk. *Journal of Applied Microbiology* 90: 201–207.

Marth, E.H. and Steele, J.L. 2005. *Applied Dairy Microbiology*. New York: Marcel Dekker, Inc.

Mead, P.S., Slutsker, L., Dietz, V., McCaig, L.F., Bresee, P.M., Shapiro, C., Griffin, T.M., and Tauxe, V. 1999. Food-related illness and death in the United States. *Emerging Infectious Diseases* 5: 607–625.

Mena, C., Almeida, G., Carneiro, L., Teixeira, P., Hogg, T., and Gibbs, P.A. 2004. Incidence of *Listeria monocytogenes* in different food products commercialized in Portugal. *Food Microbiology* 21: 213–216.

Morgan, D., Newman, C.P., Hutchinson, D.N., Walker, A.M., Rowe, B., and Majid, F. 1993. Verotoxin producing *E. coli* 0157 infections associated with the consumption of yoghurt. *Epidemiology Infection* 111: 181–187.

Mortimore, S. and Wallace, C. 2013. *HACCP: A Practical Approach* (3rd Edition). New York: Springer.

Mostert, J.F. and Jooste, P.J. 2002. Quality control in the dairy industry. In: R.K. Robinson (Ed.) *Dairy Microbiology Handbook* (pp. 655–736). 3rd edn., New York: John Wiley.

Mucklow, R. 1998. *Guidelines for Developing Good Manufacturing Practices (GMPs) and Standard Operating Procedures (SOPs) for Raw Ground Products.* Institute of Food Science and Engineering, Texas A&M University College Station, Texas, for National Meat Association Oakland, California.

Myhr, A.N., Irvine, D.M., and Arora, S.K. 1982. Late gas defect in film-wrapped cheese. *XXI International Dairy Congress* (Vol. 1, Book 1, pp. 431–432). Moscow, Russia: Mir Publishers.

Nath, K.R. and Kostak, B.J. 1985. Etiology of white spot defect in Swiss cheese made from pasteurized milk. *Journal of Food Protection* 49: 718–723.

Nichols, G., Greenwood, M., and de Louvois, J. 1996. The microbiological quality of soft cheese. *PHLS Microbiology Digest* 13: 68–75.

Nout, M.J.R. 2001. Fermented foods and their production. In: M.R. Adams and M.J.R. Nout (Eds.), *Fermentation and Food Safety* (pp. 1–38). Gaithersburg, MD: Aspen Publishers.

Nout, M.J.R., Rombouts, F.M., and Havelaar, A. 1989. Effect of accelerated natural lactic fermentation of infant food ingredients on some pathogenic organisms. *International Journal of Food Microbiology* 8: 351–361.

Olasupo, N.A., Smith, S.I., and Akinsinde, K.A. 2002. Examination of the microbial status of selected indigenous fermented foods in Nigeria. *Journal of Food Safety* 22: 85–94.

Pacheco, F.P. and Galindo, A.B. 2010. Microbial safety of raw milk cheeses traditionally made at a pH below 4.7 and with other hurdles limiting pathogens growth. In: *Current Research, Technology and Education Topics in Applied Microbiology and Microbial Biotechnology*, 2, 1205–1216.

Pazakova, J., Turek, P., and Laciakova, A. 1997. The survival of *S. aureus* during the fermentation and storage of yoghurt. *Journal of Applied Microbiology* 82: 659–662.

Perez, P.F., Minnaard, J., Rouvet, M., Knabenhans, C., Brassart, D., De Antoni, G., and Schiffrin, E.J. 2001. Inhibition of Giardia intestinalis by extracellular factors from lactobacilli: An in vitro study. *Applied and Environmental Microbiology* 67: 5037–5042.

Pitt, J.I. and Hocking, A.D. 1999. Spoilage of stored, processed, and preserved foods. In: J.I. Pitt and A.D. Hocking (Eds.), *Fungi and Food Spoilage* (p. 506). Gaithersburg, MD: Aspen Publishers.

Ramírez Vela, A. and Martín Fernández, J. (2003). Barriers for the developing and implementation of HACCP plans: results from a Spanish regional survey. *Food Control* 14(5): 333–337.

Ramsaran, H., Chen, J., Brunke, B., Hill, A., and Griffiths, M.W. 1998. Survival of bioluminescent *L. monocytogenes* and *E. coli* O157:H7 in soft cheese. *Journal of Dairy Science* 81: 1810–1817.

Reid, G. 2001. *Regulatory and Clinical Aspects of Dairy Probiotics*. Background paper. FAO/WHO expert consultation on evaluation of health and nutritional properties of powder milk with live lactic acid bacteria. Lawson Health Research Institute, Ontario, Canada.

Reitsma, C.J. and Henning, D.R. 1996. Survival of enterohaemorrhagic *E. coli* O157:H7 during the manufacture and curing of cheddar cheese. *Journal of Food Protection* 59: 460–464.

Robinson, R.K. 2002. *Dairy Microbiology Handbook: The Microbiology of Milk and Milk Products*. New York: Wiley-Interscience.

Rocourt, J., BenEmbarek, P., Toyofuku, H., and Schlundt, J. 2003. Quantitative risk assessment of *Listeria monocytogenes* in ready-to-eat foods: The FAO/WHO approach. *FEMS Immunology & Medical Microbiology* 35(3): 263–267.

Rohm, H., Eliskasses, F., and Bräuer, M. 1992. Diversity of yeasts in selected dairy products. *Journal of Applied Bacteriology* 72: 370–376.

Rubin, H.E., Nerad, T., and Vaughan, F. 1982. Lactate acid inhibition of *S. typhimurium* in yoghurt. *Journal of Dairy Science* 65: 197–203.

Rudolf, M. and Scherer, S. 2001. High incidence of *Listeria monocytogenes* in European red smear cheese. *International Journal of Food Microbiology* 63: 91–98.

Saad, S.M.I., Vanzin, C., Oliveira, M.N., and Franco, B.D.G.M. 2001. Influence of lactic acid bacteria on survival of *E. coli* O157:H7 in inoculated Minas cheese during storage at 8.5°C. *Journal of Food Protection* 64: 1151–1155.

Schaak, M.M. and Marth, E.H. 1988. Survival of *L. monocytogenes* in refrigerated cultured milks and yoghurt. *Journal of Food Protection* 51: 848–852.

Shadbolt, C., Ross, T., and McMeekin, T.A. 2001. Differentiation of the effects of lethal pH and water activity: Food safety implications. *Letters in Applied Microbiology* 32: 99–102.

Silverglade, B.R. 2000. The WTO agreement on sanitary and phytosanitary measures: weakening food safety regulations to facilitate trade? *Food and Drug Law Journal*. 55: 517.

Simon, F., Emmanuelle, E., and Vaillant, V. 2002. *An Outbreak of Salmonellosis Serotype Infantis Linked to the Consumption of Brie Cheese*, France, 2001. http:/www.epinet.org/seminar/2002/SimonSorial1.htm.

Singh, P. and Prakash, A. 2008. Isolation of *E. coli*, *S. aureus* and *L. monocytogenes* from milk products sold under market conditions at Agra region. *Acta Agriculturae Slovenica* 92(1): 83–88.

Sperber, W.H. and Doyle, M.P. 2009. *Compendium of the Microbiological Spoilage of Food and Beverages*. New York: Springer.

STTA Report. 2005. *New Cheese Manufacturing Processes for Dairies in Kosovo Margarita Petrova*, December 2005. Chemonics International Inc. United States Agency for International Development or the United States Government.

Tamime, A.Y. and Robinson, R.K. 1985. *Yogurt: Science and Technology*. Oxford: Pergamon Press.

Tamime, A.Y. and Robinson, R.K. 1999. *Yoghurt: Science and Technology* (2nd Edition). Cambridge: Wood head Publication Limited.

Tatini, S.R. and Kauppi, K.L. 2003. Microbiological analyses. In: H. Roginski, J.W. Fuquay, and P.F. Fox (Eds.) *Encyclopedia of Dairy Sciences* (Vol. 1, pp. 74–79). Amsterdam, the Netherlands: Academic Press and Elsevier Science.

Tekinşen, K.K. and Özdemir, Z. 2006. Prevalence of food borne pathogens in turkish Van otlu (Herb) cheese. *Food Control* 17: 707–711.

USDA. 1999. *Guidebook for the Preparation of HACCP Plans*. Washington, DC: Food Safety and Inspection Service Office of Policy, Program Development, and Evaluation (OPPDE). http://www.fsis.usda.gov/index.htm.

Van Gerwen, S.J.C. and Zwietering, M.H. 1998. Growth and inactivation models to be used in quantitative risk assessments. *Journal of Food Protection* 61: 1541–1549.

Varnam, A.H. and Sutherland, J.P. 1994. *Milk and Milk Products*. London: Chapman & Hall.

Warnekulasuriya, M.R., Johnson, J.D., and Holliman, R.E. 1998. Detection of *Toxoplasma gondii* in cured meats. *International Journal of Food Microbiology* 45: 211–215.

Wood, B.J.B. 1998. *Microbiology of Fermented Foods* (2nd Edition, 2 Vols, 852 pp). London: Blackie Academic & Professional.

Wood, G.W. and Adams, M.R. 1992. Effects of acidification, bacterial fermentation and temperature on the survival of rotavirus in a model weaning food. *Journal of Food Protection* 55: 52–55.

Yabaya, A. and Idris, A. 2012. Bacteriological quality assessment of some yoghurt brands sold in Kaduna metropolis. *The Journal of Research in National Development* 10(2): 1596–8308.

Yadav, J.S., Grover, S., and Batish, V.K. 1993. *A Comprehensive Dairy Microbiology* (1st Edition). New Delhi, India: Metropolitan Book Co. Pvt., Ltd.

Yusof, R.M., Morgan, J.B., and Adams, M.R. 1993. Bacteriological safety of a fermented weaning food containing L-lactate and nisin. *Journal of Food Protection* 56: 414–417.

24

Packaging of Fermented Milks and Dairy Products

P. Narender Raju and Ashish Kumar Singh

Contents

24.1	Introduction	638
24.2	Packaging Materials for Fermented Milks	642
	24.2.1 Paper and Paper-Based Materials	642
	24.2.2 Glass-Packaging Materials	643
	24.2.3 Metal-Packaging Materials	644
	24.2.3.1 Tin and Chromium-Coated Steel	644
	24.2.3.2 Aluminum	645
	24.2.4 Plastic-Packaging Materials	646
	24.2.4.1 Polyolefins	646
	24.2.4.2 Polystyrene	647
	24.2.4.3 Polyvinyl Chloride	647
	24.2.4.4 Polyvinylidene Chloride	648
	24.2.4.5 Polyester	648
	24.2.4.6 Polyamides	648
24.3	Packaging Techniques for Fermented Milks	649
	24.3.1 Aseptic Packaging	649
	24.3.2 Modified Atmosphere Packaging	649
	24.3.3 Active Packaging	651
	24.3.4 Antimicrobial Packaging	652
	24.3.5 Edible Films and Coatings	653
	24.3.6 Nanocomposite Packaging	653
24.4	Packaging of Fermented Milks and Dairy Products	654
	24.4.1 Dahi and Yogurt	654

24.4.2 Cheese 656
24.4.2.1 Whey Cheeses 662
24.4.3 Packaging of Probiotic Fermented Dairy Products 664
24.5 Conclusions 665
References 666

24.1 Introduction

Food packaging like any other packaging is an external means of preservation of food during storage, transportation, and distribution. Hence, it forms an integral part of the product manufacturing. Food packaging performs four different major functions (Figure 24.1). Protection is considered as the primary function of packaging, and it includes protecting the contents of the package from environmental effects such as water, water vapor, gases, odor, microbes, dust, shocks, vibrations, and compressive forces. Convenience is becoming the key function of a package, especially with the modern lifestyle-related changes among men and women. Convenience, as desired by consumers, includes easy to hold, open and pour as appropriate, and ready to eat. The ability of supermarket consumers and alike to instantly recognize products through distinctive branding and labeling is accomplished by the communication function of a package, where it acts as a "silent salesman." Containment function makes a great contribution to protect the environment from the myriad of products, which are moved from one place to another on different occasions. Faulty packaging could also result in a major pollution of the environment. In pursuit of achieving these goals, many materials have been discovered by man for use as food-packaging materials, namely, wood and paper, glass, metals, plastics, and composites. Packaging materials used in the food industry are classified into three different forms, namely, flexible, semirigid, and rigid depending on their response to applied external force. A packaging material that is able to be deformed with little force and generally able to return to original shape on deformation is termed as flexible. A material is termed as rigid when it is able to be deformed with strong force and not able to return to original shape. However, a material is termed semirigid when it is able to be deformed with moderate force and may or may not return to original shape. Packaging is classified into four levels, namely, primary, secondary, tertiary, and quaternary. When a packaging material is in direct contact with the product, it is called a primary package, while a package is termed secondary when it helps to

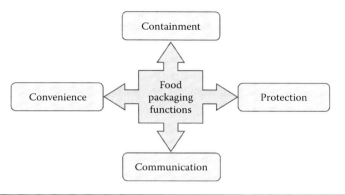

Figure 24.1 Basic functions of food packaging.

contain the primary package. Tertiary and quaternary levels are distribution or shipping packages. The inherent properties of these packaging materials make them either highly suitable or unsuitable for a particular food product or level of packaging.

Milk and milk products not only contain almost all the essential and nonessential nutrients required for the growth of humans but also serve as a very good media for the growth of many spoilage-causing microbes, making these a highly perishable commodity. Hence, there is an obvious need to preserve these. Many food-processing techniques (i.e., pasteurization, sterilization, ultrahigh temperature treatment, condensing, drying, fermentation) have been applied to preserve milk. Fermentation is one of the oldest methods used to extend the shelf life of milk and has been practiced for thousands of years (Tamime 2002). A large number of fermented milk products are manufactured worldwide. However, these could be classified into (1) lactic fermentations including mesophilic, thermophilic, and therapeutic or probiotic types; (2) yeast-lactic fermentations; and (3) mold-lactic fermentations (Robinson and Tamime 1990). The physical, biochemical, and technofunctional properties of different fermented milks and milk products vary and as such require different packaging interventions to preserve and extend their shelf life. The characteristics of broadly classified fermented milk products and their packaging requirements are given in Table 24.1. Hence, a packaging material has to be thoughtfully chosen depending upon the nature of the product, availability, machinability, cost, and so on. In this chapter, an overview of the packaging materials and techniques suitable for the fermented dairy product packaging is discussed.

Table 24.1 Characteristics of Fermented Dairy Products and Their Packaging Requirements

CLASSIFICATION OF FERMENTED MILKS	SELECTED PRODUCTS	SOME CHARACTERISTIC ATTRIBUTES OR CONTENTS	PACKAGING REQUIREMENTS
Cheese	Cheddar Gouda Mozzarella Camembert Roquefort Processed cheese Shredded cheese	Acid Alcohol Milk fat Milk protein Carbon dioxide Flavor Surface molds	• Protection against chemical and microbial contamination especially surface mold growth • Reduction of loss of moisture from the surface • Protection of flavor • Protection from light • Prevention of oxidation of fat • Prevention of physical deformation of cheese blocks • Convenience in terms of use • Provision for product labeling and brand identification
Lactic fermented milks	Cultured buttermilk Bulgarian buttermilk Lassi Shrikhand Stirred yogurt Set yogurt Dahi Misti dahi Labneh/Zabadi Filmjolk Langfil	Acid Tartness Flavor Diacetyl Acetoin Acetaldehyde Firm body in set products Free from syneresis in set products May contain fruits	• Protection of flavor • Protection from light • Prevention of oxidation of fat • Protection against post-process microbial contamination, especially surface mold growth on set-type products • Protection against physical abuse and chemical contamination • Provision for product labeling and brand identification

(Continued)

Table 24.1 (*Continued*) Characteristics of Fermented Dairy Products and Their Packaging Requirements

CLASSIFICATION OF FERMENTED MILKS	SELECTED PRODUCTS	SOME CHARACTERISTIC ATTRIBUTES OR CONTENTS	PACKAGING REQUIREMENTS
Long-life products	Aseptically packaged lassi Aseptically packaged yogurt-based drinks	Acid Tartness Flavor Acetoin Acetaldehyde Fruits	• Protection against chemical and microbial contamination • Protection of flavor • Protection from light • Convenience in terms of use • Provision for product labeling and brand identification
Therapeutic- or probiotic-type fermented products	Acidophilus milk Yakult Probiotic dahi Probiotic lassi	Probiotic organisms Acid Flavor Firm body in set products Free from syneresis in set products	• Low oxygen for maximum viability of probiotic organisms • Protection of flavor • Protection from light • Prevention of oxidation of fat • Protection against post-process microbial and chemical contamination and physical abuse especially for set-type products • Provision for product labeling and brand identification
Yeast-lactic and mold-lactic fermented milks	Kefir Koumiss Acidophilus yeast milk Viili	Acid Alcohol Carbon dioxide Flavor Viscous Gelatinous Ropiness	• Protection of flavor • Protection from light • Prevention of oxidation of fat • Protection against post-process microbial and chemical contamination • Provision for product labeling and brand identification

24.2 Packaging Materials for Fermented Milks

In order to have a significant impact on the shelf life and consumer acceptability of packaged fermented dairy products, a heterogeneous group of materials such as paper, plastics, metals, and composite materials with a corresponding range of performance characteristics is used in packaging. However, the largest share of global packaging market is accounted for paper and board followed by plastic-packaging materials (WPO 2008). The nature and basic characteristics of commonly used packaging materials are as follows.

24.2.1 Paper and Paper-Based Materials

Paper and packaging is a natural pair as paper is the most important packaging material and packaging is an important use of paper. It is the oldest packaging material and the most versatile one. *Paper* is a generic term encompassing a broad spectrum of materials derived from cellulosic fibers used for the production of paper, paperboard, corrugated board, and other similar products. Depending on the processing treatments given to pulp, type, quality, and properties of paper vary greatly. The pulp may be bleached or unbleached and may be subjected to different finishing treatments such as calendaring, surface sizing, or coating. As a result, a wide range of papers is obtained such as Kraft, waxed, greaseproof, and so on. Further, almost all paper is converted by undergoing further treatments after manufacture such as embossing, coating, laminating, and forming into special shapes and sizes such as bags and boxes (Robertson 2006). Paper can acquire barrier properties and extended functional performance, such as heat seal ability, heat resistance, grease resistance, and product release, by coating, lamination, and impregnation. Materials used for these purposes in these ways include extrusion coating with plastics such as polyethylene (PE), polypropylene (PP), PE terephthalate, and ethylene vinyl alcohol; lamination with plastic films or aluminum foil; and by treatment with wax, silicone, or fluorocarbon. Papers can be impregnated with a vapor-phase metal-corrosion inhibitor, mold inhibitor, or coated with an insect repellent (Kirwan 2005).

The distinction between paper and paperboard is not always very clear, but paperboard is heavier and more rigid than paper. Usually all paper sheets that are greater than 12 divisions thick, when measured by a micrometer (300 μm or 1200 gauge or 0.012 inch) or when its grammage (i.e., grams per square meter or gsm) exceeds 250 g/m², are classified as paperboard. Multiply boards are produced by the consolidation of one or more web piles into a single sheet of paperboard that is subsequently used to manufacture rigid boxes, folding cartons, and similar products (Robertson 2006). Folding cartons are containers made from sheets of paperboard, which have been cut and scored for bending into desired shapes; they are delivered in a collapsed state for assembly or erection at the packaging point. A number of steps are involved in converting paperboard into cartons. Where special barrier properties are required, coating and laminating are carried out. Corrugated fiberboard is characterized by its cellular structure that imparts high compressive strength at a relatively low weight and is manufactured from three basic sheets—two liner boards and a central corrugated sheet or medium (flute). These materials can be varied as to weight, type, and number and/or height of the corrugations in the fluting medium. Outstanding rigidity and structural strength as well as its exceptional cushioning properties of corrugated fiberboard are due to its sandwich-type construction. The properties of a corrugated board depend largely on the type, number, and position of the corrugations. Different types of corrugated fiber boards include single face, single wall, double wall, triple wall, and so on. Mostly corrugated fiber boards are used as secondary or tertiary packaging in the distribution and marketing of dairy products.

24.2.2 Glass-Packaging Materials

Glass is a substance that is hard, brittle, and usually transparent and one of the most important packaging materials because of its high barrier and see-through properties. Compositionally, it is an inorganic substance formed from a mixture of sand, sodium oxide, and calcium oxide with a proportion of broken glass. Recycled broken glass (cullet) may account for as much as 60% of all raw materials (Marsh and Bugusu 2007). Soda-lime glass, which accounts for nearly 90% of all glass produced (Robertson 2006), is made from silica ($\approx 75\%$), soda ($\approx 15\%$), and limestone ($\approx 10\%$). Lesser percentages of aluminum, potassium, and magnesium oxides may be included. When the components are melted together, these fuse into a clear glass, which can be readily shaped while in a semimolten state. The melting of glass is usually accomplished at a temperature of more than 1500°C in a very large furnace. For obtaining different colored glasses, various chemicals are added to the melt such as iron oxide for amber color, cadmium compounds for yellow color, and so on. Two types of glass containers include those with narrow necks (i.e., bottles) and wide necks (i.e., jars) are frequently used for food-packaging applications. Different types of closures are used for closing the containers, namely, simple wrapping without sealing, heat seals, and threaded screw closures.

24.2.3 Metal-Packaging Materials

Metal is the most versatile of all packaging forms, as it offers a combination of excellent physical protection and barrier properties, formability and decorative potential, recyclability, and consumer acceptance (Marsh and Bugusu 2007). Steel, aluminum, tin, and chromium are commonly used metals for food-packaging applications. Tin and steel and chromium and steel are used as composite materials in the form of tinplate and electrolytic chromium-coated steel, sometimes referred to as tin-free steel. Aluminum is used in the form of purified alloys containing small amounts of magnesium and manganese (Robertson 2006).

24.2.3.1 Tin and Chromium-Coated Steel Low-carbon mild steel sheet (0.15–0.50 mm) with a coating of tin (0.4–2.5 µm) on each surface of the material is termed as *tinplate*. The combination of tin and steel produces a material that has good strength and possesses qualities such as ductility and drawability as well as good solderability, weldability, nontoxicity, lubricity, lacquerability, and a corrosion-resistant surface of bright appearance (Robertson 2006). Tinplate is commonly used for the manufacture of three-piece cans for canned foods. Low-carbon steel electrolytic coated with chromium is referred to as the electrolytic chromium-coated steel (ECCS). Usually, the ECCS consists of a duplex coating of metallic chromium (0.07–0.15 gsm) and trivalent chromium present as oxide (0.03–0.06 gsm) giving a total coating weight of approximately 0.15 gsm. Although metals are important packaging materials, they are also susceptible to corrosion.

Corrosion, a universal process affecting most metals, is described as the chemical reaction between a metal and its environment to form useless corrosive compounds. Metal containers are usually coated with lacquers or enamels to avoid interaction with food contents, which otherwise would cause taints and off-flavors and reduce the shelf life or quality of the product to an unacceptable level. By simple interposition of a barrier between the food and container, the lacquer reduces or eliminates the area available for anodic dissolution and migration of metal. It also reduces or eliminates the cathode area available for reduction of plant pigments leading to color changes. Many types of internal enamel are available for food containers including oleoresinous, vinyl, acrylic, phenolic, and epoxyphenolic (Simal-Gandara 1999). ECCS is less resistant to corrosion than tinplate; therefore, it is enameled on both sides (Robertson 2006).

24.2.3.2 Aluminum Aluminum is a lightweight, silvery white metal derived from bauxite ore, where it exists in combination with oxygen as alumina. It is commonly used to make cans, foil and laminated paper, or plastic packaging besides providing an excellent barrier to moisture, air, odors, light, and microbes; aluminum has good flexibility and surface resilience, excellent malleability and formability, and outstanding embossing potential (Marsh and Bugusu 2007). Aluminum foil is a thin-rolled sheet of alloyed aluminum varying in thickness from about 4 to 150 μm. It is essentially impermeable to gases and water vapor, when it is thicker than 25.4 μm. It can also be converted into a wide range of shapes and products including semirigid containers with formed foil lids, caps and cap liners, composite cans, laminates containing plastic, and sometimes paper or paperboard. Aluminum foil has an important role as light and gas barrier in the five- to seven-layered laminate used for packaging of UHT products. The corrosion resistance of aluminum is excellent in the neutral range of pH 4–9 but will corrode in acid solutions (pH below 4) and alkali solutions (pH above 9). Further, aluminum is liable to undergo severe corrosion, if brought into metallic contact with copper, iron, or other more positive metals in the presence of an electrolyte (Robertson 2006). The main disadvantages of aluminum are its high cost compared to other metals such as steel and its inability to be welded that renders it useful only for making seamless containers.

24.2.4 Plastic-Packaging Materials

Food industry is a major user of plastic-packaging materials. Plastics are usually divided into two broad categories, namely, thermosets and thermoplastics. Mostly thermoplastics are being used for packaging of dairy products. Brief information about some common thermoplastics used in dairy industry such as PE, PP, ethyl vinyl acetate, polystyrene (PS), polyvinyl chloride (PVC), polyvinylidene chloride (PVDC), polyethylene terephthalate (PET), and polyamides (nylon) is given next.

24.2.4.1 Polyolefins Polyolefins form an important class of thermoplastics and include low, linear, and high-density PE and PP. Industry commonly divides PEs into two broad categories based on their densities as high-density polyethylene (HDPE) and low-density polyethylene (LDPE). The properties that make PE film a popular packaging medium in dairy industry are its low price, nontoxic, excellent heat-sealing property, flexibility, pleasing appearance, softness, and chemical inertness. LDPE accounts for the biggest proportion (>50%) of the plastics used in food packaging due to its versatility. It can be extruded into film, blown into bottles, injection molded into closures, extruded as a coating on paper, aluminum foil, or cellulose film, and made into large tanks and other containers, and it can be easily heat sealable. Linear LDPE (LLDPE) is also used commonly in place of LDPE. HDPE gives greater rigidity to bottles than LDPE at an equal wall thickness. HDPE films have a cloudy appearance, occasionally utilized as a liner for the bulk packaging of skim milk powder and carton liner due to their high strength and barrier properties. However, PE is not suitable for packing foods with strong aromas or the products, which have to be packed under vacuum (Robertson 2006). PP is nontoxic, is harder than either LDPE or HDPE, and has a less waxy feel and an excellent grease resistance. It can be converted into film and sheet or can be thermoformed to give thin-walled trays of excellent stiffness. It has better mechanical strength and is less prone to stress cracking than PE. The major factor that makes PP one of the most widely used clear plastics is its resistance to deformation at the temperatures used for sterilization or retort processing. However, it tends to be brittle at low temperatures, which could be addressed by co-polymerizing it with a maximum ethylene content of 20% (Kadoya 1990). The physical and mechanical properties of PP can be improved by orienting the film. The oriented PP has better tensile strength and low permeability to water vapor and oxygen gas.

24.2.4.2 Polystyrene PS is an amorphous, crystal-clear, hard, brittle, low-strength material with a relatively low melting point of 90°C and poor impact strength. It is crystal clear, sparkling, and relatively cheap and has high water vapor and gas transmission rate, hence also called as "breathing film." Properties of PS could be improved by incorporating synthetic rubber during its manufacture. Rubber-modified PS is known as toughened, impact, high-impact, and super-high-impact PS, depending upon the proportion of rubber to styrene (Sacharow 1976). Co-polymerization with polybutadiene or styrene butadiene rubber (10%) makes PS tough and improves impact resistance. Such PS is called as high-impact polystyrene (HIPS). PS is also used for thermoformed cups, trays, and glasses for yogurt, ice cream, meat, soft drinks, and so on. However, it is unsuitable for heat sterilization and deteriorates on exposure to sunlight. HIPS is an excellent material for thermoforming (Robertson 2006). PS can be foamed by adding foaming agents and as such is called as foamed PS or expanded PS (EPS). EPS is characterized by excellent resistance to bacterial and mold growth and almost nil water absorption (Sacharow 1976). It has lightweight and excellent cushioning properties.

24.2.4.3 Polyvinyl Chloride PVC, also called as vinyl, has several overall balanced properties required by food-packaging systems such as glass-like clarity; good mechanical strength; resistance to water vapor, gases, and chemicals; retention of flavor; excellent printability; and lower weight-to-volume ratio (Robertson 2006). However, pure PVC polymer is a tough and naturally brittle material and requires addition of large amounts of other chemical compounds called "plasticizers" (Sacharow 1976). Of the total plasticizers used, 80% of them are used with PVC alone with phthalic acid esters being the commonly used (Robertson 2006). One of the most important properties of PVC films is their ability to be made into shrink film. PVC is commonly used for food packaging in various forms such as films and sheets, bags, liners, shrinkable tubes/films, skin/blister packs, film laminates, bottles, and sachets.

24.2.4.4 Polyvinylidene Chloride PVDC is a homopolymer of vinylidene chloride. Pure PVDC homopolymer yields a rather stiff film, which is hardly suited for packaging use. However, when polymerized with 5%–50% vinyl chloride, a soft, tough, and relatively impermeable film results. These copolymers were first introduced as "Sarans" during World War II (Sacharow 1976). The film has excellent water vapor and gas barrier properties, besides outstanding characteristics such as retention of odors and flavors, resistance to oils and fats, heat-sealing characteristics, nontoxicity, abrasion, and chemical resistance. Although PVD shrink films have good impact resistance, once punctured they tear readily. As part of a flexible laminate, PVDC can be used in its film form or as a coating on paper, film, or foil. PVDC may also be coextruded with polyolefins to form either film or semirigid sheets with improved barrier properties. The laminates comprising of PVDC are widely used for the packaging of baby foods, snacks, chips, powdered soups and sauces, powdered coffee, coffee beans, sweets, biscuits and crackers, and powdered sugar (Robertson 2006).

24.2.4.5 Polyester PET, the important polyester used in food industry, is formed by condensation polymerization of ethylene glycol and terephthalic acid. It is a brittle opaque film in its crystalline state and weak in its amorphous state. The polymer is usually biaxially oriented to obtain superior film properties. PET films are characterized by exceptional strength, excellent chemical resistance, and transparency and can be used over a wide range of temperature from frozen food applications to boil-in-bag pouches (Sacharow 1976). However, they are not readily heat sealed. It has wide uses outside packaging as well including polyester clothing and carpets (Hernandez et al. 2000). Its main applications include in metallized films, vacuum and gas packaging, shrink packaging, cured meat, and boil-in-bag applications. This film is of great interest to food packagers as it contains no plasticizers and is nontoxic.

24.2.4.6 Polyamides Polyamides, commonly referred to as nylon, are condensation polymers made from monomers with amine and carboxylic acid functional groups resulting in amide linkages. Two different types of polyamide films are available based on their resin manufacture. One variety is made by condensation of mixture of diamines and dibasic acids, and the other type is formed by condensation of omega amino acids (Sacharow 1976). First type is identified by the number of carbon atoms in the diamine followed by the number of carbon atoms in the diacid, for example, nylon 6,6 and nylon 6,10. Second type is identified by using a single number signifying the total amount of carbon atoms in the amino acid, for example, nylon 6 and nylon 11. The unique properties of nylon film are that it has high mechanical strength, high elongation capability, excellent resistance to cutting, perforation, abrasion and bursting, high chemical resistance to oils and fats, outstanding impermeability to gases and vapors, easy printability, and easy metallizing (Hernandez et al. 2000). The film can be biaxially oriented, and its properties remain unchanged between −30° and 120°C. Due to its excellent elongation and ability to stretch without breaking, nylon can be successfully used in applications involving irregular shapes (Sacharow 1976).

24.3 Packaging Techniques for Fermented Milks

24.3.1 Aseptic Packaging

Aseptic packaging is the filling of sterile containers with a commercially sterile product under aseptic conditions, and it is then hermetically sealed so that reinfection is prevented. Aseptic packaging has advantages in terms of being a high temperature–short time sterilization process, enabling use of containers that are unsuitable for in-package sterilization and extending the shelf life of products at normal temperatures by packaging them aseptically (Robertson 2006). The shelf life of cultured milk products can be extended in two ways: first, production and packaging under aseptic conditions and, second, heat treatment of the finished product, either immediately before packing or in the package (Bylund 1995).

24.3.2 Modified Atmosphere Packaging

The primary causes of quality deterioration in dairy products are microbial growth and development of rancidity. The intrinsic properties of individual dairy products are responsible for these changes.

For example, hard cheeses with relatively low water activity are normally affected by the growth of molds, whereas products with high water activity such as cream and soft cheeses are more susceptible to fermentation and rancidity. It is evident that the shelf life of various dairy products is limited in the presence of normal air. There are two principal factors that are responsible for the deterioration of dairy products in the presence of normal air. The first being the chemical effect of atmospheric oxygen and second the growth of aerobic spoilage microflora. These factors bring about changes in odor, flavor, color, and texture, leading to an overall deterioration in quality either individually or in association with one another. The normal composition of air is 78% N_2, 21% O_2, 0.03% CO_2, and traces of noble gases. The modification of the atmosphere within the package by reducing the oxygen content while increasing the levels of CO_2 and/or N_2 has been shown to significantly extend the shelf life of perishable foods at chill temperatures. The modified atmosphere packaging (MAP) is the technology of producing a gas atmosphere around a product that deviates from ambient air composition by either passive or active methods in order to achieve improved quality and prolonged shelf life (Ucherek 2004). Four packaging techniques are included within this definition, namely, vacuum packaging, passive MAP, gas-flush MAP, and MAP by active packaging (O'Connor-Shaw and Reyes 1999). MAP has also been referred to as "protective atmosphere packaging" or as "packaged in a protective atmosphere" when used in labeling.

Removal of air from the pack and its replacement with a single gas or mixture of gases by either passive or active methods is the key principle in MAP. The gas mixture used is dependent on the type of product. The three major gases used in MAP of foods are O_2, N_2, and CO_2. Different combinations of two or three of these gases are used to meet the needs of a specific product. Usually for nonrespiring products such as dairy products, where the microbial growth is the main spoilage parameter, about 30%–60% CO_2 is used, rest being pure N_2 or pure O_2 for O_2-sensitive foods or combinations of N_2 and O_2. In order to minimize the respiration rate of respiring food products, around 5% CO_2 and O_2 each are usually used, with the remainder being N_2 (Sivertsvik et al. 2002). Although several other gases such as carbon monoxide, ozone, ethylene oxide, nitrous oxide, helium, neon, argon, propylene oxide, ethanol vapor, hydrogen, sulfur dioxide, and chlorine have been tried, factors such as regulatory restrictions, safety concerns, cost, and negative effects on sensory quality restrict their use. Passive modification is a slow process and requires interactions between the food and its surrounding gases with the package to play the role of a regulator. On the contrary, active modification is faster and can be achieved by gas flushing, by vacuum application, or by using gas scavengers/emitters (Floros et al. 2000). Gas-flush MAP involves the establishment of a specific gas composition within the package in a single stage during the packaging operation, by flushing with the selected gas mixture before sealing. Depending on the desired residual O_2, a vacuum operation may be needed prior to gas flushing. The instant modification of the package atmosphere is the main advantage of gas-flush MAP over the relatively slow passive process (O'Connor-Shaw and Reyes 1999). Vacuum packaging involves the removal of air from a package without replacement by another gas.

24.3.3 Active Packaging

A package may be termed as active, if it performs some role other than performing the basic functions as given in Figure 24.1. Active packaging is defined as a type of packaging that changes the condition of the packaging to extend shelf life or improve safety or sensory properties, while maintaining the quality of food (Vermeiren et al. 1999). Active packaging should not be confused with "intelligent packaging," which informs or communicates with the consumer regarding the present properties of the food or records aspects of its history (Rooney 2005). It is generally specific to the situation and essentially influences the environment of the food. Important active packaging techniques include those that are concerned with substances that absorb oxygen, ethylene, moisture, carbon dioxide, flavors, and odors and those that release CO_2, antimicrobial agents, antioxidants, and flavors. Among all, O_2 scavengers (50%), desiccants (34%), and removal of undesirable elements (purge absorbers) (14%) account for almost all active packaging technologies presently used (Brody 2006). Quality changes of O_2-sensitive foods can be minimized by using O_2 scavengers. The technique includes iron powder oxidation, ascorbic acid oxidation, photosensitive dye oxidation, and enzymatic oxidation (e.g., glucose oxidase and alcohol oxidase).

Although the nature of the food and packaging favors the choice of a particular technique, the majority of the commercial success was achieved by iron oxidation only. Ethylene is a plant hormone responsible for accelerating the ripening process and leading to maturity and senescence. Although it has positive effects on fruits and vegetables postharvest physiology, it is often the cause for deteriorating the quality and shelf life. To prolong the shelf life and maintain acceptable quality of fresh fruits and vegetables, ethylene-scavenging techniques based on potassium permanganate, activated carbon, and so on have been successfully applied. As a result of deterioration and respiration reactions, in some food products CO_2 is generated. Removal of such CO_2 is important to extend the shelf life of such food products for which CO_2 scavengers are used. Moisture control is an essential process for food products susceptible to moisture damage. Excessively high moisture levels cause the softening of crispy products, caking of milk and other food powders, and so on. On the other hand, excessive moisture evaporation through packaging material might result in desiccation or may trigger chemical changes such as lipid oxidation. Use of packaging films with appropriate water vapor permeability or moisture-controlling pads or sachets would help in establishing the required relative humidity in the headspace of the packaged food.

24.3.4 Antimicrobial Packaging

Control of food spoilage microorganisms on foods can be achieved by coating antimicrobial compounds on food-packaging material, usually referred to as antimicrobial packaging. It is also regarded as a component of active packaging. The direct incorporation of antimicrobial additives in packaging films is a convenient methodology by which antimicrobial activity can be achieved (Perez-Perez et al. 2006). Antimicrobial packaging is a system that can kill or inhibit the growth of microbes and, thus, extend the shelf life of perishable products and enhance the safety of packaged products (Han 2005). Various antimicrobial agents such as organic acids, acid salts, bacteriocins, antioxidants, phenolics, and plant/spice extracts have been tried to incorporate into conventional food-packaging systems.

24.3.5 Edible Films and Coatings

Films are generally defined as stand-alone thin layers of materials, which usually consist of polymers that provide mechanical strength to the stand-alone thin structure. Coatings are a particular form of films directly applied to the surface of materials, which could be removed but not intended for disposal separately from coated materials (Han and Gennadios 2005). Edible films and coatings are produced from edible biopolymers, namely, proteins, polysaccharides or lipids, and food-grade additives (e.g., plasticizers). Modification of the physical and functional properties of films is achieved by combining film-forming biopolymers with additives. The use of edible films and coatings as carriers of active substances, especially antimicrobial agents, is a promising application of active food-packaging technique for enhancing shelf life of perishable foods.

24.3.6 Nanocomposite Packaging

Polymer composites are a mixture of polymers with inorganic or organic additives having certain geometries such as fibers, flakes, spheres, and particulates. These new-generation composites exhibit significant improvements in modulus, dimensional stability, and solvent or gas resistance with respect to the pristine polymer. Nanoparticles are generally accepted as those with a particle size below 100 nm. The use of nanoscale fillers is leading to the development of polymer nanocomposites and represents a major alternative to these conventional polymer composites. Various nanoparticles have been recognized and used as possible additives to enhance the polymer performance such as nanoclay and carbon nanotubes. The first successful example of a polymer-clay hybrid was a nylon-clay hybrid (Sorrentino et al. 2007). Natural biopolymers have the advantage over synthetic polymers in that they are biodegradable, renewable, as well as edible. However, relatively poor mechanical and water vapor barrier properties of those films are causing a major limitation for their industrial use. Efforts have been focused on the property modification of natural biopolymer-based films to improve their mechanical and water vapor barrier properties (Rhim and Ng 2007). The attention of readers is drawn toward the edible or biodegradable films incorporated with nano-sized antimicrobial materials such as nano-silver that could enhance the shelf life of fermented dairy products or synthetic and conventional packaging materials incorporated with nanoparticles such as clay to obtain improved barrier properties such as LLDPE with improved O_2 barrier property (Gokkurt et al. 2012).

24.4 Packaging of Fermented Milks and Dairy Products

The requirements for packaging of fermented dairy products are similar to those of many other dairy and food products. Important considerations about materials used to package dairy products are toxicity and compatibility with the product, resistance to impact, maintenance of sanitation, odor and light protection, tamper resistance, size specifications, shape and weight requirements, marketing appeal, printability, and cost (Alvarez and Pascall 2011). General information on packaging materials and packaging techniques commonly used in dairy and food processing industries was discussed to give a background to novice readers. Packaging of specific fermented milks and dairy products being practiced by the industry and R&D is discussed here.

24.4.1 Dahi and Yogurt

Yogurt is the best known of all cultured milk products and is the most popular worldwide. Typically it is classified as set-type, stirred-type, drinking-type, frozen-type, and concentrated. Concentrated yogurt is sometimes called as strained yogurt, labneh, or labaneh (Bylund 1995). Dahi and yogurt are highly perishable products, and packaging protects these during handling and helps to maintain their physicochemical, nutritional, and sensory characteristics. The packaging materials for yogurt and dahi must be acid resistant, prevent the loss of volatile flavors, and be impermeable to oxygen as yeasts and molds grow actively in the presence of oxygen. These products are soft in nature and generally require rigid or semirigid containers such as glass or plastic containers for set-type products, while flexible packaging materials in the form of paperboard cartons and plastic pouches could be used for stirred or pourable products (Alvarez and Pascall 2011; Zhao 2004). Plastic materials such as HDPE, PP, PS, PVC, and PVDC provide desired strength for yogurt and dahi. These are inert, low cost, and contribute no off-flavors to the products. With a view to meet the product requirements and attract consumer attention, dahi and yogurt are packaged in materials of various shapes and designs from previously discussed materials, but cups, tubs, or paperboard laminates are packaging forms usually found in the market. PP containers have become the best option for yogurt packaging as these can be made with thinner walls while maintaining the same structural integrity resulting in significantly less plastic usage (Zhao 2004). The PS containers, especially the high-impact PS (HIPS), which are now extensively used by dairy industry for packaging of dahi and yogurt, are very clean in appearance, give good shining look, are light in weight, and are unbreakable, but the problem associated with PS and PP cups is that of whey off in set-type products such as dahi during storage. Preformed and cut aluminum foil and PE or PS laminates of about 100 mm diameter with a pull tab for easy opening are usually used as lids to seal dahi and yogurt cups or tubs. Among the rigid containers, glass bottles are still used in some countries to package yogurt (Alvarez and Pascall 2011).

Packaging materials available for yogurt in the developed countries are PS, PP, or glass (Lafougere 1998). The influence of PP and PS on the sensory and physicochemical characteristics of flavored stirred yogurts was studied by Saint-Eve et al. (2008). It was reported that PS seems to be preferable, especially for yogurts with 4% milk fat, for avoiding loss of fruity notes and for limiting the development of odor and aroma defects. In a study on sweetened low-fat (1%) plain yogurt, filling of CO_2 (0.08–0.09 kg/cm^2) into yogurt cups stored at 4°C had been shown to be beneficial in extending the shelf life up to 21 days and was acceptable to consumers with no alterations in the sensory characteristics (Karagül-Yüseer et al. 1999). Nitrogen-flushed plain yogurt has showed no contamination after 8 months, but the products filled under ambient conditions lasted for only 14 weeks. In India, dahi is packaged and sold in local market by *halwais* (milk-based confectioners) and mini dairies in earthen pots or cups and dried leaf-based cups (Paltani and Goyal 2007), which has a firm body. Earthenware vessels or pots are produced from clay, and part of the container coming in contact with the product is glazed. However, earthenware has the drawback of being heavy, is susceptible to break, and above all leads to excessive shrinkage of the product during storage due to moisture seepage through the pores of earthenware. Shelf-stable products such as long-life drinking yogurt and stirred or set-type yogurts are manufactured by employing aseptic packaging techniques using six layered laminates containing PE, aluminum foil, and paperboard (Bylund 1995; Bockelmann and Bockelmann 1998). Paper-based cartons made from laminates are usually used to package dehydrated yogurt. Kumar and Mishra (2004) reported that the shelf life of mango soy-fortified yogurt powder packaged in aluminum foil-laminated PE was found to be better compared to high-density polypropylene pouches. Metal cans are also used for packaging of some type of dried yogurts (Alvarez and Pascall 2011).

Adhesion of foods and the resulting residues in packaging cause economic loss and poor product appearance, and in this context, approximately 10% of all fermented milk products remain on the inside of packaging. Hansson et al. (2012) reported that adhesion of fermented milks and yogurt on packaging materials depends on the product contact time to the surface and among different packaging materials studied, namely, PS, PE, PET, and glass, and the adhered amount of fat was lowest (4%–8%) with PE contact layer package. Further, it was also reported that the product with higher fat content will adhere more to the surface after disruption or shaking of package prior to consumption.

24.4.2 Cheese

Cheese is a generic name of a group of fermented milk-based food products, produced in a wide range of flavors and forms throughout the world, and cheese making is a complex system with different reactions taking place during manufacturing, maturation, and storage stages (Fox and McSweeney 2004). Packaging of cheese is an important aspect in cheese making as it plays a vital role during curing and storage, in the final cheese shape and appearance, and in its protection from the environment. Cheese packaging can be broadly grouped as (1) wrapping of cheese for storage and ripening and (2) retail packaging for consumers. Quality and water loss of finished product depend on the chemical properties and on the storage conditions in unpackaged cheese. However, in packaged cheeses, along with the storage conditions, these are dependent on the permeability and protection provided by the packaging material. Light-induced degradation of lipids, proteins, and vitamins in cheeses causes both formation of off-flavors and color changes, which may swiftly damage product quality and marketability and ultimately may lead to loss in nutritional value and also formation of toxic products such as cholesterol oxides. Dalsgaard et al. (2010) reported that the contents of protein (dityrosine and dimethyl disulfide) and lipid (lipid hydroperoxides, pentanal, hexanal, and heptanal) oxidation products were significantly low in vacuum-packaged cheeses compared to cheeses packed in normal air. For a detailed discussion on light-induced changes in packaged cheeses, it is suggested to refer to an excellent review carried out by Mortensen et al. (2004). Light initiates the oxidation of fats, even at refrigerated temperatures, and unripened cheeses result in "cardboardy" or "metallic" off-flavors (Robertson 2006). Such photooxidation changes may be prevented by either minimizing exposure to light or controlling light barrier properties of packaging material. Depending on the reflectance, transmittance, and oxygen permeability, packaging materials provide protection against light-induced changes in cheeses. Metals, paper and paperboard, various plastics, and finally glass transmit light in increasing order (Bosset et al. 1994). Incorporation of titanium dioxide into plastic materials as an additive increases the light scattering, especially light at wavelengths of less than 400 nm, thereby reducing light transmittance (Nelson and Cathcart 1984). However, the reduction of light transmittance is dependent on the amount of added titanium dioxide. Hence, in selecting an appropriate packaging material for a particular type of cheese, factors such as the type of cheese and its textural characteristics (hard or soft); the presence of characteristic microorganisms; the type of package (wholesale or retail); permeability to different fluids, namely, water vapor, oxygen, carbon dioxide, and ammonia, and light; and provision for printing and labeling have to be considered.

Bulk cheeses are either paraffin or vacuum packed in flexible film. For waxing, the cheeses can be lifted by means of suction and half-immersed in wax and then the other half can be immersed. Generally, mineral waxes (paraffin) are usually applied to low-moisture fresh cheese immediately after its manufacture as these are cheap, are easily applied, and do not act as a nutrient for ordinary molds and other microbes. However, a waxed coat will not prevent mold growth if it cracks or if air gets into cheese in any way (Davis 1965). Currently, a latex emulsion is being used for semihard cheese varieties in place of paraffin (Alvarez and Pascall 2011). Also, a variety of vacuum-packaging machines, gas-flushing machines, overwrapping machines, and vacuum skin–packaging machines are available for bulk packaging using high barrier plastic films. Coextruded pouches with low-oxygen permeability or similar materials have good barrier properties that can be used for vacuum packaging of cheese (Sayer 1998). However, during storage and associated ripening changes, the properties of ethylene vinyl alcohol (EVOH) and nylon change as moisture uptake occurs. In-bag curing under vacuum using barrier bag technology is now possible for Emmental and Parmesan cheese without loss of vital flavor and texture characteristics (Sayer 1998). In case of surface mold-ripened cheeses such as Camembert and Brie, packing does not take place until the mold has grown to a desirable extent. In such cases, packaging material must have a limited permeability to oxygen so as to minimize the growth of anaerobic proteolytic bacteria, which can also develop if the permeability to water vapor is too low resulting in condensation inside the package. Further, the material should not adhere to the surface mold of the cheese. Oriented PP with carefully designed perforations that allow controlled quantities of water vapor permeability or paper coated with wax or laminated-to-perforated film are suitable for surface mold-ripened cheeses (Robertson 2006). The packaging requirements for fresh cheeses such as cottage and quark cheeses are similar to those of other types of cheeses. Thermoformed HIPS coextruded or extrusion coated with PVC or PVDC and pigmented with titanium dioxide are used to improve the barrier properties and protection against light, while thermoformed nylon LDPE or injection-molded HDPE containers with slits in the side to allow drainage of whey are usually used for fresh cheese packaging (Robertson 2006).

For retail consumers, blocks of cheese are cut into small random weight size; vacuum packed in barrier bags; hot water shrunk; and then weighed, labeled, and reassembled as a large block and put into the carton for distribution. Also, about 8–10 pieces of rectangular or triangular chiplets of cheese or processed cheese of convenient sizes (25 g) are individually wrapped with aluminum foil and placed in an outer plastic or paperboard container. Cheese spreads may also be packed in plastic tubs and sealed with aluminum foil laminates and closed with a press on plastic lid or plastic-laminated or coextruded squeezable tubes or bottles. Cheese is also available in slices individually wrapped in plastic films. About 10 such slices are placed and sealed in an outer high barrier plastic pouch. Some of these packaging forms can be seen in Figure 24.2.

Although used for a wide range of products in dairy sector, MAP has been mainly applied to the packaging of cheese. As hard cheese quality is most likely to be affected by mold growth, an atmosphere consisting solely of CO_2 is most appropriate for MAP. Acceptable eating quality with shelf life trebled from around 4 weeks to up to 3 months can be achieved by using MAP. Hard cheeses such as cheddar are commonly packed in 100% CO_2 using horizontal form-fill-seal pillow pack machines (Damske 1990). The packaging materials used include PVDC-coated cellophane or PET-PE. Modified atmosphere packaged cheese in PP film has a shelf life of up to 4 weeks, compared with only 14–15 days when packaged under normal conditions (Hampton 1982). Cheeses are sold as cuts/blocks, sliced, or

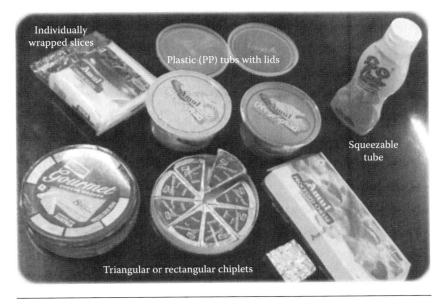

Figure 24.2 Different packaging materials and forms used for cheeses in retail packaging.

grated in order to accommodate consumer demands with respect to convenience. Reports reveal that sliced and grated cheeses are also packaged under MAP. Grated cheese is packaged using flexible films similar to those used for hard cheese blocks. However, the films may be metallized. Cheese slices are packaged in similar gas mixtures but interleaved with paper for easy separation of slices. For these products, it is not possible to use 100% CO_2 as the absorption of the gas by the product causes the packaging to collapse, crushing the product, thus interfering with the ease of separation. Therefore, N_2 is used as part of the gas mixture to stop the total collapse of the film around the product. The gas mixture typically used for these value-added products is 70% N_2/30% CO_2. Transparent films transmitting more than 80% of the incident light in the wavelength range of 400–800 nm are generally used for packaging of semihard cheeses (Kristensen et al. 2000). Sliced or grated cheeses have a large surface area exposed to light and the surrounding atmosphere and are, thus, more susceptible to oxidative-induced color and flavor changes (Juric et al. 2003). Mortensen et al. (2004) studied the impact of MAP and storage conditions on the photooxidation of sliced Havarti cheese and reported that when exposed to light and 0.6% residual O_2, the photooxidation, as quantified by the formation of 1-pentanol and 1-hexanol, significantly increased during 168 hours of storage. Therefore, it was reported that for the better storage of sliced Havarti cheese with minimal photooxidation, the O_2 content should be less than 0.6%.

The packaging of mold-ripened products under modified atmosphere is more complicated than the MAP of hard cheeses owing to the presence of live molds. As mentioned earlier, the requirement of packaging used for these products is that the mold growth is allowed to continue but at a controlled rate without causing the whole surface of the cheese to be covered with mold. The presence of air within the pack causes the mold to spread uncontrollably on the product. On the other hand, if the O_2 is totally excluded from the pack, the mold dies off, making the product unacceptable. Because of the mold-inhibiting properties of CO_2 and N_2, cheese that incorporate natural molds are not normally packed using modified atmosphere technology (Subramaniam 1993). By using a packaging material with the correct permeability and gas flushing, some manufacturers have found an acceptable solution for packaging of mold-ripened cheeses. Therefore, a compromise is required in the packaging of these cheeses.

Softer cheeses are more successfully packaged under MAP because of the cushioning effect of the gas. Gas flushing with CO_2/N_2 mixtures has been successful in doubling the shelf life to 21 days (Subramaniam 1993). The effectiveness of flushing commercial packages of cottage cheese, stored at 8°C, by pure CO_2 on the shelf life extension was evaluated by Mannheim and Soffer (1996) who reported that the shelf life could be extended by about 150% without altering the sensory properties or causing any negative effect. Study on direct-set cottage cheese by Maniar et al. (1994) reported that among various atmospheres, the quality of direct-set cottage cheese packaged in 100% CO_2 was satisfactory even after 28 days of storage at 4°C. Mozzarella, a fresh, soft unripened cheese with high moisture (50%–60%), is susceptible to microbial spoilage. Alam and Goyal (2006a, 2006b) studied the effect of five different atmospheres (air, vacuum, 100% CO_2, 100% N_2, and a mixture of 50% N_2 and 50% CO_2) on the chemical quality of Mozzarella cheese made from mixed milk and stored at 7°C ± 1°C and deep freeze conditions (–10°C to –15°C) and reported that the Mozzarella cheese packed with 100% CO_2, among others, could be stored for 12 weeks and 12 months at refrigerated (7 + 1°C) and deep freeze conditions (–10°C to –15°C), respectively, with least chemical changes. The stability of sliced Mozzarella cheese packed under three different atmospheres was studied by Alves et al. (1996) who reported that the shelf life was 63 and 45 days when packed in 100% CO_2 and 50% CO_2/50% N_2, respectively, while it was only 13 days when packed in normal air. Similarly, Eliot et al. (1998) studied the effect of MAP on microbial growth in shredded Mozzarella cheese and reported that CO_2 and N_2 in the ratio of 75:25 was optimal to suppress undesirable organisms and reduced gas formation. Cameros cheese, another soft cheese that takes its name from the Cameros geographical area in the province of La Rioja (Spain), is made from pasteurized goat's milk. It has a short shelf life of only 7 days owing to the presence of oxygen, high water activity, and high pH that allows the growth of psychrotrophs, molds, and yeasts (Olarte et al. 1999). With a view to extend the shelf life, Gonzalez-Fandos et al. (2000) have studied the effect of MAP on Cameros cheese and reported that packaging in

50% CO_2/50% N_2 and 40% CO_2/60% N_2 is the most effective conditions for extending the shelf life with good sensory characteristics.

24.4.2.1 Whey Cheeses Whey is the largest by-product of the dairy industry and is obtained during the manufacture of cheese, casein, paneer, chhana, and shrikhand. In India, milk products such as paneer, shrikhand, and chhana are very popular and are in great market demand, while cheese consumption is steadily increasing due to changing food habits. With the increase in their production levels, there is a corresponding increase in the whey as a by-product (Raju et al. 2005). Whey contains lactose, proteins, minerals, and traces of fat and organic acids, which add up to about 7% of total solids, 75% of which is lactose and 10% of which is whey protein (Mulvihill 1991). The preparation of whey cheeses, of the Ricotta type, is a better alternative for recovery of whey proteins, because denatured whey proteins are consumed as such (Mathur and Shahani 1981). Fresh whey cheeses have high pH (>6.0), high moisture content, and low salt content, and these dairy products are very susceptible to microbial spoilage, by molds, yeasts, and Enterobacteriaceae, especially under improper temperatures. Storage of fresh whey cheeses under aerobic conditions results in rapid spoilage, usually in less than 7 days, while vacuum packaging may extend the storage life to 20–30 days (Papaioannou et al. 2007). Portuguese whey cheese, also called as Requeijão, contains high contents of protein, water, and lactose. But due to poor handling practices, it is easily contaminated by environmental microbes that bring about physicochemical changes and has only 2–3 days of shelf life. Pintado and Malcata (2000) had studied the effect of MAP on physicochemical and sensorial characteristics in Requeijão cheese following the response to surface methodology using storage time, storage temperature, and fractions of CO_2 as manipulated variables and reported that plain CO_2 as a flushing gas ensures more constant composition of the cheese until 15 days at 4°C storage.

In Greece, the most popular whey cheeses are Myzithra, Anthotyros, and Manouri produced from the whey of Feta cheese or hard cheeses such as Kefalotyri or Graviera. Anthotyros has relatively high sugar content (i.e., lactose), a fat content of at least 18%, and moisture content not greater than 70%. Papaioannou et al. (2007) had studied the effect of MAP on Anthotyros cheese during storage under vacuum and MAP conditions at 4°C and reported that MAP extended the shelf life by about 10 and 20 days when stored at 4°C in 30% CO_2/70% N_2 and 70% CO_2/30% N_2, respectively, compared with vacuum packaging. However, under the same conditions, when stored at 12°C, it was reported that the shelf life was 2 and 4 days in 30% CO_2/70% N_2 and 70% CO_2/30% N_2, respectively. Myzithra Kalathaki, another typical whey cheese, has a shelf life of about 11 days when packaged aerobically due to its high moisture content (75%), low salt content (1%–1.5%), and high pH (6.8). Dermiki et al. (2008) studied the effect of MAP on the quality characteristics of Myzithra cheese when packed in four different atmospheres, namely, vacuum, 20% CO_2/80% N_2, 40% CO_2/60% N_2, and 60% CO_2/40% N_2 at 4°C ± 1°C and reported that Myzithra cheese packaged in 40% CO_2/60% N_2 and 60% CO_2/40% N_2 had a shelf life of about 14–16 and 18–20 days, respectively. Further, these reported that the lipolysis, proteolysis, and lipid oxidation were inhibited due to the presence of CO_2-containing atmospheres.

In a study on MAP Danbo cheese (30% CO_2 and 70% N_2) blocks packaged in thermoformed, transparent polylactic acid (PLA) trays with a lid made of PLA film (500 μm) and stored in a display cabinet with exposure to fluorescent light under similar conditions to those in retail stores, Holm et al. (2006) revealed that PLA packages can be used for storage of Danbo cheese for a period of 56 days where moisture loss and lipid oxidation were limited. The oxidative stability of cream cheese stored in thermoformed trays made of amorphous PET-PE, PS-EVOH-PE, and PP-PE with a transparent lid made of PET-aluminum oxide-PE film with different depth and color was studied by Pettersen et al. (2005). It was reported that cream cheese stored in trays made of amorphous PET-PE offered best protection against oxidation with respect to sensory flavor, and color of the packaging material was an important factor regarding the formation of volatile oxidation products during exposure to light.

Interest in edibles with or without antimicrobial coatings for extending the shelf life of food products including fermented dairy products is increasingly growing (Han 2000; Cagri et al. 2004; Bourtoom 2008; Kuorwel et al. 2011). Cerqueira et al. (2009) reported that among three biopolymers, namely, chitosan, galactomannan from *Gleditsia triacanthos*, and agar from *Gracilaria birdiae*, galactomannan presented the best properties to coat the *Saloio* cheese with decreased respiration rates and mold growth at the surface. Ricotta cheese, a fresh-type cheese, coated with chitosan-whey protein composite film and stored at 4°C under 40% CO_2 and 60% N_2 MAP conditions was found to have reduced microbial growth at the end of 30 days, suggesting potential for shelf-life extension (Pierro et al. 2011). Guldas et al. (2010) investigated the shelf life of single-baked mustafakemalpasa cheese-based sweets coated with different edible biopolymers, namely, κ-carrageenan, chitosan, corn zein, and whey protein concentrate, and reported that the latter two prolonged the shelf life of sweets from 3 to 10 days. The investigations on the antimicrobial effectiveness of sodium caseinate, chitosan, and sodium caseinate-chitosan coatings on the native microflora of cheese revealed that chitosan and sodium caseinate-chitosan coatings exerted significant bactericidal effect on mesophilic, psychrotropic, and yeast and mold counts (Moreira et al. 2011).

24.4.3 Packaging of Probiotic Fermented Dairy Products

As per the widely accepted definition, probiotics are defined as "live microorganisms which when administered in adequate amounts confer a health benefit on the host" (FAO/WHO 2001, p. 5). The challenge for food industry is to provide probiotic dairy products with viable probiotics to consumers by adopting appropriate processing and packaging techniques including the selection of packaging material. Being anaerobic and microaerophilic in nature, probiotics require lowest possible oxygen in the container. This would ensure viability of requirement number of cells in the finished product with desired functionality and avoid toxicity and death of the microbes. Exposure to dissolved oxygen during processing and storage is highly detrimental to *Bifidobacterium bifidum* and *Lactobacillus acidophilus* (Cruz et al. 2007). Dave and Shah (1997) reported that the dissolved oxygen contents of yogurts containing *Lb. acidophilus* filled into glass remained low, whereas those filled into high-density PE containers significantly increased. The influence of high oxygen barrier packaging materials such as PS containers (300–350 μm) and a multilayered structure (HIPS-EVOH-PE) on the dissolved oxygen in probiotic yogurt was studied by Miller et al. (2002), and it was reported that the oxygen levels in the PS containers varied from 20 to 40 ppm, while they decreased to 10 ppm at the end of 42 days of refrigerated storage period. The same researchers worked on incorporating oxygen absorbers into PS (oxygen-scavenging packaging system) and reported that such oxygen scavenger–integrated packaging materials created best conditions for favorable growth of anaerobic probiotics (Miller et al. 2003). Talwalkar et al. (2004) investigated the effect of HIPS, HIPS-EVOH-PE, and HIPS-EVOH-PE integrated with oxygen-scavenging material on survival of *Lb. acidophilus* and *Bifidobacterium* ssp. in yogurt by monitoring the oxygen concentration during storage and reported that the dissolved oxygen depended on the type of packaging material used. At the end of 42 days of storage period, the dissolved oxygen content increased (30%–38%) steadily in HIPS-packaged yogurt. The oxygen levels in the HIPS-EVOH-PE declined to values lower than 4.29 ppm, whereas the oxygen scavenger–integrated HIPS-EVOH-PE dropped to 0.44 ppm.

The effect of packaging material and storage temperature (4°C and 25°C) on the viability of microencapsulated *B. longum* B6 and *B. infantis* CCRC 14633 was studied by Hsiao et al. (2004). It was reported that survival of bifidobacteria was enhanced, when they were stored at 4°C in glass bottles. The effect of glass bottles, plastic cups, and clay pots on the survival of *B. longum* NCTC11818 incorporated into probiotic buffalo milk curd was investigated by Jayamanne and Adams (2004). The study concluded that bacteria survived best in glass bottles, followed by plastic packages and clay pots, when stored at 29°C ± 2°C. Further, it was reported that curds packaged in glass bottles and plastic packages exhibited values of 10^6 CFU/ml for up to 8 days. The effect of PP, PS, PE, and glass containers on the progress of acid development in probiotic yogurts made from goat and cow milk during 21 days of refrigerated storage was studied by Kudelka (2005). The study reported that the lowest acidity values of yogurt was in yogurt stored in PS containers, and glass packages favored the survival of probiotics due to its extremely low oxygen permeability.

24.5 Conclusions

Dairy products are highly perishable, and hence, there is an obvious need to preserve these. Although fermentation itself is a way of preservation, food packaging complements and further enhances the shelf life of fermented dairy products. In view of the availability of large number and variety of fermented milk products, packaging materials have to be judiciously chosen depending on the nature and requirement of the fermented milk product, availability, machinability, cost, and so on. Most fermented milks are packaged in rigid containers such as cups, whereas cheeses of convenient sizes (8–10 g) are individually wrapped with aluminum foil and placed in an outer plastic or paperboard container, in plastic pouches or tubs with a press on plastic lids, or in squeezable tubes or bottles. However, flexible materials are often used in emerging packaging techniques such as MAP and antimicrobial packaging. A lot of work is presently being carried out on the antimicrobial coatings of cheeses using edible polymers to enhance food safety and improve shelf life.

References

Alam, T. and Goyal, G.K. 2006a. Influence of modified atmosphere packaging on the chemical quality of Mozzarella cheese during refrigerated storage. *Journal of Food Science and Technology* 43(6): 662–666.

Alam, T. and Goyal, G.K. 2006b. Influence of modified atmosphere packaging (MAP) on the chemical quality of Mozzarella cheese stored in different packages at deep freeze conditions. *Indian Journal of Dairy Science* 59(3): 139–143.

Alvarez, V.B. and Pascall, M.A. 2011. Packaging. In: *Encyclopedia of Dairy Sciences*, 2nd edition, eds. J.W. Fuquay, P.F. Fox, and P.L.H. McSweeney, 16–23. London: Academic Press.

Alves, R.M.V., Luca-Sarantopoulos, C.I.G.De, Van Dender, A.G.F., and Assis-Fonseca-Faria, J.De. 1996. Stability of sliced Mozzarella cheese in modified atmosphere packaging. *Journal of Food Protection* 59(8): 838–844.

Bockelmann, B.V. and Bockelmann, I.V. 1998. Long-life acidified products. In: *Long-Life Products: Heat Treated, Aseptically Packaged: A Guide to Quality*, 193–196. Akarp, Sweden: B.V. Bockelmann.

Bosset, J.O., Gallmann, P.U., and Sieber, R. 1994. Influence on light transmittance of packaging materials on the shelf life of milk and dairy products—A review. In: *Food Packaging and Preservation*, 222–268. Glasgow: Blackie Academic & Professional.

Bourtoom, T. 2008. Edible films and coatings: Characteristics and properties. *International Food Research Journal* 15(3): 237–248.

Brody, A.L. 2006. Commercial uses of active food packaging and modified atmosphere packaging systems. In: *Innovations in Food Packaging*, ed. J.H. Han, 457–474. London: Elsevier Academic Press.

Bylund, G. 1995. Cultured milk products. In: *Dairy Processing Handbook*, 255–276. Lund, Sweden: Tetra Pak Processing Systems AB.

Cagri, A., Ustunol, Z., and Ryser, E.T. 2004. Antimicrobial edible films and coatings. *Journal of Food Protection* 67(4): 833–848.

Cerqueira, M.A., Lima, A.M., Souza, B.W.S., Teixeira, J.A., Moreira, R.A., and Vicente, A.A. 2009. Functional polysaccharides as edible coatings for cheese. *Journal of Agricultural and Food Chemistry* 57: 1456–1462.

Cruz, A.G.D., Faria, J.D.A.F., and Dender, A.G.F.V. 2007. Packaging system and probiotic dairy foods. *Food Research International* 40: 951–956.

Dalsgaard, T.K., Sorensen, J., Bakman, M., Vognsen, L., Nebel, C., Albrechtsen, R., and Nielsen, J.H. 2010. Light-induced protein and lipid oxidation in cheese: Dependence on fat content and packaging conditions. *Dairy Science and Technology* 90: 565–577.

Damske, L.A. 1990. Modified atmosphere packaging of dairy products—Machinery and materials. In: *Proceedings of Pack Alimentaire, 4th Annual Food and Beverage Packaging Expo & Conference*, May 1990, San Francisco, CA.

Dave, R.I. and Shah, N.P. 1997. Viability of yoghurt and probiotic bacteria in yoghurt made from commercial starter cultures. *International Dairy Journal* 7: 31–41.

Davis, J.G. 1965. The after-treatment of cheese: Packaging. In: *Cheese. Vol. I. Basic Technology*, 388–431. New York: American Elsevier Publishing Company.

Dermiki, M., Ntzimani, A., Badeka, A., Savvaidis, I.N., and Kontominas, M.G. 2008. Shelf-life extension and quality attributes of the whey cheese "Myzithra Kalathaki" using modified atmosphere packaging. *LWT—Food Science and Technology* 41(2): 284–294.

Eliot, S.C., Vuilleumard, J.C., and Emond, J.P. 1998. Stability of shredded Mozarella cheese under modified atmospheres. *Journal of Food Science* 63(6): 1075–1080.

FAO/WHO. 2001. *Joint FAO/WHO Expert Consultation on Evaluation of Health and Nutritional Properties of Probiotics in Food Including Powder Milk with Live Lactic Acid Bacteria*, 5. Cordoba, Argentina.

Floros, J.D., Nielsen, P.V., and Farkas, J.K. 2000. Advances in modified atmosphere and active packaging with applications in the dairy industry. *International Dairy Federation Bulletin* 346: 22–28.

Fox, P.F. and McSweeney, P.L.H. 2004. Cheese: An overview. In: *Cheese: Chemistry, Physics and Microbiology*, ed. P.F. Fox, P.L.H. McSweeney, T.M. Cogan, and T.P. Guinee, 1–18. London: Elsevier Academic Press.

Gokkurt, T., Findik, F., Unal, H., and Mimaroglu, A. 2012. Extension in shelf life of fresh food using nanomaterials food packages. *Polymer-Plastics Technology and Engineering* 51: 701–706.

Gonzalez-Fandos, E., Sanz, S., and Olarte, C. 2000. Microbiological, physical and sensory charcateristics of Cameros cheese packaged under modified atmospheres. *Food Microbiology* 17: 407–414.

Guldas, M., Akpinar-Bayizit, A., Ozean, T., and Yilmaz-Ersan, L. 2010. Effects of edible film coatings on shelf-life of mustafakemalpasa sweet, a cheese based dessert. *Journal of Food Science and Technology* 47(5): 476–481.

Hampton, J. 1982. Mould controlled. *Packaging Reviews* 8(10): 61.

Han, J.H. 2000. Antimicrobial food packaging. *Food Technology* 54(3): 56–65.
Han, J.H. 2005. Antimicrobial packaging systems. In: *Innovations in Food Packaging*, ed. J.H. Han, 80–107. London: Elsevier Academic Press.
Han, J.H. and Gennadios, A. 2005. Edible films and coatings: A review. In: *Innovations in Food Packaging*, ed. J.H. Han, 239–262. London: Elsevier Academic Press.
Hansson, K., Andersson, T., and Skepo, M. 2012. Adhesion of fermented dairy products to packaging materials. Effect of material functionality, storage time and fat content of the product. An empirical study. *Journal of Food Engineering* 111: 318–325.
Hernandez, R.J., Selke, S.E.M., and Culter, J.D. 2000. *Plastic Packaging: Properties, Processing, Applications and Regulations*. Munich, Germany: Hanser Publishers.
Holm, V.K., Mortensen, G., and Risbo, J. 2006. Quality changes in semi-hard cheese packaged in a poly (lactic acid) material. *Food Chemistry* 97: 401–410.
Hsiao, H.C., Lian, W.C., and Chou, C.C. 2004. Effect of packaging conditions and temperature on viability of microencapsulated bifidobacteria during storage. *Journal of the Science of Food and Agriculture* 84(2): 134–139.
Jayamanne, V.S. and Adams, M.R. 2004. Survival of probiotic bifidobacteria in buffalo curd and their effect on sensory properties. *International Journal of Food Science and Technology* 39(7): 719–725.
Juric, M., Bertelsen, G., Mortensen, G., and Petersen, M.A. 2003. Light-induced colour and aroma changes in sliced, modified atmosphere packaged semi-hard cheeses. *International Dairy Journal* 13: 239–249.
Kadoya, T. 1990. *Food Packaging*, 131–137. New York: Academic Press.
Karagül-Yüseer, Y., Coggins, P.C., Wilson, J.C., and White, C.H. 1999. Carbonated yogurt-sensory properties and consumer acceptance. *Journal of Dairy Science* 82: 1394–1398.
Kirwan, M.J. 2005. Paper and paperboard—Raw materials, processing and properties. In: *Paper and Paperboard Packaging Technology*, ed. M.J. Kirwan, 1–49. Oxford: Blackwell Publishing Ltd.
Kristensen, D., Orlien, V., Mortensen, G., Brockhoff, P., and Skibsted, L.H. 2000. Light-induced oxidation in sliced Havarati cheese packaged in modified atmosphere. *International Dairy Journal* 10(1/2): 95–103.
Kudelka, W. 2005. Changes in the acidity of fermented milk products during their storage as exemplified by natural bio-yoghurts. *Milchwissenschaft* 60(3): 294–296.
Kumar, P. and Mishra, H.N. 2004. Storage stability of mango soy fortified yoghurt powder in two different packaging materials: HDPP and ALP. *Journal of Food Engineering* 65(4): 569–576.
Kuorwel, K.K., Cran, M.J., Sonneveld, K., Miltz, J., and Bigger, S.W. 2011. Antimicrobial activity of biodegradable polysaccharide and protein-based films containing active agents. *Journal of Food Science* 76(3): R90–R102.
Lafougere, C. 1998. Packaging of milk products in Europe. *Latte* 23(10): 121–125.
Maniar, A.B., Marcy, J.E., Bishop, R., and Duncan, S.E. 1994. Modified atmosphere packaging to maintain direct-set cottage cheese quality. *Journal of Food Science* 59(6): 1305–1308.

Mannheim, C.H. and Soffer, T. 1996. Shelf-life extension of cottage cheese by modified atmosphere packaging. *Lebensmittel Wissenschaft und Technologie* 29: 767–771.
Marsh, K. and Bugusu, B. 2007. Food packaging—Roles, materials, and environmental issues. *Journal of Food Science* 72(3): R39–R55.
Mathur, B.N. and Shahani, K.M. 1981. Ricotta cheese could be your best vehicle for whey. *Dairy Field* 164(11): 110–112, 114.
Miller, C.W., Nguyen, M.H., Rooney, M., and Kailasapathy, K. 2002. The influence of packaging materials on the dissolved oxygen content in probiotic yoghurt. *Packaging Technology and Science* 15: 133–138.
Miller, C.W., Nguyen, M.H., Rooney, M., and Kailasapathy, K. 2003. The control of dissolved oxygen content in probiotic yoghurts by alternative packaging materials. *Packaging Technology and Science* 16: 61–67.
Moreira, M.D.R., Pereda, M., Marcovich, N.E., and Roura, S.I. 2011. Antimicrobial effectiveness of bioactive packaging materials from edible chitosan and casein polymers: Assessment on carrot, cheese and salami. *Journal of Food Science* 76(1): M54–M63.
Mortensen, G., Bertelsen, G., Mortensen, B.K., and Stapelfeldt, H. 2004. Light-induced changes in packaged cheeses—A review. *International Dairy Journal* 14: 85–102.
Mulvihill, D.M. 1991. Trends in the production and utilization of dairy protein products: Production. *CSIRO Food Research Quarterly* 51(3–4): 145–157.
Nelson, K.H. and Cathcart, W.M. 1984. Transmission of light through pigmented polyethylene milk bottles. *Journal of Food Science* 47: 346–348.
O'Connor-Shaw, R.E. and Reyes, V.G. 1999. Use of modified atmosphere packaging. In: *Encyclopedia of Food Microbiology*, eds. C. Batt, P. Patel, and R. Robinson, 410–416. Amsterdam, the Netherlands: Elsevier Science Publications.
Olarte, C., Sanz, S., Gonzalez-Fandos, E., and Torre, P. 1999. Microbiological and physiocochemical characteristics of Cameros cheese. *Food Microbiology* 16: 615–621.
Paltani, I. and Goyal, G.K. 2007. Packaging of dahi and yoghurt—A review. *Indian Journal of Dairy Science* 60(1): 1–11.
Papaioannou, G., Chouliara, I., Karatapanis, A.E., Kontominas, M.G., and Savvaidis, I.N. 2007. Shelf-life of a Greek whey cheese under modified atmosphere packaging. *International Dairy Journal* 17: 358–364.
Perez-Perez, C., Regalado-Gonzalez, C., Rodriguez-Rodriguez, C.A., Barbosa-Rodriguez, J.R., and Villasenor-Ortega, F. 2006. Incorporation of antimicrobial agents in food packaging films and coatings. In: *Advances in Agricultural and Food Biotechnology*, eds. R.G. Guevara-Gonzalez and I. Torres-Pacheco, 193–216. Trivandrum, India: Research Signpost.
Pettersen, M.K., Eie, T., and Nilsson, A. 2005. Oxidative stability of cream cheese stored in thermoformed trays as affected by packaging material, drawing depth and light. *International Dairy Journal* 15: 355–362.
Pierro, P.D., Sorrentino, A., Mariniello, L., Giosafatto, C.V.L., and Porta, R. 2011. Chitosan/whey protein film as active coating to extend Ricotta cheese shelf-life. *LWT—Food Science and Technology* 44(10): 2324–2327.

Pintado, M.E. and Malcata, F.X. 2000. Optimization of modified atmosphere packaging with respect to physicochemical characteristics of Requeijao. *Food Research International* 33: 821–832.

Raju, P.N., Rao, K.H., and Devi, N.L. 2005. Whey proteins and their uses in food industry. *Indian Food Industry* 24(5): 19–27, 45.

Rhim, J.W. and Ng, P.K.W. 2007. Natural biopolymer-based nanocomposite films for packaging applications. *Critical Reviews in Food Science and Nutrition* 47(4): 411–433.

Robertson, G.L. 2006. *Food Packaging: Principles and Practice*, 2nd edition. Boca Raton, FL: Taylor & Francis Group.

Robinson, R.K. and Tamime, A.Y. 1990. Microbiology of fermented milks. In: *Dairy Microbiology*, 2nd edition, ed. R.K. Robinson, Vol. 2, 291–343. London: Elsevier Applied Science.

Rooney, M.L. 2005. Introduction to active food packaging technologies. In: *Innovations in Food Packaging*, ed. J.H. Han, 63–79. London: Elsevier Academic Press.

Sacharow, S. 1976. *Handbook of Package Materials*. Westport, CT: The AVI Publishing Company Inc.

Saint-Eve, A., Levy, C., Moigne, M.L., Ducruet, V., and Souchon, I. 2008. Quality changes in yoghurt during storage in different packaging materials. *Food Chemistry* 11: 285–293.

Sayer, G. 1998. Packaging trends for cheese and other dairy products. In: *Modern Food Packaging*, 451–454. Mumbai, India: Indian Institute of Packaging.

Simal-Gandara, J. 1999. Selection of can coatings for different applications. *Food Reviews International* 15(1): 121–137.

Sivertsvik, M., Rosnes, J.T., and Bergslien, H. 2002. Modified atmosphere packaging. In: *Minimal Processing Technologies in the Food Industry*, eds. T. Ohlsson and N. Bengtsson, 61–86. Cambridge: CRC Press and Woodhead publishing limited.

Sorrentino, A., Gorrasi, G., and Vittoria, V. 2007. Potential perspectives of bionanocomposites for food packaging applications. *Trends in Food Science and Technology*, 18: 84–95.

Subramaniam, P.J. 1993. Miscellaneous applications. In: *Principles and Applications of Modified Atmosphere Packaging of Food*, ed. R.T. Parry, 170–188. Glasgow: Blackie Academic & Professional.

Talwalkar, A., Miller, C.W., Kailasapathy, K., and Nguyen, M.H. 2004. Effect of packaging materials and dissolved oxygen on the survival of probiotic bacteria in yoghurt. *International Journal of Food Science and Technology* 39(6): 605–611.

Tamime, A.Y. 2002. Fermented milks: a historical food with modern applications—A review. *European Journal of Clinical Nutrition* 56(Suppl 4): S2–S15.

Ucherek, M. 2004. An integrated approach to factors affecting the shelf life of products in modified atmosphere packaging (MAP). *Food Reviews International* 20(3): 297–307.

Vermeiren, L., Devlieghere, F., van-Beest, M., de-Kruijf, N., and Debevere, J. 1999. Developments in active packaging of foods. *Trends in Food Science and Technology* 10(3): 77–86.

WPO. 2008. *Market Statistics and Future Trends in Global Packaging*. World Packaging Organization. http://www.worldpackaging.org/publications/documents/market-statistics.pdf. Accessed on January 4, 2013.

Zhao, Y. 2004. Semisolid cultured dairy products: Packaging, quality assurance and sanitation. In: *Handbook of Food and Beverage Fermentation Technology*, eds. Y.H. Hui, L. Meunier-Goddik, A.S. Hansen, J. Josephsen, W.K. Nip, P.S. Stanfield, and F. Toldra, 133–146. New York: Marcel Dekker.

25

Xylitol

An Alternative Sweetener for Dairy Products

MAMATHA POTU, SREENIVAS RAO RAVELLA, JOE GALLAGHER, DAVID WALKER, AND DAVID BRYANT

Contents

25.1	Introduction	673
25.2	Xylitol Applications in Weight Control, Diabetes, and Dentistry	675
25.3	Xylitol in the Dairy Industry	678
	25.3.1 Sweetened Milk	678
	25.3.2 Cheese	678
	25.3.3 Ice Cream	679
	25.3.4 Yogurt	679
	25.3.5 Dadih	680
	25.3.6 Milk Chocolate	680
25.4	Pharmaceutical Applications and Confectionery Ingredient	680
	25.4.1 Xylitol Is Effective	681
	25.4.2 Xylitol Is Safe	682
25.5	Conclusion	683
References		683

25.1 Introduction

In recent years, the production of xylitol, a naturally occurring five-carbon sugar alcohol (alditol), has increased owing to its commercial application in a range of food, dental, and pharmaceutical products. The demand for xylitol is ever increasing, and the current global market size for xylitol is around 125,000 tons per annum, which is met through the chemical hydrogenation of xylose, obtained from the hemicellulosic fraction of corn cobs and/or other lignocellulosic feedstock (Rao et al. 2012). The sustained market growth, value, and versatility contribute to xylitol being one of the top-10 molecules derived from lignocellulosic biorefineries (Bozell 2008).

Several epidemiological studies have linked the consumption of sugary products with increased risk of obesity in humans, and the American Calorie Control Council found that as many as 90% of adult Americans (173 million people) eat sugar-free and other diet foods on a daily basis. Excessive sugar consumption leads to multiple major health problems such as hypoglycemia, diabetes, mineral imbalance in the body, and elevated triglyceride levels in addition to tooth decay, periodontal diseases, and obesity. Therefore, there are health benefits for people who consume foods that are low in both calorific value and glycemic index (GI) and contain a tooth-friendly sugar. Having a similar taste, lower calorific value [2.4 kcal (10.08 kJ) versus 4 kcal (16.8 kJ)], and sweetness equivalence to sucrose, xylitol is also anticariogenic and inhibits growth of *Streptococcus mutans*, the causal agent of dental caries—tooth decay. These unique properties of this sugar alcohol make it a suitable bulk, sugar-free substitute sweetener for sucrose. As such, the European Union along with the U.S. Food and Drug Administration has approved xylitol as a sweetening agent, providing it "generally recognized as safe" status for use as a food additive.

Xylitol is abundant in nature, with many fruits being known to contain this polyol sugar; however, the concentration is at such a low level that the technoeconomic feasibility of industrial-scale extraction is rendered unviable. Current commercial production processes are chemical based, whereby pure xylose is hydrogenated using acid and Raney nickel catalyst requiring high temperature and pure xylose. Cognate to these processes are the safety and environmental hazards and concerns that stem from the handling of corrosive reagents at elevated temperature and their requisite disposal following use. Biotechnological production of xylitol through microbial conversion (Kim et al., 1999, 2000, 2010; Rao et al. 2004, 2007; Prakasham et al. 2009; Aranda-Barradasa et al. 2010), by the enzymatic reduction (hydrogenation) of the functional carbonyl group of xylose by xylose reductase, offers the promise of an environmentally benign process that can use low-cost, unpurified xylose from hemicellulosic streams of several biorefining feedstocks (Granström et al. 2007a, 2007b). A variety of lignocellulosic wastes has been used as substrate for xylitol production (Rao et al. 2006; Nair and Zhao 2010; Branco et al. 2011; Li et al. 2013; Misra et al. 2013; Ping et al. 2013; Wang et al. 2011, 2012). Xylitol can be produced through fermentation using microorganisms (Cortez and Roberto 2010; Kim et al. 2012; Wang et al. 2012; Zhang et al. 2012; Zhoua et al. 2012; Guo et al. 2013; Kamat et al. 2013; Pérez-Bibbins et al. 2013).

The use of xylitol as an alternative sweetener has become quite wide as it is used in many food products, for example, juices, jams, cold drinks, cakes, and confectionary. In the past few years, the application of xylitol in dairy sectors has received attention, attracting researchers to study its incorporation into low-calorie dairy products. The positive effects of xylitol are well documented and the effect of xylitol on appetite control has been seen when formulated in yogurt (King et al. 2005). This chapter includes various applications of xylitol in different foodstuffs with special reference to dairy-based foods.

25.2 Xylitol Applications in Weight Control, Diabetes, and Dentistry

Because of its unique properties, xylitol has the potential to be used as an alternative bulk sweetener (Rao et al. 2012), and some of the chemical characteristics are listed in Table 25.1. Xylitol readily dissolves in water, tastes like sucrose, and is three times sweeter than mannitol and two times sweeter than sorbitol. As xylitol has a high negative heat of solution, it has the additional advantage of conferring a fresh and cool pleasant sensation during oral consumption (Alminoff et al. 1978; Pepper and Olinger 1988; Pepper 1989), which may be beneficial in yogurt and related dairy products. An overview of xylitol applications is given in Figure 25.1.

Xylitol is a natural wood sugar, and due to its unique properties, its global demand is constantly increasing. It is widely used in chewing gums, confectionary, jams, low-calorie drinks, ice creams, and dental products to reduce tooth decay and to cure ear infections, especially in children (Pepper and Olinger 1988; Mäkinen 2000a; Lynch and Milgrom 2003; Tange et al. 2004; Inagaki et al. 2011; Elsalhy et al. 2012). It is also used as a sugar substitute for diabetic patients (Gare 2003), and the xylitol market continues to experience a high

Table 25.1 Characteristics of Xylitol

PROPERTIES	
Molecular formula	$C_5H_{12}O_5$
Molar mass	152.15 g mol^{-1}
Boiling point	216°C
Density	1.52 g/cm³
Melting point	92°C–96°C
Solubility in water	~1.5 g/ml

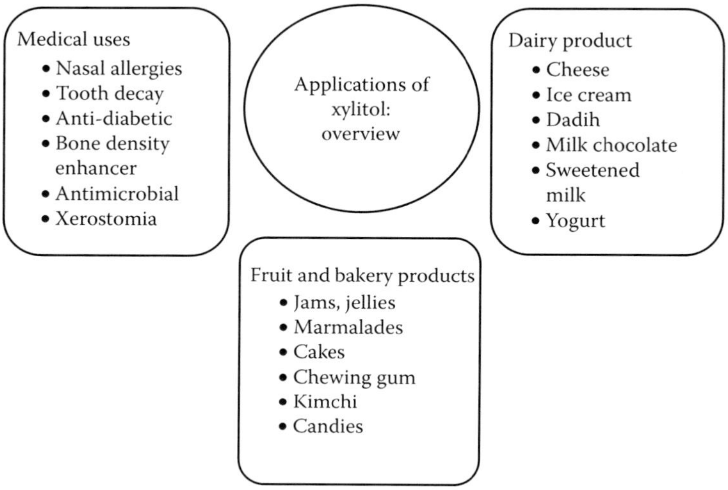

Applications of xylitol: overview

- **Medical uses**
  - Nasal allergies
  - Tooth decay
  - Anti-diabetic
  - Bone density enhancer
  - Antimicrobial
  - Xerostomia

- **Dairy product**
  - Cheese
  - Ice cream
  - Dadih
  - Milk chocolate
  - Sweetened milk
  - Yogurt

- **Fruit and bakery products**
  - Jams, jellies
  - Marmalades
  - Cakes
  - Chewing gum
  - Kimchi
  - Candies

Figure 25.1 Overview of xylitol applications.

demand and rapid growth worldwide due to increasingly health conscious consumerism.

Xylitol is also used in many different processes such as controlling the economically important mite *Acarus farris* on Cabrales cheese (Sanchez-Ramos and Castanera 2009), pasteurization at low pH (Klewicki 2007), interacting with other sweeteners in aqueous systems (Gliemmo et al. 2008), tooth-friendly soft drinks (Kolahi et al. 2009), starch industry (Sun et al. 2014), plasticized bio-nanocomposites (Liu and Chaudhary 2013), preparation of low-calorie sugar-free sponge cakes (Ronda et al. 2005), digestive stability of green tea (Shim et al. 2012), juices (Türkmen and Eksi 2011), and tablets (Inagaki et al. 2011).

Acid-producing cariogenic bacteria cannot utilize xylitol as a carbon source for their growth, which is one mode of action, among others, whereby xylitol significantly contributes to reducing tooth decay

(Rolla et al. 1987; Mir 1988; Mäkinen 2000a; Miake et al. 2003; Kakuta et al. 2003). In the early 1970s at Turku University, scientists conducted studies on xylitol and its effect on dental caries. They demonstrated significant and convincing results that helped food companies in developing xylitol-based products (Makinen 2000b). Xylitol is specific in its inhibition of the *mutans streptococci* group, the bacteria that causes caries, and it can also inhibit these bacteria in the presence of other sugars (Mäkinen 2000a). Multiple field trials have demonstrated that a small amount of xylitol can reduce dental caries (Isokangas et al. 1988; Mir 1988; Gary et al. 2000), and the dental health benefits afforded by the use of this molecule are regarded superior compared to all other polyol sweeteners (Bar 1991; Mäkinene 1992, 2000a; Maguire and Rugg-Gunn 2003; Miake et al. 2003; Moynihan 2005). Following these convincing data in terms of caries prevention, a number of xylitol-based medical and nutritional products have now been developed (Makinen 2000b; Hayes 2001).

Jannesson et al. (2002) studied the effect of toothpaste on the growth of *S. mutans* that is present in saliva and dental plaque. Three types of dentifrice were tested: (1) "Colgate total" with 10% xylitol called total xylitol; (2) Colgate total, and (3) Colgate total without xylitol and triclosan. The 6-month study demonstrated that the addition of 10% xylitol to a triclosan-containing dentifrice reduced the number of *S. mutans* in saliva and dental plaque. Miake et al. (2003) studied the re-mineralization effect of xylitol on artificially demineralized enamel. The artificially demineralized samples were treated with 20% xylitol for 2 weeks at 37°C. Contact microradiography and a high-resolution electron microscopy were used to study re-mineralization, with re-mineralization being observed in samples treated with xylitol.

On its own or in combination with other sweeteners, xylitol can be used to prepare a wide variety of products for a sugarless confectionary industry (Pepper 1989; Mussatto 2012). Pure xylitol is now available in supermarkets, in addition to several products that contain xylitol as an ingredient. The use of xylitol in the preparation of low-calorie bakery products, jams, jellies, marmalades and desserts, and other dairy products are potential applications of xylitol in a number of food sectors (Emidi 1978). Food products have a GI rating between 1 and 100, with pure glucose and sucrose rating a GI of 100 and 110, respectively, while the healthiest food stuffs have a low GI value (Gare 2003). Diabetics can tolerate xylitol as it has a low GI of 7, based on the effect on blood sugar after immediate intake of xylitol-containing foods, and its metabolism is non-insulin dependent. As a result, xylitol has been advertised as "safe for diabetics and individuals with hyperglycaemia." Gare (2003) discussed the healing effects of xylitol for different diseases.

25.3 Xylitol in the Dairy Industry

Presently, nutraceuticals and therapeutic dairy foods (e.g., yogurt, cheese, and ice creams) are getting popular due to increased lifestyle diseases (e.g., diabetes, cancer, high cholesterol). Different remedies are available for these diseases, for example, diabetes can be controlled using alternative sweeteners. The role of xylitol as a sweetener is well documented, and in this section, some of the applications of xylitol in dairy products are explained.

25.3.1 Sweetened Milk

Generally, the sweet taste of milk is the result of available sucrose, which imparts both high GI and calorie intake. Therefore, xylitol can be used along with a low amount of sucrose to prepare a low-calorie dairy product. In Peru, a study was conducted by Castillo et al. (2005) to determine the taste acceptability of milk that contained xylitol. For this, Peruvian children ($n = 75$) aged from 4 to 7 were chosen, where the children tasted two cups of milk containing xylitol (0.021 and 0.042 g/ml). The results demonstrated that milk sweetened with xylitol was well accepted by the children, thereby indicating its potential application in dairy.

25.3.2 Cheese

The effects of xylitol on the functional properties of low-fat processed cheese were studied, and it was observed that xylitol significantly decreased the hardness of low-fat processed cheese, and improved the functional properties of low-fat processed cheese. Kommineni et al. (2012) demonstrated that xylitol can maintain the soft texture of cheese that enables the reduction of 80% of the fat during cheese manufacturing. Xylitol improved the functional properties of low-fat cheese processing by maintaining the key desirable characteristics of texture and melting property, which are technically difficult issues in low-fat cheese processing. Use of 2%–4% of xylitol can maintain a good texture and melting characteristics.

25.3.3 Ice Cream

Ice cream is a high-calorie product, and sucrose is the basic sweetening component. In order to reduce the calorie measure of ice cream, sucrose was replaced by xylitol, resulting in a significant reduction in calorie intake without any adverse effect. The acceptability or sensory evaluations of this ice cream were well received. However, the consistency of ice cream was altered, as the ice cream containing xylitol was soft as compared to sucrose-based ice cream (Kracher 1975). Soukoulis et al. (2010) observed that xylitol, when added in ice cream, had the most prominent viscosity-enhancing behavior and concluded that the hygroscopic character of xylitol is the major reason for such rheological properties of the product.

25.3.4 Yogurt

Yogurt is a very popular dairy product worldwide, and its therapeutic attributes are also well understood. However, its sweetened taste that increases its calorie value may be overcome by reducing sucrose levels and supplementing it with alternative sweeteners such as xylitol. Hyvönen and Slotte (1983) evaluated the impact of xylitol in yogurt. Thier study did not show any change in pH upon supplementation; sensory evaluations were as good as with sucrose; and it was found to be a good yogurt sweetener for preincubation sweetening at a level of 8%. The retarding effect of xylitol was noticed as negligible. Additionally, xylitol-consisting yogurt imparts functional food attributes to the product (King et al. 2005) the benefits of which have been discussed earlier. Different yogurt formulations prepared with xylitol, polydextrose, and sucrose demonstrated that xylitol and polydextrose could be useful in preventing weight gain due to their low-calorific value. In general, sucrose is widely used in yogurt industry as a sweetening agent (Ozer 2010), but during the past two decades, consumer demands for low-calorie fruit yogurt have increased vastly.

25.3.5 Dadih

Dadih is popular among Southeast Asian people; however, it has a high-calorific value due to its sweetened ingredients, and trials to formulate a low-calorie dadih have recently been performed (Mohd Thani et al. 2014). These data again demonstrated that the presence of xylitol in *dadih* did not affect the sensory qualities; however, it did impart a softer texture, as compared to *dadih* containing sucrose.

25.3.6 Milk Chocolate

It is well known that chocolates are high-calorie products. Chocolates consisting of milk as an ingredient also contain a proportion of butter fat that impacts the eutectic effect and prevents bloom formation, eventually resulting in a lower melting point and soft texture. Xylitol is used for the manufacture of low-calorie or sugar-free products. The replacement of sucrose with xylitol affects the rheological properties, other processing conditions, and finally the quality of chocolates. Chocolate sweetened with isomalt also resulted in higher plastic viscosity, while xylitol provided a higher flow behavior index (Afoakwa et al. 2007).

25.4 Pharmaceutical Applications and Confectionery Ingredient

Xylitol has a wide variety of applications to reduce caries, sinus, and acute otitis media (Table 25.2). It can be used to deliver health benefits, high-quality texture, an extended shelf life, and a range of processing benefits (Table 25.3). Xylitol delivers clean sweetness with no aftertaste and an intense cooling effect. In dental protection products such as sugar-free gum, it helps protect against plaque to reduce the development of cavities. Danisco® is one of the world's largest producers of xylitol using hardwoods and maize as the feedstocks, and the pharmaceutical industry uses xylitol as a sweetening agent in its products. DuPont™ produces xylitol to satisfy both industry requirements and health-conscious consumers, offering innovative solutions

Table 25.2 Xylitol Applications in Different Industries

FUNCTIONS	APPLICATIONS
Easily metabolized in the human body	Diabetic food (da Silva et al. 2012)
Granular quality improver, which prevents denaturation by freezing and gives firm texture and maintains whiteness	Frozen fish and meat products
Improved shelf life of hard-coated chewing gum in respect of atmospheric moisture	Hard-coated chewing gum (Maguire and Rugg-Gunn 2003)
Improved stability	Improved digestive stability and absorption of green tea polyphenols (Shim et al. 2012)
Nonsugar sweetener	Chewing gums, sugarless chocolates, puddings, jams, baked products, and ice creams (Gare 2003; Mussatto 2012)
Provide calorific value	Preparation of low-calorie artificial honey
Starch industry	Change physiochemical and morphological properties of wheat starch (Sun et al. 2014)
Sweetening agent	Soft drinks (Kolahi et al. 2009), sugar-free cakes (Ronda et al. 2005), juice (Türkmen and Eksi 2011)
Sweetening property of xylitol matches that of sucrose	Beverages (Mäkinen 2000b)

for no/low-sugar confectionery ingredients and for reducing the fat content in confectionery products. DuPont™ and Danisco® sweeteners include XIVIA™ xylitol, which is proposed to be a sustainable solution, using unique processes that consume fewer materials and 85% less energy than in conventional processes for the production of xylitol. New, innovative, scalable, industrial biotechnology fermentation processes for xylitol production using xylitol-producing yeast are being developed in Europe. These activities are funded by the European Institute for Innovation Technology through Climate-KIC alongside biorefining activities in BEACON, funded by the Welsh European Funding Office.

25.4.1 Xylitol Is Effective

Food and confectionary products that are high in sugars promote caries. Xylitol is tooth friendly and reduces caries. Many clinical trials conducted using xylitol products have shown the health benefits of xylitol. The past three decades of testing xylitol-containing products confirm xylitol to be the best molecule or bulk sweetener for healthier

Table 25.3 Medical Applications of Xylitol

FUNCTIONS	APPLICATIONS
Allow the sugar combination to penetrate the blood brain barrier	Pharmaceutical preparations (da Silva et al. 2012; Levi et al. 2013)
Cavity fighter	Chewing gum, which curbs bacteria and reverses early decay (Mäkinen 2000a)
Medicament to lower the intraocular pressure	Topical application in the eye for treatment of increased intraocular pressure
Non-carcinogenic humectant sweetener to inhibit plaque formation and reduce gingivitis, caries, and fissures	Manufacture of dentifrices
Nonsugar sweetener	Chewing gum, tablets, cough syrups, mouthwashes, toothpastes, and medical chewing gum for oral hygiene and treatment of oral disease (Mäkinen 2000a, 2000b)
Reduce the sinusitis	Nasal spray (Levi et al. 2013)
Remove dental plaque and act as antimicrobial agent	Mouthwash of pleasant taste and toothpaste
AS A NUTRIENT	
Adhesive property	Coatings
Applications in miscellaneous sector	Preparation of oral and intravenous nutrients
Heat storage material	Reusable heat device
Restore the balance of skin flora while inhibiting pathogenic germs	Cosmetic preparation

products that reduce dental caries, and several dental associations have released endorsements for further use of xylitol (Isokangas et al. 1988; Gary et al. 2000; Makinen 2000b; Hayes 2001; Jannesson et al. 2002; Maguire and Rugg-Gunn 2003; da Silva et al. 2012).

25.4.2 Xylitol Is Safe

Based on the Federation of American Societies for Experimental Biology report, xylitol is safe if used properly. It is safe for all ages and can be delivered through a variety of products. Different brands of chewing gums that are 100% sweetened with xylitol are available in markets all over the world. Due to high demand for food in low calories, the xylitol market is expanding, and several new formulations are currently in development for yogurt, ice cream, and other dairy and milk products sweetened with xylitol.

25.5 Conclusion

Xylitol is a natural sweetener and has additional health benefits beyond that of simply replacing sucrose and fructose sugars. During the last two decades, consumer demand for low-calorie fruit yogurts has increased tremendously, and xylitol can be used in combination with low levels of sucrose for the preparation of low-calorie yogurt and other dairy products. Future advances in fermentation technology will enable xylitol production through industrial biotechnology processes.

References

Afoakwa, E.O., Paterson, A., and Fowler, M. 2007. Factors influencing rheological and textural qualities in chocolate—A review. *Trends in Food Science & Technology* 18: 290–298.

Alminoff, C., Vanninen, E., and Doty, T.E. 1978. Xylitol-occurrence, manufacture and properties. *Oral Health* 68(4): 28–29.

Aranda-Barradasa, J.S., Garibay-Orijela, C., Badillo-Coronab, J.A., and Salgado-Manjarreza, E. 2010. A stoichiometric analysis of biological xylitol production. *Biochemical Engineering Journal* 50: 1–9.

Bar, A. 1991. Xylitol. In: *Alternative Sweetener*, eds. L.O. Nabors, and R.C. Gelardi, 351. New York: Marcel Dekker.

Bozell, J.J. 2008. Feedstocks for the future-biorefinery production of chemicals from renewablecarbon. *Clean* 36: 625.

Branco, R.F., Santos, J.C., and Silva, S.S. 2011. A novel use for sugarcane bagasse hemicellulosic fraction: Xylitol enzymatic production. *Biomass Bioenergy* 35: 3241–3246.

Castillo, J.L., Milgrom, P., Coldwell, S.E., Castillo, R., and Lazo, R. 2005. Children's acceptance of milk with xylitol or sorbitol for dental caries prevention. *BMC Oral Health* 5: 6.

Cortez, D.V. and Roberto, I.C. 2010. Individual and interaction effects of vanillin and syringaldehyde on the xylitol formation by *Candida guilliermondii*. *Bioresource Technology* 101: 1858–1865.

Da Silva, A.F., Ferreira, A.S., da Silva, S.S., and Raposo, N.R.B. 2012. Medical application of xylitol: An appraisal. In: *D-Xylitol: Fermentative Production, Application and Commercialization*, eds. S.S. da Silva, and A.K. Chandal, 325–342. Berlin, Germany: Springer-Verlag.

ElSalhy, M., Sayed Zahid, I., and Honkala, E. 2012. Effects of xylitol mouthrinse on *Streptococcus mutans*. *Journal of Dentistry* 40: 1151–1154.

Emidi, A. 1978. Xylitol, its properties and food applications. *Food Technology* 32: 20–32.

Gare, F. 2003. *The Sweet Miracle of Xylitol*. Laguna Beach, CA: Basic Health Publications.

Gary, H., Hildebrandt, D.D.S., and Brandon, S.S. 2000. Maintaining mutans streptococci suppression with xylitol chewing gum. *The Journal of the American Dental Association* 131: 909–916.

Gliemmo, M.F., Calvino, A.M., Tamasi, O., Gerschenson, L.N., and Campos, C.A. 2008. Interactions between aspartame, glucose and xylitol in aqueous systems containing potassium sorbate. *LWT—Food Science and Technology* 41: 611–619.

Granström, T.B., Izumori, K., and Leisola, M. 2007a. A rare sugar xylitol. Part I: the biochemistry and biosynthesis of xylitol. *Applied Microbiology and Biotechnology* 74: 277–281.

Granström, T.B., Izumori, K., and Leisola, M. 2007b. A rare sugar xylitol. Part II: biotechnological production and future applications of xylitol. *Applied Microbiology and Biotechnology* 74: 273–276.

Guo, X., Zhang, R., Li, Z., Dai, D., Li, C., and Zhou, X. 2013. A novel pathway construction in *Candida tropicalis* for direct xylitol conversion from corncob xylan. *Bioresource Technology* 128: 547–552.

Hayes, C. 2001. The effect of non-cariogenic sweeteners on the prevention of dental caries: A review of the evidence. *Journal of Dental Education* 65: 1106–1109.

Hyvönen, L. and Slotte, M. 1983, Alternative sweetening of yoghurt. *International Journal of Food Science and Technology* 18: 97–112.

Inagaki, S., Saeki, Y., and Ishihara, K. 2011. Funoran containing xylitol gum and tablets inhibit adherence of oral *Streptococci*. *Journal of Oral Biosciences* 53: 82–86.

Isokangas, P., Alanen, P., Tiekso, J., and Makinen, K.K. 1988. Xylitol chewing gum in caries prevention: A field study in children. *The Journal of the American Dental Association* 117: 315–320.

Jannesson, L., Renvert, S., Kjellsdotter, P., Gaffar, A., Nabi, N., and Birkhed, D. 2002. Effect of a triclosan-containing toothpaste supplemented with 10% xylitol on mutans streptococci in saliva and dental plaque. A 6-month clinical study. *Caries Research* 36: 36–39.

Kakuta, H., Iwami, Y., Mayanagi, H., and Takahashi, N. 2003. Xylitol inhibition of acid production and growth of mutans *Streptococci* in the presence of various dietary sugars under strictly anaerobic conditions. *Caries Research* 37: 404–409.

Kamat, S., Khot, M., Zinjarde, S., Kumar, A.R., and Gade, W.N. 2013. Coupled production of single cell oil as biodiesel feedstock, xylitol and xylanase from sugarcane bagasse in a biorefinery concept using fungi from the tropical mangrove wetlands. *Bioresource Technology* 135: 246–253.

Kim, J.H., Ryu, V.W., and Seo, J.H. 1999. Analysis and optimization of a two substrate fermentation for xylitol production using *Candida tropicalis*. *Journal of Industrial Microbiology and Biotechnology* 22: 181–186.

Kim, M.S., Kim, C., Seo, J.H., and Ryu, Y.W. 2000. Enhancement of xylitol yields by xylitol dehydrogenase defective mutant of *Pichia stipitis*. *Korean Journal Biotechnology and Bioengineering* 15: 113–119.

Kim, S.R., Ha, S.J., Kong, I.I., and Jin, Y.S. 2012. High expression of XYL2 coding for xylitoldehydrogenase is necessary for efficient xylose fermentation by engineered *Saccharomyces cerevisiae*. *Metabolic Enggneering* 14: 336–343.

Kim, S.Y., Yun, J.Y., Kim, S.G., Seo, J.H., and Park, J.B. 2010. Production of xylitol from d-xylose and glucose with recombinant *Corynebacterium glutamicum*. *Enzyme and Microbial Technology* 46: 366–371.

King, N.A., Craig, S.A., Pepper, T., and Blundell, J.E. 2005. Evaluation of the independent and combined effects of xylitol and polydextrose consumed as a snack on hunger and energy intake over 10 d. *British Journal of Nutrition* 93: 911–915.

Klewicki, R. 2007. The stability of gal-polyols and oligosaccharides during pasteurization at a low pH. *LWT—Food Science and Technology* 40: 1259–1265.

Kolahi, J., Fazilati, M., and Kadivar, M. 2009. Towards tooth friendly soft drinks. *Medical Hypotheses* 73: 524–525.

Kommineni, A., Amamcharla, J., and Metzger, L.E. 2012. Effect of xylitol on the functional properties of low-fat process cheese. *Journal of Dairy Science* 95: 6252–6259.

Kracher, F. 1975. Xylit, Bedeutung-Wirkung-Anwendung. *Kakao & Zucker* 27(3): 68–70, 72–74.

Levi, J.R., Brody, R.M., McKee-Cole, K., Pribitkin, E., and O'Reilly, R. 2013. Complementary and alternative medicine for pediatric otitis media. *International Journal of Pediatric Otorhinolaryngology* 77: 926–931.

Li, Z., Qu, H., Li, C., and Zhou, X. 2013. Direct and efficient xylitol production from xylan by *Saccharomyces cerevisiae* through transcriptional level and fermentation processing optimizations. *Bioresource Technology* 149: 413–419.

Liu, H. and Chaudhary, D. 2013. Effect of montmorillonite on morphology, glass transition and crystallinity of the xylitol-plasticized bionanocomposites. *Carbohydrates Polymers* 98: 391–396.

Lynch, H. and Milgrom, P. 2003. Xylitol and dental caries: An overview for clinicians. *Journal of the California Dental Association* 31: 205–209.

Maguire, A. and Rugg-Gunn, A.J. 2003. Xylitol and caries prevention—Is it a magic bullet? *British Dental Journal* 194: 429–436.

Mäkinen, K.K. 1992. Latest dental studies on xylitol and mechanism of action of xylitol in caries limitation. In: *Progress in Sweeteners*, ed. T.H. Greenby, 331–362. London: Elsevier Applied Science.

Mäkinen, K.K. 2000a. Can the pentitol-hexitol theory explain the clinical observations made with xylitol? *Medical Hypothesis* 54: 603–613.

Mäkinen, K.K. 2000b. The rocky road of xylitol to its clinical application. *Journal of Dental Research* 79: 1352–1355.

Miake, Y., Saeki, Y., Takahashi, M., and Yanagisawa, T. 2003. Remineralization effects of xylitol on demineralized enamel. *Journal of Electron Microscopy (Tokyo)* 52: 471–476.

Mir, A. 1988. Caries prevention with xylitol. A review of the scientific evidence. *World Review of Nutrition and Dietician* 55: 183–209.

Misra, S., Raghuwanshi, S., and Saxena, R.K. 2013. Evaluation of corncob hemicellulosic hydrolysate for xylitol production by adapted strain of *Candida tropicalis*. *Carbohydrate Polymers* 92: 1596–1601.

Mohd Thani, N., Mustapa Kamal, S.M., Taip, F.S., and Awang Biak, D.R. 2014. Assessment on rheological and texture properties of xylitol-substituted Dadih. *Journal of Food Process Engineering*. doi:10.1111/jfpe.12100.

Moynihan, P.J. 2005. The role of diet and nutrition in the etiology and prevention of oral diseases. *Bulletin of the World Health Organization* 83: 694–699.

Mussatto, S.I. 2012. Application of xylitol in food formulations and benefits for health. In: *D-Xylitol: Fermentative Production, Application and Commercialization*, eds. S.S. da Silva and A.K. Chandal, 309–323. Berlin, Heidelberg, Germany: Springer-Verlag.

Nair, N.U. and Zhao, H. 2010. Selective reduction of xylose to xylitol from a mixture of hemicellulosic sugars. *Metabolic Engineering* 12: 462–468.

Ozer, B. 2010. Strategies for yogurt manufacturing. In: *Development and Manufacture of Yogurt and Functional Dairy Products*, ed. F. Yildiz, 47–46. Boca Raton, FL: Taylor & Francis Group.

Pepper, T. 1989. The use of xylitol in confectionery production. *Confectionery Production* 3: 253–256.

Pepper, T. and Olinger, P.M. 1988. Xylitol in sugar free confections. *Food Technology* 10: 98–106.

Pérez-Bibbins, B., Salgado, J.M., Torrado, A., María Guadalupe Aguilar-Uscanga, M.G., and Domínguez, J.M. 2013. Culture parameters affecting xylitol production by *Debaryomyces hansenii* immobilized in alginate beads. *Process Biochemistry* 48: 387–397.

Ping, Y., Ling, H.Z., Song, G., and Ge, J.P. 2013. Xylitol production from nondetoxified corncob hemicellulose acid hydrolysate by *Candida tropicalis*. *Biochemical Engineering Journal* 75: 86–91.

Prakasham, R.S., Rao, R.S., and Hobbs, P.J. 2009. Current trends in biotechnological production of xylitol and future prospects. *Current Trends in Biotechnology and Pharmacy* 3: 8–36.

Rao, R.S., Bhadra, B., and Shivaji, S. 2007. Isolation and characterization of xylitol producing yeasts from the gut of colleopteran insects. *Current Microbiology* 55: 441–446.

Rao, R.S., Gallagher, J., Fish, S., and Prakasham, R.S. 2012. Overview on commercial production of xylitol, economica analysis and market trends. In: *D-Xylitol: Fermentative Production, Application and Commercialization*, eds. S.S. da Silva and A.K. Chandal, 291–306. Berlin, Heidelberg, Germany: Springer-Verlag.

Rao, R.S., Jyothi, Ch., Prakasham, R.S., Sharma, P.N., and Rao, L.V. 2006. Xylitol production from corn fiber and sugarcane bagasse hydrolysates by *Candida tropicalis*. *Bioresource Technology* 97: 1974–1978.

Rao, R.S., Prakasham, R.S., Krishna, K.P., Rajesham, S., Sharma, P.N., and Rao, L.V. 2004. Xylitol production by *Candida* sp.: Parameter optimization using Taguchi approach. *Process Biochemistry* 39: 951–956.

Rolla, G., Schele, A.A., and Assev, S. 1987. Plaque formation and plaque inhibition. *Deutsche Zahnärztliche Zeitschrift Z* 42: 39–41.

Ronda, F., Gómez, M., Blanco, C.A., and Caballero, P.A. 2005. Effects of polyols and nondigestible oligosaccharides on the quality of sugar-free sponge cakes. *Food Chemistry* 90: 549–555.

Sanchez-Ramos, I. and Castanera, P. 2009. Chemical and physical methods for the control of the mite *Acarus farris* on Cabrales cheese. *Journal of Stored Product Research* 45: 61–66.

Shim, S.M., Yoo, S.H., Ra, C.S., Kim, Y.K., Chung, J.O., and Lee, S.J. 2012. Digestive stability and absorption of green tea polyphenols: Influence of acid and xylitol addition. *Food Research International* 45: 204–210.

Soukoulis, C., Rontogianni, E., and Tzia, C. 2010. Contribution of thermal, rheological and physical measurements to the determination of sensorially perceived quality of ice cream containing bulk sweeteners. *Journal of Food Engineering* 100: 634–641.

Sun, Q., Dai, L., Nan, C., and Xiong, L. 2014. Effect of heat moisture treatment on physicochemical and morphological properties of wheat starch and xylitol mixture. *Food Chemistry* 143: 54–59.

Tange, T., Sakurai, Y., Hirose, M., Noro, D., and Igarashi, S. 2004. The effect of xylitol and fluoride on remineralization for primary tooth enamel caries in vitro. *Pediatric Dental Journal* 14: 55–59.

Türkmen, I. and Eksi, A. 2011. Brix degree and sorbitol/xylitol level of authentic pomegranate (*Punica granatum*) juice. *Food Chemistry* 127: 1404–1407.

Wang, L., Wu, D., Tang, P., Fan, X., and Yuan, Q. 2012. Xylitol production from corncob hydrolysate using polyurethane foam with immobilized *Candida tropicalis*. *Carbohydrate Polymers* 90: 1106–1113.

Wang, L., Yang, M., Fan, X., Zhu, X., Xu, T., and Yuan, Q. 2011. An environmentally friendly and efficient method for xylitol bioconversion with high-temperature-steaming corncob hydrolysate by adapted *Candida tropicalis*. *Process Biochemistry* 46: 1619–1626.

Zhang, J., Geng, A., Yao, C., Lu, Y., and Li, Q. 2012. Effects of lignin-derived phenolic compounds on xylitol production and key enzyme activities by a xylose utilizing yeast *Candida athensensis* SB18. *Bioresource Technology* 121: 369–378.

Zhoua, P., Li, S., Xu, H., Feng, X., and Ouyang, P. 2012. Construction and co-expression of plasmid encoding xylitol dehydrogenase and a cofactor regeneration enzyme for the production of xylitol from d-arabitol. *Enzyme and Microbial Technology* 51: 119–124.

Index

Note: Locators followed by "*f*" and "*t*" denote figures and tables in the text

6-phosphogluconate/
    phosphoketolase pathway
    (6-PG/PK), 134–135

A

ACE inhibitory activity of viili, 510
Acetaldehyde, 120
Acetic acid bacteria (AAB), 466
    kefir beverages, 468
    kefir grains, 104, 466–467
Acetone, 121
Acid cheese whey, 429, 430*t*, 435
Acid curd cottage cheese, 331
Acidophilus milk, 103
    biochemical effects of, 258
    biological value of, 250
    carbonated, 235
    containing *B. bifidum*, 290–291
    feeding of, 256
    inulin, 250
    manufacturing
        economical method of, 232
        of infant milk powder, 243, 243*f*
        of spray dried banana powder, 243, 244*f*
    new process of, 238
    nutritional benefits of, 250–253
        improved bioavailability of minerals, 251–252
        improved vitamin supplementation, 252–253
    products available in global market, 230, 231*t*
    technology and microbiology of, 229–250
        acidophilin, 245–246
        cheese, 241–242
        cream culture, 238–239
        ice cream, 239–241
        paste, 242
        powder, 242–245
        sour milk, 230–236, 233*f*
        sweet milk, 236
        whey, 238

Acidophilus milk (*Continued*)
  yeast milk, 237–238
  yogurt, 246–250
    therapeutic benefits of, 253–258
Acido yeast milk, 253
Acids and alcoholic compounds, 121–122
Active packaging, 651–652
Activia, 288
Adenosine triphosphate (ATP), 145
Adjunct probiotic cultures added to yogurt, 321, 321*t*
Agaran fermented cream, 11
Alcoholic whey beverages, cheese, 434–436, 436*t*
Allergy/autoimmune disease, probiotic health benefits, 565
α-keto acid decarboxylases, 116
American Calorie Control Council, 674
Amino acid metabolism, 115
Amino acids, 28, 115, 123, 314–315
Ammonium salt, 171
Ampiang dadih, 378
Angiotensin converting enzyme (ACE), 504
Antibiotic-associated diarrhea (AAD), probiotic health benefits, 563
Anticancer activities, 37–38, 38*f*
Anti-HIV activity, 39
Antimicrobial packaging, 652
Antimicrobial potential, 37
Antioxidative effect of viili, 509
Antitumor activity of viili, 511
Appetite suppressors, 38–39
Arkhi (high-alcoholic drink), 485
Aroma compounds, 117
Aromatic esters, 116, 123
Ascorbic acid, 186
Aseptic packaging, 649
Astringent, 206
ATP-binding cassette (ABC), 140–141

Ayran, 406–408
  average composition of, 406
  categories of, 406
  flavor compound of, 407
  manufacturing steps for, 406, 407*f*
  total production in Turkey (2012), 406

B

Bacillus cereus bacteria, 463
Backslopping method, 402
Bacterial cultures, 334
Bacteriocins, 100, 101*t*, 125
B-actin, 9
Bamboo ater, 386
Bamboo gombong, 386
Bamboo tubes as container of dadih product, 386–387
Belgian Co-ordinated Collections of Microorganisms (BCCM/LMG), 576
Bergey's Manual of Determinative Bacteriology, 274
Bergey's Manual of Systematic Bacteriology, 273
Betabacterium, 89, 89*t*
β-galactosidase, 248, 438, 446–447
Beyaz cheese, 63
Bifidobacteria (*Bifidobacterium*), 90–91, 105, 272–278, 485, 577
  ability of, 275, 279
  as biotherapeutic agent, 283
  in carbohydrate fermentation, 275
  consumption of, 270
  distribution pattern, 275
  human strains of, 275
  important milestones in history of, 273, 273*t*
  isolated from human and milk, 274, 274*t*
  measures to improve viability in bifidus products, 292–295

microorganism, 53t, 55, 59
products containing, 283–292
    acidophilus milk containing
        B. bifidum, 290–291
    bifider, 291
    bifidogene, 291
    bifidus active, 287
    bifidus baby foods, 287–288
    bifidus milk, 284–285, 286f
    bifidus milk with yogurt
        flavor, 286, 287t
    bifidus yogurt, 288
    bio-spread, 291
    BRA sweet milk, 289
    diphilus milk, 291
    milk- and water-based cereal
        puddings, 291–292
    Mil-Mil, 288–289
    progurt, 289–290, 290f
    sweet acidophilus bifidus
        milk, 289
    sweet bifidus milk, 289
    tarag, 289
safety aspect of, 295
strains, 280, 280t, 297
technologies to overcome
    limitations, 278
Bifidobacterium animalis, 558
Bifidobacterium bifidum, 90, 521
Bifidobacterium infantis, 564
Bifidobacterium longum, 59, 189
Bifidogenic factors, 275–276, 276t–277t
Bifidus milk, 105
Bifidus pathway, 275
Bifidus products
    classification of, 278, 279f
    in developed countries, 270
    history, 271–272
    measures to improve viability of
        bifidobacteria in, 292–295
        encapsulation, 292–293
        exploiting cellular stress
            responses, 293

        genetic characterization,
            293–295, 294t
        use of prebiotics, 293
    milk-based, 284, 284t–285t
    production limitations, 278–283
        acid tolerance, 282–283
        cultures, 280–281
        growth of bifidobacteria in
            milk, 279–280
        interaction with starter
            bacteria, 281–282
        presence of oxygen, 283
        requirements for marketing,
            299–300
    therapeutic aspects of, 295–299
Bifiline, 287
Bile salt hydrolase (bsh) genes, 146
Bile salts, 145–146
Bioactive molecules, 32
Bioactive peptides, 9, 40, 156
Biological oxygen demand
    (BOD), 428
Bio-yogurts, 320
Blaand, 13
Blood pressure, effects on, 35
Blood sugar management, role in,
    36–37
Blue mold, 92
Bovine milk yogurt, 63
Branched-chain amino acids
    (BCAAs), 141, 155–156
Branched chain aminotransferases
    (BcAT), 115
Breathing film, 647
Brevibacterium casei, 91
Brevibacterium linens, 91
Brie cheese, 607–608
Budapest Treaty, 575
Buffalo milk
    acidification of, 359
    in bamboo tube, 381f
    butterfat in, 380t, 388
    carbohydrate in, 387

Buffalo milk (*Continued*)
  coagulation of, 386
  dadih, 379
  fermentation, biochemical changes during, 387
  fresh raw, 383
  gelation of, 386
  mistidoi prepared from, 360
  nutritive value of, 380
  protein in, 382
  SCBM in, 217
  transformation of, 387
  in West Sumatra, 383
Buttermilk, 15–16, 204, 207–210. *See also* Lassi
  in cheese formulations, 14
  chemical composition of, 209
  cultured, 102, 205, 206f, 210–213
    production of, 213
    shelf life, 217
    starter cultures used for, 210–213
  effects of, 220
  fat content in, 204
  in food formulations, 215
  glycol-phospholipids in, 208
  health benefits of, 220–221
  hydrolyzed proteins of, 220
  large-scale production scheme of, 213, 214f
  microorganisms as starter culture for preparation of, 211f
  preparation of, 207
  processing and drying of, 210
  reduce blood pressure, 221
  spray drying of, 210
  types of, 208t
  uses of, 215–220
  utility of, 207
Butyric acids, 121–122

C

Cacık, 403
Calcium in milk, 29
Camel cheese, 10–11
Canestrato Pugliese hard cheese, 60
Carbohydrate in buffalo milk, 387–388
Carbon dioxide, yogurt, 315
Carbonyl compounds, 120–121
Casein micelles, 66
Casein proteins, 28
Catabolite control protein (CcpA) gene, 148–149
Catabolite repression process, 438
Cell(s)
  concentration, 178
  envelope proteinases, 153–154
  harvest of, 172–173
  mass, 95, 388–389
  pellet, 95
Chakka, 15, 367
Chal. *See* Shubat
Cheddar cheese, 58, 70–71, 528–529
Cheese(s), 10, 330–331. *See also specific cheese*
  acidophilus, 241–242
  camel, 10–11
  cause of late blowing, 606
  churkham, 13
  churpi, 4
  classification of, 330
  French, 103
  fresh, 330
  Gammelost, 103
  gassiness defects in, 606
  Gjetost, 349–350
  molds, 92–93
  packaging, 656–664
    Cameros cheese, 661
    classification, 656
    direct-set cottage cheese, 661

forms, 659, 659f
hard cheese, 659
light-induced changes, 657
machines, 658
MAP, 659–660
materials, 657, 659f
mozzarella cheese, 661
quality and water loss, 656
requirements, 658
for retail consumers, 658
softer cheese, 661
waxing, 657–658
whey cheese, 662–664
paneer, 344–348
probiotics in, 528–530, 529t
proteolysis, 605
qula, 12
raw materials in, 26
spoilage of, 605
Swiss-type, 54, 60, 62–63, 69, 91
Taleggio, 103
Turkish, 409–420
    çökelek, 418–419
    herby, 419–420
    kashar, 414–416
    production (2012), 409
    tulum, 416–418
    types of, 409, 410t
    white cheese, 411–414
with viili, industrial manufacturing of stretching style, 504, 506f
xylitol in, 678–679
yeast in, 605
Chemical composition of kefir, 469–476
    effect of pH, 475
    flavor, 469–473
    grain activity, 473–474
    grain-to-milk ratio, 474
    impact of environmental factors on, 473–476

    incubation temperature, 474–475
    starter, 474
    storage conditions, 475–476
Chemical oxygen demand (COD), 428
Chhana, 217–218
Chile, progurt, 289
Cholesterol, sterol, 27
Cholic acid, 146
Churkham cheese, 13
Churpi cheese, 4, 12–13
Citrate lyase, 114–115
Citrate permease (CitP), 151–152
Codex Alimentarius Commission (CAC), 407–408, 608
    exposure assessment, 611
    hazard characterization, 610
    microbiological risk assessment by, 609, 610f
Çökelek cheese, 418–419, 419t
Commercial starters, 390
Concentrated buttermilk (CBM), 217
Concentrated yogurt, 654
Conjugated bile acid hydrolase (CBAH), 145–146
Conjugated linoleic acids (CLAs), 31, 241, 252
CONSORT checklist, 579
Cottage cheese, 331–336
    classification of, 331–332
    creamed, 331
    curd for, 334
    direct acidification technique, 332
    dry curd, 331
    manufacture method of, 332, 333f
    production, 612
    shelf life of, 336
    technology of, 332

Cow's milk, 29
  bifidobacteria, 279
  paneer from, 347–348
Cream cheese, 342–343
  manufacture of, 342, 343f
  in North America, 342
Cream dressing, 331
Crohn's disease, probiotic health benefits, 564–565
Cross-flow membrane filtration, 173
Cryoprotectants, 173–174
Cultured buttermilk, 102, 205, 206f, 210–213
  production of, 213
  shelf life, 217
  starter cultures used for, 210–213
Curd
  cutting and cooking of, 335
  formation, 334–335
  preparation of, 217
Czechoslovakia, Femilact, 287–288

D

Dadih
  ampiang, 378
  in bamboo tubes, 380
  Datuk in West Sumatra, 378
  fermentation, 382–389
  in home industry, 381
  homolactic starter cultures in, 387–388
  hypocholesterolemic activity of, 392
  Indonesia, 377–395
  making in Padang Panjang West Sumatra, 381f
  manufacture of, 378–379
  microbial ecology of, 379
  nutrient density in, 382
  nutrition profile of, 381–382
  probiotic for human health promotion, 393–395
  probiotic properties of LAB isolated from
    novel probiotics, 392
    strains from dadih, 389–392
  product description of, 379–381
  proteolytic system in fermentation, 388
  served at weddings, 378
  traditional food, 378
  transformation of buffalo milk into, 387
  xylitol in, 680
Dahi, 13, 102, 356–365
  antibiotic-resistant lactis in, 360
  in Bangladesh, 359
  buffalo milk, 360–361
  demand in Bhaktapur, 361
  halwais (sweets makers), 357
  homemade, 358
  industry in India, 357
  lactobacilli species in, 358
  market size for, 357
  mistidoi, 360–361
  in Nepal, 357
  nutraceutical quality of, 361
  nutritional and therapeutic properties, 361–365
  packaging, 654–656
    in India, 655
    materials, 654–655
  preparation of, 356–357, 356f
  probiotics, 361–365, 535–536, 536f
  problem in, 359
  quality of, 359
  sensory properties of, 359
Dairy drinks, 15–17
  buttermilk, 15–16
  fruit lassi, 16
  honey lassi, 16
  lassi, 15–16
  leben, 16–17
  ymer, 17

Dairy industry, xylitol in, 678–680
　cheese, 678–679
　dadih, 680
　ice cream, 679
　milk chocolate, 680
　sweetened milk, 678
　yogurt, 679–680
Dairy plants hygiene, 615–619
　air hygiene, 619
　building and facilities, 616
　chemicals and cleaning, 617
　drainage and sewage system, 616–617
　equipment hygiene, 617–618
　floors, walls, and ceiling materials, 616
　location of, 616
　personnel hygiene, 617
　pest control program, 617
　plant hygiene, 616–617
　production and process controls, 618–619
　sanitation program, 617
　standards for prior to pasteurization/heat treatment, 620, 620t
　washrooms, lunch halls, and change rooms, 616
　water hygiene, 619
Dairy products, 4, 31, 357. *See also* Fermented milks and dairy products
　approaches for development of fermented, 8–9
　with associated microorganisms, 10t
　cheese, 10
　commonly used prebiotics, 9
　companies, 8
　dental benefits, 36
　diversified fermented, 10–17
　health benefits, 17–18, 34f
　kefir, 13–14
　koumiss, 14
　matsoni, 14
　piima, 14
　popularity, 4
　probiotics
　　challenges for, 541
　　in cheese, 528–530, 529t
　　as delivery vehicle of live, 525–526
　　in fresh fermented milks, 530–531, 531t
　　manufacturing of, 533–541
　　in non-fermented dairy products, 531–533, 533t
　　overview, 526–527
　　strains for functional, selection of, 522–524
　production of fermented milk, 6–8
　quark, 14–15
　ryazhenka, 15
　starter cultures, 6
　traditional diet, 4
　yeasts, 92
Danisco®, 680–681
Danone, 288
Datuk, 378
Dehydration protectants, 174–175
Dental benefits from dairy foods, 36
Dentistry, xylitol in, 675–678
Diabetes
　in human population, 361–362
　reduce risk of, 362
　xylitol, 675–678
Diacetyl, 120–121
Diacetyl reductase, 475–476
Direct acidification method, 334
　B. bifidum, 532
　food-grade acid or acid whey, 332, 333f
　for mozzarella cheese manufacturing, 341f

Direct vat inoculation (DVI) cultures, 98, 169
Direct vat set (DVS) cultures, 95, 169, 280
Dose–response model, 611
Dried concentrated cultures, 95
Dried starter cultures, 95
Dried yogurt, 318–319
Dry curd cottage cheese, 331
Drying process, culture
  concentrated and DVS lactic starter cultures, 183–184
  lactic starter cultures, 182–183
  long-term preservation, 181–182
  spray, 183
  vacuum drying, 184
DuPont™, 680–681

E

Electrolytic chromium–coated steel (ECCS), 644
Embden–Meyerhof pathway (EMP), 111
Encapsulation
  during preservation of lactic starter cultures, 188–189
  to improve viability of bifidobacteria in bifidus products, 292–293
Endomyces lactis, 384
Enriched and fortified yogurts, 319
Enteric coating technology, 547
Enterococcus species, 52t, 88, 91
Esterase, 116, 123
Ester compounds, 122–123
Ethanol, 122
  bioethanol, 436–437
  concentration of kefir, 472
  fermentation of whey, 450
  production from cheese whey, 439–441, 442t–443t, 449

European Food Safety Authority (EFSA), 562, 575–576
  for fermented milk products, bacteria in petitions to, 585, 585t–588t
  lactose malabsorption, 583
  reasons for rejected claims, 580
  cause and effect, 581t
  experimental design and methodology, 580t
  statistics, 581t
Ewe's milk yogurt, 64
Exopolysaccharide (EPS), 56, 142, 172, 358
  addition of, 360
  production of, 359
  secreting ropy LAB classification, 64–65
  texture of fermented milk products, 64–67
  viili, 498
    biosynthesis, 508–509
    composition and sugar components of, 507–508
Expanded PS (EPS), 647

F

Fatty acids, 27
Femilact, 287–288
Fermented foods
  approaches for development of dairy, 8–9
  milks and characteristics, 5–9
  traditional diet, 4
Fermented milks and dairy products, 4, 7t, 355, 387, 620–628, 639
  advantages of intake of, 490–491
  bacteria in petitions to EFSA, 585, 585t–588t
  beneficial bacteria, 573

characterizing microorganism(s), 574–579, 604
  defining dose for fermented milk product, 577–578
  dose response effects, 578
  food matrix, 578–579
  methodology to, 576–579
  nomenclature, 575–576
  spoilage-causing group, 606
dahi, 102, 356–365
diversified, 10–17
experiments designed to validate efficacy, 579–584
  demonstrating cause and effect, 582
  food vs. drug effects, 582–583
  learning from past experience, 579–580
  reasons for rejected claims, 580, 580t, 581t, 582
  yogurt and lactose digestion, 583–584
flavor compounds, 117–124
foodborne outbreaks, 607–608, 609t
FSMS, 626
HACCP, 620–622, 626
  description for fermented milk products, 624t–625t
  plan for management of pathogens, 623t
  reassessment of plan, 627
health benefits of, 17–18
on heart health, 35–36
human-feeding trials, 573
ISAPP suggestions to test prebiotics/probiotics, 583–584, 584t
lassi, 365–367
manufacture, 96
materials for packaging, 642–649
  considerations, 654
  effect of, 665

  glass, 643–644
  metal, 644–645
  paper and paper-based, 642–643
  plastic, 646–649
microbiological issues
  quality, 605–606
  safety, 606–608
microbiological risk assessment, 608–613
  antimicrobial activity of, 611
  by CAC, 609, 610f
  exposure assessment, 611–612
  hazard characterization, 610–611
  hazard identification, 609–610
  risk characterization, 612–613
  risk communication, 613
  risk management, 613
milk production and, 6–8
packaging of, 654–665
  cheese, 656–664
  dahi and yogurt, 654–656
  probiotic fermented dairy products, 664–665
  requirements and characteristics, 639, 640t–641t
pathogens, 608, 610, 613
possible health claims for, 584–585, 589
production of, 206f
properties of, 639
quality assurance of, 613–620
  GMP, 614–619
  hygiene measures, 613–614
  monitoring process plant, 619–620
quality problems, 604
regulatory aspects of, 628–629
shrikhand, 367–369
sources of contamination, 604

Fermented milks and dairy products (*Continued*)
  techniques for packaging, 649–654
    active packaging, 651–652
    antimicrobial packaging, 652
    aseptic packaging, 649
    edible films and coatings, 653
    modified atmosphere packaging, 649–651
    nanocomposite packaging, 653–654
  verification procedure, 626–628
  viili, 105
  yakult, 105
  yeasts, 60
Fermented Milks and Lactic Acid Bacteria Beverages Association of Japan, 578
Fermented whey-based beverages, cheese, 430–437
  alcoholic, 434–436
  bioethanol, 436–437
  functional, 431–433
  Kefir grains in, 433
  by LAB, 431–432
  lactic fermentations of, 430
Fibronectin-binding protein (fnb), 141
Films in food packaging, 653
Foamed PS, 647
Food and Agriculture Organization (FAO), 558
Food and Drug Administration (FDA), 359
Food(s)
  fermentation, 4
  functional, 30–32, 40–42
    Japan, 30
    market, 41
  health and wellness through, 20
  industry, 3
  packaging, 638
    convenience, 638
    functions, 638, 639*f*
    levels, 638–639
    materials, 638
  therapeutic benefits, 3–4
Food Safety and Standards Regulations (FSSR 2011), 331
Food safety management system (FSMS), 614
Foods for specific health use (FOSHU), 30
Fortification of yogurt with different molecules, 319*t*–320*t*
France
  bifidogene, 291
  bifidus active, 287
  bifidus baby foods, 287
  botulism, 607
  cheeses, 103
  listeriosis, 608
Freeze-dried starter cultures, 181, 183, 185
Freeze-drying process, 95, 181, 547
Freezing in long-term preservation, 176–181
French National Collection of Cultures of Microorganisms, 576
Friendly bacteria, 32
Frozen concentrated starter culture, 95
Frozen starter culture, 95
Frozen storage and thawing, 178–181
Frozen yogurt, 318
Fructooligosaccharides (FOS), 277
Fruit lassi, 16
Fruit yogurt, 318
Functional beverages, cheese whey, 431–433
Functional Food Science in Europe, 30

G

Galactose, 113
Gammelost cheese, 103
Gariss, 11
Gastrointestinal (GI) disorders, 31
Gastrointestinal health benefits,
    probiotic dairy products,
    562–565
  AAD, 563
  Helicobacter pylori infection,
    563–564
  IBD, 564–565
  IBS, 564
  infectious diarrhea, 563
  intestinal transit, 564
  NEC, 563
Gastrointestinal tract (GIT),
    134, 142
  prebiotics, 293, 523, 526
  stress-tolerance by LAB, 144–146
  warm-blooded animals, 271
Generally recognized as safe
    (GRAS), 138, 144
Genetically modified
    microorganisms, 33
Genetic modification, 189
Genome characteristics of LAB,
    135–144
  common LAB, 136t
  lactobacilli, 138–143
  lactococci, 137–138
  leuconostoc, 143–144
  Streptococcus thermophilus, 144
Genomics, 126
Geotrichum candidum, 92–93, 105,
    500–501
Germany
  bifidus baby foods, 287
  bifidus milk, 284
  bifidus yogurt, 288
  sweet bifidus milk, 289
Gjetost cheese, 349–350

Glass-packaging materials, 643–644
  melting of, 644
  soda-lime, 643
  types of containers/closures, 644
Glucose, 28
Glycerol, 174
Glycerol teichoic acid (GTA), 184
Glycolysis, 141, 148–149
  and citrate metabolism, 113–115
  LAB strains, 121
  and sugar transport in LAB,
    148–149
Glycosyltransferase gene, 65
Good hygiene practices (GHP), 614
Good manufacturing practices
    (GMP), 614–615
  defined by FDA, 615
  necessity, 616
  precautions to assure product
    quality and safety, 615
  at processing condition, 615–619
    air hygiene, 619
    equipment hygiene, 617–618
    personnel hygiene, 617
    plant hygiene, 616–617
    production and process
      controls, 618–619
    water hygiene, 619
Grain activity, kefir, 473–474
Grain-to-milk ratio, kefir, 474

H

HACCP-based FSMS, 620–628
  FSMS, 626
  HACCP, 620–622, 626
    description for fermented milk
      products, 624t–625t
    plan for management of
      pathogens, 623t
    reassessment of plan, 627
    verification procedure, 626–628
Halwais (sweets makers), 357

Haydari, 403
Health Canada (HC), 577
  dose for fermented milk product, 578
  probiotic (bacteria), 577
Heat-treated yogurt, 318
Herby cheese, 419–420
Heterofermentative pathways, 112, 212*f*
Heterolactic starter cultures, 388
High-density polyethylene (HDPE), 646
High-impact polystyrene (HIPS), 647
HIV-1 enzymes, 39
Homemade viili, 502, 502*f*
Home method, suusac, 12
Homofermentative bacteria, 470
Homofermentative pathways, 111–112, 212*f*
Homolactic starter cultures, 387–388
Honey lassi, 16
Horizontal gene transfer (HGT) mechanism, 135, 147, 315
Hypertension, milk-based food health benefits, 35
Hypocholesterolemic effect, 256
Hypocholesteromic components, 36

I

Ice cream
  acidophilus, 239–241
  frozen yogurt, 318
  probiotic, 538
  production of, 219
  xylitol in, 679
Immune health benefits, probiotic dairy products
  allergy/autoimmune disease, 565
  respiratory tract infections, 565–566

Immunomodulatory effects of viili, 510–511
Incubation temperature of kefir, 474–475
India
  buttermilk, 207
  dadih, 378
  dahi, 102, 355–356, 402
  lassi, 355, 365, 402
  mistidoi, 360
  ricotta, 350
  shrikhand, 355, 367
Indian cheese. *See* Paneer
Indian probiotic industry, 361
Indian sweet
  chhana, 217–218
  rasgulla, 218
  sandesh, 218
Indigenous enzymes, 385–386
Indonesian dadih, 377–395
Inflammatory bowel disease (IBD), probiotic health benefits, 564–565
Insertion sequence (IS) elements, 138
International Dairy Federation (IDF) standard, 300
International Life Sciences Institute (ILSI), 30
International Scientific Association for Probiotics and Prebiotics (ISAPP), 583–584, 584*t*
Intracellular ice formation (IIF), 176
Intrinsic factors, 187
Irritable bowel syndrome (IBS), probiotic health benefits, 564
Iso-malto-oligosaccharides, 58–59

J

Japan, 30
  bifider, 291
  bifidus yogurt, 288

INDEX

Fermented Milks and Lactic Acid Beverages Association, 300
sweet acidophilus bifidus milk, 289
sweet bifidus milk, 289
Yakult Honsha Company, 288

K

Kargi tulum cheese, 418
Kashar cheese, 414–416
  categories of, 414–415
  chemical composition of, 415
  commercial production of, 415
  LAB counts in, 415
  mineral composition of, 415
  production, 415, 416f
Kefir, 408
  acetic acid in, 472–473
  acetoin in, 472
  acetone in, 472
  beverage production, milk, 476, 476f
  chemical composition, 469–476
    effect of pH, 475
    flavor, 469–473
    grain activity, 473–474
    grain-to-milk ratio, 474
    impact of environmental factors on, 473–476
    incubation temperature, 474–475
    starter, 474
    storage conditions, 475–476
  culture, 104
  dairy products, 13–14
  ethanol concentration of, 472
  fermentation, characteristics of LAB in milk, 470, 471t
  grains, 60–62, 433, 463–466
    fluorescence staining and confocal laser scanning microscopy imaging, 463–464
    internal surface of, 464, 465f
    naked-eye examination, 463, 464f
    SEM, 464–466
    viewed with naked eye, 464, 465f
  lactic acid in, 470
  microbial composition (beverages)
    AAB, 468
    LAB, 467
    yeasts, 468
  microbial composition (grains)
    AAB, 466–467
    LAB, 466
    yeasts, 467
  production and uses of, 476–477
  species of
    bacteria in, 468, 468t
    yeasts in, 468, 469t
  taste, 463
  types of
    fermentation in, 462–463
    yeasts in, 472
Kefiran, 104, 463
Kluyveromyces microorganism, 53t
Koumiss, 408–409
  ACE-inhibitory peptides in, 492
  beautifying agent, 493
  in clinical practices, 493–494
  culture, 104–105
  dairy products, 13–14
  historical background, 484–485
  microbial diversity in, 485, 486t
  microflora of, 485–486
  in Mongolia, 485
  nutritional properties of, 489–490
    amino acid, 489
    fat, 490
    lactose, 490

Koumiss (*Continued*)
   minerals, 489
   protein, 490
   vitamins, 489
   overview, 483–484
   physiochemical and biological characteristics of finished, 487, 488*t*
   to produce good-quality, 485
   production method, 486–489
      light, 489
      moderate, 487
      strong, 487
   regular consumption of, 492
   steps in preparation of, 487, 488*f*
   therapeutic values of, 490–494
Kulfi, probiotics, 539–540, 540*f*
Kurut, 404–406
   composition of, 405
   LAB levels, 406
   method of preparing, 405
   milk-fermented product, 13
   pH of, 405

L

Labneh. *See* Strained yogurt
Lactic acid bacteria (LAB), 6, 48, 84, 169, 171, 204, 270
   acids and alcoholic compounds, 121–122
   citrate utilization and flavor generation, 151–152
   distinct acidification properties, 54–55
   fermentation of milk products, 117*t*, 205
      aromatic compounds produced by, 118*t*–119*t*
      cheese whey, 431–432
      dairy products, 357
      flavor development, 63–64

   metabolic pathways, 110–112, 111*f*
   microorganism, 48*t*–54*t*
   in milk kefir, 470, 471*t*
   preparation of, 604
   genome characteristics, 135–144
      Lactobacillus, 138–143
      lactococci, 137–138
      Leuconostoc, 143–144
      Str. thermophilus, 144
   heterofermentative/homofermentative pathways of, 212*f*
   indigenous, 378
   inhibitory spectra, bacteriocins of, 101*t*
   in manufacture of yogurt and dadih, 385*t*
   microbial composition
      kefir beverages, 467
      kefir grains, 466
   microorganisms, 402
   milk environment, adaptation of, 146–148
   milk sugars, 28–29
   natural, 383–384
   probiotics
      properties of, 252, 389–392
      vs. starter cultures, 559–560
   proteolytic system, 55, 115, 140, 152–156, 388
      bioactive peptides, 156
      cell envelope proteinases, 153–154
      peptide uptake systems, 154–155
      regulation, 155–156
   respiration in, 150
   R/M systems *vs.* bacteriophage attack, 156–157
   roles of, 388
   strains, 499

stress-tolerance in GIT, 144–146
sugar transport and glycolysis, 148–149
Lactic starter cultures, 94, 168
  cultivation, 169–172
  preservation, 175–186
    drying, 182–183
    drying of concentrated and DVS, 183–184
    freezing, 177–178
    freezing of concentrated and DVS, 178
    frozen, 180
    pellet-frozen, 178
    preserved forms, 170*f*
  process for production, 168*f*
Lactobacillales, 137
Lactobacilli (*Lactobacillus*), 8, 89*t*, 138–143
  characteristic, 90
  classification, 89
  co-cultures, characteristics, 55–56
  cultures, 88–89, 102
  sour product, 101
Lactobacillus acidophilus
  characteristics of, 229
  overview, 227–229
  probiotics, 521
Lactobacillus bulgaricus, 171
Lactococci bacteria, 51*t*–52*t*, 87, 137–138
Lactococcus lactis, 56, 87, 150
Lactoferrin, 37, 39
Lactose, 9–10, 18, 112
  carbohydrate, 387
  digestion and yogurt, 583–584
  genes involved in, 149*f*
  improving intolerance, 34
  intolerance, 18, 220
  koumiss, 490
  lactic starters, 94
  in milk products, 113
  in whey, 439–446
  yeast, 468
Lactose-fermenting yeasts, 472
Laldahi, 360–361
Large-scale production scheme of buttermilk, 213, 214*f*
Lassi, 15–16, 102, 365–367
  churning dahi, 365
  instant, 366
  preparation of, 356*f*
  probiotics, 536–537, 537*f*
  quality of, 367
  shelf life of, 366
Leben, dairy drinks, 16–17
Leuconostoc bacteria, 52*t*, 58, 143–144, 151
Leuconostoc lactis, 87–88
Life style island, 148
Linear low-density polyethylene (LLDPE), 646
Lipase, 116
Lipid metabolism, 116–117
Lipids, 388
Lipolysis, 116, 126
Liquid cultures, short-term preservation, 175–176
Liquid starter, 94–95
LiveBac process, 547
Long-chain fatty acids (LCFA), 316
Long set method, 334*t*
Low-density polyethylene (LDPE), 646
Lubuluqi Journey to the East, 484
Lubuluqi, William, 484

M

Mahi. *See* Lassi
Maillard reaction, 88
Malutka, 287
Mammals milks, composition of, 380, 380*t*

Mango lassi, 16
Mare's milk, 104
 energy value of, 486
 fermentation, 491
 fresh, 492
 lactose in, 491
 raw, 492
Matsoni, dairy products, 14
Mesophilic starters, 94, 102
Metal-packaging materials
 aluminum, 645
 corrosion, 644–645
 tin and chromium-coated steel, 644–645
Metchnikoff, Elie, 227, 521
Methanethiol (MTL), 124
Microbial composition of kefir
 beverages, 467–469
  AAB, 468
  LAB, 467
  yeasts, 468
 grains, 466–467
  AAB, 466–467
  LAB, 466
  yeasts, 467
Microbial fermentation, 48
Microbiological issues, fermented milk products
 quality issues, 605–606
 safety issues, 606–608
Microbiological risk assessment, fermented milk products, 608–613
 antimicrobial activity of, 611
 by CAC, 609, 610f
 exposure assessment, 611–612
 hazard characterization, 610–611
 hazard identification, 609–610
 risk characterization, 612–613
 risk communication, 613
 risk management, 613
Microencapsulation technology, 548
Microflora of koumiss, 485–486

Microorganism(s), 6, 7t, 604
 in fermented milk product, characterizing, 574–579
  defining dose, 577–578
  dose response effects, 578
  food matrix, 578–579
  nomenclature, 575–576
 in viili starter cultures, 498–501
  characteristics during culture, 500–501
  fungi and yeasts, 500
  LAB, 499
Milk
 amino acids, 5
 -based foods, health benefits of, 33–42
  anticancer activities, 37–38, 38f
  anti-HIV activity, 39
  antimicrobial potential, 37
  appetite suppressors, 38–39
  in blood sugar management, 36–37
  dairy foods and obesity, 36
  dental benefits, 36
  effects on blood pressure, 35
  fermented dairy products on heart health, 35–36
  future perspectives, 41–42
  improving lactose tolerance, 34–35
  milk-derived bioactive peptides, 40–41
  relief against rotaviral diarrhea, 39
 bifidus, 105
 bioactive components, 5
 bulk inoculum, 168
 chocolate
  probiotics, 532, 540–541, 541f
  xylitol in, 680
 coagulation of, 346
 components, 26f

different product names, 29
drawn from mammals, 4–5
fat, 27
    fatty acids, 5
    lipase, 116
    lipolysis, 116
fermentation, 203, 215t
fermented milks and
    characteristics, 5–9
foods with health-promoting
    properties, 29
heat treatment, 345
homogenization of, 205
kefir. *See* Kefir
lipids, 27
minerals, 5
pasteurization of, 205, 332
proteins, 28, 156
raw materials in, 26
salts, 29
sterols, 27–28
sugars, 28–29, 34
therapeutic potential of
    functional, 30–33
treatment, 332–334
type of, 317, 345
ultra-high-temperature-sterilized, 205
and varieties, 4–5
and water-based cereal puddings, 291–292
whole food for adults, 5
Mil-Mil, 288–289
Minangkabau ethnic group, 378
Misti dahi, 11
Mistidoi, 11, 360–361
Misti yogurt, 11
Mixed strain starter, 93
Modified atmosphere packaging
    (MAP), 649–651
    gases used in, 650
    gas-flush, 651
    passive modification, 651
    techniques, 650
Modified suusac, 12
Molds, cheese, 92–93
Mongolia
    koumiss in, 485
    tarag, 289
Moro, Ernst, 227
Mother culture, 97
Mozzarella cheese, 57–58, 338–342
    chemistry of stretch of, 342
    classification of, 339t
    direct acidification method
        for, 341f
    Italy, 338
    production method, 339
    traditional manufacturing
        method for, 340f
Mucosal epithelia, 32
Mucus-binding protein (mub), 141, 143–144
Multiple strain starters, 93–94
Mysost cheese, 349–350

N

Nanocomposite packaging, 653–654
Natural buttermilk, production of, 206, 206f
Natural killer (NK) cells, 258
Necrotizing enterocolitis
    (NEC), probiotic health
    benefits, 563
Neslac, 288
Nestle, 9
Neurodegenerative diseases, 28
Nicotinamide adenine dinucleotide
    (NAD+), 113
Nisin-controlled expression
    system, 189
Non-lactic starters, 94
Non-starter lactobacilli (NSLAB), 54, 62–63
Non-volatile acids, 120

Nox (NADH oxidase), 316–317
Nutritional benefits of acidophilus milks, 250–253
Nutritional properties of koumiss, 489–490
  amino acid, 489
  fat, 490
  lactose, 490
  minerals, 489
  protein, 490
  vitamins, 489
Nylon in packaging, 648

O

Obesity, 36
Open reading frames (ORF), 141
Oxaloacetate, 113
Oxidative stress, 316–317

P

Packaged in protective atmosphere, MAP, 650
Packaging materials for fermented milks, 642–649
  considerations, 654
  effect of, 665
  glass, 643–644
  metal
    aluminum, 645
    tin and chromium-coated steel, 644–645
  paper and paper-based, 642–643
  plastic, 646–649
    polyamides, 648–649
    polyester, 648
    polyolefins, 646–647
    polystyrene, 647
    PVC, 647–648
    PVDC, 648
Packaging techniques for fermented milks, 649–654

active packaging, 651–652
antimicrobial packaging, 652
aseptic packaging, 649
edible films and coatings, 653
modified atmosphere packaging, 649–651
nanocomposite packaging, 653–654
Paired compatible starters, 93
p-aminobenzoic acid (PABA), 316
Pancreatic hormones, 38–39
Paneer, 217, 344–348
  from buffalo milk, 344
  chilling of, 347
  from cow milk, 347–348
  manufacturing method for, 346f
  package, 347
  yield of, 347
Paper and paper-based materials in packaging, 642–643
Pasta-filata cheese, 338
Pathogenic bacteria, 42, 363
Payodhi, 360–361
Pediococci (*Pediococcus*) microorganism, 53t, 90, 90t
Pediococcus pentosaceus, 90
Penicillium camemberti, 92, 103
Penicillium roquefortii, 92, 103
Peptides
  and amino acids, 314–315
  uptake systems, 154–155
Phage-inhibitory media, 98
Phage-resistant media, 98
Phosphoenolpyruvate (PEP)-dependent PTS (PEP-PTS), 148
Phospho-ketolase pathway, 112
Phosphorus, 29
Phosphotransferase systems (PTSs), 139
Pichia microorganism, 53t
Piima, dairy products, 14
Pineapple lassi, 16

Plants hygiene, dairy, 615–619
   air hygiene, 619
   building and facilities, 616
   chemicals and cleaning, 617
   drainage and sewage system, 616–617
   equipment hygiene, 617–618
   floors, walls, and ceiling materials, 616
   location of, 616
   personnel hygiene, 617
   pest control program, 617
   plant hygiene, 616–617
   production and process controls, 618–619
   sanitation program, 617
   standards for prior to pasteurization/heat treatment, 620, 620t
   washrooms, lunch halls, and change rooms, 616
   water hygiene, 619
Plasticizers in packaging, 647
Plastic-packaging materials, 646–649
   polyamides, 648–649
   polyester, 648
   polyolefins, 646–647
   polystyrene, 647
   PVC, 647–648
   PVDC, 648
Polyethylene terephthalate (PET) in packaging, 648
Polymerase chain reaction (PCR), 358
Polystyrene (PS) in packaging, 647
Polyvinyl chloride (PVC) in packaging, 647–648
Polyvinylidene chloride (PVDC) in packaging, 648
Powdered yogurt, 319
Primary package, 638
Primost cheese, 349–350

Probiogenomics, 293–294
Probio-Tec process, 547
Probiotic dahi, 13, 361–365
Probiotic dairy products, 270, 369
   challenges for, 541
   in cheese, 528–530, 529t
   as delivery vehicle of live, 525–526
   differences between strains from same species, 559, 559f
   in fresh fermented milks, 530–531, 531t
   functional dairy products, manufacturing of, 533–540
      dahi, 535–536, 536f
      ice cream, 532, 538–539
      kulfi, 539–540, 540f
      lassi, 536–537, 537f
      milk chocolate, 532, 540–541, 541f
      selection of strains for, 522–524
      shrikhand, 537–538, 538f
      yogurt, 533–534, 534f
      yogurt drink, 534–535, 535f
   health benefits of, 562–567
      gastrointestinal health, 562–565
      immune health, 565–566
      metabolic syndrome, 566
      other, 566–567
      reduction of serum cholesterol, 566–567
      weight management, 566
      yogurt, 562
   influence of food matrix on functionality, 560–562
      matrix on probiotic health efficacy, 561–562
      of other cultures, 560
      of other ingredients/components, 560–561
      storage conditions, 561

Probiotic dairy products (*Continued*)
  in non-fermented dairy products, 531–533, 533*t*
  overview, 526–527
  strains used in, 559, 559*t*
  vs. starter cultures, 559–560
Probiotics (bacteria), 8–9, 31, 577, 664
  commonly used, 521
  currently used, 524–525
  enteric coating technology, 547
  foods, criteria for effective, 541–548
    addition of micronutrients, 546
    ascorbic acid, 544
    cysteine, 544–545
    effective dose, 541–542
    hurdles affecting survival during processing and storage, 542–543
    improving survival in dairy products, 543
    protection against oxygen, 544–545
    selection of appropriate strains, 543–544
    stress adaptation, 546–547
    technological interventions, 547–548
    two-stage fermentation, 545–546
    use of air-tight containers, 545
  freeze drying method, 547
  functional food ingredients, 521–522
  general criteria, 32–33
  health benefits
    effects of, 522
    of functional foods prepared with, 271, 272*t*
  human, 525
  market status of, 527–528, 528*t*
  microencapsulation technology, 548
  and non-strain-specific health claims, 577, 577*t*
  positive effects of, 521
  selection of, 389
  starters to produce yogurt, 530
  used in formulation of functional products, 33
  use in bifidus products, 293
  yogurt, 320, 322*t*
Progurt, 289–290, 290*f*
Propionibacterium freudenreichii, 91
Propionic acid bacteria (PAB), 60
Protective atmosphere packaging, 650
Protein, 152, 388
  in dadih fermentation, 388
  koumiss, 490
  kurut, 405
  metabolism, 115–116
  whey, 348
Proteolytic system, LAB, 55, 115, 140, 152–156
  bioactive peptides, 156
  cell envelope proteinases, 153–154
  in milk fermentation, 67–69
  peptide uptake systems, 154–155
  regulation, 155–156
Proton motive force (PMF), 145
Pull-seal method, 182
Putative peptidases (PepX), 141
Pyruvate formate lyase (pfl), 316
Pyruvate metabolism, 113, 120

Q

Quality assurance of fermented milk products, 613–620
  GMP at processing condition, 614–619
    air hygiene, 619

equipment hygiene, 617–618
personnel hygiene, 617
plant hygiene, 616–617
production and process controls, 618–619
water hygiene, 619
hygiene measures, 613–614
monitoring process plant, 619–620
Quarg cheese, 336–338
characteristics of, 336
manufacturing method for, 337f
mesophilic bacterial culture, 336
method of production, 336
shelf life of, 338
used as, 338
Quark dairy products, 14–15
Queso Blanco, 343–344
Qula cheese, 12

R

Rasgulla, formulation of, 218
Rennet curd, 331
Requeijão, 662
Respiration in LAB, 150
Respiratory tract infections, probiotic health benefits, 565–566
Restriction/modification (r/m) systems vs. bacteriophage attack, 156–157
Ricotta cheese, 349
in gulabjaman, 350
manufacture from cheddar cheese, 349
used in, 350
Ricotta whey cheese, 445
Ripened cheeses, 330–331
Rotaviral diarrhea, relief against, 39
Rubber-modified PS, 647
Ryazhenka, dairy products, 15

S

Saccharomyces cerevisiae, 437, 446–447
Saccharomyces microorganism, 53t–54t
Scanning electron microscopy (SEM), kefir, 464–466
Second-generation probiotics, 33
Serum cholesterol reduction, probiotic health benefits, 566–567
Shelf life, 179, 182
Shigella, colonization of, 365
Shmen, 11
Shoptu, 12
Short set method, 334, 334t
Shrikhand, 15, 218, 367–369
fruit-flavored preparation, 367–368
large-scale production of, 368
preparation method of, 356f
probiotics, 369, 537–538, 538f
Shubat, 11–12
Silicone gasket, 178
Single strain starter, 93
Skim milk, 170, 174–176, 189, 332
composition and physiochemical properties of, 208t
homogenization of, 334
Skim milk powder (SMP), 210
blending of, 216
difference between SCBM and, 210
Sodium ascorbate, 186
Soft cell pellets, 173
Solid not fat (SNF), 332
Soy yogurt, 318
Spray drying, 183
Standard National Indonesia 2981:2009, 381

Starter cultures, 6, 7*t*, 559
  application, 99–105
  characteristics of dairy, 85–93
    bacteria, 87–91
    molds, 92–93
    yeasts, 92
  in cottage cheese, 334
  cultivation of lactic, 169–172
  disadvantages, 168
  enterococci, 91
  EPS-producing lactics as, 359
  exopolysaccharides in fermented milk products, 100*t*
  functions, 85
  growth temperature, 94
  in kefir, 474
  microorganisms in viili, 498–501
    characteristics of, 500–501
    fungi and yeasts, 500
    LAB, 499
  microorganisms used, 85*t*–86*t*
  named as per intended use, 96*t*
  NSLAB, 62–63
  physical forms, 94–95
  for preparation of buttermilk, 211*t*
  preservation, 175–186
    drying, 182–183
      drying of concentrated and DVS, 183–184
    freezing, 177–178
    freezing of concentrated and DVS, 178
    frozen, 180
    pellet-frozen, 178
    preserved forms, 170*f*
  probiotics *vs.*, 559–560
  problems associated with production of, 98–99, 99*t*
  production, 94, 98*f*
  propagation of, 96–98
  role of, 211
  types, 93–96
  used for cultured buttermilk, 210–213
  used in, 84
Starter flora, composition of, 93–94
  mixed strain starter, 93
  multiple strain starters, 93–94
  paired compatible starters, 93
  single strain starter, 93
Sterols, 27
Strained yogurt, 318, 403, 654
Streptococcus microorganism, 52*t*–53*t*, 89*t*
Streptococcus thermophilus, 55, 65, 69, 85*t*, 88, 171, 572
  carbon dioxide, 315
  in LAB, 144
  metabolic peculiarities in yogurt, 57–58
  symbiotic cultures, 406
  yogurt, 100, 246
Streptozotocin (STZ), 362
Stress induction, 187–188
Stress proteins, 546
Sugars, 174–175
  sucrose, 58–59
  transport, 148–149
Sulfur compounds, 123–124
Suusac, 12
Sweet acidophilus bifidus milk, 289
Sweet cheese whey, 429, 430*t*, 434–435
Sweet cream buttermilk (SCBM), 207–208
  blending of, 216
  in buffalo milk, 217
  composition and physiochemical properties of, 208*t*
  difference between SMP and, 210
Sweetened milk, xylitol in, 678
Swiss Food Regulation, 300
Swiss-type cheese, 54, 60, 62–63, 69, 91, 606

Switzerland
  botulism, 607
  foodborne diseases, 608
Synbiotics, 9, 293, 526, 529

T

Taleggio cheese, 103
Tarag, 289
Therapeutic benefits
  of acidophilus milks, 253–258
  of bifidus products, 295–299
  of functional milk foods, 30–33
  of koumiss, 490–494
  probiotic dahi, 361–365
Thermobacterium, 89*t*
Thermophilic starters, 94
Thermophilic yogurt culture, 402
Tin-free steel in packaging, 644
Tissier, Henry, 272
Torba yogurt, 403–404, 404*f*
Transferase reaction, 122
Triglycerides, 116
Tripeptidase (PepT), 68
Tulum cheese, 416–418
  composition of, 417–418
  Kargi, 418
  LAB in, 418
  production of, 417
Turkish cheeses, 409–420
  çökelek, 418–419, 419*t*
  herby, 419–420
  kashar, 414–416
    categories of, 414–415
    chemical composition of, 415
    commercial production of, 415
    LAB counts in, 415
    mineral composition of, 415
    production, 415, 416*f*
  production (2012), 409
  tulum. *See Tulum* cheese
  types of, 409, 410*t*
  white cheese, 411–414
    bacteria used in, 414
    categories of, 411–412
    chemical composition of, 413
    mineral compositions, 413
    production, 412, 412*f*
    salt in, 414
Turkish Statistical Institute, 409
Turkish yogurts, 402–403

U

United States
  bifidus milk, 270
  bifidus yogurt, 288
  dietary supplements in, 228
  Queso Blanco, 411
  sweet acidophilus, 236
  yogurt in, 4
Uruguay Round of Multilateral
    Trade Negotiations in
    Marrakesh, 628

V

Vacuum drying system, 184
Van Herby cheese. *See Herby*
    cheese
Viili, 15, 105
  annual consumption, 498
  diacetyl in, 498
  functionality of, 509–511
    ACE inhibitory activity, 510
    antioxidative effect, 509
    antitumor activity, 511
    immunomodulatory effects,
        510–511
  homemade, 502, 502*f*
  industrial manufacturing
    of stretching style cheese with,
        504, 506*f*
    of yogurt and, 503, 503*f*

Viili (*Continued*)
  slime properties of, 507–509
    composition and sugar components of EPS, 507–508
    EPS biosynthesis, 508–509
  starter cultures, microorganisms in, 498–501
    characteristics during culture, 500–501
    fungi and yeasts, 500
    LAB, 499
  starters, 62
  and viili products, production of, 501–506
    manufacture of viili, 502–504
    other, 504–505
    sources of milk, 505–506
  yeast, 500–501
Volatile acids, 120
Volatile sulfur compounds (VSC), 123–124

W

Weight control
  management, probiotic health benefits, 566
  xylitol in, 675–678
Whey cheese composition, 428–430
  acid whey, 429, 430*t*
  ethanol production from, 439–441, 442*t*–443*t*, 449
  fermented whey-based beverages, 430–437
    alcoholic, 430–431, 434–436, 436*t*, 446
    bioethanol, 436–437
    functional, 431–433
    Kefir grains in, 433
    by LAB, 431–432
    lactic fermentations of, 430
  lactose in, 429, 433–435, 439, 444–448
  long-term projections of worldwide, 427, 428*f*
  production of ethanol by, 437
  response surface methodology, 445
  sweet whey, 429, 430*t*
  whey fermentation, 437–450
    by *Kluyveromyces* ssp., 439–446
    by recombinant *S. cerevisiae* strains, 446–450, 448*f*
    by *S. cerevisiae*, 437–439
Whey cheeses, 13, 348–351
  bioethanol, 436–437
  drainage of, 347
  Gjetost, 348
  manufacture of, 348–349
  packaging, 662–664
    Greece, 662
    Portuguese, 662
  principle for making, 348–349
  proteins, 28, 39
  shelf life of, 350
  types of, 348
  use of, 350–351
White mold, 92
Winter cacık, 403
Wistar rats, 363
World Health Organization (WHO), 558

X

XIVIA™ xylitol, 681
Xylitol
  applications, 675, 676*f*
    confectionery ingredient and pharmaceutical, 680–682, 682*t*
    in different industries, 680, 681*t*
    effectiveness, 681–682

INDEX

safety, 682
  in weight control, diabetes, and dentistry, 675–678
biotechnological production, 674
characteristics of, 675, 676t
in dairy industry, 678–680
  cheese, 678–679
  dadih, 680
  ice cream, 679
  milk chocolate, 680
  sweetened milk, 678
  yogurt, 679–680
demand for, 673
FDA approved as sweetening agent, 674
positive effects of, 675
production of, 673
total, 677
use of, 675, 677

Y

Yak milk products, 12–13
  churkham, 13
  churpi, 12–13
  kurut, 13
  qula, 12
Yakult, 30, 105
Yakult Honsha Company, 288–289
Yarrowia microorganism, 54t
Yeasts, 605
  acidophilus milk, 237–238
  in cheeses, 605
  dairy products, 92
  fermented dairy products, 60
  in kefir
    beverages, 468
    grains, 467
    and similar milk products, 60–62
    species of, 468, 469t
    types of, 472

species from viili cultures, 500–501
  use, 92
Yeasts, *Candida* microorganism, 53t
Ymer, dairy drinks, 17
Yogurt, 4, 12, 312
  acidophilus, 246–250
  advantages of, 321
  bacteria, 402
  bifidus, 288
  bovine milk, 63
  as carrier to probiotics, 320–321
  classification of, 317–319
  cow milk, 382t
  culture, 100–102
  ewe's milk, 64
  exopolysaccharides (EPS), 56
  fortification of, 319t–320t
  health benefits of probiotic dairy products, 562
  hypocholesterolemic effect of, 299
  and lactose digestion, 583–584
  lactose sugar, 28
  manufacture and storage of, 246–247
  metabolic peculiarities of *Str. thermophilus*, 57–58
  microbiology, 313–317
    carbon dioxide, 315
    folate and *p*-aminobenzoic acid, 316
    formate, 315–316
    genetic material exchange, 317
    LCFA, 316
    oxidative stress, 316–317
    peptides and amino acids, 314–315
  milk with yogurt flavor, bifidus, 286, 287t
  nomadic Turkish people, 312
  packaging
    materials, 654–656
    nitrogen-flushed plain, 655

Yogurt (*Continued*)
   prepared with microencapsulated
      cultures, 249
   probiotic, 533–534
   production of, 312–313
   raw materials in, 26
   shelf life of, 319
   torba, 403–404, 404f
   Turkish, 402–403
      production (2012), 402–403
      types of, 403
   variants and styles, 317–319
      as carrier to probiotics, 320–321
      concentrated/strained
         yogurt, 318
      consistency, 317
      dried yogurt, 318–319
      enriched and fortified
         yogurts, 319
      frozen yogurt, 318
      fruit yogurt, 318
      heat-treated yogurt, 318
      non-dairy yogurt, 318
      type of milk, 317
   xylitol in, 679–680

Z

Zygosaccharomyces
   microorganism, 54t